Table of Abbreviations

A.2d	Atlantic Reporter, Second Series[3]
AAA	American Arbitration Association
ABA	American Bar Association
ABAJ	American Bar Association Journal[6]
A.C.	Appeal Cases (English)
A.D.2d	Appellate Division, Second Series (New York)
ADA	Americans with Disabilities Act (federal)
ADR	Alternative Dispute Resolution
AGC	Associated General Contractors of America
AIA	American Institute of Architects
AIA Doc.	American Institute of Architects Document
A.L.R.2d	American Law Reports, Second Series[7]
A.L.R.3d	American Law Reports, Third Series[7]
A.L.R.4th	American Law Reports, Fourth Series[7]
A.L.R.5th	American Law Reports, Fifth Series[7]
A.L.R.6th	American Law Reports, Sixth Series[7]
ASBCA	Armed Services Board of Contract Appeals[10]
Ala.	Alabama Reports[1]
Ala.App.	Alabama Appellate Court Reports
Am.Jur.2d	American Jurisprudence, Second Series[9]
Annot.	Annotation
Ariz.	Arizona Reports[1]
Ariz.App.	Arizona Appeals Reports
Ariz.L.Rev.	Arizona Law Review[6]
Ariz.Rev.Stat.Ann.	Arizona Revised Statutes Annotated[8]
Ariz.Sess.Laws	Arizona Session Laws[8]
Ark.	Arkansas Reports[1]
Ark.App.	Arkansas Appellate Reports
BCA	Board of Contract Appeals (federal)
B.C.Int'l & Comp.L.Rev.	Boston College International and Comparative Law Review[6]
B.C.L.Rev.	Boston College Law Review[6]
BNA	Bureau of National Affairs (publisher)
BOT	Build-Operate-Transfer
B.R.	Bankruptcy Reporter
B.U.L.Rev.	Boston University Law Review[6]
Baylor L.Rev.	Baylor Law Review[6]
Brooklyn L.Rev.	Brooklyn Law Review[6]
C.A.	Court of Appeals (England)
CBCA	Civilian Board of Contract Appeals[10]

Key/Footnotes:
[1] Highest Court of State
[2] Highest Court of United States
[3] Regional Reporter
[4] All New York cases
[5] All California cases
[6] Periodical
[7] Annotation to important cases
[8] Statute
[9] Encyclopedia
[10] Federal Agency Appeals Board

CCD	Construction Change Directive
CCH	Commerce Clearing House (publisher)
C.D.	Central District (federal trial court)
CERCLA	Comprehensive Environmental Response, Compensation and Liability Act (federal)
CFR	Code of Federal Regulations
CGL	Commercial General Liability
CIA rules	American Arbitration Association, Construction Industry Arbitration Rules
CM	Construction Manager
CMa	Construction Manager: adviser
CMAA	Construction Management Association of America
CMc	Construction Manager: constructor
CO	change order
CPM	critical path method
Cal.	California Reports[1]
Cal.2d	California Reports, Second Series[1]
Cal.3d	California Reports, Third Series[1]
Cal.4d	California Reports, Fourth Series[1]
Cal.App.2d	California Appellate Reports, Second Series
Cal.App.3d	California Appellate Reports, Third Series
Cal.App.4th	California Appellate Reports, Fourth Series
Cal.Bus. & Prof.Code	California Business and Professional Code[8]
Cal.Civ.Code	California Civil Code[8]
Cal.Civ.Proc.Code	California Code of Civil Procedures[8]
Cal.Code Regs.	California Code of Regulations
Cal.Com.Code	California Commercial Code[8]
Cal.Exec.Order	California Executive Order
Cal.Govt.Code	California Government Code[8]
Cal.Health & Safety Code	California Health and Safety Code[8]
Cal.Lab.Code	California Labor Code[8]
Cal.Penal Code	California Penal Code[8]
Cal.Pub.Res.Code	California Public Resources Code[8]
Cal.Rptr.	California Reporter[5]
Cal.Rptr.2d	California Reporter, Second Series[5]
Cal.Stat.	Statutes of California[8]
Calif.L.Rev.	California Law Review[6]
Cir.	United States Circuit Court
Cl.Ct.	Claims Court (federal)
Colo.	Colorado Reports[1]
Colo.App.	Colorado Court of Appeals Reports
Colo.Rev.Stat.	Colorado Revised Statutes[8]
Columb.L.Rev.	Columbia Law Review[6]
Comp.Gen.Dec.	Comptroller General Decisions (federal)
Conn.	Connecticut Reports[1]
Conn.App.	Connecticut Appellate Reports
Conn.Gen.Stat.Ann.	Connecticut General Statutes Annotated[8]
Conn.Supp.	Connecticut Supplement
Constr.Contractor	Construction Contractor[6]
Constr.Eng'g. & Mgmt.	Construction Engineering and Management[6]
Constr.L.J.	Construction Law Journal (U.K.)[6]

Constr.Lawyer	Construction Lawyer[6]	F.Supp.	Federal Supplement
Constr.Litig.Rep.	Construction Litigation Reporter[6]	F.Supp.2d	Federal Supplement, Second Series
Constr.Mgmt. & Econ.	Construction Management and Economics[6]	GAAP	generally accepted accounting principles
Cornell L.Rev.	Cornell Law Review[6]	GAO	General Accounting Office (federal)
Ct.Cl.	Court of Claims (federal)	GMC	guaranteed maximum cost
		GMP	guaranteed maximum price
D.Ariz.	District Court of Arizona (federal trial court)	GSA	General Services Administration (federal)
		GSBCA	General Services Board of Contract Appeals[10]
DB	Design-Build	Ga.	Georgia Reports[1]
DBIA	Design-Build Institute of America	Ga.App.	Georgia Appeals Reports
D.Cal.	District Court of California (federal trial court)	Ga.Code Ann.	Georgia Code Annotated[8]
		Ga.L.Rev.	Georgia Law Review[6]
D.C.Cir.	District of Columbia Circuit Court of Appeals (federal)	Geo.Wash.L.Rev.	George Washington Law Review[6]
D.D.C.	District Court of the District of Columbia (federal trial court)	Gonz.L.Rev.	Gonzaga Law Review[6]
D.Me.	District Court of Maine (federal trial court)	HUD	Department of Housing and Urban Development (federal)
D.N.H.	District Court of New Hampshire (federal trial court)	Harv.Bus.Rev.	Harvard Business Review[6]
		Harv.L.Rev.	Harvard Law Review[6]
DOTBCA	Department of Transportation Board of Contract Appeals[10]	Hastings L.J.	Hastings Law Journal[6]
		Haw.Adm.Rules	Hawaii Administrative Rules
DOTCAB	Department of Transportation Civil Aeronautics Appeal Board[10]	Hawaii	Hawaii Reports[1]
D.R.I.	District Court of Rhode Island (federal trial court)	IBCA	Department of Interior Board of Contract Appeals[10]
DSC	differing site condition	Idaho	Idaho Reports[1]
Del. Super.	Delaware Superior Court	IDM	Initial decision maker
Duke L.J.	Duke Law Journal	Ill.	Illinois Reports[1]
Duq.L.Rev.	Duquesne Law Review[6]	Ill.2d	Illinois Reports, Second Series[1]
		Ill.App.2d	Illinois Appellate Court Reports, Second Series
E.D.N.C.	Eastern District of North Carolina (federal trial court)	Ill.App.3d	Illinois Appellate Court Reports, Third Series
E.D.Va.	Eastern District of Virginia (federal trial court)	Ill.Comp.Stat.Ann.	Illinois Compiled Statutes Annotated[8]
EEO	Equal Employment Opportunity (federal)	Ill.Rev.Stat.	Illinois Revised Statutes[8]
EJCDC	Engineers Joint Contract Documents Committee	Ind.	Indiana Reports[1]
		Ind.Code	Indiana Code[8]
EPA	Environmental Protection Agency (federal)	Ind.Ct.App.	Indiana Appellate Court Reports
		Ins.Couns.J.	Insurance Counsel Journal[6]
Emory L.J.	Emory Law Journal[6]	Int'l Constr.L.Rev.	International Construction Law Review (U.K.)[6]
ENG BCA	Corps of Engineers, Board of Contract Appeals[10]	Iowa	Iowa Reports[1]
Eng.Rep.	English Reports	Iowa Code Ann.	Iowa Code Annotated[8]
Exec.Order	Executive Order (federal)	Iowa L.Rev.	Iowa Law Review[8]
		IRMI	International Risk Management Institute, Inc.
F.	Federal Reporter		
F.2d	Federal Reporter, Second Series	J ACCL	Journal of the American College of Construction Lawyers[6]
F.3d	Federal Reporter, Third Series		
FAR	Federal Acquisition Regulations	J.of Constr. Eng'g & Mgmt.	Journal of Construction Engineering and Management[6]
FHA	Federal Housing Authority		
FIDIC	Federal Internationale Des Ingenieurs-Conseils	J.Mar.L.Rev.	John Marshall Law Review[6]
Fed.Cir.	United States Circuit Court, Federal Circuit	K.B.	King's Bench (England)
		Kan.	Kansas Reports[1]
Fed.Reg.	Federal Register	Kan.App.2d	Kansas Appellate Reports, Second Series
Fed.Rules	Federal Rules	Kan.L.Rev.	Kansas Law Review[6]
Fla.	Florida Supreme Court[1]	Kan.Stat.Ann	Kansas Statutes Annotated[8]
Fla.Dist.Ct.App.	Florida District Court of Appeals	Ky.	Kentucky Reports[1]
Fla.Stat.	Florida Statutes		

(continued on back endsheets)

CONSTRUCTION LAW
FOR DESIGN PROFESSIONALS, CONSTRUCTION MANAGERS, AND CONTRACTORS

JUSTIN SWEET
John H. Boalt Professor of Law Emeritus
University of California (Berkeley)

MARC M. SCHNEIER
Attorney Editor, Construction Litigation Reporter

with
BLAKE WENTZ
Associate Professor and Director of the Construction
    Management Program
Milwaukee School of Engineering

Australia • Brazil • Canada • Mexico • Singapore • United Kingdom • United States

Construction Law for Design Professionals, Construction Managers, and Contractors
Justin Sweet, Marc M. Schneier, with Blake Wentz

Publisher: Timothy Anderson

Senior Developmental Editor: Hilda Gowans

Senior Editorial Assistant: Tanya Altieri

Senior Content Project Manager: Kim Kusnerak

Production Director: Sharon Smith

Team Assistant: Ashley Kaupert

Rights Acquisition Director: Audrey Pettengill

Rights Acquisition Specialist, Text and Image: Amber Hosea

Text and Image Researcher: Kristiina Paul

Manufacturing Planner: Doug Wilke

Copyeditor: Shelly Gerger-Knechtl

Proofreader: Harlan James

Indexer: Shelly Gerger-Knechtl

Compositor: MPS Limited

Senior Art Director: Michelle Kunkler

Cover Designer: Tin Box Studio

Internal Designer: C. Miller Design

Cover Image: © Oxford/iStockphoto

© 2015 Cengage Learning, Inc.

WCN: 01-100-101

ALL RIGHTS RESERVED. No part of this work covered by the copyright herein may be reproduced, transmitted, stored, or used in any form or by any means graphic, electronic, or mechanical, including but not limited to photocopying, recording, scanning, digitizing, taping, web distribution, information networks, or information storage and retrieval systems, except as permitted under Section 107 or 108 of the 1976 United States Copyright Act, without the prior written permission of the publisher.

> For product information and technology assistance, contact us at
> **Cengage Customer & Sales Support, 1-800-354-9706.**
> For permission to use material from this text or product,
> submit all requests online at **www.cengage.com/permissions.**
> Further permissions questions can be e-mailed to
> **permissionrequest@cengage.com.**

Library of Congress Control Number: 2013957166

ISBN-13: 978-0-357-67138-2

ISBN-10: 0-357-67138-4

Cengage
200 Pier 4 Boulevard
Boston, MA 02210
USA

Cengage is a leading provider of customized learning solutions with office locations around the globe, including Singapore, the United Kingdom, Australia, Mexico, Brazil, and Japan. Locate your local office at: **international.cengage.com/region.**

To learn more about Cengage platforms and services, register or access your online learning solution, or purchase materials for your course, visit **www.cengage.com.**

Printed in the United States of America
1 2 3 4 5 6 7   24 23 22 21 20

To my wife, Sheba
　—*Justin Sweet*

For Lisa, my favorite designer
　—*Marc M. Schneier*

To my fiancée Nadya
　—*Blake Wentz*

Contents in Brief

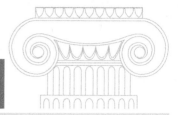

Preface
About the Authors

Part A: A Legal Framework of the Design and Construction Process

- CHAPTER 1 American Legal System
- CHAPTER 2 Forms of Business Association and Employment
- CHAPTER 3 Contracts: From Formation to Breach
- CHAPTER 4 Remedies for Contract Breach: Focus on Construction Disputes
- CHAPTER 5 Torts: Legal Relations Not Arising from Contract
- CHAPTER 6 Regulation of Land and the Construction Process

Part B: The Main Actors: The Owner, Constructor, and Design Professional

- CHAPTER 7 The Project Participants: Focus on the Owner, Prime Contractor, and Construction Manager
- CHAPTER 8 Licensing of the Design Professional, Contractor, and Construction Manager
- CHAPTER 9 The Design Professional-Client Relationship
- CHAPTER 10 Professional Design Services: The Sensitive Client Issues and Copyright
- CHAPTER 11 Design Professional Liability
- CHAPTER 12 Defenses to Claims of Design Professional Liability
- CHAPTER 13 Ethics

Part C: Project Delivery Methods

- CHAPTER 14 Project Organization, Pricing, and Delivery Methods
- CHAPTER 15 Public Contracts

Part D: Performance Disputes

- CHAPTER 16 Performance Disputes Overview: Claims and Defenses to Liability for the Owner and Prime Contractor
- CHAPTER 17 Contractor Payment and Project Completion
- CHAPTER 18 Changes: Complex Construction Centerpiece
- CHAPTER 19 Subsurface Problems: Predictable Uncertainty
- CHAPTER 20 Time: Project Scheduling and Delay Claims
- CHAPTER 21 The Subcontracting Process: An "Achilles Heel"
- CHAPTER 22 Terminating a Construction Contract: Sometimes Necessary but Always Costly

Part E: Risk Management and Dispute Resolution

- CHAPTER 23 Apportioning or Shifting Losses: Contribution and Indemnity
- CHAPTER 24 Insurance
- CHAPTER 25 Surety Bonds: Backstopping Contractors
- CHAPTER 26 Claims and Disputes: Emphasis on Arbitration

Appendices
Glossary
Reproduced Case Index
Subject Index

Contents

Preface   xviii
About the Authors   xxiii

Part A: A Legal Framework of the Design and Construction Process   1

CHAPTER 1

AMERICAN LEGAL SYSTEM   3
Scenario   3
1.1 The Importance of Law   4
1.2 The Federal System   4
1.3 Constitutions   5
1.4 Legislation   5
1.5 The Executive Branch and Administrative Agencies   6
1.6 Courts: Common Law and the Concept of Precedent   7
1.7 Courts: The Judicial System   8
    Federalism and the Court System   8
    Trial Courts   8
    Appellate Courts   8
    Public Contracts   9
1.8 The Litigation Process   10
    Hiring an Attorney: Role and Compensation   10
    The Parties   10
    Pleadings   10
    Pretrial Activities: Discovery   11
    Pretrial Dismissal   11
    Trial   11
    The Cost of Litigation   12
    Enforcement of Judgments   12
1.9 Non-Public Sources of Law   12
    Contracting Parties   12
    Publishers of Standardized Documents   13
    Restatements of the Law   13
1.10 International Contracts   13
Review Questions   14

CHAPTER 2

FORMS OF BUSINESS ASSOCIATION AND EMPLOYMENT   17
Scenario   17
2.1 Forms of Business Associations: Overview   18
2.2 Sole Proprietorships   18
2.3 Partnerships   18
    Introduction   18
    Fiduciary Duties   19
    Authority of Partner   19
    Liability of Partnership and Individual Partners   19
2.4 Corporations   20
    Introduction   20
    Personal Liability of Directors and Officers: Piercing the Corporate Veil   20
2.5 Professional Corporations, Limited Liability Companies (LLC), and Limited Liability Partnerships (LLP)   21
2.6 Joint Ventures   22
2.7 Unincorporated Associations   22
2.8 Loose Associations: Share-Office Arrangement   22
2.9 Professional Associations   23
2.10 Forms of Employment: Introduction   23
2.11 The Agency Relationship   24
    2.11A Introduction   24
    2.11B Policies Behind Agency Concept   25
    2.11C The Principal–Agent Relationship   25
    2.11D Agent Binding the Principal: Actual Versus Apparent Authority   25
        Actual Authority   26
        Apparent Authority   26
    2.11E Disputes Between Principal and Third Party   26
    2.11F Disputes Between Agent and Third Party   26

2.12 The Employment Relationship	27
2.12A Workers' Compensation	27
2.12B Employee Leasing Companies	28
2.12C Unions and Labor Disruptions: The Picket Line and Project Labor Agreements	28
2.13 Independent Contractors	30
Review Questions	31
Endnotes	32

CHAPTER 3

Contracts: From Formation to Breach   35

Scenario	35
3.1 Relevance	36
3.2 The Function of Enforcing Contracts: Freedom of Contract	36
3.3 Contract Formation	37
3.3A Definitions	37
3.3B Mutual Assent	37
Objective Theory of Contracts	37
Offer and Acceptance	38
3.3C Consideration and Its Substitute: Promissory Estoppel	39
3.3D Reasonable Certainty of Terms	40
3.4 Defects in Contract Formation	40
3.4A Defects Affecting Mutual Assent	40
Fraud, Misrepresentation, and Duty to Disclose	40
Mistake	40
Unconscionability	41
Economic Duress	41
3.4B Defect in Formal Requirements: Need for a Writing	41
Statute of Frauds	42
Homeowner Contracts	42
3.5 What Is the Contract?	43
3.5A Contract Completeness and the Parol Evidence Rule	43
3.5B Judicially Determined Terms	44
Implied Terms	44
Custom	45
Legal Requirements: Building Codes	45
Good Faith and Fair Dealing	45
Summary	46
3.6 Contract Interpretation	46
Basic Objectives	46
Language Interpretation	46
3.7 Contract Breach	48
3.7A Definition	48
3.7B Material Breach	48
3.7C Future Breach: Prospective Inability and Anticipatory Repudiation	49
Review Questions	49
Endnotes	50

CHAPTER 4

Remedies for Contract Breach: Focus on Construction Disputes   53

Scenario	53
4.1 Introduction	54
4.2 Types of Remedies for Contract Breach: Overview	54
4.3 Specific Decrees and Declaratory Judgments	55
4.4 Money Awards	55
Compensatory Damages	55
Benefit of the Bargain	55
Reimbursement	55
Restitution	56
Interest	56
Attorneys' Fees	56
Litigation Costs	56
Emotional Distress and Punitive Damages	56
4.5 Limits on Recovery	57
Causation	57
Certainty	57
Foreseeability of Loss	57
Mitigation	57
Lost Profits	58
Contract-Specified Remedies	58
4.6 Damages Specific to Construction Disputes: Introduction	59
4.7 Contractor Versus Owner	59
4.7A Project Never Commenced	59
4.7B Project Partially Completed	59
4.7C Project Completed	60
Site Chaos and Productivity	60
Actual Cost and Recordkeeping	61
Extended Home Office Overhead: Eichleay Formula	62
Productivity Loss Preferred Formulas: Measured Mile and Industry Productivity Studies	63
Total Cost	63
Jury Verdict	64

4.8 Owner Versus Contractor	64
4.8A Project Never Begun	64
4.8B Project Partially Completed	64
4.8C Defective Performance: Correction Cost or Diminished Value?	64
4.8D Delay	66
4.9 Single Recovery Rule: Claims Against Multiple Defendants	66
4.10 Owner/Design Professional Claims	67
Review Questions	67
Endnotes	68

CHAPTER 5

Torts: Legal Relations Not Arising from Contract — 71

Scenario	71
5.1 Relevance to the Construction Process	72
5.2 Tort Law: Introduction	72
5.2A Definition	72
5.2B Function	73
5.2C Threshold Classifications	74
5.3 Negligence: The Societal Concept of Reasonable Conduct	75
5.3A Elements of Negligence Claim	75
5.3B Standard of Conduct: The Reasonable Person	75
Explicating the Standard	75
Violation of Statutes and Administrative Regulations	76
Factual Versus Legal Issue	76
5.3C Duty	77
Changing Tests for Finding a Duty of Care	77
Limits on Duty	77
5.3D Causation	79
5.3E Protected Interests and Emotional Distress	79
5.3F Defenses to Negligence: Assumption of the Risk and Contributory Negligence	80
Assumption of the Risk	80
Contributory and Comparative Negligence	80
5.3G Claims Against Government Defendants	80
5.4 Misrepresentation	81
Intentional or Negligent	81
Representation or Opinion	81
Reliance	81
Person Suffering the Loss	82
5.5 Interference With Contract or Prospective Advantage	82
5.6 Premises Liability: Duty of the Possessor of Land	82
5.6A Relevance	82
5.6B Traditional Rule Based on Status	83
Passersby	83
Trespassing Adults	83
Trespassing Children	83
Licensees	83
Invitees	83
5.6C Modern Rule and Restatement (Third) of Torts	84
5.6D Defenses to Premises Liability	84
5.7 Employment of Independent Contractor	85
Generally	85
Independent Contractor Rule and Its Exceptions	85
5.8 Products Liability	86
Strict Liability	86
Third Restatement	86
Defenses to Products Liability	87
Services Contracts	88
5.9 Tort Remedies	88
Compensatory Damages	88
Punitive Damages	89
5.10 Limits on Tort Liability for Pecuniary Losses: The Economic Loss Rule	89
5.10A Introduction	89
5.10B Development of the Economic Loss Rule	90
5.10C Permutations of the Economic Loss Rule	91
Varying Tests	91
Design Professionals	91
Hazardous Defects and Prophylactic Repairs	92
Property Damage	92
Statutory Developments	93
Review Questions	93
Endnotes	93

CHAPTER 6

Regulation of Land and the Construction Process — 97

6.1 Regulation: A Pervasive Presence	97
6.2 Limits by Private Action: Restrictive Covenants in Subdivision Developments	98
6.3 Protection of Neighboring Landowners	98
Soil Support	98
Drainage and Surface Waters	98
Easements for Light, Air, and View	99

6.4 Zoning: The Mechanism for Land Use Control 99
   6.4A Euclidean Zoning: The Premise of Local Control 99
   6.4B Societal and Environmental Concerns 99
   6.4C Constitutional Limits on Land Use Controls 100
6.5 Environmental Law 102
   6.5A National Environmental Policy Act (NEPA) 102
   6.5B Superfund: Comprehensive Environmental Response, Compensation, and Liability Act (CERCLA) 102
      Overview 102
      Construction Industry Participants 103
      Brownfields 103
   6.5C State Law 103
6.6 Regulation of the Design Content and Construction Process 104
   6.6A Permits 104
   6.6B Building Codes 104
   6.6C Building Inspections 105
6.7 Safety 106
   6.7A State Safe Workplace Statutes 106
   6.7B Occupational Safety and Health Laws 107
   6.7C Tort Law 108
      Independent Contractor Rule 108
      Premises Liability 108
Review Questions 109
Endnotes 109

Part B: The Main Actors: The Owner, Constructor, and Design Professional 111

CHAPTER 7

THE PROJECT PARTICIPANTS: FOCUS ON THE OWNER, PRIME CONTRACTOR, AND CONSTRUCTION MANAGER 113

Scenario 113
7.1 The Basic Contracts: Private Ordering of the Parties' Relationships 114
7.2 The Owner 114
   7.2A Public Versus Private Owners 115
   7.2B Experience 116
   7.2C Authority Problems: Private Owners 117
      Sole Proprietors 118
      Partnerships 118
      Corporations 118
      Unincorporated Associations 119
      Spouses or Unmarried Cohabitants 119
   7.2D Authority Problems: Public Owners 119
   7.2E Foreign Owners 119
7.3 The Design Professional 119
7.4 The Prime Contractor 120
   7.4A An Industry Overview 120
   7.4B Independent Contractor Status 121
   7.4C Soil Conditions: The Common Law Rule 122
   7.4D Responsibility for Subcontractors 123
   7.4E Acceptance Doctrine 124
7.5 Construction Management 124
   7.5A Reasons for Construction Management 125
   7.5B Types of Construction Management 125
   7.5C Construction Management: Legal Issues 126
      The Owner/CM Relationship 127
      Rights and Liabilities of the CM 128
   7.5D Impact on Project Organization 129
7.6 Project Funding: Spotlight on Lender 130
7.7 Insurers 132
7.8 Sureties 132
Review Questions 132
Endnotes 133

CHAPTER 8

LICENSING OF THE DESIGN PROFESSIONAL, CONTRACTOR, AND CONSTRUCTION MANAGER 137

Scenario 137
8.1 Introduction 137
8.2 Public Regulation: Justifications and Criticisms 138
8.3 Administration of Licensing Laws: Agencies, Admission, and Discipline 139
   8.3A Licensing Agencies 139
   8.3B Admission to Practice 139
   8.3C Postadmission Discipline: *Duncan v. Missouri Board for Architects, Professional Engineers and Land Surveyors* 139
8.4 Design Professional Licensing 148
   8.4A Preliminary Issue: Constitutionality 148
   8.4B "Holding Out" Statutes 148
   8.4C "Practicing" Statutes 149
   8.4D Architecture and Engineering Compared 149
   8.4E Architect or Engineer Applying Seal to Work of Another 151
   8.4F Statutory Exemptions 152

8.4G Possessor of License: Practice by a Business Entity	152
8.4H Out-of-State Practice	153
8.5 Contractor Licensing Laws	155
8.5A Overview	155
8.5B Possessor of License: Business Organization	155
8.6 The Unlicensed Design Professional or Contractor	156
8.6A Criminal and Quasi-Criminal Sanctions	156
8.6B Recovery for Work Performed and Payment Reimbursement	156
Statute Specifies Contractor's Right to Compensation	157
Statute Silent on Contractor's Right to Compensation	157
Statute Bars Contractor's Right to Compensation	157
Payment Reimbursement	158
Summary	159
8.6C Substantial Compliance	159
8.7 Construction Manager Licensing	159
8.8 The Trained but Unregistered Design Professional: Moonlighting	160
8.8A Unlicensed Persons: A Differentiation	160
8.8B Ethical and Legal Questions	161
8.8C Recovery for Services Performed	161
8.8D Liability Problems	162
Review Questions	162
Endnotes	163

CHAPTER 9

THE DESIGN PROFESSIONAL–CLIENT RELATIONSHIP 167

Scenario	167
9.1 Traditional Phases of Architectural Services: AIA B101-2007	168
9.2 Characterizing the Client–Design Professional Relationship	169
9.3 Compensation of the Design Professional: Methods of Compensation	172
9.3A Introduction	172
9.3B Percentage of Construction Costs: Basic Versus Additional Services	173
Percentage of Construction Costs	173
Basic Versus Additional Services	174
AIA Additional Services	175
EJCDC Additional Services	175
9.3C Multiple of Direct Personnel Expense: Daily or Hourly Rates	175

9.3D Professional Fee Plus Expenses	176
9.3E Fixed Fee	176
9.3F Reasonable Value of Services or a Fee to Be Agreed On	177
9.3G Reimbursables	177
9.3H Fee Ceilings	178
9.3I Adjustment of Fee	178
9.3J Deductions from the Fee	178
9.3K The Fee as a Limitation of Liability	178
9.4 Compensation of the Design Professional: Timing of Payment	179
9.4A Service Contracts and the Right to Be Paid as One Performs	179
9.4B Interim Fee Payments	179
9.4C Monthly Billings	179
9.4D Late Payments	180
9.5 Payment Although Project Never Built	180
9.6 Client Obligations to the Design Professional Not Involving Compensation	181
9.7 Design Professional's Role in Contract Completion	181
9.8 Design Professional Suspension of Performance	182
9.9 Termination of Design Contract	182
9.10 Death or Unavailability of Design Professional	184
9.11 Breach of the Design Agreement: Remedies	185
9.11A Introduction	185
9.11B Client Claims and the "Betterment" Rule Defense	185
9.11C Design Professional Claims	187
Review Questions	187
Endnotes	188

CHAPTER 10

PROFESSIONAL DESIGN SERVICES: THE SENSITIVE CLIENT ISSUES AND COPYRIGHT 191

Scenario	191
10.1 Professional Services: Prior to Receipt of Bids	191
10.1A Assistance in Obtaining Financing	191
10.1B Economic Feasibility of Project	192
10.1C Approval of Public Authorities	192
10.1D Services of a Legal Nature	193
10.2 Cost Predictions	193
10.2A Inaccurate Cost Prediction: A Source of Misunderstanding	193
10.2B Two Models of Cost Predictions	194

10.2C A Cost Condition Scenario: *Griswold & Rauma v. Aesculapius* — 194
10.2D Cost Predictions: Legal Issues — 199
  Client's Damages: Breach of Condition — 199
  Client's Damages: Breach of Promise — 199
  Parol Evidence Rule — 199
  Cost Prediction Tolerance — 199
  Cost Condition Waived — 200
  Cost Predictions on Cost-Plus Contracts — 200
10.2E Standard Contracts and Disclaimers: AIA and EJCDC — 201
10.2F Advice to Design Professionals — 202
10.3 Site Services: Observing the Contractor's Work — 202
  10.3A Introduction — 202
  10.3B From Supervision to Observation: *Watson, Watson, Rutland/Architects v. Montgomery County Board of Education* — 203
    Summary — 207
  10.3C Site Inspections: Certification of Contractor's Progress Payments — 209
10.4 Review of Contractor Submittals — 209
10.5 Who Actually Performs Services: Use of and Responsibility for Consultants — 211
  10.5A Within Design Professional's Organization — 211
  10.5B Outside Design Professional's Organization: Use of Consultants — 212
10.6 Ownership of Drawings and Specifications — 212
  10.6A Introduction — 212
  10.6B Describing the Ownership Interest — 213
  10.6C AIA and EJCDC Approaches: License to Use — 213
  10.6D AIA and EJCDC Approaches: Electronic Data — 214
  10.6E Ownership on Design-Build — 215
10.7 Copyright of Drawings and Specifications — 215
  10.7A Introduction — 215
  10.7B Relevance to Design Professional — 215
  10.7C Work for Hire Doctrine — 216
  10.7D Owner Use of Copyrighted Work — 217
  10.7E Infringement of Copyright — 218
    Registration — 218
    Substantial Similarity — 218
    Remedies for Infringement — 218
    Copyright Duration — 219
  10.7F Copyright Transfer to Former Employee — 219
  10.7G Advice to Design Professionals: Obtaining a Copyright — 219
Review Questions — 222
Endnotes — 222

CHAPTER 11

DESIGN PROFESSIONAL LIABILITY — 225

Scenario — 225

11.1 Design Professional Liability: The Professional Standard as the Measure of Reasonable Care: *City of Mounds View v. Walijarvi* — 225
11.2 Defining the Professional Standard — 227
  Violation of Statutes and Administrative Regulations — 227
  Industry Standards — 228
  Professional Ethics — 228
11.3 Proving the Professional Standard: Expert Testimony — 229
  11.3A Need for Expert Testimony — 229
  11.3B Advice to Expert Witnesses — 230
11.4 Alternative Standards as Measure of Conduct — 230
  11.4A Contract Standard — 230
  11.4B Consumer Protection Statutes — 231
11.5 Client Claims Against Design Professionals — 231
  11.5A Suing in Contract, Tort, or Both? — 231
  11.5B "Green" or Sustainable Design — 232
    Introduction — 232
    Historical Background — 233
    Standard Form Contracts — 233
    Definition of Terms — 234
    Green Building Claims — 234
11.6 Third-Party Claims Against Design Professionals — 234
  11.6A Potential Third Parties — 234
  11.6B Contract Duty: Contracts for Benefit of Third Parties — 235
  11.6C The Design Contract and the Tort Duty of Care — 235
  11.6D Negligence Liability for Economic Losses: *Bilt-Rite Contractors, Inc. v. The Architectural Studio and SME Industries, Inc. v. Thompson, Ventulett, Stainback & Assocs., Inc.* — 237
    Negligence Versus Negligent Misrepresentation — 237
    Negligent Misrepresentation and the Economic Loss Rule — 238
  11.6E Intentional Torts: Interference with Contract or Prospective Advantage — 243
  11.6F Safety and the Design Professional — 244
    Common Law Liability — 245
    Liability under Occupational Safety and Health Act (OSH Act) — 246
  11.6G Duty to Warn — 246

Review Questions	247
Endnotes	248

CHAPTER 12

DEFENSES TO CLAIMS OF DESIGN PROFESSIONAL LIABILITY — 251

Scenario	251
12.1 Legal Defenses to Claims of Design Professional Liability: Introduction	252
12.2 Acceptance of the Project	252
12.3 Passage of Time: Statutes of Limitations	252
12.4 Design Professional Decisions and Immunity	253
12.5 Statutory Defenses	254
Certificate of Merit Statutes	254
"Good Samaritan" Laws	255
Workers' Compensation	255
12.6 Apportionment for Fault of Others	255
12.7 Contractual Risk Control	256
12.7A Scope of Services	256
12.7B Standard of Performance	256
12.7C Exclusion of Consequential Damages	256
12.7D "Limitation of Liability" Clauses	256
12.7E Immunity: Decision Making	258
12.7F Contractual Statute of Limitations	258
12.7G Third-Party Claims	259
12.7H Dispute Resolution	259
12.7I Some Suggestions	259
Review Questions	259
Endnotes	260

CHAPTER 13

ETHICS — 263

Scenario	263
13.1 Three Distinct Concepts: Law, Morality, and Professional Ethics	263
13.2 Ethics for Architects	264
13.2A Introduction	264
13.2B Historical Background: Competing for the Commission	265
13.2C The AIA Ethics Canons	265
13.3 Ethics for Engineers	268
13.4 Ethics for Contractors	269
13.5 Ethics for Design/Builders	270
13.6 Ethics for Construction Managers	271
13.7 Conclusion	272
Review Questions	273
Endnotes	273
Part C: Project Delivery Methods	275

CHAPTER 14

PROJECT ORGANIZATION, PRICING, AND DELIVERY METHODS — 277

Scenario	277
14.1 Project Planning: Owner's Choices	278
14.2 Compensating Contractor Work	278
14.3 Fixed-Price Contracts	278
14.4 Cost Contracts	280
Introduction	280
Owner Concerns and Protections	280
Cost Plus Fee	281
Guaranteed Maximum Price	281
Owner Administration	281
14.5 Unit Pricing	282
14.6 Value Engineering Change Proposal (VECP)	283
14.7 Design Responsibility: "Design" Versus "Performance" Specifications	283
14.8 Administrative Problems	284
14.8A Overview	284
14.8B Authority: Special Problems of Construction Projects	284
Introduction	284
Architect's Authority	285
14.8C Communications	285
14.9 Traditional Project Delivery: Design–Bid–Build (DBB)	286
14.9A Traditional System Reviewed	286
14.9B Weaknesses	287
14.10 Modern Variations in Project Delivery: Introductory Remarks	288
14.11 Phased Construction (Fast-Tracking)	289
14.12 Multiple Primes	290
14.13 Turnkey Contracts	293
14.14 Design–Build (DB): Combining Design and Construction	293
14.14A Reasons for Design–Build	293
14.14B Nature of Design–Build	294
14.14C Licensing and Insurance	295
14.14D Advantages and Disadvantages of Design–Build	296
14.15 Partnering	297
14.16 Teaming Agreements	298
14.17 Lean Project Delivery	298

14.18 Project Alliance	299	
14.19 Program Management	299	
14.20 Public–Private Collaboration: Public–Private Partnerships (PPP) and Build–Operate–Transfer (BOT)	300	
Public–Private Partnerships	300	
Build–Operate–Transfer (BOT)	300	
14.21 Building Information Modeling (BIM)	301	
14.22 Summary	302	
Review Questions	302	
Endnotes	303	

CHAPTER 15

PUBLIC CONTRACTS   307

Scenario   307

15.1 Public Contracts: Introduction   308
15.2 Design Contracts   308
    15.2A Hiring the Design Professional: The Brooks Act   308
    15.2B Public Works Specifications   309
15.3 Competitive Bidding: Objectives and the Pitfall of Corruption   310
15.4 The Competitive Bidding Process: Overview   311
15.5 Invitation for Bids (IFB)   311
15.6 Bid Opening and Award   312
15.7 Bid Protests   312
    Bid Responsiveness   312
    Bidder Responsibility   312
    Rejection of All Bids   313
    Bid Protest Remedies   313
15.8 Construction Management, Design–Build, and Public Contracts   313
15.9 Claims Against Public Entities: Federal False Claims Act   314
Review Questions   315
Endnotes   315

Part D: Performance Disputes   317

CHAPTER 16

PERFORMANCE DISPUTES OVERVIEW: CLAIMS AND DEFENSES TO LIABILITY FOR THE OWNER AND PRIME CONTRACTOR   319

Scenario   319

16.1 Claims: Theories of Legal Liability   320
    Contract Law Claims—Generally   320
    Contract Law Claims—Express Warranty   320
    Contract Law Claims—Implied Warranty   320
    Tort Law Claims   320
    Statutory Claims   320
16.2 Principles Underlying Claims   321
    16.2A Basic Principle: Responsibility Follows Control   321
        Design Control   321
        Defective Materials   321
    16.2B Supplemental Principle: Good Faith and Fair Dealing   322
        Introduction   322
        Deductive Changes   322
        Contractor Aware of Design Defects   323
16.3 Contractor Claims   323
    16.3A Introduction   323
    16.3B Misrepresentation Through Defective Specifications: *United States v. Spearin*   324
        Warranty of Design Specifications   326
        Misrepresentation and Defect   326
        Injury   327
    16.3C Owner Nonpayment   327
16.4 Owner Claims Against Contractor   327
    16.4A Introduction   327
    16.4B Shared Responsibility for Construction Defects   328
        Stating the Problem   328
        Defect Traced to Owner and Contractor   328
        Shared Design Responsibilities   328
16.5 Contractor's Warranty (Guarantee) Clause   329
16.6 Contractor's Implied Warranty of Workmanship   331
16.7 Homeowner Claims   331
    16.7A Overview   331
    16.7B Implied Warranties in the Sale of Homes   331
    16.7C Statutory Protections of Homeowners   331
        Consumer Protection Acts   331
        Consumer Warranties   332
        New Home Warranty Acts   332
        Right to Repair Acts   332
16.8 Financial Problems   333
    Contractor Options   333
    Owner Options   333
16.9 Contractor Defenses to Claims   333
    16.9A General Contract Defenses   333
        Mutual Mistake   333
        Impossibility   333
        Commercial Impracticability   334
        Frustration   334
        Unconscionability   334
    16.9B Contractor Followed the Design   334

16.9C Subrogation: Destruction of Project Under Construction, Covered by Owner's Insurance ... 334
16.9D Passage of Time: Statutes of Limitation and Repose ... 335
    Statutes of Limitation and Repose ... 335
    Commencement of Statute of Limitations ... 336
    Contractual Limitations ... 336
    Laches ... 336
Review Questions ... 337
Endnotes ... 337

CHAPTER 17

CONTRACTOR PAYMENT AND PROJECT COMPLETION ... 341

Scenario ... 341
17.1 The Legal Framework: Common Law Payment Rules ... 342
17.2 Progress Payments ... 342
    Rationale ... 342
    Mechanism ... 342
17.3 Retainage ... 343
17.4 Payment of Subcontractors and Suppliers ... 344
17.5 Design Professional's Certification of Payment Applications ... 345
    Observation of Work ... 345
    Design Professional Liability ... 346
    Defenses to Design Professional Liability ... 346
17.6 Construction Lender's Interest in the Payment Process ... 347
    Loan to Contractor ... 347
    Loan to Owner ... 347
    Subcontractor Claims ... 348
17.7 Surety Requests to Public Owner that Payment Be Withheld ... 348
17.8 Late Payment and Nonpayment During Performance ... 349
    Prompt Payment Acts ... 349
    Common Law Rights ... 349
    AIA Documents ... 350
17.9 The Completion Process and Payment ... 350
    Substantial Completion ... 350
    Final Completion ... 351
    Effect on Future Claims ... 352
17.10 Substantial Performance Doctrine ... 352
17.11 Work Not Substantially Complete ... 353
Review Questions ... 354
Endnotes ... 354

CHAPTER 18

CHANGES: COMPLEX CONSTRUCTION CENTERPIECE ... 357

Scenario ... 357
18.1 Changes: Differing Perspectives for Owner and Contractor ... 358
    Owner's Perspective ... 358
    Contractor's Perspective ... 358
18.2 The Common Law ... 359
18.3 The Changes Clause: Introduction ... 360
    Terminology ... 360
    AIA and EJCDC Documents ... 360
    Shifts in Bargaining Power ... 360
    An Illustrative Case ... 361
18.4 Change Order Mechanism ... 364
    18.4A Components of the Changes Process ... 364
    18.4B Limitation on Power to Order Changes ... 364
        Work ... 364
        Time ... 365
        Public Contracts ... 365
    18.4C Authority to Order Change ... 365
    18.4D Misrepresentation of Authority ... 366
    18.4E Pricing Changed Work ... 366
    18.4F Deductive Change (Deletion) ... 367
    18.4G Waiver: Excusing Formal Requirements ... 367
18.5 Contractor's "Changes" Claims ... 368
    18.5A Cardinal Change ... 369
    18.5B Constructive Change ... 369
18.6 Effect of Changes on Performance Bonds ... 370
Review Questions ... 370
Endnotes ... 370

CHAPTER 19

SUBSURFACE PROBLEMS: PREDICTABLE UNCERTAINTY ... 373

Scenario ... 373
19.1 Discovery of Unforeseen Conditions ... 374
    Effect on Performance ... 374
    Role of Geotechnical Engineer ... 374
    Risk Allocation: DB Versus DBB ... 374
19.2 Common Law Rule ... 375
19.3 Information Furnished by Owner ... 375
19.4 Risk Allocation Plans: Benefits and Drawbacks ... 376
19.5 Disclaimers—Putting Risk on Contractor ... 378

19.6 Contractual Protection to Contractor: The Federal Approach	379		20.14 Owner Acceleration of Contractor Performance	404
Type I DSC	380		20.14A The Changes Clause	404
Type II DSC	381		20.14B Constructive Acceleration	404
Notice Requirement	382		20.14C Voluntary Acceleration: Early Completion	405
19.7 AIA Approach: Concealed Conditions	382		20.15 Bonus/Penalty Clauses: An Owner Carrot	405
19.8 EJCDC Approach	384		*Review Questions*	406
19.9 The FIDIC Approach	384		*Endnotes*	407
Review Questions	385			
Endnotes	386			

CHAPTER 20

TIME: PROJECT SCHEDULING AND DELAY CLAIMS 389

CHAPTER 21

THE SUBCONTRACTING PROCESS: AN "ACHILLES HEEL" 411

Scenario	389
20.1 The Law's View of Time: Overview	390
20.2 Commencement	390
20.3 Completion	391
20.4 Categorizing Causes of and Remedies for Delay	391
Causes of Delay	391
Remedies for Delay	392
20.5 Common Law Allocation of Delay Risks	392
20.6 Contract Allocation of Fault	392
Force Majeure Clause	392
Weather	393
"Time Is of the Essence" Clause	393
20.7 Measuring the Impact of Delay: Project Schedules	394
20.8 AIA and EJCDC Approaches to Scheduling	395
20.9 The Critical Path Method (CPM) and Float	395
Description of CPM	395
Critical Path and Float	396
CPM Use in Delay Claims	398
20.10 Causation: Concurrent Causes	398
20.11 Time Extensions	399
Role of Design Professional	399
Duration of Extension	399
Notices	399
20.12 Contractor-Caused Delay: Owner Remedies	400
Actual Damages	400
Liquidated Damages Clauses	400
20.13 Owner-Caused Delay	401
20.13A Sources of Owner Delays and Some Contract Defenses	401
20.13B No-Damages-for-Delay Clauses	402
20.13C Subcontractor Claims	404
20.13D Records	404

Scenario	411
21.1 Subcontracting: An Introduction	412
21.2 The Subcontract: Source of Rights and Duties	412
Basic Structure	412
Flow-Through or Conduit Clauses	412
21.3 The Subcontractor Bidding Process	413
21.3A Statement of the Problem	413
21.3B Irrevocable Sub-Bids: Promissory Estoppel	414
21.3C Bargaining Situation: Bid Shopping and Peddling	414
21.3D Avoiding *Drennan*	416
21.4 Subcontractor Selection and Approval: The Private Owner's Perspective	416
21.5 Subcontractor Payment Claims Against Prime Contractor: "Pay When Paid" Clause	417
21.6 Subcontractor Payment Claims Against Property, Funds, or Entities Other Than Prime Contractor	418
21.7 Mechanics' Liens	418
Legal Complexity	418
Overview	419
Claimants and Lienable Work	419
Lien Priority	420
Claimants' Entitlement to Compensation	421
No-Lien Contracts	421
Criticism	421
21.8 Payment Bonds	422
21.9 Stop Notices	422
21.10 Trust Fund Legislation: Criminal and Civil Penalties	422
21.11 Nonstatutory Claims Against Third Parties	423
Owners	423
Design Professionals	423

21.12 Joint Checks	423
21.13 Performance-Related Claims Against Prime Contractor	423
21.14 Pass-Through Claims Against Owner: Liquidating Agreements	424
21.15 Owner Claims Against Subcontractors	425
21.16 Public Contracts	425
Review Questions	426
Endnotes	426

CHAPTER 22

Terminating a Construction Contract: Sometimes Necessary but Always Costly — 431

Scenario	431
22.1 Termination: A Drastic Step	432
22.2 Termination by Agreement of the Parties	433
22.3 Contractual Power to Terminate: Introduction	433
22.4 Default Termination	433
Termination by Owner	433
Termination by Contractor	434
Wrongful Termination for Default	435
22.5 Termination or Suspension for Convenience	435
Owner Suspension	435
Contractor Suspension	435
Termination for Convenience	436
22.6 Role of Design Professional	436
22.7 Waiver of Termination and Reinstatement of Completion Date	437
22.8 Notice of Termination	437
22.9 Termination Under Common Law	438
Material Breach	438
Anticipatory Repudiation	439
22.10 Keeping Subcontractors After Termination	439
Review Questions	439
Endnotes	440

Part E: Risk Management and Dispute Resolution — 443

CHAPTER 23

Apportioning or Shifting Losses: Contribution and Indemnity — 445

Scenario	445
23.1 Loss Shifting, Responsibility Apportionment, and Risk Management: An Overview	446
23.2 First Instance and Ultimate Responsibility Compared	446
23.3 Responsibility Apportionment Among Multiple Wrongdoers: Stating the Problem	447
Introduction	447
Direct Versus Third-Party Action	448
23.4 Loss Shifting and Liability Apportionment by Operation of Law: Three Devices	449
Noncontractual Indemnity	449
Contribution	450
Comparative Negligence	451
New Regime of Liability Apportionment	451
23.5 Contractual Indemnity Compared to Exculpation, Liability Limitation, and Liquidated Damages	451
23.6 Parsing Indemnity Clauses	453
Terminology	453
Components	453
23.7 Functions of Indemnity Clauses	454
23.8 Statutory Regulation	455
23.9 Common Law Regulation: Specificity Requirements	456
23.10 Indemnitor Required to Procure Insurance	457
Review Questions	457
Endnotes	458

CHAPTER 24

Insurance — 461

Scenario	461
24.1 Insurance: Risk Spreading	461
24.2 Construction Insurance: An Overview	462
24.3 Introduction to Insurance Industry and Policy	463
Standardized Insurance Policies	463
Regulation	463
Premiums	464
Deductible Policies	464
Policy Limits	464
Notice of Claim: Cooperation	464
Duty to Defend	465
Settlement	465
Multiparty Policies	465
24.4 Property Insurance	466
24.4A Introduction	466
24.4B Coverage for Project Destruction During Contract Performance	466
24.5 CGL Insurance and Defective Construction Claims	466

24.6 Professional Liability Insurance — 467
    24.6A Requirements of Professional Liability Insurance — 467
    24.6B Policy Types: Occurrence or Claims-Made — 468
    24.6C Coverage and Exclusions: Professional Services — 468
    24.6D Preparing to Face Claims — 470
24.7 Insurance and Alternative Project Delivery Methods — 471
Review Questions — 472
Endnotes — 472

CHAPTER 25

SURETY BONDS: BACKSTOPPING CONTRACTORS — 475

Scenario — 475
25.1 Introduction — 476
    Overview — 476
    Terminology — 476
    Mechanics of Suretyship — 476
25.2 Need for Bonds in Construction Industry — 477
25.3 Function of Surety: Insurer Compared — 477
25.4 Ancillary Bonds — 478
    Bid Bonds — 478
    License Bonds — 478
    Lien Release Bonds — 478
    Subdivision Bonds — 478
25.5 Performance Bonds: Surety's Promise to Owner — 478
25.6 Triggering the Performance Bond Obligation — 479
    Conference — 479
    Declaration of Default — 479
    Tender of Contract Balance — 480
25.7 Performance Bond Surety's Options — 480
25.8 Surety's Defenses — 482
25.9 Performance Bond Surety's Liabilities — 482
25.10 Payment Bonds: Functions — 483
25.11 Who Can Sue on the Payment Bond? — 484
25.12 Payment Bond Liability — 484
25.13 Asserting Claims: Time Requirements — 485
25.14 Reimbursement of Surety — 485
25.15 Regulation: Bad Faith Claims — 486
25.16 Bankruptcy of Contractor — 487
25.17 International Contracts — 487
Review Questions — 487
Endnotes — 488

CHAPTER 26

CLAIMS AND DISPUTES: EMPHASIS ON ARBITRATION — 491

Scenario — 491
26.1 Claims Resolution: Two- or Three-Step Process — 492
    Methods of Dispute Resolution — 492
    Initial Administrative Review — 492
    EJCDC — 493
    AIA — 493
    No Initial Review — 493
26.2 Reasons for Initial Design Professional Decision — 493
26.3 Procedural Matters Concerning the Initial Decision — 494
    Requirements of Elemental Fairness — 494
    Standard of Interpretation — 495
    Form of Decision — 496
    Costs — 496
26.4 Finality of Initial Decision — 497
    Range of Finality — 497
    Subject Matter of Dispute — 498
    Review Process — 498
    Other Considerations — 498
26.5 The Initial Decision Maker: Some Observations — 498
26.6 Statutory Framework of Arbitration — 499
    FAA and UAA/RUAA — 499
    Enforcement and Limited Judicial Review — 499
26.7 Abuse of Arbitration, State Regulation, and Federal Preemption — 500
    Introduction — 500
    State Judicial Regulation — 500
    State Statutes — 501
    Federal Preemption — 501
26.8 Common Law Contract-Based Defenses to Arbitration — 501
    Unconscionability — 502
    Other Defenses: Mutuality, Termination of the Contract, and Conditions Precedent — 503
    Waiver of Arbitration — 503
26.9 The Arbitration Process: Introduction — 503
26.10 Prehearing Activities: Discovery — 504
26.11 Selecting Arbitrators and Arbitrator Neutrality — 504
26.12 Multiple-Party Arbitrations: Joinder and Consolidation — 505
26.13 Award — 506
26.14 Other Dispute Resolution Mechanisms — 506
26.15 Adjuncts of Judicial System — 507
26.16 Public Contracts — 507

26.16A Federal Procurement Contracts	508
26.16B State and Local Contracts	508

26.17 International Arbitration — 508
Review Questions — 509
Endnotes — 509

APPENDICES

APPENDIX A: Standard Form of Agreement Between Owner and Contractor (AIA Document A101-2007) — A-1

APPENDIX B: General Conditions of the Contract for Construction (AIA Document A201-2007) — B-1

APPENDIX C: Standard General Conditions of the Construction Contract (EJCDC Document C-700 (2007)) — C-1

APPENDIX D: Standard Form of Agreement Between Owner and Architect (AIA Document B101-2007) — D-1

APPENDIX E: 2012 Code of Ethics & Professional Conduct (AIA) — E-1

APPENDIX F: Code of Ethics for Engineers (NSPE (2007)) — F-1

APPENDIX G: DBIA Code of Professional Conduct (2008) — G-1

APPENDICES H TO P ON WEBSITE

APPENDIX H: Standard Form of Agreement Between Contractor and Subcontractor (AIA Document A401-2007) — H-1

APPENDIX I: Performance and Payment Bonds (AIA Document A312-2010) — I-1

APPENDIX J: Standard Form of Agreement Between Owner and Engineer for Professional Services (EJCDC Document E-500 (2008)) — J-1

APPENDIX K: Suggested Form of Agreement Between Owner and Contractor for Construction Contract (EJCDC Document C-520 (2007)) — K-1

APPENDIX L: Standard Form of Agreement Between Owner and Construction Manager—Construction Manager as Owner's Agent (CMAA Document A-1 (2013)) — L-1

APPENDIX M: Standard Form of Agreement Between Owner and Construction Manager (Construction Manager At-Risk) (CMAA Document CMAR-1 (2013)) — M-1

APPENDIX N: General Conditions of the Contract for Construction, Construction Manager as Advisor (AIA Document A232-2009) — N-1

APPENDIX O: Standard Form of Agreement Between Owner and Construction Manager as Constructor where the basis of payment is the Cost of the Work Plus a Fee with a Guaranteed Maximum Price (AIA Document A133-2009) — O-1

APPENDIX P: Frank Lloyd Wright and the Johnson Building: A Case Study — P-1

GLOSSARY — GL-1

REPRODUCED CASE INDEX — CI-1

SUBJECT INDEX — SI-1

Preface

Construction Law for Design Professionals, Construction Managers, and Contractors (CLDPCM&C) is a textbook aimed primarily at architecture, professional engineering and construction management students, whether undergraduates or in graduate school. The material is adapted from *Legal Aspects of Architecture, Engineering and the Construction Process*—an icon of construction law teaching since 1970.

There is much overlap between CLDPCM&C and *Legal Aspects of Architecture, Engineering and the Construction Process* (Ninth Edition 2013). Both books are co-authored by Justin Sweet and Marc M. Schneier, with Blake Wentz contributing a new chapter on ethics for CLDPCM&C. Both books share the same educational philosophy for introducing the student to construction law. That philosophy is concisely stated in the opening two paragraphs to the Preface to *Legal Aspects of Architecture, Engineering and the Construction Process* and is reprinted here:

> The primary focus of this edition, as in editions that preceded it, is to provide a bridge for students, mainly architectural and engineering students, but increasingly, those in business schools and law schools, between the academic world and the real world. We hope to provide a cushion for the inevitable shock such a transition generates. The world of the classroom, with its teachers and its books, is not the same as the world of construction with its developers, owners, design professionals, and public officials that regulate the construction process.
>
> This cushion requires that readers understand what is law, how it is created, how it affects almost every activity of human conduct, and how legal institutions operate. This cannot be accomplished through simply stating "the law." It requires clear, concise, jargon-free text that probes beneath the surface of legal rules to uncover why these rules developed as they did, outline arguments for and against these rules, and examine how they work in practice.

In short, the authors' intention is to instruct students as to the "why" of the law, not merely the "what."

These two books differ in complexity and length. *Legal Aspects* is a much longer book. It contains detailed explanation and analysis, and reproduces several court opinions. The sophistication of its analysis of the law is such that *Legal Aspects* has repeatedly been cited in judicial opinions and law treatises.

Construction Law for Design Professionals, Construction Managers, and Contractors is tailored to be a teaching tool specifically for students in architecture, engineering and construction management schools or departments. The content is designed to satisfy not only the topical requirements of the American Council for Construction Education (ACCE) in terms of construction law, but Chapter 13 also satisfies the ACCE's ethics requirement. As compared to *Legal Aspects*, the text has been shortened, so that the subject matter may be covered in a class length of 30 to 40 lecture hours. Unduly complex legal analysis has been removed so that the student may focus on the "big picture."

The material in this book is accessible to students in both undergraduate and graduate level courses for construction management, civil engineering, and architecture. The basic concepts and real-world examples used in the text will give undergraduate students a solid grasp of the legal issues they will face in the real world. At the same time, there is ample room for graduate students to explore and further research many of these topics, with footnotes pointing to other cases or secondary sources. This combined approach makes CLDPCM&C one of the most versatile legal books on the market for construction and engineering education.

One example illustrates the changes—in both depth of legal analysis and length of discussion—from *Legal Aspects* to CLDPCM&C. *Legal Aspects* contains an entire chapter on intellectual property: copyright, patents and trade secrets. After a general introduction to these concepts, the chapter explores the interplay of intellectual property and construction law, and includes a reproduced case. There is more information—and of far greater complexity—than non-law students necessarily need.

By contrast, CLDPCM&C extracts from this chapter the vital issues of importance to future architects and engineers. The student is explained the ideas underlying copyright—including the statutory right itself, registration, infringement and the "work for hire" doctrine—and concludes with a sub-section on "advice for the design professional." There is no reproduced case. This discussion is much shorter than in *Legal Aspects* and is limited to one section (Section 10.7) within a chapter.

Speaking more generally, the change in focus between *Legal Aspects* and CLDPCM&C manifests itself in four ways: subject matter, organization, writing style, and technology. In addition, a complimentary "Instructor Solutions Manual" is included with each order of the textbook for a class.

Subject Matter

- Ethics: CLDPCM&C includes a new Chapter 13 on ethics. This chapter reviews not only the codes of professional conduct of architects and professional engineers, but also the ethics of contractors, design/builders and construction managers. In addition to codes of conduct reproduced in the text, three new appendices contain the longer codes of ethics or professional conduct.

- Focus on construction management: Construction management is introduced earlier in the book (Chapter 7) together with two other major project participants—the owner and prime contractor. The chapter on licensing has a new section (Section 8.7) on licensing of construction managers. The chapter on public contracts has a new section (Section 15.8) addressing the unique issues of the use of construction management (and design/build) on such projects. Standard form documents applicable to projects using construction managers—specifically, CMAA A-1 (2013), CMAR-1 (2013), AIA A232-2009, and AIA A133-2009—are reproduced in the book's website appendices.

- Chapter questions: Each chapter ends with ten questions. These questions help the student focus on the subject matter of the chapter, to ensure he or she understood the chapter's content.

- Glossary: A glossary at the end of the book allows students to quickly determine the definition of legal terms. Terms that are defined in the glossary are in bold type when they first appear in the text.

- Table of Abbreviations: Located on the inside of the front and back covers, this table lists the abbreviations used in the book. Any student puzzled by a reference to "N.E.2d," "Cal. Civ. Code," "J ACCL," or the like, can simply consult the table.

Organization

- All twenty-six chapters are grouped into five Parts. This format gives the students the "big picture" of how the material is organized, so they can better understand how the different chapters related to each other. The five Parts are:
    * Part A: Legal Framework of the Design and Construction Process
    * Part B: The Main Actors: The Owner, Constructor, and Design Professional

- Part C: Project Delivery Methods
  - Part D: Performance Disputes
  - Part E: Risk Management and Dispute Resolution
- Chapter scenarios: Almost all chapters begin with a *scenario*: a hypothetical fact pattern of a construction dispute. Throughout the chapter, reference is made back to the scenario so as to provide concrete examples of the legal concepts being discussed. This approach is intended to facilitate the student's understanding of these concepts.
- Chapter organization: CLDPCM&C simplifies the presentation of the material. In contrast to *Legal Aspects*—with its breakdown of the material into sections, subsections and sometimes sub-subsections—CLDPCM&C provides a more straightforward organization of the material.

Writing Style

- Shortened presentation: Even with the addition of a new chapter on ethics, CLDPCM&C has one chapter fewer than *Legal Aspects*.
- Reproduced judicial opinions: CLDPCM&C has far fewer reproduced judicial opinions than does *Legal Aspects*. More commonly, the facts and legal issues are concisely summarized, and just a paragraph or two of the court's opinion is quoted. This change means that the student is less required (than in *Legal Aspects*) to grapple with a legal opinion normally read only by attorneys or law students.
- Gender neutrality: We avoid the awkward "he or she" format. In many instances, the actor is a corporation or government entity and the neutral pronoun ("it") is used. This is particularly so for contractors, who are invariably corporations. Otherwise, the gender of the design professional often alternates with the chapters.

Technological Change: MindTap Online Course and Reader

In addition to the print version, this textbook will also be available online through MindTap, a personalized learning program. Students who purchase the MindTap version will have access to the book's MindTap Reader and will be able to complete homework and assessment material online, through their desktop, laptop, or iPad. If your class is using a Learning Management System (such as Blackboard, Moodle, or Angel) for tracking course content, assignments, and grading, you can seamlessly access the MindTap suite of content and assessments for this course.

In MindTap, instructors can:

- Personalize the Learning Path to match the course syllabus by rearranging content, hiding sections, or appending original material to the textbook content
- Connect a Learning Management System portal to the online course and Reader
- Customize online assessments and assignments
- Track student progress and comprehension with the Progress app
- Promote student engagement through interactivity and exercises

Additionally, students can listen to the text through ReadSpeaker, take notes and highlight content for easy reference, and check their understanding of the material.

Dedicated Website Support

This book has a dedicated website containing Appendices H through P. Shifting these appendices to the website means that this book is slimmer, and hence easier for the student to use and carry. To access this website, go to www.cengagebrain.com. At the home page, type in the ISBN of this book (found on the book's back cover) in the search box at the top of the page. This will take you to the product page where these resources can be found.

Instructor Solutions Manual

Each order of *Construction Law for Design Professionals, Construction Managers, and Contractors* for a classroom course includes a complimentary "Instructor Solutions Manual." The ISM fulfills two functions. First, it provides the answers to the ten questions found at the end of each chapter. Second, it includes teaching recommendations for each chapter: the focus and purpose of the chapter; the key ideas to be taught; secondary issues which may be skipped due to time constraints; and so on. Of course, these are only suggestions, for each professor to choose or ignore based on individual preference and circumstances.

Legal Citations Format

For non-law students, a brief explanation of legal citations is provided. Although legal citations seem complicated, they are, in reality, quite straightforward. A simple citation would be *Hollerbach v. United States,* 233 U.S. 165 (1911). First, the name of the case is given, usually with the plaintiff (the person starting the lawsuit) listed first, followed by the defendant. Here, an individual (Hollerbach) is suing the federal government.

"U.S." is an abbreviation of the reporter system in which the case is published, in this case the United States Supreme Court Reports. The number preceding the abbreviation of the reporter system (233) is the volume in which the case is located. The number following the abbreviation of the reporter system (165) is the page on which the judicial opinion begins. The citation ends with the year that the court announced the decision (1911).

Subsequent reference to this judicial opinion is usually by party name; in this example, the *Hollerbach* decision.

As explained in Chapter 1, the United States has two, parallel court systems: federal and state. The federal courts of appeal are printed in the Federal Reporter (now in its third series). The abbreviation of the Federal Reporter is "F." and based on the series will be F., F.2d or F.3d. The courts of appeal are called circuit courts. There are eleven federal circuits. Thus, the court will be identified as: 1st Cir., 2d Cir. and so on. As an example, *Moorehead Construction Co. v. City of Grand Forks,* 508 F.2d 1008 (8th Cir.1975), was issued by the Eighth Circuit Court of Appeals in 1975. The federal trial courts (called district courts) are published in the Federal Supplement, abbreviated as either F.Supp. or F.Supp.2d.

To understand state court citations, it is important to understand that one publishing company (West) created the West Regional Reporter System. As implied by the name, the West reporters are divided into different regions of the country: the Northeast Reporter (abbreviated N.E.); the Atlantic Reporter (abbreviated A.) and so on. State court opinions are published in the West Regional Reporter System, and sometimes may also have an official cite. As an example, *Anco Construction Co., Ltd. v. City of Wichita,* 233 Kan. 132, 660 P.2d 560 (1983), indicates that the case came from the Kansas Supreme Court and is also collected in the Pacific Reporter (P.2d).

The hierarchical nature of the court system means that parties receiving a decision by a lower-level court may try to appeal the decision. The reviewing court has several options. The two most basic options are to affirm or reverse the lower-court ruling. This is indicated after the citation to the lower-court opinion with the abbreviations "aff'd" (affirmed) or "rev'd" (reversed). If the higher court refuses to hear the appeal, this may be indicated by "review denied," "appeal denied" or "cert. denied" where "cert." is short for the term "certiorari."

An example is *McDowell-Purcell, Inc. v. Manhattan Construction Co.*, 383 F.Supp. 802 (N.D.Ala.1974), aff'd, 515 F.2d 1181 (5th Cir.1975), cert. denied, 424 U.S. 915 (1976). In this case, a corporate subcontractor (McDowell-Purcell) sued a corporate prime contractor (Manhattan). The U.S. District Court for the Northern District of Alabama issued a judgment in 1974. That judgment was affirmed on appeal by the U.S. Court of Appeals for the Fifth Circuit in 1975. The following year, the U.S. Supreme Court refused to hear the case by denying certiorari. (Had all these events happened in the same year, that year would have been placed only together with the last event.)

Appreciations

Deeply felt thanks are given to several professors who reviewed individual chapters and made suggestions for improvement: Nestor Bustamante, Florida International University; Neil Eldin, University of Houston; Stephen G. Hauser, University of Minnesota; Howard F. Haugh, Minnesota State University, Mankato; Amr Kandil, Purdue University; Ibrahim (Brian) Oenga, University of Wisconsin; Gregory F. Starzyk, California Polytechnic State University at San Luis Obispo; Peter K. Sweeney, Manhattan College; Athan Tramountanas, University of Washington; and Cassiana Aaronson Wright, Stanford University. These reviewers contributed ideas, suggestions, and new perspectives. Although not a reviewer, the contribution and support of Frank Mahuta, Milwaukee School of Engineering, was greatly appreciated.

A book of this size could not come to fruition without the tireless efforts of many parties. Tim Anderson provided assurance and guidance at the highest level of the publisher. The publisher's staff exhibited generous patience and supplied on-going, professional support; in particular, Hilda Gowans, Rose Kernan, Tanya Altieri, Shelly Gerger-Knechtl, and Harlan James. Also deserving thanks are Kelly Lowery and Jeannine Lawless as outside consultants. Working with all of these people was a distinct pleasure.

JUSTIN SWEET
MARC M. SCHNEIER
BLAKE WENTZ

About the Authors

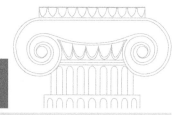

Justin Sweet

Justin Sweet received his BA (Literature) and LLB degrees from the University of Wisconsin in 1951 and 1953 respectively. He was elected to Phi Beta Kappa and the Order of the Coif (honorary law). He served as executive editor of the Wisconsin Law Review. He was admitted to the bar in Wisconsin and California.

From 1953 to 1954 he served in the Office of the Wisconsin Attorney General. From 1954 to 1957 he was an officer in the Army Judge-Advocate Corps. He engaged in private practice in Milwaukee from 1957–1958. In 1958 he was invited to join the Faculty of Law at the University of California at Berkeley. He received tenure in 1963. He taught Contracts, Torts, Insurance, Marital Property, Restitution, Remedies, Family Law and Contract Drafting. He was appointed John H. Boalt Professor of Law. He took early retirement in 1994.

After retirement he was appointed Professor of Law at the Academic College of Netanya in Israel from 1995–2006 after which he received the title of Professor Emeritus.

He taught at law schools in Italy, Belgium, Norway, Switzerland, Singapore, and Hong Kong.

He has written articles in many legal journals, many of which were collected in Sweet, *Anthology of Construction Law Writings* by the American Bar Association in 2010. He also wrote *Sweet on Construction Law* for the American Bar Association in 1997.

Beginning in 2010, he has been writing on historic cases in each issue of the Journal of Legal Affairs & Dispute Resolution in Engineering and Construction.

He was the co-founder and director of the International Construction Conference. It met in Fribourg, Switzerland, Berkeley, California and Washington D.C.

He has acted as an arbitrator and as an expert witness.

He authored the first six editions of *Legal Aspects of Architecture, Engineering and the Construction Process* and has been co-author with Marc Schneier for the Seventh, Eighth and Ninth Editions.

Marc M. Schneier

Marc M. Schneier earned his B.A. in Philosophy at the University of California, Berkeley in 1978. He earned his J.D. from the University of California, Davis (King Hall) in 1981. Marc published in the U.C. Davis Law Review and was awarded the Bureau of National Affairs Award for academic performance. He practiced law in San Francisco prior to being hired as Attorney Editor of *Construction Litigation Reporter*.

For over thirty years, Marc has been the Editor of CLR, a national reporter analyzing legal developments in the construction industry. The reporter is published by Thomson Reuters/West in print and reproduced on the Westlaw database CONLITR. In addition to writing CLR, Marc was an Adjunct Professor of Construction Law at the University of San Francisco School of Law, has published numerous articles in various legal journals, and provides consulting services.

Marc's first book was *Construction Accident Law: A Comprehensive Guide to Legal Liability and Insurance Claims* (American Bar Association 1999). He wrote *Understanding and Applying*

Construction Law as part of the "Essentials in Construction Management" series published by the Informa Center for Professional Development in 2006. Marc also authored Annotation, *Tort Liability of Project Architect Or Engineer For Economic Damages Suffered By Contractor Or Subcontractor*, 61 A.L.R.6th 445 (2011).

Marc co-authored with Justin Sweet three editions of *Legal Aspects of Architecture, Engineering and the Construction Process*—the Seventh Edition published by Thomson/Nelson in 2004, the Eight Edition published by Cengage Learning in 2009, and the Ninth Edition published by Cengage Learning in 2013.

Marc's website is buildinglaw.org.

Blake Wentz

Dr. Blake Wentz is an Associate Professor and the Director of the Construction Management Program at the Milwaukee School of Engineering, a small, private university located in Milwaukee, WI. He is a Certified Professional Constructor (CPC) as well as a LEED® Accredited Professional. His specialty areas in Construction Management are mechanical systems and construction law. Dr. Wentz has been retained and given opinion testimony as an expert witness in several construction litigation cases since 2005. Before entering academia he was the Vice-President of Wentz Plumbing and Heating in Lincoln, NE for over 5 years. Since 2009 he is the instructor for ethics in construction for the Association of General Contractors (AGC) of Wisconsin's Leadership Training Program. He is currently the Director of Region 3 of the Associated Schools of Construction (ASC) and sits on their national board of directors. He also is a trustee of the National Association of Women in Construction (NAWIC) National Education Foundation. Dr. Wentz has been the recipient of the Mechanical Contractors Association of America's (MCAA) Educator of the Year for both 2011 and 2012. He was also the recipient of the ASC National Teaching Award in 2010 and the Educator of the Year award in 2008 by Wisconsin Builder Magazine. Most recently he was named as the Associate of the Year in 2013 by the AGC of Greater Milwaukee.

PART

A Legal Framework of the Design and Construction Process

CHAPTER 1
American Legal System

CHAPTER 2
Forms of Business Association and Employment

CHAPTER 3
Contracts: From Formation to Breach

CHAPTER 4
Remedies for Contract Breach: Focus on Construction Disputes

CHAPTER 5
Torts: legal Relations Not Arising from Contracts

CHAPTER 6
Regulation of Land and the Construction Process

CHAPTER 1

AMERICAN LEGAL SYSTEM

SCENARIO

An owner wishes to have a new commercial facility built. The owner hires an engineering firm to design and oversee construction of the project. The engineering firm creates a design, which the owner uses to obtain bids from different contractors. One contractor is selected and the work begins.

The project does not go smoothly. The contractor complains the design is unclear, and it is costing more than anticipated to build the facility. The engineering firm, examining the work to authorize partial payment of the contractor, advises the owner to withhold full payment. Meanwhile, a trench collapse injures several workers. Eventually, the contractor abandons the work.

Litigation erupts. The owner, engineering firm, contractor, and subcontractor bring claims against each other, seeking money. Numerous subcontractors file mechanics' liens. (A mechanics' lien is a security interest in the improved property, meaning the owner loses its free-and-clear ownership of the property until the liens are dismissed or paid.) The injured workers also bring claims against the owner, engineering firm, and contractor, seeking to hold these persons responsible for their personal injuries.

How are these claims to be resolved? This chapter addresses the *framework* of the dispute-resolution mechanism in the United States: what is the law, which law applies, where and how are these legal disputes resolved, and so on.

The remainder of this book is *substantive* in nature, answering the question: Which claimant should prevail and why?

1.1 The Importance of Law

Law consists of coercive rules created and enforced by the state to regulate the citizens of the state and provide for the general welfare of the state and its citizens. Law is an integral part of modern society and plays a major role in the construction process. Because this text examines the intersection between law and the construction process, it is important to be aware of the various sources of law and the characteristics and functions of the law.

Many illustrations can be provided. Suppose a man who owns property wishes to build a house. Without assurance that stronger people will not use force to seize the materials with which he is building or throw him out of the house after it is built, it would take an adventurous or powerful person to invest time and materials to build the house. Similarly, workers would be reluctant to pound nails or pour concrete if they were fearful of being attacked by armed gangs. Here, *criminal law* protects both the property owner from those who might take away his property and the workers from those who might harm them.

Similarly, contractors would hesitate to invest their time or money to build houses if they did not believe they could use the civil courts to enforce their contracts and help them collect payment for their work (if owners did not pay them). Workers would be less inclined to work on a house if they were not confident they could use the civil courts to collect payment for work they had done or to compensate them if they were injured on the job. *Civil law* provides rules for resolving disputes that may arise between the participants in the construction process.

Some would be unwilling to engage in construction activity if they were not confident that an impartial forum would be available if disputes arose over performance. Were the state not to provide such a forum, participants might settle their disputes by force. The court system is a dispute-resolution mechanism available to the participants.

Society at large also has an interest in the quality of the built environment. Buildings must be safe to inhabit—not only on a daily basis—but against predictable extremes of nature, such as hurricanes or earthquakes. The building materials should not give off noxious or unhealthy fumes.

More broadly, society cannot function efficiently if there are no limits on the types of buildings that may be constructed in a particular location. The quality of life would be impacted if a large coal-driven power plant was placed in the middle of a residential community. For society to achieve these goals, it must have some control over (1) the competence of the persons performing construction work, (2) the design of the building, and (3) the type of building allowed in a particular area. Societal control over these variables is the subject of **regulation** of the construction industry.

All of these examples illustrate life in a civil society. This text focuses on civil society in the United States and its relation to the construction process. In short, civil society is the framework under which the construction process may exist and flourish.

1.2 The Federal System

Very large countries, such as the United States, Canada, Australia, and India, employ a federal system of government. Even smaller countries, particularly those with distinct religious, linguistic, ethnic, or national communities (for example, Switzerland), may choose to live under a federal system.

In a federal system, power is shared; the exact division of power between the central government and constituent members varies. For example, Canada has a looser federal system than the United States, with Canadian provinces having more autonomy than U.S. states.

Regardless of the precise relationship between the federal government and the individual states or provinces, the federal system recognizes the need for a central government to deal with certain issues on behalf of all citizens. For example, the U.S. federal government regulates currency, foreign relations, and defense—to mention a few of the areas it controls exclusively. Therefore, the states are not allowed to have their own currency, foreign policy, or military forces.

Other functions, such as the enactment of tax laws, can be shared by the federal government and state governments. Similarly, federal and state laws deal with crime, labor relations, and—to use an illustration more germane to construction—worksite safety. To avoid duplication of enforcement efforts and to relieve construction contractors from inconsistent regulations, the federal government, though dominant, can delegate workplace safety standards and their enforcement to the states so long as the states meet federal standards. Similar delegations are found in environmental protection laws.

Despite general federal supremacy over the states, the U.S. Constitution reserved some authority to the states. For example, except for contracts made by the federal government or those affected with a strong federal interest, state law determines which contracts will be enforced, the remedies granted for breach of contract, the conduct that gives rise to civil liability, and the laws that relate to the ownership of property—all core legal concepts in the construction process. Regulation of who may act as architect, engineer, and builder (through the licensing laws) is also a matter exclusive to the states. As a result, most law that regulates construction is determined by the state in which the project is located or in which the activities in question are performed.

This can and sometimes does lead to variations from state to state in legal rules that relate to construction. However, the dominance of standard contract forms that are created by national associations for nationwide use, the willingness of courts in one state to look at and often follow the decisions from another state, and the unification of areas of private law (such as the sale of goods) have all minimized the actual variation in state laws that relate to construction. Some laws vary greatly, such as mechanics' lien laws and licensing—both of which are regulated by state statutes.

1.3 Constitutions

A constitution is the fundamental source of law. No law may violate the constitution.

For more than two centuries, the U.S. Constitution has regulated the power among (1) the branches of the federal government, (2) the federal government and the states, and (3) all governments and their citizens. In brief, the federal government consists of three branches: the legislative (Congress), executive (the President and administrative agencies), and judicial (the courts). The states are sovereign but must function within the national constitutional framework, and state laws are "preempted" (or trumped) by conflicting federal laws. Finally, the U.S. Constitution protects citizens from abuse of power by the federal government and sometimes also by the states.

On the whole, state constitutions are longer than the federal Constitution and are changed more frequently. Although not identical, they share common characteristics with the U.S. Constitution, mainly the separation of powers among the legislative, executive, and judicial branches of government and the importance of protecting citizens from abuse of power by the states.

Constitutions establish the legal framework of the sovereign, but normally do not address day-to-day problems of society. These problems are addressed by the legislative branch by passing statutes and by the courts sometimes creating law when ruling on disputes.

1.4 Legislation

Legislation is the predominant source of law. A law passed by the legislative branch and signed by the executive (or passed over the executive's veto) becomes a **statute**. Statutes establish the law of the land.

Legislation is the most democratic lawmaking process. Ideally, it expresses the will of the majority of the citizens of the state. Yet legislatures are political instrumentalities. Persons and organizations seek to influence lawmakers. Although often denigrated as an area where power and money control, legislation often reflects popular attitudes. Lawmakers who vote in ways not popular with their constituents are likely to be removed from office. Also, legislation can be enacted quickly to respond to what the legislature believes to be important social and economic needs of its citizens.

The legislative process functions at different governmental levels, such as federal, state, county, and city legislative bodies.

Legislators do not have absolute freedom to enact legislation. The checks and balances so central to the American political process bar legislators from absolute power even though the legislatures do represent the political will of the citizens. Laws must be constitutionally enacted, and the constitutionality of a law is determined by the courts. In their role of interpreting legislation, the courts can indirectly control the legislature and the political desires of the majority. Similarly, local units—such as counties, cities, or special districts—are limited by the legislation that has created them.

Statutes, mainly at the state level, play an increasingly large role in the construction process, beginning with regulation of what type of building may be built and by whom. Local legislative bodies, exercising their power to protect the public and regulate land use, determine whether and which types of projects may be built. Housing and building codes control the quality of construction. Licensing and registration laws determine who may design and who may build.

Even though American law values freedom of contract, legislation increasingly regulates the content of construction contracts. Statutes preclude certain types of indemnification provisions in construction contracts. The payment process is also regulated with statutes addressing payment conditions, payment promptness, and retainage.

Statutes' may bestow certain rights on participants in the construction process. Mechanics' lien laws help subcontractors and suppliers obtain payment. Safety statutes and regulations protect workers and seek to ensure compensation in the event they are killed or injured on the job. Consumer protection and related laws protect homeowners from being swindled by an unethical contractor. More generally, statutes affect the litigation process by encouraging arbitration. In short, legislation is an integral part of the legal framework governing the building industry.

One body of statutes—the **Uniform Commercial Code** (UCC)—deserves special mention. The UCC was developed by the American Law Institute and the Commissioners on Uniform State Laws that are both private organizations devoted to unification of private law (so that the law would be the same in all states). Currently, the UCC (in whole or in part) has been "enacted"—adopted by state legislatures as statutes—in all 50 states and the District of Columbia. Article 2 of the UCC regulates the sale of materials and supplies used in a construction project. Although the UCC does not regulate construction itself (inasmuch as it does not govern services), it can be influential in a construction dispute.

1.5 The Executive Branch and Administrative Agencies

The separation of powers central to the U.S. political system is designed to prohibit any of the three branches—executive, legislative, or judicial—from wielding dominant power. Although laws come out of the legislature, the executive and judicial branches participate in lawmaking. The executive branches—the president at the federal level, the governor at the state level, and the mayor in some municipal systems—are elected by the citizens. Federal and state chief executives and sometimes local chief executives can veto legislation. To override a veto requires more than a simple majority of the legislature, usually two-thirds.

The chief executive can issue executive orders to those under his control. But more important, the chief executive, such as the president or the governor, has the power to appoint the heads of administrative agencies or administrative boards that play a significant role in construction. Examples of such regulatory instrumentalities are those that deal with workplace safety, registration of design professionals, and licensing of contractors, to name only some of the most important.

Indeed, one great change in the U.S. governmental structure since the Second World War has been the emergence of administrative regulatory agencies at every level of government, particularly at federal and state levels. The "constitution" for a regulatory agency is the legislation creating it. In that sense, the agency is a creation of the legislature. However, the chief executive usually appoints the key agency officials, often with the advice and consent of the legislature. Legislatures, particularly Congress, monitor agency performance by exercising an oversight function. This is accomplished through committee hearings during which complaints are heard and agency officials are asked to give explanations.

Agencies operate through issuance of *regulations* and through disciplinary actions. Regulations are an additional source of law. One of the most important and influential regulatory schemes is the Federal Acquisition Regulation (FAR), which governs the federal government's procurement of goods and services, including construction. Other important regulations are those governing worksite safety issued by the federal Occupational Safety and Health Administration (OSHA), and hazardous materials and environmental protection issued by the federal Environmental Protection Agency (EPA). At local levels, administrators play important roles in land use control and construction quality. Regulations are subject to judicial review.

1.6 Courts: Common Law and the Concept of Precedent

Courts play a variety of roles in the lawmaking process. First, the courts evaluate and determine the legality of the actions of the executive and legislative branches of the government. Exercising their power of judicial review, they determine whether legislation is constitutional and interpret the statutes. They have the principal responsibility for granting remedies under legislative systems. Courts also pass on and interpret administrative regulations and grant remedies for violations.

Second, court decisions, through the concept of **precedent**, establish judge-made laws, also known as the **common law**. A trial court decision is binding only upon the litigants who have appeared before the court. However, if one or both litigants appeal the trial court decision, a new ruling will be made by an appellate court (first the court of appeals, possibly followed by a ruling by the supreme court) on the correctness of the trial court. Decisions by the appellate courts become precedent, which establishes the law for all persons within that court's jurisdiction.

"Traditional" construction law was almost entirely the product of the common law. By courts ruling upon thousands of disputes between owners, contractors, subcontractors, design professionals, and other related project participants over the centuries, a common law of construction disputes developed. For example, a Pennsylvania Supreme Court ruling addressing architect liability determines the law of (is "binding precedent" in) Pennsylvania, but it also becomes a "persuasive authority" in those states in which the precise legal issue has not been litigated to that state's highest court. (Courts outside of Pennsylvania are free to adopt or reject the reasoning of the Pennsylvania high court.) While today legislatures are increasingly regulating different aspects of construction law, the common law still plays a key role in the continuing development of legal aspects of the construction process.

1.7 Courts: The Judicial System

Federalism and the Court System. Under America's system of federalism (Section 1.2), the U.S. judicial system includes two systems—federal and state—which to a degree exist side by side. Although each state has its own judicial system, the federal courts also operate in each state.

Jurisdiction is a court's authority to hear a particular dispute. State courts are courts of general jurisdiction, meaning they have authority to hear any dispute other than those the U.S. Constitution reserves exclusively for federal courts. By contrast, federal courts are courts of limited jurisdiction, meaning they have no jurisdiction over a case unless authority is explicitly granted either by the U.S. Constitution or by a federal statute.

When a particular claim may be heard in either state or federal court, there is *concurrent jurisdiction*. Generally, the court (whether federal or state) will apply the law of the state where the construction project is located unless the contracting parties have agreed that the law of a different state will be applied.

When a claim may only be heard in federal court, that court has *exclusive jurisdiction*. The principal areas of exclusive federal jurisdiction are:

1. admiralty
2. bankruptcy
3. patent and copyright
4. actions involving the United States
5. violations of federal criminal statutes.

Trial Courts. Courts are divided into the basic categories of trial and appellate courts.

Litigation—the dispute-resolution process that uses the court system—begins in the *trial court*. The federal trial court is called the district court. State trial courts go by various names: superior, district, or circuit court.

The legal claim or *lawsuit* is assigned to a *judge*. A judge is the presiding official in a courtroom. The judge can and often does question the witnesses. If no jury is involved, the judge decides both legal and factual matters.

The parties to the litigation may also ask for a *jury* trial. Although certain types of lawsuits (called suits in equity) are decided only by a judge, in most litigation the right to a jury is constitutionally guaranteed. Historically, juries have consisted of twelve laypersons selected from the community; however, some states have experimented with using smaller juries. In criminal cases, they decide guilt and sometimes also the sentence. In civil (non-criminal) cases (the vast majority of construction cases), they decide disputed factual matters (while the judge decides legal matters).

Once the trial is over, the jury makes a decision (called a **verdict**). The judge's ruling (including adoption of the verdict) is called a **judgment** or *decision*. A judgment is an order by the court stating that one of the parties is entitled to a specified amount of money or to another type of remedy.

Appellate Courts. A party who disagrees with the judgment may seek review before the *court of appeals*. The party who appeals the trial court's decision is called the **appellant**. The party seeking to uphold the trial court decision is called either the appellee or the **respondent**.

The court of appeals usually consists of a panel of three judges (also called justices). They read the parties' briefs, which are the written, legal arguments. The judges may also agree to hear oral arguments from the parties' lawyers, during which the justices may ask questions. The court of appeals reviews the trial court judgment for accuracy of both the factual basis of the decision and the lower court's legal reasoning. However, new evidence is not permitted in appeals.

The court of appeals issues a decision or **opinion** stating the reasons for the way the case was decided. Some decisions are unanimous. Sometimes the justices split between the majority opinion and the *dissent* (which disagrees with some aspect of the majority opinion). Sometimes one or more justices (but less than a majority) will agree with the majority opinion's conclusion, but for different reasons; these justices will issue a *concurrence*. The majority opinion establishes the law.

A party dissatisfied with the court of appeals decision may appeal to the supreme court. While the supreme court is mandated to take some appeals (such as death penalty cases), it has discretion as to whether it will hear the great majority of appeals. If the supreme court refuses to hear the appeal and no appeal is attempted to the U.S. Supreme Court, this is the end of the litigation. The court of appeals decision establishes the law for the parties to that dispute.

If the supreme court agrees to hear the appeal, the appellate process (written briefs, followed by oral arguments before the panel of justices) is repeated. Most state supreme courts consist of a panel of nine justices, while some have fewer. As with the court of appeals, a supreme court decision may be unanimous, include one or more dissents, or include one or more concurrences. Again, the majority opinion establishes the law.

A party dissatisfied with a state supreme court decision may attempt to appeal to the U.S. Supreme Court, but very few of those appeals are accepted. If no party attempts to appeal or that appeal is rejected, the state supreme court decision establishes the law and finally resolves the parties' dispute.

Public Contracts. Special rules may apply to contractors on public contracts. Contractors wishing to sue the public entity may be required to bring a lawsuit in special courts, often called a **court of claims**. The court of claims is the trial court for public works contractors.

Within the federal system, the U.S. Court of Federal Claims (formerly the U.S. Claims Court) hears claims against the U.S. government arising out of a contract. Appeals from this court (or from the **board of contract appeals**, not shown in Figure 1.1 below) are taken to the U.S. Court of Appeals for the Federal Circuit. Usually these cases relate to disputes between government contractors and government agencies that award government contracts.

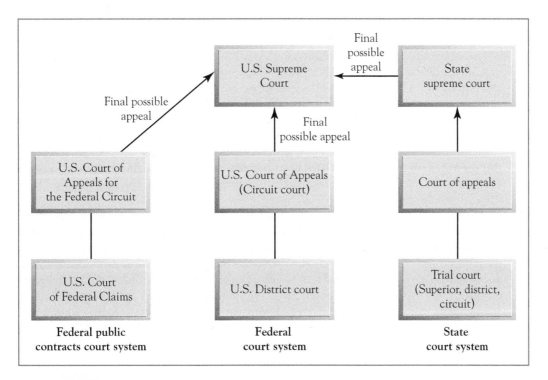

FIGURE 1.1 The court system.

1.8 The Litigation Process

Hiring an Attorney: Role and Compensation. Usually, a person with problems that may involve the law or the legal system consults an attorney (or lawyer). After an inquiry into the facts and a study of the law, the attorney advises a client of her legal rights and responsibilities. Although the attorney may also give an opinion on the desirability of instituting legal action or defending against any action that has been asserted against a client, the litigation choice is usually made by the client.

Most contracts for legal services use one of two forms of compensation. One method is for the attorney to charge an hourly fee. Hourly rates vary greatly based upon the attorney's skill, the demands on the attorney's time, the amount at stake, the complexity of the case, what the client can afford, the locality in which the attorney practices, and the outcome in the event of litigation.

Commonly in personal injury or wrongful death cases and occasionally in commercial disputes, attorneys use contingent fee contracts. The lawyers are not paid for legal services if they do not obtain a recovery for the client. Usually, a client agrees to reimburse the attorney for out-of-pocket costs, such as deposition expenses, filing fees, and witness fees. For taking the risk of collecting nothing for time spent, the attorney will receive a specified percentage of any recovery.

In the United States, each party normally is responsible for its own legal costs, including attorney fees. However, legal expense may be a major impediment to the bringing of a lawsuit. For this reason, some statutes specify that the successful party is entitled to an award of its attorney fees. For example, prompt payment statutes allow a subcontractor, who has not been timely paid, an award not only of the amount due but also of the subcontractor's legal costs in bringing the claim.

Statutes permitting award of attorney fees to the "prevailing party" may also have the purpose of discouraging frivolous litigation. If the plaintiff's claim is unsuccessful, in addition to incurring the cost of her own attorney (assuming a contingent fee arrangement is not made), the plaintiff would be required to pay the reasonable cost of the defendant's lawyer.

The Parties. Generally, the **party** (whether a person or a business entity) commencing the action (or lawsuit) is called the **plaintiff**, and the party against whom the action is commenced is called the **defendant**. Suppose an owner is unhappy with the quality of the contractor's work and refuses to make a payment. If the contractor sues the owner for breach of contract, the contractor is the plaintiff and the owner is the defendant.

In construction disputes, it is common for the defendant to sue the plaintiff (this is a *counterclaim*) or to make claims against third parties arising from the same transaction. For example, suppose an employee of a subcontractor sues the prime contractor based on a claim that the prime contractor has not lived up to the legal standard of conduct and this caused the employee to be injured on the job. The subcontractor's employee is the plaintiff and the prime contractor is the defendant.

The prime contractor, in addition to defending the claim made by the employee of the subcontractor, may sue the architect and the owner, claiming that the former was negligent and the latter was responsible for failure of the architect to live up to the legal standard, or may sue the subcontractor employer of the claimant based on indemnification. The prime contractor in asserting these *crossclaims* is called a cross-complainant in most state courts and a third-party plaintiff in the federal courts. Those against whom these claims are made would be called cross-complainants in most state courts and third-party defendants in the federal courts.

Pleadings. The lawsuit is usually begun by the plaintiff handing to the defendant a summons and **complaint**. The complaint contains allegations of the facts that gave rise to the claim. The complaint then lists the legal theories (such as breach of contract or negligence) under which the plaintiff contends the defendant is liable.

The defendant has a specified time to answer. The defendant's **answer** may assert that the plaintiff has not stated the facts correctly or that defenses exist even if the allegations of the plaintiff's complaint are true.

As noted previously, in addition to answering the plaintiff's complaint, the defendant may file a counterclaim against the plaintiff and a cross-claim against other parties.

Pretrial Activities: Discovery. The parties and their attorneys wish to learn all facts relevant to the lawsuit. **Discovery** is the legal process to uncover information in the hands of the other party. One method of discovery is written interrogatories. One party's attorney sends a series of written questions to the other party, who has a specified period of time in which to respond.

The other method is to take a deposition; that is, to compel the other party or its agent to appear at a certain time and place and answer questions asked by the attorney seeking discovery. The attorney is also likely to demand that the person being questioned bring all relevant documents relating to the matter in dispute. Discovery, although necessary for each party to learn the facts of the case, is often enormously expensive.

Pretrial Dismissal. Most cases never make it to trial. Most are settled by agreement of the parties. Sometimes, however, the trial judge will dismiss the case over the objection of one of the parties. For example, the judge may decide that, even if the facts as alleged in the complaint by the plaintiff are true, the plaintiff has no valid claim. In some states, this action is called a *demurrer*.

Suppose the facts are in dispute and whether the plaintiff has a viable claim depends upon what the facts are. In this situation, the parties may conduct discovery. The defendant is attempting to establish facts favorable to her position. Once discovery is over, the defendant may file a **motion** (a request) to the court to dismiss the lawsuit based upon the factual record as developed in discovery. This is called a motion for **summary judgment**. Based upon the facts as established, the plaintiff has no viable claim.

Yet another reason to have an action dismissed before trial is if the dispute must first be addressed by a non-judicial dispute resolution method, such as arbitration.

Dismissal of the case prior to a trial is grounds for an appeal.

Trial. The trial is usually conducted in public and is begun with an opening statement by the plaintiff's attorney. Sometimes the defendant's attorney also makes an opening statement. In the opening statement, the attorney usually states what she intends to prove and also seeks to convince the jury that the client's case is meritorious. Opening statements are less common in trials without a jury.

After opening statements, the plaintiff's attorney may call witnesses who give testimony. Also during this phase, physical exhibits and documents can be offered into evidence. In civil actions, the other party can be called as a witness.

Witnesses are supposed to testify only to matters they have perceived through their own senses. Witnesses usually cannot express opinions on technical questions unless they are qualified as experts. Claims against professionals (including architects and professional engineers) or involving technical issues (such as the nature and cause of a construction defects) often require the opinion of experts.

After the attorney has finished questioning the witness, the other party's attorney will cross-examine and try to bring out additional facts favorable to her client or to discredit the testimony given on direct examination. Cross-examination can be an effective tool to catch a perjurer or show that a witness is mistaken. However, used improperly, it can create sympathy for the witness or reinforce testimony of the witness.

Each side may introduce into evidence documents for the court or jury to consider. Documents often play a large role in construction disputes, particularly if a contractor's claim is that, because of some fault of the owner or the owner's architect, the project was more expensive or took longer to build than anticipated.

After the plaintiff's attorney has presented her client's case, the defendant's attorney will present the defendant's case. Then the plaintiff is given the opportunity to present rebuttal evidence. After all of the evidence has been presented, the judge submits most disputed matters to the jury (when one is used). The judge instructs the jury on the law, using instructions sometimes difficult for a juror to understand. The jury meets in private, discusses the case, takes ballots, and decides who prevails and how much should be awarded to the winning litigant.

The adversary system, although much criticized because of its expense and often needless consumption of time as well as the hostility it can engender, is central to the American judicial process. Each party, through its attorney, determines how its case is to be presented and can present its case vigorously and persuasively. Also, each party, mainly through cross-examination and oral argument, has considerable freedom to attack the other's case. The judge generally acts as an umpire to see that procedural rules are followed. The assumption underlying the adversary system is that, when the smoke has cleared, the truth will emerge.

The Cost of Litigation. Litigation is an expensive way to settle disputes. Even if the matter does not go to trial, the costs include attorneys' fees, witness fees, court costs, and stenographic expenses (at depositions). However, there are less obvious but no less significant expenses to the litigant. Much time must be spent preparing for and attending to the litigation, particularly if there is a trial. The litigant may have to disrupt business operations by searching through records. For these and other reasons, most lawsuits are settled out of court. The huge costs generated by the litigation process has led to the search for and the creation of the alternative dispute resolution, such as arbitration.

Enforcement of Judgments. If the plaintiff obtains a money award, the defendant should pay the amount of money specified in the judgment. However, if the defendant does not pay voluntarily, the plaintiff's attorney will deliver the judgment to a sheriff and ask that property of the defendant be seized and sold to pay the judgment. It is often difficult to find a defendant's property. In some states, defendants can be compelled to answer questions about their assets.

Even if assets can be found, exemption laws may mean that certain assets may not be taken by the plaintiff to satisfy the judgment. Statutes usually contain a long list of items of property that cannot be seized by the sheriff to satisfy a judgment. For example, homestead laws exempt the house in which the defendant lives from execution to satisfy judgments.

The chances of collection are significantly better if an insurance company or a surety backs up a losing party. This is why standard form contracts universally require either an insurance policy or a bond to cover predictable perils attendant to a construction project; however, not all risks can be insured against or bonded.

1.9 Non-Public Sources of Law

The discussion to this point has pointed to public sources of law: a constitution, statutes, executive orders, administrative agency decisions, and court decisions. Private parties or entities may also be sources of law, as examined here.

Contracting Parties. Contracting parties create their own private law governing their relationship. The terms of the contract define the conduct that must be met, and a failure to meet that standard (if not excused) results in legal liability. Of course, the autonomy granted the contracting parties to determine the terms of their exchange is broad, but not unlimited. Even though autonomy is seen as an inherent liberty in a free society, the state still determines who can contract, creates the formal requirements, and provides remedies when contracts are not performed.

Publishers of Standardized Documents. Contracts for design and construction services often are based largely on documents published by professional associations: the American Institute of Architects (AIA); the Engineers Joint Contract Documents Committee (EJCDC), a consortium of professional engineering associations; the Associated General Contractors of America (AGC) and associated trade contractors (as publishers of the ConsensusDOCS standard contracts); and the Construction Management Association of America (CMAA), to mention some of the most important. Some of the most important AIA and EJCDC standard documents are included as appendices to this book found either at the end of the book or on its website: www.cengage.com/engineering.

These associations have no official status, and some would contend they cannot be considered sources of law. Certainly, contracting parties need not use the standardized documents or, if they do use them, are free to make changes. But the frequency with which these documents are used largely unchanged justifies classifying these associations as sources of law for the contracting parties.

Restatements of the Law. Many of the common law rules, together with comments and examples, have been compiled and restated in a series of volumes called **Restatements of the Law**. They are published by the American Law Institute (ALI), which is a private organization made up of lawyers, judges, and legal scholars. The ALI's function is to collect case law from all of the states and distill it into rules. Because the common law is a dynamic and robust disputes resolution system, the legal rules generated change over time. As a result, the Restatements of the Law have been updated; for example, the Restatement (Second) of Contracts was published in 1981, while a Restatement (Third) of Torts is currently being issued.

Unless adopted by a legislature or followed by a court, Restatement rules are not law. The principal restatements for the purposes of this book are those of torts, contracts, and agency.

1.10 International Contracts

Increased worldwide competition exists for building and engineering contracts. As a result, those who design and construct may be engaged by a foreign national for work to be performed in a foreign country. These international contracts raise special problems.

Legal systems vary not only as to substantive law but also as to the independence of their judiciary and dispute resolution processes. In some countries, the judiciary is simply an arm of the state and follows (to a large degree) the will of the head of state or the agency with whom the contract has been made.

A foreigner may not have confidence in the impartiality of such a dispute resolution process. This problem can be particularly difficult when the designer or contractor is a company from an industrialized country doing business in a less developed country.

Lack of confidence in the judiciary system often necessitates a contract clause under which disputes will be resolved by some international arbitration process. Even where there is more confidence in the independence of the judiciary, such as in a transaction where disputes would normally be held before a court in the United States or a western European country, unfamiliarity with the processes and the possible application of unfamiliar law may lead to contractual provisions dealing with disputes. For example, European legal systems do not use the adversary system. They use a more inquisitorial system under which the judge is likely to be a professional civil servant who plays a more active role in resolving disputes. Similarly, laws in western European countries are often based on brief yet comprehensive civil codes quite different from those an American lawyer may encounter in domestic practice. Under such conditions, it is common for the parties to provide within the contract that disputes will be resolved by international arbitration with the applicable law of a neutral and respected legal system.

REVIEW QUESTIONS

1. Name at least three areas of the law that have been delegated to the states by the U.S. Constitution.

2. What are the three things that the U.S. Constitution regulates power among?

3. How does a law become a statute?

4. What is the purpose of the separation of powers in the U.S. political system?

5. What are the five principal areas of exclusive federal jurisdiction?

6. What is the difference between courts of general jurisdiction and those of limited jurisdiction?

7. What are the two primary forms of compensation for contracts for legal services?

8. List three examples of professional associations that publish standardized documents for construction services.

9. What are the Restatements of the Law?

10. Which legal process is used to uncover information in the hands of the other party?

CHAPTER 2

Forms of Business Association and Employment

Scenario

DD Development bought for $1 a condemned urban building, intending to convert it into apartment housing. Mr. D of DD Development telephoned ABC Engineering, Inc. and spoke to Ms. A. After brief negotiations, Mr. D and Ms. A signed a contract for ABC Engineering, Inc. to provide engineering services.

While Mr. E, an employee of ABC Engineering, Inc., was inspecting the building, the floor gave away and Mr. E fell and was injured. Mr. E sought compensation from (1) his employer (ABC Engineering, Inc.), his boss (Ms. A), (3) DD Development, and (4) Mr. D personally.

Ms. A now decides agreeing to the job was an unwise decision and calls Mr. D to cancel the contract. In response, DD Development and Mr. D sue ABC Engineering, Inc. and Ms. A for breach of contract.

Some of the legal issues this scenario uncovers include the following:

- While we know ABC Engineering, Inc. is a corporation (because of the "Inc." in its title), what kind of business entity is DD Development?
- Did either Mr. D or Ms. A have authority to sign the contract?
- What right to compensation does Mr. E have against ABC Engineering, Inc., his employer? What right to compensation does he have against DD Development and Mr. D personally?
- Can either Mr. D or Ms. A be personally liable for breach of contract?

The answers to these questions depend on many variables, including (1) the form of business association of DD Development and ABC Engineering, Inc. and (2) the employment status of Mr. D, Ms. A, and Mr. E. These matters—forms of business association and types of employment relationships—are the topics of this chapter. Understanding these concepts is of fundamental importance in the business world, including the construction industry.

2.1 Forms of Business Associations: Overview

While a construction project participant may consist of an individual, the participant commonly is a business association. This chapter reviews the basic forms of business associations that are most relevant to the construction industry. The three major forms of business associations are **sole proprietorships**, **partnerships**, and **corporations**. Although one might think that these three forms of an association are listed by size—from smallest to largest—that would be a misconception. A single person may form a corporation, and two individuals as well as two large corporations may form a partnership.

People engaged in design or construction should have a basic understanding of the ways in which individuals associate to accomplish particular objectives. The professional in private practice should know the basic elements of the forms of associations that are professionally available. Anyone who becomes an employee or executive of a large organization should understand the basic legal structure of that organization. The design professional who deals with a corporate contractor on a project should understand corporate organization. Those who are shareholders in or transact business with a corporation should understand the concept that insulates shareholders from almost all liabilities of the corporation.

The following are the most important organizational forms:

1. sole proprietorship
2. partnership
3. corporation
4. limited liability entity
5. joint venture
6. unincorporated association
7. loose association

2.2 Sole Proprietorships

A sole proprietorship is, colloquially, a one-person business. Although not a form of association, it is the logical place to begin the discussion. It is the simplest form of business and is used by many private practicing design professionals or contractors.

The creation and operation of the sole proprietorship is informal. By the nature of the business, the sole proprietor need not arrange with anyone else for operating the business. Generally, no state regulations apply except for those requiring registration of fictitious names or for having a license in certain businesses or professions. Sole proprietors need not maintain records on the business operation except those necessary for tax purposes. Sole proprietors have complete control over the business operation, taking the profits and absorbing the losses. They rent or buy space for operating the business, hire employees, and may buy or rent personal property used in the business. Capital must be raised by the proprietor or by obtaining someone to guarantee the indebtedness. (This differs from a corporation, which can issue shares as a method of raising capital.) The proprietorship generally continues until abandonment or death of the sole proprietor.

> In this chapter's scenario, if DD Development is a sole proprietorship, then Mr. D could be held personally liable to Employee E for his injury while inspecting the building.

2.3 Partnerships

Introduction. A partnership is an association of two or more persons to carry on a business for profit as co-owners. Unlike a corporation, it is not a legal entity. All states except for Louisiana have a Uniform Partnership Act, although there are variations between these different state statutes. A new form of partnership, limited liability partnership (LLP), is discussed in Section 2.5.

A partnership agreement should be in writing. Generally, the partners decide who is to exercise control. Most matters can be decided by majority vote. In a large partnership, a small group of partners may be given authority to decide certain matters without requiring that a majority of the partners approve such decisions.

In the absence of any agreement to the contrary, the partners share equally in the partnership's profits and losses.

Fiduciary Duties. Partnership is an important illustration of the **fiduciary relationship**, which is a relationship based on trust. As fiduciaries, each partner must be able to have confidence in the other. Neither must take advantage of this trust for selfish reasons or in any way betray the partnership. A partner must not harm the partnership because of any undisclosed conflict of interest. For example, where a partnership composed of design professionals represents the owner in dealing with a construction company, it would be improper for one of the partners to have a financial interest in the construction company unless it was disclosed in advance to the partners and to the owner. Generally speaking, any activity that raises a conflict of interest or competes in any way with the partnership is not proper.

Authority of Partner. Each partner is an agent of the partnership. (The concept of agency is discussed in Section 2.11.) As an agent of the partnership, the partner has the authority to make representations that are binding on the partnership.

> In this chapter's scenario, if DD Development is a partnership and Mr. D is a partner, then Mr. D as DD Development's agent would have the authority to commit DD Development to a binding contract with ABC Engineering, Inc. (unless the partnership agreement requires approval of all the partners or some other condition which was not met).

Authority may be actual or apparent. Clearly, a partner who has actual authority to make a representation binds the partnership.

"Apparent authority" means that a partner without actual authority appears to a third party to be authorized by the partnership to act. The doctrine of "apparent authority" protects the third party by charging the partnership with unauthorized acts by a partner. The partnership can create apparent authority by making it appear to third parties that a partner has authority he does not actually possess. The extreme situations are not controversial. Certain extraordinary acts, such as criminal conduct, are not charged to the partnership. On the other hand, acts that carry on ordinary partnership business bind the partnership. If the partner had no actual authority and the person dealing with the partner knew or was notified of this, the acts do not bind the partnership—in this case, the partner had neither actual nor apparent authority.

The distinction between actual and apparent authority is more fully discussed in Section 2.11D.

Liability of Partnership and Individual Partners. The liability of the partnership and the partners for debts of the partnership and individual debts of the partners is complicated and confusing. Clearly, creditors of the partnership can look to specific partnership property and, if this is insufficient, to the property of the individual partners. Creditors of individual partners who obtain a court judgment may satisfy the judgment out of the partner's interest in the partnership property, but the partnership can avoid losing the property by paying the judgment. In addition, because of the ease with which partnerships can be dissolved and reconstituted, considerable confusion exists as to the right of creditors of predecessor partnerships to hold successor partnerships liable for the obligations of the predecessor.

A partner who incurs liability or pays more than a proper share of a debt should receive contribution from the other partners. Similarly, a partner who incurs liability or pays a partnership debt should be paid back by the partnership. Usually these instances are expressly dealt with in the partnership agreement.

2.4 Corporations

Introduction. The corporate form is used by most large and medium-sized businesses in the United States and has become the vehicle by which many small businesses are conducted. In addition, practicing design professionals are increasingly choosing the corporate form, where possible, for their business organization. The corporate form takes on even greater significance in light of the increasing number of design professionals employed by corporations.

Although the partnership is merely an aggregate of individuals who join together for a specific purpose, the corporation is itself a legal entity. It exists as a distinct, legal person. It can take, hold, and convey property and sue or be sued in its corporate name. It may raise money through the sale of shares or the acquisition of debt. Other important corporation attributes are management centralized in the board of directors, free transferability of interests, and perpetual duration. In addition, the corporation offers the advantage of limiting shareholder liability for debts of the corporation to the obligation to pay for corporate shares purchased.

As noted, one advantage of the corporation is its perpetual life. Sole proprietorships end with the death of the sole proprietor. Often partnerships end with the death of any of the partners. The corporation will continue despite the death of shareholders. A court can dissolve a corporation in the event of a deadlock in the board of directors or under other circumstances.

Personal Liability of Directors and Officers: Piercing the Corporate Veil. A principal function of the corporation is to shield shareholders, directors, and officers from corporate obligations. For example, if a corporate construction company enters into a contract signed by the corporation's president on behalf of the company, the corporation may be liable for breach of that contract, but (absent extraordinary conditions) the president is not personally liable for breach of the contract. In this chapter's scenario, since ABC Engineering, Inc. is a corporation, Ms. A. would not be personally liable if the company is found to have breached its contract with DD Development.

This limitation of liability can operate unfairly where a third party relies on what appears to be a solvent corporation. The corporation thought to be solvent may be merely a shell. The assets of the corporation may not be sufficient to pay the corporate obligations. Someone injured by a corporation's acts or failure to act may find the corporation cannot pay for damages.

The injured person might then seek recovery from the corporation's shareholders and its "principals"—its directors and officers. The corporate form can be disregarded and, to use a picturesque phrase, the corporate veil pierced and shareholders or principals held liable. A court may do so if unjust or undesirable consequences would result by interposing the corporation as an entity between the injured party (or a creditor) and the shareholders or principals.

The wide variety of factual situations under which these claims for personal liability arise means that definitive legal rules are difficult to find. In *Greg Allen Construction Company, Inc. v. Estelle*,[1] the homeowners sued their corporate contractor, its president, and the sole owner for defective construction. The Indiana Supreme Court refused to find the corporation's president individually liable for his own negligent work on the house. The court pointed out that, because the president was working as the corporation's agent, he could be personally liable to the homeowners only if he had committed an independent tort. (Torts are civil wrongs;

they are the topic of Chapter 5.) Instead, the president was simply fulfilling the corporation's contract obligations, albeit in a negligent manner.

However, the Georgia Court of Appeals in *Christopher v. Sinyard* pierced the corporate veil and found the principals personally liable. The court noted that the owners of this two-person construction company did not follow corporate formalities; they intermingled their personal money with that of the corporation, and the corporation had no assets. Furthermore, the principals promised the homeowners that certain repairs would be done (while knowing that the company was "out of money"), and in reliance upon that promise, the owners permitted the closing of a loan. The court concluded that failure to pierce the corporate veil would result in an injustice to the owners. While the principals protested that failure to follow some formalities is not unusual in a "two-man business," the court responded that "if the individual who is the principal shareholder or owner of the corporation conducts his private and corporate business on an interchangeable or joint basis as if they were one, then he is without standing to complain when an injured party does the same. Under such circumstances, the court may disregard the corporate entity."[2]

Both *Greg Allen* and *Christopher* involved residential (not commercial) construction projects undertaken by closely held corporations owned by one or two persons. In both cases, the homeowners complained the work was either defective or not completed. Yet the courts came to opposite conclusions, finding the principal not personally liable in *Greg Allen*, but personally liable in *Christopher*. Does this mean that the two cases are inconsistent?

To answer that question, one must look beyond the bottom-line result to the reasoning and factual settings in which each case arose. While the *Christopher* court emphasized the principals' disregard of the corporate formalities, the under-funding of the corporation, and the misrepresentation made by the principals to the owners (so that the loan would close), none of these factors was mentioned in the *Greg Allen* case. These factual distinctions would lead one to conclude that the two courts are applying the same legal rules, tailored to the particular facts of each case.

2.5 Professional Corporations, Limited Liability Companies (LLC), and Limited Liability Partnerships (LLP)

Although design professionals traditionally have practiced as sole proprietors or partnerships, states have increasingly permitted them to practice through a professional corporation. The corporation form has been used mainly to take advantage of tax laws that allow employees of corporations to receive fringe benefits without having them included within taxable income.

Some states that have allowed professionals to incorporate have made clear that the professionals cannot use the corporate form as a shield to personal liability—a principal reason for using business corporations.[3]

Design professionals have begun to take advantage of two newer forms of business association: the **limited liability company** (LLC) and the **limited liability partnership** (LLP). The LLC is a hybrid of the corporate and partnership forms containing the pass-through income tax benefits of a partnership (in contrast to the tax treatment of corporations) with the limited liability protections of a corporation. The LLC is a legal entity, separate from its owners, who have no liability for the entity's debts. An LLP alters the rule of *joint and several liability* for general partners (under which each general partner is liable for the entire debt), replacing it with liability only of the LLP itself. One partner is not responsible or liable for another partner's misconduct or negligence.

State law varies in permitting design professionals to take advantage of these new business entities. Pennsylvania allows architects to form either LLCs or LLPs if certain requirements are met.[4] By contrast, California permits design professionals to form LLPs but not LLCs.[5]

2.6 Joint Ventures

Joint ventures can be created by two or more separate entities who associate, usually to engage in one specific project or transaction. Such arrangements are contractual and can be expressed by written contracts or implied by acts. In large construction projects, two contractors may find they cannot handle a particular project individually but can if they associate in a joint venture. Another type of joint venture can be created by an organization that performs design and an organization that constructs. Sometimes joint ventures are needed to bid on "design/build" projects (in which the same entity creates the design and builds the structure).

2.7 Unincorporated Associations

Individuals sometimes band together to accomplish a collective objective without using any of the forms of association described thus far, such as partnerships or corporations. Instead, they may organize an unincorporated association, such as a fraternal lodge, a social club, a labor union, or a church. Design professionals may perform professional services for these associations or may become involved in the associations themselves.

Generally, unincorporated associations are not legal entities; nonetheless, they are often permitted to contract, to hold property, and to sue or be sued in the name of the association. How is a contractor or design professional to know if a contract is binding on the association? Usually such groups have constitutions, bylaws, and other group-related rules that govern the rights and duties of the members. They elect officers who have specified authority, such as to hire employees and run the activities of the association. The safest approach may be to obtain a resolution—voted on by the members—authorizing the officers to make a particular contract. Because of the potential risk to officers and to members, many contracts contain provisions that limit liability to certain designated property held in trust for the association.

2.8 Loose Associations: Share-Office Arrangement

Sometimes design professionals use the term *association* to describe an arrangement under which they are independent sole proprietors but join together to share offices, equipment, and clerical help. They may also do work for each other. Sometimes such loose associations are known as *share-office arrangements*.

During the sharing arrangement, several legal problems can develop. First, suppose one associate performs services for another and the latter is not paid by his client. In the absence of any specific agreement or well-documented understanding dealing with this risk, the associate who performs services at the request of another should be paid.

Second, suppose the associate performing services at the other associate's request does not perform properly and causes a loss to the client. The client would have a claim against the associate who did the work and the associate with whom it dealt. If the latter settled the claim or paid a court judgment, he would have a valid claim for indemnification against the associate who did not perform properly.

Third, if those in the share-office arrangement create an impression to the outside world that they are a partnership, they will be treated as such. For example, if the letterhead, building directory, and telephone directory list them as "Smith, Brown & Jones," such listings may create an apparent partnership. If so, each member of the association can create partnership-like contracts and tort obligations.

When a loose association terminates or one participant withdraws, all associates in the first case and the withdrawing associate in the case of withdrawal can freely compete with former associates or those remaining. However, the associates in the first case and the withdrawing associate in the second cannot do any of the following:

1. Take association records that belong to another associate.
2. Represent that they still are associated with former associates if this is not the case.
3. Wrongfully interfere with contractual or stable economic relationships existing at the time the association ended or one associate withdrew.

2.9 Professional Associations

Professional associations of participants in the construction process, such as the American Institute of Architects (AIA), the National Society of Professional Engineers (NSPE), the Construction Management Association of America (CMAA), and the Associated General Contractors of America (AGC), have had a substantial impact on design and construction. Although they are not organizations like the others mentioned in this chapter created for the purposes of engaging in design and construction, some mention should be made of their activity and some of the legal restraints on them.

These professional associations engage in many activities. They speak for their members and the professions or industries associated with them, including before legislative bodies. In addition, they seek to educate their members in matters that relate to their activities. One of their most important activities is publishing standard documents for design and construction that are used not only as the basis for contracts for design and construction but also to implement the construction administrative process. Membership in these professional associations is not required for someone to design, manage or build—for that, a license is required (see Chapter 8).

2.10 Forms of Employment: Introduction

The business organizations described earlier in this chapter cannot grow and flourish without the businesses' ability to hire persons and have those persons (or at least some of them) interact with other businesses or persons. The hiring relationship and the ability of the hired person to interact with others on behalf of the hiring organization are separate yet related issues.

There are three major forms of employment by businesses involved in the construction industry. An owner, contractor, or design professional can hire an **agent**, an **employee**, or an **independent contractor**. Each form of employment has its own legal characteristics and consequences for the hirer.

An agency relationship is between a principal (the hiring party) and an agent (the hired party). An agent is one who, when acting on behalf of the principal, may bind that principal to another (a "third party"). Without the concept of agency, very little commercial activity would occur, because very few people in each organization would have the authority to bind the organization to a contract. Without employees, business organizations would remain tiny and weak.

Both agents and employees can subject the hiring business—the principal or employer—to liability to third parties. By contrast, a business that hires an independent contractor as a general rule is not liable to third parties injured by the independent contractor. A business's ability to hire independent contractors gives the business flexibility in getting work done with less risk of liability than if the same work had been done by employees.

The three types of employment—agent, employee, or independent contractor—are not mutually exclusive. All employees are agents of their employer, but not all agents are employees. Also, the same person can in some contexts be an agent and in other contexts an independent contractor. For example, an owner hires an engineer as an independent contractor, but in interfacing with the trade contractors the engineer may be acting as the owner's agent.

Finally, confusion often occurs because of the overlapping nature of terms such as principal–agent, master–servant, employer–employee, and employer-independent contractor. This book does not use the outdated terms of master–servant, and instead uses the terms: principal–agent, employer–employee, and hirer–independent contractor.

2.11 The Agency Relationship

2.11A Introduction

Agency rules and their application determine when the acts of one person (the agent) bind another (the principal). In the typical agency problem, there is a principal, an agent, and a third party. The agent is the person whose acts are asserted by the third party to bind the principal. See Figure 2.1. There can be legal problems relating to the rights and duties of the principal and agent as between themselves, as well as disputes between the agent and the third party. But because in the agency triangle the third party versus principal part is most important—and most troublesome—these problems will be used to demonstrate the relevance of agency law.

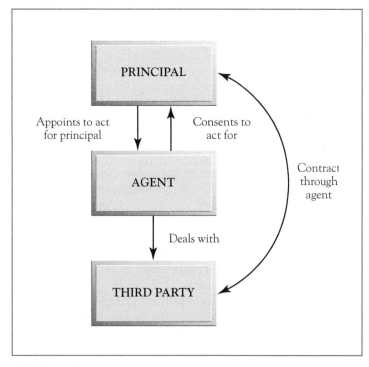

FIGURE 2.1 Agency concept illustrated.

The design professional, whether in private practice or working as an employee, may be in any of the three positions of the agency triangle. For example, suppose the design professional is a principal (partner) of a large office. The office manager orders an expensive computer. In this illustration, the design professional, as a partner, is a principal who may be responsible for the acts of the agent office manager.

Suppose the design professional is retained by an owner to design a large structure. The design professional is also engaged to perform certain functions on behalf of the owner in the construction process itself. Suppose the design professional orders certain changes in the work that will increase the cost of the project. A dispute may arise between the owner and the contractor relating to the power of the design professional to bind the owner. Here the design professional falls into the agent category, and the issue is the extent of her authority.

Suppose X approaches a design professional in private practice regarding a commission to design a structure. X states that she is the vice president of T Corporation. The design professional and X come to an agreement. Does the design professional have a contract with T Corporation? Here the design professional is in the position of the third party in the agency triangle.

The agency concept is basic to understanding the different forms by which persons conduct their business affairs, such as partnerships and corporations.

2.11B Policies Behind Agency Concept

As the commercial economy expanded beyond simple person-to-person dealings, commercial necessity required that people be able to act through others. Principals needed to employ agents with whom third parties would deal. Third parties will deal with an agent if they feel assured that they can look to the principal. The agent may be a person of doubtful financial responsibility. The concept of agency filled the need for giving third persons some assurance that they can hold the principal responsible.

Two recent cases illustrate the agency concept from the perspective of the owner. In *Shoals v. Home Depot, Inc.*,[6] a homeowner was swindled by a dishonest employee of Home Depot, a large home improvements chain. The court found that the employee's representations made to the homeowner were binding on Home Depot, the principal. In *Shoals,* the owner was the third party who personally benefited from application of the agency concept.

By contrast, in *Ciraulo v. City of Newport Beach,*[7] the contractor assured the homeowners that it would obtain a building code variance from the city, then went ahead with the construction even though it never talked to the city. After completion, the city ordered the owners to remove the unauthorized building. The court upheld the order, because the owners, as principals, were bound by the fraudulent acts of their agent, the contractor.

2.11C The Principal–Agent Relationship

As a rule, the relationship of principal and agent is created by a contract or manifested by acts in the absence of a contract. The agent may be a regular employee of the principal or an independent person hired for a specific purpose and not controlled as to the details of requested activities. Several aspects of the agency relationship distinguish it from the ordinary commercial, arm's-length relationship.

An arm's-length transaction is one wherein the parties are expected to protect themselves. In such a transaction, no general duty is imposed on one party to protect the other party, nor is any duty imposed to disclose essential facts to the other party. Although there are some exceptions, generally commercial dealings are at arm's length. On the other hand, principal and agent have a fiduciary relationship—one of trust and loyalty. In such a relationship, one person relies on the integrity and fidelity of the other. The latter must not take unfair advantage of the trust in her by benefiting at the expense of the former.

2.11D Agent Binding the Principal: Actual Versus Apparent Authority

Agency exposes the principal to risks. The principal may be liable for an unauthorized commitment made by the agent. Suppose the principal authorizes its agent to make purchases of up to $1,000, but the agent orders $5,000 worth of goods. From the third party's standpoint, this $5,000 purchase may be reasonable in light of the agent's position or what the agent had been ordering in the past. The law protects the principal from unauthorized commitments but not at the expense of reasonable expectations of the third person. That such problems can develop does not destroy the unquestioned usefulness of the agency concept. Such problems require the law to create rules and solutions to handle such questions that accord with commercial necessity and common sense. The law uses the concepts of actual and apparent authority—first mentioned in Section 2.3 with regard to partnership, but addressed here more in-depth—to answer these questions.

Actual Authority. The agent is ordinarily authorized to do only what it is reasonable to believe the principal wants done. In determining this, the agent must look at the surrounding facts and circumstances. If, for example, the principal has authorized the agent to purchase raw materials to be used in a particular manufacturing process and the agent learns that the principal has decided not to proceed with the project, it is unreasonable for the agent to believe that the authority to buy the materials still exists. To give another example: If an agent is given authority to purchase a car for the principal, it is likely that the agent also has authority to buy car insurance for the principal. Nonetheless, the agent, when possible, should seek authorization from the principal to perform acts not expressly authorized.

Suppose the agent did not have actual authority to enter into a transaction but the principal discovers the transaction. In such a case, the principal may ratify the transaction within a reasonable time and bind the third party (and itself) by notifying the third party and affirming the agent's unauthorized acts.

Apparent Authority. The most difficult principal–agency cases arise under the doctrine of apparent authority, which exists when the principal's conduct reasonably—but erroneously—leads a third party to believe that the principal consents to acts done on its behalf by the person purporting to act for it. On construction projects, the question of apparent authority often arises when an owner hires a professional person—whether design professional, construction manager, project manager, or other—to administer the contract. The professional person, although not an employee of the owner, is the owner's representative to the contractor and possesses certain designated authority. The contractor may claim that communications it had with the professional person should be treated as if coming directly from the owner, while the owner may contest that conclusion.

2.11E Disputes Between Principal and Third Party

Sometimes the principal wishes to be bound by a transaction between the principal's agent and the third party. The third party may refuse to deal with the principal, claiming it did not know it was dealing with an agent. The agent may not have informed the third party that she was an agent, and the third party may not have had reason to know that the agent was acting on behalf of someone else. If the principal discloses its status (as the agent's principal), the third party will not be bound to the principal if this would result in an injustice to the third party. Of course, if the third party accepts the principal, then the principal becomes bound to the third party.

2.11F Disputes Between Agent and Third Party

Disputes between an agent and a third party are relatively rare. If the agent is acting on behalf of an undisclosed principal, the third party has the right to sue the agent individually. The third party may lose this right if the principal discloses itself and the third party pursues its remedy against the principal.

Within the construction industry, third-party claims against the agent of a disclosed principal most commonly occur when a contractor sues the project architect or engineer.

The contractor usually claims that the design professional committed a tort for which she is personally liable. For example, the contractor may contend that the design professional negligently administered the project or intentionally interfered with the contractor's contractual relationship with the owner. A design professional who acts within the scope of her agency and is not motivated by improper motive will not be liable to the contractor.[8]

2.12 The Employment Relationship

The essence of an employment relationship is that the employer has the right to completely control what work the employee does and how the employee performs that work, and the employer may terminate the employee for disobedience. In addition to the element of control, another indication of an employment relationship is that the employee is paid net wages after the deduction of federal and state employment taxes.

For reasons of public policy, the employer is vicariously (or strictly) liable if the employee causes physical harm to others while acting within the scope of the employment services. This doctrine of an employer's **vicarious liability** is called *respondeat superior*. The issue of a hirer's liability for the torts of the hired person is a key distinction between employees and independent contractors (discussed in Section 2.13). Unlike an employer's strict liability under *respondeat superior* for the acts of its employees, the hirer of an independent contractor is presumptively not liable for the torts of that contractor.

2.12A Workers' Compensation

One of the most significant benefits of an employment relationship is workers' compensation coverage in the event the employee is injured on the job. The **workers' compensation** system is a legislative compromise. Employees are guaranteed compensation without regard to fault—meaning that they need not prove their employer was negligent and the employees are compensated even if they themselves may have been negligent. On the other hand, the workers' compensation payment is the employees' exclusive remedy—meaning that (subject to exceptions) they cannot sue their employer in tort, even if the employer had negligently caused the employee's injury.

There is considerable variation among the workers' compensation acts of different states. Some states cover occupational diseases, others do not. Although negligence by the worker will not preclude coverage, some states deny recovery if the worker was guilty of willful misconduct or intoxication and such misconduct caused the injury.

Most states require that the employer obtain private workers' compensation insurance. Early in the twentieth century, this approach proved inadequate for the construction industry, where the practice of subcontracting portions of the work is commonplace. Many subcontractors were not financially sound and may not have obtained the requisite insurance. Those who were uninsured may have been cheaper to hire, thereby creating unfair competition for subcontractors who had insured their employees against job-related injuries.

In response, most states enacted "statutory employer" statutes under which a prime contractor is viewed as the employer of a subcontractor's employees under certain circumstances. Most states impose statutory employment liability on the prime contractor if the subcontractor was uninsured.[9] Statutory employment allows an injured subcontractor's employee recovery against the prime contractor's workers' compensation insurer.

As noted, workers' compensation remedies generally supplant whatever tort remedy the worker may have had against the employer. This is accomplished by statutory immunization of the employer from tort claims by the workers. However, job-related injuries are often caused by actors other than the employer. Most states permit injured workers to bring tort claims against most third parties whose negligence caused the injury, such as other construction employers, the design professional, or the manufacturer of an allegedly defective product. (However, some workers' compensation acts have extended limited statutory immunity to design professionals, thereby protecting them from tort suits by injured construction workers.) [10] These third-party actions by injured employees against anyone they can connect with their injuries other than their employers are common, because few view workers' compensation benefits (which are set by statute) as sufficiently generous to provide them with full compensation.

> In this chapter's scenario, Mr. E would receive workers' compensation benefits from his employer, ABC Engineering, Inc. Mr. E. could also sue DD Development for maintaining unsafe premises, because DD Development is a third party not protected by workers' compensation act immunity.

Workers' compensation claims are handled by an administrative agency rather than a court. Hearings are informal and usually conducted by a hearing officer or examiner. The employee can represent himself or be represented by a layperson or lawyer. Fees for representation are usually regulated by law. Although awards by the agency can be appealed to a court, judicial review is extremely limited, and very few awards are overturned. Third-party actions, on the other hand, because they involve tort claims, are brought in court, and incredibly complicated lawsuits often result.

2.12B Employee Leasing Companies

The labor demands placed upon a construction company often fluctuate. It may be more economical for a contractor to maintain a minimum permanent staff of employees and supplement its labor pool only when awarded a large project.

One solution to the need for a flexible workforce is employee leasing companies—also called labor brokers. The labor broker maintains a list of prospective workers with expertise in different building skills, including simple manual labor. The broker then matches workers with the needs of its clients, who are construction companies. Typically, the broker employs the workers, pays them and withholds their taxes, handles the paperwork, and provides workers' compensation coverage—all for a set fee to be paid by the client. In return, the contractor–client receives a worker each day who is subject to its authority. The contractor instructs the workers and has the authority to dismiss them.

Suppose the leased employee is injured on a job while working for a client–contractor. The employee leasing company is normally subject to workers' compensation liability and so is immune from tort liability.

The more difficult question is whether the contractor may be sued in tort by the leased employee injured on the job. Some states extend workers' compensation immunity to the contractor by statute.[11] Even without such legislative direction, most courts find the contractor is a co-employer also entitled to immunity from tort liability. In sum, the leased employee has two employers: the leasing company and the client–contractor for whom the employee was working.

2.12C Unions and Labor Disruptions: The Picket Line and Project Labor Agreements

Employees of large contractors are more likely to be unionized than are employees in most other industries. The presence of unionized workforces in turn raises the likelihood of labor disputes. Labor disputes can cause lengthy work stoppages. Craft unions sometimes dispute who has the right to perform certain work. Not uncommonly, employees of unionized contractors refuse to work on a site with nonunion employees of another contractor. Workers sometimes refuse to work on a project because prefabricated units are introduced on the site in violation of a "work preservation" clause in a subcontract or collective bargaining agreement. Any of these situations can result in a strike, a picket line, or both. Workers frequently refuse to cross a picket line. Such refusals can shut down a job directly (because no workers are willing to work on the site) or indirectly (through the refusal of workers to deliver materials to a picketed site).

Pressure and even coercion are common tactics in the struggles between employers and groups that are attempting to organize the workers as well as represent them in negotiating wages, hours, job security, and working conditions. Each of the disputants—employer and union—not only seeks to persuade but also to use economic weapons to obtain a favorable outcome.

Disputants direct pressure at each other by use of economic weapons such as strikes or lockouts. Even these direct primary pressures affect neutrals not involved in the labor dispute. A strike affects nonstriking workers, the families of the strikers, and those who deal with the struck employer, such as those supplying it goods or purchasing its products or services. Likewise, a lockout by the employer affects neutrals.

Economic warfare often expands to include pressure and coercion on third parties important to the employer, such as suppliers or customers. The pressure can be direct, as with communicated coercive threats, or indirect, as with a picket line whose signs describe the union's reason for its grievance against the employer and ask that certain activity or nonactivity be taken.

As these tactics broaden the field of economic warfare, they may begin to seriously harm neutrals and become what are called *secondary boycotts*. Such impermissible activities are unfair labor practices, giving the party injured by such activity the right to damages and, more important, the right to obtain an injunction ordering that such activities cease or be modified.

The line between permitted primary and prohibited secondary boycotts is difficult to draw in industrial collective bargaining warfare. It is even more difficult in construction work because of the transient nature of the workers, the seasonal nature of the work, the proliferation of craft unions, the frequent occupation of the construction site by employees of many bargaining entities such as contractors and subcontractors, and the special rules that govern collective bargaining in the construction industry.

One of these factors merits additional comment. As indicated, the construction project usually involves work by a number of contractors, sometimes at different stages of the work, but often at the same time. Some of the employers may have collective bargaining agreements with a union, whereas others may not. Some employers may have collective bargaining agreements with one union, and others may have collective bargaining agreements with a different one. For example, a nonunion prime contractor may be working alongside union subcontractors. Sometimes union subcontractors are working on a site with employees of a subcontractor who does not have a collective bargaining agreement with any union. A union engaged in an economic struggle with an employer can maximize its bargaining power if it can shut down the entire project by putting up a picket line that no union workers will cross. The very effectiveness of common situs picketing (picketing an entrance used by all workers) is the reason it can also enmesh many neutrals and cause frequent and costly work stoppages.

The law has limited common situs (location) picketing by allowing the prime contractor to set up separate gates, one for the employees involved in the labor dispute and the other for those not involved in the dispute. This two-gate system prevents the project from being totally shut down. Picketing can be done at the first gate but not at the second.

Owners on large projects may circumvent problems of both union and nonunion trades working together *and* jurisdictional disputes among different unions by having the prime contractor and all subcontractors enter into a Project Labor Agreement (PLA). A PLA is a project-specific labor agreement. It ensures uniform terms and conditions of employment for all workers, whether union or nonunion, covering such matters as working hours, shift times, scheduling, holidays, overtime, and premium pay. All contractors, unions, and employees are subject to the same collective bargaining agreement. Nonunion contractors must hire workers through the union hiring hall (although an exception may exist for core employees), but the employees need not actually join the union. Of even greater importance for owners, the PLA establishes a dispute resolution mechanism and prohibits strikes. In short, PLAs are intended to guarantee labor peace through the life of the contract.

Legal challenges to PLAs have arisen on public works projects. In the *Boston Harbor* case, the Supreme Court ruled that PLAs are not preempted by the National Labor Relations Act, which is a federal law.[12]

Legal attacks then shifted to state laws. The most frequent complaint by nonunion contractors is that PLAs violate the state competitive bidding laws by restricting competition for public works jobs to union contractors. This narrowing of the number of contractors willing to bid on advertised projects—the opponents to PLAs contend—deprives the state from obtaining the lowest bid for the work. Most courts apply a balancing test. They will uphold use of PLAs on public projects if the size and complexity of the project are such that the danger of labor disruptions without a PLA would be very high.[13] Increasingly, courts reject the premise of opponents that PLAs discourage bidding by nonunion contractors. The legality of PLAs notwithstanding, the propriety of public agencies entering into such agreements has generated heated debate. The political contentiousness of PLAs is evident from examining the executive orders of the presidents from George H.W. Bush to Barack Obama.[14]

About half of the states have enacted "right to work" laws. These laws prohibit "union shops," which require employees to join a union as a condition for receiving or retaining a job. "Right to work" laws make it more difficult for trade unions to organize workers. These laws are one reason for a declining percentage of workers who belong to trade unions in the private sector. But public sector workers are more likely to be members of unions.

2.13 Independent Contractors

A key characteristic of an independent contractor is that the contractor, unlike an agent or employee, controls the manner in which the work is performed. An independent contractor is asked to achieve a result and is not controlled as to the means by which this is accomplished. Commonly, an independent contractor is hired on a project-specific basis. The hirer of an independent contractor is typically the owner (who hires the prime contractor), the prime contractor (who hires subcontractors), and some subcontractors (who hire sub-subcontractors).

While the element of control is the primary feature of a hirer–independent contractor relationship, it is not the sole consideration. Instead, the courts have developed a multi-factor test to determine whether a hired person or business is an agent or independent contractor:

> Control of manner work is to be done; responsibility for result only; terms of agreement between the parties; the nature of the work or occupation; skill required for performance; whether one employed is engaged in a distinct occupation or business; which party supplies the tools; whether payment is by time or by job; whether work is a part of regular business of the employer, and also the right of employer to terminate the employment at any time.[15]

Courts applying this test consistently find that a business entity or person regularly involved in construction work is an independent contractor if hired by an owner to perform building, repair, or renovation work.

Whether the person hired is an agent or an independent contractor has significant implications for the hirer. While a principal is presumed to be liable for the acts of his or her agents, under the "independent contractor rule," the hirer of an independent contractor is presumed *not* to be liable for injury to others caused by the contractor.[16] For example, suppose an owner hires a contractor, but the contractor performs the work in a negligent and unsafe manner, leading to the injury of a passer-by. The passer-by may be unable to recover an adequate financial award against the contractor and decides to sue the owner as well. The owner's first line of defense is that its lack of control over the contractor's work methods shields the owner from liability.

The injured person may contend that the hired person was an agent, not an independent contractor, and is entitled to recovery against the owner. The plaintiff may argue that the owner's right to supervise the work of the contractor was a form of control that transformed the relationship into one of agency. This argument is rarely successful. Courts normally hold that the owner's general right to supervise the work is not sufficient to establish either substantial control or an agency relationship. Supervision to ensure that the desired result is achieved is not the same thing as controlling the manner in which the work is performed.

Rather than find the contractor was an agent, courts are more likely to impose liability upon the hirer under an exception to the independent contractor rule. The independent contractor rule is discussed in more detail in Section 5.7.

REVIEW QUESTIONS

1. Which type of organization is the simplest form of business and is used by many private practicing design professionals and contractors?

2. In the court case of *Christopher v. Sinyard*, what were the factors that the court noted that lead to the ruling of piercing the corporate shield and holding the principals personally liable?

3. Which form of business organization is a hybrid of the corporate and partnership forms, containing the pass-through income tax benefits of a partnership with the limited liability protections of a corporation?

4. What type of authority exists when the principal's conduct reasonably—but erroneously—leads a third party to believe that the principal consents to acts done on its behalf by the person purporting to act for it?

5. What is the essence of an employment relationship?

6. What is an employee guaranteed under workers' compensation laws?

7. If a construction company needs a flexible workforce and finds it may be more economical to maintain a minimum permanent staff of employees and supplement its labor pool only when awarded a large project, what type of company would that contractor utilize?

8. What is a Project Labor Agreement (PLA)?

9. What are the elements developed by the courts of the multi-factor test to determine whether a hired person or business is an agent or independent contractor?

10. What is the difference between an arm's-length transaction and a fiduciary relationship?

ENDNOTES

[1] 798 N.E.2d 171 (Ind.2003).
[2] *Christopher v. Sinyard,* 313 Ga.App. 866, 723 S.E.2d 78, 81 (2012), quoting *Soerries v. Dancause,* 248 Ga.App. 374, 377, 546 S.E.2d 356, 359 (2001).
[3] *Moransais v. Heathman,* 744 So.2d 973 (Fla.1999); La.Rev.Stat. § 12:1320(D).
[4] 63 Pa.Stat.Ann. § 34.13(a)(6) & (7). The requirements pertain to the minimum number of members or managers of the LLC, or partners of the LLP, who must be licensed to practice architecture. Id., § 34.13(f) & (g).
[5] Cal.Bus. & Prof. Code § 5535 (architects may form LLPs) and Cal.Corp. Code § 17375 (LLCs may not render professional services). For a state-by-state survey, see Lurie & Anderson, *The Practice of Architecture and Engineering by Limited Liability Entities,* 24 Constr. Lawyer, No.1, Winter 2004, p. 24.
[6] 422 F.Supp.2d 1183 (E.D.Cal.2006).
[7] 147 Cal.App.4th 838, 54 Cal.Rptr.3d 515 (2007), depublished, May 23, 2007.
[8] *Wiekhorst Bros. Excavating & Equipment Co. v. Ludewig,* 247 Neb. 547, 529 N.W.2d 33 (1995). See also Section 11.6E.
[9] As examples, see Conn.Gen.Stat.Ann. § 31–291 and Mo.Rev.Stat. § 287.040.
[10] As examples, see Conn.Gen.Stat.Ann. § 31–293(c); West Fla.Stat.Ann. § 440.09(6); West Okla.Stat.Ann. Tit. 85, § 12.
[11] Fla.Stat.Ann.§ 440.11(2); Ga.Code Ann. § 34-9-11(c).
[12] *Building & Constr. Trades Council of the Metro. Dist. v. Associated Builders & Contractors of Mass./R.I., Inc.,* 507 U.S. 218 (1993) (involving a project to clean up the polluted Boston Harbor).
[13] *New York State Chapter, Inc. v. New York State Thruway Auth.,* 88 N.Y.2d 56, 666 N.E.2d 185, 643 N.Y.S.2d 480 (1996) (renovation of the Tappan Zee bridge, spanning the Hudson River in New York).
[14] The first President Bush issued Executive Order (E.O.) No. 12818 on October 23, 1992, which barred PLAs on federal projects. President Bill Clinton lifted this ban with E.O. No. 12836 in 1993. President George W. Bush renewed the ban in E.O. Nos. 13202 and 13208, both in 2001. President Obama revoked these executive orders with his E.O. No. 13502, issued on Feb. 2, 2009.
[15] *Stepp v. Renn,* 184 Pa.Super. 634, 135 A.2d 794, 796 (1957).
[16] Restatement (Second) of Torts § 409 (1965).

CHAPTER 3

Contracts: From Formation to Breach

Scenario

Jim and Jane Smith, young doctors with busy schedules, bought an empty lot on which to build a house. After an online search, they began negotiations with Mr. Builder, the president of Prime Builder, Inc., to be the prime contractor on the job.

Negotiations were protracted: face-to-face meetings, phone calls, emails, and more meetings. The Smiths were unsure of the type of house they desired, but knew they wanted the cost to be within their budget. Mr. Builder assured them he had built many houses (although he had previously built only warehouses), that he had a "good idea" of the kind of house they wanted, and that he could stay within their budget. The deal was sealed with firm handshakes.

Prime Builder, Inc. began to look around for suppliers and subcontractors. Mr. Builder first telephoned a hardware store and ordered $5000 of supplies.

Mr. Builder next entered into a signed contract with Sorry Subcontractor to perform the entire labor for the project. However, as the job was about to start, Sorry Subcontractor got an offer from another contractor to do a larger job for more money. True to its name, Sorry Subcontractor told Mr. Builder that he was sorry but he couldn't do the job. Scrambling around, Mr. Builder signed a new subcontract with Substitute Subcontractor, for $10,000 more.

Prime Builder, Inc. began performance. The city's building inspector happened to be driving by and saw the new construction. Curious, he checked and found that no building permit had been issued. He ordered the work stopped until the Smiths put in an application for a permit, which was then approved.

As the work proceeded, the Smiths were worried that it was taking too long. When they asked Mr. Builder, he pointed out that their agreement was not written down, that they had never agreed on a specified completion date, so Prime Builder, Inc. could take as long as it needed to "do the job right." In retaliation, the Smiths began to pay only half the requested interim payments. Worried that the Smiths were running out of money, Mr. Builder asked them to prove they could pay for the rest of the job. When the Smiths failed to respond, Mr. Builder stopped work and told Substitute Subcontractor to leave too.

Litigation ensued.

1. Prime Builder, Inc. sued the Smiths for breach of contract for nonpayment of the full amount the contractor claimed was due.
2. The Smiths in response sued Prime Builder, Inc. for breach of contract. The Smiths alleged that the contractor's work was defective, the project was well past the time it should have been finished, and the budget had been used up even though the house was only two-thirds finished.
3. Prime Builder, Inc. sued Sorry Subcontractor, contending that it was responsible for the prime contractor incurring $10,000 more than anticipated in labor costs because it had to hire Substitute Subcontractor. Sorry Subcontractor pointed out that they had not yet signed the subcontract (that was supposed to occur when Sorry Subcontractor showed up on the job); therefore, no contract existed and it could not be sued for breach of contract.

3.1 Relevance

This chapter introduces the law of contracts. Contracts are binding bargains, usually made between two parties. They are agreements involving an exchange of promises between the contracting parties. The promises are interdependent; each is made in exchange for the other. The bargain is "binding" in the sense that, if one party to the contract fulfills the promise but the other doesn't (and is not excused from nonperformance), the performing party may sue the nonperforming party for breach of the contract (so as to receive an award of money, as described in Chapter 4). If the breach is "material," the aggrieved party may cancel (or terminate) the contract (as explained in Section 3.7).

With rare exceptions, every person on a construction site is there by virtue of a contract. Every commercial entity is there only because it had previously entered into contracts necessary to make it a hopefully viable business concern.

A design professional makes contracts with clients, consultants, employees, landlords, and sellers of goods (among others). In addition, contracts are made between property owners and prime contractors, prime contractors and subcontractors, contractors and suppliers, employers and employees, and buyers and sellers of land.

While construction law encompasses far more than contract law, contracts provide the bedrock upon which the construction industry is based. Familiarity with the key concepts discussed in this chapter is a precondition to understanding the remainder of the book.

3.2 The Function of Enforcing Contracts: Freedom of Contract

The principal function of enforcing contracts in the commercial world is to encourage economic exchanges that in the long run (if not in particular instances) leads to greater productivity. This is accomplished by protecting the reasonable expectation of contracting parties—that each will perform as promised. Although many (perhaps most) contracts are and would be performed without resort to court enforcement, the availability of legal sanctions plays an important role in obtaining performance.

Generally, American law gives autonomy (freedom) to contracting parties to choose the substantive content of their contracts. Because most contracts are economic exchanges, giving parties autonomy allows each to value the other's performance.

To a large degree, autonomy assumes and supports a marketplace where market participants are free to pick the parties with whom they will deal and the terms on which they will deal. Not only are the parties in the best position to determine terms of exchange, but the alternative—state-prescribed rules for economic exchanges—would lead to rigidity and heavily burden the state. Also, parties are more likely to perform in accordance with their promises if they have participated freely in making the exchange and determined its terms. Finally, such autonomy—often called *freedom of contract*—fits well in a free society that encourages individual enterprise.

However, broad grants of autonomy assume contracting parties of relatively equal bargaining power, equal accessibility to information, and a relatively free marketplace. Such conditions place a check on overreaching. A party who believes the other party's terms to be unreasonable can decide instead to deal with others. If the parties do arrive at an agreement under such conditions, the give-and-take of bargaining should ensure a contract that falls within the boundaries of reasonableness.

The development of mass-produced contracts and the emergence of large blocs of economic power often dealing with parties with limited or no bargaining power have made this earlier model of the negotiated contract the exception. If the state, through its courts, enforces **adhesion contracts** (contracts presented on a take-it-or-leave-it basis), the state is according almost sovereign power to those who have the economic power to dictate contract terms. For this reason, many in-roads have been made on contractual freedom by federal and state legislation, regulations of administrative agencies, and courts through their power to interpret contracts and determine their validity.

3.3 Contract Formation

Many technical rules have been developed to determine whether a contract was formed. Some of these rules—such as the mental capacity of the parties to contract or whether one party was simply joking when purporting to agree to contract—have little real-world application to the construction industry. This discussion focuses on contract formation issues of practical relevance to the industry.

3.3A Definitions

A contract is an enforceable bargain. To be valid, a contract must meet several requirements, including the following.

1. Agreement by the parties, sometimes called manifestations of *mutual assent.*
2. *Consideration*, which means the promises are exchanged one for the other and bring to the bargain either a legal benefit or detriment to the contracting parties.
3. Formal requirements applicable to some types of contracts.
4. The *promisor* is a person who makes a promise, and the *promisee* is the person to whom the promise is made. In most two-party contracts, each party is a promisor and promisee, both making and receiving promises.
5. The *offeror* is a person who makes an offer, and the *offeree* is a person to whom the offer is made.
All other definitions will be given as the particular term is discussed.

3.3B Mutual Assent

Objective Theory of Contracts. For an agreement to exist (leaving aside for now whether it is a binding agreement), common sense says that each party (participant) to the agreement must have actually agreed on the same thing. Yet if the test for an agreement is the *subjective* perception of each party, enormous problems of proof arise and great uncertainty will be injected into the law of contracts.

Instead, the law uses an *objective* theory of contracts that protects the reasonable expectations of the parties. Under this theory, a party is bound by what it manifests to the other party; secret intentions are not relevant. If one party—innocently or otherwise—misleads the other into thinking it has serious contractual intention, a contract exists although the parties do not actually (or subjectively) agree.

Offer and Acceptance. Typically, the process by which contracting parties make an agreement involves communications—oral and written—that culminate with one party making an offer and the other party accepting it.

When parties exchange communications at the same time and place and an agreement is reached, it rarely is necessary to determine whether there has been an offer and acceptance. But when parties communicate with each other at a distance or over an extended period of time, it is often necessary to determine whether the parties have arrived at an agreement, and this requires an examination of whether there has been an offer and acceptance.

Offers are different from preliminary negotiations or proposals for offers. The effect of an offer is to create a "power of acceptance in the offeree." The offeree, by accepting the offer, creates a legally enforceable obligation without any further act of the offeror.

How is it determined whether the offeree has a reasonable belief that he has a power of acceptance? The entire written proposal is examined to see whether a hypothetical "reasonable person" receiving it would think he could "close the deal." Factors include the certainty of the terms, any indication that the offeror will not have to take further action, the past dealings between the parties, and the person to whom the offer is made. For example, if nothing is stated on essential terms such as price, quantity, or quality, the bargaining is probably in a preliminary stage.

> In this chapter's scenario, the Smiths and Prime Builder, Inc. had a handshake deal, but did they have a contract? Many terms were not finalized, including the scope of work and the completion date.

Suppose the offer is sufficiently detailed that the offeree's acceptance would "close the deal." If the offeree rejects it, then the offer is terminated. Suppose the offeree does not reject the offer but instead proposes a counteroffer. The counteroffer terminates the original offer, but at the same time gives the original offeror the power to accept the counteroffer.

One complication is devising a rule of acceptance between a buyer and seller with their own standard form contracts, which rarely agree. To remedy this problem for transactions involving the commercial sale of goods (sales between merchants), the Uniform Commercial Code adopted § 2-207. Including additional or different terms does not necessarily preclude the communication from being an acceptance if it appears the offeree is accepting the terms of the offer and does not condition his acceptance expressly on the offeror agreeing to additional or different terms. Those additional terms are considered proposals for additions to the contract that are sometimes binding without further communication of the offeror. A contract has been concluded on the terms that match with the supplemental terms sometimes being supplied by law.

The discussion at this point has examined acceptance by communication. Acceptance may also happen by conduct or performance. If a party performs in accordance with the requirements of the contract, the courts may find that the party has accepted the offeror's terms, even in the absence of a written or oral communication.

In this chapter's scenario, one could argue the parties at the time of the handshakes did not have a contract. Yet Prime Builder, Inc. began performance, and the Smiths paid the work as it was done. It may be that the parties cured any defect in the initial agreement by their own performance, so that a contract existed by the time Prime Builder, Inc. stopped work.

3.3C Consideration and Its Substitute: Promissory Estoppel

While an offer and acceptance result in a bargain, the added element of **consideration** is required to transform that bargain into an enforceable contract. Consideration is loosely defined as bargained-for exchange: one person giving up something of value in return for receiving a benefit from the other person to the bargain. Using more technical language, consideration is found to exist if the promisee (the recipient of the promise) suffers a legal detriment (for example, pays money) and this detriment induces the promisor to act (give the promisee an apple).

Of particular importance to the construction industry is the doctrine of **promissory estoppel**—also called "promised-induced reliance"—which makes a promise binding on the promisor even in the absence of consideration.

Suppose a prime contractor uses a subcontractor's bid (an offer to perform the subcontract work at a given price) to obtain the prime contract without the prime contractor and subcontractor having first entered into a subcontract (because the prime contractor does not know yet if the owner will award it the job). If the prime contractor is awarded the job, but the subcontractor then refuses to enter into a subcontract, the prime contractor cannot sue the subcontractor for breach of contract, because no consideration existed and hence no contract existed. However, if the prime contractor cannot find a substitute subcontractor at the same (or lower) price, it will have to perform the contract at a financial loss.

While no contract has been formed, many courts allow the promisee (the prime contractor in this case) to sue the subcontractor under the doctrine of promissory estoppel. The prime contractor must show that it relied upon the subcontractor's promise (by submitting to the owner a bid premised upon use of the subcontractor's offer), that it was reasonably foreseeable (or likely) to the subcontractor that the prime contractor would rely on its offer in this manner, and (because of lack of consideration) that enforcement of the promise is the only way to prevent an injustice to the prime. While promissory estoppel can be applied in a variety of circumstances, in the construction process, it primarily has been used as a means of enforcing bids made by subcontractors to prime contractors.

In this chapter's scenario, Prime Builder, Inc. is suing Sorry Subcontractor for the increased cost of having to hire Substitute Subcontractor. The first question is whether Prime Builder, Inc. and Sorry Subcontractor had a binding contract. If so, the contractor would have to sue for breach of contract, since promissory estoppel is a back-up theory used only in case no contract exists.

Assuming no subcontract exists, may Prime Builder, Inc. sue Sorry Subcontractor under a theory of promissory estoppel? Note that Prime Builder, Inc. might have difficulty proving *reliance* on Sorry Subcontractor's promise. This is because the Smiths and Prime Builder, Inc. had already agreed to a contract price (assuming there was a contract and it included a price), and only afterwards did the builder and Sorry Subcontractor begin negotiations. While Prime Builder, Inc. may have been looking for the lowest price, it did not rely on Sorry Subcontractor's offer when negotiating a contract price with the Smiths.

3.3D Reasonable Certainty of Terms

As mentioned in Section 3.3B, one test for determining whether an offer has been made is the clarity and completeness of the terms of the proposal. Even when an agreement on terms has been reached, agreed terms that lack reasonable certainty of meaning preclude a valid contract.

Why require that even agreed terms be reasonably clear in meaning before a valid contract can exist? First, vagueness or incompleteness of terms may indicate that the parties are still in the bargaining stage. Second, a third party, such as a judge or an arbitrator, should be able to determine without undue difficulty whether the contract has been performed. But reasonable certainty does not require terms so clear that there can be no doubt as to their meaning.

> In this chapter's scenario, many essential contract terms—including price, scope of work, and time—appear to have been vague or even nonexistent when the Smiths and Prime Builder, Inc. entered into their bargain. It appears unlikely an enforceable contract existed at that moment.

3.4 Defects in Contract Formation

Suppose there has been an offer and acceptance of that offer, the bargained-for exchange is supported by consideration, and the content of the contract is reasonably certain. Nonetheless, defects in the contract formation process may result in a contract that is unenforceable (usually at the request of the promisee).

3.4A Defects Affecting Mutual Assent

The acts of offer and acceptance presume the parties are acting free of constraints and with awareness or knowledge of the true circumstances. Defects in the process affecting mutual assent would mean that one or both parties could argue the resulting bargain was not binding or enforceable.

Fraud, Misrepresentation, and Duty to Disclose. The negotiation process frequently consists of promises and factual representations made by each party. If important factual representations or promises are false, the premise underlying a contract—the voluntary entering into a bargained-for exchange—is negated. Generally speaking, the deceived party will be able to cancel the contract and (depending on the nature of the misrepresentation) also be allowed to sue the deceiver and obtain a monetary award. Generally, the difference between fraud and misrepresentation turns on the degree of deception involved.

Suppose one party to a contract wishes to cancel the transaction because of a claim that the other party should have disclosed important facts that, had they been known, would have persuaded the former not to enter into the transaction. Here, the claim is not one of active (or even innocent) misrepresentation, but instead a failure to disclose information. Although early cases rarely placed a duty to disclose on contracting parties, modern courts tend to do so where the matter that is not disclosed is important and where the party knowing the facts should have realized that the other party was not likely to ascertain them.

A duty to disclose is more commonly imposed upon public entities. These public agencies must disclose to contractors information they need to know (and cannot access themselves) in order to create a viable bid.

Mistake. The doctrine used most often in attacking the formation process is mistake. Mistakes can be unilateral or bilateral (involving one or both of the contracting parties). There may be mistakes as to the terms of the contract. One party may have simply made an error in an intended response; for example, with a misplaced decimal point on a number.

This would be a unilateral mistake. In addition, parties may be mistaken as to the basic assumptions on which the contract was made. Everyone who makes a contract holds certain underlying assumptions that, if untrue, render the contract undesirable. This would involve a bilateral mistake.

A few generalizations can be made. Although there is always some carelessness in the making of a mistake, the greater the carelessness the less likely that the party making the mistake will be given relief. A comparison of the values exchanged is relevant. If one party is getting something for almost nothing, the argument of mistake is more likely to be employed to relieve the other party from the contract. The question of when the mistake is uncovered is also important. If the mistake is discovered before the other party has relied on or performed under the contract, there is a greater likelihood that the contract will not be enforced. Of course, the law also seeks to prevent parties from avoiding performance of their promises by dishonest claims of mistake that are economically advantageous.

Unconscionability. Courts may refuse to enforce contract terms that are unconscionable. For "equity" courts in "old England", these were terms that shocked the conscience of the judge. Borrowing this concept, the Uniform Commercial Code § 2-302 gave the trial judge the power to strike out all or part of an unconscionable contract for the sale of goods. Gradually, this concept is being accepted in all contracts.

An unconscionability analysis focuses on several variables.

1. Is the contract one of adhesion (in which the terms are dictated by the economically stronger party)?
2. Is there an absence of true negotiation or meaningful choice?
3. Is the objectionable term hidden in the contract; for example, located in small print?
4. Is the term unfair or one-sided?

Economic Duress. A valid contract requires consent freely given. Consent cannot be obtained by duress or compulsion. Duress in the commercial world principally involves economic duress, which can exist if one party exerts excessive pressure beyond permissible bargaining and the other party consents because it has no real choice. Courts have found economic duress to exist in very few cases. Their hesitance undoubtedly relates to the inherent pressures involved in the bargaining process and the fear that many contracts could be upset if the economic duress concept were used too frequently.

3.4B Defect in Formal Requirements: Need for a Writing

For purposes of contract jurisprudence (or legal theory), written and oral contracts are equally enforceable, although proof of the existence of an oral contract may be more difficult. However, if the parties did not intend to be bound until their agreement was reduced to writing, then the courts will honor that intent.

> In this chapter's scenario, Prime Builder, Inc. and Sorry Subcontractor intended to sign their subcontract on the first day Sorry Subcontractor showed up for work, which it never did. If a court finds the parties intended not to be bound until a written subcontract was signed, then there would be no enforceable subcontract.

In addition, statutes as a matter of public policy may require certain types of contracts to be evidenced by a written memorandum or be in writing and contain certain terms in order to be enforceable. This section reviews these statutes.

Statute of Frauds. Each state has a **Statute of Frauds**. The statute lists specific transactions and requires that a sufficient written memorandum be signed by the party to be charged before there can be judicial enforcement of that transaction. In essence, the legislature views these transactions as being so important that there must be some writing to document the agreement.

For purposes of the construction industry, the most important transactions subject to the Statute of Frauds are promises to answer for the debt or default of another (which covers suretyship agreements discussed in Chapter 25); agreements that cannot be performed within a year from their making;[1] and contracts for the sale of goods over a specified value. (The Uniform Commercial Code § 2-201 sets this amount at $500.) Under legislation passed by both Congress and the states, electronic records and signatures comply with the Statute of Frauds.

Courts have eroded the Statute of Frauds by creating exceptions. In some transactions, either partial or full performance by the parties will make an agreement enforceable notwithstanding the lack of a written memorandum.

> In this chapter's scenario, Mr. Builder telephoned a hardware store and ordered $5000 worth of supplies. Since both parties are merchants and the transaction was for the purchase of goods, the UCC would apply. Since there is no evidence of a writing, that contract would be unenforceable under the UCC § 2-201, as the amount is for more than $500. However, if the hardware store then ships the supplies and Prime Builder, Inc. pays for them, this full performance would make the contract enforceable—notwithstanding its initial defect.

Homeowner Contracts. As with the Statute of Frauds, many states require certain contracts, entered into by consumers, to be in writing. This consumer protection trend has been extended to include regulation of building contracts entered into by homeowners. Typically, they mandate that these contracts be in writing, be of a minimum type size, and include consumer protection information (such as the right of cancellation and warning of the danger of mechanics' liens). Most statutes regulate the relationship between the homeowner and contractor,[2] although some also extend to design agreements.[3]

While a contract that violates the Statute of Frauds is unenforceable, a similar result may not apply to agreements that violate consumer protection statutes. Unless the statute clearly states that the contract is invalid if not in writing, the courts will examine a variety of factors to determine the effect of the contract's illegality on its enforceability. The courts balance variables such as the evil sought to be eliminated by the statute, the seriousness of the misconduct, the parties' reasonable expectations, and the danger of forfeiture (a total loss for the contractor). Since the purpose of statutes regulating home improvement and construction contracts is to protect homeowners from fraud and manipulation by unethical contractors, courts are less likely to void contracts entered into by sophisticated homeowners who would reap a windfall if the contractor were denied recovery.

As an example, a California case involved a $4 million oral contract to build a house for Cher, the movie star. She refused to pay and defended on the ground the contract was not in writing, as required by a statute. The court of appeal ruled that although the contractor could not sue for breach of the contract (because it was not in writing), it could sue outside the contract to recover the value of labor and materials provided so as to prevent the owner from obtaining a windfall.[4] As another example, an Illinois court allowed the contractor to sue the owner—a real estate attorney—for breach of an oral contract (because the written contract violated the statute).[5]

In this chapter's scenario, the transaction is for the construction of a residence; therefore, state law must be examined to determine whether such contracts must be in writing. Assuming such a statute exists, are the Smiths—presumably intelligent people as they are both doctors—entitled to the statute's protection? The answer is surely "yes," as there is no indication they are sophisticated with regard to construction work; indeed, quite the opposite appears to be true.

3.5 What Is the Contract?

The contract-making process is often messy, whether the underlying project is a home improvement or a new commercial building. Negotiations are often protracted. Communications are done face-to-face, at a distance, and by various media—from talking, to e-mails or letters—leading to the exchange of formal proposals and counter-proposals. Even the signing of a formal contract may not end the question of "what is the contract," as the parties' subsequent conduct may shed light on what they each believe the content of the contract is (but is this something the courts should examine?).

Out of this messiness, the law must determine what is the contract. Broadly speaking, this question requires three separate inquiries. The first inquiry involves the relationship between the written contract (which includes a provision that it is the complete and exclusive agreement of the parties) and prior or contemporaneous oral agreements allegedly made between the parties. This is the subject of the parol evidence rule addressed in Section 3.5A. Second, what terms may be added to the contract by a judge? This is the subject of judicially determined terms addressed in Section 3.5B.

Once the contract's content is established, the next question is: How are the words of a written agreement to be interpreted? This question of *contract interpretation* is the subject of Section 3.6.

3.5A Contract Completeness and the Parol Evidence Rule

The **parol evidence rule** bars the introduction of evidence (prevents a party from testifying) about alleged oral agreements (1) made prior to or contemporaneous with the signing of the contract (2) that contradict or vary the terms of a written contract. For the rule to be applied, the judge must first determine that the contract is "integrated," meaning the parties intended the written contract to be the full and final embodiment of their negotiations. Normally, this is done by including an "integration" clause in the contract.

As an example of when this issue of parol evidence may arise, a client may contend that the design professional agreed to perform certain services that were not specified in the written agreement. The latter's attorney may contend that the writing expressed the entire agreement and that parol (or oral) evidence is not admissible to add to, vary, or contradict the writing.

Reducing an arrangement to writing does not necessarily protect the design professional from assertions of additional oral agreements. In general, the more detail included in the agreement and the greater the likelihood that the client understood the terms or had legal counsel, the greater the probability that the client will not be allowed to prove the claimed oral agreement. On the other hand, most courts will permit either party, especially the client, to show prior oral agreements not included in the writing. Undoubtedly, one reason why clients are usually allowed to testify as to promises not found in the written contract is the close relationship between the design professional and the client.

As noted, the parol evidence rule is often triggered through inclusion of an integration clause in the written agreement. The American Institute of Architects owner/architect agreement, Document B101-2007, § 13.1, is an example of such a clause: "This Agreement represents the entire and integrated agreement between the Owner and the Architect and supersedes all prior negotiations, representations or agreements, either written or oral." By contrast, the Engineers Joint Contract Documents Committee's owner/engineer agreement, E-500 (2008), does not contain an integration clause.

> In this chapter's scenario, could the Smiths testify that Mr. Builder had orally promised that the work would take no more than three months? The answer is "yes"—the parol evidence rule does not apply (meaning the testimony will be allowed) because the parties never entered into a written contract.

3.5B Judicially Determined Terms

Implied Terms. Written contracts only rarely address all problems that may arise. It is difficult for the parties to consider all contingencies, or they have believed the resolution of the matters to be so obvious that contract coverage would be unnecessary. Also, the parties may have discussed certain matters during negotiations but were unable to agree on a contract solution to the problem, while still intending to have a binding contract. Courts may be asked to fill in the gaps not covered by the contract or decide matters left for future agreement where the parties could not agree. Courts will be more likely to perform these functions if convinced that the parties intended to make an enforceable agreement, especially where performance has begun.

Courts exercise restraint when asked to imply terms. A contract that expressly (in writing) addresses a particular subject matter generally precludes terms on that subject being implied. Also, custom (discussed next) should take precedence over implied terms. However, judicial implication sometimes must be exercised in order to articulate the reasonable expectations of the parties. For example, one court used an implied term to make explicit an owner's duty to furnish a site, reasoning as follows:

> Each party to a contract is under an implied obligation to restrain from doing any act that would delay or prevent the other party's performance of the contract. ... A party who is engaged to do work has a right to proceed free of let or hindrance of the other party, and if such other party interferes, hinders or prevents the doing of the work to such an extent as to render the performance difficult and largely diminish the profits, the first may treat the contract as broken and is not bound to proceed under the added burdens and increased expense.[6]

This implied obligation prevents the owner from impeding the contractor's work. A similar obligation may be imposed on the contractor and in favor of a subcontractor.

Implied obligations can go further, requiring an owner to perform positive acts to assist the contractor or for a prime contractor to assist the subcontractors. As an example of the latter, a prime contractor has an implied duty to coordinate the work of the subcontractors, even if the prime contract does not explicitly impose this duty. Similarly, an owner who hires the trade contractors directly, but no prime contractor to oversee the work, has an implied obligation to coordinate the work of the trade contractors.[7]

The contractor is usually the most sophisticated party, and the contractor has an implied obligation (one not explicitly stated in the prime contract) to protect the owner. For example, a contractor has an implied obligation to warn the owner of soil conditions that, if not adequately addressed in the design, will eventually lead to failure of the project.[8]

Custom. Customary practices are important in determining the rights and duties of contracting parties, particularly in complex transactions such as construction, where not everything can be stipulated in the contract. In addition, courts recognize that parties contract with reference to existing customs. In that sense, courts can be said to be simply giving effect to the actual intention of the parties. However, a contracting party may also be held to those customs that it knew or should have known. The law places a burden on contracting parties to learn customs that apply to the type of transaction they are about to enter and in the place where they contract.

In any event, an established custom can be a more convenient and proper method of filling gaps than a court determination of what is reasonable. For example, if customarily the contractor obtains a building permit, the court is likely to place this responsibility on the contractor, but only where there is a contract gap on this point. Invoking custom is preferable to a judge's determination of who should "reasonably" obtain the permit. Custom can also help the court interpret terms or resolve conflicts in the contract documents.

> In this chapter's scenario, custom may help determine who had the obligation to obtain a building permit, as the contract never addressed this issue.

Legal Requirements: Building Codes. Building codes and land use controls play a pervasive role in construction, defining the legal parameters under which the project may be performed. (Land use is discussed in Chapter 5.) The contract frequently states expressly that the contractor will comply with applicable laws. (If the contract is silent on this matter, such an obligation would be implied.) Further analysis is needed to separate several related but distinct problems.

Where design and construction are separate, the former is the owner's responsibility and is usually done by the design professional. Suppose the design violates building code requirements. While the contractor may contend that its job is to build the design, it should not ignore violations of law. It should direct the design professional's attention to any obvious code violations in the design.

This is the approach taken by the American Institute of Architects. AIA Document A201-2007 § 3.2.3, which states that a contractor, upon receipt of the design, is not required to determine whether it is in accord with legal requirements; however, the contractor must promptly report to the architect any nonconformity discovered or made known to the contractor (presumably by a subcontractor). Under § 3.7.3, if the contractor performs work knowing it to be contrary to such laws without notifying the architect, the contractor "shall assume appropriate responsibility for such Work and shall bear the costs attributable to correction." A similar approach is taken in EJCDC Doc. C-700, § 6.09B (2007).

This solution effectuates a sound middle ground between putting an unreasonable responsibility on the contractor and allowing the contractor to close its eyes in the face of danger.

Good Faith and Fair Dealing. A hallmark of modern contract law is the implied *duty* of **good faith and fair dealing** that is imposed by law (independent of the intentions of the parties) on each contracting party. Each party should not only avoid deliberate and willful frustration of the other party's expectations but should also extend a helping hand where to do so would not be unreasonably burdensome. Contracting parties, although not partners in a legal sense, must recognize the interdependence of contractual relationships. This implied duty is generally viewed as applying to all contracts, including to transactions subject to the Uniform Commercial Code: sales of goods between merchants.[9]

Some applications of the doctrine are simply recognition of implied terms, as discussed previously. But more expansive use of the doctrine emphasizes the unspoken objectives of the contracting parties, the spirit of the contract itself, and the need for elementary fairness.

Summary. A party should look first and foremost to the written contract to determine its obligations and those of the other party. Yet a variety of legal doctrines—implied terms, custom, legal requirements, and the duty of good faith and fair dealing—expand these obligations in ways that may catch the contracting parties by surprise. Common to all of these doctrines is the law's attempt to create a binding contract (where doing so was clearly the intent of the parties) in which the *reasonable* expectations of both parties (even if not explicitly articulated in the contract) are protected. Knowing the law's *motivation* should help the contracting parties to understand the full spectrum of their obligations on the project.

3.6 Contract Interpretation

Basic Objectives. Once an enforceable bargain is formed and the content of the agreement is determined, the parties may disagree as to the *meaning* of the contract's terms and obligations. The basic objective in contract interpretation is to determine the intention of the parties. However, within this relatively simple standard lurk many problems:

1. What can be examined to ascertain the intention of the parties?
2. Once the relevant sources are examined, how is the intention of the parties determined?
3. What if each party has different intentions?
4. What if one party knows of the other party's intention?
5. What if the parties have no particular intention about the matter in question?
6. How is the intention of the parties determined when they did not write the contract language but instead used a standard form contract (e.g. by the AIA or EJCDC)?

The preceding list is given merely to indicate that the linch-pin phrase of contract interpretation—"intention of the parties"—is deceptively simple, often hiding difficult interpretation problems. Note that the term "intention of the parties" must be understood in light of the objective theory of contracts discussed in Section 3.3B. Courts will not examine any undisclosed intentions of the parties they claim existed at the time they made the contract. If these intentions are made known to the other party, they may be relevant.

Language Interpretation. Words have no inherent meaning—they develop meanings because people who use them as tools of communication attach meanings to them. Courts seek to put themselves in the position of the contracting parties and determine what the contracting parties must have meant or intended when they used the language in question.

Yet what could otherwise be a free-wheeling form of inquiry *must* be constrained by articulated approaches to contract language interpretation; otherwise, an appellate court would have no basis for reviewing the decision of the trial judge or jury. As difficult as inquiry into the meaning of contract language is, that inquiry must proceed using an objective and accepted method or methods. Some of these methods are listed here.

Courts begin with the contract wording itself. Generally, they assume the parties intended the plain and ordinary meaning of the words they use. Dictionaries may be used to ascertain possible meanings. In commercial settings, such as construction contracts, the judge may also examine trade practice or industry custom to discern the meaning of the words. However, if the meaning is clear, it cannot be contradicted by claims of trade practice. For example, if the contract calls for one coat of primer and two coats of paint, the contractor cannot do less by claiming it is industry custom to apply one coat of primer and one coat of paint.

If the contract is ambiguous—subject to two different but reasonable interpretations—the courts will admit the introduction of "extrinsic" evidence (outside the contract itself) to determine the words' meaning. The parties may testify as to their intentions or understandings of the contract's requirements.

Courts invoke other interpretation guideposts. One is to look at the conduct of the parties themselves to see if it sheds light on how they understood their contract obligations. For example, if the contract requires changes in the work to be authorized in writing, but the owner repeatedly orally orders changed work which it then pays for, the court will interpret the parties intention as to *not* require strict compliance with the contract's notice or procedural steps. Likewise, the prime contractor periodically billing the owner in accordance with certain unit prices and the owner paying these billings would indicate that the unit prices were correct. The contractor doing what the owner had directed without complaint could indicate the contractor's acquiescence in the owner's interpretation. Note that this doctrine of practical interpretation applies only to language susceptible to more than one interpretation. However, this rarely is a difficult obstacle.

Several *canons of interpretation* (all with Latin names) may be invoked to ascertain the meaning of words by looking at the context in which they are found. Suppose the contract provides a list of items (for example, that a party is excused from performance in the event of strikes, fire, explosion, storms, or war). Under one canon, the occurrence of an event not mentioned, such as a drought, would not excuse performance. Where the parties have expressed five justifiable excuses, they must have intended to exclude all other excuses for failure to perform. (This guide is sometimes harshly applied and does not give realistic recognition of the difficulty of drafting a complete list of events.)

Another important interpretation guide, **contra proferentem** "against the one who puts it forward" (meaning the drafter of the language), interprets ambiguous language against the party who selected the language or supplied the contract. Usually this guide is applied not to negotiated contracts but to those mainly prepared in advance by one party and presented to the other on a take-it-or-leave-it basis (adhesion contracts). One basis for this guide is to penalize the party who created the ambiguity. Another and perhaps more important rationale is the necessity of protecting the reasonable expectations of the party who had no choice in preparing the contract or choosing the language. The *contra proferentem* guide can break a tie when all other evidence is either inconclusive or unpersuasive.

Another canon of interpretation is that courts presented with two possible interpretations of the language often state that they will choose an interpretation which is reasonable and fair over an interpretation that is unreasonable or unfair.

Yet another canon is that an interpretation of the contract will not be adopted if it nullifies (makes irrelevant) other parts of the contract. Rather, an interpretation will be preferred if it gives meaning to all parts of the contract.

Courts will apply particular scrutiny to contract clauses that appear to be one-sided or that could encourage careless conduct. **Exculpatory clauses** relieve one party from liability for its own negligence. They are not enforced unless evidence of the other party's consent to accept the risk is sufficiently clear.

Finally, courts employ different interpretive approaches depending on whether the contracts may be classified as negotiated or adhered to (accepted without meaningful choice). The latter are referred to as *contracts of adhesion*.

A negotiated contract arises when two parties with reasonably equivalent bargaining power enter into negotiations, give and take, and jointly work out a mutually satisfactory agreement. Courts faced with negotiated contracts do not give special preference to either party.

The adhesion contract has no or minimal bargaining. The dominant party hands the contract to the weaker party on a take-it-or-leave-it basis. At its extreme, the weaker party who wishes to enter into the transaction must accept all the terms of the stronger party. In addition, in many transactions the weaker party will find the same terms used by the competitors of the party whose terms were unpalatable or will find no competitor (because the stronger party also has monopoly power).

Courts apply special interpretive rules to adhesion contracts. Ambiguous terms will be interpreted against the party who supplied the contract. (This is the canon of *contra proferentem* mentioned earlier.) In addition, it is more likely that terms in an adhesion contract than in a negotiated contract will be considered unconscionable (see Section 3.4A.)

3.7 Contract Breach

3.7A Definition

A contract is an exchange of promises which are manifested in performance. Absent a legal excuse, failure to perform in the time and manner required constitutes breach of a contract term. Stated differently, a party that fails to perform a promise is in breach of the contract. Every breach entitles the aggrieved party to a claim for damages (a monetary award).

The aggrieved party's entitlement to damages must be distinguished from its obligation to perform under the contract. Unless the breach is "material," the aggrieved party remains obligated to continue performance. Conversely, if the breach is "material" and is not cured or likely to be cured, the aggrieved party may cancel (or terminate) the contract, and its obligation to perform is ended.

3.7B Material Breach

The law examines the surrounding facts and circumstances to determine whether the breach is **material**. The Restatement (Second) of Contracts § 241 articulates factors that are significant in determining whether a particular breach is material. These factors include:

1. the extent of injury to the aggrieved person
2. the extent to which the aggrieved party can be compensated
3. the extent to which the breaching party will suffer forfeiture
4. the likelihood the breaching party will cure
5. the reason for the breach: whether the breaching party's failure to perform comports with the standards of good faith and fair dealing.

One can look at these factors from the perspective of contractor breach and owner breach. Looking first at breaches by the contractor, the first factor examines the importance of the deviation. The second factor examines whether the defect can be remedied. The third looks to whether the contractor will suffer uncompensated damages if the contract is terminated. (If so, that would suggest against a finding of a material breach.) The fourth item addresses whether the contractor will likely repair the defect. As an example of the final factor—the reason for the breach—if defective or late work was performed by a subcontractor that the prime contractor could not have reasonably prevented, this would suggest against terminating the prime contract.

An owner breach would probably be nonpayment of money, failure to furnish the site, or noncooperation. The first factor would look at the extent of the breach and the likelihood of future performance, while the second would look at whether the breach was easily compensable, such as interest for a breach consisting of not making a progress payment.

The third factor examines the harm nonperformance would cause the owner, such as lost loan commitments, liability to prospective tenants, or other lost business opportunities. The fourth factor seeks to determine whether the breaches are likely to be cured. Finally, the fifth factor examines the reasons for nonperformance. Financial reverses or a steep rise in interest rates makes it less likely that the contractor can terminate for failure to pay. If, however, there was "bad blood" between the design professional and the contractor and the latter seized on the breach by the owner to injure the design professional, termination would be less likely. Similarly, if it appears the contractor sought an excuse to end the contract, the breach would less likely be material.

As should be clear, the determination of whether a breach is material requires a balancing of several factors. Not only are these factors fact-dependent, they must be evaluated in light of the larger context of the project as a whole, the interests of the parties, and even their motivations (or, more accurately, their "bona fides"—whether they acted in good faith). Not only is an evaluation of the breach difficult on a practical level, but a "wrong" guess by the aggrieved party would mean that its termination of the contract was wrongful, and the *original* aggrieved party is now the one in material breach. In light of the importance of a decision to terminate the contract, this matter will be discussed in more depth in Chapter 22.

3.7C Future Breach: Prospective Inability and Anticipatory Repudiation

Sections 3.7A and 3.7B dealt with breaches that have occurred. This section deals with breaches that *may* occur in the future.

Each party owes the other party the duty to appear to be ready to perform when the time for performance arises. Suppose the contractor discharges some of its employees, a number of the employees quit, the contractor cancels orders for supplies, its suppliers indicate they will not perform at the time for performance. In such cases, the owner has reasonable concerns as to whether the contractor will be able to perform. Under the Restatement (Second) of Contracts § 251, the owner may demand *assurance* and, in the absence of assurance, legally terminate any obligation to use the contractor.

The contract may address the rights of parties who are concerned the other might not perform. For example, the AIA General Conditions, A201-2007, provide the contractor with specific options in the event the owner fails to make periodic (or progress) payments as required by the contract. Under § 2.2.1 of A201-2007, the contractor may ask the owner for evidence of its financial arrangements. Under § 9.7.1, the contractor may suspend performance. These contractual rights make explicit the common-law right of adequate assurance.

A breach by *anticipatory repudiation* occurs when one party indicates to the other that it cannot or will not perform. The aggrieved party may rely upon the other party's anticipatory repudiation of the contract and terminate its obligation without waiting for the other party to commit an actual breach.

> In this chapter's scenario, some of the questions raised are (1) did the Smiths' payment of less money constitute a material breach entitling Prime Builder, Inc. to leave the job? (2) did the Smiths' failure to provide proof of adequate funding constitute an anticipatory breach of the contract? (3) did the contractor have a right under the common law (since any contract was silent on the matter) to demand an assurance of funding before it continued performance?

REVIEW QUESTIONS

1. What are the three requirements that must be met in order for a contract to be valid?

2. What is the objective theory of contracts?

3. What is consideration?

4. What is promissory estoppels and when might it be used in the construction industry?

5. What are the four primary variables that an unconscionability analysis focuses on?

6. What is an implied contract term and what are the major sources of judicially determined terms?

7. What is the parol evidence rule?

8. What is a breach of contract?

9. According to the Restatement (Second) of Contracts, what are the five factors used in determining whether a particular breach is material and what is the legal consequence of a finding the breach was material?

10. How does a contractual breach by anticipatory repudiation occur?

ENDNOTES

[1] *R. Loftus, Inc. v. White*, 85 N.Y.2d 874, 649 N.E.2d 1196, 626 N.Y.S.2d 52 (1995) (oral agreement to build house *and* to provide a one-year warranty is subject to the statute).

[2] Cal.Bus. & Prof.Code §§ 7159 (home improvement contracts) and 7164 (construction of single-family homes). California's Statute of Frauds is found in Cal.Civ.Code § 1624.

[3] Cal.Bus. & Prof.Code § 5536.22. The statute extends to contracts between an architect and a client, without limitation to homeowners.

[4] *Arya Group, Inc. v. Cher*, 77 Cal.App.4th 610, 91 Cal.Rptr.2d 815 (2000).

[5] *K. Miller Construction Co., Inc. v. McGinnis*, 238 Ill.2d 284, 938 N.E.2d 471 (2010).

[6] *United States ex rel. E. & R. Constr. Co., Inc. v. Guy H. James Constr. Co.*, 390 F.Supp. 1193, 1206 (M.D.Tenn.1972), aff'd without opinion, 489 F.2d 756 (6th Cir.1974).

[7] *Broadway Maintenance Corp. v. Rutgers, State Univ.*, 90 N.J. 253, 447 A.2d 906 (1982).

[8] *Farmer v. Rickard*, 150 P.3d 1185 (Wyo.2007).

[9] UCC § 1-304 (formerly 1-203).

CHAPTER 4

Remedies for Contract Breach: Focus on Construction Disputes

Scenario

Oliver Owner wishes to build an office building. Oliver hired Original Architect, Inc. to create the design. Over several months, Original Architect, Inc. gave Oliver several drafts of increasing detail.

Uninspired by what he saw, Oliver terminated its contract with Original Architect, Inc. and hired Replacement Architect, Inc. in its place. Oliver gave Replacement Architect, Inc. the last drawing Oliver had received from Original Architect, Inc. Replacement Architect, Inc. used this last drawing to create its own final design.

Oliver then hired Prime Builder, Inc. to be the prime contractor on the job. The contract price was for a fixed amount, with work to be completed in one year.

The project did not go smoothly. Oliver failed to obtain a zoning change on time, and work had to begin one month late. Meanwhile, Prime Builder, Inc. found the design literally impossible to implement; for example, all of the specified conduits would not fit into the space allotted, forcing Replacement Architect, Inc. to redesign. This took Replacement Architect, Inc. two weeks to do.

Meanwhile, Replacement Architect, Inc. was unhappy with the quality of the contractor's work. Within a week after the foundation was poured, it began to crack. Replacement Architect, Inc. determined the soil preparation work was not properly done. It ordered Prime Builder, Inc. to remove the foundation and redo the soil preparation, which the contractor did.

The project was finished three months late. As a result, several commercial tenants who had promised to move into the new building found rentals elsewhere. Overwhelmed with all the problems on the construction project, Oliver did not even try to find replacement tenants until two months after the building was completed. Eventually, Oliver fully rented the building.

Litigation ensued

1. Original Architect, Inc. sued Oliver Owner for (a) nonpayment of the design work the architect had done and (b) emotional distress for not being chosen as the project architect.
2. Oliver sued Replacement Architect, Inc. and Prime Builder, Inc. for late completion.
3. Prime Builder, Inc. sued both Oliver and Replacement Architect, Inc. for (a) higher performance costs than anticipated, (b) home office overhead expenses, and (c) lost profits on the new contracts Prime Builder, Inc. would have signed had it not been still working on Oliver's project after the original, one-year completion date.

4.1 Introduction

Determining the remedies available for breach of contract is much more difficult than determining whether a valid contract has been made. It is exasperatingly difficult to predict before trial whether the law will require that a breaching party must pay and the amount of that payment. Even after all of the evidence has been produced at trial, it can be very difficult to determine the precise remedy that the party entitled to relief should receive.

Although the primary function of awarding a remedy for contract breach is to compensate the injured party, measuring losses incurred and gains prevented can be very difficult. There may be agreement on general objectives, yet implementing these objectives is complicated by the great variety of fact situations, the different times at which a breach can occur in the history of a contract, the variety of causes that generate the breach, and the different judicial attitudes toward breach of contract itself. In addition, courts must balance full compensation against the need for reasonable precision in tracing the loss to the breach, as well as the aggrieved party's duty to limit its losses as much as possible.

The common law has developed conventional formulas intended to remedy the effects of a contract breach. However, the complexity and nature of construction disputes means that a law of contract damages peculiar to such disputes has supplemented the common law formulas. Here too, recognition of a financial injury must be balanced with the difficulty of proof.

4.2 Types of Remedies for Contract Breach: Overview

Judicial remedies can be divided into the following:

1. Money award: judgments that simply state that the defendant owes the plaintiff a designated amount of money or "damages."
2. Specific performance: judgments that specifically order the defendant to do something.
3. Injunction: judgments that specifically order the defendant to stop doing something.
4. Declaratory judgment: judgments that declare the parties' contract rights.

Section 4.3 addresses the last three remedies together, while Section 4.4 focuses on the much more common remedy of a damages or money award.

4.3 Specific Decrees and Declaratory Judgments

The specific decree—whether specific performance or an injunction—is a personal order or command by the judge. It is a very effective remedy. A defendant who does not comply may be cited for contempt of court. The defendant is brought before the judge to explain why she has not complied with the decree. If the explanation is not satisfactory, the judge can punish the defendant by imposing a fine or imprisonment or by coercing the defendant into performing by stating that the defendant must pay a designated amount or stay in jail until she performs.

A **declaratory judgment** delineates the contract rights of the parties under stipulated facts. A party seeking a declaratory judgment is asking the court to assess the party's rights under the contract, usually as a means to preclude further litigation. Most commonly, declaratory judgment actions are sought in insurance disputes to determine whether insurance coverage exists under the terms of the insurance policy. In one construction case, an owner's request for a declaratory judgment—that its contractor's invocation of a contract clause defense was untimely—was disallowed.[1]

The great majority of claims addressed in this book will, if successful, result in money awards.

4.4 Money Awards

Money is awarded for different purposes. While virtually all money awards for the claims discussed in this book are for compensatory damages, money may be awarded in the judgment for other purposes. The major components of a money award are the following.

Compensatory Damages. The law compensates the nonbreaching contracting party for the losses it suffered because of the other party's breach of contract. The court judgment seeks to (1) put the injured party in the position it would have been in had there been no breach, (2) restore the injured party to the position it occupied before performance began, or (3) prevent the breaching party's unjust enrichment.

Benefit of the Bargain. The most common measure of breach of contract damages is to give the plaintiff the "**benefit of the bargain**," meaning the plaintiff is put in the financial position it would have been in had the defendant properly performed.

> In this chapter's scenario, suppose Oliver after a couple of years discovers the roof leaks, and the defect is traced to poor workmanship by Prime Builder, Inc.'s roofer. Protection of Oliver's expectation interest—to give him the benefit of his bargain, which is to receive a properly constructed office building in exchange for the contract price—would mean an award against Prime Builder, Inc. of enough money for Oliver to hire a new roofer to fix the problem.

Otherwise stated, Oliver would be entitled to an award of the repair cost of the contractor's defective work.

Reimbursement. On rare occasions, compensatory damages means restoring the party to the position it occupied before performance began. The plaintiff is reimbursed the losses it had suffered because of the defendant's breach, so it is placed in the financial position it would have been in had no contract been made.

> In this chapter's scenario, Original Architect, Inc. had produced multiple designs for Oliver, for which the architect was not paid. Compensation of the architect's reliance interest would reimburse the labor cost of having produced those designs.

Restitution. Suppose the contract suffers from a technical defect or work is performed outside of the scope of the contract. In these situations, compensatory damages for breach of contract is not possible. Here, the purpose of damages is to restore the status quo that existed before the contract. The plaintiff is awarded both the loss to itself and the benefit it has conferred on the defendant.

Again, plaintiff's claim is not for breach of contract but for **restitution**. A claim for restitution is phrased as one either for *quantum meruit* (in Latin, meaning as much as he deserved) or unjust enrichment. Under either theory of recovery, restitution is ordered lest the defendant be unjustly enriched by retaining the benefit of the plaintiff's services.

For an example of restitution, recall in Section 3.4B the statutory requirement that residential contracts be in writing. That section noted that refusal to enforce oral contracts would result in forfeiture to the contractor, even if the homeowner was a sophisticated party and not in need of statutory protections. While an unpaid contractor could not sue the homeowner for breach of contract (because the contract violated the statute), some courts (to avoid forfeiture) allow the contractor recovery in restitution, so the contractor could recover the cost of the labor and material it had expended. Failure to allow restitution would have resulted in unjust enrichment of the homeowner.

Interest. Considerable time may pass between when the breach occurred and when the court judgment is awarded. An award of prejudgment interest, which is calculated as a percentage rate applied to the amount of compensatory damages, compensates the plaintiff for its lost use of funds. If not specified in the contract, the interest rate is set by statute.

Some statutes mandate the award of prejudgment interest to the party who prevailed in the litigation. For example, statutes which require the payment of subcontractors within a certain time after the prime contractor has been paid by the owner (prompt payment statutes) will specify an interest rate that must be added to the compensatory award. See Section 17.8.

Attorneys' Fees. Even if a plaintiff receives the compensatory award it requested coupled with an award of prejudgment interest, it will not have achieved full compensation. Such an award does not take into consideration the transaction cost of achieving the award: Attorneys' fees and litigation costs—to say nothing of resources devoted to litigation by the businesses or persons involved.

Focusing first on attorneys' fees, the "American rule" is that each party bears the cost of its own fees. The purpose of the American rule is to encourage the bringing of claims, as the party filing the lawsuit knows it will have to pay only its own lawyers' fees—regardless of whether it wins or loses.

The American rule may be overridden by agreement, and often claims for attorneys' fees are based on a contract provision. Documents published by the AIA do not—except for indemnification—provide for attorneys' fees. Instead, each party bears its own litigation costs.

Attorneys' fees also may be mandated by statute, particularly in consumer or civil rights cases. The purpose is to promote socially important public policies by removing an economic impediment to bringing a claim. Similarly, prompt payment statutes (discussed in the earlier subsection) often mandate the award of attorneys' fees to the prevailing party.

Litigation Costs. The cost of construction litigation—particularly of complex disputes that involve defects and impact claims—cannot be ignored. These claims generate immense costs to reproduce, classify, analyze, and store documents as well as the staggering costs involved in conducting pretrial discovery. To these costs must be added the expense of preparing exhibits and retaining expert witnesses and the nonproductive costs incurred by personnel in preparing for the lawsuit. Those who draft contracts may wish to deal with these costs. Such costs should remind the parties they must make every effort to avoid litigation.

Emotional Distress and Punitive Damages. The basic purpose in awarding damages for breach of contract is to compensate a party who has suffered financial losses or been prevented from making gains. The common law did not view contract breach as an immoral act that might justify punishment or as causing emotional distress deserving of compensation. This view may have been based on the importance—in a market-oriented society—of people engaging in economic exchanges. Excessive sanctions may discourage people from making contracts.

For these reasons, **emotional distress** and **punitive damages** awards have been quite rare in construction disputes, particularly for claims of breach of contract. (These damages are more likely to be awarded in tort claims and will be addressed in Chapter 5.)

> In this chapter's scenario, Original Architect, Inc.'s claim of damages for emotional distress for not being used as the project's architect would be denied as a type of loss not properly subject to remedy for a breach of contract.

4.5 Limits on Recovery

After proving a breach of contract by the defendant, the plaintiff seeking money for its legal injury must overcome limits on financial recovery. These limits should be kept in mind when reading the remainder of this chapter, which addresses compensatory damages in different contexts.

Causation. The claimant must show that the defendant's breach has caused the loss. Losses may be caused by more than one actor, and causal factors can include other events and conditions. For example, the contractor's performance may be delayed by the owner, strikes, material shortages, and the contractor's poor planning. It may be difficult—if not impossible—to establish the amount of the loss caused by contributing causes or conditions.

The defendant will be responsible for the loss if its breach was a *substantial factor* in bringing about the loss. It need not be the sole cause of the loss.

Causation problems can be particularly difficult when claims are made for breach of a contract to design or to construct, mainly because of the number of different entities that participate in the process and also because of the variety of conditions that may affect causation.

> In this chapter's scenario, the project was completed three months late. Oliver and Prime Builder, Inc. sued each other (and both sued the architect), seeking to recover damages caused by the delay: lost income for the owner and higher performance costs for the contractor. Yet how is each plaintiff to establish that the defendant *caused* the delay when all three parties appear to have contributed to the delay? Oliver did not obtain the zoning change; prime Builder, Inc.'s soil preparation work had to be redone; and Replacement Architect, Inc.'s design had to be revised. In addition, the project was delayed by weather, which is not the fault of any party. Unless the consequences of each party's breach are apportioned, all three may be denied recovery because of their failure to prove causation.

Certainty. A claimant must prove the extent of losses with reasonable certainty. This limitation may best be understood by emphasizing the word "reasonable." If the fact of damage is conclusively established, the extent or amount need not be shown with mathematical certainty, but only with reasonable certainty. Sometimes this limitation is expressed in the negative by stating that a damages award cannot be based on mere speculation.

Foreseeability of Loss. A series of improbable events can combine such that a minor breach leads to disproportionately large losses. To prevent a defendant from being burdened with unexpectedly large and disproportionate liability, the common law limits a defendant's liability to damages that would foreseeably result from the breach as determined at the time the contract was made. For example, the law asks: What financial losses to the owner would a contractor reasonably anticipate at the time it signed the contract, in the event the contractor later breaches that contract? Unanticipated or unforeseeable losses should not be charged to the contractor, without evidence that it agreed to do so or should have known these losses may occur.

Mitigation. A plaintiff's duty of mitigation is also called the *doctrine of avoidable consequences*. However phrased, it requires the victim of a contract breach to do what is reasonable to mitigate or reduce the damages. This limiting rule relates to the requirement that the breaching party is responsible only for those losses that its breach has caused. If the loss could have been reasonably avoided or reduced by the plaintiff, the plaintiff cannot transfer that loss to the breaching party.

> In this chapter's scenario, Oliver delayed its search for replacement tenants. The income it lost from this delayed search was avoidable, meaning that Prime Builder, Inc.'s damages may be reduced by Oliver's failure to mitigate its damages.

Lost Profits. Claims for lost profits implicate all of the preceding limitations on contract recovery: causation, certainty, foreseeability, and avoidable consequences. Proof of lost profits with reasonable certainty is particularly difficult for a new business—lacking a solid history of profitable earnings it can point to as a baseline to show the consequences of the defendant's breach.

In the construction setting, an owner whose project's completion was delayed may claim lost profits in the form of lost leases from prospective tenants or lost opportunities for resale profits. A contractor who claims breach by the owner may assert that future contracts were lost because of its diminished reputation or reduction in bonding capacity. (A bond, issued by a surety, backstops the contractor; a previously terminated contractor would be more risky to guarantee with the result that any new bond, if not denied outright, would cost the contractor more.)

> In this chapter's scenario, Prime Builder, Inc.'s claim for lost profits on *other* jobs it would have obtained had the owner and architect not delayed *this* project would probably be denied as too speculative.

Lost profit claims may be barred by waiver of consequential damages clauses, discussed next.

Contract-Specified Remedies. The preceding discussion has assumed an absence of any controlling contract clauses regulating the remedy. Such clauses may displace the common law remedy. Several contract-specific remedies are of particular importance to the construction industry.

Exculpation or limitation-of-liability clauses seek to eliminate or cap one party's potential liability. Liquidated damages clauses specify an amount of money due in the event of a particular breach; most commonly, late completion by the contractor.

> In this chapter's scenario, if the prime contract contained a liquidated damages clause, Oliver's recovery of damages for late completion of the building would be determined by application of this clause. Oliver would recover the per-day amount specified in the clause, multiplied by the number of days of delay for which the contractor would be responsible. This recovery would preclude Oliver's claim for lost rental income.

A breach of contract may cause not only direct costs, but also indirect or "consequential" costs which can be large and often unquantifiable. Although consequential damages may ultimately be barred under the foreseeability requirement, waiver of consequential damages clauses seek to preclude the claim from arising in the first place. Probably the most significant consequential damage claim precluded by these clauses in one of lost profits on *other* transactions allegedly caused by the defendant's breach of *this* transaction. These clauses have been adopted by the AIA in both the "General Conditions of the Contract for Construction," A201-2007, § 15.1.6, and in the owner/architect agreement, B101-2007, § 8.1.3. Recall that standard form documents (including those of the AIA) are not compulsory, and even when used on a project may have been modified by the parties. For example, a powerful owner may delete the waiver of consequential damages clause so as to increase its recoverable damages.

4.6 Damages Specific to Construction Disputes: Introduction

The discussion up to this point has addressed the law of contract damages generally. The principles discussed previously apply equally whether the underlying transaction is a consumer's purchase of a car from a dealer, a restaurant's purchase of food from a farm, an employee's claim her employment was wrongfully terminated, or a subcontract to perform services on a construction project.

The remainder of this chapter addresses the law of damages as uniquely applied to construction disputes. Section 4.7 covers contractor claims against the owner (this analysis applies equally to subcontractor claims against the prime contractor), while Section 4.8 covers owner claims against the contractor.

4.7 Contractor Versus Owner

Claims by contractors against owners can arise in many contexts. The principal claims are based on the following:

1. refusal to award the contract to the successful bidder.
2. refusal of the owner to permit the contractor to commence performance after the contract has been awarded.
3. wrongfully terminating the contractor's right to perform during performance.
4. failing to pay the contractor for work performed under the contract.
5. committing acts that increase the contractor's cost of performance.

As noted, this discussion of contractor claims against the owner applies equally to subcontractor claims against the prime contractor. In that case, the prime contractor occupies a position similar to that of the owner, and the subcontractor occupies a position similar to that of the prime contractor.

4.7A Project Never Commenced

The contractor is entitled to protection of its expectation interest. This would put it in the position it would have been in had it been allowed to complete performance. Had it performed, it would have received the contract price. From that is subtracted the expense saved, which is the cost of performance. The difference would have been the contractor's profit on the project.

4.7B Project Partially Completed

The damages award should place the contractor in the position it would have been in had the owner performed in accordance with the contract. Different formulas can be used to calculate the award. Probably the most common is to begin with the contract price and then subtract payments received *plus* the reasonable cost of completion. The difference is the contractor's profit on the job.

For example, assume that a project has the following figures:

1. contract price—$100,000
2. expenditures in part performance—$60,000
3. cost of completion—$30,000
4. progress payments received—$50,000

The contractor would receive the following:

$100,000	(contract price)
−30,000	(cost of completion)
$70,000	(subtotal)
−50,000	(progress payments received)
$20,000	(net recovery)

This damages formula assumes a relatively straightforward project in which matters continued pretty much as planned until the point of termination.

A different formula—one based on restitution—may be used where the other party committed a serious breach so that the nature of the contractor's performance obligation was fundamentally altered. The performance bears little resemblance to the contract or the assumptions underlying the contract. Breaches sufficiently serious to justify restitution have been failure to make progress payments, excessive changes, and failure to perform those acts during performance that would allow the contractor to perform in the most expeditious way.

The contractor is entitled under restitution to receive the value that his performance has benefited the owner, generally the reasonable value of the materials and labor it contributed to improving the owner's land. Normally, this is measured by the actual costs incurred by the contractor in performing its contractual obligations. A contractor who seeks restitution may recover profit on performed work—but not on unperformed work.

Sometimes the owner will claim the project is not completed and the contractor is in breach of contract. The contractor in defense will argue that the project, although not technically complete, is *substantially complete*, and it is entitled to the payment of the contract balance less the cost to perform any minor work not yet done. Substantial completion in addressed in Section 17.9.

4.7C Project Completed

A contractor who has fully performed but has not been paid the balance of the contract price is entitled to the unpaid balance of the contract price and any other losses it can prove. This subsection will address the claims for costs in addition to the contract balance as well as the methods for proving the damages amount of such claims. But first, the setting or context from which such claims arise is explained.

Site Chaos and Productivity. Many complex construction contract disputes involve claims for higher-than-anticipated performance costs. The contractor may assert that it was required to perform in an inefficient and costly manner for a variety of reasons—separately or in combination—such as defective specifications, unexcused delays, excessive changes, and/or administrative incompetence of the owner. To set the scene, a court facing such a claim stated,

> We note parenthetically and at the outset that, except in the middle of a battlefield, nowhere must men coordinate the movement of other men and all materials in the midst of such chaos and with such limited certainty of present facts and future occurrences as in a huge construction project such as the building of this 100 million dollar hospital. Even the most painstaking planning frequently turns out to be mere conjecture and accommodation to changes must necessarily be of the rough, quick and ad hoc sort, analogous to ever-changing commands on the battlefield. Further, it is a difficult task for a court to be able to examine testimony and evidence in the quiet of a courtroom several years later concerning such confusion and then extract from them a determination of

precisely when the disorder and constant readjustment, which is to be expected by any subcontractor on a job site, become so extreme, so debilitating and so unreasonable as to constitute a breach of contract between a contractor and a subcontractor. This was the formidable undertaking faced by the trial judge in the instant case and which we now review on the record made by the parties before him.[2]

Walter Kidde Constructors, Inc. v. State involved a contractor (Kidde-Briscoe) claim in a dispute over a project that finished almost 900 days later than planned. Much of the responsibility fell on the public owner. The owner failed to make certain parts of the site available as promised. It issued many hold orders that substantially interfered with and disrupted construction while plans were being redesigned. It was late in processing change orders and approving shop drawings while still insisting that the contractors accelerate (speed up) their work under threat of imposing $250-per-day penalty for not completing within the time requirements of the contract. In describing the effect of these delays, the court stated,

> The long delay in completing construction had a devastating impact on Kidde Briscoe. They were not only required to provide labor and materials far in excess of and different from that called for in the original contract but to perform work out of sequence at an accelerated rate and in a manner not planned and not utilized under normal conditions. The unanticipated almost two and a half years of delayed construction extended through a period of increasing inflation and escalation in labor rates and material costs, through two winters with adverse building conditions and into a time when Kidde-Briscoe was faced with an 89 day regional strike by the sheet metal workers and 137 day regional strike by the plumbers union.
> A computation of the consequent damages to Kidde-Briscoe involves damages of two sorts: (a) delay damages due to the extended periods of field and home office overhead and (b) damages due to disruption, loss of productivity, inefficiency, acceleration and escalation.[3]

In addition to the items noted in the *Walter Kidde* decision, contractors often claim and are sometimes awarded the following:

1. idleness and underemployment of facilities, equipment, and labor
2. increased cost and scarcity of labor and materials
3. use of more expensive modes of operation
4. stopgap work needed to prevent deterioration
5. shutdown and restarting costs
6. maintenance
7. supervision
8. equipment and machinery rentals and cost of handling and moving
9. travel
10. bond and insurance premiums
11. interest.

Actual Cost and Recordkeeping. The preferred method for proving damages is through the introduction of actual cost data for the additional work claimed to have been caused by the owner's breach. Data of actual cost, once tied to the owner's breach so as to meet the requirement of causation (invariably through expert testimony), fulfills the certainty requirement discussed in Section 4.3. The contractor who seeks to prove the reasonable value of the labor and materials it has furnished will be in a substantially better position if it can produce accurate, detailed records of what it spent in performing the work. Records consist of supplies invoices, time cards, daily logs, performance schedules created and then modified during the project, correspondences, and other records preferably generated at the time the event was happening (not created months or years after-the-fact in the midst of litigation). The recordkeeping also should be segregated by activity, so that the cost of the claimed additional work can be shown with precision.

To lesser or greater degrees, the damages formulas discussed in the remainder of this section are a second-best method of proving damages to the use of actual cost data. Indeed, courts will not allow the use of some of these damages formulas without the contractor first proving that the use of actual cost data was impossible or impractical because of the site chaos discussed immediately above.

Extended Home Office Overhead: Eichleay Formula. A claim for extended home office overhead presents one of the more complicated questions in measuring damages for owner breach. This law was largely developed in the field of federal procurement.

When a contractor bids for construction work, it usually takes into account not only job site overhead but also home office overhead. Home office overhead includes costs that are incurred to the mutual benefit of all contracts and cannot be tied to a specific project. One commentator stated,

> Costs which are normally included are: executive and clerical salaries, outside legal and accounting expense, mortgage expense, rent, depreciation, property taxes, insurance, utilities/telephone, auto/travel, professional and trade licenses and fees, employee recruitment, relocation, training and education, photocopying, data processing, office supplies, postage, books and periodicals, miscellaneous general and administrative expenses, advertising, interest on borrowing and other financial costs, entertainment, contributions and donations, bad debts, losses on other contracts, and bid and proposal costs.[4]

Contractors seek to use the *Eichleay* formula to measure extended home office overhead when the government agency causes an indefinite suspension in the contractor's performance, so the duration of contract performance is longer than the contractor anticipated when it calculated the cost of the job. "*Eichleay*" refers to a 1960 federal administrative board ruling which has been adopted by the U.S. Court of Appeals for the Federal Circuit.[5] Under this formula, the claimant must submit proof that (1) a government-caused delay extended the contract completion date; (2) the contractor was required to be on standby for an uncertain period of time; and (3) the contractor was unable to take on replacement or substitute work during the delay. Proof of these elements is, to a large degree, in the hands of accountants employed as expert witnesses and requires careful understanding of job site overhead, home office overhead, fixed overhead costs, variable overhead costs, and unabsorbed home office overhead.

It is important to understand that unabsorbed home office overhead is an indirect cost. When work is suspended indefinitely, the contractor has been placed on "standby," meaning it must be available to return to work at a moment's notice. Though its indirect costs continue to run, it is not receiving payments against which these indirect costs can be charged. But if the contractor can use its resources to take on other work, that work can absorb the indirect costs. In that case, there was no "standby," there are no unabsorbed overhead costs and no *Eichleay* claim. But if the replacement work absorbs only part of the indirect overhead, the contractor can recover the balance of the unabsorbed overhead incurred during the standby period under the *Eichleay* formula.[6]

Although developed as part of federal procurement law, the *Eichleay* formula has been influential among many state courts.[7] Other state courts are strict in demanding accurate proof of the loss caused by the delay and are unwilling to allow formulas to be employed in place of strict proof requirements.[8] The disagreement between those courts adopting a formula-based approach to the proof of extended home office overhead and those that require instead strict proof that these losses in fact occurred reflects two views of the "reasonable certainty" requirement of proving damages noted in Section 4.5.

The AIA waiver of consequential damages clause (discussed in Section 4.5) explicitly bars a contractor from recovering extended home office overhead; see A201-2007, § 15.1.6.2.

Productivity Loss Preferred Formulas: Measured Mile and Industry Productivity Studies. In a claim for lost productivity, the contractor asserts that it was required to use more manpower than anticipated. This claim is an attempt to recover excess labor costs of performance. Two techniques are used to measure productivity losses.

The "measured mile" method compares contractor productivity in an undisputed area of work with the contractor's productivity on a similar task during a disrupted work period. The productivity in the undisputed area of work establishes the contractor's productivity: the cost of labor to perform a defined amount of work in a given period of time. That ratio is then applied to disturbed areas of work. Any labor costs in excess of the baseline productivity cost is attributed to the defendant. The "measured mile" method only may be used where the contractor is performing repetitive work under similar conditions.

Another preferred method of proving diminished productivity is industry or trade productivity studies. One federal Board of Contract Appeals (an agency board to which contractor claims against the government are brought) addressed the contractor's attempt to prove productivity damages using a Mechanical Contractors of America Association (MCAA) bulletin. The Board provides this description of the bulletin:

> The MCAA bulletin identifies sixteen productivity factors and includes a narrative description of each factor. For each factor, the bulletin assigns a percentage productivity loss for three condition categories (minor, average, severe).[9]

Some of the listed factors are morale and attitude, reassignment of manpower, concurrent operations (or "stacking" of the contractor's workforce), dilution of supervision, and learning curve.[10]

Total Cost. Some contractors seek to use an even rougher formula than those that have been noted in this section. A contractor faced with the almost insurmountable obstacle of establishing actual losses knows it has lost money and feels its losses were attributable not to its own poor estimating or inefficient performance but to acts of the owner or to those for whom the owner is responsible. As a result, the contractor will sometimes seek to use what is called the *total cost method,* which is a comparison of the actual costs of performance with what the contractor contends should have been the cost of the project.

Although the federal Court of Claims allowed this method in *WRB Corporation v. United States,* it placed limits on possible abuse by requiring proof that

> (1) the nature of the particular losses make [*sic*] it impossible or highly impracticable to determine them with a reasonable degree of accuracy; (2) the plaintiff's bid or estimate was realistic; (3) its actual costs were reasonable; and (4) it was not responsible for the added expenses.[11]

The difficulty of using the total cost method is illustrated by a federal court of appeals decision in *Propellex Corp. v. Brownlee.*[12] The contractor (Propellex) sought to use this method to recover increased costs from a moisture problem attributed to the government. The court first emphasized that a contractor's use of actual cost data is the preferred method for it to prove any claimed increased costs. The court then ruled that Propellex failed to prove the first element of *WRB* test (the impracticability of proving its actual losses directly), because the evidence showed that Propellex could have set up its accounting system to track the costs of the moisture investigation and that it had failed to do so.

The *modified* total cost method seeks to avoid the crudeness of the total cost method. It focuses on the impacted work activities and adjusts the original estimate to remove mistakes, inaccuracies, and work items not affected.

Jury Verdict. Another rough measurement sometimes employed either in conjunction with a total cost theory or by itself is the jury verdict. A judge using the jury verdict method seeks to employ the same educated guesswork used by a jury. This formula recognizes the inherently imprecise nature of the proof that can be produced in such cases. However, a jury verdict can be used only where there is a clear proof of injury, there is no more reliable method for computing damages, and the evidence is sufficient to make a fair and reasonable approximation of damages.

4.8 Owner Versus Contractor

The principal damages measurement problems relate to the contractor's unjustified failure to start or complete the project, to complete the project as specified, or to complete the project on time. Just as the analysis of contractor claims against an owner would apply equally to subcontractor claims against the prime contractor, so too analysis of owner claims against the prime contractor would apply equally to contractor claims against a subcontractor.

4.8A Project Never Begun

Damages are determined by subtracting the contract price from the market price of the work. Market price is usually based on the best competitive price that can be obtained from a successor contractor for the same work. Suppose the contract price were $100,000 and the successor cost $120,000. The owner would be entitled to $20,000. This would protect its contract bargain.

In addition to the expectation interest (loss of bargain), the owner would be entitled to any losses caused by the delay (its reliance interest).

4.8B Project Partially Completed

Suppose the contractor ceases performance unjustifiably or is properly terminated by the owner. Failure to complete most commonly generates two types of damages: excess cost of a successor contractor to complete the project and the cost to repair defective work.

First, the owner very likely will have to pay more than the balance of the contract price to have the work completed by a successor. There are several rules for the higher cost of a successor contractor. The successor will not be selected (as a rule) by competitive bidding because of time constraints. Also, a successor will need to incur expenses to bring its own equipment and personnel to the site. It may also determine that it will need a premium to take into account that it may be held liable in the end for defective work performed by the original contractor. Similarly, it may worry about working with an owner with whom the original contractor had problems—no matter who may have been at fault in causing the termination.

The law would award the owner any excess costs in using a successor for doing the work required to be performed under the contract. This would put the owner in the position it would have been had the work been completed by the original contractor (thereby satisfying the owner's expectation interest).

Second, in many cases, the original work will have been performed defectively, and this too may justify additional compensation—the topic of the following subsection.

4.8C Defective Performance: Correction Cost or Diminished Value?

The contractor may have completed the work yet performed it (or parts of it) improperly. The law seeks to place the owner in the position it would have occupied had the contractor performed properly. One way of doing so is by giving the owner the *cost of correction*—the amount necessary to correct the defective work or complete the work to bring it to the state required under the contract. This is the preferred method of compensation.

Another method of giving the owner what it was promised is to award the *diminished value*—the difference between the market value of the project as defectively built and what it would have been worth had it been completed as promised. This measure gives the owner's balance sheet what it would have had by the time of full performance. This aim is achieved by combining the value of the project as it sits with the diminished value measure of recovery because performance has been less than complete.

Each method presents problems. Suppose cost of correction is used, and the correction costs greatly exceed the value correction would add to the property. The likelihood that the owner will not use the money for this purpose presents the possibility that the money will be a windfall. The award would put the owner in a better position than if the contractor had performed properly. If the owner *did* correct defective work or complete the work when it would not be economically sound to do so, this would waste scarce societal resources.

But awarding diminished value deprives the owner of the performance bargained for and forces the owner to take an amount that is usually determined by the testimony of expert witnesses. Such testimony is expensive to procure and may not take into account the subjective expectations of the owner when it made the contract. In addition, this evidence is "softer" than the evidence of the cost of correcting the work. Finally, if the award is less than the cost of correction and the owner elects to correct or complete, the award will be inadequate.

Most courts span this divide between these two approaches to compensation by applying the cost of correction measure unless it will generate "economic waste." The concept of economic waste may be understood from an examination of the classic 1921 case of *Jacob & Youngs v. Kent*.[13] On a contract to build a custom house, one specification for the plumbing work provided, "All wrought-iron pipe must be well galvanized, lap welded pipe of the grade known as 'standard pipe' of Reading manufacture." After moving in, the owner discovered that some of the pipe, instead of being made by Reading, was the product of other manufacturers. When the contractor asked for the final payment, the owner refused until the nonconforming pipe was replaced with Reading pipe. As the court succinctly noted, this would have "meant the demolition at great expense of substantial parts of the completed structure," and rather than do so, the contractor sued to recover the contract balance.

The court noted that the use of non-Reading pipe was not intentional, but simply the result of inadvertence. It observed that the contractor sought to introduce evidence that the nonconforming pipe was of equal quality to the Reading pipe. In short, the court found it was required to choose between literal enforcement of the contract requirements—regardless of cost—with the danger of creating a forfeiture for the contractor. Stating that "[w]e must weigh the purpose to be served, the desire to be gratified, the excuse for deviation from the letter, [and] the cruelty of enforced adherence," the court continued:

> In the circumstances of this case, we think the measure of the allowance is not the cost of replacement, which would be great, but the difference in value, which would be either nominal or nothing. ... It is true that in most cases the cost of replacement is the measure. [Citation omitted.] The owner is entitled to the money which will permit him to complete, unless the cost of completion is grossly and unfairly out of proportion to the good to be attained. When that is true, the measure is the difference in value. Specifications call, let us say, for a foundation built of granite quarried in Vermont. On the completion of the building, the owner learns that through the blunder of a subcontractor part of the foundation has been built of granite of the same quality quarried in New Hampshire. The measure of allowance is not the cost of reconstruction.[14]

One commentator has explained "economic waste" as follows.

> "Economic waste" is primarily a result-oriented concept, not a fiscal one. Economic waste comes into play in those cases in which the defective building is still serviceable and useful to society. If repairs are

possible but would completely destroy a substantial portion of the work, damage or injure good parts of the building, impair the building as a whole, or involve substantial tearing down and rebuilding, then that is "economic waste."[15]

Individual cases may not fit neatly within the general rule of applying the cost of correction unless doing so may result in economic waste, in which case the owner is awarded the diminution in value. Suppose a homeowner's renovated house suffers from an aesthetic defect, and the cost of correction would exceed the diminution in value. The court may take into consideration the owner's emotional attachment to the house and award the cost of repair measure of damages.[16]

In a federal public works project, the material did not meet technical requirements of the specifications, yet it exceeded by 20 the safety factor for the overall project. The U.S. Court of Appeals for the Federal Circuit applied the diminution in value measure of damages.[17]

What if the defective construction is so pervasive that the structure is either unusable or correction would be unaffordable and the building has basically no market value? Here, an award using the cost of repair standard may not require simply repairs, but the actual removal and replacement of the structure or component. This may result in other complications. Where the remedy is replacement (rather than repair) of a component, the result may be the owner obtaining a product of better quality than originally specified.[18]

Finally, an owner may not seize upon the presence of defective construction so as to rebuild a better project and seek to shift that higher cost to the contractor. In *City of Westminster v. Centric-Jones Constructors*,[19] a city entirely rebuilt a defective structure, while adding design enhancements. The Colorado appellate court ruled that the city must apportion those costs caused by the redesign, which cannot be charged to the contractor.

4.8D Delay

The basic measurement formula used for an owner's delay damages is lost use, measured as a rule by rental value. In this chapter's scenario, Oliver had entered into leases with tenants who could not be put into possession because of delayed construction.

The uncertainties over the lost rental value and other related expenses for delay provide an incentive for owners to use liquidated damages clauses, which allocate a set amount of damages for each day of delay. These clauses are discussed in Section 20.12.

4.9 Single Recovery Rule: Claims Against Multiple Defendants

In this chapter's scenario, Oliver sued Replacement Architect, Inc. and Prime Builder, Inc. for delay damages, while Prime Builder, Inc. sued both Oliver and Replacement Architect, Inc. for increased costs. What rules apply when a claimant seeks the same or related damages from multiple defendants?

A claimant should not recover more than it has lost. The claimant would be unjustly enriched if this occurs. Under the single recovery rule, a claimant who sues more than one defendant (and has been paid by one defendant) must reduce its recovery against the other defendant, so it is not paid twice for the same loss.

For example, a contractor may sue both the owner and the owner's architect, claiming both caused the contractor to incur higher-than-anticipated costs in building the project. If the contractor settles with one of these defendants, the other defendant will be entitled to a reduction in its own liability by the amount of that earlier settlement. In effect, the owner and its architect are considered one entity for determining the scope of the contractor's recovery. Any amount received from the owner would offset any liability of the architect, while any amount received from the architect would offset any liability against the owner.[20]

However, where the owner sues both its architect and the contractor, the analysis differs. Ordinarily, a contractor is not responsible for defects in the completed structure that are traced to errors in the design. For this reason, the owner is suing the architect and contractor not for the same injury, but for different injuries. A recovery from the architect should not be subtracted from a later recovery against the contractor. Both recoveries are for different injuries; hence, the owner has not been compensated twice.

4.10 Owner/Design Professional Claims

A design professional's claims against the owner as client are usually for nonpayment of the fee, which is a topic discussed in Section 9.3.

An owner's claims against its architect or engineer are ordinarily for defective design and/or errors in administering the project. The measure of damages is similar to those the owner would bring against the contractor.

REVIEW QUESTIONS

1. What are the four types of judicial remedies for contract breach?

2. What are the three primary things a court judgment seeks to do for compensatory damages?

3. In regards to limits on recovery, what are the four primary factors that must be met or proven for a claimant to receive compensation?

4. What are the five principal claims made by contractors against owners?

5. If a contractor had an original contract price of $800,000, had already expended $210,000 in part performance, estimates that the cost of completion would be $625,000, and has already received $175,000 in progress payments, how much would that contractor be entitled to in damages if you were to use the reasonable cost of completion formula?

6. If the data is available, which is the best method of proving damages?

7. Under the *Eichleay* formula, what three things must the claimant prove in order to use this formula?

8. Assuming actual data is not available, what are the two techniques used for a contractor to measure productivity losses for a claim?

9. What is the Single Recovery Rule?

10. What is "economic waste" and what role does it play in measuring the owner's damages?

ENDNOTES

[1] *Milford Power Co., LLC v. Alstom Power, Inc.*, 263 Conn. 616, 822 A.2d 196 (2003).

[2] *Blake Constr. Co. v. C. J. Coakley Co.*, 431 A.2d 569, 575 (D.C.App.1981).

[3] 37 Conn.Supp. 50, 434 A.2d 962, 977 (1981).

[4] Long, *Extended Home Office Overhead Damages*, VI KC News, No. 2, June 1989, p. 1.

[5] *Eichleay Corp.*, ASBCA No. 5183, 60-2 BCA (CCH) ¶ 2688, aff'd on reconsideration, 61-1 BCA (CCH) ¶ 2894. The *Eichleay* formula was adopted by the Federal Circuit in *Capital Electric Co. v. United States*, 729 F.2d 743 (Fed.Cir.1984).

[6] *P.J. Dick, Inc. v. Principi*, 324 F.3d 1364, 1370 (Fed.Cir.2003).

[7] For example, the *Eichleay* formula was adopted in *Complete General Constr. Co. v. Ohio Dept. of Transp.*, 94 Ohio St.3d 54, 760 N.E.2d 364 (2002) and *Fairfax County Redevelopment & Housing Auth. v. Worcester Bros. Co., Inc.*, 257 Va. 382, 514 S.E.2d 147 (1999).

[8] *Berley Industries, Inc. v. City of New York*, 45 N.Y.2d 683, 385 N.E.2d 281, 412 N.Y.S.2d 589 (1978).

[9] *P.J. Dick Inc.*, VABCA No. 5597, etc., 01-2 BCA (CCH) ¶ 31,647, p. 156,341, aff'd in part & rev'd in part, *P.J. Dick v. Principi*, 324 F.3d 1364 (Fed.Cir.2003).

[10] Id., 01-2 BCA (CCH) pp. 156,341–156,342.

[11] 183 Ct.Cl. 409, 426 (1968).

[12] 342 F.3d 1335 (Fed.Cir.2003).

[13] 230 N.Y. 239, 129 N.E. 889 (1921).

[14] Id., 129 N.E. at 891.

[15] Abney, *Determining Damages for Breach of Implied Warranties in Construction Defect Cases*, 16 Real Est. L.J. 210, 218 (1988). Modern cases are collected in Annot., 41 A.L.R.4th 131 (1985).

[16] In *Orndorff v. Christiana Community Builders*, 217 Cal.App.3d 683, 266 Cal.Rptr. 193 (1990), homeowners were awarded repairs costs that exceeded the loss of market value by 2.5 percent.

[17] *Granite Constr. Co. v. United States*, 962 F.2d 998 (Fed.Cir.1992), cert. denied, 506 U.S. 1048 (1993).

[18] In *Hendrie v. Board of County Comm'rs*, 153 Colo. 432, 387 P.2d 266 (1963), a new pool (represented as adequate by the manufacturer/installer) was removed and replaced with a different model compatible with soil conditions. In *State of Kansas ex rel. Stovall v. Reliance Ins. Co.*, 278 Kan. 777, 107 P.3d 1219 (2005), the owner was allowed to replace an earthen trench system with a concrete one, where the cost would be less than installing a new earthen system.

[19] 100 P.3d 472 (Colo.App.2003).

[20] *Huber, Hunt & Nichols, Inc. v. Moore*, 67 Cal.App.3d 278, 136 Cal.Rptr. 603, 619 (1977); *RPR & Assocs., Inc. v. University of N.C.—Chapel Hill*, 153 N.C.App. 342, 570 S.E.2d 510, 519-20 (2002), cert. dismissed and review denied, 357 N.C. 166, 579 S.E.2d 882 (2003)

Mr Doomits/Shutterstock

CHAPTER 5

TORTS: LEGAL RELATIONS NOT ARISING FROM CONTRACT

SCENARIO

Oliver Owner decided to build a new office building. Oliver Owner hired Awesome Architect, Inc. to design and administer the project. Awesome Architect, Inc. assigned Awesome Junior to conduct occasional site visits to monitor the progress of the work.

Oliver Owner then hired Prime Builder, Inc. to be the prime contractor on the job. Prime Builder, Inc. performed some of the work itself, but subcontracted the excavation work to Trencher, Inc., the framing work to Framer, Inc., and the roofing work to Roofer, Inc.

The project suffered from a series of mishaps. Trencher, Inc.'s employees dug a deep trench in the pouring rain. Awesome Junior, the architect's representative, came by and measured the length of the trench so as to determine how much work Trencher, Inc. had done, but did not speak to either the subcontractor or to Prime Builder, Inc. about the subcontractor's employees being at the bottom of the trench. Shortly after Awesome Junior left, the trench collapsed, injuring two workers. A federal Occupational Safety and Health Administration (OSHA) official issued a citation to Prime Builder, Inc. and Trencher, Inc., because under OSHA safety regulations, the trench's depth meant it should have been shored (provided with supports) or sloped so as to prevent trench collapse.

As the building went up, Framer, Inc. began the framing work. Fred Framer, an employee of Framer, Inc., was using a staple gun, but it became jammed. Fred attempted to open the gun to undo the jam. Instead, the gun discharged, and he received a staple into his arm.

Prime Builder, Inc. needed a package delivered to the job site. Mike Messenger worked for a bicycle delivery service. He got the package and drove his bike unto the site, speeding toward the office trailer at dusk as it was becoming dark. When his front wheel struck a hole dug by Prime Builder, Inc., Mike was thrown from his bike and injured.

Roofer, Inc. began the roofing work. Randy Roofer, an employee of Roofer, Inc., accidentally dropped his hammer, which hit the roof of a passing car being driven by Dan Driver. Dan was startled, and the car hit a tree and came to a complete stop. Although not physically injured, Dan suffered an anxiety attack. His emotional distress was so great that he had to go on medication and missed a week of work.

Finally, Prime Builder, Inc. was forced to redo some of the work when Oliver Owner discovered the windows were of the wrong type. The defect was traced to an error in Awesome Architect, Inc.'s design.

Litigation ensued.

1. The two employees of Trencher, Inc. who were injured in the trench collapse received workers' compensation benefits. (See Section 2.12A.) They then sued Prime Builder, Inc. and Awesome Architect, Inc. for negligently causing their personal injuries.
2. Fred Framer sued the manufacturer of the staple gun for creating a defective product in that it did not contain a safety latch for Fred to use when the gun jammed.
3. Mike Messenger sued Oliver Owner and Prime Builder, Inc. for the injuries he sustained when he fell off his bike.
4. Dan Driver sued Prime Builder, Inc. and Roofer, Inc. for damage to his car, damages for emotional distress, and lost income at work.
5. Prime Builder, Inc. requested Oliver Owner pay for increased performance costs caused by Awesome Architect, Inc.'s specification of the wrong window. When Oliver Owner filed for bankruptcy (which would make it difficult for the contractor to obtain any reasonable recovery), Prime Builder, Inc. sued Awesome Architect, Inc. for the increased costs.

5.1 Relevance to the Construction Process

During the construction process, events can occur that might harm persons, property, or economic interests. Workers or others who enter a construction site may be injured or killed. The owner or adjacent landowners may suffer damage to land or improvements. Participants in the process may incur damage to or destruction of their equipment or machinery. The owner or other participants in the project may incur expenses greater than anticipated. Investors in the project or those who execute bonds on participants may also suffer financial losses.

After completion of a project, people who enter or live in the project might be injured or killed because of defective design, poor workmanship, or improper materials. Those who invest in the project may find investment value reduced for similar reasons.

The construction project is a complex undertaking involving many participants. It presents a high risk of physical harm to those actively engaged in it. Sometimes such harm is caused by participants failing to live up to the standards of conduct required by law. Losses sometimes occur by human error that does not constitute wrongful conduct or because of unpredictable and unavoidable events for which no one can be held accountable. Because of the varying causes of losses; the many participants in the project; and the complex network of laws, regulations, and contracts, placing responsibility is difficult.

5.2 Tort Law: Introduction

5.2A Definition

Tort law may be understood by contrasting it with contract law. Chapter 3 introduces the law of contracts. It outlines the law applicable to parties who have entered into binding agreements. To a large extent, the contract itself defines the parties' rights and obligations.

Tort law defines civil (non-criminal) rights and obligations independent of any agreement. These rights and obligations are imposed (or granted) by society as expressed in statutes, regulations, and judicial decisions. Society defines what conduct is permissible, which injuries are entitled to legal protection, and the remedies for those injuries. Virtually all tort law is state law.

A **tort** has been defined as a civil wrong—other than a breach of contract—for which the law will grant a remedy, typically a money award. This definition—though not very helpful—mirrors the difficulty of making broad generalizations about tort law in the United States. One reason is the incremental or piecemeal development of tort law necessitated by new activities causing harm. For this reason, much American tort law consists of a collection of wrongs called by particular terms that were given legal recognition in order to deal with particular problems.

Although tort law and criminal law have features in common—each regulating human conduct—they operate independently. Crimes are offenses against the public for which the state brings legal action in the form of criminal prosecution. Prosecution is designed to protect the public by punishing wrongdoers through fines or imprisonment and to deter criminal conduct. By contrast, the tort system is essentially private. Only individual victims (or sometimes a large number of people who bring a class action) can use the system.

However, criminal sanctions and tort liability may be complementary. For example, the automobile driver who violates the criminal law may be fined or imprisoned. Those who are injured because of such criminal conduct are likely to institute a civil action to transfer their losses. This civil action is part of the tort system.

Criminal sanctions and tort remedies serve similar purposes. Just as one purpose of criminal sanctions is to punish, so too tort law allows for the recovery of punitive damages to punish those who acted in an egregious (or morally reprehensible) manner. By comparison, punitive damages are not a remedy for breach of contract.

Finally, administrative regulation can be one means of establishing tort liability. A government agency with authority to regulate certain conduct may issue an administrative penalty. For example, the federal Occupational Safety and Health Administration (OSHA) may investigate a construction site after an accident and issue citations for unsafe conditions. (About half of the states have their own OSHA agencies.) The injured person (usually a construction worker) may use the citation as evidence in a tort action against the business that received the citation.

Tort law is simply too diverse and immense to cover comprehensively in this text. By and large, this chapter will cover the primary tort actions applicable to the construction industry:

- Section 5.3: Negligence
- Section 5.4: Misrepresentation
- Section 5.5: Interference with contract or prospective advantage
- Section 5.6: Premises liability
- Section 5.7: Liability of one who hires an independent contractor
- Section 5.8: Products liability

5.2B Function

Tort law has different functions. The particular function most emphasized at any given time depends on social and economic conditions in which the system operates.

The principal functions are to compensate accident victims, to deter unsafe or uneconomic behavior (to encourage investment of up to but not beyond the point at which incremental safety costs—the added cost to make the project even safer—equal incremental injury costs), to punish wrongful conduct, to protect social norms from a sense of outrage generated by perceived injustice, to provide a dispute resolution mechanism, or a combination of any of these goals.

In seeking to implement these functions, the law must often choose between goals and interests of individuals and groups. One person's desires may be filled at the expense of another. One person may wish to drive a car at a high speed, exposing others to danger. Property owners may wish the freedom to maintain their property as they wish. But this freedom may come at the expense of those who enter the land and are injured. A manufacturer may wish complete freedom to design a product that will earn the highest profit. But this freedom may come at the expense of buyers who suffer harm from using the product. Adjusting these conflicts can reflect conscious decisions to select or favor one competing interest or goal over another—a form of social engineering.

5.2C Threshold Classifications

Three important threshold concepts should be distinguished. The first concept describes the *interests* (or rights) that merit tort protection:

1. *personal,* harm to person, including death, and sometimes also psychic or emotional interests
2. *property,* tangible and intangible
3. *economic,* unconnected to harm to a person or damage to property

The second concept classifies the *conduct* of the person causing the loss:

1. *intentional,* including not only the desire to cause the harm but also the realization that the conduct will almost certainly cause the harm
2. *negligent,* usually defined as failure to live up to the standard prescribed by law
3. *nonculpable,* though in a sense wrongful, in which the actor neither intends harm nor is negligent

These threefold classifications play important roles in determining which victims will receive reparation from those causing the loss. On the whole, harm to a person is most deserving of protection, with harm to property being considered second in importance. At the conduct end, intentional conduct that causes harm is least worthy of protection, followed by negligent conduct. Finally, tort law is more likely to provide a remedy for physical harm, less likely for emotional harm (unless incident to a physical injury), and least likely allow compensation for purely financial injury.

These classifications are gross, and many subtle distinctions must be made. For example, while purely economic or financial injury is less likely to receive protection, that is not true if the defendant's conduct was intentional and with the likely result of causing the victim financial injury.

Some have examined the unruly and disparate thousands of tort cases and have attempted to articulate factors that affect tort liability. One treatise listed the following items as important:

1. a recognized need for compensation
2. historical development (meaning the doctrine of precedent)
3. the moral aspect of the defendant's conduct
4. the administrative burden of recognizing a legal right on the judicial system
5. the capacity of each party to bear or spread the loss
6. the extent to which liability will prevent future harm[1]

5.3 Negligence: The Societal Concept of Reasonable Conduct

5.3A Elements of Negligence Claim

To justify a conclusion that the defendant was negligent—to establish a claim of **negligence**—the plaintiff must establish the following elements (or components).

1. The defendant owed a *duty* to the plaintiff to conform to a certain standard of conduct in order to protect the plaintiff against unreasonable risk of harm.
2. The defendant did not conform to the *standard of conduct* required.
3. A reasonably close *causal connection* existed between the conduct of the defendant and the injury to the plaintiff.
4. The defendant invaded a *legally protected interest* of the plaintiff.

The following subsections discuss each of these elements in more detail, beginning with the second element: the standard of conduct.

5.3B Standard of Conduct: The Reasonable Person

Under negligence law, every person has a duty to use the level of care of a reasonable and prudent person. An act or an omission (a failure to act) not in conformity with this **standard of conduct** is negligence.

Explicating the Standard. Reasonable conduct is not the opposite of fault, and it is not based on morality. Rather, it is a baseline level of conduct society expects of all individuals. This community standard requires that the defendant do what the reasonable person of ordinary prudence would have done. The standard holds people who live in the community to an average community standard. For example, an inexperienced driver has a legal duty to drive as well as the average driver.

The community standard of reasonable conduct is an "objective" standard. This means a person's conduct is not measured solely by what they subjectively knew. Instead, defendants are judged by what they knew or should have known or by what they did or should have done.

Exceptions to the community standard do exist, and a person can be held to a lower or higher standard than that of the community. Usually such exceptions are created by designating special subcommunities smaller than the general community and then applying an objective subcommunity standard. For example, children generally are not held to the adult standard but are held only to the standard of children of similar age and experience. Persons with special training or innate skill are expected to do better than the average person. For example, architects and engineers are held, as a rule, to the standards of their subcommunities (discussed in greater detail in Chapter 11). Professional truck drivers are expected to drive better than ordinary drivers.

> In this chapter's scenario, the injured employees of Trencher, Inc. sued Awesome Architect, Inc. Their theory of negligence would be that Awesome Junior should have perceived that the trench was dangerous and was negligent in failing to take steps to have the work stopped until the hazard was corrected. (As an employee, Awesome Junior's negligence would be attributed to Awesome Architect, Inc., his employer; see Section 2.12A.) Awesome Architect, Inc.'s conduct would be measured against that of the subcommunity of architects: How would the reasonably prudent architect have acted under similar circumstances?

A reasonable person of ordinary prudence will adjust his or her behavior based on the risk of harm in the event of an accident. A reasonable contractor working on a roof will take extra precaution to ensure equipment or supplies do not fall off the roof, given the hazard to those below. Yet society will not ban roofing work because of the inherent danger involved. A balancing test is needed. In short, in deciding whether particular conduct is negligent, society takes into consideration not only the magnitude of the risk but also the utility of the conduct and the burden of eliminating the risk This balancing approach has been adopted in the Restatement (Third) of Torts: Liability for Physical and Emotional Harm, adopted in 2005.[2] Section 3, titled "Negligence," provides:

> A person acts negligently if the person does not exercise reasonable care under all the circumstances. Primary factors to consider in ascertaining whether the person's conduct lacks reasonable care are the foreseeable likelihood that the person's conduct will result in harm, the foreseeable severity of any harm that may ensue, and the burden of precautions to eliminate or reduce the risk of harm.

The inherent flexibility in this "balancing approach" means that negligence law can adjust to individual, factual situations.

Violation of Statutes and Administrative Regulations. The construction process is governed by a multitude of laws dealing with land use, design, construction methods, and worker safety. What effect do violations of those statutes and regulations have on civil liability?

Courts view a statute as a legislative declaration of proper community conduct. If the injured person is within the class of persons protected by the statute, and the injury was caused by a risk the statute sought to avoid, the injured person may use the statutory violation as conclusive evidence of negligence—called **negligence *per se***. The trier of fact, whether judge or jury, need not decide whether there had been negligent conduct. Examples of such statutes is one requiring safety measures in construction work or building codes establishing minimum construction standards.

Sometimes the legislature by statute designates administrative agencies to issue detailed regulations covering the activity. For example, the Occupational Safety and Health Administration (OSHA) has issued detailed regulations governing safety on employment sites, including construction projects. An injured construction worker may cite violation of OSHA safety regulations (called "standards") as evidence that the defendant was negligent. However, courts generally reserve the "negligence *per se*" doctrine to violations of statutes—not violations of regulations.

> In this chapter's scenario, OSHA had cited Prime Builder, Inc. and Trencher, Inc. because the trench did not comply with safety standards. Trencher, Inc.'s employees may not sue their employer, who is shielded from tort liability by workers' compensation immunity. (See Section 2.12B.) However, the employees may point to the OSHA citations as some evidence of negligence on the part of the Prime Builder, Inc.

Factual Versus Legal Issue. A fundamental difference between the standard of care and duty (addressed next) is that the standard of care is an issue of fact, while duty is an issue of law. A factual issue is decided by the jury, while a legal question is decided by the judge. (If there is no jury, the judge also acts as the fact-finder.) If a judge decides the defendant did not owe the plaintiff a duty of care, the plaintiff's case will not go before the jury.

Suppose the judge finds a duty of care and the jury decides whether the defendant was negligent. If the losing party appeals, an appellate court gives great deference to factual findings. The reviewing court knows the credibility of a witness is best evaluated by actually seeing that person testify. It knows that reading the "cold transcript" of witness testimony is a poor substitute for viewing live testimony. The appellate court will not second guess the jury's decision without strong evidence that this decision lacked a minimal, factual basis. In contrast, the reviewing court gives no deference to the trial court's legal reasoning. It will conduct its own independent review of the legal rulings, including whether the defendant owes the plaintiff a duty of care.

5.3C Duty

Proof that a defendant acted below the standard of conduct is not, in itself, sufficient to establish liability to the plaintiff. Rather, the plaintiff must also establish that the defendant owed the plaintiff a duty of care. Otherwise stated, if a defendant acted negligently and that negligent conduct actually caused the plaintiff an injury, the defendant still would not be liable under negligence law to the plaintiff if the defendant did not owe that specific plaintiff a duty of care. As noted previously, deciding whether a duty exists is the exclusive function of the judge.

Changing Tests for Finding a Duty of Care. Explaining the concept of duty is complicated by (1) the changing nature of the concept in Anglo-American law in the past 300 years, (2) the changing terminology used to explain the legal concepts, and (3) fluctuations in the perceived need to place an outer limit on potential liability—lest risky yet societally useful conduct (including engaging in construction work) becomes prohibitively expensive.

Fundamentally, the purpose of imposing a duty of care is to regulate *risky* behavior. As phrased by the Restatement (Third) of Torts: Liability for Physical and Emotional Harm, § 7(a): "An actor ordinarily has a duty to exercise reasonable care when the actor's conduct creates a risk of physical harm." Under this definition, a "risk of physical harm" created by an actor's conduct is "ordinarily" sufficient grounds for finding that a duty to exercise reasonable care exists.

Note that the Restatement's test for the existence of a duty addresses only conduct that creates a risk of physical harm, purposefully leaving out purely emotional and financial harm. Recall the threefold classification in Section 5.2C, stating a duty is more likely to be found if the injury is physical rather than psychic or financial. This does not mean that negligence law is entirely unconcerned with financial injury. A defendant whose risky behavior results in a plaintiff's emotional or financial injury (but no physical harm) may well owe that plaintiff a duty of care. But a finding of a duty of care in those instances usually requires a finding of a *special relationship* between the parties. Where the harm is physical, the plaintiff ordinarily need not prove a special relationship.

> In this chapter's scenario, claims based on physical harm may be brought by Trencher, Inc.'s employees and by Dan Driver. These plaintiffs can easily establish a duty of care under the Restatement test. By contrast, Prime Builder, Inc.'s suit against Awesome Architect, Inc. is for financial injury. Awesome Architect, Inc. may argue that it did not owe the contractor a duty of care.

Limits on Duty. The Third Restatement definition of duty is intended to replace the common law view that dominated the twentieth century and remains to some degree in effect today. In 1928, the New York Court of Appeals (that state's highest court) addressed the liability of a train company whose employee's negligence set off a series of freak mishaps that led to the injury of the plaintiff, a member of the public. In *Palsgraf v. Long Island Railroad Company*,[3] the court defined the duty of care using the concept of foreseeability. *Foreseeability* refers to the class of persons or property and type of injury that the defendant could have reasonably foreseen would have been injured by his negligent conduct. Otherwise stated, a victim cannot recover against the defendant under a theory of negligence unless the defendant could have reasonably anticipated that someone would be injured in *this* way by *this* careless conduct of the defendant. Negligence liability does not extend to unforeseeable plaintiffs injured by freak accidents, as had happened in *Palsgraf*.

> In this chapter's scenario, an employee of Roofer, Inc. dropped a hammer while working on the roof. Under *Palsgraf's* foreseeability test, Roofer, Inc. would owe a duty of care to all persons or property the defendant could reasonably see would be injured by the falling hammer. Under the foreseeability test, Roofer, Inc. owed a duty of care to Dan Driver.

While the *Palsgraf* decision used the concept of foreseeability to *restrict* the concept of duty, in the second half of the twentieth century a concern grew that the foreseeability test resulted in too much potential liability. Potential liability without clearly defined limits may choke off economically useful activity. The Third Restatement of Torts, § 7(b), explains when a duty of care may be modified (limited) or eliminated:

> In exceptional cases, when an articulated countervailing principle or policy warrants denying or limiting liability in a particular class of cases, a court may decide that the defendant has no duty or that the ordinary duty of reasonable care requires modification.

A judge who believes a generalized, no-duty rule is merited or that a duty of care should be limited in scope must, under § 7(b), articulate a "countervailing principle or policy" that justifies such a conclusion. Unlike under the *Palsgraf* rule, foreseeability no longer defines the limit of duty. A judge may not use lack of foreseeability to create a generalized, no-duty rule.

Three examples of modified (or limited) duty are of particular interest to the construction industry.

First, the doctrine of *premises liability* defines the legal responsibility of an owner or possessor of land to those who are injured when they come unto the land. The common law traditionally created different rules of liability for the owner or possessor, depending on the status of the entrant (the person coming unto the land). For example, the owner or possessor owes a lessor duty of care to a trespasser than to one invited onto the land for the owner's or possessor's benefit. Premises liability is discussed in Section 5.6.

Second, as explained in Section 2.13, much of the construction industry involves the employment of independent contractors. One difference between employment of an independent contractor and employment of an employee is the potential liability of the hirer if either an independent contractor or an employee injures a third party. If an employee—while acting in the scope of his employment—injures a third party, the employer is vicariously (or strictly) liable to the victim. By contrast, if an independent contractor injures a third party, the contractor's hirer is *not* vicariously liable to the victim; here, the duty of care—for public policy reasons—has been modified or limited. Employment of an independent contractor is discussed in Section 5.7.

Third, recall that under the foreseeability test potential liability may be wide-ranging. For example, a project architect might owe a duty of care to the prime contractor, the subcontractors, injured construction workers, future purchasers of the building, and maybe even more distant parties—all of whom may claim to have been injured (physically or financially) by design defects or negligence in the architect's supervision of the project under construction.

The experience of design professionals was shared by many other business enterprises. Concern grew that the foreseeability doctrine broadened potential liability too much, so that persons remote to the defendant's conduct could nonetheless sue for negligence. Many manufacturers and professionals experienced increases in insurance premiums, which they blamed on a runaway tort system.

One way courts responded to the concern with overly broad negligence liability was through creation of the *economic loss rule*, discussed in Section 5.10. This judge-made rule limits the application of negligence and strict liability law to injuries for physical harm. Under this doctrine, a plaintiff whose injuries consisted exclusively of financial losses (such as lost profits, repair or replacement costs, or higher than anticipated costs of performance) cannot recover these losses in tort. (An exception exists if the losses were caused by an intentional tort.) Instead, these economic losses may be recovered only under a theory of contract or warranty. Using the vernacular of the Third Restatement of Torts, § 7(b), the economic loss rule is an example of the courts invoking an "articulated countervailing principle or policy" in order to limit a defendant's duty of care.

5.3D Causation

Assuming the defendant owed the plaintiff a duty of care and that the defendant's conduct fell below the standard of care, there still is no liability unless a causal connection exists between that conduct and the plaintiff's injury. This causal connection has been expressed using different terms: "factual cause" (this is the Third Restatement of Torts wording), "cause in fact," and "but-for" causation. Regardless of the terminology used, the Third Restatement of Torts, § 26, provides this definition: "Conduct is a factual cause of harm when the harm would not have occurred absent the conduct." Factual cause is for a jury to decide.

The question of causation can raise difficult issues. First, the defendant's conduct need not be the sole cause of the loss. Many acts and conditions join together to produce a particular event. Suppose an employee of a subcontractor suffers a fatal fall while working on a scaffold high above the ground. Several factors may have played a role in bringing about the death: defective scaffolding supplied by a scaffolding supplier; failure of the subcontractor or prime contractor to fix the scaffolding in response to complaints; weather conditions that made the scaffold particularly slippery on the day of the accident; a low-flying plane that momentarily distracted the worker; the worker's refusal to wear a safety belt; and the subcontractor's or prime contractor's failure to enforce safety belt rules.

Notwithstanding the difficulty of establishing factual cause, it would be unfair to relieve any party whose failure to live up to the legal standard played a significant role in the injury simply because other actors or conditions also played a part in causing the fall. Liability in such a case would depend on a conclusion that any of the defendants substantially caused the injury.

5.3E Protected Interests and Emotional Distress

The fourth and final element of a negligence claim, as outlined in Section 5.3A, is that the defendant invaded a legally protected interest of the plaintiff. Even if all of the other requirements for negligence are met, the particular harm that has resulted may not receive judicial protection.

Physical harm to the person is most worthy of protection. Death not only ends one's life but also can have a severe financial and emotional impact on the deceased's dependents. Physical injury often means medical expenses and diminished earnings as well as pain and suffering. Harm to property (such as damage or destruction) is also considered worthy of protection because of the importance placed on property in modern society.

As the harm moves away from personal and property losses, the interest receives less protection. Economic harm—such as diminished commercial contractual expectations, lost profits, or additional expenses to perform contractual obligations—although sometimes protected, is considered less worthy of protection than harm to person or to property. Even less protection is accorded emotional distress and psychic harm, which are sometimes called *noneconomic losses*.

A plaintiff normally cannot recover for mental disturbance caused by the defendant's negligence in the absence of accompanying physical injury. Emotional distress and mental disturbance (the terms are synonymous) can arise in the context of construction work. For example, a severely defective house can haunt the owner with fears of toxic mold or collapse of the house in an earthquake. Without questioning the authenticity of this emotional pain, courts almost universally find it is not compensable unless it is the result of morally objectionable conduct, such as an intentional tort.

> In this chapter's scenario, Dan Driver should easily obtain an award for the damage to his car and even perhaps for the cost of medication and his lost income. However, it is least likely he will obtain damages for his emotional distress, as he was not physically injured.

5.3F Defenses to Negligence: Assumption of the Risk and Contributory Negligence

Assumption of the Risk. Assumption of risk completely bars recovery—even if the defendant has been negligent. Advance consent by the plaintiff relieves the defendant of any obligation toward the consenting party. The plaintiff has chosen to take a chance. Suppose the plaintiff voluntarily entered into a relationship with the defendant with knowledge that the defendant would not protect the plaintiff from the risk. In such cases, the plaintiff implicitly assumed the risk. Sometimes the plaintiff is aware of a risk created by the negligence of the defendant but proceeds voluntarily to encounter it.

Assumption of the risk may be "express" (by written agreement) or "implied." Courts are concerned that express assumption of the risk involves over-reaching by the more powerful party against the weaker one and will scrutinize such agreements carefully. For example, workers are frequently prohibited from signing agreements to assume the risk of physical harm.

Most cases involve implied assumption of risk. Did the plaintiff know and understand the risk? Was the choice free and voluntary? Voluntariness has generated considerable controversy where workers take risks under the threat that they will be discharged if they do not continue working.

Some comparative negligence statutes, discussed next, abolish assumption of the risk. However, the majority of statutes make no reference to assumption of the risk, leaving the matter for resolution by the courts.

Contributory and Comparative Negligence. What is the effect if the plaintiff was also negligent to some degree? Here, too, the law has evolved, but on this issue it is due in large part to legislative action. The traditional rule of contributory negligence was that the slightest negligence by a plaintiff was sufficient to bar any recovery.

While some courts changed this rule under their common law authority, the doctrine of contributory negligence has been virtually eliminated by legislative adoption of comparative negligence. The effect of comparative negligence is to allow the plaintiff to bring a claim but then to reduce any damages award proportionate to the plaintiff's percentage of fault. Comparative negligence thus shifts any consideration of the plaintiff's fault from an issue of liability to one of damages.

5.3G Claims Against Government Defendants

For various reasons—some metaphysical and some practical—English law immunized the sovereign from being sued in royal courts. This doctrine was adopted early in the nineteenth century by the U.S. federal courts, and it soon became established that the federal government could not be sued without its consent.

In 1946, the Federal Tort Claims Act (FTCA)[4] was adopted. This law gives individuals the right to sue the United States for certain wrongs it committed. The act has two important exceptions that prohibit tort liability of the federal government (1) for certain intentional torts and (2) for injuries caused by certain discretionary functions or duties performed by government officials. Under the intentional tort and "discretionary function" exceptions to the FTCA, the federal government would be immune even though a private party might be liable under state law. Virtually all states have also limited sovereign immunity through enactment of tort claims acts.

Even if no immunity protects a government defendant, the public entity may escape liability under the public duty doctrine. This is a common law doctrine that reflects the courts' sensitivity to the financial burden that tort liability may impose upon state and local governments and—by extension—taxpayers. Under this doctrine, a duty owed to the public at large is not enforceable in tort by any particular member of the public. As an example, a city whose building inspector is charged with negligence owes no duty to the specific, individual owners whose house was improperly inspected; hence, the city would not be liable to the owners in negligence.

5.4 Misrepresentation

Intentional or Negligent. Both the tort of misrepresentation and the tort of interference with contract or prospective advantage (discussed in Section 5.5) may be either intentional or negligent. While negligence is conduct below the standard of care, intentional torts involve acting with a purpose (or with substantial certainty) to effect a harmful result.

It is this element of purpose that makes intentional torts particularly blameworthy. For example, a plaintiff who suffered an injury negligently inflicted would be entitled to compensatory damages, but a plaintiff who suffered the same injury intentionally inflicted might be entitled to punitive (or punishment) damages as well. (Punitive damages are discussed in Section 5.9B.) In addition, while the economic loss rule (discussed in Section 5.10) may bar compensation in negligence for purely financial losses, this doctrine generally does not apply to intentional torts.

The more wrongful the conduct by the person making the representation, the greater the likelihood of recovery against the party.

Representation or Opinion. This section treats the liability of people whose business it is to make representations. A surveyor makes representations as to boundaries, a geotechnical engineer as to soil conditions, and a design professional as to costs and the amount of payment due a contractor.

Representations should be distinguished from opinions. For example, an architect may give her best considered judgment on what a particular project will cost. The prediction, however, may not be intended by her or understood by the client to be a factual representation that will give the client a legal claim in the event the prediction turns out to be inaccurate. If the statement is merely an opinion and not a representation of fact, it is reasonably clear that the person making the representation will not be liable simply because she is wrong.

Reliance. In addition to the representation having to be material or serious, it must be relied on reasonably by the person suffering the loss. If there is no reliance or if the reliance is not reasonable, there is no liability for the misrepresentation.

For example, often the owner makes representations as to soil conditions to the contractor and then attempts to disclaim responsibility for the accuracy of the representation. The disclaimer is an attempt to transfer the risk of loss for any inaccurate representations to the contractor. It is intended to negate the element of reliance, which is a basic requirement of misrepresentation. However, if the misrepresentation was intentional, fraudulent, or even negligent, the disclaimers may not be enforced.

Person Suffering the Loss. A distinction may be drawn based on who was harmed: the person to whom the representation was made or a third party. For example, the geotechnical engineer may make a representation of soil conditions to a client. If the representation is incorrect, the harm may be suffered by the client or—in some cases—by third parties, such as a contractor, subcontractor, subsequent purchaser, or occupant.

Liability to third parties often depends on the type of harm suffered. If personal harm such as death or injury results, the absence of a contractual relationship between the person suffering the harm and the person making the misrepresentation is not likely to constitute a defense.

Where harm is economic, lack of privity (a contractual relationship) between the claimant and the person who made the misrepresentation causes the greatest difficulty. In the construction context, misrepresentation cases may arise in a variety of situations, including against surveyors or design professionals. A surveying error often injures the client, usually the owner. Contractors or subcontractors may sue the design professional for misrepresentation, seeking to recover economic losses in the form of higher-than-anticipated performance costs, lost profits, or damages due to delay. These types of claims and possible application of the economic loss rule as a defense for the design professional are the topic of Section 11.8D.

5.5 Interference With Contract or Prospective Advantage

In the competitive market system, persons are allowed—and even encouraged—to seek commercial advantages by claiming they can do better than their competitors. But at a certain point, attempting to persuade someone to use one's services or buy one's goods will run afoul of the protection tort law accords valid contracts: a commercial relationship, such as a stable supplier–customer relationship, or a negotiation that has proceeded so far that it is very likely it will be concluded successfully. In such cases, the person who makes the claim has the reasonable expectation of commercial gain that has been frustrated wrongfully.

Whether a third party has interfered wrongfully requires balancing protection given a commercial relationship with the freedom given persons to seek to persuade others that they can perform services better than the one who has been retained to perform them.

The two issues that surface when claims are made against third parties are whether the conduct in question has risen to the necessary level of wrongdoing and whether the defendant's conduct is nonetheless protected. For example, in the construction context, a claim for intentional interference may arise if an owner—upon advice of the architect—fires the contractor. This topic is discussed in Section 11.8E.

5.6 Premises Liability: Duty of the Possessor of Land

5.6A Relevance

Tort law determines the duty owed by the possessor of land—that is, the one with operative control—to those people who pass by the land or enter on it. Before, during, or after completion of a construction project, members of the public will pass by the land or (with or without permission) enter on the land with the potential of being injured or killed by a condition on the land or by an activity engaged in by the person in control of the land. Workers also may suffer injury or death because of the condition of the land or activities on it. Liability for such harm depends on the particular nature of the obligation owed to the plaintiff by the possessor of land near which or on which the physical harm was suffered. It can depend on the injured party's permission to be near or on the land and the purpose for being there.

The term *possessor of land* is used in this section without exploring the troublesome question of whether the owner, the prime contractor, or the subcontractors fall into this category during the construction process. Each may, depending on the circumstances.

As the following discussion makes clear, the courts are nearly evenly split on the duty a possessor owes to those who enter on to the land. Under the traditional common law rules, followed by a slight majority of states, the possessor's duty varies depending on the status of the entrant. However, under the minority modern rule, adopted in the Restatement (Third) of Torts, a duty of reasonable care extends to all entrants, regardless of their common law status.

5.6B Traditional Rule Based on Status

The traditional common law **premises liability** law varied the land possessor's duty based on the status of the entrant: trespasser, child, licensee, or invitee.

Passersby. Passersby can expect that the conditions of the land and activities on it will not expose them to unreasonable risk of harm. The possessor owes passersby a duty to act with due care.

Trespassing Adults. The trespasser enters the land of another without permission. In so doing, the trespasser is invading the owner's exclusive right to possess the land. Veneration for landowner rights led to a very limited protection for trespassers by English and American law. The possessor was not liable if trespassers were injured by the possessor's failure to keep the land reasonably safe or by the possessor's activities on the land.

Exceptions developed as human rights took precedence over property rights. Despite their unfavored position, trespassers are not outlaws. Discovered trespassers are entitled to protection. The landowner must avoid exposing such trespassers to unreasonable risk of harm.

Trespassing Children. Special rules have developed for trespassing children. Under the "attractive nuisance" doctrine, found in § 339 of the Restatement (Second) of Torts, a land possessor is liable for the injury of trespassing children if the possessor knew or should have known that children are likely to trespass on land containing a hazardous condition; the children because of their youth did not discover the condition or realize the risk; and the possessor failed to exercise reasonable care to eliminate the danger. This doctrine has been extended to children trespassing on a residential construction site. Contractors should use fencing or other barriers to preclude children from trespassing on the project.

Licensees. A licensee has a privilege of entering or remaining on the land of another because of the latter's consent. Licensees come for their own purposes rather than for the interest or purposes of the possessor of the land. Examples of licensees are people who take shortcuts over property with permission, people who come into a building to avoid inclement weather or to look for their children, door-to-door salespeople, and social guests. The possessor of land must conduct activities in such a way as to avoid unreasonable risk of harm.

Invitees. Invitees are invited unto the land to further the interests of the possessor. Among the states following the common law "status" rule of possessor liability, contractors and their employees are universally viewed as invitees.

An invitee receives the greatest protection. The possessor must protect the invitee not only against dangers of which the possessor is aware but also against those that could have been discovered with reasonable care. Although not an insurer of the safety of invitees, the possessor is under an affirmative duty to inspect and take reasonable care to see that the premises are safe. The possessor then has a duty to exercise reasonable care to protect the invitee against the danger.

If the owner vacates the land and possession is assumed by the prime contractor, then the contractor owes the highest duty of care to its invitees: the subcontractors and their employees.

> In this chapter's scenario, Mike Messenger is injured when he is thrown from his bike when delivering a package for Prime Builder, Inc. Prime Builder, Inc. is the land possessor, and it invited Mike unto the premises for the contractor's own economic benefit. Hence, Mike was an invitee. Prime Builder, Inc. would owe Mike the highest duty of care under premises liability law. Mike would argue that the hole in the ground into which his bicycle fell was a hazardous premises condition created by Prime Builder, Inc. and from which Mike was entitled to reasonable protection (such as erection of a barrier).

5.6C Modern Rule and Restatement (Third) of Torts

The various categories that determined the standard of care under the traditional "status" rules were difficult to administer. Exceptions developed within the categories, and application of the categories was often uneven. For these reasons, there is clear movement toward a single rule that would require the possessor of land to avoid unreasonable risk of harm to all who enter on the land.[5]

Acceptance of the modern rule will no doubt be bolstered by its adoption in § 51 of the Restatement (Third) of Torts, which provides that a land possessor owes a duty of reasonable care to all entrants (with an exception for "flagrant" trespassers). The division between children, licensees, and invitees no longer matters. Rather, the standard of reasonable care is flexible, based on the nature of the hazard and the type of entrant. Under this approach, more care must be given to protect immature children than construction workers against the hazards posed by a construction site. The standard of conduct is the same as that found in the Third Restatement of Torts, § 7, which is discussed in Section 5.3C.

A commentary to § 51 (each Restatement contains a "commentary" or fuller explanation of the rules listed) addresses construction projects. It provides that the possessor has a duty of care in selecting agents and contractors.[6] One undertaking a construction project must exercise reasonable care in choosing a contractor who will endeavor to maintain a safe job site for entrants.

5.6D Defenses to Premises Liability

Underlying premises liability law is a balancing of responsibility between the land possessor and the entrant. Just as the possessor must take reasonable steps to protect entrants from hazardous conditions of the premises, so too must the entrant exercise prudence while on the property. It is sometimes said that the land possessor's liability is premised upon his *superior knowledge* of the hazard. If the entrant's knowledge is equal to that of the possessor—or if a reasonable person in the entrant's position would have discovered the hazard and appreciated the risk—then the possessor's liability is either eliminated altogether or at least reduced by the degree of the entrant's negligence.

This onus upon the entrant to exercise care and prudence when on the premises has congealed into a number of defenses a land possessor may assert. One defense is that the entrant was injured by a "patent" defect—one the entrant knew of or should have discovered. Another defense is that the invitee exceeded the scope of the invitation—that he wandered off into an area into which he was not authorized to enter.

> In this chapter's scenario, Prime Builder, Inc. would argue that the hole in the ground was a patent defect and that Mike Messenger had a duty to look out and protect himself from such dangers. Instead, Mike sped on his bike (even though it was getting dark), and that is why he was unable to stop before hitting the hole.

Several states have passed "recreational use" statutes, whose purpose is to encourage property owners—both private and public—to provide the public with free access to their property for recreational use. With narrow exceptions, these statutes immunize owners from liability to entrants injured while engaging in recreational activities.[7]

Of importance to the construction industry are various defenses that apply to injuries to independent contractors. An owner who turns the premises entirely over to the control of a prime contractor may argue that the contractor is now the possessor and that the landowner is therefore no longer subject to premises liability. (This defense may not be successful if the hazardous condition already existed at the site when the prime contractor arrived.)

> In this chapter's scenario, Oliver Owner would argue that he could not be liable to Mike Messenger under a theory of premises liability, because Oliver had turned possession over to Prime Builder, Inc.

Another defense is that the hazardous condition was itself the object of the contractor's work or arose from its activities.[8] The reason for this defense arises directly from the "superior knowledge" rationale of premises liability law: a contractor hired to repair or who created a hazardous condition while doing its work was obviously aware of the dangerous condition, so its knowledge was at least equal to that of the landowner.

5.7 Employment of Independent Contractor

Generally. As explained in Section 2.13, an independent contractor, unlike an agent or employee, controls the manner in which the work is performed. The contractor is an independent business hired on a project-specific basis to achieve a specified purpose. Virtually all work on a construction project is performed by independent contractors. The hirer of an independent contractor can be either the owner (who hires the prime contractor) or the prime contractor (who hires subcontractors).

Suppose an owner hires a contractor to do remodeling work and the contractor negligently drops tools from the roof, injuring a passerby. Clearly, the passerby can sue the contractor for negligence. As explained in Section 2.12A, if the person who dropped the tool were an *employee* of the owner, the owner would be vicariously (or strictly) liable. But may the passerby sue the owner for the negligence of its independent contractor?

Independent Contractor Rule and Its Exceptions. Under the **independent contractor rule**, the owner would not be liable to the passerby, because the injury was caused by an independent contractor—not an employee. The contractor's control over the method and manner by which it undertakes the work relieves the owner of responsibility for the accident. The contractor's freedom to do the work in the manner it chooses includes an obligation to perform in a safe manner, and that obligation is not shared by the owner. As articulated in the Restatement and (Second) of Torts, § 409, and subject to numerous exceptions, "the employer of an independent contractor is not liable for physical harm to another by an act or omission of the contractor or his servants."

The degree to which the independent contractor rule shielded the hirer from liability has evolved. Throughout much of the twentieth century, the common law courts carved numerous exceptions to the independent contractor rule. If the work performed was exceptionally dangerous, the contractor's performance violated a statutory duty, or the owner retained some right of control over how the work was to be performed, the courts imposed either vicarious liability (akin to strict liability) or a duty of care on the hirer. As a result of these exceptions (and others), the potential liability of a hirer of an independent contractor mirrored that of society as a whole when duty was defined by the foreseeability test—increasing and seemingly disproportionate liability.

Beginning in the final quarter of the twentieth century, a judiciary concerned with overly expansive liability has reinvigorated the independent contractor rule, thereby immunizing the hirer from liability for the harm caused by its contractor. One method used by the courts to achieve this result was by interpreting exceptions to the rule narrowly. For example, the "retained control" exception was limited to situations in which the hirer actively or affirmatively interfered in the contractor's manner of performance, causing the work to become more dangerous and ultimately leading to the underlying injury.[9]

Another significant expansion of the independent contractor rule (also beginning in the last quarter of the twentieth century) concerned the identity of the victim of the accident. When the victim was a stranger to the construction process—such as a passer-by or visitor—the overriding concern of compensating the victim justified expanded liability for the hirer, because the independent contractor might not have the resources to pay the entire award.

However, when the victim was a construction worker, the concern for compensation does not carry the same weight, since the worker will receive workers' compensation benefits. (As discussed in Section 2.12A, workers' compensation is a statutory, no-fault insurance system that compensates employees injured on the job; in return, the employer is immunized from tort liability to its employee.) Courts also point out that vicarious liability is particularly unfair to the hirer, who pays twice for the same loss. The hirer pays the first time when it hires the independent contractor, whose contract price includes the workers' compensation premium covering its employees; the hirer pays the second time when subjected to vicarious liability. For this same reason, the injured construction employee receives a windfall not available to other industrial workers whose employer is not hired by anyone else.

In addition to these considerations, there is a particular unfairness to imposing vicarious liability on the hirer. Construction work is specialized and dangerous. An unsophisticated homeowner may hire a contractor precisely so that the work would get done in a safe manner. Yet that desire for a safe project instead will subject the owner to double *and strict* (or vicarious) liability—even though the contractor itself was in all likelihood the primary cause of the accident.[10] These practical and equitable considerations of policy have persuaded many courts that any exceptions to the independent contractor rule should not extend to injured construction workers.

5.8 Products Liability

Manufacturer liability has become important in the construction process, as harm can be caused by defective equipment or materials.

Strict Liability. The historical development of legal rules relating to the liability of a manufacturer for harm caused by its products manifests a shift from protection of commercial ventures toward compensating victims and making industry bear the normal enterprise risks. Unlike the theories of tort law described earlier in this chapter—all of which are variations on either negligence law or intentional torts—a manufacturer's liability also may be based on strict liability. This means that the defendant may be liable for the harm even if it was not negligent and the parties were not in privity (did not have a contractual relationship). A plaintiff to receive a recovery need not prove that the manufacturer deviated from the standard of care.

Third Restatement. The modern law of **products liability** is found in the Restatement (Third) of Torts: Products Liability, published in 1998. The Third Restatement defines a defective product as one that is "not reasonably safe" for the user. A product may be defective in three ways: the manufacturing, design, or marketing of the product. A manufacturing defect is a production flaw; an example is the sale of contaminated food. A design defect occurs when the design of the product is needlessly dangerous, such as a car that is prone to roll over. A marketing defect means a lack of instructions or warnings explaining to the end user how to operate the product safely. All of these defects are measured against the reasonable expectations of the ultimate consumer of the product.

Under the Third Restatement, liability for some defects are in strict liability, while for other defects the plaintiff must prove negligence. Section 2, titled "Categories of Product Defect," states:

> A product is defective when, at the time of sale or distribution, it contains a manufacturing defect, is defective in design, or is defective because of inadequate instructions or warnings. A product:
>
> (a) contains a manufacturing defect when the product departs from its intended design even though all possible care was exercised in the preparation and marketing of the product;
>
> (b) is defective in design when the foreseeable risks of harm posed by the product could have been reduced or avoided by the adoption of a reasonable alternative design by the seller or other distributor, or a predecessor in the commercial chain of distribution, and the omission of the alternative design renders the product not reasonably safe;
>
> (c) is defective because of inadequate instructions or warnings when the foreseeable risks of harm posed by the product could have been reduced or avoided by the provision of reasonable instructions or warnings by the seller or other distributor, or a predecessor in the commercial chain of distribution, and the omission of the instructions or warnings renders the product not reasonably safe.

Subsection (a) imposes liability for a manufacturing defect "even though all possible care was exercised"—this is the strict liability standard.

By contrast, subsection (b) imposes liability for a design defect only if the manufacturer did not adopt "a reasonable alternative design." Use of the term "reasonable" indicates a negligence standard. The existence of a reasonable alternative design is determined using a risk/utility analysis.

Finally, subsection (c) imposes a duty to provide "reasonable instructions or warnings." Again, use of the term "reasonable" imposes a negligence standard. On this matter, the Third Restatement varies from the Second. Under the Second Restatement, a manufacturer's warning or directions rendered an otherwise defective product nondefective. The Third Restatement no longer states that instructions or warnings will make a product nondefective. In addition, the new Restatement mandates a safer design (where possible) over the use of warnings.

Defenses to Products Liability. A product is defective in design or marketing only if the harm was reasonably foreseeable. The manufacturer will not be liable if the consumer's unforeseeable misuse of the product was the sole cause of the harm. Unforeseeable misuse may show that the product was not defective, that the defect did not cause the injury, or that the plaintiff's recovery should be reduced by his comparative negligence. Under the Third Restatement, there is no duty to design or warn against unforeseeable risks and uses.

Suppose a user discovers the defect, uses the product anyway, and is injured. Both the Second and Third Restatements allow an "assumption of the risk" defense to claim of inadequate warning. Under the Third Restatement, the plaintiff's fault does not bar the claim, but it results in a reduction in his recovery.

The "sophisticated user" defense is of particular importance to the construction industry. This defense bars failure-to-warn claims by highly trained workers who use complicated— and potentially dangerous—machinery in their trade. For example, in *Johnson v. American Standard, Inc.*,[13] a certified heating, ventilation, and air conditioning (HVAC) technician was brazing (welding) a refrigerant line and was injured when the line released phosgene gas. He sued the manufacturer of the refrigerant line for failing to warn that brazing could lead to the release of the gas. The California Supreme Court rejected the claim on the ground that sophisticated users of a product are charged with knowing the product's dangers and therefore cannot sue for failure to warn.

One defense manufacturers may assert is that federal government regulations preempt (that is, displace) state product liability law. For example, the New Jersey Supreme Court has ruled that a construction worker, while injured by a forklift manufactured in conformity with federal standards, could not sue the manufacturer for failure to include in the forklift additional warning devices.[14]

Finally, a manufacturer's liability for economic losses, such as delay damages, lost profits, or injury solely to the product itself, may be disallowed under the economic loss rule. See Section 5.10.

> In this chapter's scenario, Fred Framer sued the staple gun manufacturer in products liability. If Fred claims the gun has a design defect because it lacks a safety latch, he will have to prove that an alternative, safer design could have been used. If he claims the defect was lack of instructions as to what to do in the event the gun jams, the manufacturer could invoke the sophisticated user defense.

Services Contracts. The discussion to this point addressed mass-produced products made by large-scale, commercial manufacturers. Does this paradigm extend to the construction industry, and does strict products liability also apply to individual products made by service providers?

Developers who mass-produce homes[15] or lots[16] have been held strictly liable in some states. Contractors or subcontractors are not subject to strict liability (other than incidentally, such as the furnisher of appliances or other products).

The Restatement (Third) of Torts: Products Liability, § 19(b), states that services—even when provided commercially—are not products.

5.9 Tort Remedies

Compensatory Damages. The principal function of awarding tort damages is to compensate the plaintiff for the loss. In the ordinary injury case, the plaintiff is entitled to recover economic losses and certain noneconomic losses. Examples of economic losses are lost earnings and medical expenses. The principal noneconomic loss is pain and suffering.

Recovery for emotional distress always has been given hesitantly. Where there is physical injury, however, there has been no difficulty in allowing recovery for pain and suffering. Often the plaintiff's attorney will seek to obtain a large award for pain and suffering by asking the jury to use a per diem or even per hour method to compute the pain and suffering award. Breaking down the period of pain and suffering into small units can generate a large award.

Recovery of pain and suffering damages, which are often open-ended, has been justified by the fact that a large amount of the damage award usually goes to pay the victim's attorney. Plaintiff advocates emphasize that pain and suffering are real and that placing an economic value on them—although difficult—can give victims a sense that the legal system has taken adequate account of the harm they have suffered.

> In this chapter's scenario, Dan Driver sued Prime Builder, Inc. and Roofer, Inc. for damage to his car, damages for emotional distress, and lost income at work. He should be able to recover the repair cost for his car and probably also his lost income. However, since he was uninjured, Dan Driver might not be able to receive damages for emotional distress.

Punitive Damages. Tortious conduct that is intentional and deliberate—bordering on the criminal—can be punished by awarding punitive damages. Such damages are designed not to compensate the victim but to punish and make an example of the wrongdoer and to deter others from committing similar wrongs. In some areas of tort law, such as defamation, punitive damages play an important role because compensatory damages are often difficult to measure. There has been a tendency to award punitive damages for wrongful refusal to settle claims by insurance companies with their own insureds to ensure fair dealing in claim settlement. A few courts have awarded punitive damages in claims against manufacturers of defective products where the manufacturer seemed unwilling to place a high value on human life.[17]

In 2003, the U.S. Supreme Court established a test for whether punitive damages awards are constitutional. In *State Farm Mutual Auto Insurance Co. v. Campbell*,[18] it held that the due process clause of the U.S. Constitution was violated when a state awarded 145 million dollars in punitive damages where the compensatory damages were one million dollars.

The Court found that the award was grossly excessive or arbitrary. This violated the U.S. Constitution. Such an award serves no legitimate purpose and constitutes an arbitrary deprivation of property. Also, an excessive award violates due process as the defendant is deprived of fair notice that his conduct will subject him to a defined punishment.

The amount must bear some relationship to compensatory damages. In practice, said the Court, few awards exceeding a single-digit multiplier (the amount of punitive damages times compensatory damages) will satisfy due process. More than four times is close to the line, even to the point of presumptively being invalid. But the Court created no fixed rule.

5.10 Limits on Tort Liability for Pecuniary Losses: The Economic Loss Rule

5.10A Introduction

As explained in Section 5.3C, the scope of tort liability has not remained static; it has either been restricted or expanded by the common law in response to perceived societal needs and goals. During much of the twentieth century, a duty of care was owed to all foreseeable victims who suffered a foreseeable harm. However, concern grew as to the impact of such far-reaching potential tort liability on economic activity.

Recall too that Section 5.2C distinguished between the types of interests protected by tort law. Tort law traditionally has sought to provide a remedy for physical harm to person or property—less so for injury to solely economic interests. Economic or pecuniary losses relating to defective products (or buildings) include

1. damage to the defective product itself
2. diminution in the value of the product
3. natural deterioration of the product
4. cost of repair or replacement of the product
5. loss of profits caused by use of a defective product

Beginning in the mid-1960s, these twin considerations—reigning in seemingly limitless liability and a lesser concern for protection against solely pecuniary losses (not resulting from injury to person or property)—combined with a third concern of common law courts: maintaining a proper distinction between contract and warranty law on one hand and tort law (both negligence and strict liability) on the other. Courts feared that the intrusion of tort law into contract or warranty disputes would inject a "wild card" into commercial disputes. If a dissatisfied product purchaser sought recovery of lost profits in strict products liability, what would remain of the law of warranty? If contracting parties could sue each other for negligence, would this not have the practical result of the courts rewriting the parties' contract?

In sum, courts in the latter third of the twentieth century began to view the type of damages sought—economic losses—as establishing a dividing line between contract (including warranty) law and tort law. Common law courts adopted the **economic loss rule** for these several reasons. That rule, in its simplest form, bars the recovery of solely economic losses under a theory of nonintentional tort. Adoption of the economic loss rule has had far-reaching impact on construction industry disputes—not only for litigants and the courts but increasingly for legislatures as well.

5.10B Development of the Economic Loss Rule

Modern development stems from the 1960s explosion of strict products liability. During that period, the courts had to face claims by purchasers of defective products against manufacturers when the product itself did not function properly and caused economic losses to the purchaser. In the leading case of *Seely v. White Motor Co.*,[19] the purchaser bought a truck for use in his hauling business. The truck's brakes were defective, and the truck overturned. The purchaser, who was unhurt, sued in strict products liability to recover the profits lost in his business because he was unable to use the truck. In rejecting this claim, the California Supreme Court stated that the manufacturer can protect itself from liability for physical harm by meeting the tort standard. But it should not be charged for the performance level in the purchaser's business without showing a breach of warranty (which is a promise of a specific result). The purchaser could not charge the success of his business venture to the manufacturer. The modern economic loss rule thus had its origin in strict products liability law. Today, a clear majority of courts disallow recovery of purely economic losses under a theory of strict liability.

In *East River Steamship Corp. v. Transmerical Delaval, Inc.*,[20] an admiralty case (involving the law of the high seas), the U.S. Supreme Court held that a charterer of a supertanker could not recover in tort from the turbine manufacturer who had designed and manufactured the turbines installed in the charterer's vessels for damage to the turbines themselves. (Defective turbine seals damaged the turbines but not other parts of the ship). Speaking for the Court, Justice Blackmun stated that three views had developed in the state courts. The majority would not allow tort to be used in such a claim, preserving a proper role for the law of warranty if a defective product causes purely monetary harm. At the other end of the spectrum, a minority of courts would allow recovery where the product itself was injured; those courts saw no distinction between harm to person and property and purely economic loss. Those jurisdictions are not concerned about unlimited liability, because a manufacturer can predict and ensure against product failure. Some courts sought a compromise solution, allowing recovery if the defective product creates a situation potentially dangerous to people or property.

The Supreme Court followed the majority rule. When harm to the product itself occurs, the resulting loss due to repair costs, decreased value, and lost profit is the failure of the purchaser to receive the benefit of its bargain, which is an injury compensable by contract law—not tort law. The Court rejected the intermediate position (allowing tort recovery in the case of dangerous defects) as too indeterminate to allow manufacturers to structure their business behavior. Also, the minority view did not adequately keep tort and contract law in their proper compartments and failed to maintain a realistic limitation on damages.

Justice Blackmun stated that tort law was concerned with safety but not with injury to the product itself. In such a case, the commercial user

> stands to lose the value of the product, risks the displeasure of its customers who find that the product does not meet their needs, or, as in this case, experiences increased costs in performing a service. Losses like these can be insured. Society need not presume a customer needs special protection.[21]

In these transactions, as had been suggested the *Seely* court, contract law and the law of warranty function well.

Finally, the Supreme Court was not persuaded that the limitations set forth in the minority of cases granting tort recovery for economic losses—that of foreseeability—would be an adequate brake on unlimited liability. The Court stated: "Permitting recovery for all foreseeable claims for purely economic loss could make a manufacturer liable for vast sums. It would be difficult for a manufacturer to take into account the expectations of persons downstream who may encounter its product."[22]

5.10C Permutations of the Economic Loss Rule

While having its modern origin in products liability law, the economic loss rule has subsequently spread to virtually all aspects of construction defects litigation. The rule has been applied to claims (1) brought by third parties and between contracting parties and (2) by and against both commercial parties and unsophisticated homeowners.

Varying Tests. In truth, there is no one economic loss rule. While there is consensus as to what economic losses are (see the list at the beginning of Section 5.10A), courts apply the rule (or find it inapplicable) in different situations and for different reasons. A comprehensive discussion of the economic loss rule is beyond the scope of this text, and what follows is intended simply as illustrations of this complex topic.

The most expansive use of the economic loss rule is by states that apply the rule in a literal and mechanistic manner, flatly refusing to allow the recovery of economic losses under theories of nonintentional torts. In these states, such as Illinois, the rule applies whether the parties are in privity, regardless of whether the plaintiff has a contract remedy, and to claims against design professionals.[23] Other states permit third-party claims if the relationship between the parties approximates that of privity or if the parties have a "special relationship."[24]

Other tests exist. Colorado permits recovery of economic losses between parties in privity if the defendant breached a tort duty that was independent of its contract obligations. Under this "independent duty" test, the analysis to determine when a tort duty of care exists focuses not on the type of damages suffered but on the source of the duty that formed the basis of the action.[25] Wisconsin limits the economic loss rule to claims for defective products and does not extend the defense to services contracts, including between commercial parties in privity.[26]

One approach gaining increasing acceptance is to disallow third-party claims if there exists a scheme of interrelated contracts that affords the claimant a contract remedy—even if not against the actual tortfeasor. Under this rationale; for example, a subcontractor will be denied a tort claim against the project's architect on the grounds that the subcontractor has a contract remedy against the prime contractor for the economic losses caused by the architect's negligence.[27]

Other jurisdictions are more willing to allow the recovery of economic losses in tort. These states apply the economic loss rule only under narrow circumstances. Suppose a state's pre-existing common law permitted a contractor or subcontractor to sue the project designer in tort. These courts refuse to eliminate this pre-existing duty of care through adoption of the economic loss doctrine.[28]

Design Professionals. Should the economic loss rule extend to a contractor's (or subcontractor's) negligent design or negligent administration claim against the project's architect or engineer? This problem is particularly vexing because strong arguments can be made either way. Arguments in favor of allowing the claim include the lack of a contract between the contractor and designer; the close connection between the design defect and the contractor's economic losses (usually consisting of higher than expected construction costs and delay damages); and the concern that applying the economic loss rule would eviscerate the well-established law of professional liability. The contrary view is that the contractor's claim falls squarely within the parameters of the economic loss rule; allowing a tort claim creates uncertainty for architects and engineers and so will drive up the cost of design services; and the contractor is not without a remedy because it has a misrepresentation claim against the owner based on the defective design.[29]

> In this chapter's scenario, Prime Builder, Inc. sued Awesome Architect, Inc. for increased performance costs caused by defective design. Because the contractor's damages are in the nature of purely financial losses, the architect may invoke the economic loss rule. Awesome Architect, Inc. in effect would be arguing that its tort duty of care is limited to not causing physical damage to the contractor. Prime Builder, Inc. would counter that its financial loss was caused by Awesome Architect, Inc.'s design negligence, and that—Oliver Owner having filed for bankruptcy—Awesome Architect, Inc. is the contractor's only source of a remedy.

Hazardous Defects and Prophylactic Repairs. Another difficult issue is whether the economic loss rule should bar the recovery of prophylactic remediation costs—repair costs incurred so as to prevent a perceived defect from actually inflicting damage. This question is particularly acute when the perceived defect is viewed as hazardous to the owner or those who enter upon the property (for whom the owner may be liable under premises liability law, as discussed in Section 5.6).

An increasing number of courts hold that the economic loss rule bars both homeowners and commercial owners from a tort remedy when they perceive their building contains hazardous defects which have not (yet) caused either property damage or personal injury. As indicated in the discussion of *East River Steamship*, the Supreme Court rejected the "intermediate position" (allowing recovery for hazardous defects) as too indeterminate a standard of manufacturer liability.[30]

In *Aas v. Superior Court*,[31] a divided California Supreme Court disallowed the recovery of remediation costs of hazardous defects—including improper shear walls meant to stabilize the building in an earthquake and lack of fire protection in party walls—on the ground that the homeowners had not suffered actual injury, which is a prerequisite to a tort action. The court majority rejected the dissent's view that it would be "economically efficient" to allow the homeowners to make the repairs and then charge that cost to the builders.[32] But some courts allow remediation even of nonhazardous defects under the rationale that the owner is mitigating damages.[33]

Property Damage. Recall that the economic loss rule is defined as disallowing a (nonintentional) tort claim to recover purely economic losses in the absence of physical harm to person or property. Recall also that the Supreme Court in *East River Steamship* found the economic loss rule applies if the defective product damages only itself. The *East River Steamship* Court's definition of economic loss involving damage to the product itself, was adopted in the Restatement (Third) of Torts: Products Liability § 21(c).

Under this definition, economic losses are recoverable in tort if the plaintiff suffered damage to "other property." Defining what constitutes "other property," especially in construction defect disputes, is not straightforward. Every building is made up of many separate components: the exterior stucco, the windows, the roof, and so on. A defect in any one of those components may cause physical damage (from water infiltration) to the rest of the building. Does this water damage to the building constitute damage to "other property"—in which case the economic loss rule does not apply—or is the entire building viewed as the "product," so that the damage was only to the product itself? As would be expected, courts have come to differing conclusions on this difficult question.

Statutory Developments. Legislatures are increasingly addressing the economic loss rule, particularly to safeguard the rights of homeowners. The California Supreme Court's *Aas* decision, mentioned previously, was overruled by the legislature's enactment of a Right to Repair Act governing newly built residences.[34] This act defines a construction defect and creates a dispute resolution procedure for getting the defects repaired without the need to resort to litigation. If the parties are unable to resolve their differences, the homeowner may then bring a lawsuit. The result is a statutory preemption of the common law economic loss rule as applied to owners of newly built homes in California. Other states have adopted similar legislation; see Section 16.7.

REVIEW QUESTIONS

1. What are the principal functions of tort law?

2. What are the two different threshold classifications of tort law?

3. What are the four elements a plaintiff must establish to prove a defendant was negligent?

4. What is the fundamental difference between the standard of care and a duty?

5. What is foreseeability?

6. What is the difference between contributory negligence and comparative negligence?

7. Under which doctrine is a land possessor liable for the injury of trespassing children if the possessor knew or should have known that children are likely to trespass on land containing a hazardous condition; the children because of their youth did not discover the condition or realize the risk; and the possessor failed to exercise reasonable care to eliminate the danger?

8. What may be the difference in liability of an owner if the prime contractor injures (a) a passerby and (b) a subcontractor's employee?

9. According to the Restatement (Third) of Torts: Product Liability, what are the three ways a product may be defective?

10. What are the two types of remedies for tort claims?

ENDNOTES

[1] W. P. KEETON et al, TORTS, 20–26 (5th ed. 1984).
[2] The first two Restatements of Torts were each published as a single, multivolume work. By contrast, the Restatement (Third) of Torts is being published in separate volumes, each addressing individual topics. Hence, the Restatement (Third) of Torts: Liability for Physical and Emotional Harm and the Restatement (Third) of Torts: Products Liability are stand-alone volumes with their own separate titles. The Restatements of the Law as sources of law is explained in Section 1.9.

[3] 248 N.Y. 339, 162 N.E. 99 (1928).
[4] 28 U.S.C. §§ 2671–80.
[5] *Rowland v. Christian,* 69 Cal.2d 108, 443 P.2d 561, 70 Cal.Rptr. 97 (1968) is a seminal case.
[6] Restatement (Third) of Torts: Liability for Physical and Emotional Harm § 51, comment g.
[7] Cal. Civ. Code § 846; Wash.Rev.Code Ann. § 4.24.210.
[8] *Blair v. Campbell,* 924 S.W.2d 75 (Tenn.1996) (contractor, hired to perform repairs on a decrepit building, knew the work site was unstable); *Baber v. Dill,* 531 N.W.2d 493 (Minn.1995) (worker fell on protruding reinforcing steel rods—a hazardous condition the contractor had created).
[9] *Toland v. Sunland Housing Group, Inc.,* 18 Cal.4th 253, 955 P.2d 504, 74 Cal.Rptr.2d 878 (1998).
[10] *Privette v. Superior Court,* 5 Cal.4th 689, 854 P.2d 721, 21 Cal.Rptr.2d 72 (1993) (Restatement §416); M. SCHNEIER, CONSTRUCTION ACCIDENT LAW: A COMPREHENSIVE GUIDE TO LEGAL LIABILITY AND INSURANCE CLAIMS, pp. 131–47 (1999).
[11] 217 N.Y. 382, 111 N.E. 1050 (1916).
[12] The leading case is *Greenman v. Yuba Power Products, Inc.,* 59 Cal.2d 57, 377 P.2d 897, 27 Cal.Rptr. 697 (1963).
[13] 43 Cal.4th 56, 179 P.3d 905, 74 Cal.Rptr.3d 108 (2008).
[14] *Gonzalez v. Ideal Tile Importing Co., Inc.,* 184 N.J. 415, 877 A.2d 1247 (2005), cert. denied sub nom., *Gonzalez v. Komatsu Forklift, U.S.A., Inc.,* 546 U.S. 1092 (2006).
[15] *Schipper v. Levitt & Sons, Inc.,* 44 N.J. 70, 207 A.2d 314 (1965); *Kriegler v. Eichler Homes, Inc.,* 269 Cal. App.2d 224, 74 Cal.Rptr. 749 (1969). But see *Wright v. Creative Corp.,* 30 Colo.App. 575, 498 P.2d 1179 (1972). See Section 16.07B.
[16] *Avner v. Longridges Estates,* 272 Cal.App.2d 607, 77 Cal.Rptr. 633 (1969).
[17] *Romo v. Ford Motor Co.,* 113 Cal.App.4th 738, 6 Cal.Rptr.3d 793 (2003).
[18] 538 U.S. 408 (2003).
[19] 63 Cal.2d 9, 403 P.2d 145, 45 Cal.Rptr. 17 (1965).
[20] 476 U.S. 858 (1986).
[21] Id. at pp. 871–72.
[22] Id. at p. 874.
[23] *2314 Lincoln Park West Condominium Ass'n v. Mann, Gin, Ebel & Frazier, Ltd.,* 136 Ill.2d 302, 555 N.E.2d 346 (1990) (client must sue its architect in contract); *Anderson Elec., Inc. v. Ledbetter Erection Corp.,* 115 Ill.2d 146, 503 N.E.2d 246 (1986) (claim against engineer barred despite lack of contract remedy).
[24] *Ossining Union Free School Dist. v. Anderson LaRocca Anderson,* 73 N.Y.2d 417, 539 N.E.2d 91, 541 N.Y.S.2d 335 (1989) (relationship approximating privity); *Eastern Steel Constructors, Inc. v. City of Salem,* 209 W.Va.392, 549 S.E.2d 266 (2001) ("special relationship" between architect and contractor).
[25] *Town of Alma v. AZCO Construction, Inc.,* 10 P.3d 1256 (Colo.2000).
[26] *Insurance Co. of North Am. v. Cease Elec. Inc.,* 276 Wis.2d 361, 688 N.W.2d 462 (2004).
[27] *SME Industries, Inc. v. Thompson, Ventulett, Stainback & Assocs., Inc.,* 28 P.3d 669 (Utah 2001).
[28] *Indemnity Ins. Co. of North America v. American Aviation, Inc.,* 891 So.2d 532 (Fla.2004); *Kennedy v. Columbia Lumber & Mfg. Co., Inc.,* 299 S.C. 335, 384 S.E.2d 730 (1989).
[29] The contractor would bring a *Spearin* claim against the owner. See Section 16.3B.
[30] 476 U.S. at p. 870.
[31] 24 Cal.4th 627, 12 P.3d 1125, 101 Cal.Rptr.2d 718 (2000). The *Aas* decision has been overruled by statute; see infra note 34.
[32] 12 P.3d at p. 1140.
[33] *Toll Bros., Inc. v. Dryvit Systems, Inc.,* 432 F.3d 564 (4th Cir. 2005) (Connecticut law); *Kelleher v. Marvin Lumber and Cedar Co.,* 152 N.H. 813, 891 A.2d 477, 496 (2005).
[34] Cal. Civ. Code §§ 895–924.

CHAPTER 6

REGULATION OF LAND AND THE CONSTRUCTION PROCESS

6.1 Regulation: A Pervasive Presence

Regulation—legal limitations on what can be built, where it can be built, and how and by whom it is built—permeates the construction process. Regulation is the legal framework under which land improvement takes place. It comes broadly from two sources: (1) limits on land ownership and use and (2) regulation of the construction process itself. Every owner, designer, developer, construction manager, or prime contractor should have at least a basic understanding of the concepts addressed in this chapter. This section provides an overview of this chapter.

Limits on land ownership may be created by *private* action, such as when the developer of a housing subdivision imposes in the grant deeds (or purchase agreements) limits on the future owners' ability to modify their houses. The purpose of these "restrictive covenants," monitored by homeowner associations, is to preserve the value and aesthetics of the entire subdivision development.

Most restrictions, however, are *public* land use controls. Public land use controls are of two types: judicial and legislative. Tort law imposes controls on land use designed to safeguard neighboring landowners from interference with their own basic rights of ownership.

Local governments (through powers given to them by the state) employ zoning and related laws to broadly regulate the type of buildings that may be built. Local governments may go one step further by promoting developments broadly intended to improve the community (for example, by promoting a revitalized downtown). These myriad forms of public control, in turn, implicate constitutional protections against overreach.

Regulation of hazardous materials provides an overlapping layer of land ownership limitations. This form of regulation grew out of the environmental movement that began in the late 1960s. A complex array of federal and state environmental laws impose liability upon both land owners and contractors.

This chapter then moves from regulation of land to regulation of the construction process. Three broad regulatory schemes govern: what is built, how it is built, and who builds it. First, building codes specify minimum building standards. Compliance with the building code is subject to review by local governmental entities under the permitting and building inspection processes. Second, as with all major industrial activities, actors on the construction site must comply with occupational safety and health laws, which promote worker safety. Last, regulation of who may build is the subject of the licensing laws discussed in Chapter 8.

6.2 Limits by Private Action: Restrictive Covenants in Subdivision Developments

A land owner generally has no interest in how the land is used after the deed (the contract of sale) has transferred ownership. But suppose the owner is a residential developer who is selling lots or houses to a large number of buyers, as in the case of a housing subdivision development. The developer-seller may wish to assure buyers in the development that the land will retain its residential character to protect the enjoyment of and investment in the land. To accomplish the land-planning objectives, the developer-seller can obtain an express promise from each buyer relating to land use. These promises are called *restrictive covenants*.

To be effective over a long period of time, such restrictions must be "tied to" or "run with" the land. All buyers, present and future, must be bound by the restrictive covenants. To do this, a developer's uniform, common scheme of restrictions is recorded in the land records. All owners—present and future—may obtain judicial enforcement of the recorded restrictive covenants.

Over time, private restrictive covenants can become too rigid to accommodate changing conditions. To deal with the need for flexibility, many planned developments create **homeowner associations** composed of owners in the development. These associations can modify existing restrictions under certain circumstances and pass judgment on requests to deviate from the restrictive covenants, including aesthetic and architectural questions.

Restrictive covenants are limited to multi-housing developments and are of lesser importance than land use control, which is discussed next.

6.3 Protection of Neighboring Landowners

Public land use controls broadly further two goals: protection of neighboring landowners and regulation of the construction process. Protection of neighboring landowners, addressed here, covers the topics of soil support, drainage and surface waters, and certain easements. These are originally (and remain mostly) common law doctrines.

Public land use controls regulating the construction process is addressed in Section 6.4.

Soil Support. An owner is entitled to lateral (or soil) support of his property. That lateral support may be weakened by an adjoining owner's excavation. What is the liability of the excavating owner or contractor?

A distinction is drawn between land in a natural state and developed land. An excavating owner is strictly liable (the injured owner need not prove negligence) for damage caused by depriving adjoining land in its natural state of lateral support.

By contrast, liability for damage to improved land caused by loss of lateral support is premised upon a showing of negligence. The excavator is liable only if the excavation work was done negligently. The owner is not liable for the contractor's negligence unless the owner was negligent in selecting the contractor or some other exception to the independent contractor rule applies.

State legislation frequently controls the result in excavation cases. For example, in Illinois, the excavator must notify adjacent landowners of its intent to excavate, the proposed depth of excavation, and when excavation will begin. Legal rights depend on the depth of the excavation.[1]

Drainage and Surface Waters. Construction frequently affects and is affected by drainage at the site and adjacent land. Drainage changes can cause troublesome collection of surface waters. Construction of an improvement should be undertaken so as to avoid disturbing the surface water of the adjoining property.

Easements for Light, Air, and View. Light, air, and view are important considerations in constructing buildings and residences. However, as a general rule, a landowner has no natural right to air, light, or an unobstructed view. In the absence of a common law remedy, protection of light, air, and view largely are found either in restrictive covenants (discussed in Section 6.2) or in zoning statutes or ordinances.

6.4 Zoning: The Mechanism for Land Use Control

Zoning is the division of land into distinct districts and the regulation of the uses and developments within those districts. Every construction project must comply with the zoning code in order to obtain approval by the local building authorities.

Traditionally, the states delegated land use powers to local government in the belief that local authorities were in the best position to determine land use regulation appropriate for their community. Eventually, criticism that this form of local control led to sprawl and other social and environmental ills led to the development of a new model for zoning: smart growth.

All of the zoning measures described in the following sections are subject to challenge under the United States Constitution. The operative terminology is that a zoning regulation which amounts to a "taking" is unconstitutional. However, the constitutional analysis makes little sense without an understanding of the zoning measures themselves. Therefore, different forms of land use regulation are explored first (sometimes including reference to constitutional challenges) followed by a discussion of the constitutional framework in Section 6.4C.

6.4A Euclidean Zoning: The Premise of Local Control

Zoning activities began at the turn of the twentieth century in response to the country becoming more urban. Early zoning efforts sought to segregate housing and industrial activities into separate districts. Each district (or zone) was allowed particular uses. For example, residential districts permitted only residences, commercial districts only commercial activities, and industrial districts only industrial activities. Major categories were further divided. For example, industry might be divided into heavy industry and light industry. Residential use was divided into single-family homes, two-family homes, and multiple-family uses. Such homogenous districts were based on the assumption that differing uses within a district would harm property values.

This system was sometimes called Euclidean zoning after the U.S. Supreme Court decision that first validated a zoning plan. In *Village of Euclid v. Ambler Realty Co.*,[2] the Court justified the limitation of property rights under the city's police power to provide a safe and more pleasant community.

In addition to creating districts of permitted use, the early zoning acts regulated matters such as height, bulk, and setback lines. Rigid patterns of land and development were assumed. Single-family homes were to be placed on gridlike lots.

6.4B Societal and Environmental Concerns

Over time, the societal and environmental consequences of zoning have become of broader concern. Beginning in the mid-1950s, the decline of central cities was coupled with "white flight" to the suburbs. The suburbs, in turn, sought to attract light industry that paid high property taxes but did not require many employees of low or moderate income. The latter were to be avoided. People of middle income and above wanted to live with others who shared their social and cultural values. Some zoning techniques used to accomplish these purposes were as follows:

1. large lot requirements
2. minimum house size requirements
3. exclusion of multiple dwellings
4. exclusion of mobile homes
5. unnecessarily high subdivision requirements

Judicial backlash to this type of exclusionary zoning took the form of constitutional attacks. In 1965, the Pennsylvania Supreme Court, in *National Land and Investment Co. v. Kohn*,[3] invalidated a four-acre minimum lot requirement. Noting that evaluating the constitutionality of a local zoning ordinance was not easy and declining to be "a planning commission of the last resort," the court nevertheless saw itself as a judicial overseer "drawing the limits beyond which local regulation may not go."[4]

This judicial backlash was sometimes followed by a legislative one. In 1969, the Massachusetts legislature amended its regional planning law to establish a mechanism for the approval of "low or moderate income housing."[5]

Some local governments have sought to include housing for those often excluded. Typically, such ordinances (county or municipal laws) require that developers of more than a certain number of units set aside a designated percentage for low- and moderate-income tenants. Developers challenge these ordinances as a taking or an unconstitutional deprivation of their due process rights. Courts that uphold "inclusionary" zoning measures view them as a valid exercise of the local government's police powers.[6]

Euclidean zoning has been criticized not only because of its exclusionary effect on low- and moderate-income persons, but also, beginning in the 1960s, by those concerned with the environment. The once accepted goal of growth as the proven road to prosperity and social mobility began to be questioned seriously. Growth became synonymous with urban sprawl—the often uncontrolled development of land at the outskirts of major American cities. Urban sprawl has been negatively characterized as unaesthetic, wasteful of valuable land resources, and unduly increasing the cost of providing municipal services.

A recent effort to counter urban sprawl through zoning reform is the concept of "smart growth." The Massachusetts Smart Growth Zoning and Housing Production Act encourages development of anti-sprawl, mixed-population communities. It defines smart-growth zoning:

> Smart growth is a principle of land development that emphasizes mixing land uses; increases the availability of affordable housing by creating a range of housing opportunities in neighborhoods; takes advantage of compact design; fosters distinctive and attractive communities; preserves open space, farmland, natural beauty, and critical environmental areas; strengthens existing communities; provides a variety of transportation choices; makes development decisions predictable, fair, and cost effective; and encourages community and stakeholder collaboration in development decisions.[7]

Smart growth is being adopted by an increasing number of states. Legal challenges, including constitutional limitations, remain to be resolved.

6.4C Constitutional Limits on Land Use Controls

Although action or inaction of a local governing body can be challenged politically, such as by initiative, referendum, or the election process, the opponents of an ordinance or those who wish to contest a permit decision often go to court. The primary constitutional challenge is that the ordinance is an uncompensated "taking."

The power to control public land use essentially belongs to each state with actual regulation performed by local entities (counties and cities). But exercise of state powers is limited because property rights have received special protection in the United States. Owners cannot be deprived of their property without due process of law. If public regulation has placed too great a limit on an owner's right to use his land, such a limitation can constitute a "taking" of the property, which entitles the property owner to compensation from the regulating authority.

Takings can take two forms. There can be a physical taking. The state may use its powers of eminent domain and "condemn" the property, forcing the owner to sell the property to the government. If the state uses its powers of condemnation, the Fifth Amendment of the U.S. Constitution requires that it pay the owner "just compensation." The only issues in such physical takings are whether the "takeover" is for a public purpose and the value of the land taken.

The first issue, the requirement of a public purpose, presents no difficulty if the state appropriated the land for schools, roads, or parks. But increasingly, the state, particularly cities, seeks to take away land from one private person and hand it over to another who it believes will make better economic use of the land. Even though the land is being transferred from one private owner to another, the city still contends this is being done for a "public purpose." In *Kelo v. City of New London, Connecticut*,[8] the U.S. Supreme Court interpreted the term "public purpose" broadly to encompass the destruction of nonblighted private homes as part of a city's economic development plan.

A backlash to the *Kelo* decision by property rights activists has led to local proposals to restrict municipalities from engaging in similar economic development plans. Some state courts have rejected the premise, which was accepted in *Kelo*, that economic benefit alone satisfies the "public use" requirement.[9]

Suppose the state does not use eminent domain to physically take the owner's land. Instead, the owner contends that land use or environmental rules have so reduced the value of the land—prohibited all economically beneficial use—as to create a "taking," entitling the owner to compensation. In a 2002 decision, the U.S. Supreme Court commented on the distinction between physical and regulatory "takings" cases.

In *Tahoe-Sierra Preservation Council, Inc. v. Tahoe Regional Planning Agency*,[10] the Court was faced with a moratorium or temporary regulation. The issue was whether two moratoria totaling thirty-two months on any development in the Lake Tahoe basin was a taking requiring that the landowners be compensated. The moratoria were to give the regional planning authority time to formulate a land use plan for the area.

In a six-to-three opinion, the Court distinguished between a physical taking and an alleged regulatory taking. While the government's compensation obligation has been broadly applied in cases of physical takings, a more nuanced test applies to claims of regulatory takings. In the latter situation, whether a taking occurs requires a balancing test. The Court explained the need for greater judicial deference in regulatory taking cases:

> Land-use regulations are ubiquitous and most of them impact property values in some tangential way—often in completely unanticipated ways. Treating them all as *per se* takings would transform government regulation into a luxury few governments could afford. By contrast, physical appropriations are relatively rare, easily identified, and usually represent a greater affront to individual property rights. "This case does not present the 'classi[c] taking' in which the government directly appropriates private property for its own use," *Eastern Enterprises v. Apfel*, 524 U.S. 498, 522, 118 S.Ct. 2131, 141 L.Ed.2d 451 (1998); instead the interference with property rights "arises from some public program adjusting the benefits and burdens of economic life to promote the common good," *Penn Central*, 438 U.S., at 124, 98 S.Ct. 2646.[11]

This case-by-case approach, the immense variety that regulation can take, and the different impact such regulation can have on land ownership means that "taking" questions will continue to be a complex, controversial area of constitutional law.

6.5 Environmental Law

During the late 1960s, the American public began to demand that government make strong efforts to protect the natural environment. The goals of the environmental movement are evident in the preamble of the National Environmental Policy Act (NEPA)[12] enacted by Congress in 1969:

> The Congress, recognizing the profound impact of man's activity on the interrelations of all components of the natural environment, particularly the profound influences of population growth, high-density urbanization, industrial expansion, resource exploitation, and new and expanding technological advances and recognizing further the critical importance of restoring and maintaining environmental quality to the overall welfare and development of man, declares that it is the continuing policy of the Federal Government, in cooperation with State and local governments, and other concerned public and private organizations, to use all practicable means and measures, including financial and technical assistance, in a manner calculated to foster and promote the general welfare, to create and maintain conditions under which man and nature can exist in productive harmony, and fulfill the social, economic, and other requirements of present and future generations of Americans.[13]

Congress created the Environmental Protection Agency (EPA) as the agency in charge of environmental matters. The EPA writes regulations for and enforces environmental statutes. Those statutes most strongly impacting the construction process are briefly listed below.

6.5A National Environmental Policy Act (NEPA)

NEPA is a broad-based environmental statute that imposes continuing responsibility on the federal government to protect the natural and human environment. As a means to that end, Section 102 of NEPA requires all federal agencies to create a statement—commonly known as an environmental impact statement (EIS)—that details the environmental impact of proposed federal actions "significantly affecting the quality of the human environment."[14]

An owner or developer making a proposal that requires approval by a federal agency must wait for the agency to go through the EIS-creation process. This takes time. If the agency decides based on the EIS that the project may go forward, citizens' groups or other parties may challenge the agency's determination, thereby delaying the project. This challenge takes the form of a lawsuit against the public agency, claiming the agency had not followed the statutory requirements for approving an EIS.

Since its enactment, NEPA has created considerable litigation; still, very few projects have been enjoined (stopped by court order) under the statute. When agency determinations are challenged, courts usually extend great deference to agencies based on the latter's expertise and experience. Therefore, the most significant burden of NEPA appears to be the time and expense of preparing the EIS.

6.5B Superfund: Comprehensive Environmental Response, Compensation, and Liability Act (CERCLA)

Overview. CERCLA[15], commonly known as Superfund, gives the federal government broad powers to clean up hazardous waste sites. Of greater importance to construction industry actors is the potential exposure to statutory liability.

Section 107 of CERCLA imposes liability on four categories of "potentially responsible parties" (also known as PRPs), including the "owner" or "operator" of a "facility" and anyone who disposes or arranges to dispose "hazardous substances."[16] The costs of liability under CERCLA can be extremely high. Potentially responsible parties are liable for the federal government's cost to clean up the hazardous waste and may also be subject to costly civil and criminal penalties for violating the statute or consent decrees issued under the law.

CERCLA liability is "joint and several," meaning the bad acts of one PRP may subject other PRPs on the same property to statutory liability. Traditional notions of causation carry little weight in CERCLA case law, so even *de minimis* (minimal) contributors may be held liable for the entire cost of cleaning up a hazardous waste site. Liability is strict and retroactive, and can attach regardless of when the material was deposited. However, a PRP who voluntarily remediates the site may seek to recoup those expenses from other PRPs.[17]

Construction Industry Participants. The wide-ranging scope of CERCLA's strict liability directly impacts the construction industry. Owners and developers should be aware of any hazardous substances that may exist on their property. They risk being PRPs as the "owner" or "operator" of a "facility"—the contaminated site.

More surprising, CERCLA has ensnared prime contractors who unknowingly worked on contaminated land. In a 1992 decision, a federal appeals court ruled that an excavating and grading contractor, who quite unknowingly moved some contaminated soils in the course of his work and who stopped work as soon as he noticed something amiss, was nevertheless a responsible party under CERCLA. Liability was imposed because the contractor was found to be both a "transporter" of hazardous substances and as an "operator" of a facility (the construction site), and his dispersal of the contaminated soil was a "disposal" within the meaning of the statute.[18]

In contrast, design professionals have not been found to be PRPs, because they lack control over actual disposal or movement of the hazardous materials. For example, an architect who designed a wood treatment plant that released hazardous wastes when operated was found not to be a responsible party.[19] Another court found that a consulting engineer who provided not only the design but also substantial technical assistance during the construction phase still lacked the requisite control to merit PRP status.[20]

Brownfields. CERCLA's onerous liability provisions had the unintended effect of causing developers to shy away from potentially contaminated urban sites—popularly called "brownfields"—opting instead to pursue unpolluted green space at the outskirts of towns and cities. In response, CERCLA was amended in 2001 by what is popularly known as the Brownfields Act.[21] The Brownfields Act seeks to promote development of these urban sites by creating defenses ("safe harbors") to CERCLA's strict liability.

Three significant safe harbors exist: for innocent owners, for contiguous (neighboring) owners, and for prospective purchasers. Focus here is on the final safe harbor provision. It seeks to exempt future purchasers of contaminated property from CERCLA liability so long as they do not hinder existing or future cleanup operations. This innocent prospective purchaser defense does not require the purchaser to be unaware of the contamination. Moreover, a prospective purchaser who discovers contamination during its due diligence inquiry remains protected so long as it does not hinder any federal Environmental Protection Agency (EPA) cleanup response. By allowing a prospective purchaser to develop a contaminated site without threat of CERCLA liability, this final safe harbor provision may do the most to stimulate development of brownfields.[22]

6.5C State Law

The environmental movement has also led state legislatures to enact environmental laws often based on, if not more protective than, federal environmental laws. For example, roughly three-fourths of the states have enacted NEPA-type laws. In California, the law includes detailed requirements for agency procedures, substantive requirements for environmental impact reports (EIRs) and agency review, and judicial review standards.[23]

Similarly, many states have enacted stringent hazardous waste laws modeled after CERCLA that may create additional regulatory burdens for owners and developers. For example, under Massachusetts state law, parties liable for hazardous waste cleanup are required to pay the Commonwealth of Massachusetts for costs of assessment, containment, and removal of hazardous wastes and for damages for injury to and destruction of natural resources.[24] The Massachusetts law also promotes the development of brownfields.[25]

Some states have moved beyond the federal legislation in their regulation of environmental risks. California has responded to "sick building" complaints with enactment of the Toxic Mold Protection Act of 2001.[26] The Act requires the state Department of Health Services to convene a task force to develop permissible exposure levels to mold and to devise remediation standards. A second, new law requires a different task force to study and publish its findings on fungal contamination of indoor environments.[27]

6.6 Regulation of the Design Content and Construction Process

Both the design content and construction process are subject to regulation, primarily by state and local entities. The purpose of regulation is to ensure that the design and actual construction of the structure conforms to minimal requirements.

6.6A Permits

Local governments use the permitting process to review the project's design content. Permits issued by local authorities must be obtained at different stages of the project. Building permits allow the construction to go forward, and occupancy permits authorize the owner to use the completed structure. Each permit requirement provides the government with the opportunity and power of regulatory review. For example, the city or county will not approve the building permit unless the design complies with the building code and zoning laws.

Who must obtain permits required by law? Construction documents should cover these matters. Under the AIA "General Conditions," A201-2007, § 3.7.1, the contractor obtains building permits.

If the contract is silent, the law will likely require the owner to obtain the more important permanent permits, such as pertaining to land use control and any environmental impact report.[28] The contractor would have the obligation to obtain operational permits, such as building and occupancy permits and permits that are associated with facilities and equipment, such as for utility hookup. Determining who should obtain particular construction permits may also depend on the extent of experience the particular owner has had in construction projects.

Public authorities, inundated with permit requests, are not always able to process these requests expeditiously, thereby causing significant project delays. To ameliorate this bottleneck, large urban centers (beginning with New York City in 1976) created a process of "self-certification," under which architects, engineers, or certain trade contractors (such as plumbers and electricians) certify their own work as being in compliance with the building code. Because of the obvious conflict of interest, public authorities continue to spot-check self-certifications, and violators are subject to severe sanctions, including loss of the future ability to self-certify projects and loss of licensure.[29]

6.6B Building Codes

Building codes play a pervasive role in construction. Their original purpose was to promote safety in the design of the project and its construction. Some states' modern codes address more tangential concerns, such as the health of building occupants and energy conservation. These multiple purposes are evident in the 2010 California Building Code, which states as its purpose:

> to establish the minimum requirements to safeguard the public health, safety and general welfare through structural strength, means of egress facilities, stability, access to persons with disabilities, sanitation, adequate lighting and ventilation and energy conservation; safety to life and property from fire and other hazards attributed to the built environment; and to provide safety to fire fighters and emergency responders during emergency operations.[30]

Building codes are reissued periodically to keep up with technology changes. A leading template for the states is the International Building Code (IBC). All fifty states and the District of Columbia have adopted the IBC, although not all use the most recent edition.[31] The 2010 California Building Code, for example, adopts the 2009 IBC.

As discussed in Section 11.5B, "green" or sustainable design and construction is a growing trend. While such practices were first adopted through private decision-making, adoption of a green building code would have a state-wide impact. California also recently adopted the California Green Building Standards Code, known as CALGreen.[32]

Building codes are issued by the federal, state, and local governments. The federal government is the least active participant, mandating design standards on such matters as the Americans with Disabilities Act.[33] Indeed, under the Tenth Amendment to the U.S. Constitution, building codes are a matter of state concern, and—whether for this reason or otherwise—there is no national building code. To complicate matters further, cities and counties may adopt their own codes, which supplement the state's uniform code. This means that the architect and contractor must comply with multiple layers regulating construction standards.

This section addresses building code issues applicable to owners and prime contractors. Section 11.2 deals with designs that violate building codes.

Where design and construction are separate, the former is the owner's responsibility and is usually done by a design professional. Suppose the design violates building code requirements—what are the contractor's responsibilities?

The AIA Document A201-2007 § 3.2.3, states that a contractor, upon receipt of the design, is not required to determine whether the contract documents are in accord with legal requirements. However, if the contractor discovers or is informed (presumably by a subcontractor) of a design nonconformity, it must promptly report the nonconformity to the architect. Under § 3.7.3, if the contractor performs work knowing it to be contrary to law without notifying the architect, the contractor "shall assume appropriate responsibility for such Work and shall bear the costs attributable to correction."

This solution effectuates a sound middle ground. A contractor is responsible based on a subjective standard: only if it actually knew of the design nonconformity. Under this test, the contractor (1) is not liable under a broader, objective standard (used to determine negligence) yet (2) cannot simply close its eyes in the face of known danger on the ground that the design professional—not the contractor—is responsible for the design.

Finally, the design may simply be silent on how to build to a particular level of detail. In this instance, the contractor need only comply with the building code. Essentially, the code is a gap-filling instrumentality.

6.6C Building Inspections

While local governments use the permit process to review the project design, they use the inspection process to try to ensure the construction meets code requirements. Suppose a building is inspected by the city and a certificate of occupancy is issued, but the owners upon moving in discover significant defects they believe should have been discovered during the inspections. Is the city liable to the owners for having conducted negligent inspections?

Although narrow exceptions exist, the "public duty doctrine" would protect local government from liability to the owners. Under this doctrine, the government is not liable for breach of a duty owed to the public in general. The purpose of the building codes and building inspections are to require minimum performance standards in construction—not to benefit the owners of particular buildings. Individual homeowners cannot hold the city liable for its violation of a duty owed to the public at large.

Courts may point to public policy considerations in favor of government immunity. Local government, saddled with budgetary and personnel constraints, is ill-equipped to guarantee a contractor's compliance with building codes. Indeed, neither issuance of a permit nor inspection of the building absolves the contractor of its legal obligation to build in conformity with the codes.

An exception to the public duty doctrine might be found in the case of a "special relationship" between the owners and local government. For example, the city may be liable (its defense of immunity denied) where the building official knew of a dangerous condition yet failed to order the problem corrected.

6.7 Safety

Construction is dangerous work, and even the most careful contractor cannot eliminate the possibility of accidents. Regulation of the construction process includes mandating safety measures. Regulation encompasses both legislative action (statutes and regulations) and private action through tort law.

Safety statutes regulating the construction process divide broadly into two groups. Workers' compensation laws, discussed in Section 2.12B, guarantee compensation to injured employees or the estates of employees who had died. While this section addresses statutory and common law means to promote safety, good safety practices by contractors often originate from workers' compensation insurers who may exercise the threat of increased premiums.

The purpose of the statutes discussed in this section is not to ensure compensation but to make construction projects safer. These statutes divide into two broad types: (1) occupational safety and health acts and (2) safe workplace statutes. While both types of laws seek to promote project safety, they try to achieve this common goal with two fundamentally different approaches.

Occupational safety and health laws function prospectively, attempting to prevent accidents from happening through regulation of the workplace, which is enforced through administrative inspections and citations. Under these laws, employers may be penalized for maintaining unsafe conditions—even if no accident has occurred.

In contrast, safe workplace statutes come into operation after an accident has happened. They seek to promote safety by imposing statutory liability or facilitating the imposition of common law liability.

6.7A State Safe Workplace Statutes

State safe workplace statutes came first. Legislatures grappled with the high injury and death rates associated with large-scale construction projects. New York passed the first "scaffold act" in 1885, and Illinois followed with a "structural work act" in 1907, which was largely patterned on the New York law. (The Illinois statute was repealed in 1995.)

The New York Scaffold Act protects construction workers against gravity-related hazards: falling from a height or being struck by a falling object.[34] It states that "[a]ll contractors and owners and their agents" (except for small residential owners) "shall furnish or erect" scaffolding or other devices sufficient to protect against these hazards. For purposes of the statute, the owner's "agent" does not include an architect or professional engineer, although it may include a construction manager.[35] Liability is strictly imposed, so long as the statutory violation was a proximate cause of the accident. (There is no statutory liability if the accident arose solely from the negligence of the employee.) Statutory liability is supplemented by safety regulations issued by the Labor Commissioner. As should be clear from this briefest of summaries, the Scaffold Act uses the threat of strict, statutory liability as incentive for owners and prime contractors to maintain a safe workplace.

6.7B Occupational Safety and Health Laws

Safety legislation changed significantly with enactment of the Occupational Safety and Health (OSH) Act of 1970.[36] The OSH Act seeks to increase industrial safety (including at construction sites) by imposing upon each "employer" the duty to comply with the statute and safety "standards" (or regulations) promulgated by the Secretary of Labor. The federal OSH Act permits states to adopt their own occupational safety and health acts, subject to approval by the Secretary of Labor.

The Occupational Safety and Health Administration (OSHA), an arm of the U.S. Department of Labor, has the power to issue safety standards and to enforce the OSH Act by inspecting workplaces and issuing citations for statutory violations. Compliance officers may conduct inspections without advance notice to the employer. Challenges to citations are through the administrative process. If a citation is upheld by an **administrative law judge**, appeal is made to the Occupational Safety and Health Review Commission (OSHRC). Further appeals are made from the OSHRC to the federal courts of appeal and then to the U.S. Supreme Court.

The OSH Act classifies violations discovered during an inspection as willful or repeated, serious, and nonserious.[37] The Secretary of Labor assesses civil penalties after considering an employer's good faith, business size, previous violation history, and the gravity of the violation. In addition to civil penalties, criminal sanctions in the form of a $10,000 fine and/or six months imprisonment may be imposed for first-time willful violations which result in death.[38]

The OSH Act imposes two duties on the employer:

- to provide its employees with a place of employment "free from recognized hazards that are causing or likely to cause death or serious physical harm" and
- to comply with the safety standards issued by the Occupational Safety and Health Administration (OSHA).[39]

OSHA standards are the safety regulations governing construction sites; for example, they specify what safety measures must be undertaken when excavating a trench of a minimal depth. An employer who violates this standard is subject to citation by a compliance inspector—even if the trench has not collapsed and no one was injured.

The existence of these two overlapping statutory duties has a unique impact on construction projects. Clearly, a prime contractor is responsible under the OSH Act for the safety of its own employees—just as each subcontractor is statutorily responsible for the safety of its employees. However, a construction site is characterized by multiple employers working in a small, confined site in which hazardous activities by one employer may cause injury to employees of other employers. Is an employer subject to statutory liability if its conduct threatened or injured other employers' employees but not its own?

Although not definitively decided by the U.S. Supreme Court, an increasing number of federal courts of appeals, as well as the Review Commission, say "yes." Under the "multi-employer" doctrine, a construction employer (whether prime contractor or subcontractor) is liable under the Act if its violation of an OSHA standard exposes employees of another employer to the risk of injury or death—even if there was no threat to the violating employer's own employees.[40] One federal court of appeals extended the "multi-employer" doctrine to criminal liability under the OSH Act.[41]

OSH Act liability of architects and engineers is addressed in more detail in Section 11.8F, and statutory liability of construction managers is addressed in more detail in Section 7.5C.

6.7C Tort Law

Every project participant may be liable for its own negligence that results in injury or death. That said, this section will focus on the tort liability of prime contractors for worksite accidents arising out of two sources: (1) as hirer of independent contractors and (2) as possessor or occupier of the premises (land and building).

Independent Contractor Rule. As discussed in Section 5.7, the independent contractor rule states that a hirer of an independent contractor is presumed immune from liability for harm to others caused by the contractor's negligence. For example, a prime contractor as a general rule is not liable for the negligence of its subcontractors. This rule is not without criticism. During much of the twentieth century, numerous exceptions to the rule were created, thereby expanding hirer liability. However, beginning in the final quarter of the last century, courts have restricted or eliminated many exceptions, especially where the injured person is a construction worker and so subject to the workers' compensation system.

Nonetheless, many exceptions to the independent contractor rule sprang from the prime contractor's unique ability to promote (but not guarantee) project safety. It can specify in the subcontracts that safety measures be employed. It can employ safety inspectors who can walk through the site, look for hazardous situations, and then have them remedied. It can coordinate the work of the subcontractors to make sure different trades are not working in the same area at the same time. It can have weekly safety meetings. Only the prime contractor is in a position to enforce these mandates, using the threat of termination against any subcontractor who does not comply.

Premises Liability. Where the project involves new construction, the prime contractor may be the only entity in actual possession of the entire site. It can look out for hazardous conditions (whether natural or created during the building process) and then take measures to protect others from those conditions, such as giving warning to subcontractors and blocking off access to the area. Even on renovation projects, the prime contractor has the ultimate authority to exclude entrance to the work site, especially by strangers to the project (such as passers-by or trespassers).

These considerations inform the prime contractor's liability as the possessor of the construction site. *Premises liability* refers to the legal responsibilities owners or possessors of land or buildings have toward those who enter the premises. (Think of a visitor to a building whose foot goes through a rotted floor board; the visitor would sue the building owner for premises liability.) Liability is imposed if an entrant is injured by a defective premises condition which was (1) latent or unknown to the entrant (who did not see the floor board was rotted) and (2) known to the possessor or discoverable by it with a reasonable inspection (who should have been aware of the floor's rotted condition). Premises liability is discussed in Section 5.6.

For a construction site accident caused by a premises condition, the nature of the hazard to a large extent determines whether the owner or prime contractor is liable. If the premises when turned over to the prime contractor already contained the hazardous condition, then it is the owner who would be responsible. Again, the owner would be liable only if the condition was known (or knowable) to it, but was not obvious to the construction worker. However, if the dangerous condition was created by the construction work itself, liability would more likely lie with the prime contractor.

REVIEW QUESTIONS

1. What are restrictive covenants?
2. Regarding zoning reform and urban sprawl, what is "smart growth?"
3. What is the National Environmental Policy Act and what is the purpose of the Environmental Impact Statement (EIS) that is required by this act?
4. What is the "Superfund" law and what is its relevance to members of the construction industry?
5. How does the Brownfields Act modify CERCLA and what significance does that have for the construction industry?
6. According to the AIA 201-2007 Document, which party obtains a building permit?
7. What are some of the purposes of building codes?
8. What are the two broad types of statutes for construction safety?
9. What are the two duties that are imposed on an employer according to the OSHA Act?
10. What is premises liability law and what is its role in construction site accidents?

ENDNOTES

[1] 765 ILCS 140/0.01.
[2] 272 U.S. 365 (1926).
[3] 419 Pa. 504, 215 A.2d 597 (1965). For a more recent case, see *C & M Developers, Inc. v. Bedminster Township Zoning Hearing Bd.*, 573 Pa. 2, 820 A.2d 143 (2002) (one-acre minimum lot requirement is unreasonable). The township's amended ordinance was upheld in *Piper Group, Inc. v. Bedminster Township Bd. of Supervisors*, 612 Pa. 282, 30 A.3d 1083 (2011).
[4] 215 A.2d at 607.
[5] Mass.Gen.Laws Ann. ch. 40B, §§ 20–23.
[6] *Home Builders Ass'n of Northern Cal. v. City of Napa*, 90 Cal.App.4th 188, 108 Cal.Rptr.2d 60 (2001) (upholding the ordinance); Annot., 22 A.L.R.6th 295 (2007).
[7] Mass.Gen.Laws Ann., ch. 40R, § 1 et seq., effective July 1, 2004. The law is reviewed and a town's adoption of smart growth upheld in *DiRico v. Town of Kingston*, 458 Mass. 83, 934 N.E.2d 208 (2010).
[8] 545 U.S. 469, reh'g denied, 545 U.S. 1158 (2005).
[9] *City of Norwood v. Horney*, 110 Ohio St.3d 353, 853 N.E.2d 1115 (2006); Annot., 21 A.L.R.6th 261 (2007).
[10] 535 U.S. 302 (2002).
[11] Id. at 323-24 (footnote omitted).
[12] 42 U.S.C. §§ 4321–4375.
[13] Id. § 4331(a).
[14] Id. § 4332(C).
[15] 42 U.S.C. §§ 9601–9675.
[16] Id. § 9607(a).
[17] *United States v. Atlantic Research Corp.*, 551 U.S. 128 (2007).
[18] *Kaiser Aluminum & Chemical Corp. v. Catellus Dev. Corp.*, 976 F.2d 1338 (9th Cir.1992). Other federal circuits agree, see *Tanglewood E. Homeowners v. Charles-Thomas, Inc.*, 849 F.2d 1568 (5th Cir.1988) and *Redwing Carriers, Inc. v. Saraland Apartments*, 94 F.3d 1489 (11th Cir.1996).
[19] *Edward Hines Lumber Co. v. Vulcan Materials Co.*, 861 F.2d 155 (7th Cir.1988).
[20] *City of North Miami, Fla. v. Berger*, 828 F.Supp. 401 (E.D.Va.1993). See also *Blasland, Bouck & Lee, Inc. v. City of North Miami*, 96 F.Supp.2d 1375 (S.D.Fla.2000) (environmental engineering firm, which negligently implemented a remediation plan, is not a PRP under CERCLA).

[21] Public Law No. 107-118, H.R. 2869 (2002), codified at 42 U.S.C. § 9601 et seq.
[22] Vanderberg, *The Brownfields Revitalization Act of 2001: New Hope for Urban Development*, 23 Constr. Lawyer, No. 3, Summer 2003, p. 39.
[23] California Environmental Quality Act, Cal.Pub.Res. Code §21000 et seq.
[24] Mass.Gen.Laws Ann. ch. 21E.
[25] Id. § 19.
[26] Cal. Health & Safety Code §§ 26100 et seq.
[27] Id. §§ 26200 et seq.
[28] *COAC, Inc. v. Kennedy Eng'rs*, 67 Cal.App.3d 916, 136 Cal.Rptr. 890 (1977) (owner must obtain environmental impact report).
[29] Kubes, *The Design Professional's Project Self-Certification: A Key to Efficiency or Liability?* 26 Constr. Lawyer, No. 4, Fall 2006, p. 5. Wrongful certification may also expose the architect to liability to third parties. See *27 Jefferson Avenue, Inc. v. Emergi*, 18 Misc.3d 336, 846 N.Y.S.2d 868 (2007).
[30] 24 Cal.Code Regs. § 1.1.2 (2012).
[31] See http://www.iccsafe.org/gr/Pages/adoptions.aspx, last visited March 11, 2013.
[32] 24 Cal.Code Regs. part 11 (2012).
[33] The ADA is discussed in Section 11.2.
[34] N.Y. Labor Law §§ 240–242. See *Ross v. Curtis-Palmer Hydro-Electric Co.*, 81 N.Y.2d 494, 618 N.E.2d 82, 85, 601 N.Y.S.2d 49 (1993) (statute is limited to "such specific gravity-related accidents as falling from a height or being struck by a falling object that was improperly hoisted or inadequately secured").
[35] Architects and engineers are exempt from statutory liability so long as they did not "direct or control the work for activities other than planning and design." N.Y. Labor Law § 240(1). Construction manager liability is addressed in *Walls v. Turner Constr. Co.*, 4 N.Y.3d 861, 798 N.Y.S.2d 351, 831 N.E.2d 408 (2005). Construction management is dealt with in Section 7.5.
[36] 29 U.S.C. §§ 651–658.
[37] 29 U.S.C. § 666(a)–(c). See Annot., 161 A.L.R.Fed. 561 (2000) (willful violation); Annot., 151 A.L.R.Fed. 1 (1999) (repeated violations); Annot., 45 A.L.R.Fed. 785 (1979) (serious violation).
[38] 29 U.S.C. § 666(e).
[39] Id. § 654(a).
[40] *Brennan v. Occupational Safety and Health Review Comm'n*, 513 F.2d 1032 (2d Cir.1975) (first adopting the doctrine); *Secretary of Labor v. Summit Contractors, Inc.*, 23 BNA OSHC 1196, 2010 CCH OSHD ¶ 33079 (OSHRC 2010).
[41] *United States v. Pitt-Des Moines, Inc.*, 168 F.3d 976 (7th Cir.1999).

PART

B

THE MAIN ACTORS: THE OWNER, CONSTRUCTOR, AND DESIGN PROFESSIONAL

CHAPTER 7

The Project Participants: Focus on the Owner, Prime Contractor, and Construction Manager

CHAPTER 8

Licensing of the Design Professional, Contractor, and Construction Manager

CHAPTER 9

The Design Professional–Client Relationship

CHAPTER 10

Professional Design Services: The Sensitive Client Issues and Copyright

CHAPTER 11

Design Professional Liability

CHAPTER 12

Defenses to Claims of Design Professional Liability

CHAPTER 13

Ethics

CHAPTER 7

THE PROJECT PARTICIPANTS: FOCUS ON THE OWNER, PRIME CONTRACTOR, AND CONSTRUCTION MANAGER

SCENARIO

Mega Manufacturer, Inc. (MMI) makes widgets (a generic product). Believing the widget market is about to skyrocket, it decided to build two new widget factories in opposite sides of the country.

MMI decided to turn these two projects into a practical experiment to see which is the most efficient way to undertake a major construction project. It therefore decided to build each factory using a different construction delivery method:

- Project #1: Build using competitive bidding to hire a prime contractor with the work administered by the project architect.
- Project #2: Build using a construction manager as agent (CMa) with MMI hiring a prime contractor and the work administered by both the CMa and the project architect.

MMI hired Awesome Architect, Inc. to design and administer the projects. Since both jobs were for the construction of a widget factory, Awesome Architect, Inc. created one design and told MMI it would save money by using that identical design at both locations. The architect estimated each project would cost about $5 million, which was within MMI's budget.

Project #1

MMI advertised for the project in local building trade journals and on the internet, stating that the completed design was available for examination at its office and inviting any interested contractor to bid for the job. Sealed bids were required at MMI's office by a particular date and time.

Five bids were submitted. Three were for around $5 million, but two were much lower: Prime Builder, Inc. submitted a bid for $4,500,000 and Second-to-None, Inc.

submitted the second-lowest bid for $4,501,000. Adhering to its plan to test the worth of competitive bidding, MMI hired Prime Builder, Inc. as the prime contractor, only because its bid was $1000 less.

Prime Builder, Inc. decided to do the framing work itself (its owner, Bob Builder, started work as a carpenter) and to subcontract the rest of the work. In addition, Prime Builder, Inc.'s supervisor—Sam Super—was assigned to the jobsite to coordinate the work. Each morning the different subcontractors met with Sam Super who told them which work they were to perform next; however, Sam Super never told them *how* to do their jobs.

The job did not go well for Prime Builder, Inc. The cost of lumber unexpectedly rose, and—as the entity performing the framing work—it had to pay $50,000 more for lumber than anticipated. The electrical subcontractor filed for bankruptcy halfway through, and Prime Builder, Inc. had to hire a replacement contractor at a higher subcontract price. Finally, the plumbing subcontractor did not connect the pipes properly, and a water leak caused extensive water damage to the completed cabinets, which had to be replaced.

Prime Builder, Inc. asked MMI to pay it extra money to account for these project difficulties. MMI responded that it would pay no more than the contract price.

Note: Project #2 is described at the beginning of Section 7.5C.

7.1 The Basic Contracts: Private Ordering of the Parties' Relationships

Every construction project involves three elements: a legal right to improve property, a design describing the intended improvement, and implementation of that design through the services of a constructor. These three elements are reflected in two major contracts: the design contract between the owner and design professional *and* the prime contract between the owner and constructor, assuming (for now) there is a prime contractor. As a practical matter, subcontracts are a third essential agreement for a project of any appreciable size. These private contracts create the basic law governing the parties' relationships.

There is an increasing use of standard contracts in the construction industry. It is impossible to anticipate all of the problems and deal with them properly in each individual construction contract. Standard contracts rely heavily on the experiences of the past and on the expertise of people knowledgeable about construction projects. The most commonly used standard form contracts are those published by the American Institute of Architects (AIA) and the Engineers Joint Contract Documents Committee (EJCDC). In addition, a coalition of primarily contractor associations has published a series of standard contracts under the title ConsensusDOCS.

Good standard contracts are planned carefully; however, their existence does not solve all problems. When asked to pass judgment on these contracts, lawyers may reject them completely or substantially modify them. Also, a standard contract is often unread or misunderstood by the other party if the party to whom the contract is presented is not represented by a lawyer. Finally, a possibility exists that the contract will not represent all the law regulating the relationship between the parties.

7.2 The Owner

As a rule, the owner is the entity that provides the site, procures the design, determines the organizational process, and acquires the money for the project. With some exceptions, such as real estate developers, owners tend to be "one-shot" players—not the repeat players found in the contractor, construction manager, and design professional segments of the industry. There is a wide range of those who commission construction; some differentiation between them is thus essential.

More importantly, any person contracting with an owner must have an understanding of the capabilities and limitations of that client. Some limitations are legal in nature; Section 7.2C addresses who in different business organizations has authority to enter into contracts that will bind the business. Some limitations are subjective in nature, involving such issues as the client's sophistication and "smarts" in dealing with complex business transactions. All of these variables will bear on the relationship the contracting party (whether design professional or constructor) will have with the owner and no doubt its broader experience on the project.

Note that the term "owner" is commonly used in the United States and in standard form documents utilized in this country. Other countries use that term but may also use other terms, such as "principal," "employer," and "proprietor."

7.2A Public Versus Private Owners

The most important difference in types of owners is between public and private entities. A private owner can select its design professional (by competition, competitive bid, or negotiation), its contractor (by competitive bid or by negotiation), and its contracting system (single contract or multiple prime, prime contractor, or construction manager) in any manner it chooses.

Public agencies, in contrast, are limited by statute or regulation. As a rule, they must hire their designers principally on the basis of design skill and design reputation rather than fee. Construction services generally must be awarded to the lowest responsible bidder through competitive bidding. Often a public entity must use separate or multiple prime contracts because of successful efforts in the legislatures by specialty trade contractors. Public owners who wish to use construction management or design/build require separate, legislative authorization.

> In this chapter's Project #1 scenario, MMI decided to award three projects using three different project delivery methods. As a private owner, MMI was free to make these choices.

Other important differences exist between public and private owners. Public contracts have traditionally been used to accomplish goals that go beyond simply getting the best project built at the best price in the optimal period of time. Contracts to build public projects have often been influenced by the desire to improve the status of disadvantaged citizens, to remedy past discrimination, to give preferences to small businesses, to place a floor on labor wage rates, and to improve economic conditions in depressed geographical areas. Considerations exist that are not likely to play a significant role in awarding private contracts.

Public entities are more likely than are their private counterparts to be required to deal fairly with those from whom they procure design and construction services. Yet public entities, having responsibility for public monies, usually impose tight controls on how that money is to be spent. As a result, such transactions have often generated intense monitoring by public officials and by the press to avoid the possibility that public contracts will be awarded for corrupt motives or favoritism. In addition, public projects are more controversial. To whom the project is awarded, the nature of the project, and the project's location often excite fierce public debate and occasional treks to the courthouse.

Public owners often seek to control the dispute resolution process either by contract or by statute. Some public owners require arbitration. Many experienced public owners, such as federal contracting agencies and similar agencies in large states, have developed a specialized dispute resolution mechanism using specialized regulatory, arbitral, or judicial forums.

7.2B Experience

Another differentiation is between experienced and inexperienced owners. An experienced owner engages in construction, if not routinely, at least repeatedly. This owner is familiar with common legal problems that arise, construction legal and technical terminology, and standard construction documents. She may also have a skilled internal infrastructure of attorneys, engineers, risk managers, and accountants.

> In this chapter's Project #1 scenario, MMI is clearly an experienced owner. It is both a commercial owner (not a residential one) and it knows enough about the construction industry to be aware of different project delivery methods.

An inexperienced owner, though often experienced in its business, is likely to find the construction world strange and often bewildering. Such a client lacks the internal infrastructure and—even if she has resources to hire such skill—may not even know whether she should do so and, if so, how it can be done. The prototype, of course, is an owner building a residence for his or her own use.

Many other owners are "inexperienced" yet in less obvious ways. Although as a general rule, public owners are more experienced than private ones, care must be taken to differentiate between, for instance, the U.S. Army Corps of Engineers and a small local school district. The former has experienced contracting officers, contract administrators, and legal counsel and operates through comprehensive agency regulations, standard contracts, and an internal dispute resolution mechanism. The small local school district, in contrast, may be governed by a school board composed of volunteer citizens, be run by a modest administrative staff, and have a part-time legal counsel who may be unfamiliar with construction or the complicated legislation that regulates public works projects.

Similar comparisons can be made between private owners. Compare a car company building a new plant with a group of doctors building a medical clinic or a limited partnership composed of professionals seeking tax shelters through building or renting out commercial space. The clinic or limited partnership at least can buy the skill needed to pilot through the shoals of the construction process. That may not be true for a person building a residence for personal use. This person as a rule does not obtain technical assistance because of costs.

Looking next to awarding construction contracts, inexperienced private owners will likely prefer competitive bidding—not because they must but because they will not know the construction market well enough to sit across the negotiating table from a contractor. In contrast, an experienced owner may be able to review the contractor's proposal and be aware of the market and other factors necessary to negotiate.

> In this chapter's Project #1 scenario, MMI decided to use competitive bidding for Project #1. As an experienced owner, MMI should have known to use this method with caution. For example, of the five bids received, two were about ten percent lower (about $4.5 million) than the other three (for $5 million). This would have alerted an experienced owner to the possibility of a mistake in the lower bids, especially since the higher prices agreed with the architect's estimate of cost.
>
> Only $1000 separated the two lowest bids. Even if it did not want to explore the possibility of a bid mistake by Prime Builder, Inc. and Second-to-None, Inc., MMI should at least have explored each company's experience to see which was more qualified to do this type of work. Price alone will almost never motivate a sophisticated, private owner in deciding which contractor to hire.

Much more than an experienced owner, an inexperienced owner will need to engage an architect or engineer to design and administer the construction contract. (The design professional may also protect the owner from entering into a one-sided or unfair contract offered by the contractor.) The experienced owner may have sufficient skill within its own internal organization and not need an outside advisor experienced in construction.

An inexperienced owner will more likely use standard construction contracts such as those published by the American Institute of Architects (AIA) or the Engineers Joint Contracts Documents Committee (EJCDC). It will do so because it does not wish to spend the money for an individualized contract, does not have an attorney who can draft such a contract, or wishes to acquiesce to the suggestions of its architect or engineer.

An owner's lack of experience with complex standard contract terms can generate many legal problems. If people operating under the contract are not familiar with or do not understand the contract, they are likely to disregard provisions, leading to claims that those provisions have been waived.

The client's level of experience may affect performance in other ways. Inexperienced owners may make many design changes that raise the cost of construction and increase the likelihood of disputes over additional charges. They may also refuse legitimate contractor requests for additional compensation because they are unaware of those provisions of a contract that may provide the basis for additional compensation. Inexperienced owners may not keep the careful records that are so crucial when disputes arise.

Where the language must be interpreted by a court, doubts will very likely be resolved in favor of the inexperienced owner. An experienced owner could have made the contract language clear. A claim made by a contractor that it be excused from default or be given additional compensation because of unforeseen events during performance is less likely to be successful against an inexperienced owner.

7.2C Authority Problems: Private Owners

Private owners run the gamut from individual homeowner to international corporation. Across that entire spectrum, questions may arise as to who has authority to make binding representations and enter into contracts. The general agency principles that regulate these issues are discussed in Section 2.11, especially in Section 2.11D which deals with actual and apparent authority. While the question of owner authority on private projects applies equally to contractors, construction managers, and design professionals, for sake of simplicity, this discussion will presume the relationship between design professional and client.

Design professionals will be motivated by many considerations when they decide whether to enter into a contract to perform design services. One factor will be representations made by the prospective client (for example, the project will likely be approved by appropriate authorities, adequate financing can be secured, or the design professional will work out the design with a particular representative of the client). How can the design professional know he may rely on these representations?

In the best of all worlds, the people at the top of the client's organization would be contacted to determine the authority of the person with whom the design professional is directly dealing. However, it may not be easy to determine who is at the top of the client's organization, and it may not always be politic to take this approach. The realities of negotiation mean that the design professional will have to rely largely on appearances and common sense to determine whether the person with whom he is dealing can make such representations. This may depend on the position of the person in the corporate structure, the importance of the representation, and the size of the corporation. The more important the representation and the more important the contract, the more likely it is that only a person high in the organizational structure will have authority to make representations that will bind the corporation. If the corporation is particularly large, a person in a relatively lower corporate position may have authority to make representations. For example, in a large national corporation, the head of the purchasing department may be authorized to make representations that would require a vice president's approval in a smaller organization.

One useful technique is to request that important representations and promises be incorporated in the final agreement. If the person with whom the design professional is dealing is unwilling to do so, that person may not have authority to make the representations and promises that have been made.

This approach has an additional advantage. In the event of a dispute that goes to court, the parol evidence rule (discussed in Section 3.5A) may make it difficult to introduce evidence of representations that are not included in the final agreement. Even if such evidence is admitted, failure to include such representations in the written agreement may make it difficult to persuade judge or jury that the representations were made if the client denies making them.

Next, it is important to examine the authority to contract for design services. This problem will be approached by looking at the principal forms of private organizations that are likely to commission design services.

Sole Proprietors. Generally, sole proprietorships are small business operations. The sole proprietor clearly has authority to enter into contracts for design services or construction. An agent of a sole proprietorship is not likely to have authority to enter into such significant contracts. Sole proprietorships are discussed in Section 2.2.

Partnerships. Partnerships are usually not large businesses. Generally only the partners have the authority to enter into contracts for design services or construction. The partners may have designated certain partners to enter into contracts. Unless this comes to the attention of the design professional, it is unlikely that such a division of authority between the partners will affect her. Although partners have authority to enter into most contracts, it is advisable to get all of the partners to sign the contract.

If the partnership is a large organization, agents may have authority to enter into a contract for performance of design services and construction. It is probably best, however, to have some written authorization from a partner that the agent has this authority.

Partnerships are discussed in Section 2.3.

Corporations. The articles of incorporation or the bylaws of a corporation generally specify who has the authority to make designated contracts and how such authorization is to be manifested. The more unusual the contract or the more money involved, the higher the authority needed. Contracts needing board approval are passed by resolution and entered into the corporate minutes of the board meeting.

For maximum protection, the design professional should check the articles of incorporation, bylaws, and any chain-of-authority directive of the corporation. The design professional should see who has authority to authorize the contract and whether the proper mechanism, such as the appropriate resolution and its entry into the minutes, was used. Then the design professional should determine whether the person who wants to sign or has already signed for the corporation has authority to do so. The design professional can request that the contract itself be signed by the appropriate officers of the corporation, such as the president and secretary of a smaller corporation or the vice president and secretary of a larger corporation.

Sometimes such precautions need not be taken. The project may not seem important enough to warrant this extra caution. The design professional may have dealt with this corporation before and is reasonably assured there will be no difficulty over authority to contract. However, laziness or fear of antagonizing the client is not a justifiable excuse.

In this chapter's Project #1 scenario, the owner is Mega Manufacturer, Inc. Use of the designation 'Inc.' in the company name means the company is a corporation.

Corporations are discussed in Section 2.4.

Unincorporated Associations. Dealing with an unincorporated association seems simple but may involve many legal traps. To hold the members of the association liable, it is necessary to show that the people with whom the contract was made were authorized to make the contract. In such cases, it is vital to examine the constitution or bylaws of the unincorporated association and to attach a copy of the resolution of the governing board authorizing the contract. The persons signing the contract should be the authorized officers of the association. It may be wise to obtain legal advice when dealing with an unincorporated association.

Unincorporated associations are discussed in Section 2.7.

Spouses or Unmarried Cohabitants. A design professional may deal with a married couple or an unmarried couple living together. Even with spouses, no presumption exists that one is agent of the other. But one can bind the other if the former acts to further a common purpose; for example, if both persons are on the title to the improved property. In any event, it is advisable to have both spouses or members of an unmarried couple sign as parties to the contract.

7.2D Authority Problems: Public Owners

Authority issues for public entities arise with more frequency than in private contracts because of the need to protect public funds and preclude improper activity by public officials. For these reasons, contracts (or more commonly, modifications of existing contracts) entered into by an unauthorized government employee or using improper procedures are unenforceable by the contractor. Courts will protect the public entity from liability even when this is unfair to the contractor or design professional and even if a private owner under similar circumstances might be bound by the doctrine of estoppel (which would enforce the contractor's reasonable reliance on representations made by the owner or its agents).

7.2E Foreign Owners

Crucial differences exist between foreign and domestic owners. Contracts made with foreign owners are most likely to involve the sovereign or one of its agencies. Under such circumstances, the people contracting to provide design or construction services must be aware of the sovereign's power to regulate foreign exchange rates, import or export of goods or money, local labor conditions, the necessity for bribes, the lack of an independent judiciary, and the risk of expropriation.

Even contracting with a private owner in a foreign country may involve significant risks. Problems may arise from unfamiliar laws, different subcontracting practices, different laws and customs regulating the labor market, and different legal solutions.

7.3 The Design Professional

Design professionals interact with both the owner and contractor. On a design–bid–build project, the owner hires the design professional to create a design subject to parameters set by the owner, such as budget and aesthetic considerations. The design professional may then assist the owner in selecting the contractor. Once construction begins, the design professional may assume administrative obligations, such as acting as the owner's representative on the site and interacting with the contractor as questions or problems arise. In sum, design professionals find themselves in the sometimes uncomfortable position of working for the owner, yet they are expected to make impartial decisions during the construction phase of the project.

The unique position of the design professional is clarified when the project is viewed from the perspective of agency law. Clearly, the owner is the principal and the contractor is an independent contractor. The design professional, however, is a hybrid: both an agent and independent contractor, depending on the activity he is engaged in. He is an independent contractor—exercising professional judgment—when creating the design. An error in the design will not subject the owner to liability under agency principles to third parties. (Owner liability to the contractor for misrepresentations in the design—under a theory of implied warranty—is discussed in Section 16.3B.)

During the contractor selection process, the design professional acts as the owner's representative and agent. Representations made by the design professional to potential contractors are binding on the owner (unless precautions are taken in the contract to negate any oral representations). Similarly, the design professional is the owner's representative (hence, agent) during the construction phase. At the same time, as mentioned previously, decisions made by the design professional during this phase must be impartial, favoring neither owner nor contractor. Contractors may question whether in fact the design professional is impartial, especially when they believe their performance has been made more difficult because of errors in the design. Can the design professional be trusted to be impartial when agreement with the contractor's position by implication means an error on his part when creating the design? These difficult questions are explored in greater depth in Sections 12.4 and 26.2.

7.4 The Prime Contractor

On a design–bid–build or traditional project, the prime contractor is responsible for performance of the construction work. A fundamental distinction is drawn between a prime (or general) contractor and a subcontractor (or trade contractor). Taking the latter category first, a subcontractor specializes in a particular construction discipline, such as electrical; heating, ventilation and air conditioning (HVAC); or roofing. A subcontractor is not in charge of the entire building but only a part of it. A subcontractor is hired by the contractor—not the owner.

By contrast, the prime contractor is hired by the owner with the responsibility to build the entire building. The prime contractor could do all of the work himself; however, he normally hires subcontractors to perform parts of the overall work. Regardless of how much actual construction work the prime contractor does, his responsibility is to coordinate the work of the various subcontractors, enforce work quality, and deliver to the owner a finished building on time, on budget, and in strict conformity with the design requirements. A prime contractor's responsibility to the owner for defective work performed by subcontractors is discussed in Section 7.4D. A contractor's liability for safety violations is addressed in Section 6.7.

7.4A An Industry Overview

The contracting industry is very decentralized with a large number of small and medium-sized firms. Although concentration characterizes other industries, construction is largely local, with most contractors serving a single metropolitan area. Few construction companies are even regional—let alone national or international.

Half a million companies engage in construction. The average company is family owned with five to ten permanent employees. Most workers are hired for a particular job through unions or other means. Contractors obtain their work by competitive bidding. Profit margins are usually low and the likelihood of bankruptcy can be high. Because construction requires outside sources of funds, the constructor is often at the mercy of changing monetary and fiscal policies.

Two out of three contractors are specialty contractors, and one out of two workers plies a specialized trade, such as plumbing, electrical work, masonry, carpentry, plastering, and excavation. This means that in many construction projects, the prime contractor acts principally as a coordinator rather than as a builder. The prime contractor's principal function is to select a group of specialty contractors who will do the job, schedule the work, police specialty trades for compliance with schedule and quality requirements, and act as a conduit for the money flow. Moreover, in a fixed-price contract, the contractors provide financial security to the owner by giving a fixed price.

The volatility of the construction industry adds to the high probability of construction project disputes. Because the fixed-price or lump-sum contract is so common, a few bad bids can mean financial disaster. Contractors are often underfinanced. They may not have adequate financial capability or equipment when they enter into a project. They spread their money over a number of projects. They expect to construct a project with finances furnished by the owner through progress payments and with loans obtained from lending institutions.

Labor problems, especially jurisdictional disputes, are common. Many of the trade unions have restrictive labor practices that can control construction methods. Some contractors are union; others are nonunion; still others are "double-breasted," having different entities for union and nonunion jobs.

Some contractors do not have the technological skill necessary for a successful construction project. Often the technological skill, if there is any, rests with a few key employees or officers. The skill is often spread thinly over a number of projects and can be effectively diminished by the departure of key employees or officers.

As with every industry, there are contractors of questionable integrity and honesty. These contractors will try to avoid their contractual obligations and conceal inefficient or defective performance. Such contractors are skillful at diverting funds intended for one project to a different project or (even worse) taking a down payment and absconding. For example, in *State v. Colvin*,[1] an unlicensed person who defrauded six homeowners seeking to rebuilt after Hurricane Katrina was sentenced to sixty years imprisonment at hard labor—a decade for each bilked homeowner.

7.4B Independent Contractor Status

For nearly two centuries, since the advent of the industrial age, the construction contractor has been viewed as the quintessential independent contractor. (Independent contractor status is explained in Section 2.13.) A contractor (whether prime or subcontractor) is an independent, licensed business entity. It is hired to perform a job, does so using its own resources, employs means and methods of its own choosing, and is usually paid a fixed price to deliver the completed project. The prime contractor is an independent contractor vis-à-vis the owner, and the subcontractor is an independent contractor vis-à-vis the prime contractor.

> In this chapter's Project #1 scenario, Prime Builder, Inc. is an independent contractor hired by MMI. MMI is employing Prime Builder, Inc. to deliver a completed product (the warehouse) at a fixed price. The escalation in the cost of lumber, which is an industry-wide risk and not the fault of either party, is borne by Prime Builder, Inc. as the independent contractor who priced the job.
>
> Similarly, the subcontractors hired by Prime Builder, Inc. are independent contractors. This is why, although Sam Super (Prime Builder, Inc.'s supervisor) coordinates in which order and where each subcontractor works, he does not tell them *how* to do their jobs. As independent contractors, each subcontractor generally decides the means and manner of its performance.

The contractor's status as an independent contractor is the fulcrum upon which the construction industry is founded. This status means that an owner can undertake a project knowing in advance the risks of its enterprise. Were the contractor instead viewed as an agent, it would mean that any construction site accident would subject the owner to imputed (or strict) liability, even when the contractor's own methods of performance caused the accident. As remarked by New York's highest court at the dawn of the twentieth century, were an owner strictly liable in tort for the acts of its architect or contractor, "it would be very difficult, if not impossible, for any owner of land in a great city to use it for building purposes without subjecting his entire estate to such enormous hazards as would virtually amount to a prohibition of such use."[2]

7.4C Soil Conditions: The Common Law Rule

Every structure is built on a unique plot of land. Defective soil will not support a structure as built. The traditional common law rule, which was universally followed up to the beginning of the twentieth century, placed the risk of project failure caused by defective soil squarely upon the contractor. The owner had a duty not to intentionally deceive the contractor but had no duty to disclose to the contractor soil conditions that the owner knew were relevant to the contractor. Rather, the risks of contract performance—including soil conditions and design adequacy—belonged to the contractor.

In 1918, the Supreme Court announced the modern rule in *United States v. Spearin*,[3] reproduced in Section 16.3B. Under the *Spearin* doctrine, the owner—not the contractor—assumes the risk of defects in design specifications that prevent the contractor from building an acceptable project. Because a building must necessarily be built on a particular plot of land, the owner's implied warranty of design includes the risk that the design failed to accommodate the soil conditions. As explained by one court:

> Plaintiff [owner] contends here that the court below failed to distinguish between defects inherent in the plans and specifications and defects extrinsic to such specifications, *such as a latent defect in the soil*. This argument is untenable, since plans and specifications do not exist in a vacuum; they are made for a particular building at a particular place. The defect in the plans and specifications for the building in question was the failure to make provision for adequate pilings and other support for the floor; the fact that these plans and specifications might provide for an adequate building in some other place does not render the plans and specifications less defective for the location in question.[4]

The owner's implied warranty (or representation) as to the adequacy of the design specifications is counterbalanced by its right to rely upon the contractor's expertise in building. As discussed in Section 16.6, a contractor has an implied duty to perform in a workmanlike manner (as would a reasonably competent contractor). Courts include within the contractor's implied warranty of workmanlike conduct a duty to warn the owner of adverse subsurface conditions known or reasonably known to the contractor.

As an illustration, the Mississippi Supreme Court in *George B. Gilmore Co. v. Garrett*[5] held a builder responsible for severe settlement damage to a house it built, because the builder did not warn the buyer of potential soil problems and did not recommend that soil tests be made. The court held in favor of a subsequent purchaser, based on the builder's duty to use its technical knowledge to build a house on Yazoo clay in a way that would protect users from the house settling. The dissenting judge would have exonerated the builder because he concluded that no such duty was imposed on the builder and because the builder had built according to plans and specifications required by the Veterans Administration.

> In this chapter's Project #1 scenario, suppose the completed Project #1 suffers settlement. This means the foundation does not adequately and evenly support the pressures of the building on the soil. Suppose MMI seeks to hold Prime Builder, Inc. responsible for the defective soil. The contractor in its defense will argue that Awesome Architect, Inc. had a responsibility to design the foundation to accommodate the soil conditions.

As a general proposition, then, the contractor will bear the risk of unforeseen subsurface conditions unless (1) it can establish that it has relied on information furnished by the owner (see Section 19.3, (2) the contract itself provided protection (see Section 19.6), (3) the owner did not disclose information it should have disclosed,[6] or (4) the cost of performance was extraordinarily higher than could have been anticipated, a fact that would justify applying the doctrines of mutual mistake or impossibility, as discussed in Section 16.9A.

The number of exceptions to the basic rule—contractor responsibility for defective soil—should not convey the impression that the policy of placing these risks on the performing party no longer exists or does not reflect (at least to some courts) an important legal principle in fixed-price contracts. For example, in *W. H. Lyman Construction Co. v. Village of Gurnee*,[7] a contractor had been engaged to perform a sanitary sewer project. One basis for a claim against the public entity was that the sewer had to be constructed through subsurface soil that was for the most part water-bearing sand and silt rather than clay, as indicated by the soil-boring log shown on the plans. A high groundwater table was also discovered. This required the contractor to install numerous dewatering wells.

It is important to note the judicial attitude in the *Lyman* case, which dealt with the claim by the contractor that the public entity impliedly warranted that the plans and specifications would enable it to accomplish its promised performance in the manner anticipated. In rejecting the contractor's claim, the court stated, "It is well settled that a contractor cannot claim it is entitled to additional compensation simply because the task it has undertaken turns out to be more difficult due to weather conditions, the subsidence of the soil, etc. To find otherwise would be contrary to public policy and detrimental to the public interest."[8]

The court looked on the common law rule as expressing a principle of great importance—one needed to protect public entities and public funds. It is not clear that the court would have felt as strongly as it did were the contract a private one. Yet private owners and those who supply funds for the project are also greatly concerned with the ultimate cost of the project and rely heavily on the contract price in their planning.

Different rules apply to developers and subdividers. Developers both own the land and build the structures. As discussed in Section 16.7, the developer of a residential development may be liable to the homeowners under the implied warranty of habitability or fitness or under a theory of strict products liability if the homes are damaged by defective soil.

A subdivider prepares empty land for construction but does not do the construction work itself. The subdivider grades the land and subdivides it into lots in preparation for a home development. The subdivider then sells these lots to contractors who build the houses. If the houses later fail because of defective soils, may either the homeowners or contractors shift responsibility to the subdivider? A subdivider has a duty to disclose to the contractors defects in the soil it knew or should have known through the exercise of reasonable care. Disclosure of these defects to the contractors will shield the subdivider from liability to either the contractors or home purchasers.

7.4D Responsibility for Subcontractors

The subcontracting process permits the prime contractor to fulfill its obligations by using a subcontractor. However, this power to perform through another does not relieve the prime contractor of responsibility to the owner for subcontractor nonperformance or faulty performance.[9] Most standard construction contracts make the prime contractor responsible for subcontractor defaults.[10]

Owners can exercise a variety of controls over the selection of subcontractors (see Section 21.4). That the owner *approves* a proposed subcontractor, this should not relieve the prime contractor of responsibility. However, if the owner *dictates* the use of a particular subcontractor, this should relieve the prime contractor of responsibility. In such a case, the prime contractor may be acting as an agent of the owner to engage a particular subcontractor and should not be responsible.

Similarly, if the owner starts managing the subcontractors' performance to the exclusion of the prime contractor, the contractor cannot be held accountable for the work of the subcontractors. Its obligation is conditioned on its right to manage the job without unreasonable interference.

Note that the prime contractor may not invoke the independent contractor rule (see Section 5.7) to absolve itself of contract liability to the owner for the poor performance of the subcontractors. This defense applies to tort liability to third persons. It does not relieve the employer of the independent contractor when the independent contractor's failure to perform properly has damaged a party to whom the employer of the independent contractor owed a contract right of proper performance.[11]

> In this chapter's Project #1 scenario, the electrical and plumbing subcontractors caused damage to the project while it was still being built. These losses are Prime Builder, Inc.'s responsibility, as it is the prime contractor. It cannot shift to the subcontractors its contractual obligation to the owner to build the project properly and on time.

7.4E Acceptance Doctrine

What is the liability of a contractor if a construction defect causes injury to a third party (whether an occupant, visitor, or passerby), but only after performance was completed and the building had been accepted by the owner? Under the traditional **acceptance doctrine**, the owner's acceptance of the completed building cut off any liability the injured person might have against the contractor. The plaintiff was required instead to seek relief from the owner.

The acceptance doctrine was criticized for shifting liability from a negligent contractor to an innocent owner. Critics pointed out that the rationale in favor of the doctrine was especially weak where the injury was caused by a latent construction defect, which the owner could not have been expected to discover and remedy. Passage of statutes of repose—which cut off liability for contractors a specified number of years after completion of their work—meant that the acceptance doctrine was no longer needed to prevent unlimited contractor liability. (Statutes of repose are discussed in Section 16.9D.)

Today, many courts view the acceptance doctrine as incompatible with modern tort jurisprudence, which largely rejects blanket, no-duty rules. Those who continue adherence to the doctrine usually ameliorate its harsher results by creating a variety of exceptions to it; for example, the defense would not apply to defects that were imminently dangerous to others, were latent, or where the design was so obviously defective that no reasonable contractor would follow it. The competing reasons in favor and against continued adherence to the rule mean the common law remains in a state of flux.[12]

7.5 Construction Management

While the prime contractor has traditionally been the project constructor, sizable jobs are increasingly likely to involve a different actor: the construction manager (CM). Industry acceptance of the CM is evidenced by the following:

- Establishment of an advocacy, certification, and educational industry association, the Construction Management Association of America (CMAA).[13]
- Publication of standard form documents devoted to CM projects.
- Academic acceptance of construction management as meriting undergraduate and graduate degrees with stand-alone schools or departments.
- Burgeoning scholarly literature.
- U.S. Bureau of Statistics.[14]

In sum, the construction manager is as a project actor on par with the prime contractor. Yet, as explained next the CM's role as owner advisor is of equal importance. Indeed, two basic types of CMs exist: those in charge of performance, and those whose role is solely advisory.

7.5A Reasons for Construction Management

Before seeking to define construction management, the reasons for its development must be outlined. Principally, construction management developed because of the perceived inability of design professionals and contractors in the traditional construction process to use efficient management skills. Design professionals were faulted for their casual attitude toward costs, their inability to predict costs, their ignorance of the labor and materials market, as well as the cost of implementing the design. Owners were concerned in addition about the tendency of design professionals to take less responsibility for quality control, policing schedules, and monitoring payments.

Contractors also came in for their share of blame. Some lacked skills in construction techniques and the ability to work with new materials. Others did not have the infrastructure to comply with the increasingly onerous and detailed workplace safety regulations. Construction management was touted as an *efficient* tool for obtaining higher-quality construction at the lowest possible price and in the quickest possible time.

7.5B Types of Construction Management

Various approaches may be used to achieve the objectives promised by construction management. Nonetheless, a broad division of two types of CMs has developed in response to the widespread use of standard form contracts: the CM as agent or advisor (CMa) and the CM as constructor (CMc).

The divide between CMa and CMc is illustrated in Figure 7.1. The CMa contracts only with the owner and has no contract relationship with the subcontractors. Indeed, an owner who uses a CMa often hires a prime contractor. There is rarely any contract between design professional and CM.

By contrast, the CMc contracts with both the owner and the specialty trades. The CMc displaces the need for a prime contractor and looks very much like a prime contractor in the eyes of the law.

Compensation methods vary depending on the type of CM. The CMa acting as a professional advisor can be compensated by any of the various methods used to compensate a design professional, such as percentage of construction costs, personnel multiplier, cost plus fee, or fixed price. The CMc acting as a constructor can is paid the cost of the work plus a fee, where the fee is either a lump sum, a percentage of the cost of the work, or calculated using some other formula.

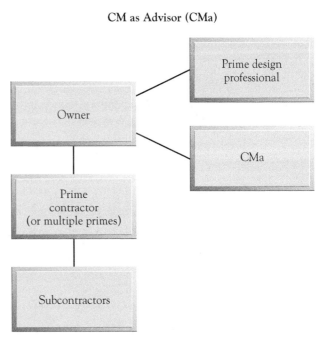
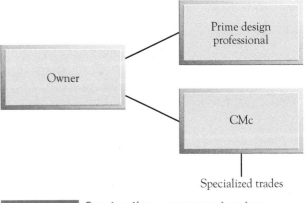

FIGURE 7.1 Construction management systems.

The CMa/CMc division is reflected in the AIA standard form documents that were last updated in 2009. The CMa documents are

- A132-2009: Owner/contractor agreement with CM as advisor
- A232-2009: General Conditions of the Contract for Construction, CMa Edition
- B132-2009: Owner/architect agreement with CM as advisor

The CMc documents are

- A133-2009: Owner/CM as constructor agreement with a guaranteed maximum price (GMP)
- A134-2009: Owner/CM as constructor agreement without a GMP

In 1993, the AIA published B144/ARCH-CM, which is an amendment to B141 (now B101), which is the standard form design agreement. B144/ARCH-CM is to be used when the architect also performs CM services. This document has not been updated.

Two of these AIA documents, A232-2009 and A133-2009, are found on this book's website as appendices.

The CMAA has published model contract documents for both agency and at-risk CMs:

- CMAA Document CMAA A-1 (2013): owner/CMa agreement
- CMAA Document CMAA CMAR-1 (2013): owner/CMc agreement

These documents are found on the book's website as appendices.

Finally, ConsensusDOCS are the standard form documents published by the Associated General Contractors of America (AGC) and numerous trade associations (not including the AIA or EJCDC). The leading ConsensusDOCS documents relating to construction management are

- ConsensusDOCS 500 (2011): CMc
- ConsensusDOCS 801 (2007, revised Nov. 2009): CMa

Also relevant is ConsensusDOCS 803 (2007), which is a standard agreement between an owner and architect/engineer when the owner has retained a CMa. These ConsensusDOCS documents are not included on this book's website.

7.5C Construction Management: Legal Issues

Before delving into legal issues involving construction management, this chapter's scenario for Project #2 is provided.

────────────── Project #2 ──────────────

After negotiations with several self-described construction management companies, MMI decided to hire CMA, Inc. (CMA) as its CMa. CMA was charged by MMI with acquiring the building permit, but CMA neglected to do so quickly enough.

After interviewing several large construction companies, CMA recommended to MMI that it hire Wonder Builder, Inc. as prime contractor, which MMI did. MMI and Wonder Builder, Inc. signed AIA Documents A132-2009 (owner/contractor agreement with CM as advisor) and A232-2009 (General Conditions of the Contract for Construction, CMa Edition).

Wonder Builder, Inc. hired a variety of subcontractors to do most of the work.

Commencement of the work was delayed a month while waiting for a building permit to be issued.

Both Awesome Architect, Inc. and CMA had employees inspecting the work. When the excavation subcontractor encountered what it believed to be a differing site condition (DSC)—a form of quicksand not predicted by the soil borings—the subcontractor informed the CMA employee who happened to be present. The CMA employee agreed this was a DSC and forwarded a recommended increase in time and cost to MMI.

The masonry subcontractor's employee fell off a scaffold and was injured. Claiming the scaffold was obviously defective and that his employer should have been prevented from using it, the employee sued Wonder Builder, Inc., Awesome Architect, Inc., and CMA. As to the architect and CM, the employee claimed they were negligent in not noticing the defective scaffold during their project inspections and then not requiring Wonder Builder, Inc. to replace it.

Legal issues surrounding construction management fall broadly into two groups. The first concerns the impact of the hiring of a CM on the owner. The second grouping of issues is the rights and liabilities of the CM itself.

The Owner/CM Relationship. While an owner may hire a CM (whether as advisor or constructor) motivated by a desire to bring greater organizational expertise and more modern construction delivery services to the project than might be expected from a prime contractor, the owner also must be aware of the legal implications of such a hiring. In general, the courts treat the CMc as akin to a prime contractor—hence, as the owner's independent contractor. This preserves the relationship existing in a design–bid–build project between the owner and prime contractor.

By contrast, a CMa is likely to be viewed as the owner's agent under certain conditions and as its independent contractor under other conditions. In this way, the CMa looks like a design professional.

> In this chapter's Project #2 scenario, commencement was delayed by one month because CMA was slow in obtaining a building permit. Under AIA A232-2009, § 3.7.1, "the Owner, through the Construction Manager, shall secure and pay for the building permit." If Wonder Builder, Inc. suffered higher performance costs caused by this delay, must it sue only the owner (as the revealed principal of an agent), or (let us assume MMI goes bankrupt) may it sue CMA directly as the party responsible for obtaining the building permit?

One drawback with this reasoning-by-analogy is that the CM ordinarily performs more services than would either the typical design professional or the typical prime contractor. For example, in *Whiting-Turner/A. L. Johnson Jt. Venture v. General Services Administration*,[15] the claimant participated as a CM during the design phase and later entered into a contract to build the project for a cost plus a fixed fee with a GMP. One of the CMc's subcontractor's made a claim based on a differing site conditions (DSC) claim (see Chapter 19). Even though the CMc was involved in assembling the information about subsurface conditions that turned out to be inaccurate, the federal board of contract appeals (an administrative agency board—not a judicial court of appeals) rejected the government's contention that the CMc—not the government—made the representations. The board held that the use of the CMc did not relieve the government of its contract obligations to compensate the contractor for DSCs. The board, while admitting that the CMc was involved in planning in a way that a normal contractor would not have been, concluded that the CMc was basically a contractor.

The role of the CM in awarding construction contracts also raises the issue of analogies. If the CM is simply a professional advisor who (unlike most design professionals) has been given the authority to award contracts, the contractor has made a contract with the owner *through* the CM. The contractor must sue the owner—not the CM who is simply the owner's agent. Thus, by using a CMa, the owner opened itself up to greater potential liability than if it had used a prime contractor. However, if the CM contracts on its own with specialty trade contractors, it looks like a prime contractor and the owner's potential liability should be less.

Must awards for CM services in public contracts be made in the same manner as other professional services? Most courts view CMs as providing primarily professional services and permit public agencies to hire them through the negotiation process, analogizing the CM to an architect or engineer.[16] Others require use of the competitive bidding statutes when hiring CMs.[17]

Rights and Liabilities of the CM. A CMc sometimes offers to include in the contract a guaranteed maximum price (GMP), which is a cost contract with a "cap". This provides the owner with similar protection to what it would get had it hired a prime contractor under a fixed-price contract. The courts have taken a strict attitude toward the GMP. Rejecting a contention by the CM that the GMP was a "target figure," an Indiana court held the CM to the GMP—despite the fact that one of the trade contractors gave a price quotation substantially higher than planned and the CM decided to do the work itself.[18]

While treatment of a GMP is a matter of contract interpretation, introduction of construction management also raises questions under statutory law. For example, in a recent case, a CM sought to assert a mechanics' lien on a property whose construction he had supervised. Were he a laborer, a subcontractor, or a design professional, he would have been entitled to a lien. But the court held that the CM had not designed or built; he had merely supervised and was not entitled to a lien under the laws of Indiana.[19] In other states, the CM might have succeeded, as mechanics' lien laws vary greatly.

As noted earlier, most legal issues have at their core whether the CM is more like a design professional or an entrepreneurial contractor. For example, must the CM be licensed by the state, and if so, which type of license is needed? The answer may depend on which form of CM is used, whether the CM is engaged solely as a professional advisor, or whether the CM undertakes to perform some construction himself. Licensing of the CM is addressed in Section 8.7.

Liability problems also involve analogies. For example, it has been held that a contractor can bring a negligence claim against the CM just as a negligence claim can be instituted against a design professional.[20] If the CM is analogized to a design professional or performs services usually performed by a design professional, will the CM be given the benefit of the professional standard described in Section 11.2?

The negligence standard applicable to CMs may well turn on the type of activity that is alleged to be negligent. (The negligence standard is discussed in Section 5.3B.) As to different project administrative services performed by the CM, the court will ask whether those services required the knowledge, training, or judgment of a professional. If so, the professional standard will apply, and expert testimony will likely be required. If not, the CM will be subject to the ordinary negligence standard.

The parties' contractual arrangements may also determine whether the CM may be subject to a negligence claim—at least for economic harm (not involving physical harm). In a California case, *The Ratcliff Architects v. Vanir Construction Management, Inc.*,[21] an *architect* was not allowed to assert a negligence claim for economic losses against the CM. (The architect contended that the CM's incompetence caused the architect to expend more services than otherwise would have been required.) The court held that the CM had no duty to the architect, that his sole duty was to the owner who had engaged him, and that the contract between owner and CM was not made for the benefit of the architect.

A CM is also exposed to claims by construction workers injured on the job. Again, liability under the common law may be based on analogies to injured worker claims against design professionals.[22] If CMs instead look like prime contractors, either because they are managing the work of the specialty trades, have overall safety responsibility, or are performing some of the construction work with their own forces, they may be found to have contractor-like liability. In *Farabaugh v. Pennsylvania Turnpike Commission*,[23] the CM had assumed responsibility in its contract for job safety, as normally does a prime contractor. The CM was found to owe a duty of care to the estate of a construction worker killed in a job site accident.

In this chapter's Project #2 scenario, the masonry subcontractor's employee would have sued Awesome Architect, Inc. and CMA for negligence. Both defendants will point to AIA A232-2009, § 4.2.5, which provides that the architect and construction manager is not responsible for job safety; that is Wonder Builder, Inc.'s obligation.

Liability may also arise under safety statutes. New York imposes a duty on all contractors and owners "and their agents" to make sure workers are furnished with safety equipment. A CM acting as the owner's advisor who has authority over the work may be liable under this statute as an agent of the owner.[24] Under federal law, a CM with a pervasive job-site presence and authority to stop unsafe work may be liable for a worker's injury under the Occupational Safety and Health Act—even if the CM did no actual construction work.[25] Safety statutes are reviewed in Section 6.7.

In some states, being an at-risk CM can be advantageous in the event of worker claims. Such a CM may be considered a "statutory employer" entitled to immunity under the workers' compensation laws, as described in Section 2.12A. This was unsuccessful where the CM did not construct anything but was simply hired to manage and inspect—more like a design professional than a contractor.[26]

The dual functions of a CM—as agent or constructor and sometimes both—raises insurance coverage difficulties. (Insurance is the topic of Chapter 24.) The CM who is principally a constructor will find its commercial general liability (CGL) coverage (which provides coverage against claims of general negligence) does not include design, whereas a CM who is essentially a design professional may find its professional liability insurance excludes any coverage relating to the construction process or the work of the contractor. For this reason, CMs should carry both CGL and professional liability insurance (as should design professionals).

7.5D Impact on Project Organization

Owners contemplating hiring CMs must consider the advantages and disadvantages of inserting this new actor into an already crowded assemblage of project participants. CMs claim to make the construction process more efficient (and hence faster and cheaper) by bringing sophistication and modern management techniques to the construction process. CMs hired as advisors, however, may be viewed by the courts as acting as the owner's agent. As explained in Section 2.11D, this could expose the owner (as principal) to liability to trade contractors who claim they were injured by the CM's actions. In addition, employing a CMa, together with an architect and prime contractor, invariably will add cost to the owner and may also cause confusion as to the chain of command. On the other hand, an owner who hires a CMc may, if problems arise, find itself in the same adversarial relationship as it would with a prime contractor, especially if a GMP is involved.

An examination of the "General Conditions" for CMa projects, AIA Doc. A232-2009, reveals the impact the hiring of a CMa may have on project organization. Only a few of the potential complications are noted.

Article 4 is titled "Architect and Construction Manager." Section 4.2—dealing with contract administration—imposes many overlapping responsibilities on the architect and CMa, which is a potential source of organizational confusion and inefficiency.

Both the architect and CMa conduct site investigations but at different times. The architect inspects at intervals appropriate to the stage of construction or as otherwise agreed with the owner (§ 4.2.2), while the CMa (under § 4.2.3) must supply at least one representative to be on-site "whenever the Work is being performed." Under § 4.2.5, the architect and CMa are not responsible for the construction means and methods.

Section 4.2.6 addresses project communications. The owner and contractor are to communicate with each other through the medium of the CMa, while "contemporaneously provid[ing] the same communications to the Architect about matters relating to the Contract Documents." What is the architect to do with these copied communications, especially if it has questions about decisions made? Must it object lest its silence be viewed as consent? The AIA document does not say.

Overlap of the CMa's and architect's responsibilities is present in three matters of great importance to the contractor: approval of payment applications, rejection of the work, and submittals. Under § 4.2.7, the CMa "and" architect review and certify applications for payment in accordance with Article 9. Under § 9.5.1, the CMa "or" architect may withhold a certificate of payment in whole or in part.

Similarly, under § 4.2.8, both the architect and CMa have authority to reject the work as nonconforming "and will notify each other about the rejection." It then goes on to say that the CMa (with no mention of the architect) may order additional inspection or testing of the work.

Review of submittals is addressed in §§ 4.2.9 (CMa) and 4.2.10 (architect). Submittals initially are submitted to the CMa, who will coordinate the information if the project involves multiple prime contractors. The CMa then transmits the submittals to the architect, who reviews them for conformance with the "design concept." The architect then returns the submittals to the CMa, who transmits them to the contractor. The submittal process involves more participants than on a traditional design–bid–build project.

The prime contractor must also adjust to a project that contains both an architect and a CMa. In the chapter's Project #2 scenario, the excavation subcontractor informed the CMA employee of a differing site condition (DSC). The CMA employee recommended to the owner that the subcontractor had encountered a DSC and was entitled to additional time and money.

Yet under AIA A232-2009, § 3.7.4, a contractor who encounters what it believes to be a DSC must inform both the architect and CM within 21 days. Under the AIA, the architect and CM must consult with each other before making a recommendation to the owner. May MMI refuse to compensate the subcontractor because (1) the subcontractor did not inform both CMA and the architect of the condition and (2) more than 21 days have passed after the subcontractor discovered the soil condition? If so, may the subcontractor sue CMA for failing to consult with the architect before making the recommendation to the owner? This scenario illustrates the complication to the contractor of a more complicated project organization.

This is a truncated review of just some of the architect's and CMa's administrative duties. The owner may view these overlapping and duplicative administrative responsibilities as increasing the likelihood of receiving a defect-free project. Yet for the contractor, the danger of miscommunications, delays, and even tensions arising out of the CMa/architect relationship must be a source of concern. The contractor must also contend with a more complicated claims process.

7.6 Project Funding: Spotlight on Lender

Because owners rarely have funds to provide for major (or even minor) construction work, an important actor along the owner chain is the entity providing financing for the project. Availability of funds is essential to the participants who provide services and materials. Often the owner's inability to obtain funds is the reason a project does not proceed, despite the design professional's having spent time on the design. Similarly, available funding may not be adequate to deal with such common events in construction as design adjustments, escalation of prices, changed conditions, and unexpected subsurface conditions. Inadequate funding can create a breakdown in the relationship among the project's participants, which in turn leads to claims.

With regard to funding, it is important again to differentiate between public and private projects. Public projects are usually funded by appropriations made by a legislature or administrative agency. In controversial projects, challenges may be made to the project itself, to the method by which the contract was awarded, or to compliance with other legal requirements. If challenges are made to the project, funds may never become available or may generate delay in payment.

Another problem generated in public projects relates to the project's being funded by a mixture of funds from various public agencies or even a combination of public agencies and private entities. For example, the interstate road system was funded 90 percent by the federal government and 10 percent by the states, while wastewater projects were funded 75 percent by the federal government, 12.5 percent by the state, and 12.5 percent by local entities. Mixed funding can generate problems, because one funding entity requires certain contract clauses that may differ from those required by another funding entity.

In private projects, funds are usually provided by construction loans made by lenders for short periods, usually only for the time necessary to complete the project. The interest rates are short term, usually floating, and are generally higher than those charged by lenders of permanent loans.

The private construction lender wants the loan to be repaid and usually takes a **security interest** in the project that can be used to obtain repayment if the loan is not repaid. The lender makes an economic evaluation to determine whether a commercial borrower will derive enough revenue to repay the loan.

In addition, the lender will wish to be certain that the borrower puts up enough of its own money, so if the project runs into problems, the borrower will not walk away from the project and avoid the loan obligation by going bankrupt—leaving the lender with only the security interest and the unpleasant prospect of taking over a defaulted project.

Similarly, the lender wants to be certain that the funds it advances will go into the project and enhance the value of the project. It does not wish to see the borrower diverting funds for other purposes.

The lender also wishes to be assured that there are no delays, which can often be caused by faulty design or poor workmanship. Costly design changes, claims for extras by the contractor, and additional expenses due to unforeseen conditions are events the lender wishes to avoid. Similarly, the lender wishes to be assured that those who provide services—such as contractors and subcontractors and those who provide materials and equipment, such as suppliers—are paid and do not assert liens against the project. (A lien is a security interest in the improved property; see Section 21.7.) If liens are asserted, it may be difficult for the construction lender to obtain a permanent "takeout" lender who would replace the short-term construction loan with a mortgage: a long-term loan that uses the completed improvement as security.

The lender usually wants to know how the project is proceeding and may, in addition to approving the construction contracts made at the outset, insist on receiving copies of contract addenda, modifications, requests for changes, change orders, claims, and even routine correspondence.

The lender's interest in knowing how the project is proceeding arises from the risk to it should the contractor abandon the job before it is completed. The lender's security in the partially completed building may be useless unless it has sufficient funds for the owner to hire another contractor to complete the project. Similarly, a contractor who has been paid the entire loan amount when the work is only half done will have an economic incentive to abscond. For these reasons, the lender will not pay the contractor all of the funds up front but will instead make periodic (usually monthly) disbursements or "progress payments" in amounts intended to equal the percentage of work completed. If the contractor abandons the project when it is half done but has received (at that point) only half of the construction funds, the lender (and borrower) should have enough funds remaining to hire another contractor to complete the work. Disbursement of accurate progress payments protects the lender's security interest.

The lender may use inspectors to verify performance of the work before paying the contractor. Suppose the inspector sees the work is incomplete or defectively done, yet the bank pays the contractor anyway? Suppose the owner warns the lender that she is unhappy with the quality of the work, but the lender still makes a disbursement? Standard language in the loan agreement will specify that any inspections made by the lender will be for its benefit alone. Nonetheless, some courts have allowed borrowers to seek recovery against the lender for wrongful disbursements. For example, in *GEM Industrial, Inc. v. Sun Trust Bank*,[27] the lender's own inspector had discovered defective construction but the lender had paid the contractor anyway; the lender was found liable to the borrowers for these wrongful payments.

Subcontractors may also claim that the lender had a duty to shield them from the contractor's misconduct. In *Hoida, Inc. v. M&I Midstate Bank*,[28] an unpaid subcontractor, unable to recover from the prime contractor, sued the project's construction lender. The subcontractor argued that the lender's disbursement of progress payments to the prime contractor was negligent, because the lender did not first verify that the prime contractor had obtained lien waivers from the subcontractors. (The lien waivers would have meant the subcontractors had been paid.) A divided Wisconsin Supreme Court instead ruled that construction lenders do not breach a common law duty of care by making disbursements, authorized by the owner, without first verifying that the subcontractors had been paid. The court feared that imposing such a duty would place too great an administrative responsibility upon lenders.[29] It also pointed out that making the lender liable would violate the public policy of the mechanics' lien act, which gives liens by construction lenders priority over liens filed by subcontractors.[30]

7.7 Insurers

Insurers are also important actors on this crowded stage. All of the major participants, such as design professionals, owners, and contractors, are likely to have various types of liability insurance. In addition, either the owner or the prime contractor will carry some form of property insurance on the work as it proceeds and on materials and equipment not yet incorporated into the project. Insurance is more fully discussed in Chapter 24.

7.8 Sureties

The owner of a construction project is, to a large degree, dependent upon the prime contractor for the project's success. The owner has two main concerns with regard to the prime contractor. First, if the contractor does not fully perform in accordance with the architect's design, the owner will be left with an unfinished or defective building. Second, if the contractor does not pay the subcontractors, the subcontractors may file mechanics' liens against the improved property. (Mechanics' liens are discussed in Section 21.7.)

A surety backstops the contractor, protecting the owner against either type of default. The surety issues a bond, which is a sum of money that can be accessed in the event the prime contractor defaults on its contract obligations. A *performance bond* backstops the prime contractor's obligation to build the project in conformity with the contract requirements. If the owner declares the contractor in default, the surety will either pay the owner the amount of the bond (so the owner can hire its own replacement contractor) or the surety will complete the project itself (by hiring a replacement contractor or even the original contractor).

A *payment bond* backstops the prime contractor's obligation to pay the subcontractors. The surety will pay the subcontractors to ensure they fulfill their contract obligations and do not file mechanics' liens against the owner's property.

Sureties are discussed in depth in Chapter 25.

REVIEW QUESTIONS

1. Every construction project involves which three elements?

2. What are the major types of owners and what are the legal consequences of those differences?

3. What are the four different situations that may occur that would allow a contractor to not bear the risk of unforeseen subsurface conditions?

4. What are the primary reasons for the development of Construction Management?

5. Explain the primary differences between the CM as Agent or Advisor (CMa) and the CM as Constructor (CMc) delivery methods.

6. What are the consequences to a CMc of agreeing to a guaranteed maximum price (GMP)?

7. What are the three matters that overlap between the CMa's and architect's responsibilities that are of great importance to the contractor?

8. In the case *Hoida, Inc. v. M&I Midstate Bank*, what did the court rule in regard to the subcontractors claim that a lender had a duty to shield them from the contractor's misconduct?

9. What is the difference between a performance bond and a payment bond?

10. What is the difference in how public projects and private projects are funded?

ENDNOTES

[1] 85 So.3d 663 (La.2012).
[2] *Burke v. Ireland*, 166 N.Y. 305, 59 N.E. 914, 916 (1901). Of course, this statement preceded widespread adoption of liability insurance.
[3] 248 U.S. 132 (1918).
[4] *Ridley Investment Co. v. Croll*, 56 Del. 209, 192 A.2d 925, 926–27 (1963) (emphasis added).
[5] 582 So.2d 387 (Miss.1991). See also Annot., 73 A.L.R.3d 1213 (1976).
[6] See *P. T. & L. Constr. Co., Inc. v. State of New Jersey Dep't of Transp.*, 108 N.J. 559, 531 A.2d 330 (1987).
[7] 84 Ill.App.3d 28, 403 N.E.2d 1325 (1980).
[8] 403 N.E.2d at 1328.
[9] *Kahn v. Prahl*, 414 S.W.2d 269 (Mo.1967); *Waterway Terminals Co. v. P. S. Lord Mechanical Contractors*, 242 Or. 1, 406 P.2d 556 (1965).
[10] E.g., AIA Document A201-2007, § 5.3.
[11] *Harold A. Newman Co. v. Nero*, 31 Cal.App.3d 490, 107 Cal.Rptr. 464 (1973); *Brooks v. Hayes*, 133 Wis.2d 228, 395 N.W.2d 167 (1986).
[12] For a compilation of the cases, see Annot., 75 A.L.R.5th 413 (2000).
[13] Its website is cmaa.com.
[14] The BLS, "Occupational Employment and Wages, May 2011," estimates a national employment of 195,000 CMs, with a mean annual wage of $93,900. See http://www.bls.gov/oes/current/oes119021.htm#nat.
[15] GSBCA No. 15401, 02-1 BCA ¶ 31,708.
[16] *Shivley v. Belleville Township High School Dist. No. 201*, 329 Ill.App.3d 1156, 769 N.E.2d 1062 (2002) and *Malloy v. Boyertown Area School Bd.*, 540 Pa. 915, 657 A.2d 915 (1995). Section 15.2A discusses public procurement of design services.
[17] *City of Inglewood-L.A. County Civic Center Auth., v. Superior Court*, 7 Cal.3d 861, 500 P.2d 601, 103 Cal. Rptr. 689 (1972).
[18] *TRW, Inc. v. Fox Dev. Corp.*, 604 N.E.2d 626 (Ind.App.1992), transfer denied May 21, 1993.
[19] *Murdock Constr. Management, Inc. v. Eastern Star Missionary Baptist Church, Inc.*, 766 N.E.2d 759 (Ind. App.), transfer denied, 783 N.E.2d 694 (Ind.2002).

[20] *Gateway Erectors Div. v. Lutheran Gen. Hosp.,* 102 Ill.App.3d 300, 430 N.E.2d 20 (1981). Contractor or subcontractor claims against a design professional is the subject of Sections 11.6D & E.

[21] 88 Cal.App.4th 595, 106 Cal.Rptr.2d 1 (2001).

[22] *Caldwell v. Bechtel, Inc.,* 631 F.2d 989 (D.C.Cir.1981), a leading case. But see *Everette v. Alyeska Pipeline Service Co.,* 614 P.2d 1341 (Alaska 1980) (CM relinquished safety responsibilities and was not liable).

[23] 590 Pa. 46, 911 A.2d 1264 (2006).

[24] *Walls v. Turner Constr. Co.,* 4 N.Y.3d 861, 831 N.E.2d 408, 798 N.Y.S.2d 351 (2005).

[25] *Bechtel Power Corp.,* 4 BNA OSHC 1005, 1975–1976 CCH OSHD ¶ 20,503 (OSHRC 1976), aff'd, 548 F.2d 248 (8th Cir.1977).

[26] *Brady v. Ralph Parsons Co.,* 308 Md. 486, 520 A.2d 717 (1987), subsequent appeal, 327 Md. 275, 609 A.2d 297 (1992).

[27] 700 F.Supp.2d 915 (N.D.Ohio 2010). See Annot., 20 A.L.R.5th 499 (1994).

[28] 291 Wis.2d 283, 717 N.W.2d 17 (2006).

[29] Id., 717 N.W.2d at 32.

[30] Id. at 33–34.

CHAPTER 8

LICENSING OF THE DESIGN PROFESSIONAL, CONTRACTOR, AND CONSTRUCTION MANAGER

SCENARIO

Mega Manufacturer, Inc. (MMI) is a Massachusetts company that makes widgets (a generic product). It decided to build a new factory in neighboring New York State.

MMI had a prior relationship with Awesome Architects, Inc. (AAI), an architectural firm located in Massachusetts. AAI is a corporation consisting of two co-owners, Andrew and Allison, both of whom are registered as architects in Massachusetts. AAI agreed to do the work, even though neither AAI, Andrew, nor Allison was licensed in New York.

Concerned about any licensing problems, AAI hired Edward Engineer, an engineer licensed in New York, to apply his seal to the design created by AAI. Edward Engineer was also to inspect the work and certify payment of the prime contractor.

Based on Edward Engineer's recommendation, MMI hired a New York prime contractor. The prime contractor wanted to use equipment manufactured by Cal-Make, Inc., a California manufacturer licensed in that state to install its equipment. Cal-Make, Inc. told the prime contractor it would sell it the equipment, but it could not install it because it was not licensed in New York. The prime contractor responded that it only trusted Cal-Make, Inc. to install the equipment properly. To induce the subcontractor to perform, the prime contractor increased the subcontract price by 20 percent and promised to pay half up front and half upon completion. Cal-Make, Inc. reluctantly agreed to install the equipment. After it did so, the prime contractor refused to pay Cal-Make, Inc. the contract balance.

8.1 Introduction

This chapter on **licensing** of design professionals, contractors, and construction managers continues the theme, begun in Chapter 6, of regulation of the construction industry. Chapter 6 addresses regulation of land use and of the construction process.

Licensing, the topic of this Chapter 8, addresses regulation of those who design and build the project. Some statutes distinguish between *registration* of design professionals and *licensing* of contractors. This chapter uses the term licensing interchangeably with registration.

This chapter deals first with professional registration laws. It then examines the licensing of contractors, followed by the licensing of construction managers, who are latecomers to the field of occupational regulation.

Licensing is a matter of state law. Typically, licensing acts (or statutes) include creation of administrative agencies that issue regulations and administer systems that determine who may legally perform these services. Because of the political nature of these laws, they often are changed. A Tennessee decision handed down in 1995 stated that the Tennessee contractor licensing law had been changed seventeen times since its original enactment in 1931.[1] This is not uncommon.

For these reasons, textual statements must be regarded as general and even tentative. Local laws must be consulted when planning to perform design and construction services or in actual cases that involve the legality of performing these services.

State licensing laws are not applied to those contractors who work exclusively for the federal government.[2] To give effect to the supremacy clause of the U.S. Constitution, the federal government must operate free of interference by the states.

8.2 Public Regulation: Justifications and Criticisms

The police and public welfare powers granted to states by their constitutions permit states to regulate who may practice professions and occupations. Such regulations are usually expressed in licensing or registration laws.

Licensing laws provide a representation to the public that the holder meets a minimal level of competence, honesty, and financial capacity. The state wishes to protect the general public from being harmed by poor construction work and seeks to accomplish this by allowing only those who meet state requirements to design and to build.

A contractor's or design professional's competence is established by having to meet specified education and experience requirements. Evidence of immoral or illegal conduct is a reason to deny an applicant a license or to suspend or revoke an existing license. A licensed person's financial integrity may include a requirement to obtain bonds for the protection of persons injured by the licensee's conduct.

Courts recognize that one purpose of licensing laws, whether they apply to architects, engineers, surveyors, or contractors, is protection both of the client and of the general public. One court explained the purpose of a contractor licensing law as to

> prevent unscrupulous or financially irresponsible contractors from deceiving and taking advantage of those who engage them to build…. It often happens that fly-by-night organizations begin a job and, standing in danger of losing money, leave it unfinished to the owner's detriment. Or they may do unsatisfactory work, failing to comply with the terms of their agreement. The licensing requirement is designed to curb these evils; the license itself is some evidence to the owner that he is dealing with an honest and qualified builder.[3]

Criticism of licensing laws exists. Educational and experience requirements and the administration of these regulatory systems increase the cost of providing these services. They also reduce the pool of contractors.

More significantly, occupational licensing laws can deny some people their only means of livelihood. (For this reason, tightening of licensing laws is often accompanied by "grandfathering in" those who already are in those occupations.) Critics attack the premise of a state determining who and how many people will be allowed to enter professions or occupations.

Critics also note that the agencies that administer licensing laws are often dominated by members of the profession being regulated. This can generate practices that keep down the number of practitioners and improve the economic status of those already in their profession.

Notwithstanding these criticisms, the social utility of licensing laws is broadly viewed as outweighing any economic drawbacks.

8.3 Administration of Licensing Laws: Agencies, Admission, and Discipline

8.3A Licensing Agencies

Modern legislatures articulate rules of conduct by statute and create administrative agencies to administer and implement the laws. Agencies created to regulate the professions can make rules and regulations to fill deliberate gaps left by the statutes and particularize the general concepts articulated by the legislature. For example, the licensing laws state that there must be examinations to determine competence. The details of the examinations, such as the type, duration, and frequency, are determined by the agency.

Agencies have both quasi-legislative power (to create rules) and quasi-judicial power (to decide if the rules have been broken). For example, agencies may decide whether a particular school's degree will qualify an applicant to take the examination or whether certain conduct merits disciplinary sanction. These agency decisions are subject to judicial review, although courts tend to defer to the agencies' expertise and knowledge of the occupation. (See Sections 8.3C and 8.4D.) Agencies also have power to seek court orders ordering that people cease violating the licensing laws.

Members of these agencies who perform quasi-judicial functions, such as deciding whether an architect or engineer should have her license suspended or revoked, traditionally are members of the professions being regulated. But there is an increasing tendency to appoint some lay persons to these boards.

8.3B Admission to Practice

Licensure requirements vary between the states. Some states require citizenship and residency. Most have minimum age requirements, usually ranging from 21 to 25. All require a designated number of years of practical experience that can be substantially reduced if the applicant has received professional training in recognized professional schools. All states require at least one examination, and some require two. Most inquire into character and honesty. Some require interviews.

8.3C Postadmission Discipline: *Duncan v. Missouri Board for Architects, Professional Engineers and Land Surveyors*

Although the regulatory emphasis has been on carefully screening those who seek to enter the professions, all states can discipline people who have been admitted. Discipline can be a reprimand or suspension or revocation of the license.

Grounds for disciplinary action vary considerably from state to state, but they are generally based on wrongful conduct in the admissions process, such as submitting inaccurate or misleading information or having conduct after admission that can be classified as unprofessional or grossly incompetent.

DUNCAN v. MISSOURI BOARD FOR ARCHITECTS, PROFESSIONAL ENGINEERS AND LAND SURVEYORS

Missouri Court of Appeals, Eastern District, 744 S.W.2d 524 (Mo.Ct.App.1988).
[Ed. note: Footnotes have been renumbered and some omitted.]

SMITH, Judge.

On July 17, 1981, the second and fourth floor walkways of the Hyatt Regency Hotel in Kansas City collapsed and fell to the floor of the main lobby. Approximately 1500 to 2000 people were in the lobby. The walkways together weighed 142,000 pounds. One hundred and fourteen people died and at least 186 were injured. In terms of loss of life and injuries, the National Bureau of Standards concluded this was the most devastating structural collapse ever to take place in this country. That Bureau conducted an investigation of the tragedy and made its report in May 1982. In February 1984, the Missouri Board for Architects, Professional Engineers and Land Surveyors filed its complaint seeking a determination that the engineering certificates of registration of Daniel Duncan and Jack Gillum and the engineering certificate of authority of G.C.E. International were subject to discipline pursuant to Sec. 327.441 RSMo 1978. The Commission, after hearing, found that such certificates were subject to suspension or revocation. Upon remand for assessment of appropriate disciplinary action, the Board ordered all three certificates revoked. Upon appeal the trial court affirmed. We do likewise.

G.C.E. is a Missouri corporation holding a certificate of authority to perform professional engineering services in Missouri. Gillum is a practicing structural engineer holding a license to practice professional engineering in Missouri. He is president of G.C.E. Duncan is a practicing structural engineer holding a license to practice professional engineering in Missouri and is an employee of G.C.E.

Gillum-Colaco, Inc., a Texas corporation, contracted with the architects of the Hyatt construction to perform structural engineering services in connection with the erection of that building. By subcontract the responsibility for performing all of such engineering services was assumed by G.C.E. The structural engineer, G.C.E., was part of the "Design Team" which also included the architect, and mechanical and electrical engineers. Gillum was identified pursuant to Sec. 327.401.2(2) RSMo 1978, as the individual personally in charge of and supervisory (sic) of professional engineering activities of G.C.E. in Missouri. His professional seal was utilized on structural engineering plans for the Hyatt. Duncan was the project engineer for the Hyatt construction in direct charge of the actual structural engineering work on the project. He was under the direct supervision of Gillum. [Ed. note: See Figure 8.1.]

* * *

We will not attempt to set forth in detail the extensive evidence before the Commission. Some review of that evidence is, however, required. The atrium of the Hyatt was located between the 40 story tower section of the hotel and the function block. Connecting the tower and function block were three walkways, suspended from the atrium ceiling above the atrium lobby. The fourth floor walkway was positioned directly above the second floor walkway. The third floor walkway was to the east of the other two walkways. As originally designed the fourth and second floor walkways were to be supported by what is referred to as a "one rod" design. This consisted of six one and one quarter inch steel rods, three on each side, connected to the atrium roof and running down through the two walkways. Under this design each walkway would receive its support from the steel rods and the second floor walkway would not be supported by the fourth floor walkway. At each junction of the rods and the walkways was a box beam–hanger rod connection. These were steel to steel connections and the design of such connections is an engineering function, because the design

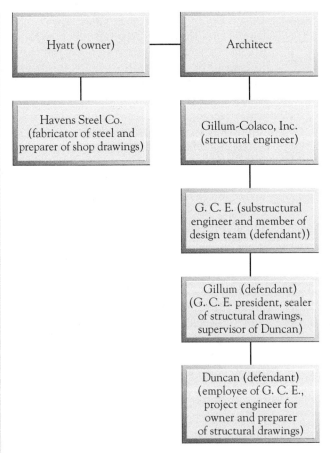

FIGURE 8.1 Participants in structural design, Hyatt Regency project.

includes the performance of engineering calculations to determine the adequacy of the connection to carry the loads for which it is designed. [Ed. note: See Figure 8.2.]

Connections are basically of three kinds, simple, complex, and special. All connections are the responsibility of the structural engineer. Simple connections are those which have no unusual loads or forces. They may be designed by looking up the design in the American Institute of Steel Construction (AISC) Manual of Steel Construction and following directions found therein. This can be done by a steel fabricator utilizing non-engineering personnel. Complex connections are those where extreme or unusual loads are exerted upon the connection or where the loads are transferred to the connection from several directions. These connections cannot be designed from the AISC manual and require engineering expertise to design.

Special connections are a hybrid having characteristics of each of the other two. A simple connection becomes special where concentrated loads are placed thereon and the AISC manual no longer provides all the information necessary to properly design the connection. Such connections may also become special where the connections are "non-redundant." A "redundant" connection is one where failure of the connection will not cause failure of the entire system because the loads will be carried by other connections. A "non-redundant" connection which fails will cause collapse of the structure. The box beam–hanger rod connections were "non-redundant." The Commission found the box beam–hanger rod connections to be special connections.

The steel fabricator on the Hyatt project, Havens Steel Company, had engineers capable of designing simple, complex or special connections. The structural engineer on a project may, as a matter of custom, elect to have connections designed by the fabricator. To do this, he communicates this information to the fabricator by the manner in which he portrays the connection on his structural drawings. The adequacy of the connection design remains the responsibility of the structural engineer. The Commission found that the structural drawings (S405.1 Secs. 10 and 11) did not communicate to the fabricator that it was to design the box beam–hanger rod connection, and did communicate to the fabricator that those connections had been designed by the engineer. Duncan testified that he intended for the fabricator to design the connections. Havens prepared its shop drawings on the basis that the connections shown on the design drawings had been designed by the structural engineer. Certain information concerning loads and other aspects of the box beam–hanger rod connections which appeared on Duncan's preliminary sketches was not included on the final structural drawings sent to the fabricator. The Commission also found that Duncan's structural drawings did not reflect the need for a special weld, did not reflect the need for stiffeners and bearing plates, and reflected that the hanger rods should be of regular strength steel rather than high strength. These factual findings are not contested. The hanger rods and the box beam–hanger rod connections shown on the structural drawings did not meet the design specifications of the Kansas City Building Code. That finding of fact by the Commission is also not contested.

Because of certain fabricating problems Havens proposed to Duncan the use of a "double rod"

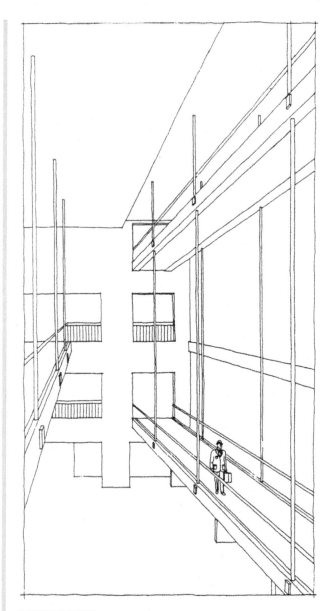

FIGURE 8.2 Schematic of walkways as viewed from north wall of atrium.

system to suspend the second and fourth floor walkways. [Ed. note: See Figure 8.3.] Under this system the original six rods would be connected only to the fourth floor walkway. A second set of rods would then connect the second floor walkway to the fourth floor walkway. The effect of this change was to double the load on the fourth floor walkway and the box beam–hanger rod connections on that walkway. There was evidence that one of the architects contacted Duncan to verify that the double rod arrangement was structurally sound and was advised by Duncan that it was. Appellants dispute that the architect's testimony clearly establishes such an inquiry and contend that the conversation dealt rather with the aesthetic nature of the change. Our review of the record causes us to conclude that the architect did testify to receiving assurances that the new design was structurally safe. It is difficult to understand why the architect would consult the structural engineer if his only concern was the aesthetics of the new design. The Commission further found that the records of G.C.E. failed to contain a record of a web shear calculation which Duncan testified he made and which would normally be a part of the G.C.E. records. Duncan's testimony reflected the need for such a calculation before approval of the double rod arrangement. The Commission also found certain additional necessary tests or calculations were not made. Appellants do not challenge that finding. It is a reasonable inference from the evidence that Duncan did not make the engineering calculations and tests necessary to determine the structural soundness of the double rod design.

Havens prepared the shop drawings of the structural steel fabrication. These drawings were returned to Duncan for review and approval. They contained the fabrication of the box beam–hanger rod connections based upon the structural drawings previously submitted by Duncan and bearing the seal of Gillum. The Commission found, and appellants do not dispute, that its own internal procedures called for a detailed check of all special connections. The primary reason for such a procedure is to provide assurance for the owner that the fabricator is conforming to the contract and that any engineering work conforms to acceptable standards. A technician employed by G.C.E. checked the sizes and materials of the structural members for compliance with design drawings. He called to Duncan's attention questions concerning the strength of the rods and the change from one rod to two. Duncan stated to the technician that the change to two rods was "basically the same as the one rod concept." Duncan did not "review" the fourth floor box beam connection shown on the Havens shop drawings nor did he, in accord with usual engineering practice, assemble its components to determine what the connection looked like in detail. The Commission found, again not disputed, that appellants did not review the shop drawings for compliance of the box beam–hanger rod connection with design specifications of the Kansas City Building Code; did not review the shop drawings for conformance with the design concept as required by the contract of G.C.E. and the specifications on the Hyatt project nor for compliance with the information given in the contract documents. Duncan and Gillum approved the shop drawings.

While construction of the Hyatt was in progress the atrium roof collapsed. Investigation into that collapse established that the cause was poor construction workmanship. During the course of their investigation of the atrium roof collapse, appellants discovered that they had made certain errors in their design drawings and that they had failed to find discrepancies in their review of shop drawings involving the atrium. These errors and discrepancies were in areas other than the walkway design. The owner and architect directed G.C.E., for an additional fee, to check the design of the entire atrium. G.C.E. undertook that review. Gillum assured the owner's representative that "he would personally look at every connection in the hotel." Appellants were also specifically requested by the construction manager to inspect the steel in the bridges including the connections. Duncan subsequently advised him that had been done. In their report to the architects, appellants advised "we then checked the suspended bridges and found them to be satisfactory." This report was a culmination of a design check of the "structural steel framing in the atrium as per the request of Crown Center [the owner]." Appellants did not do a complete check of the design of all steel in the atrium nor a complete check of the suspended bridges. Gillum reviewed the report prepared by Duncan and took no exception. At a meeting with the owner and architect, Gillum stated that his company had "run a detailed, thorough re-analysis of all of the structure. And to determine if there was any other areas that were critical or had any kind of a design deficiency or detail deficiency." Duncan reported at that meeting: "We went back, myself and another engineer, and checked all the atrium steel.... Everything in the atrium checked out very well [with one non-relevant exception]." Appellants checked only the atrium roof steel.

Approximately a year after completion of the Hyatt Regency the second and fourth floor walkways

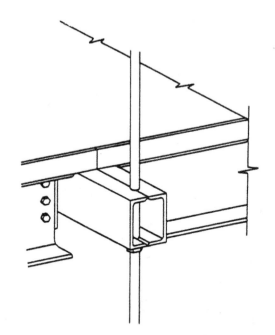

Original detail

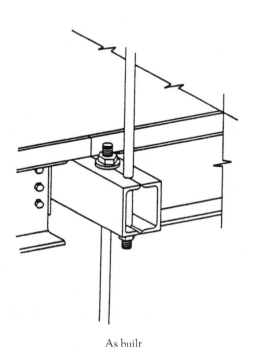

As built

FIGURE 8.3 Comparison of interrupted and continuous hanger rod details.

collapsed. The cause of the walkway collapse was the failure of the fourth floor box beam–hanger rod connections.

[Ed. note: The engineers challenged the constitutionality of the licensing laws. They contended that the "gross negligence" standard was so vague as to deny them due process under the U.S. Constitution. The court rejected this contention, concluding that the "phrase provides a guideline sufficient to preclude arbitrary and discriminatory application." See Section 8.4A.]

The Commission rejected the definition utilized in the first category of cases (difference in degree) and utilized a definition recognizing that gross negligence is different in kind from ordinary negligence. Appellants do not disagree with this selection of category. The Commission defined the phrase in the licensing context as "an act or course of conduct which demonstrates a conscious indifference to a professional duty." This definition, the Commission found, requires at least some inferred mental state, which inference may arise from the conduct of the licensee in light of all surrounding circumstances. Appellants have posited a definition purportedly different that would define the phrase as "reckless conduct done with knowledge that there is a strong probability of harm, and indifference as to that likely harm." We are not persuaded that the two definitions are in fact different. An act which demonstrates a conscious indifference to a professional duty would appear to be a reckless act or more seriously a willful and wanton abrogation of professional responsibility. The very nature of the obligations and responsibility of a professional engineer should appear to make evident to him the probability of harm from his conscious indifference to professional duty and conscious indifference includes indifference to the harm as well as to the duty. The structural engineer's duty is to determine that the structural plans which he designs or approves will provide structural safety because if they do not a strong probability of harm exists. Indifference to the duty is indifference to the harm. We find no error in the definition utilized by the Commission. It imposes discipline for more than mere inadvertence and requires a finding that the conduct is so egregious as to warrant an inference of a mental state unacceptable in a professional engineer.

* * *

Appellants . . . challenge one of the Commission's findings on the basis that the collapse of the walkways was not caused by a certain specific failure. This is raised in connection with the finding that the failure to delineate special strength steel for the rods was grossly negligent. The rods themselves did not fail. In making this assertion appellants rely on the elements of a common law cause of action for negligence, i.e., duty, breach, proximate causation, and injury or damage. They assert that proximate causation is not present. In the first place we are not dealing with a civil cause of action for negligence. We are dealing instead with a determination of whether appellants negligently breached their duty in the design of the walkways. That breach occurred at the latest when their design was incorporated into the building with their approval and they were subject to discipline whether or not any collapse subsequently occurred. It is the appellants' conscious disregard of their duty for which discipline is being imposed not the result of that breach . . . . Damage or injury is not an element of this disciplinary proceeding and proximate cause is the legal concept that authorizes civil recovery for damage resulting from negligence. It is not in and of itself an aspect of "negligence," only an aspect of a civil cause of action for negligence. Related to that concept is the fact that indeed there was damage caused by the breach. By statute and under the contract the owner of the building was entitled to a building structurally safe and sound . . . . The owner did not receive such a building because of appellants' breach of their professional responsibility. The owner received a defective building. Whether the walkways collapsed or not, the owner was damaged because it received less than it was entitled to and that damage was proximately caused by appellants' acts and omissions. Further we have previously stated that gross negligence is not required in each act of appellants; it is their overall conduct in regard to the Hyatt construction which justifies discipline.

* * *

The statutory provisions make clear that Missouri has established a stringent set of requirements for professional engineers practicing in the state. The thrust of those requirements is professional accountability by a specific individual certified engineer. These requirements establish the public policy of the state for the protection of the public. They require that plans for construction of structures in this state which require engineering expertise be prepared by or under the direct supervision of a specified certified engineer and that that engineer bear personal and professional responsibility for those plans. The affixing of his seal on the plans makes him responsible for the entire engineering project and all documents connected therewith unless he specifically disclaims responsibility for some document relating to or intended to be used for any part of the engineering project. It would be difficult to imagine statutory language more clearly evidencing the total responsibility imposed upon the engineer, and accepted by him when he contracts to provide his services. The statutory statement that the right to engage in the profession is a personal right based upon the individual's qualifications in no way impacts upon the responsibilities imposed upon an engineer. Rather the assessment of the individual "qualifications" of the engineer include his willingness and ability to accept the responsibilities imposed on him by the statutes.

[Ed. note: The court rejected the engineers' contention that the custom of relying on the fabricators to design the connections precluded any finding they were grossly negligent. The court noted that the employees of the fabricator are exempt from licensing requirements.]

* * *

The public policy of this state as it pertains to the responsibility of engineers has been established by the General Assembly in Chapter 327. That Chapter imposes upon the engineer a non-delegable duty of responsibility for projects to which he affixes his seal. ... The purpose of disciplinary action against licensed professionals is not the infliction of punishment but rather the protection of the public. ... Chapter 327 has established the responsibility a certified engineer bears when he undertakes a contract in his professional capacity. Sec. 327.191 authorizes noncertificated engineers to perform engineering work "under the direction and continuing supervision of and is checked by" a certificated engineer. It is a misdemeanor for a certified engineer to affix his seal to plans which have not been prepared "by him or under his immediate personal supervision." Sec. 327.201. A corporation may engage in engineering activities if it has assigned responsibility for proper conduct of its professional engineering to a registered professional engineer. Sec. 327.401. Gillum was the engineer designated by G.C.E. as having that responsibility. An engineer affixing his seal to plans is personally and professionally responsible therefor. Sec. 327.401. Affixing his seal to plans imposes upon the engineer responsibility for the whole engineering project unless he, under seal, disclaims such responsibility. Gillum made no such disclaimer here. The entire thrust of Chapter 327 is to place individual personal and professional responsibility upon a known and identified certificated engineer. This is the responsibility the engineer assumes in exchange for the right to practice his profession. It is the assumption of this responsibility for which he is compensated. The statutory framework is established to protect the public and to hold responsible licensed engineers who fail to afford that protection. It is clear that the statute expresses the intent to impose disciplinary sanctions on the engineer responsible for the project whether the improper conduct is that of himself or attributable to the employees or others upon whom he relies. This case differs, therefore, from the cases relied upon by Gillum and G.C.E. where the statute did not impose such non-delegable responsibility. The Commission did not err in finding that Gillum and G.C.E. were subject to discipline for the acts or omissions of Duncan.

* * *

We now turn to the sufficiency of the evidence to support the Commission's findings that discipline was warranted. In so doing we note again the concession of appellants that except for five findings the findings of fact of the Commission are supported by evidence. We have previously stated our analysis of the five disputed findings. We review the sufficiency within the legal principles and framework heretofore explicated.

We look first to Duncan. He was the project engineer for the Hyatt and as such had primary responsibility within his company for designing and approving those aspects of the Hyatt which required structural engineering expertise. The design of the connections in the walkways and the design of the walkways themselves were included in that responsibility. The walkways were intended to carry pedestrian traffic. They were suspended above the main lobby of the hotel, recognized to be the main point of congregation within the hotel. The walkways each weighed approximately 35 tons and were comprised of heavy and largely non-malleable materials such as steel, concrete, glass and wood. The connections in the walkway were nonredundant so that if any one within a single walkway failed they all would fail and the walkway would collapse. Duncan had never designed a system similar to the Hyatt walkways. It is self-evident that the walkways offered a potential of great danger to human life if defectively designed. The Commission could properly consider the potential of danger in determining the question of gross negligence. That which might constitute inadvertence where no danger exists may well rise to conscious indifference where the potential danger to human life is great. This is simply to say that the level of care required of a professional engineer is directly proportional to the potential for harm arising from his design and as we have previously stated indifference to harm and indifference to duty are closely related if not identical.

The structural drawings of Duncan furnished to the fabricator contained several serious errors. Under standard engineering practice Duncan could either design the box beam–hanger rod connections or cause the drawings to reflect his intention that they be designed by the fabricator. These drawings did neither. They appeared to be connections fully designed by the engineer and were reasonably so interpreted by the fabricator. Duncan testified that he intended the fabricator to design the connections. The drawings did not contain information indicating that the connections were to be designed by the fabricator and omitted important engineering load calculations necessary to enable the fabricator to design the connections. The drawings failed to properly identify the type of weld required, the need for bearing plates and/or stiffeners, and erroneously identified the hanger rods as standard rather than high-strength steel. The box beam–hanger rod connections and the hanger rods themselves on all three walkways, as shown by the structural drawings, did not meet the design specifications of the Kansas City Building Code. That Code is intended to provide a required level of safety for buildings within the City. It is difficult to conclude that gross failure to comply with that Code can constitute other than conscious indifference to duty by a structural engineer.

Because of certain difficulties in fabrication Havens requested a change to the double rod configuration. This request was transmitted to Duncan who approved it and verified its structural soundness and safety to the architect. He did so without having conducted all necessary engineering tests and calculations to determine the soundness and safety of the double rod arrangement. His concern was with its architectural acceptability not its structural acceptability. The result of this change was to double the load on the fourth floor walkway and impose a similar increase on the connections which were already substantially below Code requirements.

Havens supplied Duncan with its shop drawings. Under the contract, and under the statute, review and approval of the shop drawings is an engineering function. Appellants' normal in-house procedures called for detailed check of all special connections during shop drawing review. Duncan was aware of the change to the two-rod system but did not review the box beam–hanger rod connection on the fourth floor walkway. Duncan did not, as is standard practice, look for an assembled detail of the connection and did not assemble the components, either in his mind or on a sketch, to determine what the connection looked like in detail. The shop drawings did not reflect the use of stiffeners or bearing plates necessary to bring the connections within Code requirements. No review was made nor calculations performed to determine whether the box beam–hanger rod connection shown on the shop drawings met Code requirements. Shop drawing review by the engineer is contractually required, universally accepted and always done as part of the design engineer's responsibility. The box beam–hanger rod connections and the hanger rod shown on the shop drawings did not meet design specifications of the Code.

Following the atrium roof collapse appellants were requested by the architect and owner to recheck all the steel in the atrium. They reported that they had done so and included in that report was the statement "we then checked the suspended bridges and found them to be satisfactory." In fact appellants did not do a complete check of the design of all steel in the atrium and did not do a complete check of the suspended "bridges," i.e., walkways. As finally built, the hanger rods and the box beam–hanger rod connections did not meet the requirements of the Code. The walkway collapse was the result of the failure of the fourth floor box rod connections. The third floor walkway, which did not collapse, had a "high probability" of failure during the life of the building.

The determination of conscious indifference to a professional duty, i.e., gross negligence, is a determination of fact. The conduct of Duncan from initial design through shop drawing review and through the subsequent requested connection review following the atrium roof collapse fully supports the Commission's finding of conscious indifference to professional duty. The responsibility for the structural integrity and safety of the walkway connections was Duncan's and that responsibility was non-delegable. …

He breached that duty in continuing fashion. His reliance upon others to perform that duty serves as no justification for his indifference to his obligations and responsibility. The findings of the Commission as to Duncan's gross negligence are fully supported by the record.

The Commission also found Duncan subject to discipline for misconduct in misrepresenting to the architects the engineering acceptability of the double rod configuration when he performed no engineering calculations or other engineering activities to support his representation. The Commission found such representation to have been made either knowing of its falsity or without knowledge of the truth or falsity…. In either event it subjected Duncan to disciplinary action.

The Commission found Gillum subject to discipline for gross negligence under the vicarious liability theory and also personally grossly negligent in failing to assure that the Hyatt engineering designs and drawings were structurally sound from an engineering standpoint prior to impressing thereon his seal and in failing to assure adequate shop drawing review. It further found Gillum to be subject to discipline for unprofessional conduct and gross negligence in his refusal to accept his responsibility as mandated by Chapter 327 and his denial that such responsibility existed. All of these findings arise from the same basic attitude of Gillum that the responsibility imposed by Chapter 327 is not in keeping with usual and customary engineering practices and that that responsibility did not mandate his personal involvement in the design of the Hyatt. In essence he placed the responsibility for the improper design of the connections on Havens and took the position that the structural engineer was entitled to rely on Havens' expertise. What we have heretofore said in regard to the requirements of Chapter 327 and the responsibility imposed upon an engineer thereby sufficiently deals with Gillum's contentions.

His argument here that utilization of his seal without disclaimer could not impose responsibility upon him for the shop drawings of another entity prepared after impression of the seal is clearly rejected by the language of the statute. By section 327.411.2 the owner of the seal is responsible for the "whole … engineering project" when he places his seal on "any plans" unless he expressly disclaims responsibility and specifies the documents which he disclaims. The shop drawings were part of the documents comprising the engineering project and were "intended to be used for any part or parts of the… engineering project…." Gillum was by statute responsible for those drawings and he accepted such responsibility when he entered into the contract and utilized his seal. His refusal to accept a responsibility so clearly imposed by the statute manifests both the gross negligence and unprofessional conduct found by the Commission. These findings are further bolstered by the evidence of Gillum's participation in the misrepresentations concerning, and nonperformance of, a review of the atrium design upon direct request of the architect and owner. Although we have found that a specific finding of misconduct and discipline therefor cannot be based upon the atrium design review because not charged in the complaint, the evidence is relevant and persuasive on Gillum's overall mental approach to his responsibilities as an engineer and the cavalier attitude he adopted concerning the Hyatt project.

Appellant G.C.E. is, for reasons heretofore stated, subject to discipline for the conduct of its employees and particularly for the conduct of the engineer assigned the responsibility for the "proper conduct of all its … professional engineering… in this state…."

… [T]he order of the Commission and the discipline imposed by the Board is affirmed.

KAROHL, P. J., and KELLY, J., concur.

Other cases have also involved sufficiently flagrant misconduct or incompetence to justify a drastic agency decision. For example, a court affirmed an agency decision revoking the license of a professional engineer who had performed welding without being certified as required by the state administrative code.[4] In the same case, however, the court would not affirm the agency's revocation where the professional engineer designed and supervised the construction of a garage that collapsed. Although recognizing that there was an admitted error in the design of the roof supports for the garage, the court noted that the error was not obvious and that this was the first failure the engineer had experienced in eleven years of practice.

A revocation of an architect's license was affirmed based on serious design errors leading to the failure of the basement wall.[5] The architect also caused construction delay, failed to obtain a building permit, misplaced the building in reference to the lot line, and secured the owner's endorsement of payment without informing him of the facts.

Another court affirmed a six-month suspension based on a deficient ventilation plan, a superficial inspection of the premises, a superficial scanning of the architectural plan, and reliance on the judgment of two relatively inexperienced employees.[6] In addition, the engineer, after finding that the original certifications were in error and serious defects existed, did not disclose this to appropriate city officials.

Usually, attempts to discipline an architect or engineer relate to conduct after the license was issued. But California held that conduct *before* admission to practice could be misconduct that justified revocation of the license.[7] The architect in this case had difficulties with Virginia architectural board authorities that led to criminal charges. The architect did not report this when he applied for admission in California. The Supreme Court rejected the contention by the architect that the registration statutes were penal (criminal) and must be strictly construed. The court held that the registration statutes were not penal but were designed to protect the public.

The effectiveness of post-admission disciplinary powers has been limited. Attempts to suspend or revoke are almost always challenged by the design professional or contractor whose means of livelihood are being taken away. Challenges often mean costly appeals. Often administrative agencies charged with responsibility for regulating the profession or occupation are underfunded and understaffed. Suspension and revocation are unpleasant tasks. Even when action is taken, courts closely scrutinize decisions of administrative agencies. As an example, Washington held that the board findings must be based on "clear, cogent, and convincing evidence," a more demanding standard than the usual civil law standard of preponderance (majority) of the evidence.[8]

Courts *should* extend considerable deference to the agency decision. In matters as important as these, however, courts seem to redetermine what is proper. The combination of agency lethargy and overextensive judicial scrutiny when agencies do act may be one reason (among many) for increased professional liability. Civil liability supplements the regulatory licensing process as a means by which incompetent practitioners are eliminated.

8.4 Design Professional Licensing

8.4A Preliminary Issue: Constitutionality

By and large, professional regulation legislation is upheld based on the police powers granted by state constitutions and constitutional provisions permitting the state to legislate in the interest of public welfare. Sometimes constitutional attacks have been made based on statutory language or procedural problems—not the power of the state to regulate the professions and occupations. For example, in the *Duncan* decision reproduced in Section 8.3C, the court ruled that the standard of "gross negligence" was not so vague as to create an unconstitutional standard for discipline.

8.4B "Holding Out" Statutes

Licensing laws fall into two main categories. The first category is of statutes that regulate use of the professional *title* and are called "holding out" statutes.

For example, *Rodgers v. Kelley* involved an action by the plaintiff to be paid for design services on a project that was abandoned because of excessive costs. As a defense, the clients claimed the plaintiff violated the Vermont holding-out statute, which provided:

> The term "architect" or "registered architect" as used herein shall mean a person who holds himself out as able to perform, or does perform, while representing himself as an architect, any professional service such as consultation, investigation, evaluation, planning, [or] designing...

The plaintiff contended that he had never signed his name or in any way represented that he was an architect. He testified that he had twelve years of experience in the architectural field but contended that so long as he did not label himself, his plans, or his business with the title "architect" he had not violated the statute. The court rejected this narrow reading of the statute:

> But "holding oneself out as" an architect does not limit itself to avoiding the use of the label. The evidence is clear that, in the community of Stowe, this plaintiff was known as a proficient practitioner of all of the architectural arts with respect to homebuilding, at least. It was a business operation from which he received fees. He presented himself to the public as one who does the work of an architect. This constitutes holding oneself out as an architect and is part of the very activity sought to be regulated through registration.[9]

8.4C "Practicing" Statutes

In addition to "holding out" statutes, the other main category of licensing laws for design professionals focuses on allowable professional practice. Statutes define what activities fall within the practice of architecture and the practice of engineering. One who engages in these practices must be licensed accordingly. For example, a New York statute defines the practice of architecture:

> as rendering or offering to render services which require the application of the art, science, and aesthetics of design and construction of buildings.... Such services include, but are not limited to consultation, evaluation, planning, the provision of preliminary studies, designs, construction documents, construction management, and the administration of construction contracts.[10]

Another New York statute defines the practice of engineering:

> performing professional service such as consultation, investigation, evaluation, planning, design or supervision of construction or operation in connection with any utilities, structures, buildings, machines, equipment, processes, works, or projects wherein the safeguarding of life, health and property is concerned, when such service or work requires the application of engineering principles and data.[11]

The inherent vagueness of these definitions, coupled with the wide range of activities an architect or engineer may perform, means that questions remain as to which activities require a licensed practitioner. Design services can range from simply sketching a floor plan or planning to place a residence on a designated site all the way to providing construction documents with sufficient detail to obtain a bid from the contractor.

In addition to the varying design activities, a differentiation can be made between those services that are part of the design process and those services performed by a design professional during construction. Services performed during the design phase are more likely to be considered within the licensing laws than those performed during the construction phase. Nonetheless, activities during the construction process, such as interpreting documents, advising on changes, and judging performance, require specialized training and professional skills and so should be done by a licensed person.

8.4D Architecture and Engineering Compared

States regulate architecture and engineering separately. Each profession and the agencies regulating them sometimes differ over where one profession begins and the other ends. These conflicts demonstrate the economic importance of registration as well as the secondary role sometimes played by the public interest.

The two professions can be differentiated by project types and their use. One court stated,

> One prominent architect, in explaining the difference between architecture and engineering, said in effect that the entire structure and all of its component parts is architecture, if such structure is to be utilized by human beings as a place of work or assembly. He pointed out that, if the authorities were going to erect a courthouse as the building in which the [case] was being tried, they would obtain the service of an architect; but, if it was proposed to construct a power plant… they should employ an engineering firm….
> All of the architects and those who were registered as both architect and engineer agreed that the overall plan of a building and its contents and accessories is that of the architect and that he has full responsibility therefor. As one witness answered it, he is the commander in chief.[12]

In *State v. Beck*,[13] the Maine Supreme Court stated aesthetics to be the principal difference between engineering and architecture. To that court an architect was "basically an engineer with training in art." The court also stated,

> While categorically an engineer, the architect—without disparagement toward the professional engineer—is required to demonstrate that he possesses and utilizes a particular talent in his engineering, to wit, art or aesthetics, not only theoretically but practically, also, in coordination with basic engineering.[14]

The court then cited a Louisiana case that stated that an engineer "designs and supervises the construction of bridges and great buildings, tunnels, dams, reservoirs and aqueducts."[15]

As shown by the New York laws quoted in Section 8.4C, "practice" statutes do not clearly demarcate between architecture and engineering. Indeed, some statutes permit engineers to perform architectural services "incident to" engineering work and vice versa. One case allowed an engineer to seal alleged architectural drawings because, in the words of the Kansas regulation, he was qualified by education, training, and experience to prepare the drawings. (More precisely, the court found the architectural review board did not prove the engineer was not qualified by experience and training to create the drawings.)[16]

The lack of a clear dividing line between the practice of architecture and the practice of engineering means that friction is endemic to the design professions. By far, the most common scenario is for an architect to complain to the architectural licensing board that a nonarchitect (usually an engineer) is engaged in the unauthorized practice of architecture on a particular project. If the board finds against the engineer and assesses a penalty, the engineer challenges the administrative action in court.

The vast majority of courts view these complaints as a turf war between the professions and rule in favor of the accused. Little deference is given to the licensing board, which is not a disinterested examiner. In addition, courts do not defer to an agency's analysis of a legal (in contrast to factual) question, including interpretation of a "holding out" or "practice" statute. Given the overlap between the two professions, expert testimony is rarely helpful. As observed by one court:

> We are concerned that, on the testimony entered in this record, had Mr. Lomax [the architect who brought the initial complaint] been awarded the project, the Engineers' Board could have assessed civil penalties against him for the unauthorized practice of engineering. It is noted that the engineering expert witnesses testified that the project comprised 80% engineering and 20% architecture, even though the architectural expert witnesses testified that the project was 80% architectural and 20% engineering.[17]

However, the Arkansas Supreme Court came to the opposite conclusion in *Holloway v. Arkansas State Board of Architects*.[18] It rejected the argument of Mr. Holloway, the accused engineer, that the statutes distinguishing architecture from engineering were unconstitutionally vague. It then upheld the board of architecture's assessment of the maximum civil penalty, finding that Holloway's conduct threatened public health and safety.[19]

8.4E Architect or Engineer Applying Seal to Work of Another

Each licensed architect or engineer is provided a personal seal that the designer then uses to stamp documents prepared by her or by others under her personal supervision. Not all documents can be personally prepared by an engineer; however, by applying her seal, the engineer assumes personal responsibility for its contents. This principle is illustrated in the *Duncan* case, reproduced in Section 8.3C. The engineers' fault lay not in accepting for use the fabricator's (Havens's) shop drawings but in not verifying the engineering competence of those drawings. Clearly, a licensed architect or engineer has no right to rely upon the work product of one who is not licensed.

However, in many instances, a design professional will delegate to another design professional responsibility for a portion of the work. On large building projects, an architect's retained consultants can include geotechnical, civil, structural, mechanical, electrical, environmental, and hydrological engineers. If those engineering documents are faulty, what are the consequences to the architect? Two questions must be considered. First, will the architect be subject to professional discipline (as were the engineers in *Duncan*)? Second, would the architect be liable in tort for injury caused by the engineer's errors?

Before turning to the statutes for an answer, a common law perspective of these questions provides a context. As explained in Section 2.13, both the architect and the engineer are independent businesses. The engineer is the architect's independent contractor—not its agent (and certainly not its employee). As discussed in Section 5.7, the independent contractor rule, subject to exceptions, exempts the hirer of the independent contractor from liability for the torts of the contractor. Under a common law analysis, the architect generally is not responsible for the torts of the engineer. Of course, the architect's contractual responsibility to her client is not excused by her hiring of the engineer (see Section 7.4D).

Returning to the licensing laws, the first question is whether a licensee is subject to professional discipline for relying upon documents prepared by another licensee. In *Bird v. Missouri Bd. of Architects, Professional Engineers, etc.*,[20] a developer hired an architect (McInnis) to design a warehouse and office complex. After McInnis had prepared near-complete designs, a dispute arose between the developer and architect. The developer hired an engineer (Bird) to complete the design. Bird made the modifications to McInnis's design necessary to secure the planning commission's approval. Bird then signed and affixed his seal to each page of the plans. The developer submitted the plans bearing Bird's seal to the planning commission.

McInnis brought a complaint against Bird to the licensing board. McInnis contended that Bird was subject to discipline for applying his seal to McInnis's work. The board agreed, but the Missouri Supreme Court reversed. The Missouri Statutes, § 327.411, authorizes Bird to apply his seal to work prepared either by himself or by another under his "personal supervision." The court interpreted this statute as addressing responsibility for work performed by non-licensed personnel, such as draftsmen. The purpose of the regulatory scheme is to ensure that all documents were prepared under the standards of the regulated professions (both architecture and engineering). The statute does not require the licensee personally to supervise a co-licensee, such as an architect. The statute and regulations impose an affirmative duty on licensees to ensure the quality of the documents they seal (which Bird did); they do not (as the licensing board held) impose a negative duty to refrain from using plans and drawings created by another licensee not under their supervision.

The second question is the tort liability implications for the delegation of work from one licensee to another. In *Mudgett v. Marshall*,[21] the structural engineer (Megquier) subcontracted to another engineer (Marshall) the design of the steel frame. The frame collapsed while it was being erected. The estates of killed construction workers sued both engineers.

A jury traced the collapse to computational errors by Marshall, who was the subcontractor. The trial court accordingly dismissed the case against Megquier based on the independent contractor rule. The plaintiffs appealed, arguing that Megquier should be liable for Marshall's negligence.

The Maine Supreme Court found that Megquier was not liable either for its own negligence or under a theory of vicarious liability. Megquier was not liable in negligence, because it had no duty to independently review Marshall's computations. Nor was Megquier vicariously liable for Marshall's negligence under certain exceptions to the independent contractor rule.

> In this chapter's scenario, AAI created in Massachusetts the design for a New York project, then asked Edward Engineer, a licensed New York engineer, to "apply his seal to the design," without being required to conduct an independent review of the adequacy of that design. Edward Engineer will not be subjected to discipline for applying his seal to the work of another licensee. On the other hand, if there is a defect in the design which caused an accident, AAI should not be allowed to escape liability precisely because it did not hire Edward Engineer to create part of the design. A final question is whether Edward Engineer's limited contract duty to AAI, and his failure to independently review AAI's design, will subject him to personal liability.

8.4F Statutory Exemptions

Small projects do not necessarily require the services of an architect. Some statutes exempt projects based upon their size (square footage), height, and type (single-family dwellings are more likely to be exempt). For example, California Business and Professions Code § 5537 exempts in part "single-family dwellings of woodframe construction not more than two stories and basements in height" and multiple-family dwellings with no more than four units of a similar size and construction—unless public officials deem "that an undue risk to the public health, safety, or welfare is involved."

8.4G Possessor of License: Practice by a Business Entity

Design professionals traditionally functioned as sole proprietors or partners. Historically, most regulation of the design professions assumed individual practitioners.

Increasingly, for tax and liability reasons, design professionals are performing design services as members of business organizations, primarily corporations (see Section 2.4), partnerships (see Section 2.3), or limited liability companies (LLCs) (see Section 2.5). Often in this type of situation, both the individual design professionals and the business organization must be registered or licensed.

Rules and regulations may not keep up with new forms of practice. Focus on each state's law is essential, and generalizations have little reliability.

Florida permits the practice of architecture through a corporation, limited liability company, or partnership. The business organization must obtain a "certificate of authorization" which must be renewed every two years.[22] At least one principal officer of the corporation, LLC, or partner of the partnership, must be individually registered.[23] Disciplinary action by the architectural licensing board may be brought against a business organization just as it would be brought against an individual architect.[24]

In short, both the business organization and the individual architects must have their own licenses (or certificate of authorization). Legal complications arise when only some of these registration requirements are met. For example, in one Florida case, the partnership never obtained a certificate of authorization, although both of its partners were licensed architects. When the individual partners sued for nonpayment, the owner defended on the ground that its contract with the partnership was illegal because the partnership lacked a certificate of authorization. That illegality, the owner argued, meant the contract was void and so unenforceable by the architects.

The court enforced the contract. It noted that the actual work was performed by licensed persons. In addition, the owner knew the partnership was not registered but continued to accept the benefits of the contract, thereby ratifying (affirming) the contract.[25]

In an Illinois case, an architectural firm—a corporation—sought to enforce a mechanics' lien. The firm was not registered, although the vice-president and managing agent of the firm was licensed. The owner argued that the mechanics' lien was not enforceable because the underlying contract was illegal (because the firm was not registered). The court ruled that the contract underlying the mechanics' lien was valid because the architectural work was done by or under the supervision of a licensed architect.[26]

While both the Florida and Illinois courts just mentioned ruled that the underlying contract was enforceable even though the business organization was not properly registered, other courts in similar situations have come to the opposite conclusion. Ensuring the design professionals *and* the business organization in which they are employed are all registered is key to proper business practice.

> In this chapter's scenario, Awesome Architects, Inc. (AAI) is described as "an architectural firm located in Massachusetts," meaning that AAI itself is not properly registered in the state. While AAI's co-owners, Andrew and Allison, *are* registered as architects in Massachusetts, the legal question is whether a client hiring AAI to do architectural work in Massachusetts will be legally obligated to pay the corporation its fees, because it was not licensed to practice architecture. A court following the reasoning of the Florida and Illinois courts would allow recovery—but this would be achieved only after the delay, aggravation, and expense of litigation which AAI had brought upon itself through its poor practice management.

8.4H Out-of-State Practice

The practice of architecture and engineering, as well as the performance of construction services, increasingly crosses over state lines. As a result, a person licensed in one state often performs services in another.

Sometimes design professionals who perform services on a multistate basis obtain licenses in each state in which they perform services or in which projects for which they perform services are located. This has become easier because of standardized examinations and increased reciprocity.

Some state laws do not require a license for performance of services by out-of-state design professionals if they are licensed in the state where they practice principally and as long as there is a local licensed design professional in overall charge of the project. However, the licensed local design professional must not be a figurehead but must actually perform the usual design professional functions.

A review of some cases shows the ways the issue of out-of-state practice may arise and how courts respond.

In *Johnson v. Delane*,[27] the Idaho court held that a Washington engineer who performed only design services and did so in Washington was not subject to Idaho's licensing laws merely because the contract was signed and the project was built in Idaho. The court held that he was not practicing engineering in Idaho, and he recovered for his services.

In *Food Management, Inc. v. Blue Ribbon Beef Packing, Inc.*,[28] an Ohio architecture and engineering firm "associated" with an Iowa architect on an Iowa project. Nonetheless, the court ruled that the Ohio firms' design contract was not enforceable, because the local architect was not "in responsible charge" of the work, as required by Iowa's licensing laws. The court also rejected the Ohio design firm's argument that the licensing requirement should not extend to an "isolated" transaction, stating:

> One instance of untrained, unqualified, or unauthorized practice of architecture or professional engineering—be it an isolated transaction or one act in a continuing series of transactions—may be devastating to life, health or property. To exclude an isolated transaction from the proscription of the registration laws would seriously weaken the purpose of such laws at a point where their prohibition of professional services may be most needed.[29]

In *Hedla v. McCool*,[30] a Washington architect performed design services for a project to be built in Alaska and had the plans approved by an Alaskan engineer. Once again, the legal question was whether the Washington architect's use of an Alaskan engineer meant the Washington architect could recover its fee.

The court disallowed recovery of the fee. It distinguished this case from *Johnson v. Delane* (discussed earlier) by noting that the Alaska statute was broader, that this was not an isolated transaction, and that conditions in Alaska were different from those in the state of Washington. The court was more persuaded by the holding in the *Food Management* case, which had not looked kindly on out-of-state architects associating a local architect when the out-of-state architects were principally responsible for design decisions.

One important consideration permeating these cases is whether the unlicensed person seems to have performed properly. If so, denial of recovery, despite the importance of license compliance, may seem unjust. For example, in *Johnson v. Delane* the work appears to have been performed correctly. But in *Hedla v. McCool,* where recovery was denied, the costs overran considerably and the design was never used—factors that make recovery less attractive.

In this chapter's scenario, AAI agreed to create a design for a factory to be built in New York. AAI hired a New York engineer, Edward Engineer, to "apply his seal to the design created by AAI" and also to inspect the construction while it was being performed. Would AAI be allowed to recover its fees for the New York project? (Of course, the *real* answer to that question turns on a careful analysis of the New York statutes, but here answers will be explored based on the discussion in the text.)

AAI's strongest argument is that—like the architect in *Johnson v. Delane*—it was not practicing architecture in New York and so did not need that state's license. If that argument is rejected by the court, then AAI's lack of even a Massachusetts license will give a New York court pause to allow enforcement of the contract with MMI (even if MMI is a large and sophisticated company and so presumably not in need of the licensing laws for its own protection).

AAI's hiring of Edward Engineer, a New York engineer, will not be sufficient under the reasoning of *Food Management* and *Hedla v. McCool* to authorize payment on the ground that AAI had employed a local design professional. This is because Edward Engineer was not "in responsible charge" of the design work, but was instead simply asked to apply his seal to AAI's design.

Finally, a claim arising out of Edward Engineer's site responsibilities (probably brought by the contractor) should not subject AAI to liability. Edward Engineer is a separate business entity, charged with site responsibility, and AAI could invoke the defense of the independent contractor rule.

In sum, a design professional considering performing design services either in another state or for a project that will be built in another state should receive legal advice on the proper process for ensuring that he or she is not violating the laws of the state where the services are being performed or the project is located.

8.5 Contractor Licensing Laws

8.5A Overview

Along with the general proliferation in occupational licensing, there has been an increase in contractor licensing laws. At present, about half of the states have such laws. Most of these states broadly require both commercial and residential contractors to be licensed. Most of these require separate licensing of prime contractors and of the specialty trades (such as asbestos, plumbing, electrical, or roofing). Several do not require general contractors to be licensed but do require specialty trades to be licensed.[31]

Sometimes contractors are faced with multiple licensing requirements. A state-wide licensing requirement does not preclude local governments (counties and cities) from imposing their own licensing mandates. Similarly, many states require state-licensed contractors to have an additional license when undertaking specific types of work, such as residential projects.

As a rule, contractors need meet only experience requirements and not educational ones. In California, they must have four years of experience in the last ten years in the trade they wish to enter. Under certain circumstances, particular education courses taken can fulfill up to three years of this four-year experience requirement.[32] Applicants in California take a one-day written examination—half of which is devoted to law and the other half to trade practices. Even these minimal examination requirements can be waived under certain circumstances.[33]

A licensing board examines the integrity of the applicant, including financial capacity. A licensed person's financial integrity may include a requirement to obtain bonds for the protection of persons injured by the licensee's conduct. In California, an applicant for a contractor's license must post a bond for $12,500.[34] If the licensee is a limited liability company, the bond is $100,000.[35]

8.5B Possessor of License: Business Organization

As with design professionals (Section 8.4G), proper licensing of a contractor depends on the nature of the business organization in which the contractor practices. In California, contractors' licenses are issued to individuals, partnerships, corporations, and limited liability companies. Since a business cannot take an examination to obtain a license, it must obtain a license through either a responsible managing employee (RME) or a responsible managing officer (RMO).[36] The RME or RMO must exercise direct control over construction operations and see to it that the Contractors' License Law is not violated. If the firm does violate the license law, the board may suspend or revoke the license of the RME or RMO as well as the license of the firm.

A corporate contractor must obtain its own license in addition to that of its RME or RMO. In *WSS Industrial Construction, Inc. v. Great West Contractors, Inc.*,[37] the RMO was licensed, but the corporate contractor was not; hence, the contractor was denied recovery. In another California case, a physically absent RME meant the partnership contractor was not properly licensed.[38]

In California, the license must be renewed every two years.[39] There is no provision for periodic examination or continuing education.

8.6 The Unlicensed Design Professional or Contractor

States employ a combination of inducements and sanctions to encourage licensure. As an example of an inducement, California permits only a licensed contractor to hold a supplier of goods to any orally communicated price proposal made under certain circumstances.[40] (Otherwise, the supplier would only be bound if the price proposal were in writing.)

However, states primarily use the threat of sanctions to achieve licensing by design professionals and builders. The remainder of this section addresses these sanctions.

8.6A Criminal and Quasi-Criminal Sanctions

Licensing laws usually carry criminal sanctions. Violations can be punished by fine or imprisonment. Yet, use of the criminal sanction is relatively rare. Where a penal violation is found, it is likely that sentence will be suspended or probation granted. However, in a Maryland case, *Huffman v. State*,[41] the unlicensed contractor was sentenced to jail, given probation, and also fined.

8.6B Recovery for Work Performed and Payment Reimbursement

The principal sanction for the unlicensed practice of architecture, engineering, or construction work is to deny recovery for work performed. Attempts by unlicensed persons to recover for their work has been the main legal battleground. Again, generalizations should be read with caution, as the wording of the specific state's statute will govern resolution of the dispute.

The law of illegal contracts provides the analytical framework for understanding the rights of an unlicensed person and the client. For the sake of brevity, the discussion will refer only to contractors, although the law equally applies to design professionals.

Generally, neither party can enforce an illegal contract. The law provides no help if the parties are equally guilty. If a citizen bribes a public official, the citizen cannot sue to have the promised performance made nor can the public official sue to recover the promised bribe.

Courts do not view parties who hire unlicensed contractors as analogous to those who bribe public officials. In the former situation, not only are the parties not equally guilty, but they might both be ignorant of the violation (for example, if the contractor lets its license lapse through inadvertence). Also, the illegality of the two situations is not comparable in terms of intention of the parties, morality, or public policy. In legal parlance, bribing officials is a *malum in se*—intrinsically illegal—while violating a licensing law is a *malum prohibitum*—admittedly a prohibited act, but disallowing enforcement of the contract would be disproportionate to the illegality. For these reasons, the client can enforce its contract against the unlicensed contractor and maintain an action for faulty performance.

The more difficult questions are (1) may the contractor recover (get paid) for work properly performed and (2) may the client recoup (recover back) payments already made under the contract. If the contractor is not paid for work already done, or the client is allowed to recover back payments already made, the contractor will suffer *forfeiture*—the owner will retain the work without having to pay for it.

Focusing for now on the first question of the contractor's right to payment for work performed, the contractor may seek recovery (1) on the contract or (2) outside the contract. In the first situation, the contractor seeks to enforce the contract even though the contractor is not licensed. In the second situation, the contractor is not trying to enforce the contract.

An unlicensed contractor's right of recovery *on the contract* for work properly performed depends on the wording of the statute the contractor violated. Three types of statutes exist: those that specify the contractor's right to compensation, those that are silent on the contractor's right to compensation, and those that expressly prohibit the contractor's right to compensation.

Statute Specifies Contractor's Right to Compensation. Tennessee Code § 62-6-103(b) is an example of a statute that specifies a remedy for the contractor. It allows an unlicensed contractor to recover in a court actual documented expenses if established by clear and convincing proof.

Statute Silent on Contractor's Right to Compensation. Where the statute does not specify that its violation will deprive the contractor of its right to compensation, courts generally view the illegality as a *malum prohibitum* and will not impose upon the design professional or contractor the penalty of forfeiture. The following case from Idaho illustrates this reasoning.

In *Farrell v. Whiteman*,[42] Farrell, a Michigan-licensed architect, designed a condominium project in Idaho. The Idaho developer terminated the design agreement, then he successfully sold the condominium units. The developer refused to pay Farrell because he was not licensed in Idaho. Farrell sued the developer in Idaho and the developer claimed the architect was entitled to nothing because the design agreement was illegal.

The architect registration statute was silent on the effect of its violation. The Idaho Supreme Court refused to sanction forfeiture, permitting the architect to recover in quantum meruit (that is, for the value of benefit conferred upon the developer):

> This case is one in which awarding some damages is necessary to protect the public interest. Although Farrell had an obligation to ensure that he was properly licensed while he practiced architecture in Idaho, it would be contrary to the public interest to allow a developer to avoid any payment for architectural services when the work performed was not sub-standard, the building was actually constructed, and several units were sold. The architect licensing statutes aim to protect buyers and the public—who lack knowledge and expertise about architecture—from shoddy work and untrained individuals posing as architects. In this case, although Farrell was not licensed in Idaho for the entirety of his performance, he was a licensed architect in another state. The district court found that Farrell did not breach his duty of care, and that no defects existed in the building Farrell designed. Therefore, to leave the parties as the Court finds them would result in a windfall to [the developer], who obtained the benefit of a trained and capable architect at zero cost.[43]

Statute Bars Contractor's Right to Compensation. California Business and Professions Code § 7031(a) is an example of a statute that expressly prohibits an unlicensed contractor from suing the client for compensation for work performed. Section 7031(a) states that no contractor

> may bring or maintain any action, or recover in law or equity in any action, in any court of this state for the collection of compensation for the performance of any act or contract where a license is required by this chapter without alleging that he or she was a duly licensed contractor *at all times during the performance* of that act or contract, regardless of the merits of the cause of action brought by the person . . . [italics added].

The California Supreme Court has repeatedly enforced § 7031(a)—even when the result has been forfeiture for a contractor who had performed quality work for a sophisticated party. Here are some of the court's rulings:

1. A sub-subcontractor who was unlicensed when it signed the contract but licensed when it performed the work was allowed recovery because it was licensed "at all times during

the performance" of the work, as required by the statute. The law does not prohibit an unlicensed person from signing a construction contract.[44]

2. A sub-subcontractor who was unlicensed only during the first month of work was denied recovery, even though it was properly licensed when it performed the vast majority of work. The court ruled that, because the sub-subcontractor was not licensed "at all times" during performance of the contract work, it could not recover *any* compensation under that contract.[45]

3. By forbidding recovery in "any action," § 7031(a) precludes an unlicensed contractor from recovery based on restitution to be paid the value of the benefit provided by the owner by the contractor's services. That fact that the owner was unjustly enriched is not a reason to ignore the clear, statutory language.[46]

4. Section 7031(a) prohibits recovery by an unlicensed subcontractor, even though (a) the hiring party was sophisticated (a prime contractor) and so presumably not in need of the protection of the licensing laws, (b) the prime contractor knew the subcontractor was unlicensed, and (c) the prime contractor arguably committed fraud by representing that the subcontractor would be paid.[47]

The difficult public policy questions created by the licensing laws are illustrated in a 2002 Michigan Supreme Court decision. In *Stokes v. Millen Roofing Company*,[48] homeowners *knowingly* hired an unlicensed contractor to install a roof, then refused to pay for it. When the contractor filed a mechanics' lien, the owners asked the court to have it discharged because the contractor was unlicensed. The licensing statute prohibits an unlicensed contractor from suing "for the collection of compensation for the performance of an act or contract for which a license is required by this article..."[49]

The court ruled for the owners based on the contractor's lack of a license. It even denied the contractor's request for reimbursement of the cost of its supplies, finding that such a recovery would constitute "compensation" within the meaning of the statute.

But the case produced four opinions: the majority, two concurrences, and a dissent. All four opinions addressed the same issue: the fairness of preventing payment based solely on the contractor's unlicensed status, where the contractor produced quality workmanship and the owners hired him knowing he was unlicensed. Most troubling was evidence that in a *separate* case involving the same improvement, the owners had hired another unlicensed contractor and then refused to pay him because he was unlicensed. It appeared that the owners were abusing the statutory scheme: hiring unlicensed contractors and then hiding behind their lack of a license in order to deny them payment.

> In this chapter's scenario, the New York prime contractor hired Cal-Make, Inc. knowing the subcontractor was unlicensed in New York. There is no indication that the subcontract work was improperly performed, so denial of payment would result in a forfeiture.
>
> Cal-Make, Inc.'s right to recovery would depend upon the precise wording of the New York contractor licensing statute. If that statute has the same wording as California Business and Professions Code § 7031(a), then Cal-Make, Inc. will be denied recovery, either under the contract or for unjust enrichment.
>
> If the New York statute is silent on an unlicensed contractor's right to recovery, then a New York court may well allow recovery in restitution, so as to prevent forfeiture.

Payment Reimbursement. May a client recoup payments already made during performance to an unlicensed contractor? As always, analysis of this question must begin with the governing statute. In 2001, the California legislature added subsection (b) to California Business and Professions Code § 7031, which allows "a person who utilizes the services of an unlicensed contractor [to] bring an action ... to recover all compensation paid to the unlicensed contractor for performance of any act or contract." In other words, a contractor who violated the licensing laws must return to the owner payments the owner had previously made. Notwithstanding the clear hardship to the contractor, the courts have enforced the new statute as written.[50]

Where the statute is silent, some states allow the owner to recover payments it already made based on common law principles.

Summary. Taken together, these statutes and cases signal the importance of registration and licensing laws. When the statute takes a tough (even harsh) attitude toward those who perform under such illegal contracts, enforcement by the courts is likely. Where the statute is silent, the courts examine the following factors to determine whether recovery should be allowed.

1. The work for which recovery is sought conforms to the contract requirements.
2. The party seeking a defense based on a violation of the licensing law is in the same business or profession as the unlicensed person.
3. The unlicensed person apparently had the qualifications to receive the license.

8.6C Substantial Compliance

A number of states recognize the substantial compliance doctrine. In these states, excusable errors by the contractor in obtaining or renewing a license will not preclude recovery against the client. These jurisdictions are inclined to see contractor licensing laws as traps for unwary contractors who perform proper work and then face what appears to be a technicality that may bar them from recovering for their work. In some states, the substantial compliance doctrine has been codified (adopted) by statute.[51]

8.7 Construction Manager Licensing

All have states have enacted licensing statutes for design professionals, and most have enacted licensing statutes for contractors. The construction manager (CM), however, is a relatively new industry actor, and state regulation of CMs is slowly evolving.

Three states (California, Idaho, and Oklahoma) license construction management as a separate profession.[52] All three states require licensing only for public works projects, thereby leaving the private sector unregulated.

The novelty of construction management is shown by the fact that each state imposes different licensing requirements. A CM in California must be licensed as an architect, engineer, or contractor and must demonstrate to the public entity "evidence that the individual or firm and its personnel carrying out onsite responsibilities have expertise and experience in construction project design review and evaluation, construction mobilization and supervision, bid evaluation, project scheduling, cost-benefit analysis, claims review and negotiation, and general management and administration of a construction project."[53]

Idaho's licensing law allows either education—a bachelor's degree in architecture, engineering, or construction management—or a minimum of five years experience in managing construction projects as sufficient for licensure.[54] Oklahoma requires the individual be certified as a construction manager by the Construction Management Association of America.[55]

Absent licensing statutes expressly applicable to construction management, states generally require an architectural, engineering, or contractor license or a combination of these. Generally, courts look to the services offered to determine the license required.

In *Kourafas v. Basic Food Flavors, Inc.*,[56] a licensed architect agreed to design a project and then entered into a separate contract with the same owner to manage construction of the facility the architect had designed. The architect sued the owner under the construction management contract. The owner defended on the ground that the architect was not licensed to perform construction management services because he lacked a contractor's license.

The Nevada Supreme Court ruled that the architect was not required to have a contractor's license in order to perform construction management services. A statute defines the "practice of architecture" as:

> rendering services embracing the scientific, esthetic and orderly coordination of processes which enter into the production of a completed structure which has as its principal purpose human habitation or occupancy, or the utilization of space within and surrounding the structure, performed through the medium of plans, specifications, *administration of construction*, preliminary studies, *consultations, evaluations, investigations, contract documents and advice and direction*. (Italics added.)[57]

The court sent the case back to the trial court to determine whether the services performed by the architect on this project fit within the statute's list of services that constitute the practice of architecture.

In a California Court of Appeal case, *Fifth Day, LLC v. Bolotin*,[58] the plaintiff agreed to provide professional development and construction management services to a real estate company. The plaintiff acted as a CM-as-agent (a term the court did not use) and the owner separately hired the project contractor. When plaintiff sued the owner for nonpayment, the owner defended on the ground that plaintiff lacked a contractor's license.

The court ruled that because the plaintiff did not undertake any construction work either itself or through the employment of subcontractors, it did not need a contractor's license. Interestingly, there is no indication in the court's opinion that plaintiff held *any* license, whether as architect or engineer. Finally, the court reasoned that because the legislature requires licensing of construction management on public works project, but is silent on the need for licensing on private works projects, there is no such requirement on private works projects.[59]

The unsettled law of CM licensure means that one should seek advice of counsel before beginning practice.

8.8 The Trained but Unregistered Design Professional: Moonlighting

8.8A Unlicensed Persons: A Differentiation

This chapter has spoken of persons who violate the registration or licensing laws as if they all can be placed in the same category. Yet differentiation can be made between design professionals and others, such as contractors, developers, designer-builders, or self-styled handypersons.

A long period ordinarily elapses between the beginning of architecture or engineering training (architecture will be used as an illustration) and registration. As a result, there are many people with substantial education and training as design professionals, who for various reasons do not work under the supervision of a registered architect yet engage in the full spectrum of design services. The law does not differentiate these people from those without design education and training. Yet that these people often do perform these services in violation of the law demonstrates not only the financial burden involved in education and training before registration but also that a market exists for such services.

This section is directed toward some of the legal problems faced by what is referred to as the *moonlighter,* who is likely to be a student, a recent graduate, or a teacher. Some of the discussion anticipates material to be discussed in greater detail in the balance of the book, such as professional liability, exculpation, and indemnification. However, a focus here on the range of legal problems that such individuals face is also useful.

8.8B Ethical and Legal Questions

A threshold question is whether such work *should* be performed. A differentiation will be drawn between work by a moonlighter that does not violate the law because the project is exempt and the requirement that a registered designer perform the services. The former course raises no ethical or legal questions, while the latter does.

The discussion in this section should not be taken as a suggestion that moonlighters should engage in design services prohibited by law. However, two factors necessitate discussion of moonlighting. First, arrangements between clients and moonlighters demonstrate that each group feels it will benefit from such an arrangement or such arrangements would not be made. That there is a good deal of moonlighting demonstrates either that the registration laws only make conduct illegal that should not be prohibited or that there is great disrespect for the registration laws.

Second, the many cases that involve attempts by unlicensed people to recover for services that they have performed—and the periodic success of their claims—indicate some dissatisfaction with the registration laws even by those charged with the responsibility of enforcing them indirectly by denying recovery for services performed in violation of them.

Anyone tempted to moonlight and violate registration laws must take into account—in addition to the risk of criminal and quasi-criminal sanctions—the risks discussed in the balance of this section: going unpaid for services rendered or being wiped out financially because of claims by the client or third parties.

8.8C Recovery for Services Performed

Clearly, the safest path is to design only those projects exempt from the registration laws. Undertaking nonexempt projects creates substantial risks. This subsection looks at these risks and ways of minimizing the likelihood that the moonlighter will not be able to recover for services performed.

Because payments made as a general rule cannot be recovered by the client, it is best to specifically provide that interim fee payments will be made and to make certain that they are paid. If payment has not been made in full, it is possible at least in some jurisdictions to recover for services in a claim for restitution based on unjust enrichment. Such a claim is more likely to be successful if the client is aware of the moonlighter's nonregistered status, if the moonlighter has had substantial education and training in design, if the work appears to have been done properly, and if it appears that the client has benefited by a lower fee. The stated purpose of registration laws is to deter unqualified people from performing design services and to protect the public from hiring unqualified professionals. A client who is aware of the status of the designer engaged has given up protection accorded by law. To allow the client to avoid payment where the services have been properly rendered can create unjust enrichment.

If the project will expose the general public to the risk of physical harm, it is more difficult to justify either entering into such an arrangement or allowing recovery for services performed under one. But as a practical matter, this is not the type of project that moonlighters will design. The uncertainty of recovery and the high cost of litigation even if there is a recovery should deter moonlighters as a whole from doing nonexempt work.

8.8D Liability Problems

This discussion assumes that the moonlighter will not be carrying professional liability insurance. As a result, any liability that moonlighters incur will be taken out of their pockets (if there is anything in their pockets to take).

As shall be seen in Section 11.2, the professional is generally expected to perform as others in the profession would have performed. Should this standard be applied when a moonlighter is knowingly engaged by a client?

Differentiation must be made between an express agreement dealing with a standard of performance and an agreement implied by law. Suppose the parties agree that the standard of performance will be less than that usually required of a registered design professional. Although arguments for and against enforcement of such an agreement can be made, the law would probably not give effect to an express agreement under which someone performing services in violation of law would be judged by some standard less than would be used in the event that the person doing the design were registered.

If no express agreement is made regarding the standard of performing, it is even more likely the court would impose the standard of conduct required of registered architects, notwithstanding the client's benefit of a lower fee.

Another approach, related to but somewhat different from the one discussed in the preceding paragraphs, would be for the moonlighter to seek to persuade the client to exculpate (relieve) the moonlighter from the responsibility for any performance that would violate the contract or to limit the moonlighter's exposure to a designated portion of the entire fee. Because by definition the moonlighter is not insured, the moonlighter may be able to persuade the client that he would not be in a position to pay for any losses he caused, and this is taken into account in setting the fee.

Suppose the client will agree to exculpation or liability limitation. The moonlighter must take particular care to express these provisions very clearly, as courts (at the very least) will construe them against the moonlighter. Also, there is a risk that a court that feels strongly about the registration laws will not enforce such a clause if it would frustrate those laws. Keep in mind, however, that because the moonlighter is not likely to have either insurance or assets, a claim by the client is likely to be rare.

Third-party liability raises different problems, inasmuch as any arrangement for exculpation or a liability limitation between the moonlighter and the client will not affect the rights of third parties. But again, the moonlighter without insurance or assets is unlikely to be sued by a third party. A moonlighter would be concerned about the liability to third parties either because of moral considerations or because the moonlighter does have assets.

Finally, if the moonlighter is an employee of an architectural firm, will the moonlighter's negligence subject the firm to liability to the moonlighter's client? In *O'Brien v. Miller*,[60] the moonlighter was an architect in a New York firm. He provided architectural services in his individual capacity in New Jersey, although he was unregistered in that state. The New Jersey client claimed the moonlighter's work was defective, and he sued the New York firm. The court ruled that the client could not sue the architectural firm, because there was no evidence the moonlighter was acting as the firm's agent when he agreed to work for the client.

REVIEW QUESTIONS

1. What licensure requirements do all states have for engineers?
2. What were the key issues that led to the collapse of the skywalk in the *Duncan v. Missouri Board for Architects* case?

3. In the case *Duncan v. Missouri Board for Architects* the statutory provisions made it clear that Missouri has established a stringent set of requirements for professional engineers practicing in the state. What requirements were established for public policy of the state for the protection of the public?
4. What are the two main categories of licensing laws and what does each entail?
5. According to the case *State v. Beck* what did the Maine Supreme Court state was the principal difference between engineering and architecture?
6. What are some of the factors that may provide statutory exemption for some projects from needing a licensed architect to design and stamp the drawings?
7. If a design professional is considering performing design services either in another state or for a project that will be built in another state, what should that design professional do?
8. What are the licensing complications if a design professional practices in a business organization other than a sole proprietorship?
9. What is the principal sanction for the unlicensed practice of architecture, engineering, or construction work?
10. Which states currently license construction management as a separate profession?

ENDNOTES

[1] *Winter v. Smith,* 914 S.W.2d 527, 536 (Tenn.App.1995), appeal denied Dec. 18, 1995.
[2] *Gartrell Constr. Inc. v. Aubry,* 940 F.2d 437 (9th Cir.1991), following *Leslie Miller, Inc. v. Arkansas,* 352 U.S. 187 (1956).
[3] *Sobel v. Jones,* 96 Ariz. 297, 394 P.2d 415, 417 (1964).
[4] *Vivian v. Examining Bd. of Architects, etc.,* 61 Wis.2d 627, 213 N.W.2d 359 (1974).
[5] *Kuehnel v. Wisconsin Registration Bd. of Architects & Professional Engr's,* 243 Wis. 188, 9 N.W.2d 630 (1943).
[6] *Shapiro v. Bd. of Regents,* 29 A.D.2d 801, 286 N.Y.S.2d 1001 (1968). See also *Martin v. Sizemore,* 78 S.W.3d 249 (Tenn.Ct.App.2001).
[7] *Hughes v. Board of Architectural Examiners,* 17 Cal.4th 763, 952 P.2d 641, 72 Cal.Rptr.2d 624 (1998).
[8] *Nims v. Washington Bd. of Registration,* 113 Wash.App. 499, 53P.3d 52 (2002).
[9] 128 Vt. 146, 259 A.2d 784, 785 (1969).
[10] N.Y.Educ.Law § 7301.
[11] Id. § 7201.
[12] *State Bd. of Registration v. Rogers,* 239 Miss. 35, 120 So.2d 772, 774 (1960).
[13] 156 Me. 403, 165 A.2d 433 (1960).
[14] 165 A.2d at 435.
[15] Id. at 436, quoting from *Rabinowitz v. Hurwitz-Mintz Furniture Co.,* 19 La.App.811, 133 So. 498, 499 (1931).
[16] *Schmidt v. Kansas State Bd. of Technical Professions,* 271 Kan. 206, 21 P.3d 542 (2001).
[17] *Rosen v. Bureau of Professional and Occupational Affairs,* 763 A.2d 962, 969 (Pa.Commw.2000), appeal denied, 566 Pa. 654, 781 A.2d 150 (2001).
[18] 352 Ark. 427, 101 S.W.3d 805 (2003).
[19] Holloway then filed suit against the licensing board in federal court for deprivation of his due process rights. A federal court of appeals ruled against him in an unpublished decision. *Holloway v. Arkansas State Bd. of Architects,* 186 Fed.Appx. 708 (8th Cir.2006).
[20] 259 S.W.3d 516 (Mo.2008).
[21] 574 A.2d 867 (Me.1990).
[22] Fla.Stat.Ann. § 481.219(1) and (9).
[23] Id. § 481.219(7)(a).

[24] Id. § 481.219(12). Similar rules governing engineers are found at Fla.Stat.Ann. § 471.023.
[25] *District Bd. of Trustees of St. Johns River Community College v. Morgan*, 890 So.2d 1155 (Fla.Dist. Ct.App.2004), review dismissed, 918 So.2d 273 (Fla.2005).
[26] *G.M. Fedorchak and Associates, Inc. v. Chicago Title*, 355 Ill.App.3d 428, 822 N.E.2d 905 (2005), appeal denied, 215 Ill.2d 596, 833 N.E.2d 2 (2005).
[27] 77 Idaho 172, 290 P.2d 213 (1955).
[28] 413 F.2d 716 (8th Cir.1969).
[29] Id. at 723-24.
[30] 476 F.2d 1223 (9th Cir.1973).
[31] A list of contractor licensing requirements by state may be found at http://www.clsi.com/state_contractor_license_board.htm.
[32] 16 Cal.Admin.Code § 825.
[33] Cal.Bus.&Prof.Code § 7065.1.
[34] Id. § 7071.6.
[35] Id. § 7071.6.5.
[36] Id. § 7065.
[37] 162 Cal.App.4th 581, 76 Cal.Rptr.3d 8 (2008).
[38] *Oceguera v. Cohen*, 172 Cal.App.4th 783, 91 Cal.Rptr.3d 443 (2009).
[39] Cal.Bus.&Prof.Code § 7140.
[40] Cal.Comm.Code § 2205.
[41] 356 Md. 622, 741 A.2d 1088 (1999).
[42] 146 Idaho 604, 200 P.3d 1153 (2009)
[43] Id., 200 P.3d at 1161. In a subsequent appeal, the damages award in favor of the architect was affirmed, see *Farrell v. Whiteman*, 152 Idaho 190, 268 P.3d 458 (2012).
[44] *MW Erectors, Inc. v. Niederhauser Ornamental and Metal Works Co., Inc.*, 36 Cal.4th 412, 115 P.3d 41, 54-58, 30 Cal.Rptr.3d 755 (2005).
[45] Id., 115 P.3d at 48–50.
[46] *Hydrotech Systems, Ltd. v. Oasis Waterpark*, 52 Cal.3d 988, 803 P.2d 370, 277 Cal.Rptr. 517 (1991).
[47] Id. But see *Stalker Bros., Inc. v. Alcoa Concrete Masonry, Inc.*, 422 Md. 410, 30 A.3d 885 (2011) (allowing an unlicensed subcontractor recovery against the prime contractor).
[48] 466 Mich. 660, 649 N.W.2d 371, reh'g denied, 467 Mich. 1202, 651 N.W.2d 920 (2002).
[49] Mich.Comp.Laws § 339.2412(1).
[50] *White v. Cridlebaugh*, 178 Cal.App.4th 506, 100 Cal.Rptr.3d 434 (2009).
[51] E.g., Cal.Bus.&Prof.Code § 7031(e).
[52] Cal.Gov.Code § 4525(e); Idaho Code Ann. § § 54-4501 – 4514; Okla.Admin.Code § § 580:20-17-1 to -10.
[53] Cal.Gov.Code § 4529.5.
[54] Idaho Code Ann. § 54-4505.
[55] Okla.Admin.Code § 580:20-17-3(c)(1).
[56] 120 Nev. 195, 88 P.3d 822 (2004).
[57] Nev.Rev.Stat. § 623.023.
[58] 172 Cal.App.4th 939, 91 Cal.Rptr.3d 633 (2009), review denied (Cal. 2009).
[59] Supra note 52.
[60] 60 A.D.3d 555, 876 N.Y.S.2d 23 (2009).

CHAPTER 9

THE DESIGN PROFESSIONAL–CLIENT RELATIONSHIP

SCENARIO

Oliver and Olivia Owners decided to retire from life in the big city. They bought an empty lot in a quaint, New England "historic" town. Upon recommendation, they decided to hire Andrew Awesome, one of three partners in Awesome + More Architects, as their architect. The Owners drove to the architects' office and presented their ideas for a home without mentioning any budget.

Andrew Awesome created some schematic drawings for the Owners. When the Owners appeared satisfied by one drawing, they signed an AIA B101-2007 owner/architect contract with Awesome + More Architects. Andrew Awesome told the Owners that, "as a technicality," they needed to check the box next to B101-2007, § 4.1.26, titled "Historic Preservation," simply because of the historic character of the town. Trusting Andrew Awesome, the Owners checked the box without questioning its purpose in the contract.

Even though this contract called for payment based on a percentage of the cost of the work, Awesome + More Architects submitted monthly invoices to the Owners, charging them hourly rates. The Owners paid the invoices without complaint.

Andrew Awesome strongly recommended the Owners use Prime Builder, Inc. as the prime contractor, and the Owners immediately agreed. Unknown to the Owners, Andrew had a kickback scheme with Prime Builder, Inc. under which he got a small share of the contractor's payments. Andrew assuaged his feelings of guilt with the knowledge that Prime Builder, Inc. does high-quality work.

During construction, the Owners repeatedly asked for changes to the work. Andrew Awesome passed on these changes to the contractor without pointing out the cost consequences to the Owners.

As the work dragged on well past the anticipated completion date, the Owners grew suspicious of the apparently cozy relationship between their architect and contractor. On the day they were to confront Andrew Awesome with their suspicions, Betty More—another partner of Awesome + More Architects—showed up at the worksite to say that Andrew Awesome had suffered a heart attack (not too serious, fortunately) and that Betty More would take over their project. When the now-flustered Owners complained that they had specifically selected Andrew Awesome to be their architect, Betty More replied that their contract was with the partnership itself and that any partner had the legal right to provide architectural services. The Owners acquiesced, secretly hoping matters would improve.

They didn't, and the same cozy relationship between architect and contractor seemed to persist. One morning, Oliver Owner saw the contractor's president pass a thick envelope to Betty More. Convinced that the envelope was full of cash, Oliver confronted the architect and told her Awesome + More Architects' services were terminated immediately and that she had until the end of the day to remove her possessions from the job site. When Betty More opened the envelope, she was shocked to discover the envelope was, indeed, full of money, and not with the documents Andrew Awesome had asked her to get from the contractor.

9.1 Traditional Phases of Architectural Services: AIA B101-2007

The range of professional services may span from project conception to the completion of building activities. Throughout this process, in order for the relationship to be successful, both parties must fulfill their contract obligations in an atmosphere of trust.

The American Institute of Architects (AIA) "Standard Form of Agreement Between Owner and Architect," Document B101-2007, is the template for this discussion. It is reproduced in Exhibit D of this book. AIA B101-2007 describes design services provided on a design–bid–build (DBB) project (see Section 14.9). Project delivery systems which differ from DBB (see Sections 14.11–14.19) alter the relationship between the design professional and the project.

Usually the client comes to the designer with a problem it hopes the designer can solve. The client describes its needs and, as a rule, what it wishes to spend. Assuming the client hires an architect and agrees to use B101-2007, Article 3 of that document divides the architectural services into five components:

1. schematic design phase (§ 3.2)
2. design development phase (§ 3.3)
3. construction documents phase (§ 3.4)
4. bidding or negotiation phase (§ 3.4)
5. construction phase (§ 3.6)

After consulting the client, the architect first develops a schematic design and may revise the client's budget. The AIA describes the schematic design as the architect's "preliminary evaluation of the Owner's program."

> In this chapter's scenario, the Owners did not mention a budget to Andrew Awesome at the initial meeting, and there is no indication that the architect inquired into (yet alone revised) the budget. The AIA contemplates the architect taking an affirmative role—early in the parties' relationship—to become aware of the owner's financial constraints and to design accordingly.

Next, the architect studies the design and prepares drawings and possible models that illustrate the plan, site development, features of construction equipment, and appearance. The designer is also likely to prepare outline specifications and possibly again revise the predicted costs. In small projects, schematic design and design development are still designated as "preliminary studies."

After the client approves the design development, the designer prepares working drawings and specifications that cover in detail the general construction, structure, mechanical systems, materials, workmanship, site development, and responsibility of the parties. Often the designer supplies or drafts general conditions and bidding information.

The final preconstruction phase is generally called the *bidding* or *negotiation phase.* The designer helps the client obtain a construction contractor through bidding or negotiation.

During construction, the designer performs *project administration* services. The designer interprets the contract documents, checks on the progress of the work in order to issue payment certificates, participates in the change order process, and initially resolves disputes. (However, A201-2007 Article 15 introduces a new actor, the Initial Decision Maker (IDM). She makes initial decisions. If an IDM is not specifically designated, then initial decisions are made by the architect. See Sections 26.1 and 26.5.)

These five phases described in B101-2007, Article 3 constitute the architect's "basic" services under an AIA contract. The differentiation between "basic" and "additional" services (B101-2007, Article 4) relates to the architect's compensation, a topic explored in Section 9.3B.

9.2 Characterizing the Client–Design Professional Relationship

How does the law view the relationship between client and design professional (usually the architect)? Commercial (or contractual) relationships may be characterized differently based on the context.

The normal rules of conduct in the commercial world allow parties to act at *arm's length*. Each actor in that world can think mainly of its own interest.

Even in such a context, the law increasingly finds that all contracting parties owe each other the responsibility of good faith and fair dealing.[1] This is particularly so when objectives sought by the design professional and the client require close cooperation. Each party should help the other achieve its goals under the contract.

Of particular importance to the owner–design professional relationship, the law recognizes the necessity of a layperson to have trust in the judgment and actions of a professional. Not only does the layperson lack the knowledge, skill, or experience of the professional, but these deficits are precisely the reason the professional is hired. The client is not in a position to protect himself in his dealings with the professional in the same way he might be expected to protect himself in a standard, arm's length relationship.

Finally, at the opposite extreme from an arm's length relationship is a *fiduciary* one. A fiduciary relationship may be found in specific and limited contexts. Generally speaking, a fiduciary owes the other contracting party a duty of the highest loyalty and strict fidelity. It may be said that a relationship is fiduciary when one party has superior knowledge and authority and that party is in a position of trust and confidence over the weaker party. The relationship places one party in the hands of another. The fiduciary is expected to act in the best interests of the other party.

As an example, the attorney–client relationship automatically (*per se*) creates a fiduciary relationship. The attorney's advice must be based solely on the best interests of the client—not his own. There can be dealings between the attorney and client, but the law will look very carefully at any transaction to make sure it is fair to the client in light of the greater knowledge of the attorney and the trust the client places in his attorney.

The relationship of trust between a designer (more commonly the architect) and the client is not on par with the relationship between an attorney and client. In *Carlson v. SALA Architects, Inc.*,[2] the Minnesota Court of Appeals rejected the client's contention that the relationship between architect and client in itself (*per se*) creates a fiduciary relationship. However, the court left open the possibility of such a relationship between client and architect. It remanded the case (sent it back) to the trial court to determine whether the underlying facts gave rise to fiduciary duty owed by the architect.

A construction manager as advisor (CMa) is often analogized to an architect, as both act as the owner's advisor (see Section 7.5C). In *City of Meridian v. Petra, Inc.*,[3] the Idaho Supreme Court ruled that the CMa was not a fiduciary of the owner—even though the construction management agreement provided that the CM "acknowledges and accepts the relationship of trust and confidence established with Owner by this Agreement…". According to the court, a fiduciary relationship is one in which one party is in a superior position, the other party reposes a "special trust" in the superior party, and the other party reasonably believes the superior party is not acting in its own interests. The court found none of these conditions applied to an owner–construction manager relationship.

Even without a fiduciary relationship between the parties, the law would most likely construe any ambiguities against the architect both because the architect in most cases drafted the contract and because the balance of knowledge tilts toward the architect despite the owner often having superior bargaining power. The relationship between client and architect would not likely be viewed as arm's length.

This question of how to characterize the owner–architect relationship is complicated by the various roles played by the architect in both the design and the construction phases. The design professional's relationship to the owner is a combination of independent contractor and agent. While an agent owes a fiduciary duty to the principal (see Section 2.11C.), the relationship between a hirer and independent contractor can be arm's length, fiduciary, or any characterization between those extremes.

In the design phase, the architect acts as an independent contractor of the client. The architect is its own independent business. The owner cannot tell the architect who to hire or fire or how to perform its design duties. That said, the architect also can be a professional advisor, particularly in design matters that concern cost. For example, in *Getzschman v. Miller Chemical Co.*, the Nebraska Supreme Court affirmed the trial court's rejection of a jury instruction stating that the architect owes his client a fiduciary duty. At the same time, the court affirmed the following jury instruction: "An architect is bound to make full disclosures of all matters of which he had or should have knowledge and which it was important that his client should know."[4] This obligation imposed on the architect—to look out to protect the client's interests with regard to project cost—is similar to a fiduciary duty.

> In this chapter's scenario, the Owners repeatedly asked for changes to the work. Andrew Awesome passed on these changes to the contractor without pointing out the cost consequences to the Owners. Were the Owners at the end of the project to be suddenly confronted with a large pay-off demand from the contractor, they could sue their architect for negligent failure to advise them as to the cost consequence of their decisions to make changes to the work.

During the construction phase, the architect or engineer has other roles. He may be given the power by the parties to the construction contract to certify payment certificates[5] or certificates of completion.[6] In these roles, the design professional is acting as the owner's agent. A representation from the architect to the contractor made within the scope of the agency and meeting any formal requirements (for example, that the communication be in writing) is binding on the owner.

Finally, the AIA gives the architect the power to initially resolve disputes.[7] In this capacity the architect acts in a quasi-judicial role—not as agent for the client. Certainly, there is no place for any fiduciary duties even if there were such a relationship.

Two other issues that can be part of the owner–design professional (whether architect or engineer) relationship concern confidentiality and conflict of interest. (Indeed, these ancillary issues often arise in construction.) The parties may learn information that one party wishes to stay confidential. Also, any decision making can be affected by a conflict of interest.

The client should be able to trust its professional advisor and receive honest professional advice. Both parties—design professional and client—should be candid and open in their discussions, and each should feel confident that the other will not divulge confidential information to third parties.

Some aspects of the parties' relationship are obvious. Design professionals should not take kickbacks or bribes. They should not profit from professional services other than by receiving compensation from the client.

> In this chapter's scenario, Andrew was receiving kickbacks from Prime Builder, Inc. Even if the contractor's work was of a high quality, the kickback scheme still injures the Owners by depriving them of the relationship of trust that they deserve to have with their architect. The Owners could sue Andrew to recover these payments which (after all) originated from their payment of the contractor.
>
> Whether Andrew Awesome's conduct violates the AIA's Code of Professional Conduct is addressed in Chapter 13, dealing with ethics.

Funds held by the design professional that belong to the client should be kept separate. Comingling is a breach of the fiduciary obligation. In such a case, any doubts about to whom the money belongs or for whom profitable investments were made is resolved in favor of the client.

Financial opportunities that come to the attention of the design professional as a result of the services he is performing for the client should be disclosed to the client if the services would be an opportunity falling within the client's business.

One of the most troubling concepts in law relates to conflict of interest. A person cannot serve two masters. The client should be able to trust its design professional to make judgments based solely on the best interests of the client. Advice or decisions by the design professional should be untainted by any real or apparent conflict of interest. The client must believe the design professional serves it—and it alone.

The AIA recognizes conflict of interest in its B101-2007, § 2.4. It provides that, unless the owner knows and consents, "the Architect shall not engage in any activity or accept any employment, interest, or contribution that would reasonably appear to compromise the Architect's professional judgment with respect to this Project." Yet even without a contract provision like § 2.4, prohibition of the design professional from engaging in a conflict of interest would be implied by the law.

Design professionals should not have a financial interest in anyone bidding on a project for which they are furnishing professional advice. Likewise, they should not have a financial interest in any contractor or subcontractor who is engaged in a project for which the professionals have been engaged. Design professionals should not have any significant financial interest in manufacturers, suppliers, or distributors whose products might be specified by them. Products should not be endorsed that could affect specification writing, nor should designated products be specified because manufacturers or distributors of those products have furnished free engineering. The purpose of these restrictions is to avoid conflict of interest. Design professionals cannot serve their clients loyally if they might personally profit by their advice.

Generally, a client who is fully aware of a potential conflict of interest can nevertheless choose to continue to use the design professional. Consent to a conflict of interest should be binding only if it is clear that the client knows all of the facts and has sufficient understanding to make a choice. A design professional who intends to rely on consent by the client must be certain that such requirements are met.

The client clearly can avoid responsibility for any act by the design professional that is tainted by a breach of the fiduciary obligation. For example, if a contract is awarded to a bidder with whom the architect or engineer colluded, the contract can be set aside by the client. If the architect or engineer issued a certificate for payment dishonestly or in violation of his fiduciary obligation, the certificate can be set aside if the client so desires.

In sum, while the mere retention of a design professional does not automatically establish a fiduciary relationship, nor is the design professional–client relationship exclusively one at arm's length. Although neither extremity comfortably fits this relationship, the nature of the relationship has many attributes of a fiduciary relationship because of the client's lack of the specialized skills possessed by the design professional and the power the designer has to affect the interests of his client. The client must have confidence in the loyalty of his design professional, and the two must trust each other.

The undermining of that trust and confidence gives the client grounds to dismiss the design professional. Any bribes or gifts taken by the design professional can be recovered by the client. Any profit the design professional has made as a result of breaching the fiduciary obligation must be given to the client even if the profit was generated principally by the skill of the design professional. Obviously, the design professional must take his obligation of trust seriously and make every effort to avoid its breach or the appearance of such a breach.

9.3 Compensation of the Design Professional: Methods of Compensation

9.3A Introduction

A number of different fee arrangements are available to design professionals and owners. The choice among these arrangements is principally guided by criteria that are professional—not legal. Put another way, design professionals are generally in a better position to determine the type of fee structure than are their attorneys. Nevertheless, the law does play a limited role.

Principally, the law interprets any contractual terms that bear on fee computation when the contracting parties disagree. If no fee arrangement is specified in the contract and the parties cannot determine an agreed-on fee subsequent to performance, the law may be called on to make this decision.

Despite this limited role, certain legal principles must be taken into account in choosing a fee arrangement or in predicting the legal result if a dispute arises. For example, faced with the question of whether certain services come within the basic design fee, the law may choose to protect the reasonable expectation of the client if the design professional selected the language.

Likewise, any fee arrangement that measures compensation by a stated percentage of construction cost may be interpreted to favor the client. This can result from the belief that such a fee formula can be unfair to the client, given the design professional's incentive to run up costs. In rare cases, the contractual method selected will be disregarded because supervening events occur that neither party contemplated.

In its instructions to B101-2007, the AIA states there are at least ten ways to compute compensation for architectural services. Four are based on cost and time:

1. multiple of direct salary expense
2. multiple of direct personnel expense
3. professional fee plus expenses
4. hourly billing rates

In the first, the multiple represents benefits, overhead, and profit. In the second, the multiple represents overhead and profit. In the third, salaries, benefits, and overhead are expenses and the fee represents profit. In the fourth, salaries, benefits, overhead, and profit are included in the hourly rates.

The other methods, although indirectly related to time and expenses spent, are related more directly to the project costs or attributes of the project. They are

1. stipulated sum
2. percentage of the cost of the work
3. multiple of consultants' billing
4. square footage
5. unit cost (apartment, rooms, acres, etc., multiplied by a price factor)
6. royalty (share of owner income or profit)

These methods may be used in combination. This chapter comments on those methods used most often.

9.3B Percentage of Construction Costs: Basic Versus Additional Services

Percentage of Construction Costs. Although no longer universal (fixed-fee and cost methods of compensation being used increasingly), the stated percentage of construction costs is still a common method of fee computation. In such a method, the fee is determined by multiplying the construction costs by a designated percentage set forth in the contract.

Under AIA B101-2007—and sometimes a matter of surprise to the client—the percentage of construction costs method does not pay for all architectural services but only for "basic" services. Basic services are those described in Article 3 of B101-2007; they correspond to the five project phases listed in Section 9.1. "Additional services," listed in Article 4 of B101-2007, are compensated by a separate method agreed to by the parties. Distinguishing between basic and additional services is discussed more in-depth later in this section.

The construction cost (or "Cost of the Work" to use the AIA's terminology) is described in B101-2007, § 6.1. It is the entire cost to build the project, including the contractor's overhead and profit. However, it does not include the architect's compensation or costs that are the responsibility of the owner (such as buying the property, obtaining rights-of-way, and financing).

The percentage of construction cost method has been criticized. It can be a disincentive to cut costs and may reward the design professional who is less cost conscious. It can be too rigid, because projects and time spent can vary considerably. It may not reflect time spent. It also tends to subsidize the inefficient client at the expense of the efficient one.

Despite constant criticism, the fee method is still used. Clients seem accustomed to it. The method can avoid bargaining over fee. In normal projects it may accurately reflect the work performed. Although it may undercompensate on some projects and overcompensate on others, some design professionals feel the fees average out. The fee method can avoid extensive recordkeeping. It also is much less likely to generate a client demand to examine the design professional's records, a common feature of a cost type of fee formula.

> In this chapter's scenario, the parties agreed that the architect would be paid a percentage of the cost of the work. Yet the architectural firm billed the Owners based on hourly rates, which the Owners paid. If a fee dispute erupts, the court will have to determine whether the parties by their conduct had changed the manner by which Awesome + More Architects' compensation would be calculated.

Percentages vary, with the figure selected in major part reflecting the amount of work the design professional must perform and—increasingly these days—the risk of project failure. The amount of work depends on a number of factors.

If the design professional has worked for the client before, past experience may be relevant in determining the stated percentage. An inexperienced, inefficient client may require more work than an efficient client who has dealt with construction before. Sometimes the percentage is based on whether the project is residential or commercial, whether the construction contracts are single or separate, and whether the construction contract price is fixed or a cost type. Smaller projects may have a minimum fee.

To avoid criticism that the fee method encourages high costs and discourages cost reduction, some design professionals use a flexible percentage. One method is a percentage that declines as the costs increase. The actual percentage for the entire project is determined from a schedule that has variable percentages depending on the ultimate construction cost. For example, a fee schedule may provide for a 5 percent fee if the costs do not exceed $1 million, a fee of 4 percent if the ultimate cost is between $1 million and $1.5 million, and 3 percent if the cost is over $1.5 million.

Another method to encourage cost consciousness is to employ a sliding scale under which the highest percentage is applied to a cost up to a specified amount and then the percentage reduces on succeeding amounts. For example, the fee can be 8 percent on the first $1 million of cost, 7 percent on the next $4 million, and 6 percent on all amounts over $5 million.

Courts have interpreted fee provisions, but because of the different provisions that can be or are employed, generalizations are perilous. In close cases, courts are likely to favor the position of the client if the design professional selected the contract language.

Basic Versus Additional Services. As alluded to earlier, the question of basic versus additional services becomes a central issue and often a contentious one. The client may believe that the services in question are part of basic services. If there are too many additional services that require compensation in addition to the basic fee, the client may believe that it is not getting much for the basic fee and that the basic fee is only a "sticker" price unrelated to the ultimate payout.

While the purpose of distinguishing between basic and additional services is to ensure the architect is properly paid, often the procedures required by additional compensation clauses, such as written orders and notices, are ignored by the parties. This generates claims by the design professional that these formal requirements were met or waived by the client. Courts have not been sympathetic to these claims, primarily because the architect is viewed as the author of the additional-compensation agreement. For example, in *Belot v. Unified School District No. 497*,[8] a Kansas court denied the additional services claim because the architect had failed to segregate costs of additional services from costs incurred for basic services. Similarly, in *Newman Marchive Partnership, Inc. v. City of Shreveport*,[9] the Louisiana appellate court ruled that the architect's failure to get advance authorization showed that the architect viewed the services as basic; accordingly, his claim for compensation for additional services was rejected.

One difficulty with the scope of basic services is the increasing number of services that must be performed by someone—whether the design professional, the owner, or a consultant retained by either. One reason for this expansion of services is the increased public controls over construction, the emphasis on the environment, and the increasing likelihood of disputes that often end up in arbitration or litigation. These new services can be needed in the design or construction phase. Additionally, each project can be handled differently, depending upon the client's capacity to do them itself and the specific nature of the project. In this regard, AIA Document B101-2007, § 4.1, lists twenty-seven specific services that can be done by the architect, done by the owner, or not be provided.

Besides guaranteeing proper compensation for the architect, a separate listing of "additional" services benefits the owner by providing him with a wider range of possible services than are included in the basic fee. If ordered by the owner, the additional services must be performed (but the design professional earns additional compensation). In that sense, the list of additional services is similar to a "changes clause" in a contract for construction services (discussed in Chapter 18). Both grant unilateral power to the owner to demand that certain services be performed. In the case of a contract for design services, they are designated as additional services. In the case of construction services, they must be within the general scope of the work.

A federal district court case, *Plante & Moran Cresa, L.L.C. v. Kappa Enterprises, L.L.C.*,[10] demonstrates the effect of not having a clause detailing additional services. A project manager (PM) gave a fixed price without any provision for additional services. Despite an eighteen-month delay in completion and further delay after completion when a burst sprinkler head flooded the site, the PM received no additional compensation.

AIA Additional Services. As noted, AIA Document B101-2007, Article 4, is entitled "Additional Services." Article 4 has three parts.

Section 4.1 applies to additional services agreed upon at the time the contract for design services is made. Twenty-seven such services are listed. As to each, the parties must decide whether the service will be done by the architect or by the owner or will not be provided. Some of these listed services include site evaluation and planning (§ 4.1.5), tenant-related services (§ 4.1.18), and historic preservation (§ 4.1.26).

Section 4.2 is a blank section for the parties to write in additional services.

Section 4.3 applies to additional services the architect discovers needs to be performed after signing the contract. The architect must notify the owner of the situation but then not proceed with the additional service until given written authorization. These new services are usually initiated by the owner (§ 4.3.1) or involve unanticipated dealings with the contractor (§ 4.3.2).

Under the AIA system dealing with additional services, there can be a substantial difference between the basic fee and the ultimate fee payout where the compensation is not one based on cost. Inefficient or incompetent contractors, unforeseeable or even foreseeable events that affect performance adversely, and dithering or mind-changing owners can, under B101-2007, turn a percentage of construction cost or a fixed-price contract into a cost contract.

> In this chapter's scenario, Andrew Awesome told the Owners that, "as a technicality," they needed to check the box next to B101-2007, § 4.1.26, titled "Historic Preservation," simply because of the historic character of the town. In fact, § 4.1.26 references the parties use of AIA B205-2007, which is titled "Standard Form of Architect's Services: Historic Preservation." The document is intended for projects that are "historically sensitive," such as renovation of a historic building. Here, Andrew Awesome is abusing the "additional services" scheme of B101 by trying to shift services from basic to additional as a means of increasing his income.

EJCDC Additional Services. In Document E-500, Exhibit A (2008), the EJCDC lists twenty-five *optional* additional services that require the owner's written authorization (¶ A2.01) and nine *required* additional services not requiring the owner's written authorization (¶ A2.02).

9.3C Multiple of Direct Personnel Expense: Daily or Hourly Rates

Personnel multipliers determine the fee for basic and additional services by multiplying direct personnel expense by a designated multiple ranging from 2 to 4 (the average being 2.5 to 2.7). AIA Document B141-1997, ¶ 1.3.9.4, defined direct personnel expense as "the direct salaries of the Architect's personnel engaged on the Project and the portion of the cost of their mandatory and customary contributions and benefits related thereto, such as employment taxes and other statutory employee benefits, insurance, sick leave, holidays, vacations, employee retirement plans, and similar contributions." Once fringe benefits were truly on the "fringe." Today, they can constitute as much as 25 to 40 percent of the total employee cost. The contract must clearly specify personnel compensation cost beyond the actual salaries or wages.

In AIA Document B101-2007, this paragraph has been deleted. The AIA believes that this method is less significant. In its Instructions to B101-2007, it merely defines Multiple of Direct Personnel Expense as "salaries plus benefits … multiplied by a factor representing overhead and profits." Multiples are still common in contracts for engineering services.

There are obvious disadvantages to daily or hourly rates. Because a day is a more imprecise measurement than an hour, if either method is used it is likely to be the hourly rate. Such a method requires detailed cost records that set forth the following:

1. The exact amount of time spent.
2. The precise project on which the work was performed.
3. The exact nature of the work.
4. Who did the work.

Different hourly rates may be used for work by personnel of different skills.

When compensation is based on cost incurred by the design professional, the client often prescribes the records that must be kept, how long they must be kept, and that they be made available to the client. Many owners will insist on contract language that requires the architect to keep any relevant cost records based on GAAP (generally accepted accounting principles) standards.

B101-2007 does not prescribe accounting standards for records that must be kept. However, B101-2007, § 11.10.4, specifies that "Records of Reimbursable Expenses, expenses pertaining to Additional Services, and services performed on the basis of hourly rates shall be available to the Owner at mutually convenient times."

9.3D Professional Fee Plus Expenses

The form of fee arrangement which the AIA calls "professional fee plus expenses" is analogous to the cost type of contracts discussed in greater detail in Section 14.4. One advantage of a cost type of contract for design services is that the compensation is not tied to actual construction costs and there should be incentive to reduce construction costs. It can be a disincentive, however, to reduce the cost of design services.

The "professional fee plus expenses" contract necessitates careful definition of recoverable costs. Costs—direct and indirect—can be an accounting nightmare. Disputes can arise over whether certain costs were excessive or necessary. In cost contracts, advance client approval can be required on the size of the design professional staff, salaries, and other important cost factors. Cost contracts require detailed recordkeeping. (A ceiling can be placed on costs.)

Suppose the design professional estimates what the costs are likely to be in such a contract. Although the client may wish to know approximately how much the design services are likely to cost, an estimate can easily become a cost ceiling. If an estimate is given that is not intended as a ceiling, it should be accompanied by language that indicates the assumptions on which the estimate is based and that shows it is not a fixed ceiling or a promise that design costs will not exceed a designated amount.

9.3E Fixed Fee

Design professional and client can agree that compensation will be a fixed fee determined in advance and incorporated in the contract. This method reflects the desire on the part of owners and those providing financial resources for price certainty "up front" and a great fear of uncertainty caused by cost overruns and claims.

Before such a method is employed, a design professional should have a clear idea of direct cost, overhead, and profit as well as appreciate the possibility that contingencies may arise that will affect performance costs. A fixed fee should be used only where the scope of design services is clearly defined and the construction project well planned. It works best in repetitive work for the same client.

Does the fixed fee cover only basic design services? Does it include additional services and reimbursables? Standard contracts published by professional associations usually limit the fixed fee to basic design services. A design professional who intends to limit fixed fees to basic services should make this clear to the client. See Section 9.3H for discussion of fee provisions that place an absolute ceiling on compensation.

9.3F Reasonable Value of Services or a Fee to Be Agreed On

The fee will be the reasonable value of the services where the parties do not agree on a compensation method. If there is no agreed valuation method for additional services, compensation is the reasonable value of the services. The reasonable value of a design professional's services will take into account the nature of the work, the degree of risk to the design professional, the novelty of the work, the hours performed, the experience and training of the design professional, and any other factors that bear on the value of these services, including overhead and a reasonable profit. Proving the reasonable value of services requires detailed cost records. Leaving the fee open is generally inadvisable.

Where this issue does arise, each party usually introduces evidence of customary charges made by other design professionals in the locality as well as evidence that bears on factors outlined in the preceding paragraph. In many cases, great variation exists among the testimony given by the expert witnesses for each party. It is not unusual for the court or jury to make a determination that falls somewhere in between.

9.3G Reimbursables

B101-2007, § 11.8 is entitled "Compensation for Reimbursable Expenses." Subsection 11.8.1 specifies that compensation for these expenses is in addition to compensation for the basic and additional service. It then lists some of these expenses, including taxes levied on professional services and on reimbursables (§ 11.8.1.9), as well as site office expenses (§ 11.8.1.10). The catch-all "Other similar Project-related expenses" (§ 11.8.1.11) makes clear that the list is not exclusive. The list always seems to grow.

Incurring obligations for the client and paying them can impose an administrative burden on the design professional. Sometimes design professionals charge the client a markup for handling reimbursables. For example, suppose the design professional incurred expenses of $1,000 for traveling in connection with the project and long-distance calls. Under a markup system, the design professional might bill the client $1,000 plus an additional 10 percent or $100, making a total of $1,100. The markup percentage can depend on the number of reimbursables and the administrative overhead incurred in handling them. A design professional who wishes to add an overhead markup should explain this to the client in advance and obtain client approval.

B101-2007, § 11.8.2, adds to the amounts incurred by the architect a blank percentage for expenses incurred—in effect a markup.

9.3H Fee Ceilings

Owners are often concerned about the total fee, particularly if the fee is based on cost. But even in a compensation plan under which the design professional is paid a fixed fee or a percentage of compensation, the client may be concerned about additional services, reimbursables, or increases based on an unusual jump in the construction cost. Public owners with a specified appropriation for design services may seek to limit the fee to a specified amount. It may be useful to look at two cases that have dealt with fee limits.

In *Hueber Hares Glavin Partnership v. State*,[11] the contract limited the fee. It also excluded recovery for work to correct design errors. The court held the language was unambiguous and the fee limit was not an estimate. The fee limit could not be exceeded by costs attributed to design errors. By dictum, the court stated that the city would have been precluded from asserting the fee limit had it ordered extra services knowing the fee limit had been reached.

Harris County v. Howard[12] involved an AIA document. The upset price was also held to unambiguously include additional services and reimbursables. The public owner inserted a detailed recital on the fee limit and what it included. The court rejected the architect's contention that his having been paid $20,000 over the limit showed the limit did not include additional services or reimbursables.

Substantial changes in project scope should eliminate any fee ceiling that may have been established.[13]

9.3I Adjustment of Fee

Generally, the law places the risk that performance will cost more than planned on the party who has promised to perform. Careful planners with strong bargaining power build a contingency into their contract price that takes this risk into account. Suppose a client directs significant and frequent changes in the design. Suppose for any reason the design services must be performed over a substantially longer period than planned. It is unlikely that the law will give the design professional a price adjustment under a fixed-price contract. The AIA has sought to protect architects from these risks and others in B101-2007, Article 4, dealing with additional services.

9.3J Deductions from the Fee

Acts of the design professional may cause the client to incur expense or liability, and the client may wish to deduct expenses incurred or likely to be incurred from the fee to be paid to the design professional. Suppose a design professional commits design errors that cause a claim to be made by an adjacent landowner or by the contractor against the client, and the client settles the claim.

Suppose the client then wishes to deduct an amount from the architect's fee to reimburse itself for having paid the claims. Often, the cause of a claim may be difficult to determine. The AIA seeks to foreclose the owner from withholding the architect's compensation "unless the Architect agrees or has been found liable for the amounts in a binding dispute resolution proceeding." AIA Doc. B101-2007, § 11.10.3.

9.3K The Fee as a Limitation of Liability

Looking ahead to professional liability, the fee can serve another function. Some design professionals seek to limit their liability exposure to their client, to the amount of their fee. See Section 12.7D.

9.4 Compensation of the Design Professional: Timing of Payment

9.4A Service Contracts and the Right to Be Paid as One Performs

In service contracts, the promises exchanged are payment of money for performance of services. Unless such contracts specifically deal with time for payment to occur, the performance of all services must precede the payment of any money. Put another way, the promise to pay compensation is conditional on the services being performed.

Such a rule operates harshly to the person performing services. First, if the performance of services spans a lengthy time period, the party performing these services may need a source of financing to perform. Second, the greater the performance without being paid, the greater the risk of being unpaid. For these reasons, the law protects manufacturers and sellers of goods by giving them the right to payment as installments are delivered. However, this protection was not accorded to people performing services. If the hardships and risks described are to be avoided, contracts for professional services must contain provisions giving design professionals the right to be paid as they perform.

9.4B Interim Fee Payments

Design professionals commonly include contract clauses giving them the right to interim fee payments. This avoids the problems described in the preceding section. From the client's standpoint, interim fee payments can create an incentive for the design professional to begin and continue working on the project.

Usually, interim payments in design professional contracts become due as certain defined portions of the work are completed. Although AIA Document B101-2007, § 11.5, leaves blanks for interim fee payments, the 1977 B141a ("Instruction Sheet") suggested interim fee payments as

- schematic design phase 15 percent
- design development phase 35 percent
- construction documents phase 75 percent
- bidding or negotiation phase 80 percent
- construction phase 100 percent

Any schedule used should depend on the breakdown of professional services and the predicted work involved in each phase.

Dividing the design services and allocating a designated percentage of the fee to each service simplifies resolution of damages in the event the owner terminates the contract before contract completion. Assuming the termination was not for cause, the architect would be entitled to payment based on the phase of work done. The architect would be precluded from claiming actual damages in an increased amount, and the owner would be precluded from claiming actual damages in a lower amount.

9.4C Monthly Billings

In many projects, months may elapse before a particular phase is completed. To avoid overly long periods between payments, contracts for such projects should provide for monthly billings within the designated phases. See B101-2007, § 11.10.2.

9.4D Late Payments

Financing costs are an increasingly important part of performing design professional services. Late payments and reduced cash flow can compel design professionals to borrow to meet payrolls and pay expenses. The contract should provide that a specified rate tied to the actual cost of money be paid on delayed payments. For example, AIA B101-2007, § 11.10.2, provides blank spaces to be filled in for the rate of interest and when payable. If the blanks are not filled, late payments invoke the "legal rate" at the architect's principal place of business. (In most states, the "legal rate" is the amount of interest payable on court judgments.)

When clients delay payments, the design professional should make a polite and sometimes strong suggestion that payments should be made when due. A pattern of delayed payments should make the design professional seriously consider exercising any power to suspend further performance until payments are made. (See Section 9.7C.) If the suspension continues for a substantial time period, the design professional should consider terminating the contract.

In cases of suspension or termination of performance, it is desirable to notify the client of an intention to either suspend or terminate unless payment is received within a specified period of time. This gives the client an opportunity to make the payment. It also shows the client that failure to make interim fee payments as promised will not be tolerated.

9.5 Payment Although Project Never Built

A *condition* is an event that must occur or be excused before a party is obligated to begin or continue performance. Often contracting parties do not wish to begin or continue performance unless certain events occur or do not occur. For example, an owner may not wish to start construction until it has obtained a loan or permit. Design professionals may wish to condition their obligations on their ability to rent additional space or hire an adequate staff. Yet each party may wish to make a binding contract in the sense that neither can withdraw at its own discretion.

In contracts for design services, the client frequently asserts that the nonoccurrence of conditions gives the client power to end the relationship *and* (depending on the language of the condition) precludes the design professional from being compensated for services rendered before termination. *Parsons v. Bristol Development Co.*[14] is instructive in this regard. The contract provided that a condition precedent to payment by the owner was the owner's obtaining a loan to finance the project. The architect commenced work and was paid part of his fee. He then began to draft final plans and specifications for the building. However, the owner was unable to obtain a construction loan and abandoned the project.

The owner pointed to the condition in the contract as a basis for its refusal to pay for any unpaid services that the architect had rendered. The architect pointed to language in the contract stating that if any work designed by the architect is abandoned, the architect "is to be paid forthwith to the extent that his services have been rendered." Such language frequently appears in documents published by the AIA and is known as a "savings clause." (In B101-2007, see § § 9.3 and 9.6.) However, the court interpreted the two apparently inconsistent provisions in favor of the more specific one dealing with the condition of obtaining a construction loan.

The case demonstrates the tendency of courts to construe language against the parties that supplied the language and to protect clients from having to pay for design services when the project is abandoned. Such client protection is based on the understandable reluctance of the client to pay for services that it ultimately does not use. But professionals generally expect to be paid for their work. Fee risks can be taken, but any conclusion that the risk was taken should be supported by strong evidence that the design professional assumed the risk.

Courts facing claims of finance conditions often come to different results depending on the language, the surrounding circumstances, and judicial attitude toward outcomes that either deny any payment for work performed or force a party to pay for services it cannot use.

9.6 Client Obligations to the Design Professional Not Involving Compensation

Contracts for design commonly place a number of obligations on the owner. These usually relate to furnishing information that is within its knowledge as owner and obligations that it is in the best position to perform.

AIA Document B101-2007, Article 5, is captioned "Owner's Responsibilities." The owner is required to provide information through a written program (§ 5.1). It establishes and updates its budget (§ 5.2). It designates a representative to act on its behalf (§ 5.3). It provides surveys that "describe physical characteristics, legal limitations … and a written legal description of the site." (§ 5.4.) It furnishes the services of geotechnical engineers (§ 5.5), furnishes tests and inspections required by law (§ 5.7), as well as legal, insurance, and accounting services (§ 5.8). Article 5 includes other operational responsibilities, but the items listed show the significant role played by the owner.

In addition to express provisions, obligations can be implied into contracts. The client impliedly promises not to interfere with the design professional's performance and to cooperate. For example, the client should not refuse the design professional access to information necessary for performing the work. Refusal to permit the design professional to inspect the site would be prevention and a breach of the implied obligation owed by the client to the design professional.

Positive duties are owed by the client. The client should exercise good faith and reasonable promptness in passing judgment on the work of the design professional and in approving work at the various stages of the latter's performance. It should request bids from a reasonable number of contractors and should use best efforts to obtain a competent bidder who will agree to do the work at the best possible price. If conditions exist that will require acts of the client, such as obtaining a variance or obtaining financing, the client impliedly promises to use best efforts to cause the condition to occur.

9.7 Design Professional's Role in Contract Completion

Generally, the owner would like the design professional to perform until the project is "completed." At this point, the owner takes possession of the work. After that point, except for post-completion services such as furnishing or reviewing as-built drawings, the owner no longer needs the design professional's services.

In most construction contracts executed on standard documents published by the professional associations, completion has two stages: substantial and final. AIA Document B101-2007, § 3.6.1.3, states that the obligation of the architect to provide administration "terminates on the date the Architect issues the final Certificate for Payment." This end date corresponds to A201-2007, § 4.2.1, which provides that the architect will perform until he issues the final certificate for payment.

Clearly, some time will pass between the dates of substantial and final completion. It is during this time that the contractor performs a **punch list** of relatively minor items to fully complete the project. But the AIA recognizes that issuance of a final certificate may be delayed because of the contractor's unwillingness to correct the punch list items. B101-2007, § 4.3.2.6, entitles the architect to additional compensation if services extend sixty days after substantial completion. That services performed after the sixty-day period are considered additional may surprise the client.

9.8 Design Professional Suspension of Performance

Suspension of performance in contracts for design and construction is not uncommon. The owner may wish to suspend performance of either design professional or contractor if it is having financial problems or it wishes to rethink the wisdom of the project. Those who perform services in exchange for money, such as the design professional and the contractor, may wish to have the power to suspend performance if they are not paid. Suspending performance can provide a powerful weapon to obtain payment for work that has been performed. Also, suspending performance reduces the risk that further performance will go uncompensated.

The common law did not develop clear rules that allowed suspension. The performing party either must continue performing or under proper circumstances could discharge its obligation to perform.

An interim remedy—suspension—was not well recognized until states adopted the Uniform Commercial Code, beginning in the early 1950s. Section 2-609 of the code allowed a party to a contract involving goods to demand assurance under certain circumstances and withhold its performance until reasonable assurance was provided. This doctrine has carried over to contracts that do not fall within the jurisdiction of the Uniform Commercial Code, such as those that involve design or construction.[15]

Contracts for design services frequently include provisions granting the client the power to suspend the performance of the design professional. Usually such clauses give the design professional an immediate right to compensation for past services and a provision stating that suspension will become a termination if it continues beyond a certain period. If the designer has enough bargaining power, he will seek to include a provision stating that he can suspend if he is not paid and that if suspension continues for a designated period, he can terminate his performance under the contract.

Document B101-2007, § 9.1, gives the architect the power to suspend or terminate if the owner fails to make payment. The election to suspend is effective after seven days' written notice to the owner. The seven-day period allows for cure by the owner and continuation of performance. Before resuming services, the architect must be paid "all sums due." Fees and time are "equitably adjusted."

Document B101-2007, § 9.2, gives the owner the power to suspend the architect's performance. If suspension continues for more than thirty consecutive days, the architect is paid for past work. If work is resumed, he is paid for "expenses incurred during the interruption and resumption." Also, fees and schedules are "equitably adjusted." Under B101-2007, § 9.3, suspension for over ninety consecutive days gives the architect the power to terminate.

9.9 Termination of Design Contract

Contracts frequently contain provisions under which one or both parties can terminate their contractual obligations to perform. Some provide that one party can terminate if the other commits a serious breach of the contract. Some allow termination powers for any breach, which is a method intended to foreclose any inquiry into the seriousness of the breach. However, such a power can be abused. For that reason, such a clause will be interpreted against the stronger party to the contract and may be found unenforceable. These two provisions allow termination for "default." A contract may also permit termination for no cause or at the "convenience" of one party.

AIA Document B101-2007 employs both of these approaches. B101-2007, § 9.4, is a "default" termination clause: either party may terminate the other for substantial failure to perform through no fault of the terminating party. This standard of a substantial failure to perform codifies the common law concept of **material breach**. Generally, a material breach by the client is an unexcused and persistent failure to pay compensation or cooperate in creating the design. A material breach by the design professional is likely to be negligent performance or excessive delays. See Section 3.7B.

The owner has no need to invoke B101-2007, § 9.4. Under § 9.5, an owner may terminate the architect for the owner's convenience and "without cause." However, if the owner terminates for convenience, § 9.6 gives the architect payment for services performed and not yet paid for and reimbursement for certain expenses incurred prior to termination. The architect is also entitled to termination expenses under § 9.7. These include expenses for which the architect has not been compensated and profit on services not performed.

Termination clauses often require written notice of termination. For example, AIA Document B101-2007, § 9.4, allows termination "upon not less than seven days' written notice." Although it is clear that termination does not actually become effective until expiration of the notice period, it is not always clear what the rights and duties of the contracting parties are during the period between receipt of notice and the effective date of termination. This may depend on the purpose of the notice.

The notice period can serve as a cooling-off device. Termination is a serious step for both parties. If it is ultimately determined that there were insufficient grounds for termination, the terminating party has committed a serious and costly breach. A short notice period can enable the party who has terminated to obtain legal advice and to rethink its position.

If the right to terminate requires a contract breach, notice can have an additional function. It may be designed to give the breaching party time to cure past defaults and provide assurances that there will be no future defaults. If cure is the function of the notice period, actual termination should occur only if the defaults are not cured by the expiration of the notice period. During the notice period, the parties should continue performance. If it appears there is no reasonable likelihood that past defaults can be cured and reasonable assurances given, performance by the defaulting party should continue only at the option of the party terminating the contract. The latter should not be forced to receive and perhaps pay for substandard performance.

Probably the principal purpose of a notice is to wind down the work to allow the parties to plan new arrangements made necessary by the termination. A short continuation period can avoid a costly shutdown of the project or the unavoidable expenses that can result if the design professional must stop performance immediately. The notice period can enable the client to obtain a successor design professional while retaining the original professional for a short period. It can also enable the design professional to make workforce adjustments, get employees back to home base, cancel arrangements made with third parties, and allow time to line up new work for employees.

Contracting parties should decide in advance what function the notice period is to serve and what will be the rights and duties of the parties during this period. Once this determination is made, the contract should reflect the common understanding of the parties, and any standardized contracts should be modified accordingly.

Finally, a contractual notice period may play an inadvertent role in the termination process. Suppose an owner, after much patience and increasing frustration, decides to terminate the design professional. Factually, the termination may be well justified. However, in his anger the owner may overlook (assuming he was even aware of) the notice period, telling the design professional to "leave by the end of today" or "by the end of the week." By failing to comply with the notice period, the owner may have himself have breached the contract. An entirely justified termination would have been turned into a wrongful one.

In this chapter's scenario, Oliver Owner terminated Andrew Awesome for receiving an apparent kickback from the contractor. The termination was justified because, in fact, the envelope was full of cash as Oliver had suspected. However, the parties were using an AIA contract, and Oliver told the architect to leave by the end of the day, thereby violating the seven-day notice requirement.

A court would likely reject Andrew Awesome's argument that the Owners' failure to give him notice of termination seven days before the effective date of the termination meant that the termination was wrongful and that the Owners were the ones in breach of contract. Here, the reason for the termination—illegal behavior, which was also a clear conflict of interest—meant that a notice period had no function to fulfill. Moreover, the two wrongs are not equivalent: Andrew Awesome engaged in a *malum in se* while Oliver Owner committed a *malum prohibitum*. (See Section 8.6B.) The court is likely to ignore Oliver Owner's technical breach and find Andrew Awesome in material breach.

9.10 Death or Unavailability of Design Professional

Suppose a key person is no longer available to perform professional services. Is that person's continued availability so important that his inability to perform—because of either disability, death, or an employment change—will terminate the contract? The issue usually arises if the client wishes to terminate its obligation because a key design person is no longer available. That key person can be the sole proprietor, a partner, or an important employee of a partnership or professional corporation.

Contract obligations generally continue despite the death, disability, or unavailability of people who are expected to perform. Only in clear cases of highly personal services will performance be terminated.

Unavailability of key design people can frustrate contract expectations. For example, in the absence of a contrary contractual provision, the death of a design professional who is a party to the contract, such as a sole proprietor or partner, will terminate the obligation of each party. The personal performance of that particular design professional was very likely a fundamental assumption on which the contract was made. A successor to the design professional can, of course, offer to continue performance, and this may be acceptable to the client. However, continuation depends on the consent of both successor and owner. Without agreement, each party is relieved from further performance obligations.

In this chapter's scenario, Andrew Awesome suffered a heart attack and Betty More, another partner of Awesome + More Architects, insisted that the Owners must continue services with her, as their contract is with the partnership, not with Andrew personally. Nonetheless, a court is likely to find the Owners had personally selected Andrew as their architect and that imposing upon them services being provided by Betty More would violate their reasonable contract expectations. An owner's personal relationship with the architect is of particular importance in a residential project where personal taste is paramount.

Suppose the person expected to actually perform design services is an employee of a large partnership or professional corporation. That person's unavailability may still release each party, but it would take stronger showing that the unavailable design professional was crucial to the project and that her continued performance was a fundamental assumption on which the contract was made.

The parties should consider the effect of the unavailability of key design personnel and include a provision that states clearly whether the contract continues if that person dies, becomes disabled, or becomes unavailable for any other reason.

AIA Document B101-2007, § 10.3, binds owner and architect and their "successors" to the contract. Under this obscure language, it appears that the parties contemplate successors stepping in if for some reason a contracting party, such as the architect, can no longer perform. This appears to require that the client continue dealing with the partnership if the partner with whom the client had originally dealt is no longer available.

Continuity may be desirable. However, the close relationship required between design professional and client may mean that the client does not wish to continue using the partnership if the person in whom it had confidence and with whom it dealt is no longer available. Similarly, a successor may not want to work for the client. Specific language should be included dealing with this issue.

9.11 Breach of the Design Agreement: Remedies

9.11A Introduction

Basic judicial remedies for contract breach are discussed in Chapter 4. That chapter also discusses claims in the context of the construction contract (owner–contractor disputes). This section applies basic legal doctrines to special problems found in the relationship between client and design professional.

9.11B Client Claims and the "Betterment" Rule Defense

The principal claims that clients make against professionals relate to defective design. A breach of contract by the design professional entitles the client to protect its restitution, reliance, and expectation interests. (Refer to Section 4.4.)

Although clients occasionally seek to protect their restitutionary interest by demanding return of any payments made, the principal problem relates to the client's expectation interest. If the project is designed defectively, the client is entitled to be put in the same position it would have been had the design professional prepared a proper design. The first issue that can arise is whether the client can measure its expectation loss by proving the cost of correcting the defective work or whether it is limited to the difference between the project as it should have been designed and the project as it was designed.

This issue can be demonstrated by looking at *Bayuk v. Edson*.[16] In this case, the owner complained about faulty design consisting of, among other things, an improperly designed floor, closets too small, outside doors constructed for a milder climate than where the house was built and of an unusual type that could not be constructed by artisans in the area, unaesthetic kitchen tile, sliding doors that did not fit properly in their tracks, and a fireplace that became permanently cracked.

A number of witnesses testified that it would not have made economic sense to repair the defects. One witness testified that tearing out and repairing would cost more than the cost of rebuilding the house in its entirety. The plaintiff produced an expert real estate appraiser who fixed the value of the house without the defects at $50,000 to $60,000 and with the defects at $27,500 to $31,500. The trial court awarded a judgment of $18,500, the least of the possible remainders. This was affirmed by the appellate court.

Suppose, however, that it would not have been economically wasteful to correct the defective work. This would entitle the owner to the cost of correction. In claims against a design professional for improper design, application of this standard involves the **betterment rule**, which is based on the cost of correction sometimes unjustly enriching the owner. For example, in *St. Joseph Hospital v. Corbetta Construction Co., Inc.*,[17] the hospital sued its architect, the contractor, and the supplier of wall paneling that had been installed when it was disclosed that the wall paneling had a flame spread rating some seventeen times the maximum permitted under the Chicago Building Code.

After the hospital had been substantially completed, it was advised that it could not receive a license because of the improper wall paneling. The city threatened criminal action against the hospital for operating without a license. The hospital removed the paneling and installed paneling that met code standards. The jury awarded $300,000 for removal of the original paneling and its replacement by code-complying paneling and an additional $20,000 for architectural services performed in connection with removal and replacement.

In reviewing the jury award of $320,000, the appellate court noted that had the architect complied with his obligation to specify wall paneling that would have met code standards, the construction contract price for both paneling and cost of installation would have been substantially higher. The court stated that the hospital should not receive a windfall of the more expensive paneling for a contract price that assumed less expensive paneling. The paneling that should have been specified together with installation would have cost $186,000, whereas the paneling specified with installation cost $91,000. This, according to the court, should have reduced the judgment by $95,000. The court reduced the award an additional $21,000 for items installed when the panels were replaced that were not called for under the original contract. As a result, the judgment was reduced some $116,000.[18]

Another claim sometimes made by clients is unexcused delay in preparing the design or performing administrative work during construction. Delay can harm the contractor, and most delay disputes are between an owner and a contractor. Claims for delay during the design phase are difficult to establish because of the likelihood of multiple causes. But suppose the project is completed late because of negligence by the architect in passing on submittals of the contractor. There are two main damage items. First, the owner will lose the use of its project for the period of unexcused delay. Second, the contractor may make a claim against the owner or design professional for any loss suffered because of delay wrongfully caused by the design professional. (The second claim is discussed in Section 20.13A.)

The first damage item, that of loss of use caused by the delay, was before the court in *Miami Heart Institute v. Heery Architects & Engineers, Inc.*[19] This case demonstrates the liability exposure of design professionals when they have delayed completion of a project without justification. In this case, the architect had agreed to design plans for the building of a new hospital structure that would house patients. During construction, the patients were housed in older buildings on the premises. The architect's failure to comply with code requirements caused a ten-month delay in the issuance of a certificate of occupancy.

The court rejected the contention by the architect that damages for loss of use can be recovered only when there is a delay by the contractor. It also rejected the contention that lost-use value can be recovered only if the project itself was to be rented to others. The court concluded that the proper measure of recovery should be the reasonable rental value of the structure during the period of delay.

The court was faced with the fact that the patients who would have been in the new structure were housed in the old structure for the period of the delay. With this in mind, the court concluded that the proper measure of recovery was the difference between the reasonable rental value of the new structure and that of the old structure for the period of delay. The patients were undoubtedly inconvenienced in still being housed in the old structure, and the quality of the medical services may have been poorer than if they had been in the new structure. Yet to establish the actual economic value of such losses would be almost impossible. Of course, some of this may be factored into the reasonable rental value amounts. (The difficulty of proving an owner's delay damages is a primary motivation behind liquidated damages clauses, discussed in Section 20.12.)

There were other claimed damages. The court held that the betterment rule would prevent the owner from recovering for additional expenses incurred to make the building meet code requirements. It noted that the plans called for 2 feet of electrical wire, but proper plans would have required 12 feet of wire. But the owner's expense in purchasing the additional 10 feet cannot be chargeable to the architect. The owner would have incurred this expense regardless of the architect's defective plans.

Finally, the court concluded that the owner could also recover any delay damage claims it paid to the contractors and subcontractors as a result of the architect's breach.

Delay can also cause less direct, or what are sometimes called *consequential*, damages. For example, delay in completing an industrial plant may generate losses caused by the inability of the plant owner to fill the orders of its customers, exposing the plant owner to claims by the customers and causing it to lose profits on the transactions. (Lost profits are discussed in Section 4.5.)

AIA Document B101-2007, § 8.1.3, provides that each party "waives consequential damages for claims … arising out of or relating to this Agreement." In the *Miami Heart Institute* case, damages for lost use would not be recoverable nor would the losses caused by delay in completing the industrial plant were such a waiver clause in the contract. As a result, owners may seek to delete this clause from the AIA contract.

9.11C Design Professional Claims

The principal claims made by a design professional against the client relate to the latter's failure to pay for services performed. These claims have not raised difficult valuation questions. Design professionals commonly seek to protect their restitution interests and recover the reasonable value of their services.[20] Occasionally, clients have resisted this claim by contending that they did not use the plans and specifications drafted and thereby have not been enriched. Such defenses (to a claim for unjust enrichment or quantum meruit) have been generally unsuccessful.[21]

Suppose the design professional seeks to protect his expectation interest. To do so, he must be put in the position he would have been had the client performed as promised. This could be computed by what he could have earned had he fully performed less the expense he has saved or his expenditures in part performance plus his profit and less what he has already been paid.

REVIEW QUESTIONS

1. According to Article 3 of the AIA B101-2007 Document, what are the five components of architectural services?

2. How should funds that are held by the design professional that belong to the client be handled?

3. In the instructions to the B101-2007 Document, what are the four methods to compute compensation for architectural services that are based on cost and time?

4. According to the EJCDC Document E-500, Exhibit A, how many optional additional services that require the owner's written authorization and how many required additional services not requiring owner's written authorization are listed?

5. What are the three parts of Article 4 of the AIA Document B101-2007 that is entitled "Additional Services?"

6. If a design professional decides to use an hourly rate for compensation for additional services, what four detailed cost records are required?

7. The law generally places the risk that performance will cost more than planned on which party?

8. What are the suggested interim fee payments for design professionals according to the 1977 B141a ("Instruction Sheet") Document?

9. What are the two stages of completion according to most construction contracts executed on standard documents published by the professional associations?

10. What are the two types of provisions for termination of a design contract?

ENDNOTES

[1] Restatement (Second) of Contracts § 205 (1981): Uniform Commercial Code § 1–304 (formerly 1–203). See Section 3.5B.
[2] 732 N.W.2d 324 (Minn.App.2007), review denied (Minn. Aug. 21, 2007).
[3] 154 Idaho 425, 299 P.3d 232 (2013).
[4] 232 Neb. 885, 443 N.W.2d 260, 268 (1989), distinguished on a different ground, prejudgment interest, in *Folgers Arch. Ltd. v. Kerns*, 262 Neb. 530, 633 N.W.2d 114 (2001).
[5] AIA Doc. A201-2007, §§ 9.4, 9.5.
[6] Id. at § 9.8.4.
[7] AIA Doc. A201-1997, ¶ 4.4. In 2007 the AIA changed this. It now gives this power to an Initial Decision Maker, which can be the architect or someone else. AIA Doc. A201-2007, § 15.2. This is discussed in Section 26.1.
[8] 27 Kan.App.2d 367, 4 P.3d 626 (2000).
[9] 944 So.2d 703 (La.App.2006), writ denied, 949 So.2d 448, 452 (La.2007).
[10] 2006 Westlaw 1676411 (E.D.Mich.2006), aff'd, 251 Fed.Appx. 974 (6th Cir. 2007).
[11] 75 A.D.2d 464, 429 N.Y.S.2d 956 (1980).
[12] 494 S.W.2d 250 (Tex.Ct.App.1973).
[13] *Herbert Shaffer Assoc., Inc. v. First Bank of Oak Park*, 30 Ill.App.3d 647, 332 N.E.2d 703 (1975).
[14] 62 Cal.2d 861, 402 P.2d 839, 44 Cal.Rptr. 767 (1965).
[15] Restatement (Second) of Contracts, § 237, illustration 1, §§ 251, 252 (1981).
[16] 236 Cal.App.2d 309, 46 Cal.Rptr. 49 (1965).
[17] 21 Ill.App.3d 925, 316 N.E.2d 51 (1974).
[18] For further discussion of the betterment rule, see also *Lochrane Eng'g, Inc. v. Willingham Realgrowth Investment Fund, Ltd.*, 552 So.2d 228 (Fla.Dist.Ct.App.1989), review denied, 563 So.2d 631 (Fla.1990); Bales, O'Meara & Azman, *The "Betterment" or Added Benefit Defense*, 26 Constr. Lawyer, No. 2, Spring 2006, p. 14.
[19] 765 F.Supp. 1083 (S.D.Fla.1991), aff'd, 44 F.3d 1007 (11th Cir.1994).
[20] *Getzschman v. Miller Chemical Co.*, supra note 4.
[21] *Barnes v. Lozoff*, 20 Wis.2d 644, 123 N.W.2d 543 (1963) (measured by rate of pay in community); *In re Palms At Water's Edge, L.P.*, 334 B.R. 853 (Bankr.W.D.Tex.2005) (recovery in quantum meruit).

CHAPTER 10

Professional Design Services: The Sensitive Client Issues and Copyright

SCENARIO

This chapter does not include a single scenario. Instead, the chapter reproduces two cases: the *Griswold* case in Section 10.2C presents a "real world" scenario involving project cost predictions, while the *Watson* case in Section 10.3B involves a design professional charged by the owner with failing to discover defective workmanship by the contractor while the work was on-going. Sections 10.6 and 10.7, dealing with design ownership and copyright issues, reference several cases.

10.1 Professional Services: Prior to Receipt of Bids

During the project planning stage, the owner may turn to the designer to assist in matters unrelated to design. The owner may view the designer as experienced in project development and capable of providing valuable advice. This request for guidance is understandable in light of the relationship of trust between the client and designer, as noted in Section 9.2. The designer, in turn, may view the provision of non-design services as beneficial to the success of the project. Yet these good intentions by both parties must be tempered with awareness of the risk involved for the design professional. Those risks are addressed in this section.

10.1A Assistance in Obtaining Financing

A building project frequently requires lender financing. To persuade a lender that a loan should be granted, the client generally submits schematic designs or even design development, economic feasibility studies, cost estimates, and sometimes the contract documents and proposed contractor. This submission includes materials prepared by the design professional.

Must the design professional do more than permit the use of design work for such a purpose? Does the basic fee cover such services as appearing before prospective lenders, advising the client as to who might be willing to lend the client money, or helping the client prepare any information that the lender may require? Must such work be done only if requested as additional service and paid for accordingly?

Assistance in obtaining financing should not be considered part of basic design services. The professional education and training of a design professional do not include techniques for obtaining financing for a project. Nor is it likely that the design professional will be examined on this activity when seeking to become registered. The American Institute of Architects (AIA) does not consider services related to project financing to be part of basic or additional services.

An architect who agrees to provide financing advice creates a risk of liability exposure if she does the job badly. For this reason, it is better to view such a service as provided pursuant to a separate agreement—not as an additional service.

10.1B Economic Feasibility of Project

If the architect should hesitate before making representations as to the availability of funds, she should certainly avoid venturing into economic feasibility studies. These are not part of basic services. If such studies are requested and made and the venture is unsuccessful, the design professional will be exposed to claims by the client. Such predictions are treacherous and are ones for which design professionals are rarely trained by education and experience.

Reversing its prior practice, the AIA beginning with B101-2007 does not include "economic feasibility studies" as a service that the architect must provide if requested by the owner. (AIA Doc. B141-1997, ¶ 2.8.3.6 included "economic feasibility studies" as a service that the architect must provide if requested by the owner.)

10.1C Approval of Public Authorities

Greater governmental control and participation in all forms of economic activity have meant that the design professional increasingly deals with federal, state, and local agencies. Must the design professional help the client or its attorney prepare a presentation for the planning commission, zoning board, or city council? Must she appear at such a hearing and act as a witness if requested? Is the design professional who does these things entitled to compensation in addition to the basic fee?

The increased likelihood that the owner will ask for help on permit and public land use control matters has led the American Institute of Architects to seek to make clear that these services are not part of basic services and merit additional compensation. (See Section 9.3B.) AIA Document B101-2007, § 4.3.1.7 states that "preparation for, or attendance at, a public presentation, meeting or hearing" is an additional service for which the architect is entitled to additional compensation beyond the basic fee.

Reflecting the different nature of engineering projects and their funding sources, EJCDC E-500 (2008) states in Exhibit A, ¶ A2.01A(1), that the following are additional services requiring advance authorization:

> Preparation of applications and supporting documents (in addition to those furnished under Basic Services) for private or governmental grants, loans or advances in connection with the Project; preparation or review of environmental assessments and impact statements; review and evaluation of the effects on the design requirements for the Project of any such statements and documents prepared by others; and assistance in obtaining approvals of authorities having jurisdiction over the anticipated environmental impact of the Project.

Design professionals may decide it is useful or desirable to perform such services without requesting additional compensation. This section deals solely with the questions of whether they are obligated to perform the services and whether they are legally entitled to additional pay for doing so.

10.1D Services of a Legal Nature

Some design professionals volunteer or are asked to perform services for which legal education, training, and licensure may be required. This may be traceable to the high cost of legal services and the uncertainty as to which services can be performed only by a lawyer.

Undoubtedly, one of the most troublesome activities is drafting or providing the contract. It may not be wise for laypeople to draw contracts for themselves, although the law allows professional designers to supply or draft contracts for their own services. In addition, the AIA in B101-2007, § 3.4.3 contemplates the architect's assistance in creating the prime contract as an "additional service."

On the other hand, B101-2007 § 5.8 requires the owner to furnish legal services as necessary. The AIA clearly does not want architects to perform legal, insurance, and accounting services.

As a cautionary note, the AIA specifies in its standard forms of agreement that the document has important legal consequences and that "consultation with an attorney is encouraged with respect to its completion or modification."

The border between legal and nonlegal services sometimes is blurry. For example, although land use matters can and do involve legal skills, the architect may be knowledgeable about the politics and procedures in land use matters, particularly if she has had experience in dealing with land use agencies.

Suppose a client asks the design professional whether a surety bond should be required. Here the balance tips strongly in favor of considering these legal services. Sometimes bonds are required for public projects. Using surety bonds may largely depend on other legal remedies given subcontractors and suppliers, such as the right to assert a mechanics' lien or to stop payments. The design professional should simply answer specific questions of a nonlegal nature, rather than advise generally on such matters.

If pressed by the client, the design professional may decline advice on the ground that such services are excluded from professional liability insurance coverage.

10.2 Cost Predictions

10.2A Inaccurate Cost Prediction: A Source of Misunderstanding

The relative accuracy of cost predictions is crucial to both public and private clients. Clients may be limited to bond issues, appropriations, grants, loans, or other available capital. Unfortunately, many clients think that cost estimating is a scientific process by which accurate estimates can be ground out mechanically by the design professional. For this reason, design professionals should start out with the assumption that cost predictions are vital to the client and that the client does not realize the difficulty in accurately predicting costs.

The close relationship between design professional and client often deteriorates when the low bid substantially exceeds any cost figures discussed at the beginning of the relationship or even the last cost prediction made by the design professional. Not uncommonly, the project is abandoned, and the client may claim that it should not have to pay the design professional. The design professional contends that cost predictions are educated guesses and that their accuracy depends in large part on events beyond her control. Clients frequently request cost predictions before many details of the project have been worked out. They also frequently change the design specifications without realizing the impact such changes can have on earlier cost predictions. Design professionals point to unstable labor and material costs. The amount a contractor is willing to bid often depends on supply and demand factors that cannot be predicted far in advance. Design professionals contend they are willing to redesign to try to bring costs down. They assert that, unless it can be shown they have not exercised the professional skill that can be expected of people situated as they were, they should be paid for their work.

The terminology relating to *cost predictions* indicates this is a sensitive area. The most commonly used expression (although used less frequently in contracts made by professional associations) is *cost estimates*. The term *estimate* is itself troublesome. In some contexts, it means a firm proposal intended to be binding, and in others, it is only an educated guess. Probably many clients believe that cost estimates will be roughly accurate, whereas design professionals may look on such estimates as educated guesses.

The response of professional associations to the issue of cost prediction is reviewed in Section 10.2E.

10.2B Two Models of Cost Predictions

The many reported appellate cases demonstrate not only the frequency of misunderstanding but also the difficulties many design professionals have predicting costs. Recognition of this has led the sophisticated client to seek more refined ways to control and predict costs. It is important at the outset to distinguish between what can be called the *traditional method* and these more *refined methods*.

The traditional method usually involves the design professional's using rough rules of thumb based on projected square or cubic footage, modulated to some degree by a skillful design professional's sense of the types of design choices a particular client will make. Cost predictions are likely to be given throughout the development of the design, but they are usually based on these rough formulas and to be refined somewhat as the design proceeds toward completion.

As time for obtaining bids or negotiating with a contractor draws near, the cost predictions should become more accurate. In this model, there is a great deal of suspense when bids are opened or when negotiations become serious. Under this system, a greater likelihood exists of substantial (if not catastrophic) differences between the costs expected by the client and the likely costs of construction as reflected through bids or negotiations. This model gives rise to the bulk of litigation.

The other model—that of more efficient techniques—is likely to be used by sophisticated clients who are aware of the difficulties design professionals have using the model just described. Clients are likely to engage someone who can more accurately predict costs as the design evolves. Sometimes they hire an experienced cost estimator—someone close to the quantity surveyor used in England—as a separate consultant. Sometimes they hire a construction manager (CM) who is supposed to have a better understanding of the labor market, the materials and equipment market, and the construction industry, as well as of the construction process itself. (See Section 7.5.) Using a CM is intended not only to free the designer from major responsibility for cost predictions but also to keep an accurate, ongoing cost prediction. Sometimes the CM agrees to give a guaranteed maximum price (GMP) that may vary as the design evolves. To do so, the CM may obtain firm price commitments from the specialty contractors.

This second model—a fine-tuned model—is designed to avoid the devastating surprises common under the traditional model. To determine whether the design professional bears the risk of losing the fee, differentiating between the two models is vital with the risk being greater in the traditional model.

10.2C A Cost Condition Scenario: *Griswold & Rauma v. Aesculapius*

Before undertaking a legal analysis of a cost prediction dispute, it is useful to read a judicial decision. The *Griswold* case, reproduced here, involved the use of AIA documents. Note how the court carefully recounts the facts that ultimately led to this lawsuit. Pay attention to how the issue of construction cost is raised by the parties and how diligent they were (or weren't) in updating that estimate as the design concept evolved. Note next which provisions from the AIA contract the court feels are important to resolution of the parties' obligations.

To explain the legal position of the client in this, it is important to understand the concept of conditions in the law of contracts. For purposes of this case, a condition is some event that must occur before one contracting party is obligated to the other contracting party. When clients assert they have no obligation to pay the design professional because the low bid exceeded predicted costs, they are asserting a cost condition. They are claiming their obligation to pay was conditioned on the accuracy of a design professional's cost prediction. A cost condition is a gamble by the design professional that the cost prediction will be reasonably accurate. In the *Griswold* case, what factors does the court examine in deciding whether (as the trial court held) the architect has not met a cost condition and so is not entitled to its fee, or whether instead the architect's claim for its fee is not foreclosed?

GRISWOLD AND RAUMA, ARCHITECTS, INC. v. AESCULAPIUS CORP.
Supreme Court of Minnesota, 301 Minn. 121, 221 N.W.2d 556 (1974)
[Ed. note: Footnotes omitted.]

Peterson, Justice.

Plaintiff brought this action to enforce and foreclose a lien for $19,438.65 for architectural services provided defendant. Defendant answered by denying that it owed plaintiff anything and filed a counterclaim to recover $17,436.04 already paid plaintiff, defendant's theory being that plaintiff was entitled to no fee because it breached its contract by grossly underestimating the probable cost of construction of the as-yet-unbuilt building. The trial court found for defendant. We reverse, for reasons requiring an extended recital of the factual setting out of which the litigation arose.

Plaintiff is a corporation engaged in providing architectural services. Defendant is also a corporation, the principal stockholders of which are Drs. James Ponterio, P. J. Adams, and A. A. Spagnolo. Defendant owns a medical building in Shakopee which it rents to the Shakopee Clinic, which in turn is operated by Drs. Ponterio, Adams, and Spagnolo.

In early 1970 defendant decided to expand the Shakopee Clinic to allow for a larger staff of doctors. As a result the members of the corporation and their business manager, Frank Schneider, contacted various architectural firms and selected plaintiff.

At his first meeting with the doctors in February or March 1970 David Griswold, one of plaintiff's senior architects, was shown a rough draft of the proposed addition and given a very general idea of what the doctors wanted. Although the evidence is conflicting, it appears that at this meeting the doctors talked in general terms of a budget of about $300,000 to $325,000.

After a number of subsequent conferences with the doctors and Mr. Schneider, plaintiff prepared and delivered to the doctors on May 8, 1970, a document entitled "Program of Requirements." This document, which outlined and discussed the requirements of the project as then contemplated, contained the following final section:

BUDGET

The design to evolve from this program will indicate a certain construction volume that can be projected to a project cost by the application of unit (per square foot and per cubic foot) costs; and eventually, as the design is developed in detail, by an actual materials take-off. Inevitably the projected cost must be compatible with a budget determined by available funds. It is obvious that adjustment of either the program or the budget may be necessary and that possibility must be recognized.

The project budget established, as currently understood, is $300,000. It has not been stated if this is intended to include non-building costs such as furnishings, equipment and fees—which may be approximately 25% of the total expenditure—as well as construction costs. Advice in this respect will eventually be necessary.

The essential principle to be considered in the design development is as previously stated in the paragraphs of the section titled PROJECT OBJECTIVES.

The construction shall be as economical as possible within the limitations imposed by the desire to build well and provide all of the facility required for a medical service.

In spite of the suggestion at the end of the second paragraph quoted above, neither party at any time thereafter sought to define more particularly what the budget was intended to include. Mr. Schneider testified, however, that he believed the original budget figure included the cost of construction, architects' fees, and the remodeling of the old building. In contrast, Dr. Ponterio testified that it was his belief that the original budget figure did not include a communications system valued at $12,755, architects' fees, or the $20,000 remodeling of the existing building.

* * *

On June 1, 1970, plaintiff provided defendant with a "Cost Analysis" of the [preliminary] plan chosen by defendant. This cost analysis showed the dimensions of the project in square feet as then contemplated, and computed the cost of the project at two different rates per square foot. At the higher rate per square foot, the cost came to $322,140, plus an estimated $20,000 for remodeling the old building, totaling $342,140. At the lesser rate per square foot, the cost came to $284,575, plus $15,000 for the remodeling of the old building, totaling $299,575. The cost analysis memorandum also noted that "the best procedure for projecting costs is by a materials take-off" which was to be done "when sufficient information is available."

It is undisputed that subsequent to the June 1, 1970, cost estimate, no further cost estimates were ever conveyed to defendant. What is disputed is whether in the ensuing months there was any discussion as to whether the project was coming within the budget. According to the testimony of Mr. Schneider, defendant was assured at all times during the preparation of the building plans and in all discussions with Griswold that the construction would come within the budget. Dr. Ponterio also emphasized that Griswold constantly mentioned the budget figure of $300,000 at their meetings. Griswold, however, denied that he had ever assured defendant that the project was coming within the budget.

Although the architectural services began in March and the first billing was May 6, 1970, no written contract was forwarded until June 23, 1970. At that time, a standard American Institute of Architects (AIA) contract was forwarded, calling for payment at plaintiff's standard hourly rate and for reimbursement of expenses and recognizing that a lump sum fee for the construction phase would be negotiated prior to its commencement. The following provisions of the contract have relevance to this case:

[Ed. note: The court quotes the AIA provisions that the architect will submit to the owner a "Statement of Probable Construction Cost" at the schematic design, design development and construction documents phases. It also quotes a provision requiring the owner to notify the architect of any "nonconformance with the Contract Documents."]

CONSTRUCTION COST

3.4 ... Accordingly, the Architect cannot and does not guarantee that bids will not vary from any Statement of Probable Construction Cost or other cost estimate prepared by him.

3.5 When a fixed limit of Construction Cost is established as a condition of this Agreement, it shall include a bidding contingency of ten per cent unless another amount is agreed on in writing.

3.5.1 If the lowest bona fide bid ... exceeds such fixed limit of Construction Cost (including the bidding contingency) established as a condition of this Agreement, the Owner shall

1. give written approval of an increase in such fixed limit,
2. authorize rebidding the Project within a reasonable time, or
3. cooperate in revising the Project scope and quality as required to reduce the Probable Construction Cost.

In the case of (3) the Architect, without additional charge, shall modify the Drawings and Specifications as necessary to bring the Construction Cost within the fixed limit. The providing of this service shall be the limit of the Architect's responsibility in this regard, and having done so, the Architect shall be entitled to his fees in accordance with this Agreement.

PAYMENTS TO THE ARCHITECT

6.3 If the Project is suspended for more than three months or abandoned in whole or in part, the Architect shall be paid his compensation for services performed prior to receipt of written notice from the Owner of such suspension or abandonment, together with Reimbursable Expenses then due and all terminal expenses resulting from such suspension or abandonment.

Between June 1, 1970, and November 25, 1970, when the bids were opened, plaintiff worked actively with defendant both in the design development and construction documents phases of the project, plaintiff continuing to bill defendant on an hourly basis without objection by defendant. Although the project was substantially increased in size and scope during this period, plaintiff did not furnish and defendant did not request any up-to-date cost projections. Significantly important changes in the project made during this period include:

1. Replacement of offices with examining rooms necessitating additional plumbing;
2. More extensive X-ray space;
3. A doubling of the size of the laboratory;
4. Addition of a sophisticated communications system;
5. Addition of 2,100 feet of finished space in the basement (to provide facilities originally projected for the remodeled old building);
6. Enlargement of the structure as follows:

    waiting area 1710' to 1724'

    first floor 5880' to 6650'

    basement 6260' to 6814'

All of the changes were discussed, approved, and understood by defendant. Defendant alleges, however, that it was under the impression that all such changes would be included in the original budget figure.

Bids were opened on November 25, 1970. The low construction bid was Kratochvil Construction Company at $423,380. Deductive alternates agreed to by defendant would bring the total low bid cost, including carpeting, down to $413,037.

Subsequent to the opening of the bids, the doctors called a meeting with Griswold at which the doctors informed Griswold that they could not complete the building according to the cost evidenced by the bids. Thereafter, Griswold met with the doctors and offered suggestions as to how the low bid figure could be reduced. Approximately $42,000 of reductions were projected, so that the final bid as reduced totaled $370,897. This figure included construction, carpeting, and remodeling of the old building and would meet the program of requirements without reducing size in any way. Griswold also pointed out to the doctors that the project cost could be further reduced by eliminating "bays" (series of examination rooms) from the building at a saving of approximately $35,000 per bay. Defendant was willing to accept the $42,000 reduction but was not in favor of eliminating any bays from the proposed project.

From November 25, 1970, when the bids were opened, until October 12, 1971, plaintiff and defendant met and corresponded many times in connection with various possible revisions of the project. During this time defendant never indicated that the project was abandoned and in fact, in January 1971 made a payment of $12,000 on its bill....

The building, in fact, was never constructed. Plaintiff filed its lien on April 30, 1971, and commenced this action to enforce it in October 1971. [Editor: A lien is statutory procedure to force an owner to pay for improvements to its property. See Section 21.7.]

In analyzing the facts of this case we have considered Minnesota decisions as well as decisions from other jurisdictions. From these decisions we have extracted a number of factors which we believe are relevant to a determination of what the effect on compensation of an architect or building contractor should be when the actual or, as here, probable cost of construction exceeds an agreed maximum cost figure.

One very significant factor is whether the agreed maximum cost figure was expressed in terms of an approximation or estimate rather than a guarantee. Where the figure was merely an approximation or estimate and not a guarantee, courts generally permit the architect to recover compensation provided the actual or probable cost of construction does not substantially exceed the agreed figure.

Another significant factor is whether the excess of the actual or probable cost resulted from orders by the client to change the plans. Where the client ordered changes which increased the actual or probable construction costs, courts are more likely to permit the architect to recover compensation notwithstanding a cost overrun.

A third factor is whether the client has waived his right to object either by accepting the architect's performance without objecting or by failing to make a timely objection to that performance.

A fourth factor, applicable in a case such as this where the planned building was never constructed, is whether the architect, after receiving excessive bids, suggested reasonable revisions in plans which would reduce the probable cost. Courts have held that if the architect made such suggestions and the proposed revisions would not materially alter the agreed general design, then the architect is entitled to his fee, again provided that the then probable cost does not substantially exceed the agreed maximum cost figure.

Considering this case in light of these factors, we conclude that the trial court erred in denying plaintiff's motion for amended findings of fact, conclusions of law, and order for judgment.

First, it does not appear that plaintiff guaranteed the maximum cost figure. ... [T]he intention of the parties, as evidenced by the record, especially by contract provision 3.4, quoted earlier, more reasonably supports an established cost estimate than a guaranteed cost figure. Secondly, the probable cost of the project in our view did not substantially exceed the cost figure. At trial architect Griswold testified and Mr. Schneider agreed, that the project could be completed for approximately $360,000 to $370,000 without reducing the square footage of the project. A probable construction cost of $370,000 exceeds the agreed cost estimate maximum found by the trial court, $325,000, by only 13 percent. It seems difficult to classify such a degree of cost excess as substantial. A review of cases cited in Annotation, 20 A.L.R.3d 778, 804 to 805, suggests that most courts would not consider such a degree of cost excess substantial.

Thirdly, we think it relevant that defendant approved substantial changes beyond the original plans....

Finally, we think it relevant that plaintiff showed defendant how they could reduce the cost of the lowest bid below $370,000. ... Provision 3.5.1 of the contract required the parties to revise the project scope and quality if necessary to reduce the probable construction cost. While it might be against public policy to allow an architect, under such a provision, to reduce substantially the area of a proposed project, it seems that, barring a specifically guaranteed area, a reasonable reduction in the size of the project should be allowed when necessary to meet the construction cost. Reversed and remanded.

SHERAN, C. J., took no part in the consideration or decision of this case.

The *Griswold* case illuminates the human aspect of the client–design professional relationship and the law's effort to resolve rights when that relationship breaks down. It presents a "real life" scenario, as does the *Watson* case reproduced in Section 10.3B. Some issues to consider:

1. After the first cost estimate given in June 1, 1970, no further cost estimates were ever conveyed to the client. Should the architect have raised the topic on a periodic basis?
2. The architect used two different rates-per-square-foot methods to estimate construction costs. Given the results in this case—that the low bid was so much higher—are these methods deficient?
3. As often happens, a client wants changes made to the original design concept. Again, there is no indication the architect pointed out the cost consequences of these suggestions as they were being made.
4. In deciding whether the architect had provided a cost *guarantee* or instead an *estimate*, the court first looked to the contract language. This shows the importance of understanding the content of a contract—preferably before it is signed. (AIA and EJCDC contract disclaimers as to cost predictions are discussed in Section 10.2E.)
5. Although mostly edited out of this reproduced opinion, the court looked not only to Minnesota law but also to judicial decisions from other states to ascertain legal principles and how that law should be applied to these types of disputes. *Griswold* illustrates the common law process in operation.

*

Another case study involving one of the most famous of American architects—Frank Lloyd Wright—shows the human side of the architect–client relationship that is not always depicted in court cases. This case study involved Wright and one of his most famous projects, the Johnson Building in Racine, Wisconsin. While this project did *not* culminate in a written contract, a lawsuit, and a reported appellate decision, Professor Stewart Macaulay gained access to records and obtained testimony from witnesses so as to able to recount the story behind the project. His case study is published in the Wisconsin Law Review,[1] and may be accessed at the website supporting this book.

10.2D Cost Predictions: Legal Issues

Client's Damages: Breach of Condition. As noted earlier, when clients assert they have no obligation to pay the design professional's fee because the low bid exceeded predicted costs, they are asserting a cost condition. They are claiming their obligation to pay was conditioned on the accuracy of a design professional's cost prediction. Normally, as in the *Griswold* case, the architect sues to recover her fee, and the owners assert failure of the cost condition as a defense.

Client's Damages: Breach of Promise. Clients may instead view a cost prediction as a promise. Breach of a promise to design a project within budget would subject the designer to a claim for damages for breach of contract. The client's damages will depend on its response to the higher-than-anticipated construction cost. A client confronted with a wrongful cost prediction has two options: to abandon the project or to go forward with it notwithstanding the higher construction cost.

Abandoning the project may cause client losses, such as wasted expenditures in reliance on the design professional's promise to bring the project in within a designated cost. The client may also seek to recover interim fee payments to the design professional based on restitution.

Suppose the client goes ahead with the project after a redesign. The redesign very likely caused delay, and delay damages are likely recoverable if proved with reasonable certainty and were reasonably foreseeable at the time the agreement was made. AIA Doc. B101-2007, § 6.7, states that redesign is the limit of the architect's responsibility. This should not relieve the architect from responsibility for delay damages caused by negligence.

When the project is built, the architect should not be held for the excess of actual costs over predicted costs. The client benefits by ownership of property presumably of a value equal to what the client has paid for the improvement. Damages should not be awarded unless the client can prove the economic utility of the project was reduced in some ascertainable manner because of the excessive costs.

Parol Evidence Rule. The issue that has arisen most frequently in cost cases relates to the parol evidence rule (discussed in Section 3.5A). This rule determines whether the terms of a written contract may be altered by evidence of prior or contemporaneous (i.e., made at the time the contract is signed) oral (non-written) agreements or statements. If the parol evidence rule is enforced, such evidence is barred by the trial judge.

In the context of a cost condition, the issue is whether the client will be permitted to testify that it had made an earlier oral agreement or had an understanding that it could abandon the project and not pay for design services if the low construction bid substantially exceeded the cost prediction of the design professional. Permitting such testimony is based on the conclusion that the written contract is not the final and complete repository of the entire agreement between design professional and client.

Standardized contracts prepared by the AIA and EJCDC include an **integration clause** stating that the agreement is the entire agreement of the parties. Such a clause seeks to prevent the owner from using parol evidence to establish oral agreements of a cost condition.[2]

Cost Prediction Tolerance. As noted by the *Griswold* court, a design professional is allowed some tolerance in determining whether the cost prediction was accurate. It cited testimony that the project could have been built at 13 percent over the cost estimate, then added: "It seems difficult to classify such a degree of cost excess as substantial."

The degree of tolerance permitted may also depend on the language used to create the cost limitation. The more specific the amount, the more likely a small tolerance figure will be applied. The court may view the cost estimate as a "hard" number and subject to a small tolerance.

Even absent a written contract provision, a cost condition can be created by implication. Certainly, design professionals do not operate in the dark. If they know what funds are available and are aware of the remoteness of obtaining additional funds, any cost specified may be hard.[3]

Cost Condition Waived. A cost condition generally is created for the benefit of the client. If it so chooses, the client can dispense with this protection. Courts that conclude the client has dispensed with the cost condition usually say that the condition has been waived. When this occurs, the condition is excused, and the design professional is entitled to be paid even if the cost condition has not been fulfilled.

The most common basis for excusing the condition is the client's making excessive changes during the design phase. Also, the client proceeding with the project despite the awareness of a marked disparity between cost estimates and the construction contract price can excuse the condition. By proceeding, the client may be indicating it is willing to dispense with the originally created cost conditions. However, proceeding with the project should not automatically excuse the condition.

Finally, the owner may assume responsibility for cost estimating. In *Getzschman v. Miller Chemical Co., Inc.*,[4] the Nebraska Supreme Court ruled that, when the owner assumes this obligation, the architect will be entitled to her fee even though the lowest bid is substantially more than the owner could afford.

Cost Predictions on Cost-Plus Contracts. The issue of cost predictions on a cost-plus contract deserves separate mention. On these contracts, the owner is least likely to be protected against cost escalation, because the prime contractor did not enter into a fixed-price contract.

In *Williams Engineering, Inc. v. Goodyear, Inc.*[5] Williams, the engineer, had been retained to design a recreational water slide. To meet the client's desire to be open for business by the start of summer, Williams suggested that the client proceed on a "fast track" basis with the contractor hired on a cost-plus basis. The client agreed.

During design and construction, the engineer submitted written invoices for his fee based on a percentage of the estimated cost of $409,000. Three days before the slide opened, the engineer's bill still showed the cost as $409,000. But twenty days after the water slide was opened for business, the client was billed for construction costs of almost $888,000. The project was still only 82 percent complete. Even worse, the engineer submitted a new bill based on the projected cost of almost $1,000,000. The water slide turned out to be a financial failure.

After an eight-day trial, the jury decided the engineer had breached the contract, and the client should have damages of $125,000 plus expert witness fees of $3,000.

Declaring that "[i]t is generally recognized that an architect or engineer has an affirmative duty to give a definite idea of the reasonable cost of a project,"[6] the Louisiana Supreme Court affirmed a judgment in favor of the client—not based on the initial cost estimate but on the failure to update that estimate as the cost-plus construction proceeded:

> The vigorous contention of Williams … is that Williams did not guarantee the cost of the project and cannot be cast in damages for the overrun on costs. However, there was expert testimony that Williams breached the contract, not by giving an inaccurate initial estimate, but by failing to employ a professional estimator, failing to look at other water slides, failing to advise the owners about other contractual possibilities, and failing to provide revised cost estimates. The owners, while admitting a somewhat blind faith in Williams, argue that they could have modified the

design of the project or given it up entirely if they had been fully apprised of what their exposure might be.

Aside from his failure to make complete and reliable disclosure to these owners of their construction options, it is clear that Williams breached the contract by failing to re-estimate the cost during the course of construction. The breach caused the resulting damages to the owners. The owners were completely unaware that the cost had escalated and received this rude advice almost a month after the facility opened. Contrary to Williams' contention that the written contract was abrogated by the cost-plus, fast track contract, the owners were more in need of reliable estimates of construction costs under that contract than they would have been if bids had been received. Had bids been received, the architect's responsibility for estimating costs would have ended when he completed his design and bids were submitted. In this instance, the design and construction phases were pursued simultaneously and the engineer had an on-going responsibility to keep the owners advised of the costs.

Any doubt about the meaning of the contract must be considered in light of the fact that Williams, the engineering professional who drafted the instrument, was dealing with lay persons.[7]

10.2E Standard Contracts and Disclaimers: AIA and EJCDC

Professional associations have dealt with cost problems by including language in their standard contracts to protect design professionals from losing fees when cost problems develop. The protective language has ranged from the brief statement that cost estimates cannot be guaranteed to the elaborate contract language of AIA Document B101-2007.

This standard form owner/architect agreement seeks to shift responsibility for the budget squarely onto the owner. The owner's initial information must include its "budget for the Cost of the Work."[8] Under Article 6 of B101-2007, the owner must periodically update the budget with input from the architect. If the architect's estimate exceeds the owner's budget, the architect and owner will work together to adjust the project's size or quality. At the same time, the AIA emphasizes that the architect's inputs are estimates—not guarantees. Section 6.2 of the AIA document warns that "the Architect cannot and does not warrant or represent" that her cost evaluations will not vary from bids received. (This is intended to preclude a client's claim that an inaccurate estimate is a misrepresentation by the design professional.)

If the lowest bona fide bid or negotiated price exceeds the budget, the owner must do one of the following:

1. approve an increase in the fixed limit
2. authorize rebidding or renegotiating
3. terminate and pay for work performed and termination expenses if the project is abandoned
4. cooperate in project revision to reduce cost
5. implement any other mutually acceptable alternative[9]

If option 4 is chosen, under § 6.7 the architect must bear the cost of redesign (specified as the limit of the architect's responsibility), and she is to be paid for any other service rendered.

The Engineers Joint Contracts Document Committee (EJCDC) in document E-500 (2008), ¶ 5.01, refers to "opinions of probable construction costs." In optional Exhibit F, ¶ F5.02 of E-500 (2008), the owner and engineer agree to a "construction cost limit." If the bids exceed the limit, the owner will either increase the limit, rebid, or renegotiate the project or work with the engineer to revise the project's scope, extent, or character consistent with sound engineering practices.

10.2F Advice to Design Professionals

Many design professional–client relationships deteriorate because of excessive costs. Design professionals should try to make the cost prediction process accurate. The chief method chosen by the professional associations has been to use the contract to protect their members from losing their fees where their cost predictions are inaccurate. The professional associations have included provisions in their contracts that are supposed to ensure both that fees will not be lost when cost predictions are inaccurate and that fees will be lost only where the cost predictions are made negligently.

Protective language should be explained to the client. Design professionals who give a reasonable explanation to a client are not likely to incur difficulty over this problem. They should inform the client how cost predictions are made and how difficult it is to achieve accuracy when balancing uncontrollable factors. They should state that best efforts will be made but that for various specific reasons the low bids from the contractors may be substantially in excess of the statement of probable construction costs. The suggestion should be made that under such circumstances the design professional and the client should join to work toward a design solution that will satisfy the client's needs. In helping the client be realistic about desires and funds, the design professional should ask the client to be as specific as possible as to expectations about the project.

The design professional who takes these steps may lose some clients. It may be better to lose them at the outset than to work for many hours and then either not be paid or be forced to go to court to try to collect. Without honest discussion with the client at the outset, the design professional takes these risks.

Design professionals should also consider greater flexibility in fee arrangements. If the stated percentage of construction costs is used, it is advisable to reduce or eliminate any fee based on construction costs that exceed cost predictions.

During performance, the design professional should state what effect any changes made by the client will have on any existing cost predictions. It is hoped that not every change will require an increase in the cost predictions. If the client approves any design work, the request for approval should state whether any cost prediction has been revised.

10.3 Site Services: Observing the Contractor's Work

10.3A Introduction

Section 10.3 examines the design professional's role on the job site during the project's construction phase. The client may believe that the design professional has been paid to ensure that the client receives all it is entitled to receive under the construction contract. The client may expect the design professional to accomplish this by watching over the job, seeing that the contractor performs properly, and being responsible if the contractor does not so perform. As noted in Section 10.3B, a design professional may take a different view of her function. The purpose of Section 10.3 is to explore this problem by examining what the law has done when faced with resolving the often different expectations of client and design professional.

Site services are but a part of the total professional services performed by the architect or engineer. The design professional's performance is measured by either the professional standard of care or the contract obligations. Relevant to this inquiry are any contractual disclaimers which seek to make the contractor alone responsible for its defective work and to relieve the design professional from responsibility for project defects.

10.3B From Supervision to Observation: *Watson, Watson, Rutland/Architects v. Montgomery County Board of Education*

This section addresses the design professional's site visits during the construction phase. The details of site visits are often handled in the design agreement. It may require the design professional to take an *active role* in the process, performing such tasks as solving execution problems, directing how work is to be done, searching carefully to determine failure to follow the design, and seeing that safety regulations are followed. Alternatively, it can limit the design professional to a *passive role* under which she "walks the site" periodically to check to see how things are going, perhaps measures progress for payment purposes, and reports to the client about work in progress.

If the design professional takes an active role, she will be diluting the authority and responsibility of the contractor and exposing herself and her client to liability. Yet some clients may prefer this route because they see it as a means of ensuring a more successful project.

The professional associations representing design professionals have favored placing the authority and responsibility for executing the design solely on the contractor and limiting the architect or engineer to a more passive role. To implement this separation, the standard documents published by professional associations include language that clearly gives sole responsibility for executing the design to the contractor. Such language disclaims any responsibility on the part of the design professionals for how the work is to be done or for the contractor's failure to comply with the contract documents. The effect of such **disclaimers** (discussed later) can generate concern both by clients ("What am I getting for my money?") and by courts ("Is the design professional taking any responsibility?"). The AIA's approach to disclaimers is explored in the *Watson* case (reproduced in this section).

The following issues can arise as to the specific obligation to visit the site:

1. When must the design professional be on the site (continuous presence versus periodic visits)?
2. What is the intensity level of checking for compliance (intense inspection to ferret out deficiencies versus casual observation)?
3. If a deficiency or noncompliance is discovered, what must the design professional do (direct work be corrected versus report to owner)?

The following issues can arise as to the contractual disclaimers:

1. Do they totally exculpate the design professional from any responsibility for failure by the contractor to comply with the design documents? Will they entitle the design professional to a "summary judgment"—a ruling of no-liability without the judge having to hear witnesses or to submit the dispute to a jury?
2. Do they exculpate the design professional even if she discovers the noncompliance by the contractor?
3. Do they exculpate the design professional if she has failed to comply with her contractual obligation? (Is she responsible for what she would have detected had she complied with her contractual commitment?)

The exact nature of the design professional's site visits depends, of course, on the contract—both its express and implied terms. The contract may specify continuous versus noncontinuous presence, the frequency and timing of visits, the intensity of inspection, and action to be taken on discovery of deficiencies.

In addition to contract requirements, the design professional's site inspections must comply with the professional standard of care. In deciding whether the duration, frequency, and timing of site visits was adequate, courts examine such factors as the size of the project, whether the owner had its own technical staff to conduct inspections, the type of construction contract (cost contracts require more monitoring), and whether inspections took place during crucial events in the construction process.

The *Watson* case (reproduced here) examines the responsibility of the design professional for failure of the contractor to comply with the contract documents, with special reference to the effect of disclaimers.

WATSON, WATSON, RUTLAND/ARCHITECTS, INC. v. MONTGOMERY COUNTY BOARD OF EDUCATION, et al.
Supreme Court of Alabama, 559 So.2d 168 (Ala.1990)

MADDOX, Justice

[Ed. note: A school board decided to build a school. It entered into separate contracts with the prime contractor, the manufacturer of the roof membrane and the architect (Watson). The court states that the design agreement was modeled after language found in AIA contracts. The roof was installed by a roofing subcontractor hired by the prime contractor.

The roof on the completed school leaked. The school board sued the prime contractor, roof manufacturer, and Watson. These defendants brought claims against each other.

The school board settled with the prime contractor and roof membrane manufacturer. At a trial against the architect, the school board's breach of contract claim was submitted to the jury. (The trial court had ruled that the school board's negligence claim was barred by the statute of limitations; see Section 16.9D.) The jury returned a verdict of $24,813 against the Architect. The court entered a judgment based on that verdict. The architect appealed.]

At the center of this dispute is the architectural agreement between the Architect and the School Board; it included the following language:

> ARTICLE 8. Administration of the Construction Contract. The Architect will endeavor to require the Contractor to strictly adhere to the plans and specifications, to guard the Owner against defects and deficiencies in the work of Contractors, and shall promptly notify the Owner in writing of any significant departure in the quality of material or workmanship from the requirements of the plans and specifications, but he does not guarantee the performance of the contracts.

* * *

The Architect shall make periodic visits to the site and as hereinafter defined to familiarize himself generally with the progress and quality of the Work and to determine in general if the Work is proceeding in accordance with the Contract Documents. On the basis of his on-site observations as an Architect, he shall endeavor to guard the Owner against defect and deficiencies in the work of the Contractor. The Architect shall not be required to make continuous on-site inspections to check the quality of the Work. Architect shall not be responsible for construction means, methods, techniques, sequences or procedures, or for safety precautions and programs in connection with the Work, unless spelled out in the Contract Documents, and he shall not be liable for results of Contractor's failure to carry out the work in accordance with the Contract Documents.

* * *

The Architect shall not be responsible for the acts or omissions of the Contractor, or any Subcontractors, or any of the Contractor's or Subcontractor's agents or employees, or any other persons performing any of the Work.

* * *

The Architect argues that the exculpatory language in Article 8 of the architectural agreement absolves the Architect from liability to the School Board because all roof leaks involved were attributable to the faulty workmanship of the contractor. The Architect points out that Article 8 provides for two types of inspection services by the Architect; that is, the owner could elect to receive only general site inspection by the Architect, or the owner could elect to pay an additional fee for continuous on-site inspections (known as the "clerk of the works" alternative); the School Board elected not to pay for the second options.

* * *

A case analogous to this one is *Moundsview Indep. School Dist. No. 621 v. Buetow & Associates, Inc.,* 253 N.W.2d 836 (Minn.1977), where the contract language regarding inspection duties was virtually identical to that used here; the trial court entered a summary judgment for the architect, and the Minnesota Supreme Court affirmed. The supreme court found it significant that the school district had elected not to obtain continuous supervisory services from the architect through the "clerk of the works" clause. The court held that the contractual provisions "absolved [the architect] from any liability, as a matter of law, for a contractor's failure to fasten the roof to the building with washers and nuts." 253 N.W.2d at 839. [Ed. note: The term "as a matter of law" means the dispute is decided by the judge, without involvement of a jury.]

Other courts have also held that similar language absolved the architect from liability as a matter of law. [Citing cases.]

The Architect contends that imposition of liability on it here would be nothing short of a holding that the Architect was the guarantor of the contractor's work.

The School Board responds to the Architect's argument by saying that it has never contended that the Architect should be a guarantor of the contractor's work, but the School Board strongly contends that where a contract is for work and services, there is an implied duty to perform with an ordinary and reasonable degree of skill and care, citing, *C. P. Robbins & Associates v. Stevens,* 53 Ala.App. 432, 301 So.2d 196 (1974). The School Board also argues that its claim was covered under the terms of the agreement because its claim was limited to those deviations from the plans and specifications that should have been obvious to one skilled in the construction industry. According to the School Board, if the Architect's construction of the contract language is accepted, the Architect would not be accountable to anyone for its failure to make a reasonably adequate inspection so long as it made an "inspection," no matter how cursory, on a weekly basis. The School Board points out that the contract also included the following language:

> On the basis of his on-site observations as an Architect, he shall endeavor to guard the Owner against defects and deficiencies in the work of the Contractor.

> The Architect shall have authority to reject Work which does not conform to the Contract Documents.

> The administration of the contract by the Architect is not normally to be construed as meaning the furnishing of continuous inspection which may be obtained by the employment of a Clerk of the Works. However, the administration shall be consistent with the size and nature of the work and must include, at least, one inspection each week, a final inspection, and an inspection at the end of the one year guarantee period shall be required on all projects.

It is apparent that our focus must be drawn to the exculpatory language contained in Article 8 of the agreement. Most courts in other jurisdictions that have considered similar exculpatory clauses have recognized that while such clauses do not absolve an architect from all liability, an architect is under no duty to perform continuous inspections that could be obtained by the employment of a "clerk of the works."

The critical question in most of the cases we have reviewed from other jurisdictions seems to focus on the extent of the obligation owed when an architect agrees to perform the type of inspection that the architect agreed to perform in this case. Some of the courts hold, as a matter of law, that the agreement does not cover particular factual situations, and at least one court makes a distinction based on whether a failure of a contractor to follow plans and specifications is known *to the architect during the course of the construction.*

A fair reading of Article 8 obviously operates to impose certain inspection responsibilities on the architect to view the ongoing construction progress. The frequency and number of such inspections is made somewhat specific by the contractual requirement that these visits to the site be conducted at least once each week. The difficulty arises in construing the particular terminology used, such as "*endeavor* to require," "familiarize himself generally with the progress and quality of the work," and "*endeavor* to guard the owner against defect and deficiencies." When coupled with the contract language stating that "[t]he administration of the contract by the Architect is not normally to be construed as meaning the furnishing of continuous inspection which may be obtained by the employment of a Clerk of the Works," these phrases make it obvious that the Architect's duty to inspect is somewhat limited, but we cannot agree with the argument on appeal that there could never be an imposition of liability under an agreement similar to Article 8 no matter how serious the deviation of the contractor from the plans and specifications.

* * *

Although the contract here clearly made the Architect's inspection duty a limited one, we cannot hold that it absolved the Architect from all possible liability or relieved it of the duty to perform reasonably the limited contractual duties that it agreed to undertake. Otherwise, as the School Board argues, the owner would have bought nothing from the Architect. While the agreement may have absolved the Architect of liability for any negligent acts or omissions of the contractor and subcontractors, it did not absolve the Architect of liability arising out of its own failure to inspect reasonably. Nor could the Architect close its eyes on the construction site and not engage in any inspection procedure, then disclaim liability for construction defects that even the most perfunctory monitoring would have prevented, or fail to advise the owner of a known failure of the contractor to follow the plans and specifications.

The issue here is whether the Architect can be held liable for its failure to inspect and to discover the acts or omissions of the contractor or subcontractors in failing to follow the plans and specifications. Under the terms of the contract, the Architect had at least a duty to perform reasonable inspections, and the School Board had a right to a remedy for any failure to perform that duty. There is no question that the Architect performed inspections; the thrust of the School Board's argument is that these inspections were not as thorough as the Architect agreed they would be.

As we have already stated, the only evidence concerning the Architect's duty under the agreement was from an architect who stated that under his interpretation of the contract the Architect was not under a duty to inspect for the specific defect that was alleged in this case.

As was pointed out in *Moundsview*, an architect is a "professional," and we are of the opinion that expert testimony was needed in order to show whether the defects here should have been obvious to the Architect during the weekly inspections. Just as in cases dealing with an alleged breach of a duty by an attorney, a doctor, or any other professional, unless the breach is so obvious that any reasonable person would see it, then expert testimony is necessary in order to establish the alleged breach.

The nature and extent of the duty of an architect who agrees to conduct the inspection called for by the subject agreement are not matters of common knowledge. The rule of law in Alabama concerning the use of expert testimony is as follows: "[E]xpert opinion testimony should not be admitted unless it is clear that the jurors themselves are not capable, from want of experience or knowledge of the subject, to draw correct conclusions from the facts. The opinion of the expert is inadmissible on matters of common knowledge." *Wal-Mart Stores, Inc. v. White*, 476 So.2d 614, 617 (Ala.1985) (quoting—C. Gamble, *McElroy's Alabama Evidence* § 127.01(5) (3d ed.1977)).

The breach alleged in this case involved architectural matters that would not be within the common knowledge of the jurors, yet the School Board presented no expert testimony regarding the Architect's inspections and any deficiencies in those inspections. ...

In a case startlingly similar to this one, a New York court, while holding that an architect could be found liable under the AIA contract for a failure to notify the board of education about leaks in a school's roof, did so only after finding that there was evidence that the architect had knowledge during the course of the construction that the contractor was not following the plans and specifications. In *Board of Educ. of Hudson City School Dist. v. Sargent, Webster, Crenshaw & Folley,* 146 A.D.2d 190, 539 N.Y.S.2d 814, 817–18 (1989), the court stated:

> While this very clause in the standard AIA architect/owner contract [the clause absolving the architect from responsibility for the contractor's failures] has been given exculpatory effect in this State and other jurisdictions [citations omitted] *none of the cases involved defects known by the architect during the course of construction*, which he failed to apprise the owner of under the contractual duty to keep the [School District] informed of the progress of the work. We decline to extend the application of the clause in question to an instance such as this, *where the trier of facts could find that the architect was aware of the defect and failed to notify the owner of it.* (Emphasis added.)

The key distinction between that case and this one is that in *Sargent, Webster* there was evidence that the architect knew of the defects during construction and failed to notify the owner. Clearly, the architect there would have breached his contractual duty if he failed to notify the owner of a known defect. It is undisputed in this case that during construction the Architect did not know of the defects with the weep holes. In fact, the School Board does not suggest that the Architect did know, only that under the provisions of the contract the trier of fact could have found that the Architect should have conducted an inspection that would have discovered the defect.

We conclude that, although the Architect had a duty under the contract to inspect, exhaustive, continuous on-site inspections were not required. We also hold, however, that an architect has a legal duty, under such an agreement, to notify the owner of a *known* defect. Furthermore, an architect cannot close his eyes on the construction site and refuse to engage in any inspection procedure whatsoever and then disclaim liability for construction defects that even the most perfunctory monitoring would have been prevented. In this case, we hold, as a matter of law, that the School Board failed to prove that the Architect breached the agreement.

In making this judgment, we would point out the value, and in some cases, the necessity, for expert testimony to aid the court and the trier of fact in resolving conflicts that might arise.

* * *

For the above-stated reasons, that portion of the judgment based on the jury's verdict against the Architect on the School Board's contract claim is reversed, and a judgment is rendered for the Architect on that claim. That portion of the judgment based on the directed verdict in the Architect's favor on the School Board's negligence claim is affirmed.

HORNSBY, C. J., and JONES, SHORES, HOUSTON and STEAGALL, J. J., concur.

Some observations on the *Watson* decision and some discussion of other important decisions may be useful in understanding the legal aspects of the general obligation of the design professional to perform site services.

Summary. After concluding that the exculpatory language did not completely shield the architect, the court faced the question of whether the inspections were as thorough as the architect agreed they would be. The court was not able to determine whether the architect had conducted a proper inspection in the absence of expert testimony showing whether the defects should have been obvious to the architect during his weekly inspections. Because the school board had the burden of providing this expert testimony but did not, it failed to prove that the architect had breached the contract.

First, note that the AIA documents do not specifically require the architect to make inspections until the contractor claims that the work has been substantially completed. Prior to substantial completion, the architect's site visits are observations—not inspections.[10] Probably unaware of the AIA's distinction, the *Watson* decision spoke of the duty to inspect.

The *Watson* case illustrates the interrelationship that often exists between a breach of contract claim involving a design professional and the professional standard applicable to a claim of professional negligence (malpractice). (The professional standard is discussed in Section 11.2) An architect or professional engineer owes her client a duty—not only to act in conformity with the contract requirements, but also in a non-negligent manner. For this reason, an owner who has hired a design professional to inspect the project during the construction phase—but who ends up with a defective building—can sue the design professional for breach of contract and for tort. (In *Watson*, the school board's negligence claim was barred by the statute of limitations.)

Whether the claim is approached as one based on breach of contract or professional malpractice, evidence will have to be introduced that bears on whether the design professional did what she promised or what the law requires. Depending on the nature of the contractual obligation, the evidence necessary to establish breach of contract or malpractice may be identical or it may be very different.

In comparing the proof necessary to establish a breach of contract and that necessary to establish a tort, an initial distinction must be drawn between two types of contract terms. Where the standard of performance dictated by the contract is unambiguous and easily discernable, expert testimony is not necessary to establish its breach. The reason is that the contract creates its own standard of expected performance, so failure to meet that standard (absent excuse) results in a breach of contract. A party who breaches an unambiguous contract term cannot raise as a defense that it acted in a non-negligent manner. As an example, the modified AIA contract addressed in *Watson* required the architect to provide "at least one inspection each week, a final inspection, and an inspection at the end of the one year guarantee period …" An architect who fails to inspect at least weekly would be in breach of contract—even if the professional standard did not require that many inspections.

The second type of contract term establishes a standard of performance which is not easily discernable. For example, the contract addressed in *Watson* requires not only weekly inspections but also that "the administration [of the contract by the Architect] shall be consistent with the size and nature of the work" and that the architect through her inspections shall "endeavor to guard the Owner against defects and deficiencies in the work of the Contractor." These contract requirements create no objective, readily discernible, and unambiguous standards of conduct. Must the court turn to the professional standard and ask what inspections would the reasonable architect make in an effort to protect the owner from defective work by the contractor?

Here, the *Watson* court created a distinction by way of its citation to *Board of Education of the Hudson School District v. Sargent, Webster, Crenshaw & Folley*.[11] The *Sargent* case also involved an architect's duty of inspection under an AIA contract. A newly installed roof failed, and the school district sued the inspecting architect. Evidence revealed that the architect's representative *saw the roof being installed improperly*. The appellate court entered judgment for the school board, ruling that the architect breached the contract when she did not inform the owner of a defect known to the architect from her inspections.

Citing *Sargent,* the *Watson* court adopted the rule that an architect with a contractual duty of inspection breaches that contract when it fails to notify the owner of a *known* defect. The fact that the contractual duty itself does not contain a readily discernible, objective standard of conduct does not mean that in *all* cases the owner must present expert testimony to prove a breach of contract. However, excluding this narrow exception of a *known* defect in the contractor's work, the owner cannot prove the architect's breach of her duty of inspection absent expert testimony. (Even when the architect has performed the required inspections, expert testimony may be necessary to establish whether they were carried out properly.[12])

In imposing an *objective* standard on the architect's duty to inspect—the negligence standard of care—the *Watson* court refused to allow the AIA disclaimer language to exculpate the architect "no matter how serious the [contractor's] deviation of the contractor from the plans and specifications." The court was frustrated by the AIA's attempt to provide a loose standard as exemplified by phrases such as "endeavor to require," "familiarize himself generally with the progress and quality of the work," and "endeavor to guard the owner." Frustration with the imprecision of these terms undoubtedly motivated the court to employ an objective standard.

In its 2007 version of the design agreement, the AIA attempted to create a *subjective* standard. Under a subjective standard, the architect would not be liable unless she actually knew of the construction defects and did nothing. (This is what happened in the *Sargent* case.)

The AIA sought to achieve this result by making two changes to B101-2007, § 3.6.2.1. First, it eliminated the "endeavor to guard" language. Second, it imposed upon the architect a duty to report to the owner only "known deviations" from the design requirements and "defects and deficiencies observed in the Work." While these changes in language may tip the balance in favor of the architect in close cases, it is difficult to imagine a court allowing a design professional to entirely exculpate herself from liability in the face of evidence of obvious construction defects which implicate safety or structural integrity.

The cases discussed here demonstrate the inherent difficulties of standard form contracts. It is difficult to quarrel with the basic allocation expressed in AIA documents of responsibility for executing the design. That the architect checks the work periodically and issues certificates of payment should not disturb the division of authority and responsibility under which the architect designs and the contractor builds. However, general language often cannot take into account unforeseen circumstances. If the architect has not breached her obligation to observe (or inspect) and has not learned of defects, the exculpatory language should be sufficient to grant her a summary judgment. But when she has learned of defects and not reported them or when she has not complied with her contractual obligations, the exculpatory language should not provide a shield. Although it must be recognized that contracts drafted by the professional associations will inevitably be designed to favor members of those associations, such standard contracts cannot anticipate every problem.

10.3C Site Inspections: Certification of Contractor's Progress Payments

As explained in Section 17.5, the design professional inspects the contractor's work before certifying the contractor's entitlement to progress (interim) payment. If the design professional overcertified payments—so that the contractor receives during performance more than it is entitled to—the remaining funds may be insufficient to complete the project if the prime contractor defaults. The design professional's overcertification will subject it to liability to the owner and third parties—in particular a performance bond surety (who guarantees the contractor's obligation to build the project).

10.4 Review of Contractor Submittals

The contractor is usually required to submit information generally known as *submittals* to the design professional. The best-known submittals are shop drawings. (The term comes from the fabrication process: Shop drawings instruct the fabrication "shop" how to fabricate component parts.) AIA Document A201-2007, § 3.12.1, defines shop drawings as data especially prepared for the work that is "to illustrate some portion of the Work." Submittals also include product data, defined in § 3.12.2 as including "illustrations, standard schedules, performance charts, instructions, brochures, diagrams and other information" that illustrate the materials or equipment the contractor proposes to use. Finally, § 3.12.3 defines samples as physical examples that illustrate "materials, equipment or workmanship and establish standards by which the Work will be judged."

Often submittals are prepared by the subcontractor and reviewed by the contractor and the design professional or her consultant. See Figure 10.1. Although submittals should prevent defective work from being performed, they also generate additional expense.

The owner, design professional, her consultant, the prime contractor, and specialty subcontractors all view the submittal system from their own perspective.

An owner sees the submittal system as a double- and even triple-checking system that should reduce the likelihood of defects or accidents.

The design professional has several concerns about the submittal review process. Submittal review exposes the design professional to claims if people are injured or other parts of the project are damaged often the result of deficient construction methods (indicated in submittals)

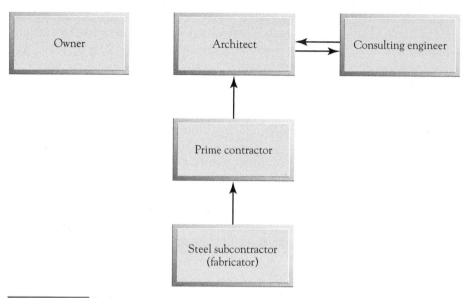

FIGURE 10.1 Path of steel fabrication submittal.

that are not within the design professional's expertise or responsibility. Project failure may also be traced to submittal review.[13]

The submittal process may expose the design professional to liability for another form of injury: claims by the client and the contractor that improper delay in reviewing submittals delayed the project or disrupted the contractor's planned sequence of performance. See Section 20.13A.

As explained in Section 10.5B, the design professional must rely heavily on her consultants who have expertise in the subject matter addressed in the submittal. Since the consultant is a subcontractor of the design professional, any fault of the consultant will leave the design professional accountable to the owner. The consultant, in turn, may have to indemnify the design professional against whom a claim has been made.

The consultant who reviews submittals, such as a structural engineer, sees the submittal process as one that exposes him to the risk of license revocation or other disciplinary action. The *Duncan* case reproduced in Section 8.3C dealt with disciplinary action against a licensed engineer based on shop drawing review, among other things.

The prime contractor often relies on subcontractors to prepare submittals. Just as the prime design professional will be held responsible for the acts of her consultant, the contractor will be held responsible for the acts of the subcontractors.

Subcontractors and prime contractors see submittals as forcing them to fill in gaps in the design created by the design professional. In effect, part of the design has been delegated to the contractors. The question of allocating risk when this is done has troubled the construction industry. Contractors are concerned the submittal process exposes them to liability. In addition, the contractor may believe that approval of its submittals transfers certain risks and responsibilities to the owner, a position often contrary to contract language.

The AIA's response to this risk allocation question is found in A201-2007, § 3.12.10. It allows the owner (and the architect) to delegate the responsibility for the design quality of specified parts of the project to the contractor. The latter must use a licensed design professional to do the design and be responsible for it.

The principal purpose of a submittal is to obtain a representation from the contractor as to how it plans to execute the aspects of the design for which submittals are required. In traditional design–bid–build construction projects, such as those for which AIA documents are used, the contractor determines how the design is to be executed. The architect does not control the contractor's methods of executing the design. Consistent with this division of responsibility, AIA Document A201-2007, § 4.2.7, describes the architect's review of contractor submittals to be only for "the limited purpose of checking for conformance with information given and the design concept." The architect does not use the submittal process to check the methods by which the contractor intends to execute the design.

In *Waggoner v. W. & W. Steel Co.*,[14] workers were injured when a scaffold collapsed. They sued the architect for approving shop drawings that did not provide for temporary connections on the scaffold's expansion joints. After a review of the AIA contract, the court stated it was the duty of the contractor, not of the architect, to see that the shop drawings included how the temporary connections were to be made.

It would have been more correct for the court to state there was no requirement that the contractor communicate its plan as to temporary connections to the architect through submittals. The architect could assume that a failure to include provisions in the shop drawings for temporary connections did not indicate that the contractor would use unsafe connections but indicated only that this was not information required by the submittal process.

Another function of the submittal process is to make more definite the contract document requirements for which submittals are required. Language and even graphic depiction have communication limitations. (Try to describe or draw even a simple house!) As the contractor plans its actual performance rather than review the design prior to making its bid, its submittal indicates its interpretation of the contract, particularly in sensitive areas such as steel detailing. Submittal review can be the process by which some potential interpretation disputes are exposed.

In addition, submittals of product data and samples may be the method by which the contractor proposes substitutions if they are permitted by the contract.

Finally, although the contractor is not entitled to technical advice from the design professional on how to execute the design, the submittal process can be the method by which the contractor seeks information from the prime design professional or her consultants, particularly as to engineering techniques. The contractor may be stating what it proposes to do and asking the design professional, "What do you think?" or "Can you help me out?" If this request relates to construction means, methods, or techniques, the design professional need not respond. But if the design professional knows that the contractor proposes to proceed in a way that will lead to project failure or accidents, the design professional must draw attention to this danger.

10.5 Who Actually Performs Services: Use of and Responsibility for Consultants

10.5A Within Design Professional's Organization

The design professional may operate through a corporation, limited liability company or a partnership, or be a sole proprietor. The actual performance of the work may be done by the sole proprietor, a principal[15] with whom the client discussed the project, another principal, employees of the contracting party, or a consultant hired by the design professional. Is the client entitled to performance being rendered by any particular person? Can certain portions of the performance be rendered by people other than the design professional without affecting the obligation of the client to pay the fee?

Services of design professionals, especially those relating to design, are generally considered personal. A client who retains a design professional usually does so because he is impressed with the professional skill of the person with whom he is dealing or the firm that person represents. The client is likely to realize that licensing or registration laws may require that certain work be done by or approved by people who have designated licenses. Yet the client is also likely to realize that some parts of the performance will be delegated to other principals in the firm with whom the client is dealing, employees of that firm, or consultants retained by the design professional organization.

As to design, the client probably expects the design professional with whom it has dealt to assemble and maintain a design "team" and to control and be responsible for the design. The fleshing out of basic design concepts, such as the construction drawings and specifications, is likely to be actually executed by other employees of the design professional's organization. The client is also likely to realize that the person who has overall responsibility will not actually perform every aspect of contract administration.

10.5B Outside Design Professional's Organization: Use of Consultants

Because of the complexity of modern construction, the high degree of specialization, and the proliferation of licensing laws, design professionals frequently retain consultants to perform certain portions of the services they have agreed to provide. Clients generally prefer that highly specialized work be performed by highly qualified specialists who are often outside the design professional's organization. Unless the client insists that all design services be performed within the design professional's organization, the client should not complain if the principal design professional turns to a consultant for specialized work.

Generally the consultant is an independent contractor, equivalent to a subcontractor. As a rule, the consultant is asked to accomplish a certain result but can control the details of how it is to be accomplished. See Section 2.13. Because of the consultant's independent contractor status, if an architect retains a structural engineer as a consultant, the latter's negligent conduct that causes injuries to third persons generally will not be chargeable to the architect. See Section 5.7.

On the other hand, delegation of part of the work to a consultant as a general rule does not relieve the principal design professional of her contractual obligations to the client. See Section 7.4D. For this reason, the designer should ensure that the consultant is obligated to perform in an identical manner to the designer's obligation to the client. The designer also must consider the financial responsibility of consultants. Having a good claim against a consultant may be meaningless if the consultant is not able to pay the claim. If the consultant is a small corporation, the designer should bind the consultant's individual shareholders to the contract so that they are personally liable. The designer probably should require that the consultant carry and maintain adequate professional liability insurance.

10.6 Ownership of Drawings and Specifications

10.6A Introduction

Suppose an architect creates a custom design of a house for a small developer. The house is built, and the architect is paid for her services.

A year later the architect is driving by a new development under construction. The developer is her former client and the houses under construction appear to have the identical layout as depicted in the architect's earlier design. If the developer is using the architect's design on a new development without the architect's prior consent, what are the architect's options?

There are two parallel answers to that question. The architect may have a *contractual* right of ownership in the **instruments of service**—the physical plans and specifications. Whether she does is a matter of contract. That is the topic of this section.

In addition to any contract rights, the architect may have a **copyright** interest in the design. Copyright is but one of a constellation of intellectual property rights, including trademarks, patents and trade secrets. However, copyright—the right to prohibit the unauthorized copying and use of the intellectual property—is the architect's primary, *statutory* arsenal. That is addressed in Section 10.7.

10.6B Describing the Ownership Interest

Who owns—or, more properly, who has use rights of—the tangible manifestations that are usually written (but increasingly computer generated) by the design professional and necessary to build the project? In other words, who may use the plans and specifications once the original project has been built?

Clients sometimes contend (and many agencies of state and local government insist) that the party who pays for the production of drawings and specifications should have exclusive right to their use, including on future projects. A Louisiana statute specifies that all plans, designs, and specifications "resulting from professional services paid for by any public entity shall remain the property of the public entity. ... [S]uch documents may be used by the public entity to construct another like project without the approval of, or additional compensation to, the design professional."[16] Under this statute, the state purchases a design once, then may use it repeatedly without having to pay the design professional additional compensation. The statute adds that, in the event of re-use, the designer is shielded from liability.

Design professionals do not agree that the client is purchasing the plans and specifications for use on other projects. They contend they are selling their ideas and that they are not selling the tangible manifestations of these ideas as reflected in drawings and specifications. This, along with the desire to avoid implied warranties to which sellers of goods are held, is the basis for calling the tangible manifestations "instruments of service."

Design professionals contend that the subsequent use of their drawings and specifications may expose them to liability claims. If another design professional completes the project, the original designer may be denied the opportunity to correct design errors as they surface during construction. Design professionals contend that most projects are one of a kind and that in reality design is a trial-and-error process. Similarly, liability exposure can result if the design is used either for an addition to the project or for a new project for which the design may not be suitable. Even if the design professional is absolved from liability, this may not come until after a lengthy and costly trial. In addition, re-use without adaptation may compromise the aesthetics or structural integrity of the original design.

Another reason a design professional may wish to retain ownership of the design is to receive credit as an author and to protect her professional reputation whenever her creative works are displayed to the public.

In addition, the design professional's fee is often calculated as a percentage of construction cost. (See Section 9.3B.) Had the design professional known that drawings and specifications would be used again, a larger fee would have been justified.

Yet the client also has legitimate reasons to re-use the drawings and specifications. A client who wishes to make additions or modifications to the completed building should (absent a contract prohibition) be allowed to do so without being obligated to hire the original architect. Modification of the existing building is very different from using the design for one project to create an entirely new building.

10.6C AIA and EJCDC Approaches: License to Use

The AIA and EJCDC in their standard form contracts have addressed some of the design re-use issues highlighted in the previous subsections.

Under AIA B101-2007, § 7.2, the architect and her consultants are deemed the "authors and owners" of their respective instruments of service. The AIA B101-2007, § 7.3, deals with the owner's concerns by granting the owner a *license*—an authorization for a limited purpose—to use the instruments of service to build, alter, or maintain the project. If the owner uses the documents without employing the architect, the owner will hold the architect harmless against any liability. (**Hold harmless** means the owner will indemnify any loss suffered by the architect from the unauthorized use of the instruments of service; indemnity is the topic of Chapter 23.) The owner cannot give the license to another party without the architect's prior written consent, and unauthorized use is at the owner's sole risk. If the owner exceeds the limited use of the license by using the architect's design on other projects without permission and payment of a re-use fee, the architect may ask the court to order the owner to stop this conduct.

The Engineers Joint Contracts Documents Committee (EJCDC) standard agreement between owner and engineer, E-500 (2008), ¶ 6.03 is entitled "Use of Documents." Paragraph 6.03E grants to the owner a "limited license" to use the documents on the project or for related uses of the owner. However, this grant is circumscribed: further use or re-use must include "written verification or adaptation" by the engineer. The owner who uses the documents without written verification or adaptation by the engineer does so at its own risk and will indemnify and hold harmless the engineer or her consultants from claims and other expenses, including attorneys' fees relating to this use. Any re-use of the design documents entitles the engineer under ¶ 6.03F to further compensation at "rates or in an amount to be agreed upon by Owner and Engineer."

10.6D AIA and EJCDC Approaches: Electronic Data

Technological changes mean that an increasing portion of the design is created in a digital format. How have the AIA and EJCDC addressed this change? What effect does digital information have on the basic concept of ownership?

The AIA has largely skirted the question of electronic information. It simply leaves to the owner and architect to establish protocols for the use of digital information. Thus, AIA B101-2007, § 7.1 provides: "If the Owner and Architect intend to transmit Instruments of Service or any other information or documentation in digital form, they shall endeavor to establish necessary protocols governing such transmissions."

The EJCDC deals more extensively than does the AIA with electronic data. For example, Document E-500 (2008), ¶ ¶ 6.03B and C, deal with electronic data furnished by the owner to the engineer and the engineer to the owner. Such data are furnished for the convenience of the recipient, but only the hard copies may be relied upon. Any conclusion or information from such data is "at the user's sole risk." If there is a discrepancy between "electronic files and hard copies, the hard copy governs." Finally, ¶ 6.03C deals with the possible deterioration of data stored in electronic media format, and ¶ 6.03D covers transferring of documents in electronic media format.

The EJCDC's concern with the interface between electronic and written data is well merited. Digital technologies that increasingly are used to create the design may well have the effect of redefining the basic relationships between designer, contractor, and owner. For example, when subcontractors and suppliers electronically participate in the creation of the original design components, who at that point is the overall "designer" of the project? In addition, the ease with which electronic data may be altered (without leaving a "paper trail") raises concerns as to what constitutes the actual design. (The EJCDC's response is that the hard copy governs if it conflicts with the electronic version.)

These issues (and others) undermine the basic concept of ownership so vital to the AIA's idea of the design as the architect's "instruments of service." An article questioned the adequacy of the AIA A201-1997 ¶ 1.6.1 (now A201-2007, § 1.5.1., entitled "Ownership and Use of Drawings, Specifications and Other Instruments of Service") in light of these technological changes:

The "Ownership of Documents" clause will have to be rethought in the digital age. For one thing, it completely ignores intellectual property rights in those portions of project documentation that are furnished by or through the Contractor or [subcontractors]. In a collaborative design process, is the entire project database a joint work, in which the copyright is owned by many participants or are different parts of the database owned by different parties? If portions of the Architect's contribution consist of digital "objects" imported from third party sources, to what extent is the Architect the compiler of a collective work rather than the author of an original work? With respect to control of multiple copies, there will be myriad digital files maintained in multiple storage media, not simply "one record set" retained by the Contractor. The Contractor will probably not be able to "suitably account for" all of these files to the Architect at the end of the Project. What will the "statutory copyright notice" look like in a collaboratively—generated 3D or 4D electronic file?[17]

10.6E Ownership on Design-Build

The discussion to this point has assumed that the architect or engineer was an independent contractor hired by the owner. But on the increasingly popular design–build projects discussed in Section 14.14, the contractor—not the owner—retains the architect. Who owns the design documents on a design–build project: the designer or the contractor? This question is resolved under the "work for hire" doctrine. If the designer was an employee of the contractor and the design was created within the scope of the employment, the design would be a work for hire and the employer/contractor would be the owner of the documents.[18]

10.7 Copyright of Drawings and Specifications

10.7A Introduction

The hypothetical described in Section 10.6A—of the architect who discovers her former client is re-using the design—may constitute not only a contract breach, but also a violation of the federal Copyright Act of 1976.[19] Federal law is exclusive in copyright. Any conflicting state law is preempted.

Copyright law has broader application than the design agreement. Suppose the new development was being built not by the former client, but by a third party who had secretly copied the design. Here, too, the architect would have a copyright claim, although not a contract claim. Under either scenario, the architect would sue the defendant for *infringement* (violation) of her copyright in the design.

In short, copyright—a statutory right—exists entirely independent of whatever contractual protections the designer may have in the instruments of service.

10.7B Relevance to Design Professional

Copyright law bars unauthorized reproducing, preparing derivative works, or distributing copies of copyrighted work.[20] What "work" produced by the design professional receives statutory protection? A design professional's work product consists of three steps:

1. the intellectual effort by which the solution is conceived
2. communication of the solution
3. development of the end product

The first step involves the design professional's use of her training, intellect, and experience to solve design, construction, and manufacturing problems of the clients. The second step involves reducing the proposed design solution to tangible form. These forms, whether sketches, renderings, diagrams, drawings, specifications, computer software, or models, communicate the design solution to the client and others concerned. (In addition, some aspects of the design solution, such as the floor plans, sketches, diagrams, or pictures, may be used to advertise the project.) The third step is actual construction of the structure.

The Copyright Act of 1976, § 102(a), lists the categories subject to copyright protection:

1. literary works
2. musical works, including any accompanying words
3. dramatic works, including any accompanying music
4. pantomimes and choreographic works
5. pictorial, graphic, and sculptural works
6. motion pictures and other audiovisual works
7. sound recordings
8. architectural works

Drawings or plans fall in category 5, whereas specifications fall under category 1. In that regard, literary works need not be "literary" as long as they express concepts in words, numbers, or other symbols of expression.

"Architectural works" (category 8) refers to copyright protection of the built structure. This category was added with passage of the Architectural Works Copyright Protection Act of 1990.[21] The AWCPA provides copyright protection for original design elements of buildings. Injunctive relief can be granted to enforce a copyright even if it means destruction of buildings.

What about the first step of the design professional's efforts: intellectual conceptualization of solutions to the client's needs and wishes? Ideas themselves do not receive legal protection.[22] Copyright protects *tangible manifestations* of ideas—not the ideas themselves. While no sharp line distinguishes ideas from their tangible manifestations, it can be said that the more detailed the architectural plans and drawings, the more likely they are to receive copyright protection.

In two decisions, the federal Second Circuit Court of Appeals illustrates this distinction between ideas not subject to copyright protection and copyrightable tangible manifestations. In *Attia v. Society of the New York Hospital*,[23] the court ruled that highly preliminary design concepts do not receive copyright protection. However, in *Sparaco v. Lawler, Matusky & Skelly, Engineers LLP*,[24] the same court ruled that a site plan containing detailed proposed physical improvements may receive copyright protection.

10.7C Work for Hire Doctrine

Before the 1976 Copyright Act, it was assumed, in the absence of an agreement to the contrary, that the person commissioning copyrightable works was entitled to the copyright. This was one of the reasons for the frequent inclusion of clauses in contracts between design professionals and their clients that give ownership rights to the former.

Section 201 of the 1976 act gives copyright protection to the author of the work. However, if the work is made for hire, the employer or other person for whom the work was prepared is considered the author unless the parties have agreed otherwise in a signed written agreement. Section 101 defines a **"work for hire"** as prepared by an employee or a work specially ordered or commissioned.

The design professional who operates independently and is retained by a client to prepare the design, owns the copyright in the absence of an agreement to the contrary. The standard that applies is the common law of agency. Was the work prepared by an employee or by an independent contractor? If the design professional is an independent contractor, she owns the copyright. For example, in *Kunycia v. Melville Realty Co.*[25] a federal district court ruled that an architect retained by a client was entitled to a copyright, as the copyrightable drawings were not made pursuant to a work for hire.

10.7D Owner Use of Copyrighted Work

The design professional is the owner and copyright holder of the instruments of service. Section 10.6 of this book discusses the designer giving the owner a *contract* right—using the device of a license—to use the design on the completed project or on other projects under certain circumstances.

Suppose the owner wishes to use the copyrighted work without a contract license. This may be done two ways.

First, the owner may buy the copyright to the design from the designer. The Copyright Act § 204(a) permits the transfer of copyright ownership—but only in writing and if signed by the copyright owner.

Second, the owner may contend that the architect had granted the owner an "implied nonexclusive license" to use the copyrighted materials. The term "implied" simply means that there is no writing granting the owner the license; instead, the grant is implied from the circumstances. A "nonexclusive license" is a grant by the copyright holder allowing a third person (including the owner) to use the copyrighted materials for a limited purpose.

The issue of whether a nonexclusive license was granted typically arises in the context of an architect who performed preliminary design work for an owner—only to be replaced by another architect who completes the design or administers the project's construction. May an owner who purchases the preliminary design work from the *first* architect use those documents for further development (by a *second* architect) and actual construction—in which case the owner has a nonexclusive right to the design—or must the owner compensate the original architect for the later use of the preliminary design—in which case a nonexclusive license has not been granted?

The federal Fourth Circuit Court of Appeals in *Nelson-Salabes, Inc. v. Morningside Development, LLC*[26] addressed that question as applied to architectural drawings. In that case, a developer (Strutt) hired an architect (NSI) to create a schematic building footprint, attend a zoning hearing to request an exception (which the zoning board granted), and create architectural drawings, which Strutt used to obtain zoning approval. Although Strutt paid for all this work, it never signed the AIA contract which NSI had sent it.

Strutt then pulled out of the project for financial reasons and was replaced by a second developer (Morningside). Morningside used the approved development plan created by Strutt to go forward with the project using a new architect.

NSI sued Morningside for infringement (violation) of the architect's copyright in the work it had prepared for Strutt. Morningside countered that it had an implied nonexclusive license to use the materials and therefore did not violate the copyright. The federal district (trial) court's ruled in favor of NSI.

The court of appeals stated that a key factor in determining the existence and scope of a nonexclusive license is whether the original architect (here, NSI) intended the owner to copy, distribute, or modify its work without further involvement of the architect. Here, the evidence showed that Strutt and NSI were engaged in an ongoing relationship that contemplated NSI's future involvement in the development. NSI created architectural drawings with the understanding that it would participate in the further development of the project. After all, NSI had sent Strutt an AIA contract to sign. There was no indication NSI intended a second developer (Morningside) to use those documents using a new architect.

10.7E Infringement of Copyright

In *Nelson-Salabes*, the developer/owner's claim of an implied nonexclusive license was a defense to the architect's copyright infringement claim. Even if the architect defeats that defense (as NSI did), she still must establish that the defendants—usually both the owner and the replacement architect—infringed her copyright.

To establish a copyright infringement claim, the plaintiff architect must prove (1) its ownership and registration of the copyright and (2) unauthorized copying by the defendant of original (copyright-protected) parts of the design. In the absence of direct evidence of copying, the plaintiff may prove copying by showing her design and the defendant's design are "substantially similar." "Substantial similarity" is addressed later in this section.

Registration. The Copyright Act § 411(a) requires registration with the Copyright Office before commencement of an infringement action. In addition, § 412 precludes recovery of statutory damages or attorneys' fees if the copyrighted work is not registered within three months after first publication of the work. Registration need not be difficult or expensive. The regulations permit substitutions for the original materials themselves if they were bulky or if it would be expensive to require deposit.

Substantial Similarity. A federal court of appeals case, *Sturdza v. United Arab Emirates*,[27] analyzed the requirement of substantial similarity.

In 1993, the United Arab Emirates (UAE) held a competition for the architectural design of a new embassy in Washington D.C.. The UAE informed the competitors that the building should express the "richness and variety of traditional Arab motifs." Plaintiff Elena Sturdza and defendant Angelos Demetriou both submitted designs. The UAE informed plaintiff that she had won.

Over the next two years, plaintiff and the UAE exchanged eight contract proposals. Although plaintiff agreed to the final proposal, the UAE ceased all communications with her. In 1997, plaintiff learned that the UAE had hired Demetriou to design the embassy. Plaintiff sued the UAE and Demetriou for copyright infringement.

The UAE and Demetriou did not contest that plaintiff held a valid copyright. Nonetheless, the federal district court granted summary judgment (a pre-trial ruling) in favor of the defendants on the ground that Sturdza could not prove the two designs were substantially similar. Plaintiff appealed.

The United States Court of Appeals for the District of Columbia reversed, ruling that the two designs were sufficiently similar that the issue of "substantial similarity" should be decided by a jury. The court explained that a substantial similarity inquiry consists of two steps.

The first step requires identifying which aspects of the plaintiff's work are protectable by copyright. Copyright does not extend to facts or ideas. It only protects the author's "expression" of ideas which "display the stamp of the author's originality." In the context of this case, domes, wind-towers, arches and Islamic patterns were "ideas" not subject to copyright protection.

Once the unprotected elements of the design are excluded, the second step is to determine whether the remaining protectable elements are substantially similar. This involves looking not only at the individual components of the two designs but also their "overall look and feel" of the two designs.

[Ed. note: The student can make his or her own evaluation of the "overall look and feel" of the two designs by examining the appendices included from the court's opinion and reproduced at the end of this chapter.]

Remedies for Infringement. The Copyright Act provides copyright authors with two broad remedies: injunctive relief (ordering the performance of a specific act) and damages.

The architect's entitlement to injunctive relief is illustrated in *Christopher Phelps & Associates, LLC v. Galloway*.[28] An owner built his house using infringed plans. The architect established the infringement and was awarded damages (the cost he would have charged to create the plans). The architect then sought a permanent injunction to prevent the owner from profiting from his infringement by either renting or selling the house within the period of plaintiff's copyright—95 years. The court refused on the ground that an injunction would be unduly harsh for the owner and would also undermine the public policy of allowing land to be productively used (including selling it).

The Copyright Act § 504 allows recovery of actual damages and profits made by the infringer attributable to the infringement. As an alternative to actual damages, the copyright owner may seek statutory damages of up to $30,000, with $150,000 for willful infringement.

Copyright Duration. The *Christopher Phelps* court referred to a copyright of 95 years; that is for works for hire.[29] Otherwise, the duration begins with creation of the work and continues for the life of the author, plus 70 years after the author's death.[30] These time periods are the same as in the European Union.

10.7F Copyright Transfer to Former Employee

A transfer of copyright ownership also may occur between an architectural firm and the employee (architect) who created a design. The design would be owned by the firm under the works-for-hire doctrine, which is discussed in Section 10.7C. If the employee then leaves to start her own business and the owners wish to keep working with that architect, the firm can use the Copyright Act § 204(a) to transfer ownership of the copyright to the architect. The transfer of ownership must be in writing.

10.7G Advice to Design Professionals: Obtaining a Copyright

If copyright protection is to be sought, design professionals should comply with the statutory copyright requirements. First, they should be certain they have not assigned to others their right to copyright ownership. There is still utility, as suggested in Section 10.6, in including a contract clause giving ownership rights to the plans and specifications to the design professional.

Second, design professionals should comply with the copyright notice requirements. The word *Copyright* can be written out, or the notice can be communicated by abbreviation or symbol. The authorized abbreviation is "Copr.," and the authorized symbol is the letter "C" enclosed within a circle (©). The year of first publication should be given, and the name of the copyright owner or an abbreviation by which the name can be recognized or generally known can be used. The notice must be affixed to the copies in such a manner and location as to give reasonable notice of the copyright claim. The Register of Copyrights is given authority to specify methods by which copyright notice can be given.[31]

If the design professional wishes to take advantage of the statutory damage award and to recover attorneys' fees, she should register the copyrighted work within three months of publication. Methods are available to minimize this burden. The Copyright Office should be consulted.

Finally, an architect should be cautious if an owner comes to her with plans that have already been created (or substantially completed). An architect who uses those designs may be at risk of incurring liability under the copyright laws. If possible, the architect should ascertain the source of the plans and that the owner has copyright ownership. The architect should also demand that the owner hold her harmless against any copyright liability arising out of use of the plans.

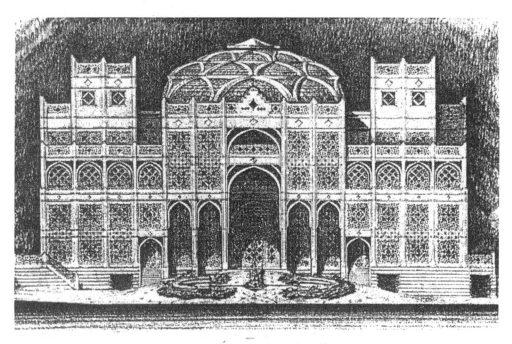

APPENDIX A Elena Sturdza

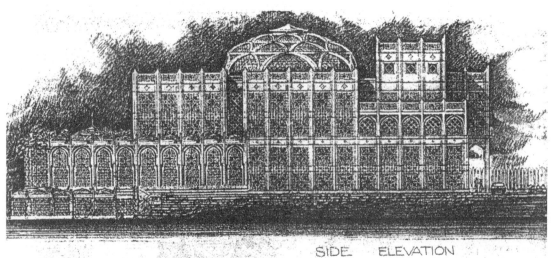

APPENDIX B Elena Sturdza

COPYRIGHT OF DRAWINGS AND SPECIFICATIONS

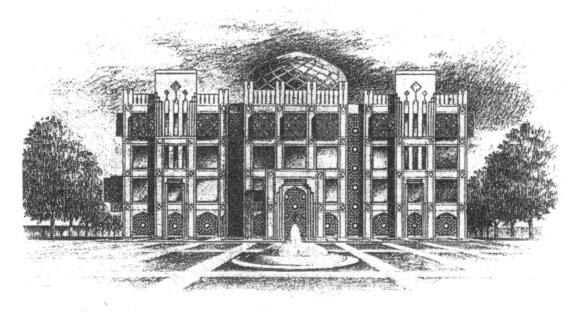

FRONT ELEVATION

APPENDIX C Angelos Demetriou (1997)

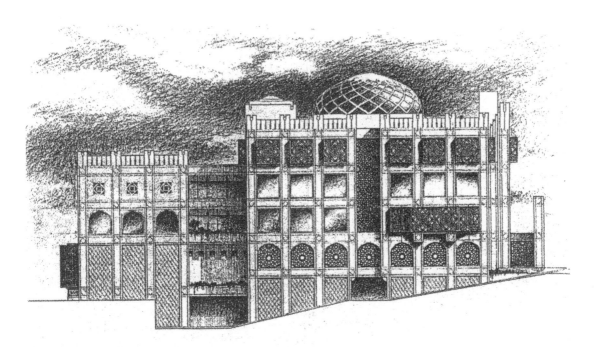

SIDE ELEVATION

APPENDIX D Angelos Demetriou (1997)

REVIEW QUESTIONS

1. Why should assistance in obtaining financing and economic feasibility studies not be considered part of basic design services?

2. What are the two models of cost predictions a design professional can use and how do they differ?

3. In the case *Griswold and Rauma, Architects, Inc. v. Aesculapius Corp.*, what were the four factors that were identified by the court as being relevant to a determination of what the effect on compensation of an architect or building contractor should be when the actual or probable cost of construction exceeds an agreed maximum cost figure?

4. What is the most common basis for waiving a cost condition?

5. What are the five choices an owner has if the lowest bona fide bid or negotiated price exceeds the budget of the project?

6. What changes did the AIA make to B101-2007, § 3.6.2.1, in an attempt to create a subjective standard in terms of an architect's duty to inspect the work?

7. In the case of *Board of Education of the Hudson School District v. Sargent, Webster, Crenshaw & Folley*, what was the primary reason the court ruled the architect breached their contract?

8. What is the principal purpose of a submittal?

9. How does the Engineers Joint Contracts Documents Committee (EJCDC) Document E-500 (2008) address the owner's use of documents generated by the design professional?

10. What does the Copyright Act protect, how does the design professional protect its copyright, and what conditions will lead a design professional to believe its copyright has been infringed?

ENDNOTES

[1] Macaulay, *Organic Transactions: Contract, Frank Lloyd Wright, and the Johnson Building*, [1996] Wis.L.Rev. 75. This article is reproduced in part in Appendix P of this book, located on the website.
[2] AIA Doc. B101-2007, § 13.1; EJCDC E-500 ¶ 7.01A(5) (2008).
[3] Sweet & Sweet, *Architectural Cost Predictions: A Legal and Institutional Analysis*, 56 Calif.L.Rev. 996 (1968).
[4] 232 Neb. 885, 443 N.W.2d 260 (1989).
[5] 496 So.2d 1012 (La.1986).
[6] Id. at 1017.
[7] Id. at 1017–18 (footnotes omitted).
[8] AIA Doc. B101-2007, § 1.1, cross referencing to B101-2007, Exhibit A, § A.1.3.
[9] Id. at § 6.6.
[10] AIA Doc. B101-2007, § § 3.6.2.1 and 3.6.6.1.

[11] 146 A.D.2d 190, 539 N.Y.S.2d 814, appeal denied, 75 N.Y.2d 702, 551 N.E.2d 107, 551 N.Y.S.2d 906 (1989).
[12] *City of York v. Turner-Murphy Co., Inc.,* 317 S.C. 194, 452 S.E.2d 615 (App.1994).
[13] Rubin & Ressler, *"To Build a Better Mousetrap"—The Search to Define Responsibility for Shop Drawing Review,* 5 Constr. Lawyer, No. 4, April 1985, p. 1. See also Circo, *When Specialty Designs Cause Building Disasters: Responsibility for Shared Architectural and Engineering Services,* 84 Neb. L. Rev. 162, 200–06 (2005).
[14] 657 P.2d 147 (Okla.1982).
[15] The term *principal* is defined as a partner in a partnership or a person with equivalent training, experience, and managerial control in a corporation.
[16] La.Rev.Stat. § 38:2317(A).
[17] Stein, Alexander & Noble, *AIA General Conditions in the Digital Age: Does the Square "New Technology" Peg Fit Into the Round A201 Hole?* Construction Contracts Law Rep., vol. 25, no. 25 ¶ 367 at p. 7 (Dec. 14, 2001).
[18] *Trek Leasing, Inc. v. United States,* 62 Fed.Cl. 673 (2004). See Section 10.7C for discussion of the work for hire doctrine.
[19] 17 U.S.C. § 101, et seq.
[20] 17 U.S.C. § 106.
[21] Title VII of Pub.L. No. 101-650, 104 Stat.5089, codified at 17 U.S.C. §§ 101, 102, 106, 120, and 301.
[22] 17 U.S.C. § 102.
[23] 201 F.3d 50 (2d Cir.1999), cert. denied, 531 U.S. 843 (2000).
[24] 303 F.3d 460 (2d Cir.2002), cert. denied, 538 U.S. 945 (2003).
[25] 755 F.Supp. 566 (S.D.N.Y.1990).
[26] 284 F.3d 505 (4th Cir.2002).
[27] 350 U.S.App.D.C. 154, 281 F.3d 1287 (D.C.Cir.2002), petition for certiorari filed, No. 02-5218, July 6, 2002, case considered closed Dec. 27, 2002.
[28] 492 F.3d 532 (4th Cir.2007).
[29] 17 U.S.C. § 302(c).
[30] Id. § 302(a).
[31] 17 U.S.C. §§ 405 and 406 provide relief if the copyright notice is omitted or erroneously made.

CHAPTER

11

DESIGN PROFESSIONAL LIABILITY

SCENARIO

As with Chapter 10, this chapter does not include a single scenario. Chapter 11 addresses design professional liability. Liability may be either to the client or to a third party. A third party claimant is usually a contractor or subcontractor in the case of economic (financial) loss or to a construction worker or passerby for personal injury.

With regard to liability to the client, Section 11.1 reproduces *City of Mounds View v. Walijarvi*, in which the client claimed the architect had promised a moisture-free basement but that the completed building did not stay dry.

Two cases address design professional liability to a contractor or subcontractor for economic losses allegedly caused by the design professional. These cases, *Bilt-Rite Contractors, Inc. v. The Architectural Studio* and *SME Industries, Inc. v. Thompson, Ventulett, Stainback & Assocs., Inc.*, are reproduced in Section 11.6D. They come to opposite conclusions on the question of liability.

These three reproduced decisions provide "real world" scenarios under which claims for design professional liability may and do arise.

11.1 Design Professional Liability: The Professional Standard as the Measure of Reasonable Care: *City of Mounds View v. Walijarvi*

As explained in Section 5.3B, society imposes upon individuals an obligation to act with due care, which is a standard defined by what the reasonable person would do under similar circumstances. However, the conduct of persons with special training or knowledge is compared to how a reasonable person with such training or knowledge would act. For professionals, including design professionals, that higher bar against which their conduct is measured is called the **professional standard**. An architect is held to the standard of conduct of a reasonable, prudent architect under similar circumstances.

In the reproduced case here, the client asserted that its architect's liability should be under a theory of implied warranty: a strict liability standard that would not require proof of the architect's fault. The Minnesota Supreme Court rejected the client's request and reaffirmed its use of the professional standard to measure an architect's liability. In so doing, the court explained the reasons underlying the professional standard.

CITY OF MOUNDS VIEW v. WALIJARVI
Supreme Court of Minnesota, 263 N.W.2d 420 (Minn.1978)
[Ed. note: Footnotes omitted.]

TODD, Justice.
[Ed. note: The city became apprehensive because of dampness in the basement of an addition to a city building. The architect wrote to the city that its design would, if executed properly, generate a "watertight and damp-free" basement. But problems grew worse, and corrective work was needed.

The city sued the architect based on claims of negligence and implied warranty. The trial court granted the architect's motion for summary judgment (a hearing before a judge with no jury present) on the implied warranty claim. The city appealed this ruling and the Minnesota Supreme Court's response to the city's argument is reproduced below.]

... As an alternative basis for recovering damages from the architects, the city urges that we adopt a rule of implied warranty of fitness when architectural services are provided. Under this rule, as articulated in the city's brief, an architect who contracts to design a building of any sort is deemed to impliedly warrant that the structure which is completed in accordance with his plans will be fit for its intended purpose.

As the city candidly observes, the theory of liability which it proposes is clearly contrary to the prevailing rule in a solid majority of jurisdictions. The majority position limits the liability of architects and others rendering "professional" services to those situations in which the professional is negligent in the provision of his or her services. With respect to architects, the rule was stated as early as 1896 by the Supreme Court of Maine (*Coombs v. Beede*, 89 Me. 187, 188, 36 A.104 [1896]):

In an examination of the merits of the controversy between these parties, we must bear in mind that the [architect] was not a contractor who had entered into an agreement to construct a house for the [owner], but was merely an agent of the [owner] to assist him in building one. The responsibility resting on an architect is essentially the same as that which rests on the lawyer to his client, or on the physician to his patient, or which rests on anyone to another where such person pretends to possess some skill and ability in some special employment, and offers his services to the public on account of his fitness to act in the line of business for which he may be employed. The undertaking of an architect implies that he possesses skill and ability, including taste, sufficient to enable him to perform the required services at least ordinarily and reasonably well; and that he will exercise and apply in the given case his skill and ability, his judgment and taste, reasonably and without neglect. But the undertaking does not imply or warrant a satisfactory result.

The reasoning underlying the general rule as it applies both to architects and other vendors of professional services is relatively straightforward. Architects, doctors, engineers, attorneys, and others deal in somewhat inexact sciences and are continually called on to exercise their skilled judgment in order to anticipate and provide for random factors which are incapable of precise measurement. The indeterminate nature of these factors makes it impossible for professional service people to gauge them with complete accuracy in every instance. Thus, doctors cannot promise that every operation will be successful; a lawyer can never be certain that a contract he drafts is without latent ambiguity; and an architect cannot be certain that a structural design will interact with natural forces as anticipated. Because of the inescapable possibility of error which inheres in these services, the law has

> traditionally required, not perfect results, but rather the exercise of that skill and judgment which can be reasonably expected from similarly situated professionals. As we stated in *City of Eveleth v. Ruble*, 302 Minn. 249, 253, 225 N.W.2d 521, 524 (1974): "One who undertakes to render professional services is under a duty to the person for whom the service is to be performed to exercise such care, skill, and diligence as men in that profession ordinarily exercise under like circumstances." ...
>
> We have reexamined our case law on the subject of professional services and are not persuaded that the time has yet arrived for the abrogation of the traditional rule. ...
>
> * * *
>
> For these reasons, we decline to extend the implied warranty/strict liability doctrine to cover vendors of professional services. Our conclusion does not, of course, preclude the city from pursuing its standard malpractice action against the architects and proving that the basement area of the new addition was negligently designed. That issue remains for the trier of fact in the district court.
>
> Affirmed.

Against whom is the conduct of the professional measured? Usually it is assumed that such conduct is measured against the conduct of others in the professional's locality. For example, if the professional practices in a small town, it is usually assumed that he should not be compared with professionals who practice in a large city. To some degree, this is based on the reasonable expectations of the client, and to some degree, on the likelihood that advances in the profession come first to large urban areas.

That said, technological changes have made it increasingly possible for design professionals to live in the suburbs or even small towns yet to be involved in a large, urban practice. The locality standard should be based on the location of the practice—not the residence of the design professional.

(For sake of completeness it should be noted that, while the vast majority of states impose upon design professionals a negligence standard of liability, South Carolina and Alabama use an implied warranty standard under which liability is based on outcome—not fault. This minority rule will not be further addressed.)

11.2 Defining the Professional Standard

The professional standard is defined as what a reasonable professional (sometimes added: in the same locality) would do under similar circumstances. The content of that standard—*what the defendant should have done*—is primarily established through the testimony of experts who have the education, training, and experience necessary to give an opinion as to the reasonable and prudent course of conduct that would have prevented the plaintiff's injury. However, plaintiffs sometimes point to other sources of proper conduct.

Violation of Statutes and Administrative Regulations. As explained in Section 5.3B, conduct that is in violation of a statute or regulation may either conclusively establish negligence (in which case violation is said to constitute negligence *per se*) or provide some (but not conclusive) evidence of negligent conduct. (Violation of a regulation is generally evidence of negligence but does not establish negligence *per se*; only statutes may do that.) If the design professional's conduct is regulated by a statute, the injured person is within the class of persons protected by the statute, and the injury was caused by a risk the statute sought to avoid, the design professional's violation of the statute would also be a violation of the professional standard.

As explained in Section 6.6B, building codes can determine design, materials, and construction methods. Design specifications, performance standards, or a combination of the two are used. Codes provide minimum standards to protect against structural failures, fire, and unsanitary conditions. Code compliance is one aspect of meeting the professional standard imposed by law on design professionals.

Housing codes are of relatively recent vintage. They have gone beyond structure, fire, and basic sanitation to include light, air, modern sanitation facilities, maintenance standards, and occupancy density rules.

Typically, a dispute as to whether the design complied with the building or housing code arises between the architect and client. The client points to an alleged violation of the code as grounds for a claim of negligence by the design professional. However, codes are complex and technical. Deciding whether a design violates the code may require a jury to sift through competing expert testimony.

The Americans with Disabilities Act (ADA),[1] enacted in 1990, generated building standards as part of an overall policy to protect disabled persons, which is a group that has long been subjected to discrimination in in many areas, including employment, housing, and public accommodations. The act comprehensively defines the treatment of disabled persons, imposes significant new obligations on employers, and (most importantly for design professionals) mandates that commercial and public accommodations be accessible to people with disabilities.

In effect, the ADA has created a form of national building code, but no entity is obligated under the ADA to review and approve drawings and specifications for compliance. Thus, parties seeking to comply may not know whether they have discharged their obligation until a complaint is filed. As with any statutory requirement, an owner sued for violation of the ADA may cite the statute's violation as evidence of negligence in a claim against the design professional.

(May disabled persons who believe a building is not in compliance with the ADA sue the architect for violation of the act? That straightforward question does not yield a simple answer because the statute is unclear as to the category of persons potentially liable under the act, the types of buildings subject to the act, and the types of activities for which one may be liable. However, in *Lonberg v. Sanborn Theaters Inc.*,[2] a federal court of appeals ruled that an architect cannot be liable under the ADA, because the architet does not have control of the finished facility as do owners and operators.)

As shown in *Duncan v. Missouri Board for Architects Professional Engineers and Land Surveyors*,[3] reproduced in Section 8.3C, licensing statutes may also establish a minimum level of professional conduct. In that case, the engineer's failure to independently review the fabricator's drawings—an exercise mandated by the licensing law—was (under the circumstances) evidence of gross negligence.

Industry Standards. Plaintiffs may argue that industry standards should inform the content of the professional standard. For example, in a Montana case, *Taylor, Thon, Thompson & Peterson v. Cannaday*,[4] the owner argued that the architect was not entitled to its fee because it had violated the American Institute of Architects' *Architects' Handbook of Professional Practice*. The court pointed out that, unless adopted by a statute, a professional practice guide is not evidence of negligence. (In general, courts reject the notion that industry norms establish the standard of care. The standard of care is decided by the courts, and what is normal within an industry may fall below that standard.)

Professional Ethics. In a Texas case, *Dukes v. Philip Johnson/Alan Ritchie Architects, P.C.*,[5] a city hired architects to design improvements to an outdoor urban park, that included a pool. After the renovations were completed, four individuals drowned in the pool. The families of the victims sued the architects for negligence. They claimed that the architects' failure to report the hazardous condition of the pool violated the ethical code governing architects.

The court ruled that the architects did not owe the victims a tort duty of care. The architects had not been hired to evaluate the pool's safety. The court cited a state supreme court decision which held that the rules of professional conduct governing attorneys do not establish civil liability standards. Applying that earlier decision, the *Dukes* court held that architects' professional ethical standards cannot establish a duty of care requiring the architects to report safety hazards they may have discovered.

Professional ethics is the topic of Chapter 13.

11.3 Proving the Professional Standard: Expert Testimony

11.3A Need for Expert Testimony

In the judicial system, judges and juries make decisions. To do so, they may have to hear, evaluate, and judge testimony and exhibits that relate to technical matters unfamiliar to them. If these matters are too difficult for a judge or jury to decide without technical assistance, a person accepted by the judge as an "expert" may give an opinion to help the judge or jury understand these technical matters. A witness may be designated as an expert based on that person's education, experience, or a combination of the two. The expert is generally not allowed to give opinions that could determine the outcome of the case. This power is still retained by the judge and (when used) the jury.

As design professionals are generally held to the professional standard of care, the judge or jury must be able to understand and evaluate (1) what is the professional standard and (2) did the defendant's conduct fall short of that standard. A design professional will take the position that both of these questions are too technical for the judge or jury to understand and so cannot be proved without the use of expert testimony. Often, the plaintiff cannot or chooses not to offer expert testimony and claims it is not needed. If expert testimony is necessary to establish the design professional's liability, and such testimony is not provided by the plaintiff, the design professional is entitled to dismissal of the lawsuit.

Not all conduct of design professionals requires expert opinion to be understood by the judge or jury. Sometimes the claimed negligence is so pervasive as to be understood using common sense, even if technical issues are involved. In a Minnesota case, *Zontelli & Sons, Inc. v. City of Nashwauk*,[6] the contractor sued the engineer (Wallace) for failing to accurately describe the site conditions, thereby leading the contractor to incur performance costs well in excess of its bid. The court ruled that the errors were so massive as to support a finding of negligence without the use of expert testimony:

> [W]e reject Wallace's claim that it has no liability because it has not been proved negligent. Wallace drastically underestimated the quantity of concrete and unsuitable materials to be removed; failed to indicate the unusual characteristics of these materials; misrepresented Highway No. 65 as a municipal, instead of state, highway, misleading bidders as to the thickness and strength of the concrete shown on the plans to be under it; and failed to indicate that construction was governed by unexpected state requirements. Knowledge of these factors would have greatly influenced amounts bid for the project and, in fact, the factors increased the project cost almost $300,000 from Wallace's own pre-project estimate of approximately $359,000. The record indicates, moreover, that had Wallace made even preliminary inquiries, it could have avoided these inaccuracies. In view of these considerations, the trial court was justified in concluding, without expert testimony, that the care, skill, and diligence exercised by Wallace in this project was less than that normally exercised by a member of the engineering profession and that Wallace therefore was negligent.[7]

Another issue regarding expert testimony is who can testify. The clearest example of a proper expert is one who is a licensed member of the same profession as the defendant. A licensed archiect could testify as an expert against a defendant architect, and so on. However, this ideal situation is not the legal standard. First, a witness who is not licensed in accordance with the registration laws of the state very likely will be permitted to testify.[8]

More importantly, recent cases have been liberal in admitting expert testimony from disciplines other than those of the defendant architect or engineer. An architect may testify as to the performance of an engineer and vice versa.[9] The Oklahoma Supreme Court allowed a physics professor to testify as to adequacy of the design of a solar system.[10] So long as the witness is qualified by education, training, or experience to provide an expert opinion, that opinion is based on scientifically valid methods,[11] and the witness's opinion will aid the jury, that witness may be allowed to testify as an expert.

Finally, the use of experts is especially contentious where scientific concensus does not exist. Within the construction industry, the reliability of expert testimony is most likely to be challenged when the claim is *not* that defective construction caused immediate damage to the building, but rather that it sickened the building's occupants years later. These "sick building" or toxic-mold claims are prone to "junk science" charges in the absence of federal or state standards on permissible levels of mold within habitable spaces and where the causal link between exposure to mold and personal injury has yet to be proven conclusively.[12]

11.3B Advice to Expert Witnesses

Space does not permit a detailed discussion of all the problems seen from the perspective of a design professional asked to be an expert witness. Some brief comments can be made, however. A clear, written understanding should precede any services being performed. Such a writing should include the following:

1. Specific language making clear that the expert will give his or her best professional opinion.
2. Language that covers all aspects of compensation for time to prepare to testify, travel time, and actual time testifying before a court, board, or commission. (Many experts use an hourly rate for preparation time and a daily rate for travel and testimony time.)
3. Specification of expenses to be reimbursed, using a clear and administratively convenient formula for reimbursing costs of accommodations, meals, and transportation.
4. A minimum fee if the expert is not asked to testify. Some attorneys retain the best experts, use the experts whose opinions best suit their case, and (by having retained the others) preclude them from testifying for the other parties.

The attorney calling the expert to testify usually provides details as to appearance, description of qualifications, methods of answering questions, explanations for opinions, and defending opinions on cross-examination. It is important to recognize who is being addressed and the reason for seeking expert opinions. The expert should help the judge or jury—people often inexpert in evaluating technical material—so opinions and explanations are understood by the people who must evaluate them. The expert should never speak down to judge and jurors.

A client who loses its case or who recovers less damages than anticipated may place the blame on the experts—whether its own or those of the opposing party. An expert sued by a client for malpractice may invoke "witness immunity," which is a defense designed to ensure that witnesses testify freely and without fear of civil liability.[13]

11.4 Alternative Standards as Measure of Conduct

11.4A Contract Standard

The professional standard is the default measure of the design professional's performance. It applies if no other standard has been specified. Sometimes, the contract itself specifies the professional standard—as do AIA B101-2007, § 2.2 and the Engineers Joint Contracts Documents Committee (EJCDC) E-500, ¶ 6.01A (2008).

One contractual standard is that the design professional promises to satisfy the client. There are two standards of satisfaction. If performance can be measured objectively, the standard is reasonable satisfaction. Would a reasonable person have been satisfied? Objective standards are more likely to be applied where performance can be measured mechanically.

A client may desire a subjective standard—that the client personally is satisfied. The subjective satisfaction standard is more likely to arise in the design phase, particularly in aesthetic matters. However, client must exercise a good-faith judgment and must be genuinely dissatisfied before she is relieved of the obligation to pay.

Sometimes the contract between design professional and client contains a specific performance standard. For example, an engineer may make a contract with a manufacturer under which it was specifically agreed that the machine designed by the engineer would produce a designated number of units of a particular quality within a designated period of time.

As discussed in Section 11.5B, a client who wants a "green" design and construction may expect specific dividends, such as lower energy costs for prospective tenants. The client may seek a claim from the design professional if the functioning building fails to meet those hoped-for goals.

In general, the design professional should avoid assuring the client that particular objectives will be achieved unless he is willing to risk the possibility of being held accountable if they are not.

Finally, can the client and design professional agree that the designer's standard of performance will be less than the professional standard?

The courts will not enforce an agreement to perform an illegal act. If the building codes require product X, the client and designer cannot agree to use product Y because it is cheaper.

Suppose the design professional would have selected X, a material that would have been designated by other professional designers because of its combination of durability, low maintenance, cost, and appearance. But the client—being aware of the tradeoffs—orders that Y be used simply because it is less expensive (but still code compliant). Suppose the material selected needed replacement earlier than the client expected. The client should not be able to recover any replacement costs against the designer. The client and the designer agreed to a standard different from the professional standard—in this case, a lower standard—which is enforceable so long as the building did not violate the building codes.

Note that the client was aware of the tradeoffs in choosing product Y. The implied obligation of good faith and fair dealing in contractual relations would impose upon the designer a duty to explain the tradeoffs. The client's consent in choosing between the two products must have been an informed consent.

11.4B Consumer Protection Statutes

The previous discussion focused on the common law. However, clients may invoke consumer protection laws when suing their design professionals, thereby benefiting from the relaxed standard for recovery (and increased penalties) specified in those laws. The Kansas Supreme Court ruled that its state's consumer protection law applies to a client's claim against his engineer,[14] and a New York court applied that state's consumer protection law to a homeowner's contract with an architect.[15]

11.5 Client Claims Against Design Professionals

11.5A Suing in Contract, Tort, or Both?

The existence of a contract between the client and design professional normally means the client has the option of bringing a claim either for breach of contract or under tort law. Both theories of liability have their advantages and disadvantages.

Tort law may provide the client with more potent remedies. For example, if the claim is based on breach of contract, it is very difficult to recover for *emotional distress* and almost impossible to recover *punitive damages* (a money award meant to punish the defendant). However, a tort claim may justify recovery of damages for emotional distress and, if a breach of contract is also considered a tort, may justify recovery of punitive damages.

Yet there are advantages to bringing the claim for breach of contract. The time limit for commencing a lawsuit is longer—often very much longer—when the claim is based on a breach of contract rather than on a tort theory. However, the limitations period—the time within which the plaintiff must begin a lawsuit—begins to "run" at different events.

Normally, the time within which a plaintiff may sue for breach of contract begins to run when the project is completed. If a construction defect is hidden from the owner, the owner may not learn of the defect until after the limitations period has run. (Some states have created a **discovery rule** to give the owner time to sue after the owner either discovered or with reasonable diligence would have discovered the defect.)

By contrast, the period to start a tort claim does not begin until the claimant discovers it has a claim against the defendant. The difficulty posed by a latent defect is usually eliminated, because the time to sue does not begin to run until the defect caused the plaintiff injury.

Yet should a client always be allowed to choose the theory of recovery? Claims by clients against their design professionals are more likely to involve specific provisions of a contract. If the client is permitted to bring its claim in tort, it may be able to bypass certain contract defenses, such as the parol evidence rule, the statute of frauds, or exculpatory provisions (protective of the design professional) found in the contract between the client and its design professional.

In addition, claims based on commission of a tort sometimes run afoul of the economic loss rule. Although this is more commonly a problem when claims are made by third parties (Section 11.6D), it can also be a problem when the claim is by one party to a contract against the other, as discussed in Section 5.10.

Normally, whether to sue in contract or tort is not an "either-or" question. The client will likely be able to assert a claim based on the commission of a tort or the breach of a contract—whichever is more advantageous to the client. Only at the stage when the client must choose a remedy must it choose the theory on which it is basing its claim.

11.5B "Green" or Sustainable Design

Introduction. From a relatively recent beginning, "green" or sustainable design and construction has become a well-entrenched and increasingly significant presence in public and large commercial construction. As often happens in a rapidly changing economy, the law lags behind in its response. This is entirely understandable, as commercial or technological changes may escape the notice of legislatures, leading to a paucity of statutory regulation, and lawsuits only slowly wind their way through the trial court and on to the courts of appeal, where a common law response with precedential value can be formulated.

"Green" or sustainable design and construction (emphasis here is on the design component) both follows and breaks with this standard paradigm. As with other rapid commercial changes, the common law awaits its first appellate court (or even trial court) rulings. By contrast, legislatures seem to have jumped on the "green" bandwagon, increasingly mandating that new public-works buildings and large commercial projects meet "green" certification.[16] Yet, here too, any legal challenges have not been the subject of judicial review and resolution (so far).

As a result, the "liability law" of "green" or sustainable design currently lies largely in the speculative province of commentators. These commentators—virtually all practitioners rather than from academia—have chronicled the statutory and regulatory changes and tried to understand the liability consequences for participants in the construction industry. However, only judicial decisions can establish a "green building" liability law.

Historical Background. 'Green," "sustainable," or "environmentally friendly" design and construction grew out of the environmental movement beginning in the 1960s and the oil embargo by the Organization of Petroleum Exporting Countries (OPEC) in the early 1970s. Statutes in the late 1960s and 1970s focused on the clean-up of hazardous waste, regulation of point sources of pollution, and increasing energy conservation (largely through regulation of consumer products). See Section 6.5.

A newer approach, begun in the 1990s, seeks to promote energy efficiency and to decrease the negative environmental impact of the construction process through manipulation of the design itself. Various industry organizations have sought to promote environmentally friendly designs. The most important such organization is the United States Green Building Council (USGBC), which has created a Leadership in Energy and Environmental Design (LEED) rating system. There are four levels of LEED certification: certified, silver, gold, and platinum. The USGBC website provides: "At its core, LEED is a program that provides third-party verification of green buildings." Under the caption "What can LEED do for you?" the website lists the following:

- Lower operating costs and increase asset value.
- Conserve energy, water, and other resources.
- Be healthier and safer for occupants.
- Qualify for money-saving incentives, like tax rebates and zoning allowances.[17]

In addition to focusing on design, this new approach also seeks to reduce the negative environmental impact of the built environment through changed building practices. Contractors are encouraged to use recycled materials, to use paints that don't give off fumes, and to recycle their waste materials. While important, these changes in "best practices" raise few, if any, legal questions and will not be further addressed here.

Standard Form Contracts. The American Institute of Architects (AIA) has embraced the "green" design movement. At its 2007 national convention (celebrating its 150th anniversary), the chosen theme was: "Growing beyond green: How you can green your projects, educate your clients, and reduce the impact buildings have on the environment." As would be expected, the 2007 edition of standard form contracts expressly addresses sustainability.

Section 3.2 of B101-2007, which covers the schematic design phase, mandates the architect to bring to the owner's attention sustainability issues. In addition, two "additional services" address "green" issues. Section 4.1.23 covers "extensive environmentally responsible design," but it is left to the parties to explain what "extensive" means. Section 4.1.24 covers "LEED® Certification" and references to B214-2007, titled "Standard Form of Architect's Services: LEED® Certification."

AIA B214-2007 allocates the responsibilities of the different project participants in achieving a building available for LEED certification. The architect is the coordinator of the effort; however, the AIA document does not guarantee that the architect's efforts will actually result in certification.

Like the AIA, ConsensusDOCS has an industry form contract specifically dealing with sustainable design and construction. ConsensusDOCS 310: "Green Building Addendum" presumes the project participants (owner, design professional, and contractor and subcontractors) will act collaboratively to undertake "Green Measures" to meet the "Green Status" for the project. "Green Status" is defined as meeting the criteria of a third-party organization, such as the government or LEED. It calls for the use of a Green Building Facilitator (GBF), who coordinates and facilitates obtaining Green Status. Unlike the architect under the AIA, ConsensusDOCS 310 makes the owner responsible for choosing a qualified GBF.

Definition of Terms. There is a lack of precision in terminology describing the "green" design and construction movement. One view is that a green building is one built with the aim of minimizing negative impacts on the environment where recycled materials are used and water- and energy-saving devices and processes are installed.

"High-performance buildings" provide measurable and verifiable improved outcomes of energy and water use when compared to a baseline of comparable buildings.

Probably the most common definition of a "green building" is one that meets a third-party certification standard, usually LEED certification. As noted, that is the definition adopted by the AIA.

The amorphous nature of green design and construction may lead to mismatched expectations between owners and designers, resulting in claims. This issue is addressed in the next section.

Green Building Claims. As green building continues its expansion, green building claims invariably will follow. One claim may arise out of the completed project's failure to achieve green building certification.

The other possible source of green liability is if the completed and operational building fails to achieve the energy savings expected by the owner. As noted earlier in this section, the parties (primarily the owner and architect) may be using terms such as "green" and "high-performance" interchangeably or without clear differentiation. An owner who invests higher construction costs so as to achieve a "green" or LEED certified building may expect a functioning building with reduced energy-consumption costs. If the building does not in fact meet these reduced energy-use expectations, the owner may seek to hold the architect responsible.

This situation will be more difficult for the architect if the design agreement guarantees minimum energy savings. (Any guarantee would open the architect to a claim for breach of warranty.) Even if there is no guarantee, one defense the architect may raise is that he has no control over the owner's operation of the building. For example, occupants may not turn off the lights when they leave the building at night, and complicated mechanical systems may not be used properly to optimize energy efficiency.

Design professionals need to be aware of the risks attendant with the "green" design movement. The convergence of imprecise terminology, design professional marketing, industry "hype," untested technologies, the ideological sentiment underlying sustainability of the built environment, owner desires for quick payback in terms of lower energy costs, and the design professional's lack of control over the building's day-to-day operations and maintenance all present the challenge of "managing green expectations."

11.6 Third-Party Claims Against Design Professionals

Third-party claims raise problems that do not arise when the claimant is connected to the design professional by contract. The proliferation of third-party claims has generated more litigation, varying state rules, and judicial opinions of divided courts than claims by clients against design professionals. This reflects rules in transition with inevitable strains and contradictions.

11.6A Potential Third Parties

The central position of the design professional in construction—both designing and monitoring performance—generates a wide range of potential third-party claimants. At the inner core are the other major direct participants in the process who work on or enter the site itself, such as contractors, construction workers, and delivery people. Around the core cluster are those who supply money, materials, or equipment, such as lenders and suppliers. Next are those who "backstop" direct participants, such as sureties (persons who promise to perform another's obligations; see Chapter 25) and insurers. Still farther from the core are those claimants who will ultimately take possession of the project, such as subsequent owners, tenants, and their employees. Situated farthest from the core are people who may enter or pass by the project during construction or after completion, such as members of the public or patrons. See Figure 11.1 and please review Figures 7.1 and 7.2.

The wide variety of potential claimants, the varying distance from core participants, the type of harm, and the difference between those who have other sources of compensation (such as workers) all combine to ensure complexity.

The analysis of third-party claims in this section focuses on four groups:

1. Contract claims under a theory of third-party beneficiary
2. Contractor or subcontractor claims for economic losses
3. Worker safety claims that arise during the construction phase of the project
4. Duty-to-warn claims brought by members of the general public

11.6B Contract Duty: Contracts for Benefit of Third Parties

Some claims by third parties are based on the assertion that the claimants are intended beneficiaries of contracts to which they are not parties. Intended beneficiaries may sue the promisor (the person making the promise) for breach of contract.

The use of this doctrine has found its way into construction claims. In the context of claims against the design professional, the doctrine is sometimes employed by contractors and subcontractors who assert that they are intended beneficiaries of the contract made between the owner and the design professional.

Such claims have had little success with most cases concluding that contracting parties usually intend to benefit only themselves. Yet the possibility of such a theory being invoked successfully has led contracting parties to seek to use the contract itself to bar claims by third parties as intended beneficiaries. See AIA Documents B101-2007, § 10.5, and A201-2007, § 1.1.2.

11.6C The Design Contract and the Tort Duty of Care

A claimant employing tort law as a basis of transferring its loss to another must show that the latter owed a duty to protect the claimant from an unreasonable risk of harm. In third-party claims against design professionals, the claimants are asserting that the architect's or engineer's breach of contract was the wrongful conduct which caused the claimant's loss. As a preliminary matter, these claimants must prove the design professional had a contractual obligation to perform the act the claimant asserts caused the loss.

A recent Illinois decision, *Thompson v. Gordon*,[18] illustrates the importance of the design contract in circumscribing (or defining) the design professional's potential liability. The plaintiff's husband and daughter died in an automobile accident. They were driving over a bridge when a car going in the opposite direction spun out of control, jumped over the bridge's median barrier, and struck the decedents' car.

The deck of the bridge had been replaced a few years earlier. The plaintiff sued the engineering firms involved in that work. According to plaintiff's expert, the engineers were negligent when they specified replacement of the median barrier with another barrier of the same type. It was the expert's opinion that, had the engineers instead specified use of an improved "Jersey" barrier, the car would not have jumped that barrier and the accident would not have happened.

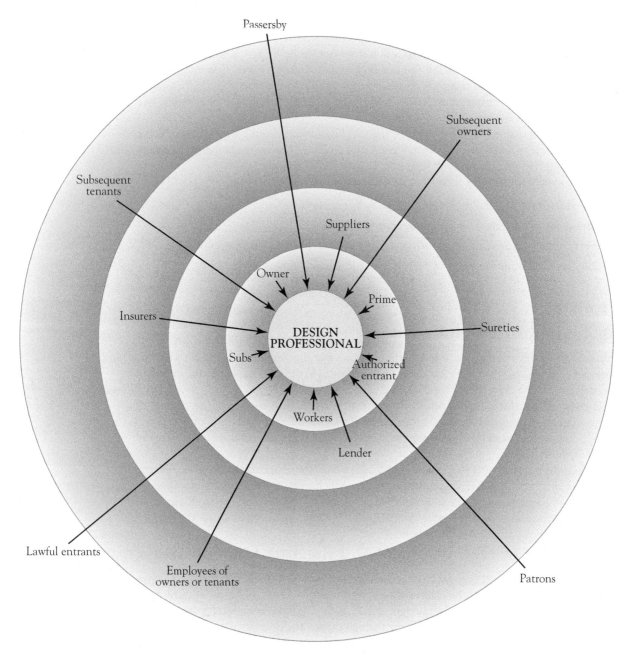

FIGURE 11.1 Potential third-party claimants against design professional.

The Illinois Supreme Court ruled that the engineers did not owe the plaintiff a duty of care. The court explained that since the engineers did not have a contractual duty to design an improved median barrier, they could not owe a duty of care to the plaintiff for failing to do so:

> [T]he scope of defendants' duty is defined by the contract between defendants and [the owner]. The plain language of that contract required defendants to replace the bridge deck, and in doing so, required defendants to use the degree of skill and diligence normally employed by professional engineers performing the same or similar services. . . .

... The trial court in this case correctly found that defendants' duty to plaintiff, and the scope of that duty, was circumscribed by the terms of defendants' contract with [the owner], which did not require defendants to consider and design an improved median barrier. Because defendants owed no duty to plaintiff to consider and design an improved median barrier, the trial court properly granted summary judgment in favor of defendants.[19]

11.6D Negligence Liability for Economic Losses: *Bilt-Rite Contractors, Inc. v. The Architectural Studio* and *SME Industries, Inc. v. Thompson, Ventulett, Stainback & Assocs., Inc.*

Tort law is not limited to claims for personal harm or damage to property. In the construction process, claimants often suffer economic (or financial) losses unconnected with personal harm or damage to property. Often, they seek to recover such losses from participants with whom they do not have a contract. For example, contractors and sureties may assert claims against design professionals by invoking tort law. Many states invoke the *economic loss rule* as a barrier to tort (especially negligence) liability. This section addresses third-party claims for economic losses brought against design professionals.

Negligence Versus Negligent Misrepresentation. For the prime contractor, economic success of the project to a significant degree depends on the actions of the design professional. Ambiguities, omissions, or outright errors in the design may delay the contractor (while corrections are made) and even may force it to tear out and redo work. During the administrative phase of the project, the contractor is dependent upon the design professional to approve submittals, approve progress payment requests, and resolve disputes fairly—to name but a few administrative duties. The design professional's negligent performance of these duties and others will impact the contractor's scheduling of the work as well as its ability to pay itself and its subcontractors and suppliers. These same claimed acts of negligence by the design professional also may impact subcontractors. (For ease of discussion, the remaining section will refer only to the contractor with the understanding that the same analysis generally applies to subcontractors.)

Lacking **privity** (a contract relationship) with the design professional, the contractor who believes its financial loss was caused by the architect or engineer must sue the design professional in tort. These tort claims are brought under theories of negligence or negligent misrepresentation. When is each theory invoked?

True to its name, a *negligent misrepresentation* claim requires the contractor to show a representation from the design professional to the contractor. The contractor must further prove that the representation was false or erroneous, that the contractor foreseeably and reasonably relied upon the representation, and that it suffered the claimed financial injury as a result. The predominant misrepresentation claim is that of negligent design. For example, the contractor may claim the following:

1. The design misrepresented the site's subsurface conditions.
2. This error was negligently made, and
3. The contractor was forced by the actual site conditions to incur higher performance costs than anticipated.

Negligence is a broader concept than negligent misrepresentation, with almost limitless applications. A contractor's claim of negligence may encompass allegations that the design professional:

1. was slow in approving submittals,
2. was negligent in rejecting submittals,
3. imposed on the contractor higher performance standards than were required by the contract, or
4. wrongly refused to approval progress payments.

Many other examples exist.

A leading case, *United States v. Rogers & Rogers*,[20] pointed to the economic life-or-death power the architect has over the contractor in order to justify imposing on the architect a duty of care owed to the contractor for economic losses. Courts that follow this decision and allow contractors to sue design professionals for negligence see no reason to distinguish whether the claimant is seeking purely economic losses or physical harm to person or property.[21]

Not all states permit contractors or subcontractors to sue design professionals in negligence for economic losses. Some states find the design professional owes a contract duty solely to the owner and does not owe a tort duty of care to the contractor or subcontractor. Other states still require privity in claims for purely economic losses. They point to tort law's historic function of protecting personal and property rights and the need to draw some lines beyond which the law should not compel a person to pay for another's loss. Also, decisions barring such claims frequently pointed to the claimant's often having a contractual right against the party with whom it contracted (for example, the contractor may look to the owner for losses caused by the design professional).

Negligent Misrepresentation and the Economic Loss Rule. As noted, a contractor may bring a claim for economic losses against the design professional under a broad theory of negligence or the narrower claim of negligent misrepresentation. In the latter case, the contractor's claim often is brought under the Restatement (Second) of Torts § 552, which imposes negligence liability for financial losses caused by the supply of false information by those in the business of supplying such information. Section 552(1) provides:

> One who, in the course of his business, profession or employment, or in any other transaction in which he has a pecuniary interest, supplies false information for the guidance of others in their business transactions, is subject to liability for pecuniary loss caused to them by their justifiable reliance upon the information, if he fails to exercise reasonable care or competence in obtaining or communicating the information.

As with negligence claims, design professionals may raise two defenses to a claim of negligent misrepresentation: (1) no duty of care absent privity of contract and (2) the economic loss rule. Whether either defense would apply to the contractor's claim is a matter of state common law. Different states will have different rules. There is no "right" or "wrong" answer on the fundamental question of whether a design professional owes a duty of care to third parties for negligently caused economic or fiscal losses.

However, it is important to understand the policy and practical considerations invoked by courts in reaching the conclusions they do. For this reason, two decisions are partially reproduced here. In *Bilt-Rite Contractors, Inc. v. The Architectural Studio*, the Pennsylvania Supreme Court adopted the Restatement (Second) of Torts § 552 as the vehicle by which a contrator may sue the design professional for negligent misrepresentation. It then rejected application of the economic loss rule to bar such a claim. By contrast, in *SME Industries, Inc. v. Thompson, Ventulett, Stainback & Assocs., Inc.*, the Utah Supreme Court ruled that the economic loss doctrine barred the subcontractor's claim. (The fact that *Bilt-Rite* involved a contractor claim, while *SME* was a subcontractor lawsuit, makes no difference to the legal analysis.)

BILT-RITE CONTRACTORS, INC. v. THE ARCHITECTURAL STUDIO
Supreme Court of Pennsylvania, 581 Pa. 454, 866 A.2d 270 (2005)

Opinion
Castille, Justice.

[Ed. note: Footnotes omitted or renumbered.]

[Ed. note: A school district hired an architectural firm ("TAS") to design a new school. The district used the plans and specifications to solicit bids for the work. The successful bidder ("Bilt-Rite") asserted that it incurred higher than anticipated construction costs because—contrary to TAS's representations—the aluminum curtain wall system, sloped glazing system and metal support systems could not be installed using standard construction techniques.

Bilt-Rite sued TAS on a theory of negligent misrepresentation under Section 552 of the Restatement (Second) of Torts, claiming that TAS's specifications were false and misleading, and seeking damages for its increased construction costs. TAS asked the trial court to dismiss the lawsuit, arguing that (1) Bilt-Rite's action was barred by the economic loss rule, which holds that a tort plaintiff cannot recover for purely economic losses; and (2) TAS owed no duty to Bilt-Rite—with whom it had no contractual relationship. The trial court dismissed the lawsuit and the Pennsylvania Superior Court (the court of appeals) affirmed.

After reviewing cases in other states analyzing application of Section 552 to contractor claims against design professional, the Pennsylvania Supreme Court provided its analysis.]

* * *

We are persuaded by these decisions from our sister jurisdictions that: (1) this Court should formally adopt Section 552 of the Restatement (Second), which we have cited with approval in the past, as applied by those jurisdictions in the architect/contractor scenario; (2) there is no requirement of privity in order to recover under Section 552; and (3) the economic loss rule does not bar recovery in such a case. Recognizing such a cause of action, with such contours, is consistent with Pennsylvania's traditional common law formulation of the tort of negligent misrepresentation.

Section 552 sets forth the parameters of a duty owed when one supplies information to others, for one's own pecuniary gain, where one intends or knows that the information will be used by others in the course of their own business activities. The tort is narrowly tailored, as it applies only to those businesses which provide services and/or information that they know will be relied upon by third parties in their business endeavors, and it includes a foreseeability requirement, thereby reasonably restricting the class of potential plaintiffs. The Section imposes a simple reasonable man standard upon the supplier of the information. As is demonstrated by the existing case law from Pennsylvania and other jurisdictions, and given the tenor of modern business practices with fewer generalists and more experts operating in the business world, business persons have found themselves in a position of increasing reliance upon the guidance of those possessing special expertise. Oftentimes, the party ultimately relying upon the specialized expertise has no direct contractual relationship with the expert supplier of information, and therefore, no contractual recourse if the supplier negligently misrepresents the information to another in privity. And yet, the supplier of the information is well aware that this third party exists (even if the supplier is unaware of his specific identity) and well knows that the information it has provided was to be relied upon by that party. Section 552 is not radical or revolutionary; reflecting modern business realities, it merely recognizes that it is reasonable to hold such professionals to a traditional duty of care for foreseeable harm.

. . . we see no reason why Section 552 should not apply to architects and other design professionals. The rationale for this application was persuasively set forth by the Court of Appeals of North Carolina in [*Davidson & Jones, Inc. v. County of New Hanover*, 41 N.C.App. 661, 255 S.E.2d 580 (1979)] where the court stated that such a duty to foreseeable third parties flows from the architect's contractual duties to the party retaining the architect, an approach we embrace:

> An architect, in the performance of his contract with his employer, is required to exercise the ability, skill,

and care customarily used by architects upon such projects. . . .Where breach of such contract results in foreseeable injury, economic or otherwise, to persons so situated by their economic relations, and community of interests as to impose a duty of due care, we know of no reason why an architect cannot be held liable for such injury. Liability arises from the negligent breach of a common law duty of care flowing from the parties' working relationship. Accordingly, we hold that an architect in the absence of privity of contract may be sued by a general contractor or the subcontractors working on a construction project for economic loss foreseeably resulting from breach of an architect's common law duty of due care in the performance of his contract with the owner.

* * *

Davidson, 255 S.E.2d at 584 (citation omitted).

Accordingly, we hereby adopt Section 552 as the law in Pennsylvania in cases where information is negligently supplied by one in the business of supplying information, such as an architect or design professional, and where it is foreseeable that the information will be used and relied upon by third persons, even if the third parties have no direct contractual relationship with the supplier of information. In so doing, we emphasize that we do not view Section 552 as supplanting the common law tort of negligent misrepresentation, but rather, as clarifying the contours of the tort as it applies to those in the business of providing information to others.

On the question of privity, . . . Section 552 imposes a duty of reasonable care upon the supplier of professional information for use by others. Both on its face and as a matter of logic, Section 552 negates any requirement of privity; therefore, we find that the absence of privity does not defeat a Section 552 claim, and does not negate Bilt-Rite's claim in the case *sub judice*.

As to the economic loss rule, the South Carolina Supreme Court took a reasoned approach to the rule in [*Tommy L. Griffin Plumbing & Heating Co. v. Jordan, Jones & Goulding, Inc.*, 320 S.C. 49, 463 S.E.2d 85 (1995)], as it discussed the realities of tort law versus contract law:

> [Our earlier] application of the "economic loss" rule maintains the dividing line between tort and contract while recognizing the realities of modern tort law. Purely "economic loss" may be recoverable under a variety of tort theories. The question, thus, is not whether the damages are physical or economic. Rather, the question of whether the plaintiff may maintain an action in tort for purely economic loss turns on the determination of the source of the duty plaintiff claims the defendant owed. A breach of a duty which arises under the provisions of a contract between the parties must be redressed under contract, and a tort action will not lie. A breach of duty arising independently of any contract duties between the parties, however, may support a tort action.

Tommy L. Griffin Plumbing, 463 S.E.2d at 88 (footnote and citation omitted). The court noted that economic losses are routinely allowed in tort actions in other contexts such as legal malpractice, accountant malpractice, and architect liability.

Like South Carolina, Pennsylvania has long recognized that purely economic losses are recoverable in a variety of tort actions including the professional malpractice actions noted by the South Carolina Supreme Court. We agree with that court that a plaintiff is not barred from recovering economic losses simply because the action sounds in tort rather than contract law. Here, Bilt-Rite had no contractual relationship with TAS; thus, recovery under a contract is not available to Bilt-Rite. Having found that Bilt-Rite states a viable claim for negligent misrepresentation under Section 552, and that privity is not a prerequisite for maintaining such an action, logic dictates that Bilt-Rite not be barred from recovering the damages it incurred, if proven. Indeed, to apply the economic loss doctrine in the context of a Section 552 claim would be nonsensical: it would allow a party to pursue an action only to hold that, once the elements of the cause of action are shown, the party is unable to recover for its losses. Thus, we hold that the economic loss rule does not apply to claims of negligent misrepresentation sounding under Section 552.

* * *

Accordingly, the order of the Superior Court is reversed and the matter is remanded to the trial court for further proceedings.

Former Chief Justice ZAPPALA did not participate in the consideration or decision of this case.

Chief Justice CAPPY files a dissenting opinion.

Justice SAYLOR files a dissenting opinion.

In *Bilt-Rite*, the prime contractor's claim of misrepresentation was that the architect's design expressly represented that the aluminum curtain wall system, sloped glazing system, and metal support systems could be installed and constructed through the use of normal and reasonable construction means and methods. This claim illustrates the flexibility of a negligent misrepresentation claim. Ordinarily, the prime contractor (as an independent contractor) is left to its own devices in deciding how to implement the design. Yet, nothing in the court's opinion intimated that the substance of the contractor's claim was somehow improper or even unusual.

Under *Bilt-Rite*, to whom and under what conditions does the architect owe a duty of care under Section 552? The court says (in a part of the opinion not reproduced here) that Section 552 imposes on design professionals nothing more "than is required by the traditional reasonable man and foreseeability tort paradigm applicable to others." There is no liability unless: (1) the architect was negligent and (2) the plaintiff was foreseeably injured by that negligence.

In short, the *Bilt-Rite* court found adoption of Section 552 to be nothing more than an extension of Pennsylvania's general negligence law, but as applied to claims of misrepresentation brought against those in the business of supplying information and where the plaintiff suffered only financial losses. The court refused to allow the economic loss rule to undermine well-established negligence principles of liability.

While *Pennsylvania is an* example of a successful third-party negligent misrepresentation claim against a design professional, other courts find such a claim barred by the economic loss rule. These courts emphasize the private ordering of the parties under which the risk of design professional negligence may be allocated in the contract—even by parties not in privity with the architect or engineer. One such case, *SME Industries, Inc. v. Thompson, Ventulett, Stainback & Assocs., Inc.*, is reproduced here.

SME INDUSTRIES, INC. v. THOMPSON, VENTULETT, STAINBACK AND ASSOCIATES, INC.
Supreme Court of Utah, 28 P.3d 669 (Utah 2001)

Opinion
RUSSON, Associate Chief Justice
[Ed note: Footnotes omitted or renumbered.]
Plaintiff SME Industries, Inc., brought this action against numerous defendants, seeking delay damages and other economic losses it allegedly incurred while working on a construction project. The trial court granted defendants' motions for summary judgment, and SME appeals.

Background
[Ed. note: SME was the structural steel subcontractor on a convention center project in Salt Lake City. SME asserted that design problems plagued its performance and caused it to incur higher performance costs and expenses from delay. For example, plaintiff alleged that it was required to submit more than 450 requests for information, and the design team's responses were allegedly late, inconsistent and otherwise unhelpful.]

* * *

IV. SME's Tort Claims Against [The Design Team]
... SME's fourth claim alleges professional negligence against all members of the design team. The trial court dismissed all such direct and assigned claims, reasoning that the economic loss rule prevented SME from recovering purely economic damages in tort.

The economic loss rule is a judicially created doctrine that marks the fundamental boundary between contract law, which protects expectancy interests created through agreement between the parties, and tort law, which protects individuals and their property from physical harm by imposing a duty of reasonable care. *See American Towers Owners Ass'n,*

Inc. v. CCI Mech., Inc., 930 P.2d 1182, 1190 (Utah 1996).22 Simply put, the economic loss rule holds that "economic damages are not recoverable in negligence absent physical property damage or bodily injury."23 *Id.* at 1189; *see also* W. Page Keeton et al., *Prosser & Keeton on the Law of Torts* § 92, at 657 (5th ed. 1984); 86 C.J.S. *Torts* § 26 (1997) . . . .

* * *

In the instant case, SME does not deny that its tort claims against the design team seek to recover what we have termed economic loss. Nevertheless, SME argues that its tort claims should be allowed because the economic loss rule is rooted in products liability law, and therefore should not be extended to bar professional negligence claims brought by contractors or subcontractors against design professionals.

Although SME correctly notes that the genesis of the economic loss rule is found in the law of products liability, it ignores the fact that the economic loss rule has been applied in other contexts . . . .

In extending the economic loss rule outside the products liability context, *American Towers* explained that the rationales underlying the doctrine are particularly applicable in the construction setting:

> Construction projects are characterized by detailed and comprehensive contracts that form the foundation of the industry's operations. Contracting parties are free to adjust their respective obligations to satisfy their mutual expectations. For example, a developer can contract for low-grade materials that meet only minimum requirements of the building code. When the developer sells those units, a buyer should not be able to turn around and sue the builder for the poor quality of construction. Presumably the buyer received what he paid for or he can bring a contract claim against his seller. Meanwhile, if the developer has a problem with the builder, he too will have a contract remedy. A buyer can avoid economic loss resulting from defective construction by obtaining a thorough inspection of the property prior to purchase and then by either obtaining insurance or by negotiating a warranty or reduction in price to reflect the risk of any hidden defects.

Id. (citations omitted). Recognizing these realities, we concluded in *American Towers* that relief for defeated economic expectations under a design or construction contract was to come from the contract itself, not from third parties. *Id.* We reasoned that to conclude otherwise would essentially impose the plaintiffs' "economic expectations upon parties whom the [plaintiffs] did not know and with whom they did not deal and upon contracts to which they were not a party." *Id.* at 1192 . . . .

Despite the above, SME argues that the rationale enunciated in *American Towers* for extending the economic loss rule outside the products liability context is inapplicable in this case because *American Towers* involved remote purchasers' claims against an architect, not, like the instant case, a subcontractor's professional malpractice claim against an architect. However, all parties to a construction project, not just the buyers and developers at issue in *American Towers*, resort to contracts and contract law to protect their economic expectations. Indeed, this is particularly true with contractors and subcontractors whose fees are founded upon their "expected liability exposure as bargained and provided for in the[ir] contract[s]." *Berschauer/Phillips Constr. Co. v. Seattle Sch. Dist. No. 1*, 124 Wash.2d 816, 881 P.2d 986, 992 (1994) (en banc). Protection against economic losses caused by another's failure to properly perform, including an architect or design professional, is but one provision a contractor, subcontractor, or sub-subcontractor may require in striking his or her bargain. Accordingly, contractors' negligence claims against architects—like the owners' negligence claims against architects in *American Towers*—are akin to the types of commercial situations to which the economic loss rule was meant to apply. *See id.* at 990 (noting that the "economic loss rule was developed to prevent disproportionate liability and *allow parties to allocate risk by contract*" (emphasis added)).

Moreover, in view of the contractual foundation of the construction industry, and the ability of contractors and subcontractors to negotiate toward the risk distribution that is desired or customary, other jurisdictions have specifically applied the economic loss doctrine to bar contractors' and subcontractors' malpractice claims against architects and design professionals. [Ed note: Citations omitted.]

Therefore, consistent with our prior analysis in *American Towers*, and the foregoing authority from other jurisdictions, we hold that the general rule in this jurisdiction prohibiting the recovery of purely economic loss in negligence is applicable to a contractor's or subcontractor's negligence claim against a design professional (e.g., an architect or engineer).

Alternatively, assuming the economic loss rule does extend to tort suits against design professionals,

> SME requests that this court apply section 552 of the Restatement (Second) of Torts to permit a subcontractor to bring a tort cause of action alleging purely economic damages against a design professional for negligent misrepresentation.
>
> * * *
>
> ... Were we to recognize a cause of action under section 552, however, parties could essentially sidestep contractual duties by bringing a cause of action in tort to recover the very benefits they were unable to obtain in contractual negotiations. Moreover, we see no principled reason why the application of section 552 would not extend liability beyond contractors and subcontractors to an unlimited number of materialmen and workmen who suffer economic injury as a result of a design professional's alleged negligence, which is precisely the type of situation the economic loss rule was designed to prevent. Therefore, to maintain the fundamental boundary between tort and contract law, we hold that when parties have contracted, as in the construction industry, to protect against economic liability, contract principles override the tort principles enunciated in section 552 of the Restatement (Second) of Torts and, thus, economic losses are not recoverable.
>
> Turning to the facts of this case, the gravamen of SME's negligence claims is dissatisfaction with the plans and specifications prepared by the design team. Indeed, SME acknowledges that its tort claims seek purely economic damages, unaccompanied by any claim of personal injury or damage to other property. Moreover, although SME did not contract with [the architect or engineers] for the design of the project and therefore had no opportunity to negotiate directly with the design team regarding the limits of liability, it did have the opportunity to allocate the risks associated with the costs of the work when it entered into a subcontract agreement with [the prime contractor], which proved to be an adequate contractual remedy considering the fact that SME settled with [the prime contractor] for $150,000 and the assignment of [the prime contractor's] claims. Therefore, we conclude that the trial court correctly dismissed SME's direct and assigned negligence claims against the design team under the economic loss rule.
>
> * * *
>
> We affirm the trial court's conclusions that ... (2) the economic loss rule bars SME's direct and assigned negligence claims against the design team ...
>
> Justice DURHAM, Justice DURRANT, Justice WILKINS, and Judge TAYLOR concur in Associate Chief Justice RUSSON's opinion.
>
> Having disqualified himself, Chief Justice HOWE does not participate herein; District Judge TAYLOR sat.

The *SME* court has an expansive view of the economic loss rule. Although the doctrine arose in the context of products liability law, it extends to services contracts as well. In addition, the court found the economic loss rule is particularly appropriate for the construction industry. Every actor on the construction site is there by virtue of a contract. Project participants should anticipate and negotiate performance risks through contractual allocation of responsibility. A contractor may negotiate with the owner (and the subcontractor may negotiate with the prime contractor) the risk of increased and/or delayed performance caused by the design professional's negligence. The fact that SME's tort claim against the design team involves professional, rather than ordinary, negligence made no difference. Finally, adoption of § 552 in the context of a construction project would simply allow the parties to "sidestep contractual duties by bringing a cause of action in tort."

11.6E Intentional Torts: Interference with Contract or Prospective Advantage

Negligence and negligent misrepresentation, which are the topics of Section 11.6D, are claims premised upon liability in negligence. As noted in Section 5.3, a defendant is negligent if he or she did not act as would the reasonable person exercising due care under the same or similar circumstances. Whether conduct (or a failure to act) is negligence is determined by a comparison to an objective, societal standard. Negligence law is largely silent on the question of the defendant's subjective state of mind.

This section addresses intentional torts, in which the defendant's state of mind is relevant. There are advantages for contractors to bring an intentional tort claim against a design professional. Courts generally refuse to apply the economic loss rule to claims of intentional torts. In addition, an intentional tort may subject the defendant to punitive damages. One disadvantage for bringing such a claim is the difficulty of proving the defendant's subjective state of mind.

Most often, a contractor may assert one of two intentional tort claims against a design professional: (1) interference with contract or (2) interference with prospective business advantage. A contractor may use either of these interrelated claims spanning the time from before a contract is awarded until when the contract is terminated.

A claim for interference with prospective business advantage is brought if the contract has not yet been awarded. Here, the contractor is asserting it would have been awarded the contract had it not been for the design professional's interference. The design professional's intentional interference caused the contractor to lose the business advantage of being awarded the contract.

A claim for interference with contract may be brought only after the contractor has entered into a contract with the owner. In this situation, the contractor is asserting that the design professional's interference made contract performance more difficult or led to the wrongful termination of the contract by the owner.

The reason such claims may be brought lies in the pivotal yet varied roles of a design professional in the construction process. In the award stage, he may advise the owner as to which bid should be accepted when a contract is to be awarded by competitive bidding. He may also give advice when a contract is awarded by negotiation.

During performance, the design professional plays a significant role both in advising the owner and in making decisions. The design professional may be asked to advise the owner on subcontractor selection or removal. He may be asked to advise the owner during a negotiation with the contractor over compensation for changes. He may also be required to certify, as in AIA Document A201-2007, § 14.2.2, that there is sufficient cause to terminate the contractor. Other illustrations can be given, but these are enough to show the design professional in his advising role may have a significant impact on the owner's perception of the contractor.

By and large, these claims have not been successful. Some defendants have been able to invoke the Restatement (Second) of Torts, § 772, which states that interference has not been improper if the person against whom the claim has been made has in good faith given honest advice when requested to do so. A design professional who provides honest advice to the owner will not be liable in tort to the contractor—even if that advice led the owner to take adverse action against the contractor.

Despite the relative lack of success of such claims, they still expose design professionals to liability for an intentional tort with the possibility of punitive damages.[24]

11.6F Safety and the Design Professional

Workplace safety is first and foremost a matter for the prime contractor and subcontractors. Section 6.7C of this textbook discusses the prime contractor's responsibility for project safety. Sections 6.7A and B review safety legislation, including the federal Occupational Safety and Health Act (OSH Act). Section 5.7 addresses the independent contractor rule, which is the primary common law doctrine applicable to project personal injury claims.

By contrast, the responsibility of design professionals for safety is a contentious issue. Does the design professional engaged to perform the normal site services have any responsibility for harm to people or to property when the principal cause of the harm has been negligence by the contractor or a subcontractor? This problem has surfaced principally in two forums. The first is the judicial system, where the problem is triggered by a claim by or on behalf of an injured worker, a member of the public, or an adjacent landowner that seeks to transfer to the design professional losses it has suffered by claiming negligence. This negligence usually consists of not taking reasonable steps to prevent contractors from performing work in a way that unreasonably exposes the claimant to personal harm or property damage.

The other forum in which this issue arises is workplace safety laws, such as the OSH Act, or equivalent state regulatory agencies. May agencies charged with administering such statutes charge design professionals with responsibility for unsafe workplaces?

Common Law Liability. Suppose a construction worker is injured on the job and brings a personal injury claim against the project's design professional. If (as is most common) the claim is for negligence, the worker must establish that the design professional owed him a duty of care. How may the design professional respond to such a claim?

Most courts find that contracts that: (1) reserve to the architect or engineer responsibility only to "observe" (not "supervise") the work, (2) impose upon the contractor responsibility for the manner or method of performance and for job-site safety, and (3) grant the owner (not the architect) the authority to stop the work do not impose a duty of care upon the design professional for the safety of the construction workers. The American Institute of Architects standard documents contain all these elements.[25]

Krieger v. J. E. Greiner Co.,[26] is an example of a claim by a construction worker injured on the job who brings a claim against the project's design professional. A construction worker was injured when a steel beam collapsed. He claimed that the erection subcontractor had not properly supported the steel column. He asserted that the work was performed under the supervision of defendant Greiner, the prime engineer, and defendant Zollman, who was a consulting engineer. He sued Greiner and Zollman, claiming that they *should have known* that the subcontractor was performing in a defective and dangerous manner and that the defendants previously had stopped the work when they perceived it being performed in a dangerous manner. The Maryland Court of Appeals (the state's highest court) rejected the worker's claim, stating:

> We have carefully examined each of the contracts in question. We find no provisions in these contracts imposing any duty on the engineers to supervise the methods of construction. . . .
> We likewise find nothing in the contracts imposing any duty on the engineers to supervise safety in connection with construction.
> The duty of the engineers under their contracts is to assure a certain end result, a completed bridge which complies with the plans and specifications previously prepared by Greiner. It will be observed that many of the cases which have held architects and engineers responsible for safety have done so on the basis of the construction by the courts of the contracts existing between the engineer or architect and the owner. We hold that a fair interpretation of the contracts between the Commission and Greiner and Zollman is that the duties of those engineers do not include supervision of construction methods or supervision of work for compliance with safety laws and regulations. Hence, the Kriegers may not recover from the engineers under the contracts between the owner and its engineers.[27]

In *Krieger*, the construction worker argued that the engineers "should have known" of the erection subcontractor's unsafe manner of performance. He essentially was saying that, while the engineers did not actually see the unsafe practice of the subcontractor, they were negligent in not having discovered it (and then stopped it) and should be liable for that negligence.

Suppose, however, that the inspecting architect or engineer *knew* of the hazardous condition beforehand and did not act to have that condition ameliorated. Some courts even here find the design professional not liable because he had no safety duty under the contract (and did not assume a duty by his conduct). The designer's mere knowledge of the hazardous condition, without more, does not create a duty of care where previously none had existed.[28] Other courts are more likely to impose a duty of care in "actual knowledge" cases.[29]

To this point, the court decisions all involved a design professional's liability to construction workers arising out of the architect's or engineer's presence on the job site. An architect who only provided the design and had no involvement in the actual construction would not be liable under such a theory. However, workers may also allege liability arising out of the architect's design duties.[30]

Liability under Occupational Safety and Health Act (OSH Act). As discussed in Section 6.7, the common law worker injury rules are supplemented with federal and state safety statutes. The most important of these is the federal Occupational Safety and Health Act. The OSH Act authorizes the Secretary of Labor to create safety regulations (called "standards") that apply to different occupations, including the construction industry. The Occupational Safety and Health Review Commission (OSHRC) is the highest administrative body to interpret the standards. (Further appeals are made from the OSHRC to the federal courts of appeal and then to the U.S. Supreme Court.)

The construction standards apply to "construction work," which is defined as "work for construction, alteration, and/or repair, including painting and decorating."[31] Clearly the standards apply to contractors and subcontractors, who are licensed and employed to perform the actual construction work. Do the standards also apply to architects or engineers?

The OSHRC first addressed this issue in 1977 when reviewing an administrative citation against a large architectural and engineering firm. In *Secretary of Labor v. Skidmore, Owings & Merrill*,[32] (SOM), the Commission found the designer's roles and authority over the actual construction work was too limited to subject the firm to the construction standards. The Commission stated: "an employer must perform actual construction work or exercise substantial supervision over actual construction. Although SOM exercises some supervision over construction, we would not characterize it as substantial in the sense that supervision by a construction manager is substantial."[33]

This has come to be known as the "substantial supervision" test. Unless the design professional's involvement on the project was "inextricably intertwined" with the actual physical labor of the construction, it is not subject to citation for violation of the construction standards. The courts will examine disclaimers of safety responsibility in the design agreement as one measure of whether the *SOM* test has been met.[34]

(Construction manager liability under the OSH Act is discussed in Section 7.5C.)

11.6G Duty to Warn

Buildings with structural defects may be in danger of collapse. Cliffs overlooking buildings also may be unstable. Suppose a design professional (usually an engineer) hired by an owner to investigate the building or topography uncovers these hazardous conditions. To whom does the engineer owe a duty to warn of the hazardous conditions, other than the owner?

Section 324A of the Second Restatement of Torts addresses a duty to warn third parties against hazards of physical harm. It provides:

> One who undertakes, gratuitously or for consideration [Ed. note: compensation], to render services to another which he should recognize as necessary for the protection of a third person or his things, is subject to liability to the third person for physical harm resulting from his failure to exercise reasonable care to protect [sic: perform] his undertaking, if
>
> (a)   his failure to exercise reasonable care increases the risk of such harm, or
>
> (b)   he has undertaken to perform a duty owed by the other to the third person, or
>
> (c)   the harm is suffered because of reliance of the other or the third person upon the undertaking.[35]

In *Burg v. Shannon & Wilson, Inc.*,[36] the City of Seattle owned bluff property overlooking houses. There was a history of landslides in the area. The city hired a geotechnical engineering firm to evaluate the cause of the latest landslides and to make recommendations to improve stability. The engineer did so. The city agreed to implement the engineer's recommendations, but before work began, a severe storm resulted in landslides that damaged the homeowners' properties. The homeowners sued the engineer for not warning them of the danger. The court of appeals ruled that the engineer had no duty to warn the homeowners of its recommendations to the city. The engineer did not owe the owners a tort duty of care and did not gratuitously assume a duty of care under the Restatement § 324A by engaging in the investigation.

Clearly, duty-to-warn cases raise difficult questions. A design professional's performance obligations are defined by his contract with the owner, and he has no obligation to go outside of those strictures and "improve" upon what the owner has agreed to.[37] Nor do the licensing laws or the code of professional ethics impose such a duty.[38] Yet, the hazardous nature of the defect makes it a matter of concern for tort law with its emphasis on safety.

Perhaps the best explanation for the courts' reluctance to use § 324A is the danger of exposing design professionals to liability to an indefinite class of persons. Knowing a building is unstable is not the same as knowing which neighboring properties are at risk. The practicalities of such a duty are also of concern: Must the engineer go door-to-door urging residents to flee? Since the obvious answer is "no," creating a duty which implies such a course of action is unjustified.

REVIEW QUESTIONS

1. Since design professionals are generally held to the professional standard of care, what are the two things a judge or jury must be able to understand and evaluate?

2. According to the Oklahoma Supreme Court, what conditions must be met in order for a person to testify as an expert witness?

3. If a design professional is considering becoming an expert witness, what four things should be present in the written understanding that precedes any services being performed?

4. What are the advantages and disadvantages of bringing a claim based on breach of contract versus bringing a claim under tort law?

5. According to the AIA B101-2007 Document, what responsibility does the architect have in regard to green design, and what additional services are addressed?

6. Who is likely to bring a third-party beneficiary claim against the design professional, what is the nature of that claim, and what is its likelihood of success?

7. What is a contractor required to show in a claim of negligent misrepresentation?

8. What are the two defenses a design professional may raise against a claim of negligent misrepresentation?

9. What are the competing approaches by the *Bilt-Rite* and *SME Industries* courts that led them to come to opposite conclusions on whether a contractor or subcontractor claim for financial losses, brought against a design professional, is barred by the economic loss rule?

10. Under what legal theories would an injured construction worker seek to impose liability upon the design professional?

ENDNOTES

[1] 42 U.S.C. § 12101 et seq.
[2] 259 F.3d 1029 (9th Cir.2001).
[3] 744 S.W.2d 524 (Mo.Ct.App.1988).
[4] 230 Mont. 151, 749 P.2d 63 (1988).
[5] 252 S.W.3d 586 (Tex.App.Ct.2008), cert. denied, 555 U.S. 1138 (2009).
[6] 373 N.W.2d 744 (Minn. 1985).
[7] Id. at 754-55.
[8] *Thompson v. Gordon*, 221 Ill.2d 414, 851 N.E.2d 1231 (2006).
[9] *Edgewater Apartments, Inc. v. Flynn*, 216 A.D.2d 53, 627 N.Y.S.2d 385 (1995) (architect may testify in claim against engineer for water intrusion), subsequent appeal, 268 A.D.2d 227, 701 N.Y.S.2d 357 (2000); *Tomberlin Assoc., Architects, Inc. v. Free*, 174 Ga.App. 167, 329 S.E.2d 296 (1985), cert. denied May 1, 1985 (civil engineer allowed to testify in claim against architect in claim of soil erosion).
[10] *Keel v. Titan Constr. Corp.*, 639 P.2d 1228 (Okla.1981).
[11] This matter of scientific reliability was addressed by a trilogy of Supreme Court decisions: *Daubert v. Merrill Dow Pharmaceuticals, Inc.*, 509 U.S. 579 (1993); *General Electric Co. v. Joiner*, 522 U.S. 136 (1997); *Kumho Tire Co. Ltd. v. Carmichael*, 526 U.S. 137 (1999).
[12] *Mondelli v. Kendel Homes Corp.*, 262 Neb. 263, 631 N.W.2d 846, opinion modified on denial of rehearing, 262 Neb. 663, 641 N.W.2d 624 (2001); O'Neal, *Sick Building Claims,* 20 Constr. Lawyer, No. 1, Jan. 2000, p.16.
[13] *Western Technologies, Inc. v. Sverdrup & Parcel, Inc.*, 154 Ariz. 1, 739 P.2d 1318 (App.1986) (claim against adverse witness); *Bruce v. Byrne-Stevens & Assocs. Engineers, Inc.*, 113 Wash.2d 123, 776 P.2d 666 (1989) (claim against own witness, an engineer, for negligence in testifying).
[14] *Moore v. Bird Eng'g Co. P.A.*, 273 Kan. 2, 41 P.3d 755 (2002).
[15] *Ragucci v. Professional Constr. Services*, 25 A.D.3d 43, 803 N.Y.S.2d 139 (2005).
[16] E.g., N.J.S.A. §§ 52:32–5.3 and 52:32–5.4. For a listing of many states' requirements, see O'Connor, Jr., *Legal Considerations in Sustainable Design and Construction*, 5 J ACCL, No. 1, Winter 2011, pp. 137, 151–54.
[17] http://www.usgbc.org/leed, last examined June 23, 2013. For a comprehensive historical review of the evolution of sustainable design, see Hurtado, *Emerging Standards of Care for Sustainable Design & Construction*, 5 J ACCL, No. 1, Winter 2011, pp. 193, 197–209.
[18] 241 Ill.2d 428, 948 N.E.2d 39 (2011).
[19] Id., 948 N.E.2d at 51–52.
[20] 161 F.Supp. 132 (S.D.Cal.1958).
[21] Annot., 61 A.L.R. 6th 445 (2011).
[22] [Ed. note. Abrogated on other grounds, *Davencourt at Pilgrims Landing Homeowners Ass'n v. Davencourt at Pilgrims Landing, LC*, 221 P.3d 234 (Utah 2009).]
[23] However, plaintiffs may recover purely economic losses in cases involving intentional torts such as fraud, business disparagement, and intentional interference with contract. *American Towers*, 930 P.2d at 1190 n. 11.
[24] Schneier, *Tortious Interference with Contract Claims Against Architects and Engineers*, 10 Constr. Lawyer, No. 2, May 1990, p. 3; Annot., supra note 21.
[25] For example, AIA Doc. A201-2007, § 2.3.1 gives the owner alone the power to stop the work.

[26] 282 Md. 50, 382 A.2d 1069 (1978).
[27] 382 A.2d at 1079.
[28] *Yow v. Hussey, Gay, Bell & DeYoung Internat'l, Inc.*, 201 Ga.App. 857, 412 S.E.2d 565, 568 (1991), cert denied Jan. 30, 1992; *Herczeg v. Hampton Municipal Auth.*, 2001 PA Super 10, 766 A.2d 866, 873–74, appeal denied, 567 Pa. 742, 788 A.2d 376 (2001).
[29] *Carvalho v. Toll Bros. & Developers*, 143 N.J. 565, 675 A.2d 209 (1996). Scholars are in accord, see Sweet, *Site Architects and Construction Workers: Brothers and Keepers or Strangers?* 28 Emory L.J. 291 (1979).
[30] See generally, M. SCHNEIER, CONSTRUCTION ACCIDENT LAW: A COMPREHENSIVE GUIDE TO LEGAL LIABILITY AND INSURANCE CLAIMS, Chapter 5 (Am.Bar Assn.1999).
[31] 29 C.F.R. § 1910.12(b) (2012).
[32] 5 O.S.H.Cas. (BNA) 1762, 1977-1978 O.S.H.D. (CCH) ¶ 22101 (O.S.H.Rev.Comm'n.1977).
[33] 5 O.S.H.Cas. (BNA) at 1764.
[34] *CH2M Hill, Inc. v. Herman*, 192 F.3d 711 (7th Cir.1999).
[35] Restatement (Second) of Torts § 324A (1965).
[36] 110 Wash.App. 798, 43 P.3d 526 (2002). See also *Gooch v. Bethel A.M.E. Church*, 246 Kan. 663, 793 P.2d 993 (1990) (collapse of church; no liability).
[37] See the discussion concerning *Thompson v. Gordon*, supra note 18, in Section 11.6C.
[38] See the discussion concerning *Dukes v. Philip Johnson/Alan Ritchie Architects, P.C.*, supra note 5, in Section 11.2.

CHAPTER 12

DEFENSES TO CLAIMS OF DESIGN PROFESSIONAL LIABILITY

SCENARIO

Mega Manufacturer, Inc. (MMI) hired Awesome Architect, Inc. (AAI) to design and administer a project to build a new factory. AAI hired a geotechnical engineering firm, Geo-R-Us, Inc., to perform a soil report. AAI used that soil report to design the excavation and foundation specifications.

Three and one-half years after completion, a one-hundred-year rainstorm struck the area where the new factory was located. Several months later, the factory began to suffer cracks and settlement damage.

MMI hired AAI to investigate the cause of the problem. After examining the situation, AAI concluded that the settlement was caused by improper construction practices by the prime contractor. The contractor disagreed and blamed AAI's design.

In response, MMI hired an engineering firm, Investigate-4-U, Inc., to examine the cause of the settlement. Investigate-4-U concluded that Geo-R-Us, Inc. was greatly deficient in its examination of soil conditions, and as a result reported soil conditions to be firmer than they in fact were. As a result, AAI's excavation and foundation specifications were inadequate. Investigate-4-U stated that, while the problem was exacerbated by the unusual storm, settlement would have happened under normal conditions, probably within a decade after project completion.

Within a few weeks of receiving Investigate-4-U's report, MMI sued AAI for breach of contract and negligence. The lawsuit was filed four years and one month after acceptance of the project.

AAI brought a third-party complaint against Geo-R-Us for breach of contract and negligence.

12.1 Legal Defenses to Claims of Design Professional Liability: Introduction

The principal issue in claims against design professionals relates to the standard of performance and whether there has been compliance. The first part of this chapter briefly outlines the most common *legal* defenses that have been used by the design professional when claims are made by the client or third parties.

In addition to the special legal defenses discussed here, a design professional has available standard defenses to *any* claim. For example, any claimant must prove causation: that a defendant's faulty conduct caused the claimant's loss.

Section 12.7 discusses contractual risk transfer strategies a design professional may employ.

12.2 Acceptance of the Project

Acceptance of the project usually involves the owner's taking possession of the completed project. Does taking possession imply that the owner is satisfied with the work and is giving up (waiving) any claim for existing defects or defects that may be discovered in the future?

If the defect was known to the owner at the time of acceptance and the owner did not object or reduce payment due the contractor or designer, a court may find that the owner waived any claim. However, certainly no such assumption should be made concerning hidden or latent defects. Moreover, standard form documents explicitly reject equating acceptance of the project with waiver of claims. For example, AIA A201-2007, § 9.10.4, makes clear that final payment does not bar most owner claims, including that the work did not conform to the design.

> In this chapter's scenario, any defects in the factory's design or construction were latent—not something that MMI (as owner) would have noticed absent an extensive investigation. MMI's acceptance of the factory was not a waiver of any claims it had against the architect or contractor.

Normally, acceptance does not bar third-party claims against design professionals. If acceptance bars claims, it is because the intervening act of the owner—that of acceptance—breaks the causal link between the design professional's negligence and the claimant's injury. This rationale is particularly weak when the claim is asserted against the design professional—the person who often decides whether the project has been completed and should be accepted. (While the design professional's decision may be appealed to arbitration, if it is not appealed, it will be final and binding on the parties.)

12.3 Passage of Time: Statutes of Limitations

Sometimes the design professional can defend by establishing that legal action was not started within the time required by law. This is usually accomplished by invoking the statute of limitations as a bar to the claim. Because this bar can apply to all claims—not only to those against design professionals—and because most construction-related claims are brought against other participants such as the owner or contractors, full discussion of this defense is postponed until Section 16.9D. A few observations concerning claims against design professionals, will be made in this section.

Statutes of limitations usually prescribe a designated period of time within which certain claims must be brought. One of the troublesome areas in construction claims relates to the point at which that period begins.

This question is exacerbated when claims are made against the design professional. While the statute of limitations begins to run on virtually all owner claims when the project is completed, suppose the owner and design professional work together to deal with defects discovered after project completion. Owners often turn to their architects and contractors for initial help in determining the source of the problem. Yet an owner faced with a defective building must know that the architect's design may be the cause of the defect—even if the architect places the blame elsewhere. A court which finds that the owner reasonably relied upon the architect's advice so as to delay filing suit against her may invoke the **continuous treatment doctrine** to **toll** (temporarily suspend) running of the statute of limitations.[1]

> In this chapter's scenario, suppose the statute of limitations for breach of contract is four years, commencing with project completion. MMI saw settlement damage after three and one-half years, but did not bring its lawsuit until four years and one month after project completion. AAI will defend against the lawsuit on the ground that the claim is barred by the statute of limitations. MMI will counter that the limitations period was tolled while it investigated the cause of the settlement, using AAI as its consultant.

When does the limitations period begin when a contractor sues the design professional for economic losses caused by a defective design? The design professional would argue that the limitations period commenced when the contractor signed the contract and began to use the design. However, courts instead find that the limitations period begins when the contractor incurs economic loss with certainty.[2]

Generally, a defense based on the passage of time—although still of value to the design professional—has provided limited protection. This has led design professionals to seek enhanced protection by contract. These contract protections are discussed in Section 12.2F.

12.4 Design Professional Decisions and Immunity

The design professional is frequently given the power to interpret the contract documents, resolve disputes, and monitor performance. When claims are brought against a design professional by the client or others for activity that can be said to resemble judicial dispute resolution (such as deciding disputes or issuing certificates),[3] design professionals sometimes assert they should receive "quasi-judicial" immunity.

This is based on the analogy sometimes drawn between judges and the design professionals performing judge-like functions. For example, the judge is given absolute immunity (legal protection) from civil action—even for fraudulent or corrupt decisions. The corrupt judge may be removed from office or subject to criminal sanctions. But a disappointed party cannot institute civil action against a judge. Immunity protects judges from being harassed by vexatious litigants and encourages them to decide cases without fear of civil action being brought against them.

Where quasi-judicial immunity is given, the design professional cannot be sued for decisions made unless they were made corruptly, dishonestly, or fraudulently. Yet the peculiar position of the design professional—independent contractor acting as designer, agent of the owner, decider of disputes, and an individual participating in construction projects—has caused difficulty. In *C. Ernst, Inc. v. Manhattan Constr. Co. of Texas*,[4] a federal court of appeals applied a limited immunity light of the many roles played by the designer:

> The arbitrator's "quasijudicial" immunity arises from his resemblance to a judge. [Ed.: The court here is speaking of the architect as arbitrator.] The scope of his immunity should be no broader than this resemblance. The arbitrator serves as a private vehicle for the ordering of economic relationships. He is

a creature of contract, paid by the parties to perform a duty, and his decision binds the parties because they make a specific, private decision to be bound. His decision is not socially momentous except to those who pay him to decide. The judge, however, is an official governmental instrumentality for resolving societal disputes. The parties submit their disputes to him through the structure of the judicial system, at mostly public expense. His decisions may be glossed with public policy considerations and fraught with the consequences of stare decisis [a Latin phrase for precedent]. When in discharging his function the arbitrator resembles a judge, we protect the integrity of his decision-making by guarding his fear of being mulcted in damages.... But he should be immune from liability only to the extent that his action is functionally judge-like. Otherwise we become mesmerized by words.[5]

The court then concluded that such immunity, as possessed by the architect as arbitrator, did not extend to unexcused delay or failure to decide. The immunity was limited to "judging"—not for failing to judge in a timely manner.

Contractors question the partiality of the design professional. They point out that the designer is hired and paid by the owner. In addition, they are concerned that the design professional will likely blame the contractor rather than their own design as the cause of project problems. For these reasons, the design professional's decision is usually preliminary and is subject to review by arbitration or litigation. See Sections 26.1 and 26.2.

12.5 Statutory Defenses

Legislative protections for design professionals fall into two broad categories. First, some statutes do not limit liability but impose special procedural or evidentiary requirements on the claimant with the intended goal of preventing frivolous claims. The second category of statutes limits the design professional's liability.

Certificate of Merit Statutes. These laws are the primary device for weeding out frivolous lawsuits through the imposition of special pleading burdens on claimants. These laws generally require the plaintiff to attach to the complaint a certificate from her attorney declaring that the attorney has consulted with an expert and that the attorney or expert has concluded that the suit is meritorious.[6]

For example, California Code of Civil Procedure § 411.35 requires the attorney for a claimant who files a claim for damages or indemnity arising out of the professional negligence of a licensed architect, a licensed engineer, or a licensed land surveyor to file a certificate stating that the attorney has reviewed the facts of the case; that she has consulted and received an opinion from a licensed design professional in the same discipline as the defendant who she reasonably believes is knowledgeable in the relevant issues; and that she has concluded "that there is reasonable and meritorious cause for the filing of this action and that the person consulted gave an opinion that the person against whom the claim was made was professionally negligent." Provisions seek to protect the identity of the design professional who has been consulted. In *Ponderosa Center Partners v. McLellan/Cruz/Gaylord & Assoc.*,[7] the California Court of Appeal approved a certificate by a structural engineer in a claim made against an architect in a roof collapse.

Georgia has gone farther, enacting legislation requiring that a complaint charging professional malpractice be accompanied by an affidavit of an expert setting forth at least one negligent act or omission.[8] Texas further restricts the expert affiant to one "holding the same professional license as, and practicing in the same area of practice as the defendant."[9] This Texas requirement may be the most restrictive in the country.

Although the statutory intent of eliminating frivolous claims against design professionals is laudable, critics complain that the certificate of merit requirement presents technical hurdles that can stop valid claims from reaching a decision on the merits.[10] Interpreting the statutory requirements also has generated a large quantity of appellate case law, which adds to the cost and delay of clients seeking a remedy.[11]

In this chapter's scenario, it is not stated whether the state where the project was located and the lawsuit brought has a certificate of merit statute. If it does, and MMI did not file an expert affidavit with the complaint, AAI will ask the court to dismiss the lawsuit. If the court finds the statute was violated and that no exception applies to excuse that omission, the court must grant AAI's request, even though the substantive questions—such as whether AAI was negligent and that negligence caused MMI's injury—are never reached.

"Good Samaritan" Laws. These laws protect design professionals (at least to a limited degree) from potential liability when they provide expert services in emergencies. The statutes vary as to the type of emergency that will invoke the protective legislation and who can benefit from such legislation. As an illustration, under California law, an architect or engineer who, at the request of a public official but without compensation or expectation of compensation, voluntarily provides structural inspection services at the scene of a declared emergency "caused by a major earthquake, flood, riot, or fire" is not liable for any harm to person or property caused by her good faith but negligent inspection of the structure. Protection does not extend to gross negligence or willful misconduct. The inspection must occur within thirty days of the "declared emergency."[12] Some states have extended this protection to those who provide inspection or other services following any natural disaster.[13]

Workers' Compensation. As explained in Section 2.12B, workers' compensation laws immunize employers from claims by their employees to workplace injuries or deaths. Beginning in the mid-1980s, associations of design professionals were successful in persuading a number of states to grant immunity to design professionals from third-party claims by injured workers (who are covered by workers' compensation) for site services unless the design professional has contracted to oversee safety or undertook to do so. None of these statutes grant immunity for negligent design.[14]

12.6 Apportionment for Fault of Others

In construction, losses are commonly caused by a number of participants and conditions. For example, an owner, believing its new building suffers from construction defects, may sue the contractor for negligent workmanship and the design professional for failure to discover the defects while inspecting the work. However, the owner may not recover twice for the same loss. (See Section 4.9.) Therefore, when more than one defendant is liable for the same injury, apportionment of damages may be possible. A few generalizations can be made about this type of situation.

Suppose the client asserts a claim against the design professional, but the loss has been caused in part by the client's negligence. If the client is allowed to pursue a tort claim, generally the negligence of the client and that of the design professional will be compared, and the amount recoverable by the client will be reduced by the amount attributed to its own negligence, as noted in Section 5.3F. For example, if the loss was $100,000 and a determination was made that 30 percent of the loss should be chargeable to the client, the client would recover $70,000. A few states apply the contributory negligence rule to bar the entire claim.

In this chapter's scenario, if MMI sues both AAI and Geo-R-Us for negligently causing damage to the facility, any recovery the owner obtains against one defendant will be subtracted from a recovery against the second. MMI will not be allowed to recover twice for the same injury.

In any claim by AAI against Geo-R-Us, Geo-R-Us will apportion its own fault with that of the architect. For example, Geo-R-Us may contend that the settlement damage was also caused by the contractor's defective workmanship, which AAI should have discovered and ordered corrected.

12.7 Contractual Risk Control

This section looks at specific contract clauses that can be useful in risk management.

12.7A Scope of Services

The contract should make clear exactly what the design professional is expected to do. It is important that the client and design professional have the same understanding of what services the designer will provide. Perhaps most important is the design professional's role in determining how the work is being performed and the responsibility of the design professional for the contractor not complying with the contract documents.

12.7B Standard of Performance

Usually the design professional wishes to be held to the professional standard discussed in Chapter 11. This is demonstrated by the AIA's inserting the professional standard for the first time in AIA Document B101-2007, § 2.2. In addition, it is important that the contract not use words such as *assure, ensure, guarantee, achieve, accomplish, fitness,* or *suitability* or any language that appears to promise a specific result or achievement of the client's objectives.

12.7C Exclusion of Consequential Damages

A breach of contract may cause not only direct costs but also indirect or "consequential" costs that can be large and often unquantifiable. Although consequential damages may ultimately be barred under the foreseeability requirement, waiver of consequential damages clauses seek to preclude the claim from arising in the first place. These contractual waivers are discussed in Section 4.5 in the subsection titled "Contract-Specified Remedies."

12.7D "Limitation of Liability" Clauses

As discussed in Section 10.5, the primary design professional (often an architect) will subcontract much of the design work to specialists, usually engineers. For example, the architect will hire a geotechnical engineer to analyze the soil conditions and recommend design of the building's foundation. Given the limited role it played in creating the design, the geotechnical engineer's fee is typically modest. At the same time, an error in the engineer's work may have catastrophic consequences to the project. In short, the engineer's potential liability greatly exceeds its compensation. This is especially so when considering the engineer's risk exposure is to both the client and third parties.

Beginning in the 1970s, professional associations of engineers, particularly the American Society of Foundation Engineers[15] and the National Society of Professional Engineers, advocated that their members seek to get their clients to agree to provisions in their contracts that would limit the liability of the engineer. Efforts to obtain such provisions were also spurred on by professional liability insurers that offered a reduced premium for those who obtained these limitations in their contracts.

A **limitation of liability** clause seeks to cap the design professional's potential liability to a specified, maximum amount.[16] Typically, this maximum amount is a particular sum or the designer's fee, whichever is greater. For example, a contract between an owner and an architectural and engineering firm, entered into in the late 1980s, contained this language:

> The OWNER agrees to limit the Design Professional's liability to the OWNER and to all construction Contractors and Subcontractors on the project, due to the Design Professional's professional

negligent acts, errors or omissions, such that the total aggregate liability of each Design Professional shall not exceed $50,000 or the Design Professional's total fee for services rendered on this project. Should the OWNER find the above terms unacceptable, an equitable surcharge to absorb the Architect's increase in insurance premiums will be negotiated.[17]

Looking at the second paragraph first, note that the designer will agree to waive the liability limitation only if the owner agrees to pay the designer's increased premium for higher insurance coverage. In effect the designer's fee remains the same, but the owner is buying increased insurance coverage. Not all limitation of liability clauses include this option.

As for the first paragraph, the owner agrees to cap the designer's "total aggregate liability" to (1) the owner and (2) the contractor and subcontractors. No matter who sues it, the designer's damages will not exceed either its fee or $50,000, whichever figure is higher. If the design professional is sued only by the client, then this limitation is easy to understand: no matter what the owner's proven damages, it cannot recover more than the monetary limit. This is a matter of agreement between the litigating parties.

But suppose the design professional is sued by third parties and they recover against the designer more than the stipulated maximum. These plaintiffs were not parties to the design agreement and did not agree to be bound by the liability limitation. In this situation, the designer enforces the limitation clause by requiring the owner to indemnify the designer against the third party damages awards in excess of the cap. ("Indemnify" means that the owner promises to pay for any claim in excess of the contractual cap. See Chapter 23.)

Limitation of liability clauses can take many forms. For example, EJCDC Doc. E-500 (2008), Exhibit I, ¶ A(1), includes three alternative liability limitation clauses for the engineer to consider. Each seeks to cover both client claims and third-party claims by specifying that the provision applies to the engineer's "total liability, in the aggregate." The first limits the engineer's liability to the amount of its compensation, the second to the amount of insurance proceeds, and the third to a specified sum that the parties agree upon.

Early opponents to limitation of liability clauses argued that they are a form of *exculpatory* clause, which essentially shields the protected party from the consequences of its own negligence. Because they may be viewed as encouraging dangerous or reckless behavior, exculpatory clauses are not enforced as a matter of public policy. However, in commercial transactions, courts usually find that the stipulated maximum liability is sufficiently high to give the designer an incentive to act with due care.[18] Stated otherwise, a clause is enforced if made by business people in a commercial setting unless the amount is so low that it removes any incentive to perform with due care.[19]

The use of limitation of liability clauses in consumer transactions is more likely to get judicial scrutiny. For example, in *Estey v. Mackenzie Engineering, Inc.*,[20] a structural engineer was retained to make a limited, visual review of a house his client was considering purchasing. His fee was $200, and the contract limited his liability for negligence to that amount. The client claimed the engineer had been negligent. At the time of trial, the client proved losses of $190,000 due to needed repairs with an estimated cost of $150,000 yet to be incurred. The engineer pointed to the limit of liability. The Oregon Supreme Court refused to enforce the clause. It stated that the client probably did not intend to take this risk.

Although it would seem unfair for the innocent client to suffer this loss, refusing to enforce such a clause in this context may mean that fewer engineers will perform this service even if they are insured, or they will drastically increase the fee. Without enforcement, a quick, cheap inspection is very likely not possible.

As noted earlier, where third-party claims exceed the cap, the client is required to indemnify the design professional for the amount of its liability in excess of the cap. However, as discussed in Section 23.8, many states have passed anti-indemnity statutes which restrict permissible indemnity agreements on construction projects, particularly if they indemnify the indemnitee (here, the design professional) when its negligence was the sole cause of the loss. Those who seek to bar enforcement of a limitation of liability often claim that the anti-indemnity statute precludes enforcement of a limitation of liability clause. While such a claim has had some success, most courts distinguish between liability limitation clauses and true indemnity clauses and hold that the statute does not affect the limitation of liability.[21]

Liability limitation clauses will continue to play a significant role in risk management for design professionals. If the specified amount does not result in complete exculpation, if the parties appear to know what they are doing and if the context is truly commercial, such clauses will very likely be enforced.[22]

> In this chapter's scenario, suppose the AAI/Geo-R-Us contract contains a limitation of liability clause, which provides that Geo-R-Us's liability under the contract was limited to its professional fee (which was $20,000) or $25,000, whichever was more. Geo-R-Us will argue that it cannot be liable to AAI for more than $25,000. AAI will contend that the clause is an improper exculpatory clause because, even if Geo-R-Us is found to be negligent, it will suffer no more than $5000 over its fee of $20,000. Under these facts, AAI will argue that Geo-R-Us would not have an incentive to act with due care.

12.7E Immunity: Decision Making

As noted in Section 12.1C, some American courts grant design professionals quasi-judicial immunity when they decide disputes under the terms of the construction contract. Many standard agreements published by the professional associations incorporate language that relieves the design professional from any responsibility if decisions are made in good faith.[23] They attempt to incorporate into the contract limited quasi-judicial immunity. To be effective, such clauses must be incorporated in the contracts both for design services and for construction services.

12.7F Contractual Statute of Limitations

As noted earlier, judicial claims can be lost simply by the passage of time via statutes of limitations. However, it may be difficult to determine which statute of limitations applies and when the limitations period (the time permitted for bringing a lawsuit) begins to run. One response is to create a private statute of limitations by contract.

The parties may agree to a shorter limitations period than that specified in the statute of limitations. Such an agreement will be enforceable if the shorter period is "reasonable," meaning it cannot be so short as to deprive the potential claimant (here, the owner) from a realistic opportunity to discover defects.

In addition, the contract may specify when the statute of limitations begins. For example, AIA Documents B141-1997, ¶ 1.3.7.3, and A201-1997, ¶ 13.7 both started the limitations period from the date of either substantial or final completion. The effect of such a clause was to deprive the claimant of the benefits of tolling doctrines such as the "continuous treatment" rule or the discovery rule.

> In this chapter's scenario, suppose the MMI/AAI contract contains a clause similar to B141-1997, ¶ 1.3.7.3. AAI will argue that, because under the contract the four-year limitations period began to run upon project completion, neither the "continuous treatment" doctrine nor the discovery rule precluded the lawsuit from being barred by the statute of limitations. The contract provision, AAI will argue, defeats application of the judicial tolling doctrines otherwise applicable to the statute of limitations. Because the claim was brought more than four years after completion of the project, the claim would be barred.

These 1997 AIA provisions, which started the limitations period from the date of either substantial or final completion, were dropped in the 2007 AIA documents. Under the new AIA Documents B101-2007, § 8.1.1 and A201-2007, § 13.7, suit must begin within the period specified by applicable law (usually the law of the state where the project is located) but, in any case, not more than 10 years after the date of substantial completion. This creates a private statute of repose.

12.7G Third-Party Claims

Third-party claims are sometimes based on the assertion that the claimant is an intended beneficiary of the contract. This contention can be negated by appropriate contract language. For example, AIA Documents B101-2007, § 10.5, and A201-2007, § 1.1.2 (with a limited exception for the architect) seek to bar those not parties to the contract from asserting rights as intended beneficiaries.

While these clauses usually are successful, they are unlikely to provide a defense to a tort claim.

12.7H Dispute Resolution

Some believe that the most important risk management tool is to control the process by which disputes will be resolved. Many American standardized construction contracts give first-instance dispute resolution to the design professional and frequently provide for an appeal to arbitration. Because arbitration has become such an important feature of construction contract dispute resolution, it is covered in detail in Chapter 26.

12.7I Some Suggestions

Many other contract protections may be attempted, such as an exculpation for consultants, a favorable choice of applicable law, waiver of a jury trial, a power to suspend work for nonpayment, insurance premiums as a reimbursable, and a stiff late payment formula (to mention a few). However, the realities of contract bargaining necessitate that emphasis be placed on those tools most useful in risk management. It makes little sense to spend precious negotiation time and bargaining power on seeking to obtain concessions that in actuality mean very little.

Finally, the clauses noted in this section are more likely to be enforced if they are drafted clearly and express specific reasons for their inclusion, if the client's attention is directed to them, and if suggestions are made to the client to seek legal advice if he has doubts or questions about them.

REVIEW QUESTIONS

1. If a defect is known to the owner at the time of acceptance, and the owner did not object or reduce payment due the contractor or designer, has the owner waived any claim against the contractor or designer?

2. If a contractor sues a design professional for economic losses caused by a defective design, when does the statue of limitations period begin?

3. What is the difference between the immunity a judge receives versus the quasi-judicial immunity that is given to a design professional who acts as an arbitrator?

4. What are the two broad categories of legislative protections for design professionals?

5. What are certificate of merit statutes and how do they function?

6. What are some words that should be avoided in the contract of a design professional in regard to their standard of performance?

7. What is the purpose of a "limitation of liability" clause and how is the maximum amount of damages determined?

8. If a client sues both the contractor and the design professional, and settles or obtains a judgment against the contractor, how might this be of benefit to the design professional?

9. What are two methods that parties can use to define the statute of limitations in a contract?

10. Under what conditions are exculpatory clauses enforced?

ENDNOTES

[1] E.g., *Lake Superior Center Auth. v. Hammel, Green & Abrahamson, Inc.*, 715 N.W.2d 458 (Minn.App.2006), review denied Aug 23, 2006.

[2] *Hardaway Co. v. Parsons, Brinckerhoff, Quade & Douglas, Inc.*, 267 Ga. 424, 479 S.E.2d 727 (1997); *MBA Commercial Constr., Inc. v. Roy J. Hannaford Co.*, 818 P.2d 469 (Okla.1991).

[3] See Section 17.5; Annot., 43 A.L.R.2d 1227 (1955).

[4] 551 F.2d 1026, rehearing denied in part and granted in part, 559 F.2d 268 (5th Cir.1977), cert. denied sub nom. *Providence Hosp. v. Manhattan Constr. Co. of Texas*, 434 U.S. 1067 (1978).

[5] 551 F.2d at 1033.

[6] Montez, *Certificate of Merit Statutes*, 24 Constr. Lawyer, No. 4, Fall 2004, p. 40.

[7] 45 Cal.App.4th 913, 53 Cal.Rptr.2d 64 (1996), review denied Aug. 14, 1996.

[8] Ga.Code Ann., § 9-11-9.1.

[9] Tex.Civ.Prac. & Rem.Code Ann. § 150.002(a) (2009).

[10] *Sharp Eng'g v. Luis*, 321 S.W.3d 748, 754 (Tex.App.Ct.2010) (concurrence by Justice Sullivan) (characterizing the statute as "trap for the unwary rather than a screen for meritless claims") (footnote omitted).

[11] Id. at 753, n. 2 (listing numerous Texas court of appeals decisions interpreting the statute).

[12] Bus.&Prof.Code §§ 5536.27 and 6706.

[13] Tenn.Code Ann. § 62-2-109; Mo.Rev.Stat. § 44.023.

[14] E.g., Conn.Gen.Stat.Ann. § 31-293(c); West Fla.Stat.Ann. § 440.09(6); Kan.Stat.Ann. § 44-501(d); West Okla.Stat.Ann.Tit. 85, § 12; Wash.Rev.Code Ann. § 51.24.035.

[15] This is the association's original name; it is now AFSE/The Geoprofessional Business Association.

[16] A more accurate name would be "limitation of damages," because these clauses do not, in fact, limit liability, but instead specify maximum damages. Still, "limitation of liability" is the commonly accepted name.

[17] *Valhal Corp. v. Sullivan Assocs. Inc.*, 44 F.3d 195, 198 (3d Cir.1995).

[18] E.g., *Valhal Corp. v. Sullivan Assocs. Inc.*, supra note 17 (cap seven times the fee).

[19] *Mistry Prabhudas Manji Eng. Pvt. Ltd v. Raytheon Engineers & Constructors, Inc.*, 213 F.Supp.2d 20 (D.Mass.2002).

[20] 324 Or. 373, 927 P.2d 86 (1997).

[21] *1800 Ocotillo, LLC v. WLB Group, Inc.*, 219 Ariz. 200, 196 P.3d 222 (2008); *Fort Knox Self Storage, Inc. v. Western Technologies, Inc.*, 140 N.M. 233, 142 P.3d 1 (App.Ct.2006).

[22] Beltzer & Orien, *Are Courts Limiting Design Professionals' Ability to Limit Liability?* 30 Constr. Lawyer, No. 2, Spring 2010, p. 17; Heley, *Professional or Not: Should Courts Preclude Contract Limitations of Liability Solely Because of the Architect's or Engineer's Status as a Licensed Professional?* 4 J. ACCL, No. 1, Winter 2010, p. 23.

[23] AIA Doc. B101-2007, § 3.6.2.4 grants the architect immunity for interpretations and decisions made in good faith.

CHAPTER 13

Ethics

SCENARIO

Please review the Scenario in Chapter 9.

13.1 Three Distinct Concepts: Law, Morality, and Professional Ethics

Three overlapping yet distinct concepts provide an understanding of professional ethics. Those three concepts are distilled in the form of the following questions.

1. What is legal?
2. What is moral (or ethical)?
3. What is one's professional, ethical duty?

The law imposes a minimal standard of conduct demanded by society. Certainly some laws—such as prohibition of murder—conform to our societal concept of what is the right (or moral) thing to do. Yet sometimes the law may seem at odds with our notion of right and wrong. Is stealing bread morally wrong if done to feed a starving family?

As shown by this last example, morality or ethics is a difficult, philosophical concept. (As used here, morality and ethics are interchangeable terms; however, only the term "morality" will be used to prevent confusion between "ethics" and "professional ethics.") A moral system is one that seeks to articulate underlying principles of proper conduct.

This chapter addresses the **professional ethics** of construction industry actors: architects, engineers, contractors, and construction managers. The nature of professional ethics may be understood by contrasting it with law and morality.

While law and morality create rules that apply universally (in the case of law, universally where that law applies), professional ethics are self-policing rules governing only members of a business association. Professional ethics define appropriate behavior for the association's members as they pursue their business (or professional) activities. The purpose of professional ethics is to seek to prevent individual members of the association from undermining the good standing and reputation of the association. Violation of professional ethics may ultimately result in a denial of continued membership in the organization. It is a way of kicking out the "black sheep" so that the association's remaining members may retain a high standing of proper conduct within the society.

The discussion to this point has emphasized distinctions between professional ethics and both law and morality. However, the nature of professional ethics may also be understood by emphasizing its *similarity* to law and morality. Just as law regulates the behavior of individuals in society, so too compliance with professional ethics is a condition for remaining a member in good standing of the association. In addition, an ethical rule may never violate the law.

Professional ethics are also similar to morality: just as a moral person does the right and appropriate act, so too does the architect, engineer, contractor, or construction manager who comports with his association's professional ethics. Just as a moral person (in most cases) has a good reputation as a member of society,[1] so too an architect, engineer, or contractor who acts in conformity with her professional, ethical obligations has a good reputation with her peers.

In sum, professional ethical rules do not have the force of law,[2] although they will be consistent with the law. They are not statements of moral philosophy, although professional ethics and morality share the goal of promoting appropriate behavior. At their core, professional ethics are rules of conduct created by business or professional associations that apply to members of that association.

13.2 Ethics for Architects

13.2A Introduction

The largest professional society in the United States for architects is the American Institute of Architects (AIA). Architects have an ethical duty to perform their services to the best of their ability and to provide their clients with the best possible buildings for their clients' wants and needs. To ensure a base level of ethical standards among its members, the AIA has developed a Code of Ethics. This code is reproduced in Appendix E.

On projects in which the lead design professional is an architect, the role of the architect is unique compared to that of the other design professionals and constructors on the same project. Architects deal directly with the owner and coordinate the design with engineers. They are the administrator for the construction contract and monitor the progress of the contractor. Architects have many different obligations in a construction project, and the AIA Code of Ethics describes the different ethical issues relating to these obligations.

The AIA Code of Ethics is divided into six "canons:"

- Canon I: General obligations
- Canon II: Obligations to the public
- Canon III: Obligations to the client
- Canon IV: Obligations to the profession
- Canon V: Obligations to colleagues
- Canon VI: Obligations to the environment

Each canon is further subdivided into Ethical Standards (E.S.) and Ethical Rules. The Ethical Standards are the specific ethical goals to which members of the AIA should aspire.

In contrast to the aspirational nature of the Ethical Standards, the Ethical Rules are mandatory and absolute minima that must be adhered to. If a member of the AIA is found to have broken a rule, that member would be immediately subject for disciplinary action by the AIA. Each rule is followed by a commentary explaining the meaning and intent of the rule.

The AIA Code of Ethics begins with a Preamble describing the goals of the code as well as the structure of the document. Statements regarding the application of the code and the enforcement of the code are also found in the Preamble. A "Statement in Compliance With Antitrust Law," also found in the Preamble, addresses the ethical practices of competitive bidding and pricing.

In this chapter's scenario, Andrew Awesome was bound by this ethical code. We are assuming he is a member of the AIA in good standing. With Andrew's kickback scheme, he is ignoring his obligations to the client as found in the AIA Code of Ethics and is instead putting his own financial gain ahead of the needs of the owners.

13.2B Historical Background: Competing for the Commission

Associations of design professionals historically have sought to discipline their members for ungentlemanly competition both in obtaining a commission for design services and in replacing a fellow member who was been performing design services for a client.

As an illustration, AIA Document J330 (1958) listed among its mandatory standards of ethics:

9. An Architect shall not attempt to supplant another Architect after definite steps have been taken by a client toward the latter's employment.
10. An Architect shall not undertake a commission for which he knows another Architect has been employed until he has notified such other Architect of the fact in writing and has conclusively determined that the original employment has been terminated.

In addition, J330 stated that the architect would not compete on the basis of professional charges.

These restraints were considered part of ethics. Undoubtedly, they are traceable to the desire to preserve the professions *as professions* and to avoid certain practices that would be accepted in the common marketplace. Although the associations must have recognized that there will be competition for work, it was hoped that competition would be conducted in a gentlemanly fashion and would emphasize professional skill rather than price.

Beginning in the 1970s, attacks were made on these ethical standards by public officials charged with the responsibility of preserving competition. One by one, association activities that were thought to impede competition came under attack, particularly those that dealt with fees. Fee schedules, whether required or suggested, were found to be illegal.[3] Disciplinary actions for competitive bidding were also forbidden.[4]

Standard 9 in J330 attempted to inhibit competition *before* any valid contract had been formed.[5] Standard 10, although less susceptible to the charge that it is anticompetitive, could have had the effect of limiting the client's power to replace one architect with another. Because of these legal challenges, the subsequent code of ethics from the AIA have been changed to include the Statement in Compliance With Antitrust Law in the Preamble, which provides that submitting competitive bids, providing discounts, or providing free services are not unethical practices.

13.2C The AIA Ethics Canons

The first canon covers the general obligations of the architect. This canon lays the basic ethical foundation for a practicing architect. These rules are very broad and cover some of the philosophical issues with being an architect. Canon I consists of five Ethical Standards. These standards show that architects should strive to always improve their knowledge and skills, engage in lifelong learning, and look to improve themselves. The Ethical Standards also state that architects should uphold human rights as well as respect and conserve their natural and cultural heritage while striving to improve the environment and the quality of life within it.

Many people in the construction and engineering trades would probably not consider some of these things in the course of their jobs. However, since architects are responsible for the overall design of the building (especially the aesthetic) they will have larger concerns than just the functionality of the building. Architects also must be concerned with how these structures fit into the larger culture of the area. It is also interesting to see the phrase "striving to improve the environment and the quality of life within it" in Ethical Standard 1.3. This phrase, found in the 2004 Code of Ethics, presages the AIA's concern with sustainable design, which came to fruition with the adoption of a new Canon VI in the 2007 edition and continued in the 2012 edition.[6]

Canon I consists of two Ethical Rules. The first mirrors the professional standard and the second is a non-discrimination mandate.

The second canon deals with the obligations to the public. Because of the seriousness of this obligation, there are seven rules and only three ethical standards in this section. Most of the rules in this section discuss how the architect will follow all legal statutes, especially those dealing with bribery such as offering or accepting gifts in exchange for judgments. These are intended to ensure that the architect follows the letter of the law and conducts business in a professional manner.

Rule 2.105 is the most important rule in this section. Because architects' designs may lead to safety concerns, such as structural stability problems and failure in the building, they must take great care to make proper decisions that ensure these types of issues do not happen. If there were to be a failure of a structural system in the building and it were to collapse, it could cost millions of dollars in damage and potentially thousands of people their lives. In light of these risks, Rule 2.105 includes very specific guidelines as to what actions an architect must take if he believes that his supervisor's or client's decisions could result in a structural failure. In that event, the AIA Code requires the architect to advise that person against the decision and refuse to consent to it, and then to report it to the building inspector "unless the Members are able to cause the matter to be satisfactorily resolved by other means."

This last phrase gives the architect some flexibility to use other means to get the client or the contractor to change their decision. Suppose an architect was inspecting a job site and observes the contractor pouring concrete for structural columns that did not meet the specified yield strength, which could lead to a failure of the concrete and of the structure itself. According to Rule 2.105, the architect should then immediately inform the owner and contractor that this is noncompliant with the contract. If both the owner and contractor refuse to change the concrete mixture to comply, the architect under Rule 2.105 should alert the local building inspector, which would cause a strain on the relationship of all involved. But Rule 2.105 allows the architect to use his authority under his contract to make sure the correct mixture is being used; for example the architect could send a letter to both parties stating that the mixture does not conform to the specifications, and as per their contract, the architect would withhold payment until this was remedied. This would be one example of how the architect could achieve compliance with Rule 2.105 without involving a local government official.

Canon III of the AIA Code of Ethics addresses the architect's obligations to his client. It reiterates that the architect will obey all of the laws in the course of his work, which shows the importance the AIA places on how they view following the law as an ethical absolute. Because architects deal directly with project owners and do so before any design work has commenced, this section of the code makes explicit that the architect needs to be honest with the client about his level of skill and what results are actually obtainable for the project. Many owners will have expectations for the completed project that are unrealistic and sometimes even physically impossible. The architect will be placed in the position of telling the owner that what the owner wants is not possible. Some may fear that the owner may go to another firm and that the architect will lose a potential client by saying no, but the code of ethics is clear that one must be upfront and honest about what the possible results of the project will be.

Another key issue that architects must be aware of when dealing with clients is privacy and confidentiality. Some owners make architects sign a formal confidentiality agreement. In the absence of such an agreement, Ethical Rule 3.401 addresses this issue. In order to perform their duties as designer of the project, architects need access to sensitive material from owners, such as financial information or manufacturing technologies that would be harmful to owners if they were made public. Architects need to keep this in mind when dealing with owners and be sure to safeguard this information and keep it confidential.

Of importance to contractors, Rule 3.202 addresses the architect's role as administrator, specifically when acting as "interpreter of building contract documents and the judge of contract performance." The rule requires the member "to render judgments impartially." If an objective review of the specifications shows that delayed performance was caused by lack of clarity in the design, the architect has an ethical obligation to accept responsibility for the delay—rather than to attempt to shift it to the contractor.

The architect's obligations to the profession are covered under Canon IV of the Code of Ethics. This is a key area of the code because architects spend a large amount of time implementing and managing the design process. The other aspect in this section of the code is the responsibility of architects to oversee and control the professionals working for them at their firm.

Rule 4.103 states that "Members speaking in their professional capacity shall not knowingly make false statements of material fact." While this means that architects should not lie, as with all things in life, there are areas of gray even in this statement. Suppose you have an architect who was brought in by a client to do an inspection of a building that was going to be sold. Assume that the architect was only instructed to certify the condition of the exterior doors and windows for the client. But during his inspection, the architect notices that the roof system shows signs of water leakage and associated damage. Should the architect put this in his report even though it was outside the scope of his work? Is it still making a false statement by omitting the observed conditions? An omission is still withholding the material facts, and it could be interpreted as a breach of the code of ethics. These types of ambiguities mean that most any claim against an architect can include an alleged violation of Rule 4.103.[7]

The other important job of an architect is to train future architects that work for the architect's firm. Many times these junior members will do much of the design work and the senior architect will then review the design to ensure its accuracy and then sign and stamp the drawings with his or her own seal. Rule 4.102 is clear that the architect must be in direct responsible control of these individuals as a condition to signing or sealing their work. This portion of the code is in place both to ensure an architect has responsible control of junior members and to prevent the architect from either using designs found online or designs that are contracted out to an unlicensed firm and then summarily signing and stamping them.

The fifth Canon of the AIA Code of Ethics deals with the architect's ethical obligations to other architects and colleagues in their profession. Most times on a construction project a single architect or architectural firm will be the only entity working on the project. This prompts the question: Why would architects have obligations to other architects? Most often this situation would occur within one firm that employs several architects or architects in training. The Ethical Statements in this canon speak to how architects must maintain a professional environment as well as help develop future professionals that would be working beneath them in the firm as well as granting those individuals credit when credit is due for their work product. These are not only good ethical statements but sound business practice.

The other portion of Canon V consists of three rules, all addressing an architect's change of employment from one firm to another. This situation has many ethical implications, especially when that architect had dealings with clients from his original employer. In this situation, the architect could try to convince these clients to switch to the architect's new firm based on their past working relationship. The architect also would have access to all of the previous design work and design data and possibly use this when working for the new employer. This could be viewed as stealing of confidential work product and would be viewed as highly unethical. It is good business practice for firms to have a written policy—signed by each architect as a condition to hiring—which prohibits this practice and spells out the consequences for breaching the policy.

The 2007 Code of Ethics has a new Canon VI (continued in the 2012 code), addressing an architect's obligation to the environment. This new canon was added in direct response to the new green building movement that has occurred in the United States during the past decade. The AIA now believes that architects need to be environmentally responsible and should advocate for sustainable buildings as part of their professional activities. Canon VI also states that architects should implement sustainable practices within their firms and should encourage their clients to do the same. Notably, Canon VI consists of three Ethical Standards but no Ethical Rule. A violation of Canon VI will not subject an architect to discipline.

The final portion of the AIA Code of Ethics (not a canon) consists of the rules of application and enforcement. These are administrative procedures of how a claim would be filed and what the procedures would be if a member were accused of violating the Code of Ethics.

This Code of Ethics provides the fundamental groundwork of what type of behavior is expected from a practicing architect that is a member of the AIA. As with all ethical questions, there are many areas subject to competing interpretations. Architects should look at the overall intent of the code to help them make ethical decisions in their practice.

13.3 Ethics for Engineers

One of the main professional societies for engineers in the United States is the National Society of Professional Engineers (NSPE). The NSPE Code of Ethics for Engineers addresses the ethical concerns that occur relative to the practice of engineering. The NSPE Code of Ethics is reproduced in Appendix F.

The NSPE Code of Ethics lists six "fundamental" canons:

- Hold paramount the safety, health, and welfare of the public
- Perform services only in areas of their competence
- Issue public statements only in an objective and truthful manner
- Act for each employer or client as faithful agents or trustees
- Avoid deceptive acts
- Conduct themselves honorably, responsibly, ethically, and lawfully so as to enhance the honor, reputation, and usefulness of the profession

The format of the NSPE Code is different from that of the AIA Code of Ethics. The six canons are listed in the beginning of the code, followed by Rules of Practice. All of these rules address all of the six canons and are not listed below each related canon as in the AIA Code. The third section of the NSPE Code lists the Professional Obligations.

Since an engineer's primary role is to design the systems of a building or other structure, the first rule addresses the health and safety of the public. If an engineer were to make an error on a design it could have catastrophic consequences: the building could collapse, the electrical system could short out and start a fire, or an airborne contaminant could infiltrate the HVAC system. All of these could lead to serious injury or even death of those people inside or near the building. The Code of Ethics outlines how engineers should conduct themselves in regard to their work product as well as what to do if their client makes a decision against the engineer's recommendation that may endanger life or property.

The remainder of the Rules of Practice is very similar to the AIA Code of Ethics. Since both engineers and architects are design professionals, it stands to reason that they would have similar ethical guidelines in how they deal with their clients. Like architects, engineers should be mindful of what is in the best interests of the public. One difference between the two professions is that the architect is more involved with aesthetics and how the building fits into the surrounding area, while the engineer is focused more on the systems of that building and ensuring that they function at a safe level.

Besides this difference in job capacity, the main difference between the two codes is the level of detail. For example, the AIA Code of Ethics Rule 4.103 states "members speaking in their professional capacity shall not knowingly make false statements of material facts." By contrast, the NSPE Code Rule 5 states that "Engineers shall avoid deceptive acts" and then lists significantly more detail to this rule in two sub-parts. Part a) gives further description of how engineers may not falsify their qualifications, misrepresent or exaggerate their responsibilities, or misrepresent pertinent facts in any brochures or presentations. Part b) stipulates that engineers shall not offer or receive any contribution to influence the award of contract from a public authority.

This code of ethics provides the fundamental groundwork of what type of behavior is expected from a practicing engineer that is a member of the NSPE. The level of detail in the NSPE Code should provide more guidance in interpreting the canons and Rules of Practice. Engineers should look at the overall intent of the code to help them make ethical decisions in their practice.

13.4 Ethics for Contractors

The role of the contractor contrasts with that of an architect or engineer. The contractor's main responsibility is to build the structure—not to design it. Their concerns are of sequence and logistics: the methods and timetable of when and how to get the project completed. The code of ethics for contractors reflects these differences in responsibilities. The American Institute of Constructors (AIC) is one of the larger organizations for contractors in the United States. Its Code of Conduct is reproduced in full in Figure 13.1.

Differences between the Contractors Code of Conduct and the AIA and NSPE ethical codes are evident. The AIC Contractors Code of Conduct is seven sentences long—compared to other multi-page codes. But the content within this code has similarities to the architect's and engineer's code of ethics.

As the actual builders, contractors' work carries a more immediate danger to workers and to members of the public near the jobsite. A contractor unconcerned with site safety could lead to serious injury and even death. The AIC Code's first rule is for the contractor to maintain full regard for the public interest. This can be

---

CONSTRUCTOR CODE OF CONDUCT

I. The CONSTRUCTOR shall maintain full regard to the public interest in fulfilling his or her professional responsibilities to the construction industry.

II. A CONSTRUCTOR shall not engage in any deceptive practice, or in any practice which creates an unfair advantage for the constructor or another.

III. A CONSTRUCTOR shall not maliciously or recklessly injure or attempt to injure the professional reputation of others.

IV. A CONSTRUCTOR shall ensure that when providing a service that includes advice, such advice shall be fair and unbiased.

V. A CONSTRUCTOR shall not divulge to any person, firm or company, information of a confidential nature acquired during the course of professional activities.

VI. A CONSTRUCTOR shall carry out responsibilities in accordance with current professional practice.

VII. A CONSTRUCTOR shall keep informed of new concepts and developments in the construction process relative to his or her responsibilities.

American Institute of Constructors
Constructor Certification Commission

FIGURE 13.1 Constructor Code of Conduct.

interpreted in different ways: with regard to safety to provide a job fence to keep people away from the jobsite activities or to have a recycling program on site in order to minimize the local environmental impact. No code could possibly document every type of issue, therefore the code must be written in a way to guide the contractor in a variety of situations.

The primary ethical issue for contractors is found in Rule II. This rule states that the contractor should not engage in any deceptive practice or in any practice that creates an unfair advantage for the contractor or another. The two main unethical practices that are found in the construction industry are bid rigging and bid shopping, and both clearly violate this rule.

Bid rigging is an illegal agreement to effectively raise prices on the owner. Competitors meet in advance of the bid and agree on who will win the bid and at what price. Often this type of scheme will have each of the conspirators take turns on who will win the bids while the others inflate their bids to keep the prices up and the profit margins unusually high. The owner would not realize this was happening, because he would go through the competitive bidding process and just see one low bidder and the rest of the bids would be higher (but at a similar range in values) leading him believe the bid was open and fair. Bid rigging is a "deceptive practice" prohibited by Rule II.

Another serious issue in the construction industry is bid shopping. Most general contractors do not perform all of the work but instead rely on subcontractors to do most (if not all) of the work. During the bid phase, the general contractors will solicit bids from several subcontractors—most often selecting the one that submits the lowest number. Once the general contractor is awarded the contract from the owner, an unethical general contractor may then shop the lowest number to other companies or demand the low bidder lower its number in order to keep the project. The general contractor does not pass any of these cost savings to the owner, but keeps it for itself. Bid shopping creates an "unfair advantage" barred by Rule II.

Bid rigging and bid shopping are not only ethical issues in the construction industry, but they are also legal issues. More detailed information and review of the legal consequences of bid rigging and bid shopping can be found in Section 21.3C.

13.5 Ethics for Design/Builders

With the rise in popularity of projects being constructed using the design–build project delivery system, the ethical considerations for these types of entities is important to note. In the design/build environment, both designers and constructors are still present and their types of services are still the same, but the contractual relationship is different. Architects are still bound by the AIA Code of Ethics, the engineers would still follow the NSPE Code of Ethics, and the contractors would still be bound by the AIC Constructor Code of Conduct. But because these different professionals are now bound together by a design/builder contract, their obligations to each other are different than on a design–bid–build project.

One of the largest organizations in the United States that deals with design/build is the Design Build Institute of America (DBIA). DBIA has published their own ethical guidelines to help members of the design-build industry deal with these new relationships and situations. The DBIA Code of Professional Conduct is found in Appendix G.

Because the individuals in the design-build process are still bound by their respective codes of ethics, the DBIA Code of Professional Conduct is quite brief compared to the other codes discussed in this chapter. This code lists only three obligations of its members:

- Obligations to the public
- Obligations to the owner
- Obligations to the design–build team

Under each of these obligations, there are a few bullet points describing the extent of the obligation. Again, there are familiar themes, such as the obligation to protect the public as well as making decisions that are in the best interests of the owner.

The primary difference is the obligation to the design–build team. Here, the code states the members should respect the talents and point of view of the other team members as well as select teammates on the basis of qualifications and best value. More importantly, all team members must respect the obligations of the design professional to protect the safety, health, and welfare of the public. The fact that the design–build team as a whole is profit-driven to reduce its construction costs cannot override the design professional's licensing and ethical requirements to make cost subordinate to safety and welfare.

In this sense, the DBIA's Code of Professional Conduct may be viewed as responding to the dissenting judge in a New York case, *Charlebois v. J.M. Weller Associates*.[8] In that case, James Weller owned a construction company (Weller Associates) and was a licensed professional engineer. The owners entered into a design–build agreement with Weller Associates (the construction company), which required the company to employ James Weller to create the design. The legal issue before the court concerned licensing: whether Weller Associates was engaged in the unlawful practice of engineering. Of more relevance here was the dissent's concern as to the professional independence of James Weller, P.E. to fulfill his duty to the client and public while employed by a construction company "beholden to that profit-motivated commercial enterprise." The dissent further explained his objections as follows (the "licensee" is James Weller):

> It is unrealistic to assume that the head of a construction company who, in managing that company's construction contract, must be concerned with time and cost restraints, allocation of resources, and profit margins, will somehow remain unaffected by these concerns simply by donning his other hat and assuming the role of professional engineer. Indeed, the interrelated nature of the "design-build" contract at issue makes this kind of distancing impossible; the fees paid by the owner for architectural and engineering services are defined as a percentage of the total project cost, thus making the financial interests of the licensee inseparably wedded to those of the contractor. Therefore, it cannot be said with any confidence that the licensee will not subordinate the owner's interests to those of his corporate employer.[9]

It appears that the DBIA (by means of its Code of Professional Conduct) is seeking to counter precisely these types of public policy concerns.

13.6 Ethics for Construction Managers

Construction managers offer consulting services at the beginning of a project to help assist with the design of the project. These same construction managers may also then assist with the actual construction of the project, depending on project delivery method. One of the largest professional organizations in the United States for construction managers is the Construction Management Association of America (CMAA). Since 1982, the CMAA has taken a leadership role in regard to critical issues impacting the construction and program management industry, including the setting of ethical standards of practice for the Professional Construction Manager.

The Board of Directors of CMAA has adopted the following Code of Professional Ethics of the Construction Manager, which applies to CMAA members in performance of their services as Construction and Program Managers. All members of the CMAA commit to conduct themselves and their practice in accordance with this code, which is reproduced in full here.

> As a professional engaged in the business of providing construction and program management services, and as a member of CMAA, I agree to conduct myself and my business in accordance with the following:

1. **Client Service**. I will serve my clients with honesty, integrity, candor, and objectivity. I will provide my services with competence, using reasonable care, skill and diligence consistent with the interests of my client and the applicable standard of care.
2. **Representation of Qualifications and Availability**. I will only accept assignments for which I am qualified by my education, training, professional experience and technical competence, and I will assign staff to projects in accordance with their qualifications and commensurate with the services to be provided, and I will only make representations concerning my qualifications and availability which are truthful and accurate.
3. **Standards of Practice**. I will furnish my services in a manner consistent with the established and accepted standards of the profession and with the laws and regulations which govern its practice.
4. **Fair Competition**. I will represent my project experience accurately to my prospective clients and offer services and staff that I am capable of delivering. I will develop my professional reputation on the basis of my direct experience and service provided, and I will only engage in fair competition for assignments.
5. **Conflicts of Interest**. I will endeavor to avoid conflicts of interest; and will disclose conflicts which in my opinion may impair my objectivity or integrity.
6. **Fair Compensation**. I will negotiate fairly and openly with my clients in establishing a basis for compensation, and I will charge fees and expenses that are reasonable and commensurate with the services to be provided and the responsibilities and risks to be assumed.
7. **Release of Information**. I will only make statements that are truthful, and I will keep information and records confidential when appropriate and protect the proprietary interests of my clients and professional colleagues.
8. **Public Welfare**. I will not discriminate in the performance of my Services on the basis of race, religion, national origin, age, disability, gender or sexual orientation. I will not knowingly violate any law, statute, or regulation in the performance of my professional services.
9. **Professional Development**. I will continue to develop my professional knowledge and competency as Construction Manager, and I will contribute to the advancement of the construction and program management practice as a profession by fostering research and education and through the encouragement of fellow practitioners.
10. **Integrity of the Profession**. I will avoid actions which promote my own self-interest at the expense of the profession, and I will uphold the standards of the construction management profession with honor and dignity.

The CMAA Code of Professional Ethics of the Construction Manager includes themes found in the other ethical codes discussed in this chapter. Construction managers must serve their clients with honesty and integrity, and should represent their qualifications in a truthful manner. They should avoid any conflicts of interest and shall not knowingly violate any law or statute.

13.7 Conclusion

While the purpose of professional ethics is to safeguard the reputation of the profession, ethical conduct by an architect, engineer, contractor, or construction manager is vital to that person's business success. Design professionals, in particular, do not advertise in mass media. Job promotion is done primarily by word of mouth. The construction industry is to a large extent local in nature. An industry actor who acquires a reputation for unethical conduct will soon find its job prospects diminished. Ethical behavior is not only mandated by the profession or association of which the actor is a member but is a precondition for long-term economic success.

REVIEW QUESTIONS

1. What is the purpose of professional ethics and what is the ultimate penalty for violating a code of ethics of a professional society?

2. What are the six canons of the American Institute of Architects Code of Ethics?

3. In the American Institute of Architects Code of Ethics, what is the difference between Ethical Standards and Ethical Rules?

4. According to the American Institute of Architects Code of Ethics Rule 2.105, what must an architect do if he believes that his supervisor's or client's decisions could endanger the safety of the public?

5. If an architect does an objective review of the specifications and it shows that delayed performance was caused by lack of clarity in the design, what does the American Institute of Architects Code of Ethics require that architect to do?

6. What are the six fundamental canons of the National Society of Professional Engineers Code of Ethics?

7. What are the seven rules of the American Institute of Constructors Code of Conduct?

8. What is the difference between bid rigging and bid shopping?

9. What are the ten categories of the Construction Management Association of America's Code of Professional Ethics?

10. According to the Construction Management Association of America's Code of Professional Ethics, what are the factors that a construction manager promises to never discriminate in the performance of their services?

ENDNOTES

[1] Socrates in ancient Athens devoted his life to understanding what is proper, ethical conduct, yet precisely that pursuit of knowledge made him a "gadfly" to most of Athenian society. See Plato, *Apology* 30e, found in THE COLLECTED DIALOGUES OF PLATO, pp. 16–17 (Hamilton & Cairns eds. 1973).

[2] For example, unethical conduct by an architect is not evidence of breach of the duty of care. See Section 11.2.

[3] *Goldfarb v. Virginia State Bar*, 421 U.S. 773 (1975).

[4] *National Society of Professional Engineers v. United States*, 435 U.S. 679 (1978).
[5] *United States v. American Society of Civil Eng'rs*, 446 F.Supp. 803 (S.D.N.Y.1977) (barred society from disciplining members who supplanted another member).
[6] See Section 11.5B.
[7] An architect's legal duty to warn is addressed in Section 11.8G.
[8] 72 N.Y. 2d 587, 531 N.E. 2d 1288, 535 N.Y. S. 2d 356 (1988).
[9] 531 N.E. 2d at 1294.

PART

PROJECT DELIVERY METHODS

CHAPTER 14

Project Organization, Pricing, and Delivery Methods

CHAPTER 15

Public Contracts

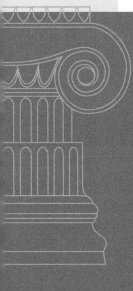

CHAPTER 14

Project Organization, Pricing, and Delivery Methods

Scenario

Mega Manufacturer, Inc. (MMI) sought to build a new factory. MMI hired Awesome Architect, Inc. (AAI) to design and administer the projects. After AAI had created a complete design, MMI solicited bids from various contractors, ultimately deciding to hire Prime Builder, Inc.

The project suffered a series of problems.

1. AAI was late in completing the design, which meant the prime contract was signed and the construction work was started several months later than MMI had planned.
2. As a result, construction took place during an unusually harsh winter, which slowed down progress.
3. During performance, Prime Builder, Inc. discovered that the design features were so complex that it could not use standard construction methods to carry out the work, which also delayed performance and greatly increased its labor costs.
4. While inspecting the work to approve a payment request, AAI discovered defective workmanship by a subcontractor. Prime Builder, Inc. was required to remove and redo part of the work.
5. The completed facility soon suffered from both settlement damage and leaks through the roof. AAI blamed Prime Builder, Inc. for improper workmanship; however, the contractor blamed defects in the design.

The president of MMI fumed: "For the next factory I build, something has got to be different!"

14.1 Project Planning: Owner's Choices

Section 7.2 addresses the project owner and notes distinctions between types of owners—primarily private versus public owners and sophisticated versus inexperienced owners. A review of Figure 7.1 at this point will refresh the basic organizational make-up of a large construction project.

This section examines the owner's objectives when making organizational planning decisions. In the design phase, quality, price, and completion date are interrelated. An owner who wishes the highest quality may have to make tradeoffs among quantity, price, and completion date. If early completion is crucial, the owner may need to sacrifice price and probably quality and quantity.

After these choices have been made, the owner may have to decide how to select a contractor, which contractor or contractors to select, and the type of construction contract or contracts to use. These choices should be made in a way that maximizes the likelihood that the owner will receive quality that complies with the contract documents and on-time completion at the lowest ultimate cost. Compromises may be needed. The contractor who will do the highest-quality work is not likely to be cheapest and quickest. The quickest contractor may not be the one who will provide the best quality. The importance the owner attaches to these objectives will affect the process it uses to select the contractors and how construction contracts are organized.

14.2 Compensating Contractor Work

One organizational decision an owner must make is how the contractor will be compensated. Different compensation systems exist. Although many payment variations are available, each major category involves risk allocation.

14.3 Fixed-Price Contracts

In American usage, fixed-price and lump-sum contracts are used interchangeably. Under such contracts, the contractor agrees to do the work for a fixed price. Almost all performance risks—events that make performance more costly than planned—fall on the contractor. For example, when the construction industry experienced a sudden and sharp increase in the price of steel after many years of price stability, contractors who had entered into fixed-price contracts were unable to pass these unanticipated costs on to the owner.[1] Similarly, as discussed in Section 7.4C, the contractor on a fixed-price contract must absorb higher performance costs caused by siting the project on defective soil.

The contract may include mechanisms for increasing the contract price. Although not common in ordinary American contracts (English and international contracts often use fluctuations clauses), some contracts have price escalation clauses under which the contract price is adjusted upward (and sometimes downward), depending on market or actual costs of labor, equipment, or materials. Such a provision protects the contractor from the risks of any unusual costs that play a major part in its performance. The provision usually requires the contractor to use its best efforts to obtain the best prices.

Many construction contracts contain differing site conditions (formerly called changed conditions) clauses that allow a price increase if conditions under the ground or in existing structures are discovered that are substantially different from those anticipated by the parties or represented by the contract. These clauses are discussed in Chapter 19.

Large public (and some private) contracts are expected to span many months and even years and may include historical weather data. The contractor would calculate its price taking into consideration the impact of likely weather conditions. If actual weather conditions are significantly worse than anticipated, the contract may specify a remedy, usually in the form of increased time to perform (more rarely, increased costs of performance).

> In this chapter's scenario, MMI sought to have the work started early enough in the year that performance would not be impacted by the winter weather. Nonetheless, Prime Builder, Inc. contracted knowing the work would be done during the winter. If the prime contract was for a fixed price, the contractor would bear the risk of higher performance costs caused by the unusually bad winter, unless a contract mechanism provided a means by which the contractor could receive additional time or compensation for extreme weather events.

The contractor's performance costs also may increase because of defects in the design. Or an owner may simply decide to make design changes midway through performance. A changes clause, addressed in Chapter 18, creates a compensation method to deal with these events.

Sophisticated procurement systems sometimes combine fixed-price contracts with incentives. For example, a fixed-price contract may include an incentive payment if the contractor completes the project early.

The discussion to this point has focused on contract mechanisms relieving contractors from the compensation ceiling of a fixed-price contract. The legal doctrine of misrepresentation may also entitle the contractor to additional compensation for defects in the design. This *"Spearin"* doctrine is addressed in Section 16.3B.

> In this chapter's scenario, Prime Builder, Inc. experienced higher performance costs than anticipated because of the complexity of the design and its inability to use standard construction methods. If the *Spearin* doctrine applies, it will provide a legal (not a contractual) method by which the contractor may obtain additional compensation.

A fixed-price contract has the obvious advantage of letting the owner and those providing funds for the project know in advance what the project will cost. It works best when clear and complete plans and specifications are drawn. Incomplete contract documents are likely to cause interpretation questions that can lead to cost increases. The fixed-price contract is used most efficiently when a reasonable number of experienced contractors are willing to bid for the work. This is less likely where the design is experimental or where construction projects are abundant.

Another advantage to the fixed-price contract is that the owner need not be particularly concerned with the contractor's recordkeeping. If changed work or extra work is priced on a cost basis, there may have to be some inquiry into the contractor's cost. On the whole, the fixed-price contract avoids excessive owner concern with the contractor's cost records. Conversely, such a contract is attractive to the contractor, which need not expose its cost records—something that occurs in a cost contract, as noted in the next subsection.

The fixed-price contract has come under severe attack for its inherently adversarial nature. A contractor that reduces costs increases profits. As long as the cost is not reduced at the expense of the owner's right to receive performance specified in the construction contract, the owner cannot object.

In construction, however, the performance required and whether such performance has been rendered are often difficult to establish. In this gray area of compliance, the interests of owner and contractor can clash. Unless the contractor seeks a reputation for quality work or values goodwill, it is likely to perform no more than the contract demands.

Another disadvantage to the fixed-price contract is that the risk of almost all performance cost increases falls on the contractor. A prudent contractor will price these risks and include them in its bid. However, in a highly competitive industry, the prudent contractor may not receive the award because others may be more willing to gamble with a low price and either hope that problems will not develop or recoup any losses by asserting claims for extra work and delays. Even if a prudent contractor does take these risks into account and does receive the award, if the risks do not materialize, the owner may be paying more than if the risk had been taken out of the contractor's bid.

14.4 Cost Contracts

Introduction. When prospective contractors cannot be relatively certain of what they will be expected to perform or are uncertain as to the techniques needed to accomplish contractual requirements, they are likely to prefer to contract on a cost basis. For either reason, projects that involve experimental design, new materials, work at an unusual site, or in which the design has not been thoroughly worked out are likely to be contracted for on a cost basis.

Usually, a cost contract allows the contractor to be paid costs plus an additional amount for overhead and profit. This arrangement should be distinguished from what is sometimes called a "time and materials" contract, which at least in one case was held to preclude recovery by the contractor of overhead on direct labor costs.[2]

From the contractor's perspective, a fixed-price contract creates an incentive to use its ingenuity and efficiencies to maximize its profit. A "cost" contract gives the contractor "downside" protection but also creates a cap on profits.

Owner Concerns and Protections. From the owner's perspective, the cost contract has two principal disadvantages. First, at the time it engages the contractor, the owner does not know what the work will cost. Second, as a general rule, a cost contract does not give enough incentive for the contractor to reduce costs. In *Smith v. Preston Gates Ellis, LLP*,[3] an owner who entered into a "cost plus" contract to build his dream home had such disastrous results that he (unsuccessfully) sued his lawyer for malpractice for allowing him to enter into the contract. Cost contracts are not used in federal procurement.

This is not to say an owner entirely lacks protections. Cost contracts often contain provisions that require the contractor to use its best efforts to perform the work at the lowest reasonable cost. Provisions are often included requiring the contractor who has reason to believe that the cost will overrun any projected costs to notify the owner or its representative and give a revised estimate of the total cost. Sometimes these provisions state that failure to give notice of prospective cost overruns will bar recovery of any amounts higher than any cost estimates given.

> In this chapter's scenario, Prime Builder, Inc. experienced higher performance costs than anticipated because of the complexity of the design and its inability to use standard construction methods. Under a cost contract, the higher than anticipated cost of performance will be borne by the owner, not the contractor. However, Prime Builder, Inc. may have a duty based on the covenant of good faith and fair dealing to warn MMI of the likely increased costs.

Another method of keeping costs down is to include provisions in the contract stating that a fiduciary relationship has been created between owner and contractor that requires that each use its best efforts to accomplish the objectives of the other and to disclose any relevant information to the other. (A fiduciary relationship is discussed in Section 9.2.) For example, certain American Institute of Architects (AIA) documents, applicable to "cost plus" projects, specify that "the Contractor accepts the relationship of trust and confidence" and will seek to do the work "in an expeditious and economical manner consistent with the interests of the Owner."[4]

Attempts by owners to use this or similar language to make the contractor on a cost project a fiduciary have been unsuccessful.

On the other hand, a contractor's obligation under the duty of good faith and fair dealing to keep the owner informed of costs has been recognized. Thus, in *Jones v. J.H. Hiser Construction Co., Inc.*,[5] a contractor on a cost contract was found to have a duty to inform the owner of prospective overruns. In addition, in *Williams Engineering, Inc. v. Goodyear, Inc.*,[6] an engineer was held responsible to the owner for cost overrun because he did not update estimates, did not advise of other types of contracts, and did not hire a cost estimator. Again, this duty arose out of the covenant of good faith and fair dealing (noted in Section 9.2C).

Cost Plus Fee. The discussion to this point has assumed a "cost plus" contract in which the "plus" refers to a percentage markup (on the direct costs) for overhead and profit. A different type of "cost" contract is the "cost plus a fixed fee" contract. The parties agree that the contractor will be reimbursed for allowable costs and paid a fee that is fixed at the time the contract is made. The fee is normally not affected when actual cost exceeds or is less than the estimated cost. However, if the scope of the work is substantially changed, sometimes the fee is renegotiated. Because the contractor's fee is not affected by cost savings, the contractor has no compensation incentive to reduce costs. For this reason, in federal procurement this type of contract has largely been superseded by cost contracts that create incentives to reduce costs.

Guaranteed Maximum Price. These methods of seeking to keep costs down, although sometimes successful, still do not accomplish the objective of letting the owner or anyone supplying funds for the project know that the costs will not exceed the particular designated amount. To deal with this problem, owners sometimes insist that the contractor give a guaranteed maximum price (GMP). The GMP should be differentiated from any cost estimates given by the contractor, although there is always a risk that any cost figures discussed will end up being a GMP.

Construction management contracts in which the CM acts as constructor may include a GMP. For example, AIA Document A133-2009, is a standard form owner/CMc agreement with a GMP. When CMs give a GMP, they usually agree to give it after the design has been worked out to a designated portion of finality, say 75 percent. They protect themselves by getting as many fixed-price contracts from specialty trades as they can and usually build enough into their GMP to take into account not only their fees but also contingencies that may arise from both design changes and unforeseen circumstances. They may also employ project delivery techniques (especially fast-track, discussed in Section 14.11) to maximize efficiencies and minimize work site overhead expenses. From the owner's perspective, a CMc with a GMP (as compared to use of a prime contractor on a fixed-price contract) combines the benefits of using a CM with a cost protection akin to that on a fixed-price contract.

A GMP may not be worth much if the design is quite incomplete at the time the GMP is given. If costs exceed the GMP, the contractor is likely to claim that the scope of the work has so changed that the GMP no longer applies.[7]

Owner Administration. The owner has additional administrative costs in a cost contract. Usually the design professional will seek a higher fee than for a fixed-price contract because many more changes are made in a cost contract as the work progresses. The design professional may have additional responsibilities for checking on the amount of costs incurred by the contractor and for ensuring that the costs claimed actually went into the project and were required under the contract. Determining costs involves not only the often exasperating problems of cost accounting but also the creation of record management and management techniques for determining just what costs have been incurred.

Innumerable variations of allowable costs exist. Usually no question arises on certain items, such as material, labor, rental of equipment, transportation, and items of the contractor's overhead directly related to the project. However, sometimes disputes arise over such matters as whether the cost of visits to the project by the contractor's administrative officials, the cost of supervisory personnel employed by the contractor, and the preparatory expenses or delay claims by subcontractors are allowable costs.

The drafter must try to anticipate all types of costs that can relate directly or indirectly to the project. A determination should be made as to which will be allowable costs for the purposes of the contract. Some troublesome areas can be highlighted by comparing Articles 7 and 8 of AIA Document A102, which deal with cost-plus arrangements. Article 7 lists reimbursable costs, and Article 8 specifies certain costs that are not to be reimbursed. EJCDC C-525, an agreement form for a cost-type contract, also illustrates cost issues. The years of experience of federal procurement have generated complicated allowable cost rules, yet problems still arise in this troublesome area.

14.5 Unit Pricing

One risk of a fixed-price contract relates to the number of work units to be performed. This risk can be removed by unit pricing. The contractor is paid a designated amount for each work unit performed.

A number of factors must be taken into account in planning unit pricing. First, the unit should be clearly described. The cost of a unit should be capable of accurate estimation. Best unit pricing involves repetitive work in which the contractor has achieved skill in cost predicting. Second, it must be clearly specified whether the unit prices include preparatory work such as cost of mobilizing and demobilizing apparatus needed to perform the particular work unit.

The legal issue that arises most frequently relates to inaccurate estimates of units to be performed. These usually involve excavation cases in which the actual units substantially overrun or underrun the estimates. Pricing the unit work usually assumes that there will not be a substantial deviation from the estimates. If the actual units substantially underrun, the cost cannot be spread over the number of units planned and will amount to more per unit than the contractor expected. If the unit is overrun, the contractor may be expected to perform more unit work in the same period of time, another factor that can increase planned costs.

Often sophisticated contracts specifically grant price changes if the number of units overrun or underrun more than a designated amount. For example, the federal procurement system grants a price adjustment if the quantity of units is 15 percent above or below the estimated quantity.[8]

AIA Document A201-2007 deals with this problem in a more limited way. Section 7.3.4 grants an equitable adjustment to either party if the quantities "are materially changed in a proposed Change Order or Constructive Change Directive" and applying the unit prices will cause "substantial inequity." This does not grant an automatic adjustment for overruns or underruns.

Suppose the owner simply gave an estimate that turned out to be inaccurate—even grossly so—and the contract had no adjustment method. As a rule, the mere existence of a variation in quantity of work does not entitle the contractor to additional compensation or a change in the unit price. Such a result would be based on the contractor having assumed the risk or, occasionally, on an express provision in the contract stating that estimates cannot be relied on.

For example, in *Costanza Construction Corp. v. City of Rochester*,[9] the specifications estimated 100 cubic yards of rock to be removed; the actual amount of rock was 600 cubic yards. The contractor claimed it bid below cost because it expected to encounter only a small amount of rock. It tried to avoid the unit price by claiming that the actual amount of rock found constituted a cardinal change that altered the essence of the contract. However, it was not granted relief, because the contract included the city's disclaimer of responsibility as to the accuracy of the estimate and required the contractor to make its own inspection. The court brushed off a claim that there was no time for an independent inspection and that the cost would have been excessive.

The dissent stated that normal variations in quantities can be dealt with by a unit price. Here, however, the difference was so great (the contract price was $936,000, the actual cost for excavation was alleged to be $800,000, and the contractor's estimate was $2,500) that the estimates became meaningless. Applying the unit price, the dissent argued, could economically ruin a good-faith bidder through no fault of its own.

14.6 Value Engineering Change Proposal (VECP)

Owners are always seeking methods of reducing costs in both fixed-price and cost contracts. Value engineering, a method developed by the federal procurement system, attempts to provide an incentive to the contractor to analyze each contract item or task to ensure that its essential function is provided at the lowest overall lifetime cost. The federal method states that an owner who accepts a **value engineering change proposal** (VECP) initiated and developed by the contractor grants the contractor a share in any decrease in the cost of performing the contract and in any reduced costs of ownership. As an example, in *ICSD Corp. v. United States*,[10] the VECP was to replace mercury batteries for the Army's night-vision gun sights with cheaper alkaline batteries. The VECP was accepted, meaning the contractor shared in the cost saving.

Beginning in 1990, the federal government adopted value engineering principles and procedures for architect/engineer contracts. When authorized by the contracting officer, such studies will be made after completion of 35 percent of the design stage or such other stages as the contracting officer determines. However, unlike value engineering for contractors, the government and the contractor do not share savings of costs. Instead, the design professional provides a fee breakdown schedule for the services relating to value engineering activities, and when approved, these services are compensated.[11]

14.7 Design Responsibility: "Design" Versus "Performance" Specifications

Design, consisting of the plans and specifications, is the link between the owner's aspirations for the completed structure and the actual, physical means to achieve that result. A fundamental organizational decision the owner must make is who creates the plans and specifications.

Generally, the owner wishes to retain control. He does so by hiring an independent design professional. The owner then uses the completed design to seek bids (prices) from contractors offering to execute the design.

An alternative approach is for the owner to list the parameters the structure would contain, leaving it up to the contractor to both create and execute the design.

Broadly speaking, this division between who creates the design corresponds to the distinction between "design" and "performance" specifications. **Design specifications** state precise measurements, tolerances, materials, construction methods, sequences, quality control, inspection requirements, and other information. They tell the contractor in detail the material it must furnish and how to perform the work.

Performance specifications state the performance characteristics required; for example, the pump will deliver fifty units per minute, a heating system will heat to 70°F within a designated time, or a wall will resist flames for a designated period. As long as performance requirements are met, how that result is achieved is up to the contractor. Under a pure performance specification, the contractor accepts responsibility for design, engineering, and performance requirements with general discretion as to how to accomplish the goal.

From an organizational perspective, the design specifications are used on a design–bid–build (DBB) project (addressed in Section 14.9), and performance specifications are used on a design–build (DB) project (discussed in Section 14.14).

14.8 Administrative Problems

14.8A Overview

This chapter addresses how a construction project may be organized. The previous discussion examines contract types (fixed-price or cost, for example). This section deals with project administration; in particular, issues of authority and communication. The remainder of this chapter is devoted to the project delivery system—whether design–bid–build or any number of variations.

Returning to the topic of this section, a successful construction project recognizes the importance of clear and efficient lines of communication among the main participants. It is particularly important to know who has authority to bind the owner (with the hope that such a person can be identified and found), how communications (and there are many) are to be made and to whom they should be directed, and what the rules are for determining their effectiveness (effective when posted or when received?).

14.8B Authority: Special Problems of Construction Projects

Introduction. On a small home renovation project under which the owner hires the prime contractor and deals with any questions that arise, the owner clearly has authority to make binding decisions. If changes involve a subcontractor's work, the prime contractor is charged with directing the subcontractor as what to do. The lines of authority are clear.

Complications arise when the owner hires an independent entity—whether architect, construction manager (CM, see Section 7.5), program manager (PM, see Section 14.19), or some other entity—to assist the owner in contract administration. For purposes of this discussion, it will be assumed the owner has engaged an architect to design and monitor performance, to act as agent of the owner, to interpret and judge performance, and (in general) to be the central hub around which the administration process revolves.

In construction projects that justify doing so, the owner may have a full-time site representative who observes, records, and reports the progress of the work. The representative (called a *clerk of the works, project representative, resident engineer,* or otherwise) can be either a regular employee of the owner or an employee of the design professional or CM and is paid for either by the employer, by the owner, or by an independent entity retained to perform this service. The representative's permanent presence invites difficulties, with the contractor often contending that the representative directed the work, knew of deviations, accepted defective work, or knew of events that would be the basis for a contractor's claim for a time extension or additional compensation.

Architect's Authority. The architect, although an agent of the owner, has limited authority. The authority of a project representative is even more limited. Which acts of the design professional or project representative will be chargeable to the owner? This question is governed by principles of agency (see Section 2.11).

Generally, the owner retains a design professional to provide professional advice and services—not to make or modify contracts or issue change orders. Although design professionals may enter into discussions with prospective contractors or even assist in administering a competitive bid, participants in the construction industry recognize that design professionals do not have authority to make contracts, modify them, or issue change orders on behalf of owners.

This does not mean that a design professional's statement or decision will never bind the owner. First, a design professional may be given express authority to perform any of these functions. For example, the design professional is frequently given authority to make minor changes in the work (see AIA A201-2007, § 7.4).

In addition, the owner, by his acts, may cloak the architect with apparent authority. For example, suppose the architect directs major changes for which he does not have express authority. If the owner stands by and does not intervene to deny such authority or pays for the changes, either may represent to the contractor that the architect has apparent authority to order that work be changed.

Care must also be taken to determine whether the principal—that is, the owner—has authority that can be exercised through its agent—the design professional. For example, suppose the engineer directs the contractor as to how the work is to be performed when this choice by contract is given to the contractor. The owner would not have the authority to direct the construction methods; therefore, neither would the engineer. This is not an agency issue but one of contract interpretation.

Another issue involves the design professional's receipt of communications from the contractor. As indicated earlier in this section, the construction process requires many communications among the participants. Clearly, a contract can provide that a notice can or must be given to the design professional. If this is the case, the notice requirement has been met if notice has been given to the design professional. The notice need not come to the owner's attention.

In the absence of specific contract language of this type, will a notice that should go to the owner be effective if it is delivered to the design professional? Because owner and design professional are closely related and the law tends to be impatient with notice requirements if they appear to bar a meritorious claim, a notice given to the design professional will very likely be treated as if given to the owner.

A separate issue—the architect's acceptance of defective work—is addressed in Chapter 17, which deals with project completion. However, it may be noted here that many contracts provide that acts of the design professional, such as issuing certificates, will not be construed as accepting defective work.[12]

As noted earlier in this section, if the design professional has limited authority, the project representative has even less. Usually the project representative is simply authorized to observe, keep records, and report.

14.8C Communications

A communication can be made personally, by telephone, or by a written communication. It is advisable to specify in the contract that a facsimile (fax) or electronic mail (e-mail) is a written communication. Generally, written communications should be required where possible. If it is not possible for a written communication to be made, at the very least a person making the oral communication should give a written confirmation as soon as possible. If the contract provision that deals with communications of notices does not state how communications are to be made, the communication can be made in any reasonable manner.

Commonly, the construction contract specifies time requirements. For example, AIA Document A201-2007, § 15.1.2, requires a notice "within 21 days after the occurrence of the event giving rise to such Claim or within 21 days after the claimant first recognizes the condition giving rise to the Claim, whichever is later." AIA Document A201-2007, § 8.1.4, specifies that it uses calendar days (as distinguished from working days).

These types of notice requirements may act as contractual statutes of limitations. A party who misses the deadline may be barred from bringing an otherwise meritorious claim or at least will be required to devote considerable efforts to having this notice period excused. Where the parties by their conduct ignore the contract's formal requirements (for example, that changes be put in writing), the court will likely excuse a failure to comply with these contractual limitations periods. But achieving that result is expensive, time consuming, and precarious.

14.9 Traditional Project Delivery: Design–Bid–Build (DBB)

The remainder of this chapter addresses different project delivery systems. The baseline against which any alternative delivery system must be understood is the traditional design–bid–build (DBB) method. This section addresses that traditional method.

14.9A Traditional System Reviewed

The traditional system separates design and construction with the former usually performed by an independent design professional and the latter by a contractor or contractors. The sequencing used in this system is creation of the design followed by contract award and execution. Two processes are available to the owner when choosing a contractor. Under **design–bid–build** (DBB), the owner provides the design to several contractors and they competitively bid to perform the work. The owner usually hires the contractor who offers to do the job at the lowest price. (Competitive bidding is discussed in detail in Chapter 15.) If the owner wants a particular contractor to perform the work, the owner will use the design–award–build (DAB) approach and give the design to a contractor it has already chosen.

The contractor to whom the contract is awarded is usually referred to as the *prime* or *general contractor*. It will perform some of the work with its forces but is likely to use specialized trades to perform other portions of the work. The prime contractor is both a manager of those whom it engages to perform work and a producer in the sense that it is likely to perform much or some of the work with its own forces. The traditional system used in the United States usually gives certain site responsibilities to the design professional who has created the design (discussed in detail in Section 10.3).

The principal advantages of this system are

1. The owner can select from a wide range of design professionals.
2. For inexperienced owners, an independent professional monitoring the work with the owner's interest in mind can protect the owner's contractual rights.
3. Not awarding the contract until the design is complete should enable the contractor to bid more accurately, making a fixed-price contract less likely to be adjusted upward except for design changes.
4. Subcontracting should produce highly skilled workers with a specialization of labor and should create a more competitive market because it takes less capital to enter.

14.9B Weaknesses

The modern variations to the traditional system developed because—despite its strengths—the traditional system developed weaknesses. It is important to see these weaknesses as a backdrop to these alternative delivery methods.

Separating design and construction deprives the owner of contractor skill during the design process, such as sensitivity to the labor and material markets and knowledge of construction techniques and their advantages, disadvantages, and costs. A contractor also would have the ability to evaluate the coherence and completeness of the design and, most important, the likely costs of any design proposed.

The frequent use of subcontractors selected and managed by the prime contractor causes difficult problems. Subcontractors complain that their profit margins are unjustifiably squeezed by prime contractors demanding they reduce their bids and, more important, reducing the price for which they will do the work after the prime contractor has been awarded the contract. Subcontractors also complain that contractually they are not connected to the owner—the source of authority and money. They also complain that prime contractors do not make enough effort to move along the money flow from owner to those who have performed work and that they withhold excessive amounts of money through retainage.

The traditional system tends to keep down the number of prospective prime contractors who could bid for work and thereby reduces the pool of competitors. The DBB method, with its emphasis on a fixed-price contract and competitive bidding, also can create an adversarial relationship between owner and contractor, ranging from lack of cooperation to the contractor trying to take advantage of the less sophisticated owner.

Also, the traditional system with its linked set of contracts—owner–design professional, owner–contractor, contractor–subcontractors—does not generate a collegial team joining together with a view toward accomplishing the objectives of all the parties. Some note that the designer and the contractor often generate a semi-adversarial mood, which can generate accusatory positions when trouble develops. The designer acting as the owner's representative can generate administrative costs with papers having to pass through many hands. All this can increase the likelihood of disputes. Some of the modern variations emphasize better organization, the creation of a construction team, and the need for all involved to pull together to accomplish the objectives.

Another weakness of the traditional system relates to the role of the design professional during construction. For the reasons outlined in Section 10.3B, modern design professionals seek to exculpate themselves from responsibility for the contractor's work and to limit their liability exposure. Perhaps even more important, many design professionals lack the skill necessary to perform these services properly.

Under the traditional method, the managerial functions of the prime contractor may not be performed properly. The advent of increasing, pervasive, and complex governmental controls over safety often found many contractors unable to perform in accordance with legal requirements. In addition, the managerial function of scheduling, coordinating, and policing took on greater significance as pressure mounted to complete construction as early as possible and as claims for delay by those who participate in the project proliferated. A good prime contractor should be able to manage these functions, as its fee is paid to a large degree for performance of these services. Not all prime contractors were able to do this managerial work efficiently. Some owners believed the managerial fee included in the cost of the prime contract could be reduced.

The division between design and construction, although at least in theory creating better design and more efficient construction, had the unfortunate result of dividing responsibility. When defects develop, the design professional frequently contends that such defects were caused by the contractor's failure to execute the design properly, whereas the contractor asserts that the design was defective. This led to bewildered owners not being certain who was responsible for the defect as well as to complex litigation.

In this chapter's scenario, MMI used a DBB project delivery method and found the experience frustrating. The division between design and construction caused frictions on the job site. For example, Prime Builder, Inc. found the design could not be implemented using standard construction techniques and was forced to use more labor than anticipated. Were the design results so unique (on a factory project) as to require such an unconventional design, or could contractor feedback to the design as it was being created have prevented this dispute from arising?

The division between design and construction also manifested itself in the dispute between the architect and contractor as to the cause of the settlement damage. In a zero-sum game, each party viewed itself as coming out ahead only at the expense of the other. The owner is left in the middle with no clear resolution in sight.

On the other hand, in a DBB setting, the architect's independent review of the payment process revealed problems with a subcontractor's work. Prime Builder, Inc. as the prime contractor, was contractually responsible for that work. MMI did not have to mediate a dispute between the contractor or subcontractor in order to get a satisfactory resolution. Prime Builder, Inc.'s responsibility to MMI was clear cut.

14.10 Modern Variations in Project Delivery: Introductory Remarks

Modern project delivery systems are becoming "the norm," especially in large commercial projects, but increasingly in public contracting as well.

Modern delivery methods may be grouped by the following variables.

1. *New project actors.* The clearest example is introduction of the construction manager, an actor similar to the prime contractor and so addressed in Chapter 7. Another arguably new actor, the design–builder, is discussed in Section 14.14.

2. *New roles for existing actors.* The vast majority of modern project delivery systems involve the same actors found on a DBB project, but with new or enhanced functions. For example, on a multi-prime project (see Section 14.12), the owner contracts directly with multiple contractors (one of whom may be designated the general contractor), rather than have a contract only with one prime contractor.

3. *New technology.* Electronic documents (including computer aided design (CAD)) and building information modeling (BIM, see Section 14.21) seek to introduce efficiencies through the use of new technologies. (This is not a new project delivery technique, but rather technology that can be used on most any project.)

On different projects, these variables may exist separately or in combination. For example, a construction manager as advisor (CMa) may be used on an otherwise traditional DBB project with the role of the CMa being to provide the owner with additional protection and efficiencies.

Or, in a "combination" situation, the CMa may be used on a design–build (DB) project. In this situation, the CMa's argument is that "owners lacking in-house technical help due to downsizing of construction staffs need independent experts to prepare requests for proposals and to ensure that plans and expectations match by reviewing drawings, submittals and quality control."[13]

Before briefly describing these new methods, some general reflections are necessary on the law's response.

Many legal rules are premised on the traditional system. Modern variations change this system. In analyzing legal disputes involving modern systems, the courts of course will look at the parties' contracts, but they may also reason by analogy to what is familiar. For example, in Section 7.5B, it was noted that the courts compare a CM as advisor to a design professional and a CM as constructor to a prime contractor.

The traditional system, which separates design from construction, has relatively clear lines of responsibility. Risk allocation methods—insurance, indemnity, and contract disclaimers—are easy to devise. By contrast, all of the modern variations have as a common denominator a blurring of the lines of responsibility. This blurring has led (and will continue to lead) to the creation of new risk allocation methods, including new insurance products and new forms of indemnity.

Sometimes legislation is necessary to adapt to these changing commercial realities. As an example, some states have passed new laws allowing the use of CM or DB on public contracts.[14]

Much standardization in the traditional system is attributable to general acceptance of construction contract documents. As alternative project delivery systems have become more accepted, industry groups have begun issuing standard form documents for these newer systems as well. The AIA publishes standard form documents for CM and DB. The Associated General Contractors of America (AGC), together with other professional and trade organizations, publishes the ConsensusDOCS, which include standard form contracts for project management, project alliance, and BIM. The Construction Management Association of America (CMAA) has also created standard form contract documents.

Even with these developments, descriptions of newer project delivery methods must be viewed as nothing more than generalizations. The reason is that these methods are almost always undertaken by sophisticated owners who are likely to tailor the agreement to meet their individual needs.

In sum, the common law, which develops by accruing judicial precedents, changes slowly. It functions to a large degree through crude categories aided by analogies, largely for administrative convenience. Law often lags behind organizational and functional shifts in the real world. New project delivery systems described in this section inevitably create temporary disharmony in the law. Over time, predictable legal rules emerge.

14.11 Phased Construction (Fast-Tracking)

Phased construction, or what has come to be known as "fast-tracking," is not an organizational variation. It differs from the traditional method in that construction can begin before design is completed. There is no reason it cannot be used in a traditional single contracting system, although it is likely to mean that the contract price will be tied to costs.

Construction can begin while design is still being worked out. Ideally, this means that the project should be completed sooner. If a contractor is engaged during the design phase and knows which material and equipment to use, it can make purchases or obtain future commitments earlier and often cut costs while saving time.

> In this chapter's scenario, the work began later than planned because the design phase took so long. This meant the work was started at the beginning of a severe winter, which impacted the project. Under fast-tracking, it might have been possible to have the foundation and body frame work completed before the winter weather.

Another advantage to the owner is the early activation of the construction loan. In traditional construction, an owner—particularly a developer—does not receive loan disbursements until the construction begins. Yet it must pay for investigation and acquisition of the project site. It may have had to pay either the purchase price for the land or installments of ground rent, as well as insurance premiums, real estate taxes, design service fees, or legal expenses. If it must make these expenditures before it starts to receive construction loan funds, it must draw on its personal unsecured credit line, necessitating monthly interest payments and reducing its credit line, which could be used for other purposes. If construction can begin through fast-tracking, the owner need not use its personal funds but can employ loan disbursements.

The principal disadvantage of fast-tracking is the incomplete design. The contractor will be asked to give some price—usually after the design has reached a certain stage of completion.

Very likely the contract price will be cost plus overhead and profit with the owner usually obtaining a guaranteed maximum price (GMP, see Section 14.4.) However, the evolution of the design through a fast-tracking system is likely to generate claims. Completing the design and redesigning can generate a claim that the contract has become one simply for cost, overhead, and profit and that any GMP has been eliminated.[15] The owner may be constrained in making design changes by the possibility that the contractor will assert that the completed drawings must be consistent with the incomplete drawings.

Two other potential disadvantages need to be mentioned here. First, there is greater likelihood that there will be design omissions—items "falling between the cracks." Needed work is not incorporated in the design given to any of the specialty trade contractors or subcontractors. Although such omissions can occur in any design, they are more likely to occur when the design is being created piecemeal rather than prepared in its entirety for submission to a prime contractor.

Second, in fast-tracking, there is a greater likelihood that one participant may not do what it has promised and thus adversely affect the work of many other participants. For example, in *Pierce Associates, Inc.,* an opinion of a federal contract appeals board discussing fast-tracking (or phased construction) in the context of construction management and multiple primes stated:

> "Phased" construction has been analogized to a procession of vehicles moving along a highway. Each vehicle represents a prime contractor whose place in the procession has been pre-determined. The progress of each vehicle, except that of the lead vehicle, is dependent on the progress of the vehicle ahead. The milestone dates have been likened to mileage markers posted along the highway. Each vehicle is required to pass the mileage markers at designated times in order to insure steady progress.[16]

When any of the vehicles do not pass the mileage markers assigned to them, claims for delays and complex causation problems are likely to result.

14.12 Multiple Primes

Multi-prime projects consist of the owner contracting directly with the principal subcontractor trades (see Figure 14.1). Use of separate contractors gives the owner greater control in specialty contracting and avoids the subcontractor complaints that they are a contract away from the source of power and funds. In addition, separate contracts are used if the owner is not confident of its prime contractor's management skill. Use of multi-prime projects was also spurred by successful legislative efforts by subcontractor trade associations to require that state and sometimes local construction procurement use separate contracts.[17]

Another factor, although of less significance, is the hope that separate contracts might result in lower contract prices. Breaking up the project into smaller bidding units allows more contractors to bid. Owners hope the managerial fees can be reduced by taking this function from the prime contractor. These hoped-for pricing gains can compensate for the additional expense of conducting a number of competitive bids.

In the traditional contracting system, the linked set of contracts determines communication and responsibility. If subcontractors have complaints, they can look to the prime contractor. If the prime has problems, it can look to the subcontractors or the owner.

Separate contractors are required to work in sequence or side-by-side on the site, but they do not have contracts with one another. Disputes among them must be worked out by the participant performing the coordination function—something that is not as neat in the separate contract system as in the single-contract system. To be sure, even in the traditional system, this problem can arise if disputes develop between subcontractors who have no contract with each other. The traditional system handled this with relative clarity by requiring that the prime contractor deal with these matters as part of its managerial function and that the owner remove itself from these problems.

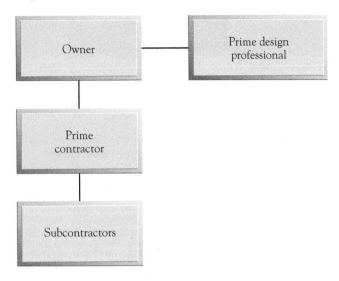
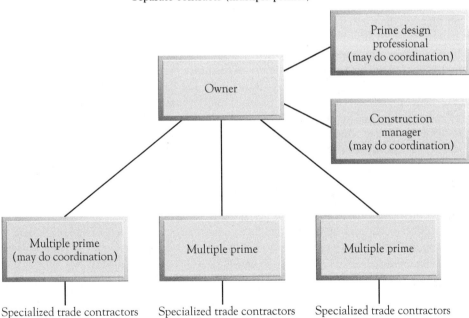

FIGURE 14.1 Traditional and separate contract systems compared.

Who performs the managerial function in a multi-prime setting? Who has legal responsibility? Under the construction management system, these often fall to the construction manager (CM). But the managerial function can be performed by the design professional if he has the skill and willingness to do so, by a staff representative of the owner, by a program manager, or by a managing or principal separate contractor. If it is not clear who will coordinate, police, and be responsible in any system, the project will be delayed and claims made.

These issues are illustrated by a leading New Jersey Supreme Court case. In *Broadway Maintenance Corporation v. Rutgers, State University*,[18] Rutgers University sought to build a medical school using multiple prime contractors. However, Rutgers, in its agreement with *one*

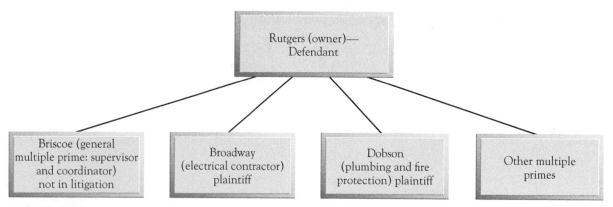

FIGURE 14.2 The relationships among parties in *Broadway Maintenance v. Rutgers*.

of those primes (Briscoe), specified that Briscoe would act as the supervisor on the job and coordinator of all the contractors. (See Figure 14.2.)

The job was finished late. Plaintiffs (two other primes: Broadway and Dobson) sued Rutgers for damages due to delays and disruptions. Plaintiffs argued Rutgers was liable for these damages because the university breached its duty to coordinate the activities of the various contractors on the site. Rutgers' defense was that Briscoe was the party responsible for coordinating the work. The trial court agreed with Rutgers and entered a judgment for the university.

The New Jersey Supreme Court affirmed. It first ruled that, in a multi-prime project, each prime contractor is a third-party beneficiary of the other prime contracts. A plaintiff prime contractor may sue a defendant prime contractor for the defendant's breach of its contract with the owner if that breach caused the plaintiff injury (commonly, delayed performance).[19]

The court then turned to the question of whether Rutgers had a duty to coordinate the work of the various primes or whether the owner could delegate that obligation to one of the primes (Briscoe). The court began its analysis with an observation:

> [A]n owner has the privilege to eliminate a general contractor and enter into several prime contracts governing the construction project. In that event the owner could engage some third party or one of the contractors to perform all the coordinating functions . . . .
>
> If no one were designated to carry on the overall supervision, the reasonable implication would be that the owner would perform those duties. In so doing, the owner impliedly assumes the duty to coordinate the various contractors to prevent unreasonable delays on the project. That is a reasonable assumption because the contracting authority has the power to use its superior position and to invoke its contractual rights to compel cooperation among contractors. The owner is impliedly obligated to act in good faith and to do that which it reasonably can to ensure that the other contractors adhere to the time schedules established for the project.[20]

The court then found that Briscoe, not Rutgers, had agreed to coordinate the work of the various prime contractors. The plaintiffs' remedies were against their co-primes as third-party beneficiaries, not against Rutgers.

How do standard form documents deal with one prime contractor being delayed or its costs increased by a co-prime? AIA Document A201-2007, Article 6, allows the owner to award separate contracts. Document A201-2007, § 6.2.3, makes the owner responsible for a contractor's costs incurred because of delayed or defective performance by another contractor. The at-fault contractor must then reimburse the owner. Under this system, an injured contractor looks to the owner for relief instead of trying to sue the co-prime contractor. This system contrasts with EJCDC Document C-700, ¶ 7.3 (2007), which gives the separate contractors rights directly against each other.

Owners contemplating a separate contract system should consider these organizational and managerial issues. Clearly, a separate contract system ought not to be used unless the managerial function can be better performed and at a lower cost than if performed by a prime contractor under a single contract system. If the separate contract system is selected, all contracts should make clear who has the administrative responsibility and who has the legal responsibility if a contractor has been unjustifiably delayed. More particularly, if this responsibility is to rest solely in the hands of the people who are given the responsibility for managing the contract, all contracts should make clear whether the owner is exculpated.

14.13 Turnkey Contracts

There are a great variety of **turnkey contracts**. At its simplest, the contract is one in which the owner gives the turnkey builder some general directions as to what is wanted and the turnkey builder is expected to provide the design and construction that will fill the client's communicated or understood needs. In theory, once having given these general instructions, the owner can return when the project is completed, get the key from the contractor, turn the key, and take over.

Many turnkey projects are not that simple. The instructions often go beyond simply giving a general indication of what is wanted. They can constitute detailed performance specifications. Also, the obligation to design and build may depend on the owner furnishing essential information or completing work on which the turnkey contractor relies to create the design and to build. Finally, the owner who has commissioned a turnkey project is not likely to remain away until the time has come to turn the key. As in design–build, the owner may decide to check on the project as it is being built and is almost certain to be making progress payments while the project is being built. The most important attribute of the turnkey project is that one entity both designs and builds, a system discussed in greater detail in Section 14.14.

Some turnkey contracts require the contractor not only to design and build the building but also to provide the land, financing, and interior equipment and furnishings.

A turnkey contract looks more like a sale than a contract for services. As a result, an owner could argue that a turnkey contract created warranties that made the seller–contractor responsible for any defects. This is something that also may be asserted in a design–build contract, which is discussed next.

14.14 Design–Build (DB): Combining Design and Construction

14.14A Reasons for Design–Build

In addition to the weakness of the traditional design-bid-build method outlined in Section 14.9B, there are other weaknesses that led the explosion of the design-build (DB) system in the 1990s. One commentator states that it can no longer be assumed that "the most advanced construction technology and knowledge of the most construction methods lie with architects and engineers.[21] He goes on to state that this knowledge lies increasingly with "specialty contractors and building-product manufacturers." He also notes that it increasingly has become difficult to prepare complete and accurate drawings and specifications, hereby exposing owners to claims based on defective specifications. He also points to increased cost and delay in determining who is responsible for defects—the designer or the contractor—and the need for "single point" responsibility. Finally, he states that the traditional method simply takes too long and costs too much in preconstruction services.

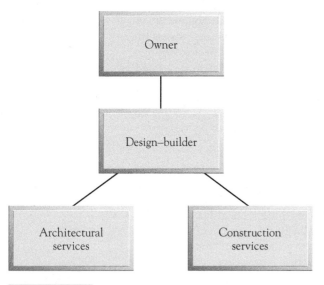

FIGURE 14.3 Design–build (DB) system.

As a result, it is increasingly popular for one entity to both design and build, which is an important variation from the traditional method of organizing for construction. See Figure 14.3.

While it is impossible to accurately compare the percentage of design–bid–build (DBB) versus DB projects, one scholar concluded that "[h]owever one measures or reports the data, the number of design–build projects has been increasing for many years and has now reached the point that it challenges the dominance of the traditional system."[22]

14.14B Nature of Design–Build

DB can encompass at one extreme the homeowner building a single-family home patterned on a house that the builder has already built and at the other a large engineering company agreeing to both design and build highly technical projects, such as petrochemical plants. The former is likely to be a builder with an in-house architect on its staff or one that engages an independent architect where it is required by law that design be accomplished by a registered architect or engineer. At the more complex extreme, the D/builder may employ a large number of construction personnel, licensed architects, and engineers in-house to offer a total package for projects such as power plants, dams, chemical processing facilities, and oil refineries. Between these extremes, DB is often used for less technical repetitive work such as warehouses or small standard commercial buildings.

Increased use of the DB system led the professional and trade associations to publish standardized documents. In addition to the American Institute of Architects (AIA) and Engineers Joint Contract Documents Committee (EJCDC), standard DB documents have been published by the Associated General Contractors of America (AGC) ConsensusDOCS, the Design–Build Institute of America (DBIA) and the Construction Management Association of America (CMAA).[23] These standard documents (or any individualized contract between the D/builder and the designer) regulate the relationship between the D/builder and the person it engages to create the design.

But what is the relationship between the owner and the designer—whether an independent design professional retained by the D/builder or one on the D/builder's staff? Even though the designer works for the D/builder and owes his primary allegiance to the one who has selected him and pays him, the owner may believe the design professional has some obligation to the owner. The owner indirectly pays him, and as a licensed design professional, he has obligations to the public and to the owner even though the owner has no contract with him. In a vague and undefined way, the owner may believe that a registered architect or engineer will make independent decisions that take his interests and those of the public into account even though the D/builder has engaged him and pays him.

In a classic DB contract, the owner gives the D/builder general parameters and the D/builder provides the design. The D/builder is responsible for a successful implementation of the owner's expectations. As explained in Section 14.7, the owner's design parameters constitute "performance" specifications. Rather than dictate to the contractor how to achieve the desired structure, the contractor decides those details for itself.

The owner may believe this "classic" DB arrangement does not provide it with an independent actor looking out for its design interests. One solution, usually called "bridging," involves the owner hiring a design professional to create the preliminary design. This design work, usually through the design development phase or approximately one-third to half of the design documents, is then used as a basis for obtaining requests for proposals (RFP) from prospective designbuilders.

This sharing of design responsibilities in a bridging/design–build setting complicates allotting liability for design defects in the completed project. No clear legal rules exist, because each bridging arrangement, as well as the locus of the design defect, is unique to each project. Nonetheless, a review of a few cases is informative.

In *Glacier Tennis Club at the Summit, LLC v. Treweek Construction Co., Inc.*,[24] a tennis club hired an architect (Thompson) to create schematic drawings showing preliminary design parameters. A design–builder then designed and built the indoor tennis facility. During construction, Thompson reviewed the contractor's work for the purpose of approving progress payments.

The completed facility suffered from design defects, which were the responsibility of the D/builder. Nonetheless, the D/builder sued Thompson, alleging the architect owed it a duty of care to bring to the contractor's attention any design errors Thompson had discovered during the payment review process. The Montana Supreme Court ruled that Thompson did not owe the contractor a duty of care.

A sophisticated owner may wish to derive the benefits of a DB structure, while retaining the option to review and even change the contractor's design. For example, in *Dillingham Construction, N.A., Inc. v. United States*,[25] the Veterans Administration (VA) used a DB contract, yet split the responsibility for designing the facility between the VA and the contractor. In the same paragraph, the VA specified that the contractor "shall provide complete construction drawings and specifications for the Clinic Building, based on the preliminary drawings and performance specifications included with this solicitation," but then added: "The government's Architects and Engineers will provide final construction plans and specifications for site improvements and the Central Mechanical Room prior to the Request for Best and Final Offer."[26] As a result, each party hired its own architectural and engineering firm. The government designer reviewed the designs created by the contractor's designer. From the D/builder's perspective, this method may create uncertainty as to where the owner's design parameters end and where the D/builder's design responsibility begins.

Similarly, in *FSEC, Inc.*,[27] the contract included drawings that required four exhaust fans and dust collectors. An amendment, in addition to describing the project as a DB contract, stated that the contractor was to submit design details for approval by the government. This labeling of the contract as a DB contract persuaded the contractor that it could design a system as long as it met performance criteria.

A specialty subcontractor determined that the performance specifications could be met with two exhaust fans. The contracting officer ordered the contractor to install four exhaust fans. The contractor's claim for additional compensation was rejected. The Armed Services Board of Contract Appeals held that all elements of the contract must be considered. It pointed to the bid documents requiring four fans and that the design was subject to the government's approval. In essence, although described as a DB contract, it was not. It was a hybrid design–award–build contract with design specifications but labeled a DB contract.

> In this chapter's scenario, MMI used a DBB method but found the results frustrating. A DB method would have united design and construction. The complexity of the design might have been avoided, because Prime Builder, Inc.'s designer would have created the design in collaboration with those responsible for the actual construction.

14.14C Licensing and Insurance

In private contracting, the principal problem in design–build again relates to licensing laws. As noted, the D/builder may be a business organization that will either furnish the design services needed in-house or retain an independent design professional to perform these design functions. In *Charlebois v. J. M. Weller Associates*,[28] New York's highest court upheld the latter type of contract against a claim of a licensing violation. The dissent worried about the effect on the public and the registration laws when the design professional is engaged by the D/builder.

Many states bar business corporations from performing architectural services. As a result, design–build contracts have been challenged—even if the design portion was to be fulfilled by the engaging of a licensed design professional. Although the trend has been toward permitting such contracts despite the absence of a registered independent design professional, the matter is still in flux.

The liability insurance of a design professional usually excludes both construction and work performed under a joint venture, whereas the liability insurance of a contractor usually excludes design. An entity planning to engage in a DB venture must check insurance carefully.

14.14D Advantages and Disadvantages of Design–Build

Despite the legal obstacles to the DB system, DB has clearly been found to be useful. The principal advantage of DB, usually associated with fast-tracking, is speed. One entity replaces different entities who design and build with the inevitable delay caused by using two entities whose work intersects. Those who design and build frequently do repetitive work and acquire specialized expertise.[29]

Perhaps most important from the standpoint of an unsophisticated owner is that the system concentrates responsibility on the D/builder. Owners are often frustrated when they look to the designer who claims that the contractor did not follow the design while the latter claims that the problem was poor design.

Another impetus for the use of design–build occurred when the AIA, at first on a provisional basis, decided that its members could engage in business without running afoul of ethical constraints. The AIA did this because it felt economic realities in the construction industry dictated that architects be given the opportunity of engaging in DB ventures.

DB has weaknesses. The absence of an independent design professional selected by the owner can deprive the owner of the widest opportunities for good design. An unsophisticated owner often lacks the skill to determine whether the contractor is doing the job well or as promised. This can reflect itself not only in substandard work but also in excessive payments being made early in the project or in slow payment or nonpayment of subcontractors. (This is one reason for an owner to employ a construction manager as advisor on a DB project.)

Owners, such as developers, do not like to make cost-plus contracts with D/builders. Clear and complete drawings and specifications protect a developer, even if it uses a cost-plus contract.

The contractor is reluctant to give a fixed price on a design–build project. It cannot know with any certainty what it will be expected to build. If the design has not been worked out at the time the contract is made, the design must be completed later. Very likely there will be redesign. As a result, the owner will prefer a fixed price, whereas the contractor would like an open-ended cost contract.

To ameliorate this possible impasse, the owner may find it helpful to prepare a set of performance specifications. Even if these specifications cannot define the elements of the building in detail, they may be able to prescribe intelligent criteria for performance in advance.

Another useful technique is to prepare a budget for each phase of the work and designate the budget estimate as a target price. If the actual cost is greater or less than the estimate, the contract price can be adjusted.

Progress payments also provide problems in DB contracts. Under traditional methods, the architect certifies progress payments. Although some complain that the architect cannot be impartial when he does this, the architect's professional stature gives hope that he will be impartial.

In DB, by contrast, payment certificates issued by an architect who has been retained by the contractor may not be reliable. Owners who use design–build methds often take control of the progress payment process—a method that can operate to the disadvantage of the contractor. As a result, it may be useful for each party to agree on an independent certifier for progress payments.

An owner building a single-residence home, a warehouse, or simple commercial building relies entirely on the contractor for design and construction. Here warranties of fitness are appropriate. The owner is not buying services but a finished product. The D/builder is held to have sold a "product" and given seller-like warranties.

The sophisticated owner who employs DB and has within its own organization the services of skilled professionals to control design or monitor performance has not bought a product but rather expert services to prepare and execute the design it chooses. Such an owner should not receive a warranty akin to a sale–of–goods (which would make the contractor strictly liable).

The differentiation between the two types also reflects itself in pricing. It is likely that the homeowner commissioning a builder to design and build a house will contract on a fixed price. The sophisticated owner building a petrochemical plant will contract on a cost basis with a GMP—at least until the design is put in final form. Such projects create immense exposure for consequential damages. It is likely that the D/builder will limit its obligation to correction of defects.

14.15 Partnering

Although it is not strictly speaking a form of organization, **partnering** is an approach that can remedy some of the stiffness and adversary tendencies that are so commonly part of the construction process.

Partnering is a team approach in which owners, contractors, architects, and engineers form harmonious relationships among themselves for the express purpose of completing a single project or group of projects. The partnering process usually is initiated when the principal project participants are assembled. It can also be used to steer a troubled project back on course. These relationships are to be structured on trust and dedication to the smooth operation of the project. Those who advocate this concept hope that it will accomplish the project successfully and minimize the likelihood of claims. Partnering, then, is a method of building a closeness that transcends organizational boundaries. This closeness should be premised on a shared culture and the desire to maximize the likelihood of attaining the common goals.

The objectives of partnering can be sought and achieved on an informal basis. In the alternative, the participants—the owner, design professional, prime contractor, and key subcontractors—may enter into an agreement to make concrete the often nebulous partnering commitments. What is important is that the parties express as clearly as possible what each party expects and what each party is willing to undertake. Of course, expressions of this sort are difficult to make sufficiently specific and concrete. But some commitments, such as confidentiality and the agreement not to profit at the other "partner's" expense, may be worth expressing in the contract.

Partnering can suffer unless the entities involved in the effort are willing to make sacrifices to keep the team intact. This is particularly a problem in dealing with design professionals, because clients often express concern when design professionals with whom they have been working are assigned to other work. But it can also be a problem in large engineering firms and the organizations that employ them. Contacts among people working at different operational levels of the various entities must be developed, and a sense of confidence and trust among these workers must be built up and maintained.

It is likely that resort to litigation by one partnering party against another will be rare in partnering arrangements. Judicial intervention will almost never take the form of ordering people to continue such arrangements or compensating one party—particularly as to acts in the future—because of the other's refusal to continue. There may, however, be resort to legal action if one party has breached confidentiality or has stolen business opportunities that should go to the other. For this reason, it is useful to deal with the possibility of litigation in the contract. Finally, legal action may be taken if one party has incurred expenses that it feels the other should bear or at least share. Again, these matters may be appropriate for contractual resolution.[30]

14.16 Teaming Agreements

Teaming agreements have been used when the federal government has embarked on programs to fulfill procurement needs, mainly in the high-tech defense industry. A teaming agreement is a contract between a potential prime contractor and another firm that agrees to act as a potential subcontractor. The prime, sometimes with the assistance of the potential subcontractor, draws up a proposal for a federal agency contract in competition with other prospective contractors. Typically, the prime promises that, if awarded the prime contract, it will reward its "teammate" with an implementing subcontract in return for the often unpaid support given in making the bid.

Most litigation has involved claims by aggrieved teammates against the prime when they have not been awarded the implementing subcontract. Such agreements generally use such phrases as "best efforts" and "reasonable efforts" as well as other nonspecific terminology. Often the question of whether the prime is actually committed to award the implementing subcontract to the potential subcontractor is left vague and uncertain. These arrangements resemble the normal prime contractor–subcontractor relationship in which the primes seek to retain as much flexibility as possible while still keeping their teammate interested in supporting the arrangement. However, the prime would like to ensure that any commitments made by the teammate can be legally enforced if necessary. Here, the twilight area between unenforceable agreements to agree and enforceable agreements to negotiate in good faith can create difficult legal questions.[31]

14.17 Lean Project Delivery

As described by one commentator,[32] **lean project** delivery involves extending the business principles of Toyota Motor Company to the construction process. The three principles underlying the "Toyota Way" are (1) allowing anyone, including factory workers, to stop production in the face of a defect; (2) just-in-time delivery of materials; and (3) subsuming the individual production units within the output of the entire project.

The organizational foundation for lean delivery is the Integrated Agreement, which is signed by the owner, architect, and construction manager (CM) or prime contractor *before* creation of the final design. These three participants are the project team. Major project-related decisions are made by consensus of these three members with the owner having the final word in the event of an impasse.

Rather than the architect's presenting a completed design to the construction participants, the Integrated Agreement calls for the core group to create a "target value design." The target value design makes explicit the value, cost, schedule, and constructability of basic components of the design criteria. This process allows the major participants to have input into the design during its creation, where problems (whether with goals, constructability, or cost) can be identified and dealt with as early into the process as possible. Also, the major trade contractors (such as mechanical, electrical, and plumbing) participate in the design of their segments of the work.

The Integrated Agreement seeks to create a system of shared risk rather than shifting it. Value is increased by emphasizing collaboration, early detection and correction of defects, and early involvement by the project team to understand the project's goals and produce a target value design.[33]

14.18 Project Alliance

Project alliancing (also called Integrated Project Delivery (IPD)) shares with partnering and lean project delivery the establishment of collaborative relationships among the owner and the major project participants. However, project alliance does not seek to mimic the Toyota Way and does not draw its theoretical framework from the manufacturing sector.

In project alliance, the owner, design team, prime contractor, and major trade contractors and suppliers all sign the same contract, called the Project Alliance Agreement (PAA). The PAA has as its goal the benefit of the project as a whole. The financial success of the project participants is contingent upon the financial success of the project. Financial goals are set forth in the PAA: each party is financially rewarded or penalized based upon whether the project as a whole meets or fails to meet these predetermined goals. In addition, the parties agree in advance to release one another from all liability arising out of the project, except for willful misconduct.[34]

The project alliance delivery system is reserved for particularly difficult, large-scale projects. It was developed by the petrochemical industry but then spread to large public works and commercial projects. The first standard form document for an alliancing project is the ConsensusDOCS 300, titled "Tri-Party Collaborative Agreement" and published in 2007 by the Associated General Contractors of America and affiliates.[35]

14.19 Program Management

The demand for program management arises not from dissatisfaction with the traditional model for project delivery, but from company downsizing and budget constraints, which have led large commercial owners to eliminate their in-house engineering and construction departments. A program manager (PM) promises to bring to commercial owners the expertise that these owners need to manage and expand already existing large facilities. The PM's role has been described as "doing for the owner what it would do for itself if it had the in-house capacity or if it chose to divert its resources to the project(s)."[36]

Akin to a CM as advisor (see Figure 7.1), the PM does not undertake responsibility for actual construction work. Instead, the PM performs organizational functions, including analysis of the owner's needs; project site evaluation and acquisition; managing a public relations program; developing a risk management and insurance program; prequalifying the design and construction team members; and administration of the construction process.[37]

Either large design firms or construction companies may market themselves to owners as qualified to provide program management services. Both AIA and EJCDC standard form documents allow either architects or engineers to offer many of these services—with the exception of assuming responsibility for site safety. Contractors, by contrast, may have more experience with constructability, pricing, and scheduling services.

Much like construction management thirty years ago, program management is still an evolving concept. One impediment to a clearer definition of PM has been the lack (until quite recently) of a standard form contract available for those providing such services. Only in 2007, with the Associated General Contractor's publication of ConsensusDOCS Document 800 (owner/PM agreement and general conditions), has a standard form contract been made available to the marketplace. The impact of this recent development remains to be seen.

14.20 Public–Private Collaboration: Public–Private Partnerships (PPP) and Build–Operate–Transfer (BOT)

As discussed in Chapter 15, public contracts normally involve a DBB format with the design professional contract procured by negotiation and the prime contractor selected through competitive bidding. This section addresses alternatives to standard public procurement.

"Public–private collaboration" refers to alternative project delivery methods for public entities. In response to a problem of inadequate funding, governments both in the United States and internationally have looked to collaboration with private industry as one solution. In the United States, these collaborations are generally called **Public–Private Partnerships** (PPP or P3), while for developing countries they are called **Build–Operate–Transfer** (BOT) projects (although alternatives to BOT also exist).

Public–Private Partnerships. PPPs differ from traditional public contracts in financing, operation, and procurement methods. Under PPPs, the private sector finances, builds, renovates, maintains, and/or operates specific public-sector activities in exchange for a contractually specified stream of future revenues generated by that activity. The PPP shifts to the private sector the cost and economic risk of a project or service which traditionally would have been provided by the government through the public procurement process (discussed in Chapter 15). As a simple example, a private company may build or maintain a road in return for the right to collects tolls. A more sophisticated example is the federal government's hiring of a contractor to fund energy-saving renovations of federal buildings with compensation of the contractor to be paid out of projected reductions in energy costs for that building.[38] A "concession agreement" spells out the predicted long-term relationship between the public entity and private-sector party.[39] A PPP differs from the concepts of "privatization" or "outsourcing," in which the government transfers actual responsibility for and title to the asset to the private sector.

Build–Operate–Transfer (BOT). In the international field, most developing countries are not able to finance large infrastructure projects. As a result, after competitive bidding, they may make an agreement with a consortium under which the latter operates the project and then ultimately transfers it back to the state. See Figure 14.4. The consortium that seeks to enter such a contract usually must provide the financial backing to build the project, get it operating,

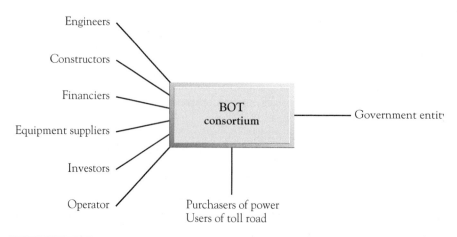

FIGURE 14.4 Build–operate–transfer (BOT) organization.

and keep it running so that it can be transferred. The consortium expects to be repaid mainly from revenues generated. BOT projects are used in large infrastructure projects such as energy generation, bridges, and transportation (roads, rails, and waterways) systems. Most BOT projects are turnkey contracts, which makes financing more secure and easier to obtain. Increasingly, these privately financed projects are being considered and used by developed countries and some American states, mainly for toll roads and high-speed rail systems. When this is done, the government may put up a part of the funds needed.

14.21 Building Information Modeling (BIM)

As noted in Section 14.10, virtually all modern variations of project delivery involve the introduction of a new participant, having existing actors assume new roles or relationships, or (more commonly) through changes in the contractual relations among the project participants (such as multiple prime, design–build, and partnering).

By contrast, building information modeling (BIM) is a technology-driven organizational model. The key to BIM is the creation of a computer model of the project that is both information-rich and information-integrative. It is information-rich in that, by clicking on any object in the model (such as a steel beam), the viewer is provided information about the object, such as the beam's size and structural characteristics (the structural forces acting on the beam and its capacity). It is information-integrative, because the object will automatically adjust to changes in other parts of the model. The design power of BIM, as compared with the traditional computer-aided design (CAD) software, was explained in a recent article:

> The difference between BIM and traditional design approaches is most striking when changes are made to the design. For example, if a steel structure is designed with traditional CAD tools, the drafted design might contain columns and beams with specific connections. If a column is removed to create a larger bay, the designer must recalculate the size of adjacent columns, resize beams, reanalyze load paths, and re-detail the connections. In object-oriented [i.e., BIM] design packages, . . . if a column is removed, the model will communicate with the remaining columns, adjust their size as necessary, change beam dimensions, and change the beam/column connections. . . . Thus, a change in the architectural requirements can ripple through the structural design without direct engineering involvement. The model can "design" itself based on rules embedded in the objects themselves. . . .Not only is this process efficient, it sharply reduces inconsistencies unforeseen when the design was modified.[40]

The goal of BIM is a paperless project in which changes to the specifications are seamlessly integrated with the rest of the design and simultaneously communicated to all the project participants. The BIM also can incorporate the project schedule (associating each phase of construction with when it is to be accomplished) and thus adjust the schedule if necessary (because of a delay in delivery of supplies or equipment, for example).

Widespread adoption of BIM by the construction industry faces serious legal impediments. BIM project delivery in many ways erases the distinctions among designer, builder, and component supplier. While the model is created in the architect's or engineer's office, it integrates contractor, vendor, and fabricator information into a seamless whole. BIM is a collaborative effort; yet the law allocates responsibility upon individuals. Who is responsible for design defects on a collaborative design? How is insurance coverage allocated?[41] Who owns the copyright in a BIM model? All of these fundamental questions are exacerbated by the absence of standard form contracts governing BIM projects.

Practical considerations also must be addressed. The level of detail in a BIM model means that the design process must start much earlier than on a design–bid–build (or even fast-track) projects with the key trades joining in the design collaboration process. Each of the many participants must have the software and know-how for dealing with the BIM model. Financial considerations also must be addressed. A BIM model creates an informational database of such detail as to eliminate the need for shop drawings, thereby eliminating the primary compensation methods of most architectural offices.

Organizations publishing standard form documents are beginning to address BIM. The AIA B101-2007, § 4.1.6, specifies that architectural services performed on a BIM project are additional services deserving compensation separate from the basic fee. A year later, the AIA added E202-2008, titled "Building Information Modeling Protocol Exhibit." Also in 2008, the Associated General Contractors of America and affiliates published ConsensusDOCS 301, titled "Building Information Modeling (BIM) Addendum."

To date, BIM projects have been limited to large owners—either private or governmental.[42] As experience grows and larger numbers of design firms adopt the technology, BIM's spread to smaller projects appears likely.

14.22 Summary

The modern variations described in this section developed because of weaknesses in the traditional system, particularly leisurely performance, divided responsibility, and complaints of subcontractors. However, time and the courts will test these new systems and reveal their deficiencies. This does not in any way diminish the utility of these variations, although it does impose a serious burden on those who wish to engage in them, to anticipate the problems in deciding whether to use a variant from the traditional process and to plan in such a way so as to minimize the difficulties—both administrative and legal.

REVIEW QUESTIONS

1. When prospective contractors cannot be relatively certain of what they will be expected to perform, or where they are uncertain as to the techniques needed to accomplish contractual requirements, what type of contract are they likely to prefer?

2. From an organizational perspective, when are design specifications used and when are performance specifications used?

3. What are the principal advantages of the design–bid–build delivery system?

4. What are the advantages and disadvantages of phased construction (fast-tracking)?

5. What can an owner do if that owner commissions a project using the design–build project delivery method and believes the arrangement does not provide it with an independent actor looking out for its design interests?

6. What are the weaknesses of the design–build project delivery method?

7. What is a teaming agreement, and when is it most commonly used?

8. What are the three principles underlying the "Toyota Way," and which are the basis for lean project delivery?

9. What is a public–private partnership and how does it differ from traditional public contracts?

10. What are some of the legal impediments caused by the widespread adoption of BIM in the construction industry?

ENDNOTES

[1] See *Holder Constr. Group v. Georgia Tech Facilities, Inc.*, 282 Ga.App. 796, 640 S.E.2d 296 (2006), reconsideration denied, Dec. 8, 2006 (*force majeure* clause inapplicable); Guidry, *The Steel Price Explosion: What Is an Owner or a Contractor to Do?* 24 Constr. Lawyer, No. 5, Summer 2004, p. 5.

[2] *Colvin v. United States*, 549 F.2d 1338 (9th Cir.1977).

[3] 135 Wash.App. 859, 147 P.3d 600 (2006), review denied, 161 Wash.2d 1011, 166 P.3d 1217 (2007).

[4] AIA A102-2007, § 3 (cost plus a fixed fee with a Guaranteed Maximum Price (GMP)) and A103-2007, § 3 (cost plus fixed fee without a GMP).

[5] 60 Md.App. 671, 484 A.2d 302 (1984), cert. denied, 303 Md. 114, 492 A.2d 616 (1985).

[6] 496 So.2d 1012 (La.1986).

[7] *C. Norman Peterson Co. v. Container Corp. of America*, 172 Cal.App.3d 628, 218 Cal.Rptr. 592 (1985).

[8] 48 CFR § 52.211-18 (2012).

[9] 147 A.D.2d 929, 537 N.Y.S.2d 394, appeal dismissed, 74 N.Y.2d 714, 541 N.E.2d 429, 543 N.Y.S.2d 400 (1989).

[10] 934 F.2d 313 (Fed.Cir.1991). The VECP regulation is 48 CFR § 52.248-1 (2012).

[11] 48 CFR § 52.248-2 (2012).

[12] AIA Doc. A201-2007, §§ 9.6.6 and 9.10.4.

[13] Engineering News Record, *CMAA Seeks to Broaden Appeal of Construction Management*, at 16 (Sept. 27, 1999).

[14] N.C Gen.Stat. § 143-128.1 (CM); Wash.Rev.Code § 47.20.780 (DB).

[15] *Marriott Corp. v. Dasta Constr.*, 26 F.3d 1057, reh'g in banc denied, 37 F.3d 639 (11th Cir. 1994); *C. Norman Peterson Co. v. Container Corp. of America*, 172 Cal.App.3d 628, 218 Cal.Rptr. 592 (1985).

[16] GSBCA No. 4163, 77-2 BCA ¶ 12,746.

[17] New York was the first state to adopt this approach with passage of the Wicks Act in 1912. The Act is scattered throughout various statutes, including N.Y.Gen.Muni.Law § 101(1) & (2).

[18] 90 N.J. 253, 447 A.2d 906 (1982).

[19] This is the majority rule, see *Moore Constr. Co., Inc. v. Clarksville Dept. of Electricity*, 707 S.W.2d 1 (Tenn. Ct.App.1985), aff'd March 24, 1986 (thorough discussion).

[20] 447 A.2d at 912 (citations omitted).

[21] Hinchey, *Karl Marx and Design-Build*, 21 Constr. Lawyer, No. 1, Winter 2001, p. 46.

[22] Circo, *Contract Theory and Contract Practice: Allocating Design Responsibility in the Construction Industry*, 58 Fla.L.Rev. 561, 567 n. 33 (2006).

[23] The most recent AIA documents are A141-2004 (owner and D/builder), A142-2004 (D/builder and contractor), B142-2004 (owner and consultant), and B143-2004 (D/builder and architect).

[24] 320 Mont. 351, 87 P.3d 431 (2004).

[25] 33 Fed.Cl. 495 (1995), aff'd, 91 F.3d 167 (Fed.Cir.1996).

[26] Id. at 497.

[27] ASBCA No. 49, 509, 99-2 BCA ¶ 30,512.

[28] 72 N.Y.2d 587, 531 N.E.2d 1288, 535 N.Y.S.2d 356 (1988).

[29] The Federal Highway Administration has reported a 14 percent reduction in project duration using DB, as well as slightly lower costs, with no reduction in quality. See *Design-Build Effectiveness Study* (Jan. 2006), found at http://www.fhwa.dot.gov/reports/designbuild/design build.htm.

[30] See Roberts & Parisi, *Partnering: Prescriptions for Success*, 13 Constr.Litig.Rep., No. 11, Nov. 1992, p. 298; Stipanowich, *The Multi-Door Contract and Other Possibilities*, 13 Ohio St. J. on Disp. Resol. 303, 378-384 (1998).

[31] For further discussion, see M. MUTEK, CONTRACTOR TEAM ARRANGEMENTS—COMPETITIVE SOLUTION OR LEGAL LIABILITY (2006); Killian & Fazio, *Creating and Enforcing Teaming Agreements*, 25 Constr. Lawyer, No. 2, Spring 2005, p. 5.

[32] Lichtig, *The Integrated Agreement for Lean Project Delivery*, 26 Constr. Lawyer, No. 3, Summer 2006, p. 25.

[33] For further discussion, see the website for the Lean Construction Institute, www.leanconstruction.org.

[34] Wilke, *Alliancing for Infrastructure Projects—Sharing Risks and Rewards with a "No Blame" Agreement,* J ACCL, May 2007, p. 211.

[35] P. O'Connor, Jr. & Bomba, *Integrated Project Delivery Part I: Collaboration Through New Contract Forms,* 3 J ACCL, No. 2, Summer 2009, pp. 71, 110–18; Peartree, *The Consensus DOCS 300 Standard Form of Agreement for Collaborative Project Delivery,* 29 Constr. Lawyer, No. 1, Winter 2009, p. 25.

[36] Terio, *Program Management: A New Role in Construction,* 16 Constr. Lawyer, No. 4, Oct. 1996, p. 4. See also Scotti, *Program Management: The Owner's Perspective,* 16 Constr. Lawyer, No. 4, Oct. 1996, p. 15.

[37] This is a partial list taken from Noble, *Program Management: The Design Professional's Perspective,* 16 Constr. Lawyer, No. 4, Oct. 1996, p. 5.

[38] Frenkil, *Energy Saving Performance Contracts: Assessing Whether to "Retrofit" an Effective Contracting Vehicle for Improving Energy Efficiency in Federal Government Facilities,* 39 Pub. Cont. L.J., No. 2, Winter 2010, p. 331; Toomey, *Energy Savings Performance Contracting: Will the Demand Remain High Despite Dropping Energy Prices?* 44 Procur. Lawyer, No. 2, Winter 2009, p. 1.

[39] Cook, *Modern Enhancements for PPP Concession Agreements,* 28 Constr. Lawyer, No. 4, Fall 2008, p. 24.

[40] Ashcraft, Jr., *Building Information Modeling: Electronic Collaboration in Conflict with Traditional Project Delivery,* 27 Constr.Litig.Rep., Nos. 7–8, July–Aug. 2006, pp. 335 and 336–37. For further discussion, see Ashcraft, *Building Information Modeling: A Framework for Collaboration,* 28 Constr. Lawyer, No. 3, Summer 2008, p. 5, Wheatley & Brown, *An Introduction to Building Information Modeling,* 27 Constr. Lawyer, No. 4, Fall 2007, p. 33.

[41] For example, a contractor's commercial general liability policy typically excludes coverage for professional services performed by the contractor. Would the contractor's input into the BIM model constitute professional services?

[42] BIM has so far been used on large, complex projects such as stadiums, entertainment venues, and large commercial facilities. P. O'Connor, *Productivity and Innovation in the Construction Industry: The Case for Building Information Modeling,* 1 J ACCL, No. 1, Winter 2007, pp. 135, 160. The federal government's use of BIM is discussed in Silberman & McKee, *GSA's Building Information Modeling Program Raises Cutting-Edge Construction, Contracting, and Procurement Issues,* 41 Procur. Lawyer, No. 4, Summer 2006, p. 4.

CHAPTER

15

PUBLIC CONTRACTS

SCENARIO

The City of Plainsville decided to extensively renovate its city hall. The city is proud of its city hall because the building is listed on the Register of Historic Places.

The city issued a Request for Proposals to various architectural and engineering firms to provide guidelines as to how the building could be renovated. Four firms responded, and the city immediately chose the cheapest proposal from Engineering-R-Us, Inc. (ERU).

ERU created plans and specifications for the work to be done. It also specified that award would be made only to a contractor who had worked on at least three buildings listed on the Register of Historic Places in the past five years.

Plainsville advertised for bids from contractors to perform the work. Three bids were submitted by the deadline by contractors A, B, and C.

However, three minutes after the noon deadline, contractor D dashed in with a sealed bid and handed it to the city clerk. The other bidders protested that this fourth bid should not be opened because D had missed the deadline. The city clerk said he would not open D's bid because it was late. D stormed off, promising to be back with his lawyer.

The city opened the three bids and discovered the following:

Bid	Bid Price	Bidder Qualification
A	$1.5 million	2 historic buildings in past 5 years
B	$2 million	4 historic buildings in past 5 years
C	$1.3 million	4 historic buildings in past 8 years

Even though it was the most expensive bid, the city award the contract to B, as the only contractor with sufficient experience to undertake the project. Contractors A and C stormed off, promising to be back with their lawyers.

15.1 Public Contracts: Introduction

Every year, hundreds of billions of dollars are spent on public construction work in the United States.[1] Much of this spending is on large infrastructure projects requiring intensive engineering input. In addition, public owners (the federal government, in particular) have been at the forefront of alternative project delivery methods, including construction management and design–build.

While private and public contracts share many characteristics, public contracts are subject to special rules. Federal, state, and local governments are inherently circumscribed in their authority and must act within statutory and constitutional limitations. In addition, government contracting programs often seek to further social and economic objectives by directing procurement of labor and goods to particular groups, such as businesses owned by minorities, women, or veterans. These affirmative action programs are subject to constitutional challenge, which is a topic outside the subject matter of this book.[2]

A major difference between public and private owners lies in the method of procurement. Design contracts are obtained through negotiation. The default or presumed method by which the government procures construction contracts is through sealed, competitive bidding. For this reason, special legislation is required to allow for the use of alternative project delivery methods (See Section 15.8).

15.2 Design Contracts

15.2A Hiring the Design Professional: The Brooks Act

The public's procurement of design agreements differs fundamentally from the process by which construction contracts are awarded. As a general rule, while construction contracts are made primarily on the basis of price when the bidders' sealed bids are opened, design agreements are entered into on the basis of negotiation, either without regard to price or with cost being of secondary importance.

Design professionals do not like to compete on the basis of price. Nonetheless, the American Institute of Architects (AIA) Code of Ethics makes clear that such competition is not unethical. (See Sections 13.2B and C.)

In 1972, Congress enacted the Brooks Act.[3] The Brooks Act determines how contracts for design services can be awarded by federal agencies. It declares that the policy of the federal government is to negotiate on the basis of "demonstrated competence and qualification for the type of professional services required and at fair and reasonable prices."

Those who perform design services submit annual statements of qualifications and performance data. The agency evaluates the statements, together with those submitted by other firms requiring a proposed product, and discusses with no fewer than three firms "anticipated concepts and the relative utility of alternative methods." The agency then ranks the three most qualified to perform the services. It attempts to negotiate a contract with the highest qualified firm and takes into account "the estimated value of the services to be rendered, the scope, complexity, and professional nature thereof." If a satisfactory contract cannot be negotiated with the most qualified, the agency will undertake negotiations with the second most qualified, and if that fails, with the third.

Many states have enacted comparable legislation regulating the award of contracts for design services by state agencies. These laws are often called mini-Brooks Acts.

> In this chapter's scenario, Plainsville "immediately" hired ERU because it offered the lowest price. If the city is subject to a mini-Brooks Act, it would have hired the engineering firm in violation of the procedures established by state law. The other design firms would have been able to protest (challenge) the award.

15.2B Public Works Specifications

A private owner may balance price and quality however it wishes in deciding which products to purchase. More importantly, a private owner is free to choose a product made by only one manufacturer, notwithstanding the higher price presumably associated with the lack of competition. The design professional has no obligation to offer private owners with different options for each product used.

Design professionals drafting specifications for public works projects do not have a similar freedom of design. To the contrary: product specifications must not be so restrictive as to result in a "sole source" solicitation. Overly restrictive specifications diminish competition and may be the result of collusion (or at least bias) between the design team creating the specifications and particular manufacturers.

For these reasons, California (as one example) requires that, if a product is described by brand name, the specifications must list at least two brands or trade names of comparable quality or utility and be followed by the words "or equal" so that bidders may furnish any equal material, product, thing, or service.[4] The decision as to whether a proffered substitute is "equal" to the brand name product is made jointly by the design professional and the awarding authority.

In *John T. Jones Construction Co. v. Hoot General Construction*,[5] the project was for the upgrade of a water treatment plant. The city hired Black & Veatch (B&V) to provide engineering services for the design, bidding, and construction of the project.

Because of the corrosive nature of the wastewater, the concrete basin must be protected by a liner. B&V's design originally specified a liner manufactured by Linabond; however, at the city's insistence, the design was changed to allow for a liner "equal" to a Linabond liner.

The liner subcontractor (Hoot) had used a liner manufactured by Ameron for a quarter of a century. Convinced the Ameron liner was "equal" to the Linabond liner, Hoot submitted to the prime contractor (Jones) a bid offering to install an Ameron liner. After the Jones bid was accepted by the city, B&V (in collaboration with the city) rejected use of the Ameron liner. As a result, Jones was required to install the more expensive Linabond liner.

To recoup this higher cost, both Jones and Hoot sued B&V for rejecting the Ameron liner as "equal" to the Linabond liner. The contractors contended that B&V had specified the Linabond liner because it had a prior relationship with that manufacturer and not with Ameron. The contractors also pointed out that B&V had relied upon Linabond for information and advice in creating the design.

In rejecting the contractors' claims, the federal district court examined the nature of the relationship between a design professional creating specifications for a public owner and of that designer's reliance upon manufacturers in deciding which products to specify. (In the following quotation from the court's opinion, Mr. Ardahl was the B&V engineer who had originally specified use of the Linabond liner.) The court stated:

> At the least, B&V had a duty to consider the Ameron submittals in good faith in the exercise of honest judgment. Jones and Hoot, particularly Jones, view this case as one in which B&V, with Linabond's connivance, drafted a "disguised" sole-source specification, one that appeared to allow for an equivalent system but in reality was all along intended by B&V and WRA [the city] to remain the sole-source specification originally drafted. . . .
>
> The Court is not convinced B&V and Linabond had any relationship other than the normal relationship between a manufacturer of construction products and an engineer with a potential use for those products. The record indicates it is common for manufacturers and vendors to pitch their products to design engineers. Ameron does it. In fact, when Mr. Hoot called Ameron's Mr. Fisher to obtain his assistance he caught Mr. Fisher just after he had made a presentation to a B&V engineer about Arrow–Lock. (Fisher Depo., Ex. 216 at 3, 15). Information from manufacturers and vendors is essential to engineers in making decisions about what products and materials to specify on a project. Mr. Ardahl's contacts with Linabond in 1995 and role as draftsman of the Linabond cut file

specification are not remarkable. . . . That Linabond was in contact with B&V while the selection of a lining system was stalled over the or-equal issue merely reflects that Linabond, as much as Ameron, wanted the job.[6]

In short, a design professional that specifies a brand product must be able to justify its decision as reasonable. Evidence of a reasonable investigation by the design professional in deciding which product brand to specify will be upheld by a reviewing court. The fact that the design professional had a prior relationship with one manufacturer or used that manufacturer as a source of information does not compel a contrary result. As stated by the *John T. Jones* court, "Information from manufacturers and vendors is essential to engineers in making decisions about what products and materials to specify on a project."

15.3 Competitive Bidding: Objectives and the Pitfall of Corruption

Just as design–bid–build (DBB) is the traditional project delivery method in the private sector and the backdrop by which other procurement methods are compared, so too competitive bidding is the standard procurement method on public contracts. Competitive bidding is similar to DBB in that the bids are not solicited until the design has been completed. Here the comparison ends. A private owner is free to accept the lowest bid, negotiate with the bidders for a better price, pass up the lowest bid in favor of a more expensive one (perhaps on the belief that the quality of the work will be higher), or reject all bids.

By contrast, the template of competitive bidding is that

- All sealed bids are received by a specified date and time.
- The bids are publicly opened at that time.
- The initial award of the public contract is made on the basis of price alone. (This initial award is subject to further review to ensure that the bid is responsive and the bidder responsible; see Section 15.7.)

As typically phrased in statutes, contract award is made to the "lowest responsible bidder" who has submitted a responsive bid. Before examining the competitive bidding process more in-depth, the objectives underlying this procurement method and the pitfall of corruption are noted.

As articulated by an Ohio court, competitive bidding "gives everyone an equal chance to bid, eliminates collusion, and saves taxpayers money. . . . It fosters honest competition in order to obtain the best work and supplies at the lowest possible price because taxpayers' money is being used. It is also necessary to guard against favoritism, imprudence, extravagance, fraud, and corruption."[7] This is a tall order.

Of relevance here is the final objective noted by the Ohio court: guarding against corruption. As corruption can play a significant role in frustrating the goal of getting the best value for public funds, it strikes at the heart of public governance.

Of course, there can be corruption even when there is competitive bidding. There can be a corruption-generated decision to exempt a project from competitive bidding. The corrupt official may reveal submitted bids to a bidder before the deadline for bid submission. She may be induced by a bribe to decide whether a bid conforms or whether to waive an irregularity in the bidding. A particular alternative can be selected because an official has been bribed.

Despite the potential for corruption in the competitive bidding process itself, competitive bidding creates transparency and openness that, along with a free and vigorous press, can create an atmosphere that discourages or exposes corruption.

15.4 The Competitive Bidding Process: Overview

The federal government provides a five-step process to the award of contracts using competitive, sealed bidding:

- preparing invitations for bid
- publicizing invitations for bid
- submission of bids
- opening and evaluation of bids
- contract award[8]

The goal of competitive bidding is for the government to enter into a firm, fixed-price contract with the lowest responsible bidder.

15.5 Invitation for Bids (IFB)

The initial step in conducting a competitive bid is the *invitation for bids* (IFB). The IFB contains a description of the project and consists of drawings, specifications, basic contract terms, general and supplementary conditions, and any other documents that will be part of the contract. The Federal Acquisition Regulation (FAR) describes a balance between the amount of detail included in the specifications and the need to preserve competition: "Invitations must describe the requirements of the Government clearly, accurately, and completely. Unnecessarily restrictive specifications or requirements that might unduly limit the number of bidders are prohibited."[9]

Sometimes soil test reports are included. In addition, the IFB may state that designated reports are available at the office of a particular geotechnical engineer or the design professional for examination by bidders.

Bidders should be given adequate opportunity to study the bidding information, to make tests, to inspect the site, and to obtain bids from subcontractors and suppliers. A contractor who does not visit the site before submitting a bid will be precluded from claiming the site conditions were different than the contractor anticipated (if the condition would have been discovered on an reasonable site visit).

Even when there is adequate time, bidders often wait to complete their bids until the bid closing deadline is imminent. Generally, this is due to reluctance on the part of subcontractors to give sub-bids to bidders until shortly before the deadline for bid submissions.

The drawings and specifications should be detailed and complete so that bidders can make an intelligent bid proposal. Imprecise contract documents and too much discretion to the design professional may discourage honest bidders from submitting bids and may encourage bidders of doubtful integrity who make low proposals in the hope that they later will be able to point to ambiguities and make large claims for extras. As noted in Section 15.2B, product specifications must not be so restrictive as to result in a "sole source" solicitation.

In order for bidders to make an intelligent bid proposal, public entities also have a duty to disclose material (important) information that the bidders are not likely to discover. In a leading case, the federal Court of Claims in *Helene Curtis Industries, Inc. v. United States*[10] held that the procuring agency must disclose technical information that it had relating to the manufacturing process that it knew the contractor intended to use. The court noted that on many occasions each party has an equal opportunity to uncover the facts. But where the government knew much more than the contractor and knew the contractor was proceeding in the wrong direction, it could not betray the contractor into a "ruinous course of action by silence."[11] State courts have also recognized a duty to disclose.[12]

The IFB should specify that any uncertainties observed by the bidder must be resolved by a written request for clarification (a "request for information" or RFI) to the design professional before bid opening. Any clarification issued should be in writing and sent to all people who may submit bids.

15.6 Bid Opening and Award

Bid proposals are sent or delivered in a sealed envelope to the owner or the person designated to administer the competitive bidding, such as a design professional or construction manager. In public contracts, bids are opened publicly at the time and place specified in the invitation to bidders. Usually, the person administering the process announces the amount of the bids and the bidders.

At this time, the public entity announces the apparent successful bidder. The second-lowest bidder (who may have missed the award by a very small margin) may review the winning bid to determine its compliance with the bidding requirements.

If the disappointed bidder finds defects in the winning bid, the public entity may reject the lowest bid and award the contract instead to the second-lowest bidder. A similar result would ensue if the disappointed bidder reveals information which would render the successful bidder nonresponsible. Since disappointed bidders have a financial stake in evaluating the correctness of the winning bid and the responsibility of the low bidder, it is sometimes said that they function as "private attorneys general," ferreting out public harm so as to reap personal benefit.[13] Protests over a bid award are addressed next.

15.7 Bid Protests

When the second-lowest bidder protests (contests) the award, the public agency must evaluate two factors: whether the bid is "responsive" to the invitation for bids and whether the bidder is "responsible." The agency may either decide to whom to award the contract, or it may reject all bids and start over.

Bid Responsiveness. When evaluating bids, the agency should follow the procedures set forth in the invitation to bidders. The bids should be checked to see whether they conform. A nonconforming bid involving a deviation is "material" (non-trivial or significant) if it gives the bidder a substantial competitive advantage and/or prevents other bidders from competing on an equal footing. Nonconformity may relate to the work, the time of completion, the bonds to be submitted, or any requirements of the contract documents.

Bidding documents sometimes allow the owner to waive bidding technicalities. This allows the owner to accept the lowest bid despite minor irregularities. However, if a bid is not in conformity and the defect is material (cannot be waived), the court will give a remedy to the protesting, second-lowest bidder.

> In this chapter's scenario, the city declared Contractor D's bid not responsive because it was submitted three minutes late. Suppose D's bid was the cheapest of the four submitted: should the city's taxpayers have to pay more for such a minor delay? On the other hand, if a three-minute delay may be waived, why not ten minutes or one hour? These difficult questions explain why courts have come to different conclusions under similar facts.[14]

Bidder Responsibility. If the bid is responsive, the agency next examines whether the apparent low bidder is "responsible." To determine the lowest responsible bidder, the owner can take into account the following factors as well as any others that bear on which bidder would be most likely to do the job properly:

1. expertise in type of work proposed
2. financial capability
3. organization, including key supervisory personnel
4. reputation for integrity
5. past performance

These examples relate to the basic objectives of obtaining a contractor who is likely to do the job properly at the lowest price and with the least administrative cost to the owner. Yet, because a "responsibility" determination necessarily involves a matter of judgment, the courts give latitude to the public owner's investigation and decision.[15]

Experience in the type of construction work being procured is an important selection criterion. However, overemphasis on experience or drawing the experience factor too narrowly carries great risk. It can result in a loss of qualified bidders that will increase the price. It can also generate a corrupt award and violate the competitive bidding laws.

> In this chapter's scenario, ERU's specifications included an experience requirement of work on at least three buildings listed on the Register of Historic Places in the past five years. Only Contractor B—the most expensive bidder—met that requirement. A court would probably hold that the experience requirements were too tightly drawn if only one bidder qualified. ERU has created a "sole source" responsible bidder, akin to a "sole source" product specification.[16]

Although the owner should evaluate criteria other than price in selecting a contractor, awarding the bid to someone other than the low bidder raises a substantial risk of a lawsuit challenging the award. Sometimes bid awards in public contracts can be judicially challenged. This should not deter even a public owner from awarding the contract to someone other than the low bidder. However, compelling reasons for not awarding it to the low bidder must exist and be documented.

Rejection of All Bids. Suppose the owner decides that all of the submitted bids are too expensive or that affordable bids were submitted by nonresponsible bidders. Typically, the invitation reserves the right to reject all bids. The owner who wishes to go forward with the project advertises (normally after first changing the design to lower the expected cost) and the bidding process is repeated.

Bid Protest Remedies. The preferred remedy by a disappointed bidder is an injunction (a court order) either barring the public entity from awarding the contract to another or (preferable for the plaintiff) ordering the contract be awarded to the protester.

Sometimes, it is too late for an injunction (because the project is completed or nearly so), or it is not available for another reason. If this is the case, the only remedy left is damages. While most courts limit damages to the bidder's bid preparation costs, others award lost profits, and some deny any monetary recovery.

15.8 Construction Management, Design–Build, and Public Contracts

As noted, public agencies as a general rule hire design professionals after a carefully scripted negotiation process set forth in the Brooks Act (or the state equivalent), while construction contracts are obtained using sealed, competitive bids.

Neither construction management nor design–build (DB) fit easily within these procurement methods. A construction manager (CM) may acts as an advisor and, if so, is akin to a design professional. Must agencies use the Brooks Act to decide which CM to hire? Suppose the agency wishes to hire a CM-as-constructor? If award is made solely on the basis of price, the purported benefits of using a CM—including greater experience and managerial capabilities than would be found in a prime contractor—would not be evaluated in the award process.

Similarly, the D/builder's combination of design and construction roles creates a dilemma as to whether a DB contract is a construction contract that must be competitively bid or one for professional services that can be negotiated. Another issue is that often a fast-track system is used for a DB project. Thus, prospective bidders cannot be given a detailed set of plans and specifications on which to make their bids.

For these reason, the use of a CM or DB on public contracts must be authorized by statute.[17] In the past decade, virtually all states have embraced alternative project delivery methods. For example, according to the Design–Build Institute of America, as of 2013, 46 states, the District of Columbia, Puerto Rico, and the Virgin Islands authorize DB on some or all state projects.[18] The federal government has also embraced DB.[19]

15.9 Claims Against Public Entities: Federal False Claims Act

Rough tactics are often part of the claims and negotiation process. The air can be filled with threats, bluster, and wild exaggerations. But the rules are different when the claim is made against a public entity—whether federal, state, or local. Such claims are governed by special statutes that seek to protect public funds. The leading statute is the Federal False Claims Act (FCA).[20] This is a civil statute creating liability for fraud against the federal government.

The purpose and effect of the FCA is demonstrated by *Daewoo Engineering & Construction Co. v. United States*.[21] Daewoo was a Korean company that contracted with the U.S. Corps of Engineers to build a 53-mile road on the Island of Palau. Daewoo presented a claim against the Corps for $64 million (the original contract price was $73 million) but ended up being ordered to pay the Corps more than $50 million. The trial judge found Daewoo had violated the Contract Disputes Act's fraud provisions and the FCA.

The government did not make its claim until Daewoo had rested its case. It claimed that there had been unexpected testimony. A number of factors persuaded the trial judge to make this order against Daewoo. He found that some of the legal arguments by Daewoo were not credible. He concluded that its witnesses testified in a vague and unreliable manner. Most important for purposes of this section, there had been testimony by one of Daewoo's executives that it made the large claim as a negotiating ploy to make the government pay attention. The judge pointed to Daewoo having misled the Corps as to its actual costs and its having lowered its claim from $50 million to $29 million. It used heightened theoretical projections instead of actual acquisition costs, claimed equipment that had been fully depreciated, and relied on an incorrect production rate to calculate its productivity loss. These can constitute fraud and an intention to deceive.

What is interesting about the *Daewoo* case is that some of its tactics would have been normal in claims against private entities. But its claim against the Corps of Engineers invoked a number of federal statutes that dealt with false claims and was the basis for a huge fine. Also, in another case, the claimant forfeited its claim for $53,534,679 under the Forfeiture of Fraudulent Claims Act because its requests for reimbursement for bond premiums were false.[22]

The FCA originated in the Civil War period. What is unusual about this and related statutes is that the person (the whistleblower) who alerts the public agency to the wrongdoing under certain circumstances can pursue the claim itself (a *qui tam* claim) and recover a generous award for bringing this matter to the attention of public authorities. Because this disclosure to the federal entity, usually by an employee, will not be appreciated by the employer, the FCA protects the whistleblower from retaliation. Most states have enacted similar statutes.

Those who engage in public projects must be aware of these statutes and their requirements. They must also emphasize to its employees that these special rules must be followed.

REVIEW QUESTIONS

1. How are both design contracts and construction contracts usually obtained in public works projects?

2. What is the primary function of the Brooks Act?

3. If a design professional is drafting specifications for public works projects, what restrictions are made on her freedom of design?

4. What are the three primary items that are the template of competitive bidding?

5. What is included in an Invitation for Bids?

6. In public contracts, how are bids opened?

7. What is a nonconforming bid?

8. What are five factors that the owner can take into account to determine whether a company is a responsible bidder?

9. If an owner decides that all bids that have been submitted for a project are too high, what may that owner do?

10. What is the purpose and the effect of the Federal False Claims Act?

ENDNOTES

[1] See http://research.stlouisfed.org/fred2/series/TLPBLCONS?rid=229, last checked July 2, 2013.
[2] E.g., Conway & Berger, *Affirmative Action for Veterans*, 43 Procur. Lawyer, No. 2, Winter 2008, p. 12; Meagher, *The Women-Owned Small Business Federal Contract Program: Ten Years in the Making*, 46 Procur. Lawyer, No. 2, Winter 2011, p. 1.
[3] 40 U.S.C. §§ 1101-04 (formerly, 40 U.S.C. §§ 541–544).
[4] Cal.Pub.Cont. Code § 3400.
[5] 543 F. Supp.2d 982 (S.D.Iowa 2008), aff'd, 613 F.3d 778 (8th Cir.2010).
[6] 543 F. Supp.2d at 1007.
[7] *United States Constructors & Consultants, Inc. v. Cuyahoga Metro. Housing Auth.*, 35 Ohio App.2d 159, 300 N.E.2d 452, 454 (1973).
[8] Federal Acquisition Regulation (FAR) 14.101, 48 CFR § 14.101 (2012).
[9] Id. § 14.101(a).
[10] 160 Ct.Cl. 437, 312 F.2d 774 (1963).
[11] 312 F.2d at, 778.
[12] *Los Angeles Unified School Dist. v. Great American Ins. Co.*, 49 Cal.4th 739, 234 P.3d 490, 112 Cal.Rptr.3d 230 (2010); Annot. 86 A.L.R.3d 182 (1978).

[13] Marshall, Meurer & Richard, *The Private Attorney General Meets Public Contract Law: Procurement Oversight by Protest*, 20 Hofstra L.Rev. 1 (1991).

[14] Waagner & Evans, *Agency Discretion in Bid Timeliness Protests: The Case for Consistency*, 29 Pub. Cont. L.J., No. 4, Summer 2000, p. 713.

[15] *Barr Inc. v. Town of Holliston*, 462 Mass. 112, 967 N.E.2d 106 (2012).

[16] See *Gerzof v. Sweeney*, 16 N.Y.2d 206, 211 N.E.2d 826, 264 N.Y.2d 376 (1965) (public contract was illegal where only one contractor fulfilled the experience requirements).

[17] E.g., N.C Gen.Stat. § 143-128.1 (CM); Wash.Rev.Code § 47.20.780 (DB).

[18] See www.dbia.org/advocacy/state; go to link "State Statute Report." Site last visited July 2, 2013.

[19] Gadbois, Heisse & Kovars, *Turning a Battleship: Design-Build on Federal Construction Projects*, 31 Constr. Lawyer, No. 1, Winter 2011, p. 6.

[20] 31 U.S.C. §§ 3729–3733.

[21] 73 Fed.Cl.547 (2006), aff'd, 557 F.3d 1332 (Fed.Cir.), cert. denied, 130 S.Ct. 490 (2009).

[22] Morse-Diesel Intern., Inc. v. United States, 74 Fed.Cl. 601 (2007). In a subsequent case, the government was awarded $7,292,213 for violations of the Anti-Kickback Act and the FCA. *Morse-Diesel Intern., Inc. v. United States*, 79 Fed.Cl. 116 (2007), reconsideration denied, 81 Fed.Cl. 311 (2008).

PART

PERFORMANCE DISPUTES

CHAPTER 16

Performance Disputes Overview: Claims and Defenses to Liability for the Owner and Prime Contractor

CHAPTER 17

Contractor Payment and Project Completion

CHAPTER 18

Changes: Complex Construction Centerpiece

CHAPTER 19

Subsurface Problems: Predictable Uncertainty

CHAPTER 20

Time: Project Scheduling and Delay Claims

CHAPTER 21

The Subcontracting Process: An "Achilles Heel"

CHAPTER 22

Terminating a Construction Contract: Sometimes Necessary but Always Costly

CHAPTER

Performance Disputes Overview: Claims and Defenses to Liability for the Owner and Prime Contractor

16

Scenario

Mega Manufacturer, Inc. (MMI) decided to build a new factory. It hired Awesome Architects, Inc. (AAI), an architectural firm, to create the design. AAI retained a geotechnical engineering firm, Geo-R-Us, Inc., to perform a soil report. AAI used that soil report to design the excavation and foundation specifications.

MMI used the completed design to solicit fixed-price bids from a variety of prime contractors. The advertisement advised that the geotechnical report was available for review at the office of Geo-R-Us.

Prime Builder, Inc. decided to bid on the job. Its president, Mr. Prime, visited the site, which was an empty lot. Mr. Prime was mostly concerned with the logistics of working in the remote location. He examined the foundation design and saw no obvious conflict with the surface soil indications (he saw no boulders or swampy areas, for example), but did not go to the office of Geo-R-Us to examine the original geotechnical report.

MMI entered into a fixed-price contract with Prime Builder, Inc. using an American Institute of Architects (AIA) General Conditions of the Contract for Construction, A201-2007. Prime Builder, Inc. then entered into several subcontracts, including one with a foundation subcontractor and one with a steel erection subcontractor.

After a mere ten months, the completed factory developed cracks and sticking doors. In addition, the roof leaked. Invoking the General Conditions' warranty (guarantee) clause, MMI demanded Prime Builder, Inc. fix the defects. Prime Builder, Inc. did so, and sued the owner on the ground that the defects were the owner's fault.

Prime Builder introduced expert testimony as to the causes of the failures. A geotechnical expert testified that the Geo-R-Us report was defective because there were too few boring logs (subsoil investigations). A roofing expert explained that the roof failed because the steel joists and the metal roof panels expanded and contracted (in response to changing temperatures) at different rates. This caused the panels to grind against each other, causing them to crack.

16.1 Claims: Theories of Legal Liability

Chapter 16 begins Part D of this book, titled "Performance Disputes." Part D is the largest part of the book, encompassing Chapters 16 through 24. It addresses the performance-related disputes arising out of a construction project with particular emphasis on disputes between the owner and prime contractor, although these concepts are also relevant to disputes between the prime contractor and subcontractors.

Chapter 16 deals with the disappointments that relate to obstacles to the contractor's performance or to insecurity about the owner's performance. These disappointments may lead the owner and prime contractor to bring claims against each other. This chapter catalogs the major claims (or theories of liability) and defenses to those claims.

This section categorizes the legal theories that can be the basis for construction-related claims between owner and contractor. Unfortunately, it has become common for claimants to assert a variety of claims based on assorted legal theories with the law generously permitting—at least up to a point in a lawsuit—the use of such a shotgun approach. This chapter deals only with claims by contract-connected parties.

Construction claims arising out of performance disputes fall into three broad categories: contract, tort, and statutory claims.

Contract Law Claims—Generally. Contract law is covered in Chapters 3 and 4. The contract consists of express (written) and implied contract terms. As explained in Section 3.5B, implied terms are added to the contract by the courts. The contract may contain express warranties; however, the law also imposes implied warranties on the owner and prime contractor.

Contract Law Claims—Express Warranty. A warranty is an assurance by one party relied on by the other party to the contract that a particular outcome will be achieved by the warrantor. A warranty is express if it is included in the written contract. An express warranty promises a performance outcome; for example, a new roof will not leak for a specified number of years. Failure to achieve the promised objectives is a breach.

Contract Law Claims—Implied Warranty. Implied warranty is a judicial doctrine that imposes certain promises on the contracting parties. An implied warranty does not require the consent or agreement of the parties. In the sale of goods, an implied warranty usually relates to the merchantability or fitness for the buyer's known purposes. In construction disputes, the owner impliedly warrants the adequacy of the design (Section 16.3B), while the contractor impliedly warrants to perform in a workmanlike manner (Section 16.6). These are "promises" the law imposes on the contracting parties, notwithstanding the absence of express terms. The closely related implied warranty of habitability applicable to residences is discussed in Section 16.7.

Tort Law Claims. Tort law is covered in Chapter 5. Negligence (Section 5.3) is the most common tort claim. Contracting parties also may assert claims of intentional torts, including misrepresentation and fraud (Section 5.4). Developers of mass housing may be sued in strict liability (Section 5.8).

Statutory Claims. Legislation increasingly protects homeowners. Homeowners may sue the prime contractor under consumer protection statutes and residential construction defects legislation (Section 16.7C).

Broadly speaking, this chapter will group claims by those brought by the contractor against the owner (Section 16.3) or those brought by the owner against the contractor (Sections 16.4 to 16.7). While either the contractor or owner may equally bring claims based in contract and tort, only homeowners may bring statutory claims.

16.2 Principles Underlying Claims

Before addressing the specific claims presented by the prime contractor or owner, it is helpful to be aware of the principles underlying these theories of liability. Of course, one principle is to look to the contract to see how the parties have allocated the risks between themselves. If in a design–build contract the contractor agrees to build a water treatment plant that will process a certain volume of water to a specified degree of purity, the parties (by agreement) have transferred to the contractor the risk that the plant will not reach those objectives. This section, however, deals with the allocation of risk that is not expressly resolved between the parties.

16.2A Basic Principle: Responsibility Follows Control

Design Control. As a basic principle, responsibility for a defect rests on the party to the construction contract who essentially controls and represents that it possesses skill in that phase of the overall construction process that substantially caused the defect. Control does not mean simply the power to make design choices. Every owner usually has the power to determine design choices. For example, an owner may require a particular type of tile to be used, which is a power within his contract rights. But the control needed to determine *who* is responsible for a design defect means a skilled choice, either one made by an owner who has professional skill in tile selection or an advisor such as an architect with those skills.

On a traditional design–bid–build (DBB) construction project where the owner separately hires the designer and contractor, certain defects caused by design are the responsibility of the owner (see Section 16.3B). As a corollary to this principle, a contractor who follows the design is not responsible for a defect resulting from the design.

By contrast, a defective design is the responsibility of a design–build contractor—one who both designs and builds.

> In this chapter's scenario, Prime Builder, Inc. sued MMI to recoup the contractor's costs to correct the defective construction. Prime Builder's theory is that the building's faults arose out of errors in the design, which are within the control of the owner and hence are the owner's responsibility.

Defective Materials. The issue of defective materials generates the most difficult problems. The huge variety of available materials (often untested), the pressure to cut costs or weight, and the inability to test or rely on manufacturers all combine to make defective materials a prime cause of defects in the project.

Materials specified may be unsuitable and may never accomplish the purpose for which they have been specified. As specification is part of design, responsibility for unsuitable materials falls on the person who controls the design or to whom the risk is transferred—the owner on a traditional project.

Materials may be suitable yet fail either because they were installed improperly—in which case the contractor clearly would be responsible for the failure—or because they were manufactured improperly. Where the materials were suitable for the purpose and installed properly, yet failed because of a manufacturing defect, who (between the owner and contractor) should be responsible?

The American Institute of Architects (AIA), A201-2007, § 3.5, specifies that the contractor warrants (promises) to the owner that the "materials and equipment . . . will be of good quality and new . . . " This is an example of the contract imposing responsibility on the contractor.

The federal boards of contract appeals too have placed responsibility for defectively manufactured materials on the contractor. They reason that, in the event a delivered product is manufactured improperly, the contractor, rather than the government, is in a better position to seek relief from the manufacturer. This is because the contractor is in privity with the manufacturer and so can protect itself contractually, while the government is not in privity with the manufacturer. The boards' reasoning follows the basic principle of responsibility following control, but in this case, "control" means the party in the best position to obtain a remedy.[1]

16.2B Supplemental Principle: Good Faith and Fair Dealing

Introduction. Parties who plan to enter into a contract, as a general rule, are not expected to look out for each other. However, a supplemental principle, begun in the mid-twentieth century, modifies this general rule. Once a contract has been made, the law also expects parties to act in good faith and to deal fairly with one another. This duty of good faith and fair dealing is an example of an implied promise, sometimes also called an implied covenant or obligation. (See Section 3.5B.)

Of course, each party should avoid deliberate and willful frustration of the other party's expectations. But in addition, the duty of good faith and fair dealing means the contracting parties should extend a helping hand where to do so would not be unreasonably burdensome. Contracting parties, although not partners in a legal sense, must recognize the interdependence of contractual relationships.

Some breaches of this implied covenant, particularly in contracts of insurance and employment, violate public policy. When they do, they can become the basis for tort remedies, including punitive damages.

Deductive Changes. The imprecise yet potent nature of a claim of breach of the implied covenant of good faith and fair dealing is illustrated by a comparison of two cases involving fundamentally the same facts: a prime contractor recommending to the owner a "deductive change" in which a subcontractor's work would be entirely or mostly eliminated. In each case, the subcontractor sued the prime contractor, contending that the recommendation of the deductive change violated the prime contractor's duty of good faith owed to the subcontractor.

In a Kansas case, *Maier's Trucking Co. v. United Construction* Co.,[2] the majority held that this did not breach the covenant of good faith and fair dealing. Two judges strongly dissented; they would have ordered a trial to determine whether the contractor's suggestion had been made in bad faith.

Yet in *Scherer Construction LLC v. Hedquist Construction, Inc.,*[3] the Supreme Court of Wyoming held that the trial court was incorrect in awarding a summary judgment (a ruling given before a trial) to the prime contractor. In the *Scherer* case, the subcontractor claimed that the obligation of good faith and fair dealing was breached when the prime suggested a change to the owner that would have wiped out 75 percent of the value of the subcontract. The subcontractor had spent a substantial amount of money to prepare to perform as required by the original contract. The court cited with approval the Restatement (Second) of Contracts § 205, which provides, "Every contract imposes upon each party a duty of good faith and fair dealing in its performance and its enforcement." The court sent the case back to the trial court to see if the subcontractor could prove its claim. (This is what the dissent in the *Maier* case would have done.)

The trial court in *Scherer* must decide whether there was a breach of the obligation. Much will depend on the prime's motive for suggesting the change. If it were done to "teach the subcontractor a lesson" or to "get even with the subcontractor," the obligation would have been breached. But if the suggestion were made in commercial good faith to help the owner cut costs, the subcontractor will not have a valid claim of breach. These will be hard claims to sustain, but the door is open.

Contractor Aware of Design Defects. As noted in Section 16.2A, the basic principle is that the owner on a traditional project is responsible for the design. This basic principle is modified by the supplemental principle—the contractor's implied obligation to act in good faith. That duty may arise when the contractor becomes aware that the design is faulty.

The implied duty of good faith requires each contracting party to warn the other when the other is proceeding in a way that will cause failure. For example, a contractor who knows the foundation design is not compatible with soil conditions should express his concerns to the owner.[4]

Sometimes the contractor's obligation to report errors is specified in the contract. The AIA General Conditions, A201-2007, § 3.2.2, states that the contractor when reviewing the design "shall promptly report to the Architect any errors, inconsistencies or omissions discovered by or made known to the Contractor." The AIA uses a subjective standard, acknowledging that "the Contractor's review [of the design] is made in the Contractor's capacity as a contractor and not as a licensed design professional." The principal purpose for the contractor's reviewing the contract documents is to prepare its bid. The owner should expect attention to be drawn only to those errors that the contractor does discover.

The contractor who *does* notify the owner of errors is not charged for defects that result despite its warnings and is entitled to be paid for what it has done. The contractor who does not report obvious errors should not be allowed to invoke the "following the plans" defense. If it performs work *knowing* that the work violates building codes or technical competence and *without* having first warned the owner, the contractor will not be able to recover for the work performed.

In this chapter's scenario, Prime Builder, Inc. is suing MMI for design defects. MMI can counter that the contractor knew or should have been aware of the design defects and had a duty under AIA A201-2007, § 3.2.2, to bring these defects to the owner's attention.

Focusing just on the issue of the foundation design, MMI would argue that Prime Builder's failure to examine the geotechnical report at Geo-R-Us's office was a violation of § 3.2.2. Prime Builder would counter that it had no legal duty to examine the geotechnical report but was entitled to rely upon AAI's design. Moreover, § 3.2.2 has a subjective standard (a duty to report only known defects), and Prime Builder's examination of the geotechnical report would likely have not given it knowledge of the design defects: it is a contractor, after all, not an engineer.

16.3 Contractor Claims

16.3A Introduction

In the course of performance, the contractor may find that its cost of performance has risen dramatically. Illustrations can be drastic increases in the cost of materials, equipment, supplies, utilities, labor, or transportation. These risks are generally assumed by the contractor under a fixed-price contract.

The common law provided little protection to a contractor who performs for a fixed price. To be sure, extraordinary and unanticipated events may occur that drastically affect the cost of the contractor's performance and may be the basis for a claim for relief. Yet the contractor's

16.3B Misrepresentation Through Defective Specifications: *United States v. Spearin*

Contractor allegations that the specifications were defective are common in construction performance disputes. Unfortunately, such allegations often produce more heat than light and obscure the real issues in the dispute. Before looking at these issues, it is useful to reproduce the fountainhead case that is often cited to support a "*Spearin*" claim.

UNITED STATES v. SPEARIN
Supreme Court of the United States, 248 U.S. 132 (1918)
[Ed. note: Footnotes renumbered.]

BRANDEIS, Justice.
Spearin brought this suit in the Court of Claims, demanding a balance alleged to be due for work done under a contract to construct a dry-dock and also damages for its annulment. Judgment was entered for him. . . .

First. The decision to be made on the Government's appeal depends on whether or not it was entitled to annul the contract. The facts essential to a determination of the question are these: Spearin contracted to build for $757,800 a dry-dock at the Brooklyn Navy Yard in accordance with plans and specifications which had been prepared by the Government. The site selected by it was intersected by a 6-foot brick sewer; and it was necessary to divert and relocate a section thereof before the work of constructing the dry-dock could begin. The plans and specifications provided that the contractor should do the work and prescribed the dimensions, material, and location of the section to be substituted. All the prescribed requirements were fully complied with by Spearin; and the substituted section was accepted by the Government as satisfactory. It was located about 37 to 50 feet from the proposed excavation for the dry-dock; but a large part of the new section was within the area set aside as space within which the contractor's operations were to be carried on. Both before and after the diversion of the 6-foot sewer, it connected, within the Navy Yard but outside the space reserved for work on the dry-dock, with a 7-foot sewer which emptied into Wallabout Basin.

About a year after this relocation of the 6-foot sewer there occurred a sudden and heavy downpour of rain coincident with a high tide. This forced the water up the sewer for a considerable distance to a depth of 2 feet or more. Internal pressure broke the 6-foot sewer as so relocated, at several places; and the excavation of the dry-dock was flooded. On investigation, it was discovered that there was a dam from 5 to 5½ feet high in the 7-foot sewer; and that dam, by diverting to the 6-foot sewer the greater part of the water, had caused the internal pressure which broke it. Both sewers were a part of the city sewerage system; but the dam was not shown either on the city's plan, nor on the Government's plans and blue-prints, which were submitted to Spearin. On them the 7-foot sewer appeared as unobstructed. The Government officials concerned with the letting of the contract and construction of the dry-dock did not know of the existence of the dam. The site selected for the dry-dock was low ground; and during some years prior to making the contract sued on, the sewers had, from time to time, overflowed to the knowledge of these Government officials and others. But the fact had not been communicated to Spearin by anyone. [Spearin] had, before entering into the contract, made a superficial examination of the premises and sought from the civil engineer's office at the Navy Yard information concerning the conditions and probable cost of the work; but he had made no special examination of the sewers nor special enquiry into the possibility of the work being flooded thereby; and had no information on the subject.

Promptly after the breaking of the sewer Spearin notified the Government that he considered the sewers under existing plans a menace to the work and that he would not resume operations unless the Government either made good or assumed responsibility for the damage that had already occurred and either made such changes in the sewer system as would remove the danger or assumed responsibility for the damage which might thereafter be occasioned by the insufficient capacity and the location and design of the existing sewers. The estimated cost of restoring the sewer was $3,875. But it was unsafe to both Spearin and the Government's property to proceed with the work with the 6-foot sewer in its then condition. The Government insisted that the responsibility for remedying existing conditions rested with the contractor. After fifteen months spent in investigation and fruitless correspondence, the Secretary of the Navy annulled the contract and took possession of the plant and materials on the site. Later the dry-dock, under radically changed and enlarged plans, was completed by other contractors, the Government having first discontinued the use of the 6-foot intersecting sewer and then reconstructed it by modifying size, shape and material so as to remove all danger of its breaking from internal pressure....

The general rules of law applicable to these facts are well settled. Where one agrees to do, for a fixed sum, a thing possible to be performed, he will not be excused, or become entitled to additional compensation, because unforeseen difficulties are encountered. *Day v. United States,* 245 U.S. 159; *Phoenix Bridge Co. v. United States,* 211 U.S. 188. Thus one who undertakes to erect a structure on a particular site, assumes ordinarily the risk of subsidence of the soil. *Simpson v. United States,* 172 U.S. 372; *Dermott v. Jones,* 2 Wall. 1. But if the contractor is bound to build according to plans and specifications prepared by the owner, the contractor will not be responsible for the consequences of defects in the plans and specifications. *MacKnight Flintic Stone Co. v. The Mayor,* 160 N.Y. 72; *Filbert v. Philadelphia,* 181 Pa.St. 530; *Bentley v. State,* 73 Wisconsin, 416. See *Sundstrom v. New York,* 213 N.Y. 68. This responsibility of the owner is not overcome by the usual clauses requiring builders to visit the site, to check the plans, and to inform themselves of the requirements of the work, as is shown by *Christie v. United States,* 237 U.S. 234; *Hollerbach v. United States,* 233 U.S. 165, and *United States v. Utah & C. Stage Co.,* 199 U.S. 414, 424, where it was held that the contractor should be relieved, if he was misled by erroneous statements in the specifications.

In the case at bar, the sewer, as well as the other structures, was to be built in accordance with the plans and specifications furnished by the Government. The construction of the sewer constituted as much an integral part of the contract as did the construction of any part of the dry-dock proper. It was as necessary as any other work in the preparation for the foundation. It involved no separate contract and no separate consideration. The contention of the Government that the present case is to be distinguished from the *Bentley Case, supra,* other similar cases, on the ground that the contract with reference to the sewer is purely collateral, is clearly without merit. The risk of the existing system proving adequate might have rested on Spearin, if the contract for the dry-dock had not contained the provision for relocation of the 6-foot sewer. But the insertion of the articles prescribing the character, dimensions and location of the sewer imported a warranty that, if the specifications were complied with, the sewer would be adequate. This implied warranty is not overcome by the general clauses requiring the contractor, to examine the site,[5] to check up the plans,[6] and to assume responsibility for the work until completion and acceptance.[7] The obligation to examine the site did not impose on [the contractor] the duty of making a diligent enquiry into the history of the locality with a view to determining, at his peril, whether the sewer specifically prescribed by the Government would prove adequate. The duty to check plans did not impose the obligation to pass on their adequacy to accomplish the purpose in view. And the provision concerning contractor's responsibility cannot be construed as abridging rights arising under specific provisions of the contract.

* * *

The judgment of the Court of Claims is, therefore, affirmed.

Warranty of Design Specifications. As explained in Section 14.7, the distinction between design and performance specifications lies in the amount of discretion given to the contractor in choosing how to build the project. Under *Spearin*, an owner who issues *design specifications*, which dictate how the contractor is to go about its work, *impliedly warrants* that following the design will lead to a result acceptable to the owner. The *Spearin* doctrine is an example of an implied warranty—imposed by the law, rather than arising from the intentions of the parties—mentioned in Section 16.1. This doctrine has been adopted in virtually every state (with Texas as perhaps the sole exception).[8]

Misrepresentation and Defect. A contractor claiming breach of an owner's implied warranty of design is asserting a *misrepresentation* claim. As the *Spearin* Court explained, "the contractor should be relieved, if he was misled by erroneous statements in the specifications."[9] Specifications can convey a wide variety of information, all of which provide the potential basis for a *Spearin* claim. According to one treatise:

> The owner's breach of its implied warranty of design has been found in factual situations such as: (1) noncompatible soils; (2) structural defects; (3) fire damage; (4) dredging difficulties; (5) highway concrete; (6) sewer design/water infiltration; (7) roof leaks; (8) survey errors; (9) concrete design mix; (10) sealant; and (11) excavation quantity error.[10]

A contractor who entered into a contract knowing the specifications were defective was not misled and so cannot assert a claim for misrepresentation. More commonly, a court denying a claim will find that the contractor had a duty to investigate the condition and failed to do so. In denying the claim, the court is in effect finding that the contractor's reliance on the misrepresentation was not justified, because it should have learned of the true conditions itself. On the other hand, a contractor's failure to investigate will not bar a *Spearin* claim if a reasonable investigation would not have uncovered the erroneous information.

The *Spearin* doctrine fits within the broader duty of an owner—and in particular a public owner—to act in good faith with a contractor it hires. It complements a public owner's duty to disclose to a contractor relevant information in its possession which the contractor is not likely to discover (see Section 15.5).

Different legal issues relate to the quality of the specifications. If building defects were caused by the design, *Spearin* stands for the proposition that the contractor who implemented the design is not liable for the cost of correction. The *Spearin* doctrine is being used as a "shield" from liability.

Contractors may also use *Spearin* as a "sword." Suppose the contractor claims that the project was more expensive to build than it anticipated because of defective specifications. If the design authorized a *means or method* of performance, the contractor will argue that the owner impliedly warranted that the contractor's use of that means will achieve an acceptable result.

For example, in *Ace Constructors, Inc. v. United States*,[11] the specifications permitted the contractor to pave an airplane runway using either slip-form or fixed-form types of pavers. However, the runway could be built only using slip-form pavers. The contractor's wasted cost of trying to build the runway using fixed-form pavers was compensable under the implied warranty.

In addition, as seen in *United States v. Spearin*, specifications can provide relevant information that the contractor needs and uses to determine whether it will enter into the contract, what will be its price, and how it expects to perform. In that case, the government did not tell the contractor there was a dam in the 7-foot sewer. Put another way, it failed to provide complete information. It described the sewer but did not inform the contractor that there was a dam in it or that there had been prior flooding problems. Information can relate to the site, its subsurface characteristics, its accesses, and any conditions that would be helpful in planning performance.

The specifications in *Spearin* involved *factual* errors. Specifications may also include *legal* errors, such as noncompliance with applicable laws such as building and housing codes and environmental regulations. Usually these kinds of legal errors are the responsibility of the owner, with ultimate responsibility belonging to the design professional. For example, in *St. Joseph Hospital v. Corbetta Construction Co.*,[12] the paneling specific in the design did not comply with the fire code. This was a legal error borne by the owner and, ultimately, the architect. As discussed in Section 16.2B, only if the contractor knew of the errors may this risk have been shifted or shared.

Where the design is *incomplete* and the contract does not clearly state who will provide the missing part of the design, it is likely to be the responsibility of the owner.[13] It was the owner who drafted the contract and hired a design professional to provide the design.

Injury. The *Spearin* Court invoked the implied warranty to relieve a contractor from liability for a failed project. In addition to this defensive use of the doctrine, contractors may also invoke the warranty in an attempt to recover higher than anticipated construction costs. However, in the latter situation, the doctrine applies only if the higher costs were incurred by the contractor in trying to implement unbuildable design specifications.

Otherwise stated, the warranty does not guarantee a perfect design and trouble-free performance. A buildable design does not violate the implied warranty—even if the construction process was more expensive and took longer than the contractor anticipated. A large number of requests for information, while perhaps evidence that the design team did a sloppy job, does not establish an injury within the meaning of the *Spearin* warranty, as long as the design is buildable.[14]

> In this chapter's scenario, Prime Builder, Inc. is invoking the *Spearin* doctrine to make MMI responsible for two design defects: in the foundation and the roof.
>
> Prime Builder's claim with regard to the foundation design defect may fail because the contractor did not examine the geotechnical report before creating its bid. There would have been no misrepresentation had Prime Builder's examination of the report revealed the foundation design's error.
>
> On the other hand, Prime Builder's claim with regard to the roof design would state a claim for breach of the owner's implied warranty of design. It is not the contractor's responsibility to double check engineering calculations of expansion and contraction of different metals. This is a latent design defect which rendered the roof unbuildable.[15]

16.3C Owner Nonpayment

As discussed in Chapter 17, the prime contractor normally is not paid one lump sum by the owner; instead, the contractor receives progress (usually monthly) payments as the work progresses. Section 17.5 discusses the contractor's rights against the owner in the event of late payment or nonpayment.

16.4 Owner Claims Against Contractor

16.4A Introduction

The sad but not uncommon discovery that the project has defects often generates a claim by the owner against parties it holds responsible for having caused the defect. Claims against the design professional were discussed in Chapter 11. This chapter concentrates on owner claims against the contractor.

Defects in a house can include a leaky roof, a sagging floor, structural instability, and an inadequate heating or plumbing system. A commercial structure can include these defects as well as an escalator that is unsafe, inefficient, or requires excessive repairs. An industrial plant can include the preceding defects as well as inadequate space to install machinery or the inability of the computer system to operate the assembly line. All three of those projects may have suffered a nonphysical defect: late completion, during which time the owner was deprived of the building's use.

When addressing defective buildings, it is important to differentiate between an owner who supplies a design—usually by an independent design professional—and one who hires a contractor to both design and build. (Design–build is discussed in Section 14.14.)

16.4B Shared Responsibility for Construction Defects

Stating the Problem. If proof establishes that the cause of the defect is clearly the responsibility of one of the parties to the construction contract, the loss is chargeable to that party. But in a venture as complicated as construction, it is not uncommon for a construction defect to be traceable to multiple causes.

For example, suppose the owner asserts a claim against the contractor based on the contractor's not having followed the plans and specifications. Suppose the contractor admits having breached the contract but points to other possible causes, such as abnormal weather conditions or third parties for whom the contractor is not responsible. Legal responsibility for breach of contract does not require the claimant to eliminate all causes except acts or omissions by the party against whom the claim has been made. The defendant's breach must be only a substantial factor in causing the harm. If other conditions or actors played a minor or trivial part in causing the loss, the contractor is responsible for the entire loss.

Defect Traced to Owner and Contractor. Multiple causation becomes more complicated when the defect is traceable both to the party against whom the claim has been made and to the claimant itself. For example, Section 16.2A stated the basic principle that, in the traditional construction project, the owner is responsible for the design and the contractor for its execution. Yet these activities are not watertight compartments. The owner plays a role in execution because of the design professional's site responsibilities. The contractor plays a role in the design because, through a specialty subcontractor, he may supply the design or may be obliged to study and compare the contract documents and report any errors observed.

Another complicating factor is that the cause of the defect may be traceable to wrongful acts of the design professional, some of which may be chargeable to the owner. This section deals with defect claims by the owner against the contractor, with the defect having been caused in whole or in part by acts of the owner or someone for whom the owner is responsible. More commonly in construction, the owner asserts a claim for defective work against both the contractor and a third party, usually the design professional.

Where defects can be traced to both the owner and the contractor, most cases have held the loss will be shared, with the owner's claim being reduced by its percentage of fault (as determined by the fact finder, either a jury or the judge). This division of damages takes into account the complexity of causation and the desire to avoid all-or-nothing outcomes.

Shared Design Responsibilities. As noted, one rough classification is to charge the owner with defects caused by design and hold the contractor accountable for defects caused by failure to follow the design or by poor workmanship. This classification can be deceptive because of the increasingly blurred roles related both to design and to its execution.

"Design delegation" is the process by which the owner shifts to the contractor creation of part of the design. Illustrations of this shared responsibility can be contracts that require the contractor to submit drawings indicating how he proposes to do the work (often prepared by specialized subcontractors) and that require the design professional to approve the submittals.

Shared design responsibility also should increase as digital technologies play a larger and larger role in design creation. Ownership of the design becomes blurred when subcontractors and suppliers contribute electronically to the formation of the original design. As discussed in Section 14.21, Building Information Modeling contemplates a collaboratively created design in which the design author cannot easily be traced or determined. Moreover, making electronic media part of the design complicates the contractor's design review role. AIA A201-2007, § 3.2.2, requires the contractor, before starting the work, to "carefully study and compare the various Contract Documents" and to report to the architect "any errors, omissions, or inconsistencies" that it discovers. An article (commenting on the same language in A201-1997, ¶ 3.2.1) questions how that review would work in a digitally based design:

> The process of "carefully studying and comparing" inter-operable object-oriented CAD files may be quite different from the process of "carefully studying and comparing" paper Drawings and Specifications. For one thing, there may be "plan-checking" software introduced to the market that the "careful study" standard would obligate the Contractor to use. And the sheer amount of information in an object-oriented interoperable file (including links to other documents and web sites) may be impossible for the Contractor to absorb in detail "before starting each portion of the Work."[16]

Sometimes bidders are asked to provide design alternates, or they do so voluntarily. Before or after award or during performance, the contractor may request to substitute different equipment or material from that specified. Although approval by the design professional is usually required, approval often is based on representations or even warranties by the contractor that the proposed substitution will be at least as good as that specified or will accomplish the desired result. The contractor, although clearly subordinate to the design professional, plays an important role in design.

Similarly, the design professional frequently monitors the contractor's performance during the work and at the end of the job. Some design professionals may even direct how the work is to be done, although this role is typically not within their power or responsibility.

As was seen in Section 16.2A, control is a key factor determining responsibility for defects. Yet the rough "partnership" between owner and contractor makes it difficult to neatly divide responsibility by design and execution.

16.5 Contractor's Warranty (Guarantee) Clause

A warranty is a guarantee. The difference between express and implied warranty pertains to the source of the guarantee.

An "express" contract term is one that arises from agreement of the parties. If that agreement has been reduced to a writing, an express warranty would refer to a warranty clause in the contract. This section addresses warranty clauses.

An "implied" contract term is one that is imposed on the parties as a matter of law. It is part of the parties' bargain—even if not explicitly agreed to by them. An implied warranty imposes certain guarantees on one of the parties. Section 16.3B discussed the owner's implied warranty of design, owed to the contractor. Section 16.6 discusses a contractor's implied warranties owed to the owner (or a subcontractor's implied warranties owed to the prime contractor).

A warranty is an assurance by one party relied on by the other party to the contract that a particular outcome will be achieved by the warrantor. Failure to accomplish the promised objectives is a breach. Fault is not required. While a breach of warranty is the same as a contract breach, warranties are less susceptible to contract doctrines that can excuse nonperformance by a contracting party. In *St. Andrew's Episcopal Day School v. Walsh Plumbing Co.*,[17] the Mississippi Supreme Court made these considerations explicit when enforcing a guarantee provision in favor of the owner (a private school, described here as the "appellant") against its contractor (Walsh):

> The appellant is in the business of running a Christian day school; it is not in the air conditioning business. Appellant does not profess to have any knowledge or expertise about air conditioning systems or equipment. That was the main reason for the provision in the Construction Specifications that the mechanical contractor "shall guarantee each and every part of all apparatus entering into this work to be *the best of its respective kind*, and he shall replace within one year from date of completion all parts which during that time prove to be defective and he must replace these parts at his own expense."
>
> Walsh was a reputable, responsible and knowledgeable contractor of mechanical work; and when . . . consulting engineers for the Day School, recommended to the appellant that the bid of Walsh Plumbing Company be accepted, the duty and responsibility was placed squarely on Walsh's shoulders to purchase and properly install the best air conditioning system on the market. Not only was Walsh to purchase and install, he was to guarantee the system and its installation for one year. This was what Walsh contracted to do, and this was what the appellant paid Walsh to do. Not knowing anything about air conditioning systems, the appellant employed experts in this field and reposed full confidence in these experts to look after its interests.[18]

The AIA General Conditions provide an example of a contractor's express warranty. A201-2007, § 3.5, lists three warranties as to the quality of the contractor's work: (1) the materials and equipment are "of good quality and new unless the Contract Documents require or permit otherwise" (the latter language permitting the owner to authorize use of used materials), (2) the work conforms to the requirements of the contract documents, and (3) the work will be free from defects. In AIA A201-2007, § 12.2.2.1, these warranties are backed up by a separate promise to make any repairs requested by the owner within one year after the project's substantial completion.

The contractor's express warranty is clearly for the owner's benefit, but it may also be viewed as benefitting the contractor. If the work is defective, the cheapest repair would be by the original contractor rather than by the owner hiring a different contractor and then charging the repair cost to the original contractor. Perhaps recognizing this, the AIA specifies that if the owner fails during the one-year period to notify the contractor of the defect and give the contractor an opportunity to repair, "the Owner waives the rights to require correction by the Contractor and to make a claim for breach of warranty." The owner may still make a claim—but it will have to be for breach of contract or negligence—not for breach of warranty, which is the easiest standard of liability for the owner to prove.

> In this chapter's scenario, the AIA warranty clause worked as intended. MMI invoked the warranty within the one-year period, and Prime Builder repaired the defects. MMI has a functioning building, while the parties litigate responsibility.

The AIA has made clear in § 12.2.5 that expiration of the period does not terminate the contractor's liability, which ends only with expiration of the appropriate statute of limitation.

What happens if defects are discovered after the one-year correction-of-work period expires? Where the same defects were discovered earlier and the contractor's repair attempts proved unsuccessful, a federal court of appeals in *Berkshire Medical Center, Inc. v. U.W. Marx, Inc.*,[19] allowed the owner to hire a new contractor to entirely replace the defective work without giving that opportunity to the original contractor.

16.6 Contractor's Implied Warranty of Workmanship

Warranty is implied so as to guarantee a result desired by society. Like express (written) warranty, no negligence need be shown to prove breach of an implied warranty. The remedy is the same as for breach of an express warranty.

Where the construction contract does not specify the quality of the contractor's services, the law usually implies a warranty that the contractor will perform its services in a workmanlike manner, free of defects of workmanship. In operation, this implied warranty will closely resemble the negligence standard. However, because warranty is more result oriented, application of a warranty is likely to place the burden on the contractor where the work has not measured up to the ordinary expectation of the owner. Although sometimes implied terms are justified as simply reflecting the intention of the parties, warranties are often implied because of "fairness," a conscious judicial allocation of risk.

Just as a prime contractor owes the owner an implied warranty of workmanlike conduct, so too a subcontractor owes the prime contractor the same warranty of workmanship.

6.7 Homeowner Claims

16.7A Overview

Owners vary greatly in sophistication and experience, from the proverbial elderly widow to a large corporation or federal agency. Homeowners as a group are viewed as unsophisticated or inexperienced. Perhaps even more importantly, a home purchase is often the largest investment owners will make in their lives. Purchase of a house with serious structural defects can have a devastating effect on the owners' finances. In addition, courts recognize that owners may have an emotional attachment to their homes, which may be deserving of special protection. For these reasons, first the courts, and increasingly legislatures, have focused attention on disputes involving defective residences.

16.7B Implied Warranties in the Sale of Homes

Before the mid-1960s, various legal doctrines meant that home buyers who discovered defects in their newly-built homes had little recourse against the sellers—or for that matter against anyone. Courts began using implied warranty to circumvent these defenses. Courts first imposed upon the builder–vendors an implied warranty of habitability, meaning that the new house was structurally secure and its basic components (plumbing, electrical, etc.) functioned properly. As an implied warranty (in contrast to an implied contract term), this benefit was soon extended by many courts to subsequent purchasers—those not in privity with the builder–vendor.

16.7C Statutory Protections of Homeowners

While the implied warranty of habitability has its origin in the common law, state legislatures increasingly are addressing the issue of defective residential construction. Depending on the wording of the statute, the existence of a statutory remedy will either displace or supplement the homeowners' common law rights.

Consumer Protection Acts. Most states have enacted consumer protection legislation meant to benefit consumers in general—not just homeowners. Such laws bar misleading or fraudulent business practices in consumer transactions. They provide a list of forbidden activities and enlarge the remedies available to consumers, such as multiplying compensatory damages (usually two-fold or three-fold) and awarding attorneys' fees. In some states, these statutes have been used against contractors who build residences.[20] Some homeowners have successfully invoked consumer protection claims against design professionals (see Section 11.4B).

Consumer Warranties. Warranties can mislead the consumer. An apparently very broad warranty may have been so hedged with restrictions that it may not amount to very much. To deal with this problem, Congress enacted the Magnuson-Moss Warranty Act, which carefully regulates consumer warranties.[21] This act also specifies remedies for breach of such a warranty and allows the consumer who prevails in a claim based on the act to recover attorneys' fees.

Although the act was thought principally to deal with consumer products, the courts have also extended it to building disputes, such as a contract to roof a house.[22] Courts looked at the legislative history and the desire to protect consumers and concluded that the act was not limited solely to everyday products, such as appliances or cars.

New Home Warranty Acts. Several states have consumer protection legislation aimed specifically at buyers of new homes. These statutes are of two types: new home warranty acts and right to repair laws.

New home warranty acts (NHWA) create statutory warranties that apply to the sale of new homes. They require contractors to guarantee that the home is free of major defects. For example, Connecticut's NHWA implies in every sale of a new residence one-year warranties that the residence is free of faulty materials, constructed according to sound engineering standards, constructed in a workmanlike manner, and fit for habitation.[23] These statutory warranties are in addition to any warranties implied by law.[24]

By contrast, Louisiana's NHWA displaces the common law rights of both the builder of a new residence and the owner of that residence. Instead, the act provides these parties with the exclusive remedies, warranties, and limitations periods.[25] For example, an owner could sue the contractor for breach of the statutory warranties but not for breach of contract and negligence.[26]

Right to Repair Acts. Right to repair (or right to cure) laws do not give homeowners new, substantive rights. Instead, they seek to prevent unnecessary litigation by giving contractors the opportunity to cure defects before the homeowner is allowed to sue. Homeowners who discover what they believe to be construction defects must comply with a statutory prelitigation procedure. They must provide the contractor with written notice of the alleged defects and then give the builder the opportunity to inspect the building and attempt repairs. The purpose of these laws is to promote repairing of defective residences without resorting to litigation.[27] According to a 2007 article, thirty states have adopted such laws.[28]

Although right to repair laws do not grant homeowners new, substantive rights, they do deprive contractors of one common law defense. By listing certain construction defects as subject to the statute, these laws preclude the builder from arguing that the economic loss rule (discussed in Section 5.10) immunizes it from liability. For example, California's right to repair law overruled the state supreme court's decision, *Aas v. Superior Court*,[29] which had applied the economic loss rule to prevent condominium owners from recovering the cost of replacing firewalls that were not code compliant.

> In this chapter's scenario, these common law and statutory rights would not apply, because the scenario involved construction of a commercial building—not a residence.

16.8 Financial Problems

After the construction contract is made and before completion of each party's performance, either party can suffer financial reverses. As a result, an owner may delay payments. A contractor may be unable to supply an adequate labor force or purchase materials. Assuming each party is suffering a cash flow problem but is not in such dire straights that it is filing for bankruptcy, what can the performing party do?

Contractor Options. Sensitive to contractor concerns, AIA A201-2007, § 2.2.1 permits them *before* beginning work to request the owner to provide "reasonable evidence that the Owner has made financial arrangements to fulfill the Owner's obligations under the Contract." Once work has begun, if the owner fails to make timely payments or enters into change orders that greatly increase the contract price, the contractor may renew that request. Section 14.1.1.4, gives the contractor the right to terminate its obligation if it has ceased performance for thirty days because the owner has not furnished the requested financial information.

Suppose the prime contract does not permit the contractor to request the owner's financial information or the information is required but not given. The common law would give the contractor the right to demand adequate assurance of financing and to suspend performance until it is furnished.[30]

Owner Options. Financial problems encountered by the contractor are not likely to give it any justification for refusal to continue performance. However, construction contracts usually grant the owner certain remedies if the contractor runs into financial difficulties, such as the ability to suspend or terminate the contract. The owner may also employ the common law doctrine of demanding assurance of continued performance.

Probably the owner's safest strategy against the risk of contractor financial problems is to obtain payment and performance bonds, as addressed in Chapter 25.

16.9 Contractor Defenses to Claims

16.9A General Contract Defenses

While certain defenses are unique to construction contracts, such contracts are subject to the same defenses as any contract. Analysis of these generalized contract defenses will be described here only briefly.

Mutual Mistake. Each contracting party, particularly the party agreeing to perform services, has fundamental assumptions often not expressed in the contract. For example, often the contracting parties assume that subsurface conditions will not vary greatly from those expected by the design professional or the contractor. Although some deviation may be expected, drastic deviation that has a tremendous effect on the contractor's performance may be beyond any mutual assumption of the parties. Such a mutual mistake would relieve the contractor from an obligation to continue performance.

Impossibility. A party to a contract does not have an obligation to perform if doing so were impossible. Suppose a contractor claims that the specifications were "impossible" to meet. The first question is who assumed the risk that the design was defective. That inquiry turns on the distinction between performance and design specifications.

In the case of performance specifications, because a design–build contractor assumed responsibility for creation of the design, it also assumed the risk of nonperformance. On a DBB project using design specifications, the contractor usually argues that the design is physically impossible to implement. In *R.P Wallace, Inc. v. United States*,[31] the contractor established that the window model was impossible to manufacture. More dramatically, in *O'Neal Engineering, Inc.*,[32] the design called for installation of a four-inch cable in an existing conduit less than four inches in diameter.

Sometimes, the specifications are impossible to implement because of time constraints. In *Triax Pacific, Inc. v. West*,[33] the specifications' painting schedule and curing requirement, when read together, made timely completion of the project impossible.

> In this chapter's scenario, the roof was physically impossible to build *properly* under the contract's design specifications. The roof can be built, but it is doomed to failure. This scenario might require an expansion of the doctrine of impossibility—one the court may find merited.

Commercial Impracticability. Because true impossibility is rare, what is usually asserted as a defense by a performing party is that performance cannot be accomplished without excessive and unreasonable cost. It is more accurate to speak of "commercial impracticability" rather than actual impossibility.

To invoke the defense of commercial impracticability, a contractor must show that something unexpected occurred, the risk of this unexpected occurrence was not allocated to the contractor, and the occurrence of the event rendered performance commercially impracticable. The contractor must show not only that it could not perform (subjective impossibility) but also that other contractors similarly situated would not have been able to do so (objective impossibility). These are formidable requirements, and relief requires extraordinary circumstances.

Frustration. Frustration looks at the effect of subsequent events not on performance but on desirability of performance. Performance is not impossible, but the underlying reason for the performance no longer makes sense. For example, suppose during construction of a racetrack a law were passed that made horse racing illegal in the state. While the race track physically can be built, the purpose of doing so has been frustrated by passage of the law.

Unconscionability. A court may void (refuse to enforce) a contract provision that is unconscionable. An unconscionability analysis has two components. The first is that the weaker party is surprised by the presence of the clause—often because the language is hidden in small print in a lengthy contract. Second, the contract provision must be one-sided and unreasonably or unfairly allocate risk in favor of the stronger party. For example, in the construction industry, an unconscionability defense might be raised to avoid an arbitration obligation included in purchase agreements in large housing subdivisions.

16.9B Contractor Followed the Design

As stated at the beginning of this chapter, a basic principle of liability for construction defects is that responsibility follows control. An owner who supplies the design has control over its creation and hence over its adequacy. Accordingly, many courts have held that the contractor who follows the design is not responsible for a defect unless it (1) knew of the defect but did not warn the owner, (2) warranted the design, or (3) was negligent. Some statutes codify this principle.[34]

16.9C Subrogation: Destruction of Project Under Construction, Covered by Owner's Insurance

During project construction, the project may be partially or totally destroyed. The work may be destroyed by causes for which neither party is chargeable, such as an earthquake or hurricane.

This section analyzes a *contractor's negligent* partial or total destruction of the project under construction in which the owner wishes to have the damage repaired and the project completed. Under the common law, the contractor would be responsible for repairing the work, or (if it did not do so) it would be liable to the owner for this cost.

The presence of insurance covering the work under progress alters application of the common law rule. While insurance is discussed in Chapter 24, a basic outline is necessary to understand this section.

Usually, each participant in the construction process obtains insurance to protect itself against certain risks. The owner acquires insurance to protect the property under construction, usually with an aptly named "builder's risk" **insurance policy**. (A policy is an insurance contract.) For example, if the project is destroyed by a fire negligently started by the contractor or one of its subcontractors, the insurance company pays the owner the cost to repair that property damage.

When the insurer pays the owner, the insurer often looks for reimbursement from a party other than the insured (the owner) who it can claim negligently caused the fire—in this case, the contractor. The common vernacular is that the insurer "steps into the shoes" of the owner and brings a **subrogation** claim against the contractor—a claim the owner could have brought had its loss not been covered by insurance.

At this point, the contractor needs a defense not to a claim by the owner, but to a subrogation claim by the owner's insurer. Construction contracts frequently use techniques to bar subrogation. One method, adopted by both the AIA and the Engineers Joint Contract Documents Committee (EJCDC), is for the owner, design professional, prime contractor, and subcontractors to waive any claims against each other to the extent the loss is covered by insurance.[35] This waiver of claims by the project participants is intended to bar as well any subrogation action by the insurance company.

To illustrate, suppose a fire has been caused by the prime contractor and the insurer pays the owner for the damage. The insurer will not be able to pursue the prime for negligently causing the fire if the owner has waived its rights of subrogation against the prime. This method seeks to place the entire loss on the insurer. It frees the participants to continue with performance of the construction project using the proceeds obtained from the owner's property insurer. Insurance replaces litigation as a means of resolving this event.

16.9D Passage of Time: Statutes of Limitation and Repose

One defense is that the claimant did not begin legal action within the time required. Time limits may be created by contract, by judge-made law, or (most commonly) by statutes. These three methods will be addressed in reverse order.

Statutes of Limitation and Repose. Statutory limits on the time available for bringing a lawsuit have generated immense complexity. Statutes barring claims based on the passage of time are of two types: statutes of limitations (discussed also in Section 12.3) and statutes of repose. Both types of statutes deal with the effect of the passage of time on whether claims may be brought. They are designed to protect defendants from false or fraudulent claims that may be difficult to disprove if not brought until relevant evidence has been lost or destroyed and witnesses become unavailable.

In addition, entirely apart from the fault aspect of delay in bringing legal action, the law can also seek to promote certainty and finality in transactions (especially commercial transactions) by terminating contingent liabilities at specific points in time. In the construction industry, this second function is accomplished by statutes of repose.

A statute of repose cuts off liability a designated number of years after substantial completion. This means the statute of repose can bar a claim before it came into existence. For example, if the statute of repose is eight years and the roof starts to leak eight years and one month after substantial completion, the owner's faulty workmanship claim against the contractor will be barred because the repose period has passed. Typically, those protected by the statute are design professionals and contractors, while protection is not accorded owners or suppliers. The period of repose varies by state. All states but New York and Vermont have enacted statutes of repose.[36]

Commencement of Statute of Limitations. One complexity is to determine when the statute of limitations begins. For example, suppose the applicable statute of limitations states that a claim for breach of contract must be begun within four years from the time the "cause of action accrued." A cause of action (a legal claim) "accrues" (and the limitations period begins to "run") at the earliest time a plaintiff may bring a lawsuit under a particular legal theory.

A breach of contract action may be brought immediately upon project completion because the owner is deemed to have suffered injury (in the form of a defectively built structure), even if she did not yet know the construction was defective. Some states have ameliorated the harsh effect of this rule. They employ the "discovery rule" to delay accrual for latent (hidden) defects until the time when a reasonable and diligent owner would have discovered the defect. Other states refuse to apply the discovery rule to breach of contract claims. (In these states, the contract statute of limitations operates like a statute of repose.)

By contrast, a negligence cause of action accrues only when a plaintiff suffers an appreciable injury. Deciding when the plaintiff suffered appreciable injury can also be difficult to determine. Suppose a contractor claims that defects in the architect's design caused the contractor to incur higher-than-anticipated performance costs. The contractor's negligence or misrepresentation action would not accrue until it suffered economic losses with certainty that were caused by the defective design.

Contractual Limitations. It is possible for the contracting parties to regulate this problem. Although contractual regulation will not affect claims of third parties, such as those who may be injured if a building collapses years after it is completed, the bulk of the exposure in this area usually involves claims by the owner against the contractor or design professional.

Parties to a construction contract can limit the time they have to bring a claim against the other. These agreed-upon limits will be enforced by the courts if the time periods are not unreasonably short. The AIA requires legal action to be brought "within the time period specified by applicable law, but in any case not more than 10 years after the date of Substantial Completion of the Work."[37] The 10-year outer limit functions as a contractual statute of repose.

Parties to a construction contract can also regulate when the limitation period begins. The 1997 AIA standard form documents specified that the limitation period began to run from the time of substantial completion.[38] Although enforced by the courts,[39] these provisions were dropped from the 2007 AIA documents in response to owner dissatisfaction.

Laches. The doctrine of laches is a judge-made limitation on the time available for bringing a lawsuit. Laches generally prevents a person who is aware of his injury from "sitting on his rights" and delaying to bring a claim—to the disadvantage of the defendant. For example, in *Chirco v. Crosswinds Communities, Inc.*,[40] a copyright holder of architectural plans waited to bring suit for two and one-half years while the defendant built a condominium using the infringed plans. The court ruled that laches barred the copyright holder's request that the infringing building be destroyed.

REVIEW QUESTIONS

1. What is the difference between an express warranty and an implied warranty?

2. What is the *Spearin* doctrine and how could a contractor use this as a "shield" or a "sword?"

3. How are the majority of claims decided when defects can be traced to both the owner and the contractor?

4. What is design delegation and what is an example of how it functions?

5. According to the American Institute of Architects A201-2007 Document, what are the three warranties as to the quality of the contractor's work?

6. What is the Magnuson-Moss Warranty Act?

7. What must a contractor show in order to invoke the defense of commercial impracticability?

8. What are the two components of an unconscionability defense?

9. Many courts have held that the contractor who follows the design is not responsible for a defect unless which things occur?

10. What are the similarities and differences between a statute of limitations and a statute of repose?

ENDNOTES

[1] *DeLaval Turbine, Inc.*, ASBCA No. 21797, 78-2 BCA ¶ 13,521; *Cascade Electric Co.*, ASBCA No. 28674, 84-1 BCA ¶ 17,210.

[2] 237 Kan. 692, 704 P.2d 2 (1985).

[3] 18 P.3d 645 (Wyo.2001).

[4] *Eichberger v. Folliard*, 169 Ill.App.3d 145, 523 N.E.2d 389, appeal denied, 122 Ill.2d 573, 530 N.E.2d 243 (1988).

[5] "271. *Examination of Site.*—Intending bidders are expected to examine the site of the proposed dry-dock and inform themselves thoroughly of the actual conditions and requirements before submitting proposals."

[6] "25. *Checking Plans and Dimensions; Lines and Levels.*—The contractor shall check all plans furnished him immediately on their receipt and promptly notify the civil engineer in charge of any discrepancies discovered therein. . . . The contractor will be held responsible for the lines and levels of his work, and he must combine all materials properly, so that the completed structure shall conform to the true intent and meaning of the plans and specifications."

[7] "21. *Contractor's Responsibility.*—The contractor shall be responsible for the entire work and every part thereof, until completion and final acceptance by the Chief of Bureau of Yards and Docks, and for all tools, appliances, and property of every description used in connection therewith."

[8] Annot., 6 A.L.R.3d 1394 (1966). In the absence of an agreement, Texas imposes on the contractor the risk of design errors. *Lonergan v. San Antonio Loan & Trust Co.*, 104 S.W. 1061 (Tex.1907).

[9] 248 U.S. at 136.
[10] 3 BRUNER & O'CONNOR ON CONSTRUCTION LAW § 9:81 at 669–70 (2002) (footnotes omitted).
[11] 70 Fed.Cl. 253 (2006), aff'd, 499 F.3d 1357 (Fed.Cir.2007).
[12] 21 Ill.App.3d 925, 316 N.E.2d 51 (1974).
[13] *Tibshraeny Bros. Constr. v. United States,* 6 Cl.Ct. 463 (1984).
[14] *Caddell Constr. Co. Inc. v. United States,* 78 Fed.Cl. 406 (2007); *Dugan & Meyers Constr. Co., Inc. v. Ohio Dept. of Adm. Servs.,* 113 Ohio St.3d 226, 864 N.E.2d 68 (2007) (delays due to excessive changes in the plans are not compensable under *Spearin*).
[15] These facts are taken from *Metric Constr. Co. v. United States,* 80 Fed.Cl. 178 (2008).
[16] Stein, Alexander & Noble, AIA *General Conditions in the Digital Age: Does the Square "New Technology" Peg Fit Into the Round A201 Hole?,* 25 Construction Contracts Law Rep., No. 25, Dec. 14, 2001, ¶ 367, pp. 3, 8. "CAD" refers to computer-assisted design.
[17] 234 So.2d 922 (Miss.1970).
[18] Id. at 924.
[19] 644 F.3d 71 (1st Cir.2011) (Massachusetts law).
[20] *Tang v. Bou-Fakhreddine,* 75 Conn.App. 334, 815 A.2d 1276 (2003); *Eastlake Constr. Co. v. Hess,* 102 Wash.2d 30, 686 P.2d 465 (1984).
[21] 15 U.S.C. § 2301 et seq.
[22] *Miller v. Herman,* 600 F.3d 726 (7th Cir.2010); *Atkinson v. Elk Corp. of Texas,* 142 Cal.App.4th 212, 48 Cal.Rptr.3d 247 (2006); and Schneier, *The Magnuson-Moss Warranty Act: Federalizing Homeowner Construction Defect Cases,* 13 Constr. Lawyer, No. 4, Oct. 1993, p. 1.
[23] Conn.Gen.Stat.Ann. § 47-118.
[24] Id. § 47-120.
[25] La.Rev.Stat. § 9:3150.
[26] *Carter v. Duhe,* 921 So.2d 963 (La.2006). New home warranty acts are reviewed in Annot., 101 A.L.R.5th 447 (2002).
[27] California's right to repair law, Cal.Civ.Code § 895 et seq., is interpreted in *Standard Pacific Corp. v. Superior Court,* 176 Cal.App.4th 828, 98 Cal.Rptr.3d 295 (2009).
[28] Quatman & Gonzalez, *Right-to-Cure Laws Try to Cool Off Condo's Hottest Claims,* 27 Constr. Lawyer, No. 3, Summer 2007, p. 13.
[29] 24 Cal.4th 627, 12 P.3d 1125, 101 Cal.Rptr.2d 718 (2000).
[30] Restatement (Second) of Contracts §§ 251, 252 (1981).
[31] 63 Fed.Cl. 402 (2004).
[32] ASBCA No. 31804, 86-2 BCA ¶ 18,906.
[33] 130 F.3d 1496 (Fed.Cir.1997).
[34] Cal.Pub.Cont. Code § 1104; La.Rev.Stat. § 9:2771.
[35] AIA Docs. A201-2007, §§ 5.3, 11.3.7; A401-2007, § 13.9; and B101-2007, § 8.1.2; EJCDC, C-700, ¶ 5.7 (2007).
[36] Tricker, Ebeler & Kortum, *Applicability of Statutes of Repose to Indemnity and Contribution Claims and 50 State Survey,* 7 J ACCL, No. 1, Winter 2013, p. 341.
[37] AIA B101-2007, § 8.1.1; A201-2007, § 13.7.
[38] AIA B141-1997, ¶ 1.3.7.3; A201-1997, ¶ 13.7.1.
[39] *Gustine Uniontown Assocs., Ltd. v. Anthony Crane Rental, Inc.,* 2006 PA Super 12, 892 A.2d 830 (2006) (interpreting identical language in B141-1987, ¶ 9.3).
[40] 474 F.3d 227 (6th Cir.2007).

CHAPTER 17

CONTRACTOR PAYMENT AND PROJECT COMPLETION

SCENARIO

Mega Manufacturer, Inc. (MMI) decided to build a new factory. It hired Awesome Architects, Inc. (AAI), an architectural firm, to create the design and administer the project.

Once the design was complete, MMI hired Prime Builder, Inc. under a fixed-price contract, using the American Institute of Architects (AIA) General Conditions of the Contract for Construction, A201-2007. Prime Builder, Inc. hired several subcontractors. Prime Builder, Inc. required all subcontractors to sign lien waivers as a condition to receiving payment.

Under the prime contract, Prime Builder, Inc. was to be paid on a monthly basis, based on AAI's certification of work done. However, 10 percent of each payment would be withheld (as "retainage") and kept in reserve by MMI.

During construction, the window supplier asked MMI to pay for the windows by writing a "joint" check made out to Prime Builder, Inc. and the window supplier. MMI agreed to do so.

Prime Builder, Inc. asked for AAI to approve payment not only for the contractor's labor and that of its subcontractors, but also for the windows. The windows had not yet been installed, but Prime Builder, Inc. told AAI that the windows were stored on-site safely, and AAI approved payment for the windows without making sure this was true.

In fact, the window supplier was not paid: it therefore took back the delivered windows. MMI was forced to buy other windows at a higher price (and better quality). As a result, Prime Builder, Inc. had insufficient money to carpet the offices, although the factory was functional. Desperate for use of the factory, MMI moved in.

MMI used the ten percent retainage to buy and install the carpeting. Meanwhile, it discovered that the wrong windows had been installed. It therefore (1) refused to pay Prime Builder, Inc. the remaining contract balance and (2) sued Prime Builder, Inc. for breach of contract, seeking the cost of replacing the windows.

17.1 The Legal Framework: Common Law Payment Rules

This chapter deals principally with compensation of the prime contractor for performing work required by the construction documents. Compensation of the design professional is addressed in Section 9.3.

A construction project involves the exchange of money (paid by the owner to the contractor) for services (delivered by the contractor). In its simplest form, this chapter addresses the *timing* of these two actions.

The common law requires that performance of services precede payment for those services. This protects the party receiving services by permitting that party to withhold payment until all of the services have been performed. Not only does this rule allow the paying party to avoid the risk of paying and then not receiving performance, but it also allowed the paying party to dangle payment before the performing party, which is a powerful incentive to deliver.

However, the "work first and then be paid" rule is disadvantageous to the performing party. The latter must finance the entire cost of performance. The party performing services must take the risk that any deviation, however trivial, will enable the paying party to withhold money greatly in excess of the damages caused by the deviation. The performing party must assume the risk that the paying party might not pay after complete performance, leaving the performing party with a legal claim.

Moreover, the "work first and then be paid" rule is simply uneconomical for anything other than the smallest construction job. Very few contractors are able to finance a large project, waiting until the smallest contract obligation is completed before payment was due. It is also inefficient for each project to be financed twice: by the contractor (to fund performance) and by the owner (to pay for the completed project).

17.2 Progress Payments

Rationale. The contract documents, most notably the basic agreement and the general conditions, change these common law rules to create a payment system tailored to the construction industry. That system is premised on *progress payments*: the contractor is paid incrementally commensurate with the percentage of work completed. The common law's "work first and then be paid" rule is divided into units of the work: as each unit is completed, payment is due for that work. The units can be either stages of completion or time elapsed (for example, monthly payments). The major standard form construction contracts all embrace similar progress payment systems.[1]

The progress payment system protects both the owner and the contractor. From the owner's perspective, if the contractor defaults (pulls out of the project) midway through performance, the owner has enough money in reserve to complete the job. From the contractor's perspective, its obligation to finance the work prior to payment is greatly reduced when compared with the common law rule. Moreover, if the owner midway through performance defaults, the contractor's claim is limited to the work performed after the last progress payment received.

Mechanism. Before requesting its first progress payment, the contractor submits to the design professional a **schedule of values**. This schedule, when approved, constitutes an agreed valuation of designated portions of the work. The aggregate of the schedule should be the contract price. Adjustments to the contract work are incorporated as the work proceeds.

After beginning work, the contractor submits to the design professional an **application for payment** a designated number of days before payment is due. This application is accompanied by documentation that supports the contractor's right to be paid the amount requested.

In addition to the work itself, the contractor may seek payment for goods purchased that are stored at the site.

The design professional inspects the work to determine whether the application should be paid in full or in part. The design professional issues a **certificate for payment**, authorizing payment of the contractor. The certificate for payment is a representation from the design professional to the owner that the work appears to have progressed to the point indicated in the application for payment and that the work is in accordance with the contract documents.

Payment to the contractor may be withheld (in full or in part) for many reasons: the work has not progressed to the point indicated in the application; the work does not conform to the design; the contractor has not paid its subcontractors or suppliers; the existence or threat of third-party claims; or concern that the project will not be completed on time, so that withholding is necessary to compensate the owner for its delay-related costs. If some of the work complies with the design requirements and some does not, the portion that complies should be paid for on time. Otherwise, the slightest nonconformity would create the threat of forfeiture for the contractor.

The amount certified for payment depends on the pricing provisions in the construction contract. In fixed-price contracts, the pricing benchmark is the contract price. In cost contracts, the principal reference point for determining payments is the allowable costs incurred by the contractor. In unit-priced contracts, the progress payments are based on the number of units of designated work performed.

Recall that the progress payments are based on the schedule of values, which is created at the very commencement of the project. However, it is not unusual for the work to be changed by agreement between the owner and contractor during performance. The contract allows the application for payment to reflect payment for changed work.

Construction contracts frequently give the design professional the power to revoke a previously issued certificate. One reason for revocation is the discovery of defective work that had been the basis of a previously issued certificate.

17.3 Retainage

In addition to the use of progress payments, a second payment device tailored for construction projects is a withholding of a portion of each progress payment. This **retainage** (or "retention") creates a financial reserve the owner can use on the project should the need arise.

Suppose the contractor is entitled to a progress payment of $50,000, and the contract retainage is 10%. The contractor is paid $45,000 and the owner retains the other $5000 in reserve, usually until contract completion.

One purpose of retainage is to provide money out of which claims that the owner has against the contractor can be collected without the necessity of a lawsuit. If a project is nearly complete but small defects remain and the contractor fails to correct the problem, the owner may use the retainage to accomplish these repairs and then refund the remainder to the contractor. Conversely, the owner's refusal to pay the retainage until these small defects are taken care of is a strong incentive for the contractor to do so.

Another purpose of the retainage is to provide the owner with money with which to pay any subcontractor or supplier liens. As explained in Section 21.7, if the owner pays the prime contractor for work performed by the subcontractors and the contractor does not pay the subcontractors, the subcontractors can file mechanics' liens against the property, thereby forcing the owner to pay twice for the same work. By paying the subcontractor liens out of the retainage, the owner ensures that the contractor—not the owner—pays the subcontractors.

> In this chapter's scenario, MMI used the retainage to buy and have installed carpeting when Prime Builder, Inc. failed to do so. By having the retention funds available, MMI was able to complete the project without having to bring a claim against Prime Builder, Inc.

One problem with retainage can develop at the end of the project. Suppose the retained amount is $50,000 and the owner is entitled to take $20,000 from it to remedy defects in the work. A solvent contractor will be paid the balance. However, in the volatile construction industry, it is not uncommon for a contractor to run into financial problems. In such a case, a horde of claimants descends on the owner and demands the money. The claimants can be lenders who have received assignments of contract payments as security for a loan, sureties who have had to discharge the obligations of the contractor, taxing authorities who claim the funds when the contractor has not paid its taxes, and in the event the contractor has gone bankrupt, the trustee in bankruptcy.

The scramble for funds usually is dealt with by the owner's initiating what is called an **interpleader action**. The owner pays the disputed funds to the court, files a lawsuit in which it names as parties all claimants to the funds, and withdraws from the fray. The court must unscramble the claims.

Contractors complain that the retainage system deprives them of earned income. They have successfully convinced some state legislatures to eliminate the retention process, replacing it with an escrow system into which the contractor may deposit securities. Some states regulate retainage on private works projects. On some federal projects, once performance has reached fifty percent, the retainage is reduced from ten to five percent.

17.4 Payment of Subcontractors and Suppliers

Much (often most) of a progress payment made to the prime contractor is to pay for work or materials provided not by the contractor but by subcontractors and suppliers. As noted, an unpaid subcontractor or supplier may be able to file a mechanics' lien against the improved property. How is the owner to ensure that these parties are actually paid by the prime? (The rest of this discussion will refer only to subcontractors, but it includes suppliers as well.)

Some contracts require the contractor to give the owner assurance or proof that the prime has paid its subcontractors and suppliers when payment applications are made. Alternatively, contracts may require the prime contractor to demand **lien waivers** from the subcontractors as a condition for payment in which the subcontractor gives up its right to a mechanics' lien for the work covered in the progress payment. Both tactics seek to protect the owner against future claims by unpaid subcontractors or suppliers.

> In this chapter's scenario, Prime Builder, Inc. required subcontractors to waive their lien rights as a condition to receiving payment. When the window supplier was not paid for the windows, it had no lien rights to enforce. It went to the job site and took the windows back. If title (ownership) to the windows had been transferred, this would have been a theft on its part.

Second, design professionals may withhold progress payments if they learn that subcontractors are not being paid or that liens have been or are likely to be filed. Usually this information is communicated quickly to the design professional.

Third, if the contractor does not give the owner proof of payment of subcontractors, some contracts may authorize the owner to contact the subcontractors and suppliers directly to ascertain whether they have been properly paid.

Fourth, the owner may seek to ensure payment of subcontractors through issuance of **joint checks** made out to the prime contractor "and" (not "or") subcontractor. The check covers work that includes work by a subcontractor or materials furnished by a supplier. The names of both payees appear on the check. Each will have to endorse the check for it to be converted into cash. This should avoid the possibility that the prime contractor will take the funds and not pay those whose services or materials have provided the basis for the payment. (Similarly, prime contractors sometimes issue payments to subcontractors by using a joint check to ensure that the subcontractor pays sub-subcontractors or suppliers.)

Suppose a joint check mechanism is used, yet the subcontractor is not in fact paid. In a California case, *Ferry v. Ohio Farmers Insurance Co.*,[2] the owner drew a check payable jointly to the contractor and a supplier after which the supplier waived its mechanics' lien rights. Both parties endorsed the check. Proceeds were paid to the prime contractor, who gave its personal check to the supplier. Unfortunately for the supplier, the check was dishonored because the contractor did not have sufficient funds in its account.

The supplier sued the contractor and received a judgment for the amount of the debt but could satisfy only a portion of it from the assets of the prime contractor. Then the supplier brought a lawsuit against the surety of the prime contractor for the balance. The court held that the unpaid supplier did not lose its right to claim on the bond despite having waived its lien and endorsed the check. Yet the joint check process clearly failed, because the supplier did not insist on being paid at the time it endorsed the check but relied on a personal check issued by the prime.

In this California case, the supplier clearly had a claim against the prime contractor and (when that failed) the contractor's surety. The basis for the claim is that the contractor owed the supplier a duty to pay it its portion of the owner's payment.

Suppose, however, that the contractor had no surety. May the unpaid supplier bring a claim against the owner? The answer is that the owner fulfills its contract obligation by writing the joint check. In legal parlance, the supplier is "deemed" paid vis-à-vis any claim it would otherwise have against the owner. The supplier's failure to protect itself from the contractor's misfeasance is not grounds for transferring that risk to the owner.[3]

> In this chapter's scenario, MMI paid the window supplier by joint check made out to Prime Builder, Inc. "and" the supplier. For reasons not explained, the supplier was not paid. However, in the eyes of the law, the supplier is "deemed" paid and would not have been able to sue MMI for payment of the windows.

17.5 Design Professional's Certification of Payment Applications

Observation of Work. Before issuing a payment certificate, the design professional visits the site to determine how far the work has progressed. This is the basis on which progress payments are made.

The inspection should uncover whatever an inspection principally designed to determine the progress of the work would have uncovered. Both the AIA and the EJCDC limit the architect's or engineer's responsibility during the progress payment certification process. Essentially, both state that the payment certificate warrants only that the work complies with the contract requirements to the best of the design professional's "knowledge, information and belief." Similarly, each standard form document warns the inspections were not "exhaustive." These disclaimers seek to shield the design professional from owner claims regarding the quantity and quality of the work.

As noted previously, the owner is also concerned with claims from unpaid subcontractors or suppliers. Professional associations also seek to disclaim responsibility for preventing such claims. The AIA B101-2007, § 3.6.3.2, excludes from the architect's responsibility in the certification process a duty to "review[] copies of requisitions received from Subcontractors and material suppliers . . . to substantiate the Contractor's right to payment." The EJCDC C-700,

¶ 14.02(B)(4)(d) (2007), specifies that the engineer's payment certification inspection does not include "any examination to ascertain how or for what purposes Contractor has used the moneys paid on account of the Contract Price[.]"

Design Professional Liability. Ideally, the amount of money a design professional certifies for a progress payment corresponds exactly to the percentage of work completed. Deviation from this ideal means the design professional has either undercertified (authorized too little) or overcertified (authorized too much) payment. Undercertification impairs the contractor's cash flow, and overcertification diminishes the retainage.

The design professional's role in the certification process can generate a variety of claims. The owner may claim that the design professional failed to discover defects during her inspections of the work (see Section 10.3). The owner also may contend that the design professional certified payment without first determining whether the contractor has paid subcontractors and suppliers or whether liens have been or will be filed.

The contractor may claim the design professional failed to issue certificates within a reasonable time. This claim may be brought against the design professional in tort or against the owner for breach of contract based on the actions of the owner's agent (the design professional). The contractor may also claim that the design professional undercertified the amount due the contractor.

The contractor's surety may also contend the design professional is liable to it for wrongful certification of progress payments. As more fully explained in Chapter 25, a surety guarantees the contractor's obligations to perform the contract and to pay subcontractors and suppliers. The surety looks to the retainage as a source of funds if the contractor defaults.

Undercertification can harm both the contractor and surety. It can adversely affect the contractor's capacity to continue or complete performance. Again, if the contractor fails its performance obligation, the owner will call upon the surety to complete the project.

Some courts have permitted third parties, such as contractors and sureties, to maintain legal action against the design professional for improper certification. An early and influential decision, *State of Mississippi, for Use of National Surety Corp. v. Malvaney*, pointed out the reliance of the construction project participants—particularly the owner and surety—on the design professional's obligation not to overcertify progress payments. In finding the architect liable to the surety notwithstanding lack of privity, the court reasoned:

> The contractual arrangement here consisted of the building contract, including the plans and specifications, and the bond which incorporated the contract as a part thereof, with the mutually interdependent obligations and rights therein contained. The duties of the architect were clearly defined in the contract. One of his very important duties was not to approve progress payments in excess of 85 percent of the contract price, and before final payment to make an inspection to find if the work was acceptable under the contract and if the contract had been fully performed, and to require the contractor to submit satisfactory evidence that all payrolls, material bills, and other indebtedness connected with the work have been paid.
> This duty was owing both to the [owner] and the surety, for whose mutual benefit and protection the retainage funds were provided. A contractual relation between the architect and the surety was not requisite to the existence of this duty. It arose out of the general contractual arrangements which contained mutually interdependent rights and obligations.[4]

Defenses to Design Professional Liability. A design professional may invoke defenses to claims of improper certification. A minority of courts point to lack of privity to bar claims brought by third parties such as contractors and sureties.[5]

One of the multifaceted aspects of the design professional's performance is to interpret the contract, judge performance, and decide disputes. As discussed in Section 12.7E, the design professional's role when performing some of these functions can be analogized to that of a judge. When the design professional acts like a judge, in some states she will be accorded quasi-judicial immunity—being liable for only corrupt or dishonest decisions or those not made in good faith. However, a design professional who is unreasonable or arbitrary in determining when and how much to withhold from payment amounts earned by the contractor will lose the defense of quasi-immunity.[6]

Without using the term "quasi-judicial immunity," both design professional associations seek to shield their members from liability for decisions made in their administration of the project. The AIA A201-2007, § 4.2.12, states that the architect "will not be liable for results of interpretations or decisions rendered in good faith." The EJCDC C-700, ¶ 9.08(D) (2007), provides that, when making decisions on the requirements of the contract documents and the acceptability of the work, the engineer "will not be liable in connection with any interpretation or decision rendered in good faith[.]"

17.6 Construction Lender's Interest in the Payment Process

Small projects may be financed by the owner and contractor using their own funds. On larger projects, both of these parties may need to borrows funds to undertake the project. (The lender's role in the construction process is described in Section 7.6.) These loans interject the lender into the payment process. Whether the owner or contractor is the recipient of the loan raises different legal questions for the lender.

Loan to Contractor. Contractors or subcontractors often must borrow funds to operate their businesses. Lenders commonly require collateral to secure them against the possibility that the borrower will not repay the loan. Sometimes collateral consists of funds to be earned under specific construction contracts. Lenders often seek information from owners regarding the construction contracts whose payments are to be used as security. The lender seeks assurances by the owner that payments will be made to the lender.

Rather than rely on a promise, lenders more commonly demand that payments to the contractor be assigned to the lender. An **assignment** transfers the right to receive payment and causes a change of ownership in the rights transferred. It is a more substantial security than a promise. As the contractor performs and becomes entitled to progress payments, those payments are made either directly to the lender or by joint check to the contractor and lender.

Loan to Owner. The lender who has made a construction loan to the owner will issue progress payments in response to the design professional's certification of payment applications. However, smaller projects (including many residential projects) may not justify the cost to the owner of employing a design professional to perform this site service.

In this situation, the construction lender may assume the role of inspecting the work before issuing the interim payments. Would these inspections subject the lender to liability to the owner for certification of payment for defective or incomplete work? This is largely a factual question that turns on variables such as the lender's representations to the owner, the sophistication of the owner, the culpability of the lender, and the wording of the loan agreement.

These many variables mean that courts come to different conclusions in different cases. In *GEM Industrial, Inc. v. Sun Trust Bank*,[7] a federal court interpreting Ohio law found a construction lender liable to the owner when the lender's inspector had discovered defective construction yet did not stop the payments from going to the contractor. In *White v. AAMG Construction Lending Center*,[8] West Virginia ruled the owner may sue the construction lender for breach of contract for making disbursements to the contractor even after the lender's inspector found the work incomplete.

Often the loan agreement specifies that inspections performed by the lender are for the benefit of the lender—not the owner. On a Colorado project, the construction loan included this Limitation of Responsibility provision:

> Inspections and approvals of the Plans and Specifications, the Improvements, the workmanship and materials used in the Improvements, and the exercise of any other right of inspection, approval, or inquiry granted to [the Bank] in this Agreement are acknowledged to be solely for the protection of [the Bank's] interests, and under no circumstances shall they be construed to impose any responsibility for liability of any nature whatsoever on [the Bank] to any party. Neither [the owners] nor any contractor, subcontractor, materialman, laborer, or any other person shall rely, or have a right to rely, upon [the Bank's] determination of the appropriateness of any Advance.

In *Alpine Bank v. Hubbell*,[9] a federal court interpreted this language as defeating the owner's right to rely on the lender to protect the owner from the contractor's improper actions.

Subcontractor Claims. Subcontractors may also claim that the lender had a duty to shield them from the contractor's misconduct. These claims have largely been unsuccessful. The lender's obligation under the contract is to the owner as borrower—not to the subcontractors. Wisconsin's supreme court, in *Hoida, Inc. v. M&I Midstate Bank*,[10] noted that imposing a tort duty in favor of subcontractors would place too great an administrative responsibility upon lenders. It also reasoned that making the lender liable would violate the public policy of the mechanics' lien act, which generally gives liens by construction lenders priority over liens filed by subcontractors.

17.7 Surety Requests to Public Owner that Payment Be Withheld

In public work, unpaid subcontractors and suppliers do not have lien rights. Instead, the contractor is required to obtain a payment bond, so the surety guarantees the contractor's obligation to pay those parties. A performance bond guarantees the contractor's performance obligation.

Sometimes the government decides to advance payments to the contractor to enable it to continue performance and complete the work. Yet in doing so, the surety's right to reimbursement from the contract balance can be affected adversely.

Another issue that may arise is that the surety believes the contractor is about to default on the contract. The surety warns the government of this and demands that future progress payments be made directly to the surety, so the surety then will pay the subcontractors and suppliers directly. Notwithstanding this demand, the government continues to pay the contractor—who then defaults.

Federal courts have sought to balance the competing interests of the government to keep performance going with protection of the surety's concern for reimbursement. While the contracting officer (the government's representative on the project) has broad discretion to pay earned progress payments to the contractor despite some minor defaults, that discretion must be exercised responsibly, and the surety's interest must be considered.

In *Balboa Insurance Co. v. United States*,[11] a summary judgment (a pretrial ruling) had been granted to the government by the trial court based solely on the contractor having been on schedule and having completed 91 percent of the project. The federal appellate court held that the summary judgment in favor of the government was improper where the surety contended that progress payments and retainage should not have been released when it notified the government of the contractor's impending default. Other factors must be examined to determine whether the government exercised reasonable discretion in disbursing the funds.

Suppose the owner accedes to the surety's request that future payments be made directly to it. In such a case, the unpaid prime contractor may, in addition to claiming a right to the payment withheld, assert a claim against the surety for wrongful interference with the prime contract.

17.8 Late Payment and Nonpayment During Performance

What are the contractor's rights should the owner delay or even stop making payments? The sources of the contractor's rights may be statutory, common law, or contractual.

Prompt Payment Acts. The federal government and state legislatures have been increasingly responsive to complaints by contractors and subcontractors that payments are not made promptly to them. Under the federal Prompt Payment Act,[12] a contractor must pay its subcontractors for satisfactory performance within seven days of receiving payment from the federal agency. Failure to pay on time subjects the contractor to an interest penalty owed to the subcontractor. Subcontractors have the same obligation to pay sub-subcontractors. The statute does not bar the use of retainage. If the contractor believes the subcontractor is not entitled to full payment, it must notify the subcontractor and the government of the reason and reduce the payment only by the amount likely to remedy the problem.

As noted, states have also adopted their own prompt payment acts. California extends a prompt payment obligation to both public and certain private contracts. Payments must be made from the contractor to the subcontractor (or from the subcontractor to the sub-subcontractor) within seven days of receipt of payment. A prime contractor who violates the law is subject to licensing disciplinary action and must pay the subcontractor a penalty of 2 percent per month in addition to normal interest. In the event of a "good faith dispute" as to a subcontractor's entitlement to payment, the contractor may withhold 150 percent of the disputed amount.[13]

Common Law Rights. For convenience, this discussion will assume an unpaid prime contractor. However, any conclusions expressed in that context are likely to apply to unpaid first- or second-tier subcontractors.

The law recognizes the importance of prompt payment. Yet under certain circumstances, drastic remedies for nonpayment may not be appropriate. Sometimes delay in payment is unavoidable. Delay may not harm the contractor. Care must be taken to avoid an unpaid party's using minor delay as an excuse to terminate an unprofitable contract and causing economic dislocation problems. Despite these possibilities, nonpayment is—and should be—considered a serious matter on the whole.

At the outset, there must be a determination that failure to make payment is a breach of contract. Because construction contracts are detailed and contract procedures are often ignored, those accused of not paying in accordance with the contract often assert that prompt payment has been waived or that full payment is not owed (because of defective or late work).

A formidable weapon in the event of nonpayment is the contractor's suspension of work. In addition to placing heavy pressure on the owner, suspension avoids the risk of further uncompensated work. The availability of a mechanics' lien is a pale substitute when there is a substantial risk of nonpayment. Until recently, the law was not willing to recognize a remedy short of termination for such breaches. Some cases now hold that the contractor can suspend work if it is not paid.[14]

Nonpayment certainly should not automatically give the right to suspend performance. Shutting down and starting up a construction project is costly. It would be unfair to allow the contractor to shut down the job simply because payment is not made absolutely on schedule.

Can an unpaid contractor terminate the contract? Suspension is temporary. Termination relieves the contractor from the legal obligation of having to perform in the future. In the absence of any contract provision dealing with this question, termination is proper if the breach is classified as "material."

The materiality—or seriousness—of a breach involves an examination of multiple factors. The most important is the effect of nonpayment on the contractor's ability to perform and the likelihood of future nonpayment. Persistent nonpayment may indicate that the problem is a serious one. A clear statement by the owner that payment will not be made is a repudiation and clearly gives the right—and probably the obligation—to stop performance. Termination is an important decision. Many cases, however, have concluded that a failure to pay (often together with other breaches) gives the contractor the right to terminate its obligation to perform under the contract.[15]

Suppose the contractor continues performance in the face of nonpayment. Continued performance may indicate that nonpayment was not sufficiently serious to constitute termination. Continued performance may manifest an intention to continue performance that is relied on by the owner. However, continued performance should not invariably preclude nonpayment from justifying termination. An unpaid contractor may choose to continue work for a short period while awaiting payment. This would be especially true if the contractor clearly indicated that, if payment were not forthcoming, work would cease.

AIA Documents. The AIA regulates the contractor's rights in the event of nonpayment. The parties should follow this contractual scheme before seeking to invoke the common law rules.

Under AIA Document A201-2007, § 9.7, seven days after a progress payment should have been made, the contractor can give a seven-day notice of an intention to stop work unless payment is made. Failure to pay after expiration of the notice period permits suspension. Section 9.7 also states that the contractor shall receive a price increase for its reasonable costs of shutdown, delay, and startup. Interest is added to these amounts. Undoubtedly, when it is not paid after a reasonable period of time, the contractor should be able to suspend work. However, the two seven-day periods may be too short.

AIA Document A201-2007, § 14.1.1.3, permits the contractor to terminate if the work has been suspended for thirty days for nonpayment. The contractor must first give a seven-day written notice that it intends to terminate.

Does this express termination provision affect any common law power to terminate for nonpayment? AIA A201-2007, § 13.4.1, states that remedies specified are in addition to any remedies otherwise imposed or available by law.

17.9 The Completion Process and Payment

Completion on small projects tends to be a simple, one-step process. The owner and contractor walk through the site together, the owner verifies that the project is apparently completed, and final payment is made.

On medium- and large-size projects where the owner employs a design professional to be the owner's representative, completion is normally a two-step process: substantial completion followed by final completion. Each event also entitles the contractor to certain rights of payment.

Substantial Completion. Substantial completion is the point at which the project may be occupied and used for its intended purpose.[16] The contractor prepares a list of final items that it acknowledges need to be completed. The design professional then inspects the project to determine whether substantial completion has been achieved and what items remain to be completed.

If the design professional finds the project is not substantially complete, she will prepare her own list of items that need to be completed before substantial completion is achieved. The contractor must correct or complete these items and then request a second inspection.

Suppose, after this second inspection, the design professional concludes the project is substantially complete. The EJCDC alone contemplates the owner being allowed seven days after receipt of a tentative certificate of substantial completion to object to the certificate. (This unique provision may simply reflect the fact that EJCDC contracts generally involve sophisticated, commercial owners.) The certificate of substantial completion again creates a *punch list*, which is a list of minor defects that need to be repaired or completed but do not interfere with the owner's occupancy and use of the project. The certificate should also specify the time by which the punch list must be completed.

While the discussion to this point has referred to completion of the project as a whole, the standard form documents contemplate partial occupancy or use by the owner.

The certificate of substantial completion specifies the owner's and contractor's responsibilities for maintenance, utilities, insurance, and security matters. Substantial completion is not project acceptance. Still, substantial completion is a major milestone for the contractor. It may (depending on the contract) entitle the contractor to payment of the retainage, less adjustments for incomplete or defective work. Warranties generally begin to run at the time of substantial completion, as does the statute of repose (discussed in Section 16.9D). Achieving substantial completion is also a prerequisite to the contractor's use of statutory remedies, such as filing a mechanic's lien. Liquidated damages may begin to run from the date of substantial completion or final completion, depending on the wording of the contract.

> In this chapter's scenario, MMI was able to use the factory but not the offices because of the lack of carpeting. The retainage funds were sufficient to pay for the carpeting, and no actual construction work needed to be done. It is likely a court would find that Prime Builder, Inc. had achieved substantial completion.

Final Completion. The contractor notifies the owner and design professional when it believes the project is ready for final completion. The design professional inspects the work to determine whether the certificate of final payment should be issued. Clearly, this inspection should be undertaken with great care.

The design professional's certification of final completion entitles the contractor to its final payment. However, this final payment is not due until the contractor submits to the owner or design professional a series of submittals, including an affidavit that all bills have been paid (or will be paid out of the final payment), a certificate of continuing insurance coverage until cancelled by the owner, consent of the surety, and releases or waivers of mechanics' liens. Of course, the owner may impose additional requirements, such as submissions of as-built drawings, copies of warranties and other close-out documents, a list of injuries that occurred on the site, and a list of all claims the contractor believes are unsettled.

The owner may be willing to accept the project even though minor work remains to be done and the project was completed late. The owner may deduct from the final payment funds needed to perform these minor repairs, as well as liquidated damages.

> In this chapter's scenario, MMI used the retainage funds to pay for carpeting. No other defect seems to have been found. MMI would be in breach of contract for failing to pay Prime Builder the remainder of the contract funds.

If the owner has not taken possession at the time of substantial completion, it will certainly do so at this point of final acceptance.

Effect on Future Claims. As a general rule, an owner's acceptance of the project or its final payment of the contractor or possession of the work does not bar it from suing the contractor for construction defects. Standard form documents make this explicit, stating that final payment does not waive the owner's claim that the work does not conform to the contract documents.[17] In addition, the universal presence of warranty (guarantee) clauses makes it quite unlikely that such a claim will be lost because of the occurrence of these acts.

Another possible claim facing the owner after final completion is in the form of mechanics' liens filed by subcontractors who have not been fully paid (even though the prime contractor has been). Standard form documents also preserve the owner's right to bring a claim against the contractor if mechanics' liens are filed.

The owner too is concerned that project completion will be followed by a contractor claim—most commonly that the contractor's cost of performance was higher than anticipated and that the owner is responsible for that economic loss. The owner can take affirmative steps to preclude such claims by requiring the contractor—as a condition for receiving final payment—to sign a "release" of future claims.

However, if a defect is obvious and the owner communicates a clear intention to relinquish a claim for the defect, he will be found to have waived his right to pursue the claim.

17.10 Substantial Performance Doctrine

Sometimes the completion process breaks down, and formal project acceptance is not achieved. The contractor is in breach of contract—some work admittedly remains to be done—and the owner could claim that breach as grounds not to make final payment. Strict enforcement of the owner's position would result in a forfeiture, as the contractor's work has become a fixture on the owner's land and cannot be removed by the contractor. Indeed, frequently the owner has taken possession of the project.

The substantial performance doctrine provides the contractor with a means of achieving at least partial payment notwithstanding its contract breach. In its simplest form, this requires the owner to pay the balance of the construction contract price if the contractor has substantially performed, leaving the owner a claim for damages based on failure of the contractor to perform strictly in accordance with the contract requirements. A prime contractor invoking the substantial performance doctrine contends that the owner has essentially what it bargained for and often points to the owner's occupation and use of the project.

The substantial performance doctrine should not be confused with substantial completion, although both generally require the owner being able to occupy and use the project. Substantial completion is a contract standard that requires certification by a designated third party (the design professional). By contrast, substantial performance is a common law doctrine intended to prevent forfeiture by an admittedly breaching party.

There is no clear-cut point at which substantial completion has been achieved. Because it is a common law doctrine, courts struggle to articulate a standard by which substantial performance may be recognized. A Missouri court defined substantial performance as "when construction has progressed to the point that the building can be put to the use for which it was intended, even though comparatively minor items remain to be furnished or performed in order to conform to the plans and specifications of the completed building."[18] A Louisiana court found the contractor had substantially performed where the building was occupied and the punch list items were mostly cosmetic.[19]

When is substantial completion not found? A New Jersey court stated that the defects must not "so pervade the whole work that a deduction in damage will not be fair compensation."[20] A California case concluded there had not been substantial performance when there was an accumulation of small defects—even though the owner was occupying the premises.[21] In a recent Iowa case, the contractor (who had left the job) claimed substantial completion even though the work was only 85 percent done; the contractor noted that subcontractors were lined up to complete the work and the owner simply had to contact them. The court found no substantial completion, pointing out that ensuring the subcontract work was performed was the contractor's job—not the owner's.[22]

The substantial performance doctrine may be invoked where remedying the contractor's breach would result in economic waste. *Jacob & Youngs, Inc. v. Kent*[23] is a leading case that found that there had been substantial performance when it was discovered after completion that one brand of pipe had been substituted for the one specified. The court noted that the pipe substituted was of equal quality to that designated. Tearing apart the completed house just to replace the piping made no economic sense.

> In this chapter's scenario, MMI sued Prime Builder, Inc. for having installed the wrong windows. The installed windows were of higher quality than what was specified. It would be economically wasteful for Prime Builder, Inc. to pay MMI to replace the windows. Indeed—and this is one rationale underlying the "economic waste" doctrine—it is likely MMI would simply pocket the awarded damages rather than actually replace the windows. The "economic waste" doctrine acts as a means to prevent an owner from obtaining a windfall from the contractor.

In *Plante v. Jacobs*,[24] the owners were living in their new home but were dissatisfied. In addition to several unfinished items (for which the owners could pay by deducting the cost from the retainage), the wall between the living room and kitchen was misplaced, narrowing the living room by more than one foot. Real estate experts testified that the smaller width of the living room would not affect the market price of the house. Tearing down and rebuilding the wall would cost $4,000 on a contract price of $26,765. The Wisconsin Supreme Court reasoned:

> Substantial performance as applied to construction of a house does not mean that every detail must be in strict compliance with the specifications and the plans. Something less than perfection is the test of specific performance. . . . There may be situations in which features or details of construction of special or of great personal importance, which if not performed, would prevent a finding of substantial performance of the contract. In this case the plan was a stock floor plan. No detailed construction of the house was shown on the plan. There were no blueprints. The specifications were standard printed forms with some modifications and additions written in by the parties. Many of the problems that arose during the construction had to be solved on the basis of practical experience. No mathematical rule relating to the percentage of the price, of cost of completion or of completeness can be laid down to determine substantial performance of a building contract. Although the defendants received a house with which they are dissatisfied in many respects, the trial court was not in error in finding the contract was substantially performed.[25]

17.11 Work Not Substantially Complete

The defaulting building contractor may not have performed sufficiently to take advantage of the substantial performance doctrine. Restitution (sometimes called *quasi-contract* or *quantum meruit*) may enable the contractor to recover for any net benefit that its performance has conferred on the owner. One purpose of restitution is to avoid unjust enrichment to the owner.

Suppose there is a construction contract in which a promise to pay $100,000 is exchanged for the promise to construct a building. Suppose the contractor unjustifiably leaves the project during the middle of performance. The defaulting contractor may have conferred a net benefit despite its having breached by abandoning the project. For example, suppose the defaulting contractor has received $20,000 in progress payments but the work has been sufficiently advanced so that it can be completed by a successor contractor for $75,000. Absent any delay damages, the defaulting contractor has conferred a net benefit of $5,000 on the owner and should be entitled to this amount. For contractors, restitution is a theory of recovery of last resort.

REVIEW QUESTIONS

1. How does the progress payment system work in the construction industry?

2. What are the purposes for retainage?

3. What are two tactics that seek to protect the owner against future claims by unpaid subcontractors or suppliers?

4. What are the implications if a design professional undercertifies or overcertifies a payment application?

5. If a contractor obtains a loan from a lender to finance construction, how might the lender use an assignment to secure paying back the loan?

6. What are the differences in rules of the courts in the *GEM Industrial, Inc. v. Sun Trust Bank* and *White v. AAMG Construction Lender Center* on one hand, and *Alpine Bank v. Hubbell* on the other?

7. What is the Prompt Payment Act, and what are the penalties if a contractor is found to break this act?

8. What is substantial completion, and what are the ramifications to the owner and contractor of the contractor achieving substantial completion?

9. What is the substantial performance doctrine, and how does it function?

10. What is restitution, what is its purpose, and when might a contractor use the doctrine in the project completion process?

ENDNOTES

[1] The payment process is found in American Institute of Architects (AIA) A201-2007, Article 9; the Engineers Joint Contracts Documents Committee (EJCDC) C-700, Article. 14 (2007); and ConsensusDocs 200, Article 9 (2012). See Nielsen, *Payment Provisions: Form Contract Approaches and Alternative Perspectives,* 24 Constr. Lawyer, No. 4, Fall 2004, p. 33, for an in-depth comparison of these three payment systems.

[2] 211 Cal.App.2d 651, 27 Cal.Rptr. 471 (1963).

[3] Barrett, Jr., *Joint Check Arrangements: A Release for the General Contractor and Its Surety,* 8 Constr. Lawyer, No. 2, April 1988, p. 7.

[4] 221 Miss. 190, 72 So. 2d 424, 431 (1954). See also Annot., 43 A.L.R.2d 1227 (1955).

[5] *Engle Acoustic & Tile, Inc. v. Grenfell,* 223 So.2d 613 (Miss.1969) (sharply limiting *State v. Malvaney,* supra note 4).

[6] *City of Mound Bayou v. Roy Collin Constr. Co.,* 499 So.2d 1354 (Miss.1986).

[7] 700 F.Supp.2d 915 (N.D.Ohio 2010).

[8] 226 W.Va. 339, 700 S.E.2d 791 (2010).

[9] 555 F.3d 1097 (10th Cir.2009). See Annot., 20 A.L.R.5th 499 (1994).

[10] 291 Wis.2d 283, 717 N.W.2d 17 (2006).

[11] 775 F.2d 1158 (Fed.Cir.1985).
[12] 31 U.S.C. § 3905.
[13] *FEI Enterprises, Inc. v. Ree Man Yoon*, 194 Cal.App.4th 790, 124 Cal.Rptr.3d 64 (2011). See also Tricker et al., *Survey of Prompt Pay Statutes*, 3 J ACCL, No. 1, Winter 2009, p. 91.
[14] *Hart & Son Hauling, Inc. v. MacHaffie*, 706 S.W.2d 586 (Mo.Ct.App.1986).
[15] *Guerini Stone Co. v. P. J. Carlin Constr. Co.*, 248 U.S. 334 (1919); *Aiello Constr., Inc. v. Nationwide Tractor Trailer, etc.*, 122 R.I. 861, 413 A.2d 85 (1980).
[16] Substantial completion is addressed with little variation in AIA A201-2007, § 9.8; EJCDC C-700, ¶ 14.04 (2007); ConsensusDocs 200, ¶ 9.6.
[17] AIA Doc. A201-2007, §§ 9.6.6 and 9.10.4; EJCDC C-700, ¶ 14.09 (2007).
[18] *Southwest Eng'g Co. v. Reorganized School Dist.* R-9, 434 S.W.2d 743, 751 (Mo.Ct.App.1968).
[19] *All Seasons Constr., Inc. v. Mansfield Housing Auth.*, 920 So.2d 413 (La.App.2006).
[20] *Jardine Estates, Inc. v. Donna Brook Corp.*, 42 N.J.Super. 332, 126 A.2d 372, 375 (App.Div.1956).
[21] *Tolstoy Constr. Co. v. Minter*, 78 Cal.App.3d 665, 143 Cal.Rptr. 570 (1978).
[22] *Flynn Builders, L.C. v. Lande*, 814 N.W.2d 542 (Iowa 2012).
[23] 230 N.Y. 239, 129 N.E. 889 (1921).
[24] 10 Wis.2d 567, 103 N.W.2d 296 (1960).
[25] 103 N.W.2d at 298.

CHAPTER 18

CHANGES: COMPLEX CONSTRUCTION CENTERPIECE

SCENARIO

Oliver and Olivia Owners bought an empty lot upon which they intended to build their "dream" retirement house. They decided to save money by buying a "stock" plans for a one-story ranch house they found on the internet.

The Owners showed the plans to Charlie Contractor and asked him to be their builder under a fixed-price contract. During negotiations, the Owners said they might want to "tweak" the design during construction. Charlie said this was not a problem, as long as the contract contained a "changes" clause which provided he would be paid a "reasonable compensation" and additional time if ordered to do work that differed from what was included in the plans. Although unclear as to what "reasonable compensation" meant, the Owners agreed.

Charlie began excavation and found a large boulder where one end of the house was supposed to be placed. The Owners and Charlie signed a change order that reconfigured the house so as to avoid the boulder.

As Charlie began to put up the house, the Owners asked for a few cosmetic changes, and Charlie made sure they signed change orders. However, as time went on this practice fell by the wayside. Moreover, the Owners became increasingly excited about the kind of house they wanted. They asked Charlie to build a swimming pool, which he agreed to do, although he warned them this would increase the price. The Owners were not deterred; indeed, their enthusiasm led them to ask for changes in the orientation of the kitchen; the kind of cabinets that would be hung there; an additional bathroom; and a pool house.

One day Charlie counted and discovered the Owners had made over fifty changes to the house, most small but some (like the pool, pool house, and kitchen orientation) quite large. He did not object, since the Owners kept paying his invoices.

One day the Owners came to the job site and told Charlie that after much thought they decided to add a second story to most of the house. Charlie said "this is not the house I agreed to build" and in response the Owners pointed to the changes clause, saying, "You have no choice but to do as we say!"

18.1 Changes: Differing Perspectives for Owner and Contractor

The generic term **changes** is used for two fundamentally different purposes. When initiated by the owner, the changes process is a means of achieving design flexibility without having to re-negotiate the contract. By contrast, a "changes" claim—brought by the contractor—seeks increased compensation and/or time within which to perform a project which now allegedly differs from what the contractor had originally agreed to build.

Owner's Perspective. After award of a construction project, the owner may find it necessary or desirable to order changes in the work. The contract documents are at best an imperfect expression of what the design professional and owner intend the contractor to execute. Circumstances develop during the construction process that may make it necessary or advisable to revise the drawings and specifications.

Design may prove inadequate. Methods specified become undesirable. Materials designated become scarce or excessively costly. From the owner's planning standpoint, the program or budget may change. Natural events may necessitate changes. For any of many reasons, it often becomes necessary to direct changes after the contract has been awarded to the contractor.

A contractual changes process is a method by which the owner may require the contractor to deviate from the design after performance has begun. Changes permit flexibility in the event modification of the original design requirements is necessary or desirable. This is done through inclusion of a "changes" clause in the prime contract.

> In this chapter's scenario, Charlie Contractor discovered a boulder where one end of the house was supposed to be placed. The parties used the changes process to avoid this obstruction.

But the owner also fears that the changes process can expose it to large cost overruns that may seriously disrupt its financial planning and capacity. In that sense, the changes process is an important element of cost control.

Owners and, perhaps more important, lenders fear a "loose" changes mechanism. They fear that bidders of questionable honesty and competence will bid low on a project with the hope that clever and skillful postaward scrutiny of the drawings and specifications will be rewarded by assertions that requested work is not required under the contract. A loose changes mechanism will generate claims for additional compensation. A changes clause, which gives either too much negotiating power to the contractor or too much discretion to the design professional, any **Initial Decision Maker** (IDM),[1] or the arbitrator, will convert a fixed-price contract into an open-ended, cost-type contract tied to a generous allowance for overhead and profit.

This is why owners and lenders want tight, complete specifications, a mechanical pricing provision (such as unit pricing), and limits on overhead and profit. Their horror is the prospect that the end of the project will witness a long list of claimed extras that, if paid, will substantially increase the ultimate construction contract payout.

Contractor's Perspective. While the owner clearly benefits from the flexibility of the changes process, the contractor needs assurance of adequate compensation for doing the new work and adequate time to fit the new work within the larger schedule of completing the project.

Although the contractor recognizes that some changes are inevitable in construction work, it fears it will not receive adequate compensation for changed work or for unchanged work that is affected by the change. It also worries that it will not receive a proper time extension or adequate compensation for delay or disruption caused by having to do work out of order. It is also concerned that poor administrative practices relating to changes will impede its cash flow and unduly burden its financial planning.

> In this chapter's scenario, Charlie Contractor insisted on including a changes clause in the contract. Contractor thus showed an awareness that such a clause can be to the builder's benefit. For example, the changes clause guaranteed not only additional compensation for Contractor for performing changed work, but also more time.

Contractors also express concern that unexpected changes may place a drain on their resources, divert capital they would like to use on other projects, and require more technical skills than they possess.

Contractors also complain about the unilateral nature of the changes clause. They must perform before they know how much they are going to receive in price and time adjustments. Although changes clauses usually provide a pricing mechanism through initial decisions by the design professional, the contractor may fear that the design professional will not grant fair compensation and time adjustment, especially when the change will reflect on the professional competence of the design professional.

The contractor may assert that a change has occurred—thus entitling it to additional compensation and/or time—notwithstanding the owner's failure to use the changes clause. In this situation, the contractor is alleging a change as a basis for a claim against the owner. More specifically, a contractor may assert either a *cardinal* change claim or a *constructive* change claim (discussed in Sections 18.5A and 18.5B, respectively).

18.2 The Common Law

Before examining a contract-based approach to pricing changed work, it is useful to see how the common law would resolve this question.

The absence of a changes clause with its pricing formulas can make pricing drastic changes difficult. This is demonstrated in *Chong v. Reebaa Construction Company*.[2] The original work was remodelling, and the contract price was $96,208. The contractor sought a written contract, but the owner-lawyer said they did not need one and should base their agreement on honor and trust. There was no written contract.

The owner made many changes and upgrades. Although warned that these changes would increase the contract price, the owner said money was no object, and he would pay what it cost. The owner made payments of $108,000. The contractor billed the owner for an additional $128,982.

The jury found for the contractor based upon a breach of the contract. But the Georgia court of appeals held that the contractor could not recover for breach of contract as there had been no agreement on price. But it did allow the contractor to recover in *quantum meruit* based on the reasonable value of the work. The owner further appealed.

The Georgia Supreme Court disagreed with the appellate court. It concluded that a breach of contract had been established by the jury, because the contract was not so uncertain as to be unenforceable.[3] Uncertainty could be cured by subsequent words or actions of the parties. When the owner requested upgrades, the contractor warned him that they would increase the project cost. The owner stated that money was no object, and he should be billed for the additional work.

The *Chong* case teaches that the absence of a changes clause can create great difficulty when, as is common, the work is changed. (Although not mentioned by the supreme court, honor and trust can plunge into bitter acrimony, which is not an uncommon event in contracts for home construction.)

18.3 The Changes Clause: Introduction

Terminology. A *change* is the term used in construction contracts that allows the owner to unilaterally direct changes to be made *without* obtaining the contractor's consent to perform the work. There are three categories of changes:

1. those that add work
2. those that delete work
3. those that substitute one item of work for another item of work

The third category might refer to replacing one piece of equipment with another by a different manufacturer. This chapter focuses on the first two categories of changes.

Changes must be contrasted with **modifications**. Modifications are two-party agreements in which owner and contractor mutually agree to change portions of the work. One way to avoid formal requirements in a changes clause is to conclude that the work in question was a modification agreed on by the parties and not a change.

Not uncommonly, the change order mechanism is disregarded by the parties. This may be because of unrealistically high expectations by the drafters, time pressures, or unwillingness by the parties to make and keep records. When this happens, the concept of *waiver* becomes paramount in the contractor's effort to be paid. The contractor argues that the owner by his actions—demanding changes but not using written change orders—manifested an intention to dispense with some of the contract's formal requirements. Waiver is discussed in Section 18.4G.

AIA and EJCDC Documents. The American Institute of Architects (AIA) has a complex changes mechanism, found in A201-2007, Article 7. Changes may be accomplished three ways: by **change order** (CO), by **construction change directive** (CCD), or by an order for a minor change in the work.

A change order happens when the owner, architect, and contractor agree in advance on the work to be changed, the price to be changed, and any adjustment in the contract time. A CO looks very much like a contract modification.

A CCD is an order to perform changed work without the agreement of the contractor as to an adjustment in the price, time, or both. A201-2007 § 7.3.3 provides a mechanism for adjusting the contract price.

The architect may order minor changes in the work that do not affect the contract price or time. The minor change must not be inconsistent with the intent of the contract documents.

The AIA uses a CCD to describe what most contracts call a change order. This chapter uses traditional terminology, such as the changes process and the changes clause. However, when the AIA documents are discussed, AIA terminology is used.

The Engineers Joint Contracts Document Committee's (EJCDC) changes mechanism is found in the General Conditions, Doc. C-700, Article 10. Article 10.1(A) allows the owner to order "additions, deletions, or revisions in the Work" using a change order. The change order is signed by the owner and contractor and spells out the change in the work, the contract price, and the contract time. If the contractor disagrees with the price or time adjustment, it files a claim with the engineer. Further appeals from the engineer's decision are in accordance with the dispute resolution process specified in Article 16.

Shifts in Bargaining Power. To appreciate the centrality of the changes process to construction, the shifts in bargaining power because of changes must be understood.

When preparing to engage a contractor, the owner, as a rule, has superior bargaining power. The hotly competitive construction industry and the frequent use of competitive bidding usually allow the owner to control many aspects of the construction contract terms.

The contractor who is performing moves into a much stronger bargaining position. This is clearly so if the owner *must,* for either practical or legal reasons, order any additional work from the contractor. It would be in an even stronger bargaining position if it could refuse to execute the change unless there were a mutually satisfactory agreement on the effect of the change on price or time.

Yet any bargaining advantages to the contractor by being in the position to refuse to do the work until there is an agreement are usually tempered by contract provisions that require the contractor to do the work even if there is no agreement on the price or time. This can be made even worse by the dominance some owners have over the changes process through their control over the purse strings. This power, exercised either directly or through the design professional to withhold payment until the contractor agrees on price and time, can exert intense pressure on the contractor to accept whatever the owner or design professional is willing to pay.

The changes mechanism can operate adversely to the contractor if the owner makes many small changes but is niggardly in her proposals for adjusting the price. (This can backfire, however, generating claims by the contractor, particularly in a losing contract, that the cumulative effect of any changes has created a cardinal change.)

To sum up, the changes mechanism on the whole favors the owner. (An exception would be if the owner is dealing with a clever, claims-conscious contractor who uses the changes mechanism to extract large amounts of money at the end of the job.) Judicial resolution of disputes that involve changes may take into account the owner's strong position, particularly if the owner seems to have abused its power.

An Illustrative Case. Before examining the changes clause in greater depth, the following case illustrates the legal issues that arise when the contract includes a changes clause which the parties followed in a haphazard fashion. How does the court sort out what changes (which the court calls "extras") were agreed to, which were not, and what are the value of those changes? More importantly, perhaps: Who has the burden of proving these matters?

WATSON LUMBER COMPANY v. GUENNEWIG
Appellate Court of Illinois, 79 Ill.App.2d 377, 226 N.E.2d 270 (1967)

EBERSPACHER, Justice.
[Ed. note: Footnotes renumbered and some ommited]

The corporate plaintiff, Watson Lumber Company, the building contractor, obtained a judgment for $22,500.0 in a suit to recover the unpaid balance due under the terms of a written building contract, and additional compensation for extras, against the defendants William and Mary Guennewig. Plaintiff is engaged in the retail lumber business, and is managed by its president and principal stockholder, Leeds Watson. It has been building several houses each year in the course of its lumber business.

* * *

[Ed. note: The project was a four-bedroom, two-bath house with air-conditioning for a contract price of $28,206. The total amount claimed as extras and awarded by the trial court was $3,840.09.]

The contractor claimed a right to extra compensation with respect to no less than 48 different and varied items of labor and/or materials. These items range all the way from $1.06 for extra plumbing pieces to $429.00 for an air-conditioner larger than plaintiff's evidence showed to be necessary, and $630.00 for extra brick work. The evidence, in support of each of these items and circumstances surrounding each being added, is pertinent to the items individually, and the evidence supporting recovery for one, does not necessarily support recovery for another.

* * *

Most of the extras claimed by the contractor were not stipulated in writing as required by the contract. The contractor claims that the requirement was waived. Prior to considering whether the parties, by agreement or conduct dispensed with the requirement that extras must be agreed to in writing, it should first be determined whether the extras claimed are genuine "extras." We believe this is an important area of dispute between these parties.

Once it is determined that the work is an "extra" and its performance is justified, the cases frequently state that a presumption arises that it is to be paid for. . . .

No such presumption arises, however, where the contractor proceeds voluntarily; nor does such a presumption arise in cases like this one, where the contract makes requirements which any claim for extras must meet.

* * *

The law assigns to the contractor, seeking to recover for "extras," the burden of proving the essential elements. . . . That is, he must establish by the evidence that (a) the work was outside the scope of his contract promises; (b) the extra items were ordered by the owner, . . . (c) the owner agreed to pay extra, either by his words or conduct, . . . (d) the extras were not furnished by the contractor as his voluntary act, and (e) the extra items were not rendered necessary by any fault of the contractor . . .

The proof that the items are extra, that the defendant ordered it as such, agreed to pay for it, and waived the necessity of a written stipulation, must be by clear and convincing evidence. The burden of establishing these matters is properly the plaintiff's. Evidence of general discussion cannot be said to supply all of these elements.

The evidence is clear that many of the items claimed as extras were not claimed as extras in advance of their being supplied. Indeed, there is little to refute the evidence that many of the extras were not the subject of any claim until after the contractor requested the balance of the contract price, and claimed the house was complete. This makes the evidence even less susceptible to the view that the owner knew ahead of time that he had ordered these as extra items and less likely that any general conversation resulted in the contractor rightly believing extras had been ordered.

In a building and construction situation, both the owner and the contractor have interests that must be kept in mind and protected. The contractor should not be required to furnish items that were clearly beyond and outside of what the parties originally agreed that he would furnish. The owner has a right to full and good faith performance of the contractor's promise, but has no right to expand the nature and extent of the contractor's obligation. On the other hand, the owner has a right to know the nature and extent of his promise, and a right to know the extent of his liabilities before they are incurred. Thus, he has a right to be protected against the contractor voluntarily going ahead with extra work at his expense. He also has a right to control his own liabilities. Therefore, the law required his consent be evidenced before he can be charged for an extra ... and here the contract provided his consent be evidenced in writing.

The amount of the judgment forces us to conclude that the plaintiff contractor was awarded most of the extra compensation he claims. We have examined the record concerning the evidence in support of each of these many items and are unable to find support for any "extras" approaching the $3,840.09 which plaintiff claims to have been awarded. In many instances the character of the item as an "extra" is assumed rather than established.[4] In order to recover for items as "extras," they must be shown to be items not required to be furnished under plaintiff's original promise as stated in the contract, including the items that the plans and specifications reasonably implied even though not mentioned. A promise to do or furnish that which the promisor is already bound to do or furnish, is not consideration for even an implied promise to pay additional for such performance or the furnishing of materials. The character of the item is one of the basic circumstances under which the owner's conduct and the contractor's conduct must be judged in determining whether or not that conduct amounts to an order for the extra.

The award obviously includes items which Watson plainly admits "there was no specific conversation." In other instances, the only evidence to supply, even by inference, the essential element that the item was furnished pursuant to the owner's request and agreement to pay is Mr. Watson's statement that Mrs. Guennewig "wanted that." No specific conversation is testified to, or fixed in time or place. Thus it cannot be said from such testimony whether she expressed this desire before or after the particular item was furnished. If she said so afterward, the item wasn't furnished on her orders. Nor can such an expression of desire imply an agreement to pay extra. The fact that Mrs. Guennewig may have "wanted" an item and said so to the contractor falls far short of proving that the contractor has a right to extra compensation.

* * *

If the construction of an entire work is called for at a fixed compensation, the hazards of the undertaking are assumed by the builder, and he cannot recover for increased cost, as extra work, on discovering that he has made a mistake on his estimate of the cost, or that the work is more difficult and expensive than he anticipated. 17A C.J.S. Contracts § 371(6), p. 413

> Some so called "extras" were furnished, and thereafter the owner's agreement was sought.[5] Such an agreement has been held to be too late. . . .
>
> The [trial court] judge, by his remarks at the time of awarding judgment, shows that the definition of extras applied in this case was, indeed, broad. He said,
>
>> substantial deviation from the drawings or specifications were made—some deviations in writing signed by the parties, some in writing delivered but not signed, but nevertheless utilized and accepted, some delivered and not signed, utilized and not accepted, some made orally and accepted, some made orally but not accepted, and some in the trade practice accepted or not accepted
>
> While the court does not state that he grants recovery for all extras claimed, he does not tell us which ones were and were not allowed. The amount of the judgment requires us to assume that most were part of the recovery. It can be said with certainty that the extras allowed exceeded those for which there is evidence in the record to establish the requirements pointed out.
>
> Mere acceptance of the work by the owner as referred to by the court does not create liability for an extra. . . . In 13 Am.Jur.2d 60, "Building & Cont." § 56, it is stated that "The position taken by most courts considering the question is that the mere occupancy and use do not constitute an acceptance of the work as complying with the contract or amount to a waiver of defects therein." Conversation and conduct showing agreement for extra work or acquiescence in its performance after it has been furnished will not create liability. . . . More than mere acceptance is required even in cases where there is no doubt that the item is an "extra" . . . .
>
> The contractor must make his position clear at the time the owner has to decide whether or not he shall incur extra liability. Fairness requires that the owner should have the chance to make such a decision. He was not given that chance in this case in connection with all of these extras. Liability for extras, like all contract liability, is essentially a matter of consent; of promise based on consideration. . . .
>
> * * *
>
> The contractor claims that the requirement of written stipulation covering extras was waived by the owner's conduct. The defendants quite agree that such a waiver is possible and common but claim this evidence fails to support a waiver of the requirement. There are many cases in which the owner's conduct has waived such a requirement. . . . In all the cases finding that such a provision had been waived thus allowing a contractor to collect for extras, the nature and character of the item clearly showed it to be extra. Also, in most cases the owner's verbal consent of request for the item was clear beyond question and was proven to have been made at the time the question first arose while the work was still to be finished. The defendants' refusal to give a written order has in itself been held to negative the idea of a waiver of the contract requirements for a written order. . . . We think the waiver of such a provision must be proved by clear and convincing evidence and the task of so proving rests on the party relying on the waiver. . . .
>
> [Ed. note: The court ordered a new trial, stating that the contractor could recover only for those extras he could prove were ordered as such by the owner in the proper form, unless he could show that the owner waived the requirements of a writing by clear and convincing evidence.]

Undoubtedly, the court is correct in emphasizing the owner's right to know whether particular work will be asserted as extra. The court's concern appears particularly justified in light of the contractor's apparent abuse of the changes mechanism. For example, in the court's (renumbered) footnote 5, it points out that the contractor had installed more expensive tile then (after the fact) got the owner to agree to pay for one-half of the cost of the more expensive material.

However, in a small construction project, the Illinois court's requirements would place an inordinate administrative burden on the contractor. It would not only have to make clear its position that particular work was extra but also obtain the written change order executed by the owner. If the contractor did not obtain a written change order, it might be able to assert the doctrine of waiver if it could persuade the court that the work was extra and that the owner was made aware of this and allowed the work to proceed.

Rather than comply with the excessively formal requirements set forth by the court in the *Watson* case, a contractor in a small project, such as that involved in *Watson*, might add a contingency in the contract price to cover small extras that are very likely to be requested.

The *Watson* case shows that courts often ignore the cost of complying with rules of law. Undoubtedly, larger projects will bear the administrative costs of doing things correctly and "according to the book." But smaller jobs may not permit "by the book" contract administration.

18.4 Change Order Mechanism

18.4A Components of the Changes Process

From a planning standpoint, a changes mechanism is essential. Design flexibility and cost control cannot be accomplished without a system for changing work.

The changes process involves the following:

1. The exercise of a power to order a change or a direction the contractor contends is a change in the work.
2. Methods for the contract or the parties to price the change and its effect on time requirements.
3. A residuary provision that controls the price in the event the parties do not agree.

18.4B Limitation on Power to Order Changes

Although a changes clause is essential to construction projects, it is—as suggested earlier in this chapter—subject to abuse. A contractor can expect some changes, but the changes clause should not be a blank check for the owner to order the contractor to do anything it wishes, to compel the contractor to perform before there has been an agreement on price, and to take care of compensation later. This behavior can be even more abusive if the owner seeks to avoid any increased cost of performing unchanged work or any cost incident to disruption of the contractor's method of performance.

Work. The California Supreme Court has addressed balancing the need for a changes mechanism with a contractor's objection to performing additional or changed work in the absence of an agreed-upon price. In *Coleman Engineering Co. v. North American Aviation, Inc.*,[6] a contract to design and manufacture equipment contained a changes clause allowing the buyer to make changes to the design which would entitle the seller to "an equitable adjustment in price and time of performance mutually satisfactory to Buyer and Seller." The court characterized the changes clause as "an agreement to agree in the future as to price." It ruled that whether a contractor (faced with changed work) is obligated to perform in the absence of an agreed price—or may instead refuse to perform and abandon the project—depends on the magnitude of the requested change:

> Undoubtedly, if the subsequent changes are minor or of not great magnitude, the contractor must perform and obtain a subsequent judicial determination as to the price of the changes. However, where the changes are of great magnitude in relation to the entire contract, the contractor must negotiate in good faith to settle the price . . ., and where he has done so, he is not required to continue performance in the absence of an agreement as to the price.[7]

In sum, in the absence of an agreed-upon price, whether the contractor must perform depends on the magnitude of the change in comparison to the contract as a whole. A contractor must perform a minor change and seek compensation later or be in breach of contract. However, if the requested change is so great as to alter the nature of the contract work, the contractor would be within its rights to refuse performance in the absence of an agreement as to price. The *Coleman* court found a cardinal change (see Section 18.5A) existed where the changed work would have cost at least $257,000 on a $527,632 contract.

In this same vein, AIA Document A201-2007, § 7.3.1 provides that the owner may "order changes in the Work within the general scope of the Contract consisting of additions, deletions or other revisions." For example, an owner may be able to order a 10 percent increase in the floor space of a residential home or that a carport be built or even a swimming pool, but not that a beach house be built twenty miles away.

> In this chapter's scenario, Charlie Contractor refused the Owners' request to add a second story to the house. Whether that additional work is outside the general scope of the contract would require a factual inquiry. The Owners would argue that Charlie was hired to build a house, and he is simply being asked to build the same house but larger.
>
> Charlie would counter that adding a second floor would require changing not only the substructure but also the plumbing, heating, ventilation, and air conditioning work. Charlie could cite the *Coleman* decision and argue that he was not obligated to perform the change because the cost would be half (if not more) of the original contract price.

Time. While the previous discussion focuses on changes in the work, may the owner use the changes clause to shorten the time of contract performance? Otherwise stated, if the owner points to the changes clause and directs that the contractor accelerate (speed up) its performance, must the contractor comply? The federal procurement system speaks directly to this point, giving the federal agency the power to order acceleration.[8]

AIA Document A201 does not address this question directly. As noted, § 7.3.1 simply provides that the owner may "order changes in the Work . . . consisting of additions, deletions or other revisions." "Work" is defined in § 1.1.3 as "the construction and services required by the Contract Documents." This definition contemplates the physical structure and not the time during which it must be completed. In light of the drastic nature of acceleration, A201 should not empower the owner to direct an acceleration. A201 should be interpreted to allow only quantity and quality changes.

Public Contracts. Special questions apply in the case of public works contracts. One initial question as to whether the additional work must be awarded by competitive bidding rather than as a change. This issue comes up frequently when there is a judicial challenge to the ordering of changed work to the contractor performing the contract.[9]

18.4C Authority to Order Change

In the absence of any contract clause dealing with the question of authority to make changes, doctrines of agency control the question of who can order a change. The owner can order changes. Members of the owner's organization may also have the authority to order changes expressly, impliedly, or by the doctrine of apparent authority. Carefully thought-out construction contracts usually specify which members of the owner's organization have the power to order changes.[10]

As an illustration, AIA Document A201-2007, § 7.2.1, states that a change order is prepared by the architect and signed by the architect, the contractor, and the owner. By contrast, § 7.3.1 states that the construction change directive is prepared by the architect but signed only by the architect and the owner. Finally, for a minor change, § 7.4 requires a written order signed only by the architect.

Absent express contract authorization, neither the design professional,[11] the construction manager, nor, surely, the project representative has inherent authority by virtue of his position to direct changes in the work. While a contractor is presumed to know the requirements for obtaining change orders, these owner representatives should not act as if they have "apparent authority" (see Section 2.11D) to order changes when they do not.

Yet facts matter. In *Tupelo Redevelopment Agency v. Gray Corporation, Inc.*,[12] the Mississippi Supreme Court pointed to the conduct of the project's construction manager (CM) to find that a public agency waived the contract's written change order requirement. The court noted that the agency's CM emphasized to the contractor that time was of the essence, the agency had already issued three change orders after the contractor had begun performance of the additional work, and the additional work stemmed from design defects and so was unanticipated.

On federal projects, the contracting officer alone has the authority to enter into and modify contracts.[13]

Sometimes there is insufficient time to obtain authorization from the owner for work needed immediately, because there is impending danger to person or property. The contractor should have authority to do such emergency work. Contracts frequently provide that extra work in emergencies can be performed without authorization; see AIA Doc. A201-2007, § 10.4. Even without express or implied authority, the contractor should be able to recover for emergency work based on the principle of unjust enrichment.

18.4D Misrepresentation of Authority

If the contractor performed the additional work at the order of the design professional and this order was beyond the latter's actual authority, what recourse does the contractor have?

First, the contractor would seek to establish that the design professional had apparent authority to order the work (see Section 2.11D). But in the absence of the owner's having led the contractor to believe that the design professional had this authority, the contractor will not be successful. Likewise, any claim that the work has unjustly enriched the owner would not be successful. The contractor would be considered a volunteer and denied recovery.

Next, the contractor would look to the design professional. The threshold question would be whether the contractor reasonably relied on misrepresented authority. In many cases, reliance would not be reasonable, because design professionals typically are not given this authority and because the contractor should have checked with the owner. But the reliance element is often minimized or even ignored. In any event, if compliance with the order had been reasonable under the circumstances, the contractor can recover the cost of performing the unauthorized work (and any cost of correction made necessary to make the work conform to the contract documents) from the design professional. The design professional would be liable for misrepresentation of her authority.

An owner who knows of the unauthorized conduct of its agent but who then expressly approves the agent's decision ratifies the decision. The owner can no longer claim that the agent's decision is not binding on it.

18.4E Pricing Changed Work

Pricing changed work is an important part of the changes mechanism. A tightly drawn pricing formula, along with clear and complete contract documents, can discourage a deliberately low bid made with the intention of asserting a long list of claims for extra work.

Where possible, work should be compensated by any unit prices specified in the agreement. If no applicable unit price is specified in the contract, compensation should be based on an analogous unit item, taking into account the difference between it and the required work. It is also important, however, to take into account the possibility of great variations in units of work requested and the effect on contractor costs (see Section 14.5). In fact, A201-2007, § 7.3.4 authorizes adjustment in the unit price in order to avoid "substantial inequity" to the owner or contractor.

The AIA provides one example of how to price changed work. When the contractor receives the construction change directive (CCD), under § 7.3.5, it must proceed with the change in the work and indicate whether it agrees or disagrees with the method set forth in the CCD of adjusting price or time. If it agrees, under § 7.3.6 it signs the CCD. If it does not respond promptly or disagrees with the proposed method of adjustment, under § 7.3.7 the price adjustment will be determined by the architect "on the basis of reasonable expenditures and savings . . . including, in the case of an increase in the Contract Sum an amount for overhead and profit."

While § 7.3.7 addresses the issue of price, § 7.3.10 makes clear that the architect may authorize adjustment to both contract price and contract time. (But this allows the architect to increase time—not decrease it—and causes the contractor to suffer an acceleration; see Section 18.4B.)

Some contracts, including A201-2007, § 7.3.7, provide that if the parties cannot agree, the contractor will be paid cost plus a designated percentage of cost in lieu of overhead and profit. Pricing extra work in this fashion can discourage the contractor from reducing costs. However, it may be difficult to arrive at a fixed fee for overhead and profit when the nature of the extra work cannot be determined until the work is ordered. For this reason, cost plus a percentage of cost will probably continue to be used to price changed work.

When a cost formula is used, the design professional must examine the cost items to see whether they are reasonable and required under the change order. The changes clause can specify—as does A201-2007, § 7.3.7—that only material, equipment, and labor or direct overhead costs are to be included as cost items.

Federal contracts frequently provide that if the government and contractor cannot agree, the contractor is entitled to an "equitable adjustment" of the contract price and time.[14] An equitable adjustment compensates the contractor for the cost impact of the change. As phrased by a Board of Contract Appeals in *Modern Foods, Inc.*, it is "the difference between what it would have reasonably cost to perform the work as originally required and what it reasonably cost to perform the work as changed."[15] Similarly, in *Hensel Phelps Construction Co. v. United States*,[16] a federal court of appeals found a subcontractor was entitled to an equitable adjustment, meaning that the subcontractor "is entitled to damages in an amount equal to the reasonable value of services and materials, including profit and overhead, for which they could be obtained under like circumstances."[17]

18.4F Deductive Change (Deletion)

Changes clauses usually permit the owner to delete a portion of the work, sometimes known as a *deductive change*. Although as a rule a deductive change is lumped together with changes that add to the contractor's contractual commitment, it raises special problems. First, the power to change the work typically is limited to those changes that fall within the scope of the work. Clearly a deductive change cannot be measured by that standard. Second, the prime contractor may recommend to the owner a deductive change that largely eliminates a subcontractor's work. As discussed in Section 16.2B, this recommendation may open up the contractor to a claim of bad faith brought by the subcontractor.

Similarly, suppose the owner wishes to delete part of the work and offer it to another contractor. Unless the owner had good reason to do so or at least acted in good faith, this would be a breach of the contract.

18.4G Waiver: Excusing Formal Requirements

Suppose the directed work comes within the scope of the changes clause and the person ordering the work had authority to do so. What happens if the parties do not comply with the contract clause's formal requirements? Most commonly, this becomes an issue when the changes clause requires the change order to be in writing, but the contractor performed based on oral instructions, and the owner is now refusing to pay for the cost of the changed work.

When the owner denies contractor claims for extras because of the absence of a written change order, contractors frequently assert the contract's formal requirements have been waived. Waiver questions can be divided into three issues:

1. Is the requirement waivable?
2. Who has the authority to waive the requirement?
3. Did the facts claimed to create waiver lead the contractor to reasonably believe that the requirements have been eliminated?

Except where the waiver concept cannot be applied to public contracts,[18] formal requirements can be waived. They are not considered an important element of the exchange and are often viewed as simply technical requirements.

As a rule, only parties who have authority to order changes have authority to waive the formal requirements. Usually only the owner or its authorized agent can waive the formal requirements, although a design professional with authority to order changes should have the authority to waive the writing requirement.

Waiver generally can be based on acts such as conduct by the owner that indicated no written change order would be required[19] or oral orders by the owner or by its authorized representative.[20] Generally, if the owner apparently gave oral orders for the changed work, the court will assume the writing requirement has been waived.

Another factor that must be taken into account is whether the owner or design professional knew of the facts that would be the basis for a the contractor's claim of a change. For example, in *Moore Construction Co. v. Clarksville Department of Electricity*,[21] one issue was whether one separate contractor could claim delay damages caused by another separate contractor. However, the claimant did not give notice of intention to claim additional compensation as required by the predecessor of AIA Document A201-2007, § 15.1.4. The court held this notice could be waived. It observed that all of the participants knew that the claimant was being delayed and that the delays were not the claimant's fault. The claimant had been given a time extension through a job site memo, but no format change order had been issued.

The court noted that the claimant could have reasonably believed that the owner would not demand that the claimant submit a written notice of a claim for additional compensation because it had been granted a time extension. In addition, throughout this project, the requirement that the contractor submit written notices had not been followed. (In fact, most of the formal requirements appear to have been abandoned.) In addition, the court concluded the owner had not shown itself to be prejudiced by the contractor's failure to give the required written notice.

The owner's payment for changed work ordered by the design professional, a resident engineer, or even project representative can be evidence of waiver. Payment can lead the contractor to reasonably believe the design professional has the authority to order changes orally.[22] However, if payment is accompanied by a notice making clear to the contractor that the formal requirements are not being waived, no waiver should be found.

A contractor's best practice is to be aware of and comply with the change order mechanism. The fallback defense of waiver is far less reliable a path to compensation for additional work.

> In this chapter's scenario, the parties initially used the changes process diligently and then stopped doing so. The Owners were actively involved in the construction, repeatedly orally ordered changes which they saw Contractor implement, and paid all invoices. A court would find that the Owners waived insistence on the contract's written change order requirement.

18.5 Contractor's "Changes" Claims

Suppose the contractor believes it had performed extra work at the owner's bequest, but the change order mechanism was not invoked by the owner. If the contractor can establish that it was directed to perform changed work and did so, it would be entitled to some form of compensation—whether applying the contract price or a "reasonable value" measurement. In this situation, the contractor is asserting a "changes" claim against the owner.

18.5A Cardinal Change

Even if a prime contract contains a changes clause, this does not mean any change can be ordered. As shown in *Coleman Engineering Co. v. North American Aviation, Inc.*,[23] an owner may not use the changes clause to order a change beyond the general scope of the contract—"where the changes are of great magnitude in relation to the entire contract," in the words of the California court. A direction that goes beyond the scope of the work need not be obeyed. However, if the contractor does perform, the cardinal change doctrine provides it with a theory of recovery.

Cardinal change is invoked in two situations: a "one-shot" scope change or (more commonly) a "nibbling" or "aggregate of changes" claim. Focusing first on a one-shot change, suppose the contractor complies. Now there has been an agreement. But if the new work does not fall within the scope of the changes clause with its procedural and pricing mechanisms and if the parties cannot agree on price, the contractor is entitled to reasonable compensation.[24]

Another type of cardinal change does not occur all at once, as in the one-shot direction, but occurs throughout the performance of the contract. It consists of many changes,[25] drastic changes,[26] or other conduct that has gone beyond the reasonable expectations of the contractor and made the transaction different from what the parties had in mind when they made the contract.[27] This is a breach.

The cardinal change doctrine was first developed in federal procurement law. Some states recognize the doctrine; others do not.[28]

18.5B Constructive Change

A constructive change occurs when the contract work is actually changed but the procedures of the changes clause were not followed. A constructive change consists of two components: an actual change in performance from what is required by the contract coupled with some order or directive from the owner to perform that work. The doctrine allows a contractor, who performed extra work in reliance upon owner directive or contract interpretation, to receive an equitable adjustment.

Both the cardinal change and constructive change doctrines originated in federal procurement. The U.S. Court of Federal Claims in *Metric Construction Co., Inc. v. United States*, recently observed that constructive change has been applied to: "(I) disputes over contract interpretation during performance; (II) Government interference or failure to cooperate; (III) defective specifications; (IV) misrepresentation and nondisclosure of superior knowledge; and (V) acceleration."[28] A federal board of contract appeals, in *Beyley Construction Group Corp. v. Department of Veterans Affairs*, ruled that a deductive change order (see Section 18.4F), which deleted the contractor's anticipated source of fill material, was a constructive change.[29]

How does the AIA handle constructive change situations? Suppose that the design professional and contractor dispute over work to be done. The design professional contends the work is called for under the contract, and the contractor claims it is not. Suppose the contractor performs the work but makes clear that he considers the position of the design professional unjustified and that he intends to claim additional compensation.

Here, as in so many other aspects of the construction process, it is important to recognize the design professional's power to interpret the document and his quasi-judicial role if he is the Initial Decision Maker (IDM). If the design professional determines that the work falls within the contract, many construction contracts make his decision final unless it is overturned by arbitration or litigation. If the contractor later requests additional compensation for the work and is met with the contention that no change order has been issued, the absence of a written change order should not bar the contractor's claim—as long as he made clear that he intended to claim additional compensation.

That said, the contractor's claim should be denied if the design professional or the IDM's decision is binding and has not been overturned. A contractor who is dissatisfied with the design professional's decision should invoke any process under which that decision can be appealed. If it later is determined that the design professional or IDM had been incorrect and the decision is overturned, the absence of a written change order should not bar the claim.

18.6 Effect of Changes on Performance Bonds

Most surety bonds provide that modifications made in the basic construction contract will not discharge the surety. Some changes clauses permit changes up to a designated percentage of the contract price without notifying the surety. EJCDC C-700, Article 10.4, makes the contractor responsible for notifying the surety of any changed work if the bond requires notice be given.

REVIEW QUESTIONS

1. What is the fundamental difference between changes that are initiated by the owner and a changes claim brought by a contractor?

2. What are the three categories of changes?

3. What are the three ways changes are accomplished according to the AIA A201-2007 Document, Article 7?

4. In the case *Watson Lumber Company v. Guennewig*, what were the essential elements the contractor had to prove to win his claim of changes?

5. What strategy might a contractor employ on a small project to cover small extras, rather than comply with the excessively formal requirements set forth by the court in the *Watson* case?

6. What three things are involved in the changes process?

7. On federal projects, which party has authority to enter into and modify contracts?

8. What are the three issues in regard to waiver questions?

9. What are the two situations in which a cardinal change can be invoked?

10. What are the two components of a constructive change?

ENDNOTES

[1] The IDM is a new position created by the AIA in A201-2007. If no separate person is designated as the IDM, then the architect is the IDM. See Section 26.1.
[2] 284 Ga.App. 830, 645 S.E.2d 47 (2007).
[3] *Reebaa Constr., Inc. v. Chong*, 283 Ga.222, 657 S.E.2d 826 (2008).
[4] We cite as some examples: An extra charge was made for kitchen and bathroom ceilings, concerning which Watson testified that he was going to give these as gifts "if she had paid her bill." According to

the testimony, the ceilings were lowered to cover the duct work. We consider it unlikely that the parties intended to build a house without duct work or with duct work exposed. Likewise an "extra" charge was made for grading, although the contract clearly specifies that grading is the contractor's duty. An "extra" charge is sought for enclosing the basement stairs, although the plans show the basement stairs enclosed. An extra charge is sought for painting, apparently on the basis that more coats were necessary than were provided in the contract.

[5] The drain tile around the foundation of the house, according to the evidence, was already in place when the owner learned that it was more expensive material. Only then did the contractor secure the owner's consent to pay for one-half the cost of the more expensive material.

[6] 65 Cal.2d 396, 420 P.2d 713, 55 Cal.Rptr. 1 (1966).

[7] 420 P.2d at 720.

[8] 48 CFR § 52.243–4(a)(4) (2012).

[9] *Bozied v. City of Brookings,* 638 N.W.2d 264 (S.D.2001); Montez, *Legislative Update: Illinois' Public Works Contract Change Order Act and Similar Statutes from Other Jurisdictions,* 25 Constr. Lawyer, No. 1, Winter 2005, p. 40.

[10] See Section 14.8B, which addresses the general question of authority on a construction project.

[11] *F. Garofalo Elec. Co., Inc. v. New York Univ.,* 270 A.D.2d 76, 705 N.Y.S.2d 327, leave to appeal dismissed, 95 N.Y.2d 825, 734 N.E.2d 762, 712 N.Y.S.2d 450 (2000).

[12] 972 So.2d 495 (Miss.2007).

[13] *Winter v. Cath-dr/Balti Joint Venture,* 497 F.3d 1339 (Fed.Cir.2007).

[14] 48 CFR § 52.243-1(b) (2012).

[15] ASBCA No. 2090, 57-1 BCA ¶ 1229.

[16] 413 F.2d 701 (10th Cir.1969).

[17] Id. at 704.

[18] *P & D Consultants, Inc. v. City of Carlsbad,* 190 Cal.App.4th 1332, 119 Cal.Rptr.3d 253 (2010); *American Safety Casualty Ins. Co. v. City of Olympia,* 162 Wash.2d 762, 174 P.3d 54 (2007); Annot., 1 A.L.R.3d 1273 (1965).

[19] *Precision Mechanical Services, Inc. v. Shelton Yacht & Cabana Club, Inc.,* 97 Conn.App. 258, 903 A.2d 692, certification denied, 280 Conn. 928, 909 A.2d 524 (2006); *Huang Int'l, Inc. v. Foose Constr. Co.,* 734 P.2d 975 (Wyo.1987); Annot., 2 A.L.R.3d 620 (1965).

[20] *City of Mound Bayou v. Roy Collins Constr. Co.,* 499 So.2d 1354 (Miss.1986).

[21] 707 S.W.2d 1 (Tenn.App.1985).

[22] *Oxford Dev. Corp. v. Rausauer Builders, Inc.,* 158 Ind.App. 622, 304 N.E.2d 211 (1973).

[23] Supra note 6. See also Silberman, *Abandonment and Cardinal Change Claims on "Projects from Hell,"* 25 Constr. Lawyer, No. 4, Fall 2005, p. 18.

[24] *Nat Harrison Assoc., Inc. v. Gulf States Util. Co.,* 491 F.2d 578, rehearing denied, 493 F.2d 1405 (5th Cir. 1974).

[25] *Wunderlich Contracting Co. v. United States,* 240 F.2d 201 (10th Cir.1957) (6,000 changes).

[26] *Saddler v. United States,* 152 Ct.Cl. 557, 287 F.2d 411 (1961) (doubling excavation in small contract).

[27] *Allied Materials & Equip. Co. v. United States,* 215 Ct.Cl. 406, 569 F.2d 562 (1978). *Wunderlich Contracting Co. v. United States,* 173 Ct.Cl. 180, 351 F.2d 956 (1965) (cumulative effect of magnitude and quality of changes).

[28] 80 Fed.Cl. 178, 185 (2008), quoting *Miller Elevator Co. v. United States,* 30 Fed.Cl. 662, 668 (1994). (A constructive acceleration, the last category listed by the *Metric* court, is discussed in Section 20.14B.)

[29] CBCA Nos. 5 & 763, 07-2 BCA ¶ 33,639.

CHAPTER 19

SUBSURFACE PROBLEMS: PREDICTABLE UNCERTAINTY

SCENARIO

Mega Manufacturer, Inc. (MMI) decided to build a new factory. It hired Awesome Architects, Inc. (AAI), an architectural firm, to create the design. AAI retained a geotechnical engineering firm, Geo-R-Us, Inc., to perform a soil report. AAI used that soil report to design the excavation and foundation specifications.

MMI used the completed design to solicit fixed-price bids from a variety of prime contractors. The design included the excavation requirements and foundation specifications. The advertisement advised that the geotechnical report was available for review at the office of Geo-R-Us and was provided "for information only."

Prime Builder, Inc. decided to bid on the job. Its president, Mr. Prime, visited the site, which was an empty lot. Mr. Prime was mostly concerned with the logistics of working in the remote location. He examined the foundation design and saw no obvious conflict with the surface soil indications (he saw no boulders or swampy areas, for example), but he did not go to the office of Geo-R-Us to examine the original geotechnical report.

MMI entered into a fixed-price contract with Prime Builder, Inc., using an American Institute of Architects (AIA) General Conditions of the Contract for Construction, A201-2007. Prime Builder began the excavation work and discovered far more boulders than anticipated. MMI refused to grant Prime Builder additional compensation or time, stating that the construction was Prime Builder's responsibility.

Within a couple of years, the completed factory developed cracks and sticking doors. When the contractor refused to make repairs, MMI sued it for breach of contract and negligence. Prime Builder then brought a third-party complaint against Geo-R-Us. Prime Builder introduced expert testimony as to the causes of the failures.

A geotechnical expert testified that the Geo-R-Us report was defective because too few boring logs were drilled. Specifically, Geo-R-Us failed to take borings from one corner of the lot which contained expansive soil. The foundation designed by AAI was inadequate to deal with this type of soil.

19.1 Discovery of Unforeseen Conditions

The discovery of unforeseen subsurface conditions is not unusual in the construction process. A subsurface condition on a renovation project refers to the existing building, such as the discovery of asbestos insulation. More commonly, a subsurface condition refers to soil conditions. This chapter addresses different legal responses to this predictable complication of the performance process, focusing on subsurface soil conditions.

Effect on Performance. When unforeseen subsurface conditions are discovered, they usually have an adverse effect on the contractor's planned performance and prediction of performance costs. Sometimes subsurface materials encountered are more difficult to excavate or extract, with many cases involving the discovery of hardpan or more rock than expected. When this occurs, performance is likely to take more time and money.

Sometimes the subsurface conditions encountered generate a great increase or decrease in the quantities to be excavated. This also can affect the time and costs, particularly if the contractor has bid a composite unit price that may be adversely affected by finding that certain work runs over and other work runs under the bid. (This is discussed in Section 14.5, dealing with unit pricing.)

Role of Geotechnical Engineer. As part of the design preparation process, the owner hires a geotechnical engineer to perform a geotechnical soil report. The engineer determines the load-bearing capacity of soil and rock and the plasticity (or expansion potential) of different soils or clays. The geotechnical engineer may use this information to recommend the design of the building's foundation.

Geotechnical engineers may be exposed to great liability if the information they report is incorrect or their suggestions do not accomplish the owner's expectations. In addition to subsurface conditions encountered during construction, problems such as structural instability, settling, or cracking can develop later. They are usually traceable to the subsurface conditions and often lead to claims against geotechnical engineers. As a result, geotechnical engineers may not perform services where liability exposure greatly exceeds anticipated profit or may sharply increase their fees. As discussed in Section 12.7D, they may use "limitation of liability" clauses to limit their liability to the owner and require indemnification of any third-party liability. These liability concerns, in turn, may lead to less extensive testing (or none at all) on smaller projects.

Use of soil borings is one technique used by geotechnical engineers to determine subsoil conditions. In *Travelers Casualty and Surety Co. of America v. United States*, the U.S. Court of Federal Claims described the soil boring process and immediately highlighted its limitations:

> A soil boring is a three inch diameter coring of the subsurface that is abstracted by drilling down with a conical drill piece and removing the earth within the coring. The narrowness of a soil boring limits the information that can be inferred from it because it is incapable of giving information of the existence of anything larger than three inches in diameter. However, soil borings can indicate cobbles, and the presence of hard materials can be detected by the resistance encountered when the drilling is pushed through the ground.[1]

In sum, the data reported may not reflect subsurface conditions throughout the entire area. Subsurface conditions even within small areas may vary greatly.

The expense of retaining a geotechnical engineer combined with the inherent limitations on the information the engineer is able to obtain helps explain why unexpected soil conditions are encountered on a regular basis.

Risk Allocation: DB Versus DBB: It is important to recognize the effect of different project delivery methods on responsibility for unexpected subsurface conditions.

First, consider construction done by a contractor who is given a site and asked both to design and to build a particular project. The case of *Stees v. Leonard*,[2] which is cited many times for the proposition that the contractor bears all risks of subsurface conditions, involved a contract under which the contractor both designed and built a house for the owner. In such cases, it is easy to place the risk of success on the contractor.

However, the traditional American construction process—design–bid–build (DBB)—divides design and construction with the owner engaging a design professional to design, a geotechnical engineer to gather data, and a contractor to build. This organizational structure makes it more difficult to determine who will bear the risk of unforeseen subsurface conditions.

Although there are abundant factual variations, typically the owner—private or public—calls for bids on a construction project that will involve subsurface excavation. Building specifications may be furnished, but methods of excavation and construction probably will be left to the contractor.

Depending on owner identity and project size, an independent geotechnical engineer makes soil tests for cost estimation, design, and scheduling. The reports are made available to the design professional and the owner. Although the information is likely to be available to the contractor, owners take different approaches as to whether they will take responsibility for the accuracy of the information given to the contractor. This topic is addressed in Section 19.3.

Whichever approach is taken, the bidder will usually be warned that it must inspect the site and, under some contracts, conduct its own soil testing. However, often the bidder will not make independent soil tests because the profit potential may be too small to justify such expenditure or because the bids are due in a relatively short period of time.

As a result, contractors frequently bid without knowledge of actual conditions that will be encountered. Bids are calculated on the basis of expected conditions, and if the unexpected is encountered, the actual cost will vary widely from that anticipated. The question then arises as to who will bear this risk.

> In this chapter's scenario, MMI used a DBB project delivery method. Prime Builder would argue that the risk of subsurface conditions is allocated to the owner.
>
> Mr. Prime visited the site and did not see obvious concern with compatibility of the project site with the design. A court would likely find that Mr. Prime's pre-bid visit was reasonable and that he had no duty to conduct a subsurface examination.

19.2 Common Law Rule

As discussed in Section 7.4C, the general rule under the common law is that the prime contractor is responsible for the consequences of encountering defective soil conditions.

19.3 Information Furnished by Owner

Owners who intend to build substantial projects often make subsurface and soil reports available to prospective contractors. One exception to the general allocation of unexpected costs to the contractor relates to the owner furnishing information relied on by the contractor. The owner may make this information available but may disclaim responsibility for its accuracy. This section emphasizes the effect of providing this information with the full realization that the contractor is likely to rely on it and without any technique to shield the owner from responsibility.

At the outset, differentiation must be made among types of information that may be given to the contractor. Reports of any tests that have been taken are usually included in such information. The reports also may contain opinions or inferences that the geotechnical engineer may have drawn from observation and tests taken. The information may include estimates as to the type and amount of material to be excavated or needed for fill or compaction. (This information also may be included in the specifications.) Although less common, the reports may recommend particular subsurface operational techniques.

Misrepresentation is the basic theory on which the contractor bases its claim. A threshold question involves what constitutes a misrepresentation. Facts and opinions must be differentiated, as a misrepresentation claim must be premised upon a false fact—not a false opinion. (Indeed, the term "false opinion" is something of an oxymoron.) Reporting the result of tests is clearly a factual representation, whereas professional judgments that seek to draw inferences from this information may be simply opinions.

A misrepresentation claim may be based on improperly selected test sites with the inference being that a contractor may believe that the test sites selected will generally represent the site. Misrepresentation can consist of a combination of providing some information but not disclosing all the information that qualifies the information given. Half-truths can be just as misleading as complete falsehoods. Misrepresentation is distinguished from simply failing to disclose any information or information that the owner knows would be valuable to the contractor and that the contractor is not likely to be able to discover for itself.

Representations may be fraudulent, that is, made with the intention of deceiving the contractor. Fraud claims, while having a heavy burden of proof, are most valuable to a contractor. Fraud may extend the time a contractor has for bringing suit. For example, in *Trafalgar House Constr., Inc. v. ZMM, Inc.*,[3] the court applied the "discovery rule" (see Section 16.9D) to extend the time available to the contractor to bring suit. Fraud also offers the contractor a variety of remedies. The contractor can rescind the contract and refuse to perform further, raise fraud as a defense if sued for nonperformance, or—most important—complete the contract and recover additional compensation in a claim for damages.

More commonly, though, misrepresentations are not made with the intention of deceiving the contractor. If the misrepresentations were made negligently, some added complexities develop. Very likely the negligence is that of the geotechnical engineer, and owner recovery against the geotechnical engineer may not always be available (or at least it will be circumscribed by a limitation of liability clause).

What is the responsibility of the owner for any negligent representations in the soil information generated by the geotechnical engineer? If the geotechnical engineer is an independent contractor—which is likely to be the case—the owner would not be chargeable with the latter's negligence. However, very likely, the owner will be chargeable either for breach of contract (that the breach is caused by a person whom it engages to perform services does not relieve it of responsibility)[4] or implied warranty of the accuracy of the information supplied by the owner that is relied on by the contractor.[5]

The least culpable conduct is that of innocent misrepresentation. In such cases, there is neither intention to deceive nor negligence. Innocent misrepresentation generally allows the contracting party who was misled to rescind the contract. However, actual rescission in these subsurface cases is rare. Many legal and factual issues may make it difficult to predict whether the right to rescind is available. Walking off the job under these conditions exposes the contractor to liability. Even if its rescission were determined to be justified, the contractor can recover only the reasonable value of its services.

19.4 Risk Allocation Plans: Benefits and Drawbacks

One of the principal planning decisions that those who prepare construction contracts must make relates to subsurface problems (broadly defined). Those who plan risk allocation for such contracts (rather than stay with the common law rules outlined in Section 7.4C) must keep in mind that generally the law will permit the contract to distribute the risks in any way chosen by the parties. That it is usually the owner who makes this determination and that the only choice the contractor has is to enter into the contract or not rarely affects the law's respect for the contract provisions.

Assuming that the owner chooses the risk distribution, the owner has essentially three options:

1. Gather no information, and let the contractor proceed based on its own evaluation.
2. Gather the necessary information, make it available to the contractor, but disclaim any responsibility for its accuracy (disclaimer system).
3. Gather the information, make it available to the contractor, and promise the contractor to equitably adjust the contract price if the actual conditions turn out to differ from those represented or anticipated (the **differing site conditions** (DSC) system).

These techniques are discussed in Sections 19.5 and 19.6. In those sections, however, the assumptions are that a particular risk distribution choice has been made and how the law deals with that choice. In this section, the emphasis is on the advantages and disadvantages of the three systems.

Emphasis in this section will be not on the first system but on the second and the third. In construction projects of any size, gathering subsurface information is crucial for design choices, and the owner rarely just allows the contractor to deal with the problem. Contractors are not always geared to gather the information and make these choices, and any attempt to use the first system runs a great risk of construction failure.

Most private and many state and local public owners make information available to the bidders but use contractual disclaimers in an attempt to relieve themselves from any responsibility for the information's accuracy. Those who adopt the disclaimer system recognize the possibility or even the likelihood that the contractor will encounter physical conditions different from those represented or anticipated. They expect the contractor to calculate the risks and include contingencies in the bid price, which takes this risk into account. Those who prefer this system also think that if the contractor will have to pay for the added expenses for corrective work, this system will encourage contractors to more carefully evaluate the information and inspect sites. Advocates of this system want the pricing of such uncertainty to go into the contract at the front end—rather than at the back end—through claims. To sum up, the principal justification for the disclaimer system is that many public entities and many private owners have to know at the outset what the project will cost.

But the disclaimer system has drawbacks. First, as shall be seen in Section 19.5, the disclaimer does not always work. The claims will still be brought, but they will be based on other contract clauses or different legal theories.

Second, if the disclaimer system does not protect the geotechnical engineer who furnishes the information, the contractor may be able to make a successful claim against the geotechnical engineer based on negligence.[6] This will frustrate the risk allocation system by giving the contractor a windfall if it has priced the contingency in its contract price and can still recover against the engineer. This uncertainty may also force the engineer to protect himself by demanding higher compensation or indemnification.

Third, the ruthlessly competitive construction market may mean that contractors do not include contingencies for subsurface conditions in their bid prices. Although this may appear beneficial to the owner, the contractor who loses money is likely to make a claim and may win it; in any event, all the parties will suffer extensive claims overhead.

Despite these undoubted disadvantages, the need for certainty, particularly when budgets are fixed and tight, motivates many owners to seek to use the disclaimer system.

The third system—the differing site conditions (DSC) method—assures the contractor that it can bid on what it believes it is likely to encounter and that it will receive an equitable adjustment if actual conditions do not turn out that way. This system is intended to generate lower bid prices because the contractor need not attempt the often difficult task of providing a contingency in its pricing for the subsurface uncertainties and need not incur extensive costs connected with its own testing if it is not certain that the information furnished by the owner under the disclaimer system is accurate. As explained by the federal Court of Claims in *Foster Construction C. A. & Williams Brothers Co. v. United States*:

> The purpose of the changed conditions clause [Ed. note: now differing site conditions clause] is thus to take at least some of the gamble on subsurface conditions out of bidding. Bidders need not weigh the cost and ease of making their own borings against the risk of encountering an adverse subsurface, and they need not consider how large a contingency should be added to the bid to cover the risk. They will have no windfalls and no disasters. The Government benefits from more accurate bidding, without inflation for risks which may not eventuate. It pays for difficult subsurface work only when it is encountered and was not indicated in the logs.[7]

Obviously, the DSC method has disadvantages. The most important is the uncertainty of the ultimate contract price. Yet over the long run—and this point is made by many owners who are repeat players—the prices will be lower. Also, the cost of administering the DSC system can be formidable, and a contractor who is not convinced it will be treated fairly may include a contingency price anyway. Determining the amount of the equitable adjustment can generate difficult problems and expensive claims overhead.

Another argument sometimes made against the DSC system is the uncertainty as to whether it actually induces lower bids. Some contend that bids are based on work load, the desirability of keeping the work force together, and the prospects of a well-administered construction project. The presence of the DSC system may be a relatively insignificant item in determining the actual bid price. Also, the bid price may not be reduced unless the contractor is confident that obtaining an equitable adjustment will not be extraordinarily difficult or administratively expensive.

Finally, another factor in favor of using the DSC system is that it should provide an incentive for the owner to furnish the best subsurface information by hiring competent soil testers and giving them enough time and compensation to develop accurate information.

Clearly, no system is perfect, and whoever must plan the risk allocation faces difficult choices. There is an increased tendency, as demonstrated in Section 19.6, to use the DSC system. But in the greater number of construction projects, the disclaimer system will likely still be employed.

19.5 Disclaimers—Putting Risk on Contractor

Commonly, the owner needs the subsurface information, commissions it, and makes it available to the contractor. Some owners use a number of techniques to avoid responsibility for the accuracy of subsurface information they have obtained and made available. Sometimes owners place the responsibility on the contractor to check the site and make its own tests. As seen in *United States v. Spearin* (reproduced in Section 16.3B), these generalized disclaimers are not always successful.

Some owners use a different approach: They do not include subsurface data in the information given bidders but state where the contractor can inspect such information available elsewhere.

Another approach is to give the information but state that it is "for information only" and is not intended to be part of the contract. This is another way of stating the contractor cannot rely on the accuracy of the information in the hope of shielding the owner from any claim based on misrepresentation or warranty. These techniques generate varied outcomes, reflecting the difficulties courts face in deciding whether a party can use the contract to shift a risk to the other party.[8]

Different judges view the disclaimer process differently. Judges who are unwilling to allow owners to furnish information and disclaim responsibility for its accuracy (the issue here is not fraud or negligent misrepresentation) seem more influenced by the apparent unfairness of allowing the owner to place the risk on a party who is often in a poorer position to distribute or shift that risk. This is even more persuasive if the owner knows that the information will be relied on and derives a benefit through lower bid prices when the contractor does not make its own tests. They are also influenced by the belief that the contracting parties owe each other the duties of good faith and fair dealing. They believe autonomy should not be used to place subsurface risks on the contractor when the contractor did as others and relied on the furnished information. For example, in *Sherman R. Smoot Co. v. Ohio Department of Administrative Services*,[9] the Ohio Court of Appeals court invoked the *Spearin* doctrine to allow the contractor to rely upon a soils report, even though the report had been provided "for information only."

Judges favoring broad autonomy are less influenced by these considerations. They seek only to determine how the risk was apportioned by the contract. They refuse to use equitable considerations to rewrite the contract agreed to by the parties.[10]

> In this chapter's scenario, MMI made the geotechnical report available for examination at the office of Geo-R-Us but also stated that the report was available "for information only." Prime Builder did not examine the report prior to bidding, relying instead on AAI's design.
>
> Here, Prime Builder is claiming two defects: (1) an unusual number of boulders to excavate and (2) an inadequate foundation design given the nature of the soil. With regard to the excavation work, Prime Builder's failure to examine the report should defeat a misrepresentation claim, as it did not rely on any alleged misrepresentation.
>
> However, the contract did contain a representation for an appropriate foundation design. Could MMI argue that Prime Builder cannot bring a misrepresentation claim because an examination of the geotechnical report would have revealed inadequacies in the foundation design?
>
> It is unlikely a court would impose upon a contractor a more discerning reading of a geotechnical report than was made by the architect. Any fault in the number and location of soil boring logs was attributable to the architect—and through the architect to the owner.

19.6 Contractual Protection to Contractor: The Federal Approach

For the reasons explained in Section 19.4, the owner may find that shifting the risk of subsurface conditions to the contractor is not in the owner's best interest. In such a case, the owner may choose to accept the risk of unforeseen site and subsurface conditions. This acceptance is accomplished in federal construction through what was formerly called the changed-conditions clause and is now known as the differing site conditions (DSC) clause.[11] The federal DSC clause states,

> (a) The Contractor shall promptly, and before the conditions are disturbed, give a written notice to the Contracting Officer of (1) subsurface or latent physical conditions at the site which differ materially from those indicated in this contract, or (2) unknown physical conditions at the site, of an unusual nature, which differ materially from those ordinarily encountered and generally recognized as inhering in work of the character provided for in the contract.

(b) The Contracting Officer shall investigate the site conditions promptly after receiving the notice. If the conditions do materially so differ and cause an increase or decrease in the Contractor's cost of, or the time required for, performing any part of the work under this contract, whether or not changed as a result of the conditions, an equitable adjustment shall be made under this clause and the contract modified in writing accordingly.

(c) No request by the Contractor for an equitable adjustment to the contract under this clause shall be allowed, unless the Contractor has given the written notice required; *provided,* that the time prescribed in (a) above for giving-written notice may be extended by the Contracting Officer.

(d) No request by the Contractor for an equitable adjustment to the contract for differing site conditions shall be allowed if made after final payment under this contract.12

A DSC clause creates two methods of obtaining an equitable adjustment: conditions different from those represented in the contract (Type I) and unanticipated conditions (Type II).

Type I DSC. A Type I DSC requires that there be an actual physical (subsurface or latent) condition encountered at the site that differs materially from the conditions "indicated" in the contract documents. First, the contract documents must be examined. In federal procurement, such documents are broadly defined and include the invitation for bids, drawings, specifications, soil-boring data, representations of the type of work to be done, and the geographical area of construction. Included among the contract documents are not only specific information of the subsurface characteristics but also the description of the nature of the project and the physical conditions that relate to the work. Physical conditions include the details of excavation or construction work, as they may include representations of the physical conditions. As an example, in *Ace Constructors, Inc. v. United States*,[13] the government's erroneous topological survey, indicating a "balanced" project when instead the project lacked adequate fill from the cut areas, was found to constitute a Type I DSC.

The contractor, while preparing its bid, must consider not only the representations contained in the physically attached documents but also information located elsewhere but made available for inspection by the bidders, and even the contractor's own experience on similar projects in the same location.[14] The contractor can draw reasonable inferences as to the physical conditions, and there need not be express representations of them. The important thing is whether the representations are sufficient to provide a reasonably prudent contractor with a sufficient basis on which to rely when the contractor prepared its bid.[15] Where the contract documents are silent as to subsurface conditions, no Type I DSC claim can be brought.[16]

The condition must be subsurface or latent. Most claims involve encountering something different than anticipated in the subsurface materials or structure—usually physical or mechanical properties, behavioral characteristics (in tunneling projects[17]), quantities, etc. But a DSC need not always be a subsurface condition; it also can be a latent physical condition at the site, such as undisclosed concrete piles, the thickness of a concrete wall, or the height of ceilings above a suspended ceiling.

The DSC must be encountered at the site, which usually is the place where construction will be undertaken, but can include off-site pits from which soil is to be borrowed for fill or disposal sites if the contractor is directed to obtain or dispose of materials off site.

The physical conditions encountered must be materially different from those indicated—an essentially factual inquiry. Was there a substantial variance from what a reasonably prudent contractor would have expected to encounter, based on its review of the contract documents and what it would have encountered had it complied with the site investigation clause in the contract? The "reasonably prudent contractor" test imposes upon the contractor not only a duty to examine all of the information stated in the contract documents or separately made available, but also to evaluate this information in a balanced manner. As explained by a federal board of contract appeals in *PCL Construction Services, Inc. v. General Services Administration*:

> For example, the contractor must consider whether the borings are numerous and well-spaced, or whether they are few and far between. In addition, the contractor must consider the general description of the site and any warnings of conditions which might be encountered. . . . [A] reasonable bidder will consider the information provided by the boring logs and then consider how other available information sheds light upon the results of the test borings and upon the extent to which the test borings are representative of conditions throughout the site.[18]

The DSC clause makes the contractor's actual reliance upon the contract documents when preparing its bid a precondition for bringing a claim. A contractor who did not rely upon certain contract documents when preparing its bid may not then use those documents as the basis for a claim.[19] Moreover, even without such a clause, a contractor would be expected to make at least a minimal inspection to familiarize itself with the site. Typically, site inspection clauses obligate the contractor to inspect and familiarize itself with the conditions at the site, but the contractor is obligated only to discover conditions apparent through a reasonable investigation. It is not expected to perform burdensome, extensive, or detailed tests or analysis. The contractor can rely on information received from the owner, particularly when it does not have adequate time or opportunity to conduct a thorough investigation. What other bidders did or what they thought they would encounter will bear heavily on the question of whether the site investigation was properly performed.

A Type I DSC clause is in essence a contractual mechanism for dealing with misrepresentation claims concerning subsurface conditions. Within federal procurement jurisprudence, the DSC clause is a subset of the *Spearin* doctrine discussed in Section 16.3B. In the case of an overlap of these two theories—such as when the specifications are allegedly defective in their misrepresentation of subsurface conditions—the contractor must pursue its claim under the more specific, contractual mechanism.[20]

The DSC clause is also akin to the *Spearin* doctrine in that courts are loath to allow disclaimers to undermine either theory of recovery. For example, a federal board of contract appeals in *Whiting-Turner/A. L. Johnson Jt. Venture v. General Services Administration*[21] refused to enforce disclaimers providing that boring logs are for information only and not part of the contract documents, as doing so would render the Type I DSC clause meaningless.

Type II DSC. A Type II DSC requires a variance between the site condition actually encountered and that which would be reasonably expected at the time the contract was made. Expectations look at the information furnished to the contractor or information acquired from other sources. Were the conditions encountered common, usual, and customary for that geographical area? The contractor is expected to be aware of conditions under which the work will be performed. For example, a contractor who is working in winter in a mountainous area where snow is common should expect to encounter wet conditions.[22]

> In this chapter's scenario, MMI and Prime Builder, Inc. used AIA A201-2007, which contains a DSC clause similar to the federal provision. Here, Prime Builder encountered an unusual number of boulders when it began excavation. Assume the contractor cannot bring a Type I DSC claim, which is a topic addressed in Section 19.7.
>
> Prime Builder's inability to bring a Type I claim would not preclude it from bringing a Type II claim. It would have to prove that the existence of boulders was an unexpected condition. Prime Builder's claim would likely be bolstered by Mr. Prime's site visit, in which he saw no boulders.

Notice Requirement. To obtain an equitable adjustment for either a Type I or Type II DSC, the contractor must notify the contracting officer promptly in writing of the alleged differing condition. The federal DSC clause subsection (a), quoted previously, begins. "The Contractor shall promptly, and before the conditions are disturbed, give a written notice to the Contracting Officer [of the DSC]."

This notice requirement generates the inevitable difficulty over substantial compliance and waiver. Waiver is the voluntary giving up of a known right. It may be demonstrated by words or by conduct—as when a party performs notwithstanding a condition excusing that performance. The contractor waives its right to assert a DSC claim by not first providing timely notice of the condition.

In *Fru-Con Construction Corp. v. United States*,[23] the U.S. Court of Federal Claims rejected waiver. On a dam renovation project, the blasting method employed by the contractor caused "overbreak" of the existing concrete walls. This delayed the project because the contractor was required to rebuild the walls. The contractor provided no DSC notice—oral or written—even though the evidence of overbreak was immediate and obvious. Indeed, the contractor did not even make a DSC Type II claim until three years after project completion and on the eve of trial.

The court rejected the contractor's arguments that late notice was excused because it was on a tight construction schedule and did not know the cause of the overbreak (that it was caused by a DSC, rather than its blasting subcontractor's improper work) until it consulted an expert. The court observed that the government was prejudiced by the late notice because it was deprived of the opportunity to consider alternative construction methods. In the court's words:

> The parties presented visual evidence of the overbreak. It was catastrophic; the damage, immediate and visible. . . . [T]he court finds that plaintiff gambled by electing to continue blasting when each successive blast wreaked havoc. While impressed with plaintiff's evidence that the Corps insisted on timely reopening of the locks and would not have welcomed—and might have resisted—a delay in the blasting in order to determine the cause of the overbreak, plaintiff cannot avoid the fact that it elected to proceed. Plaintiff did not know the cause of the overbreak, but its consequences were evident. Either plaintiff or the Corps would be responsible. Inexplicably, plaintiff's on-site personnel pressed ahead. The failure to notify the Corps of a differing site condition under these circumstances was prejudicial to the Corps, especially when considered in light of potential remedial alternatives.[24]

While earlier decisions had allowed waiver,[25] *Fru-Con* and these earlier decisions are not necessarily inconsistent. In *Fru-Con*, the contractor did not give any notice—even oral—at the time it discovered the unexpected condition; indeed, no DSC claim was brought until three years after project completion. The court was correct to point out that the role of prompt notice is to give the government an opportunity to respond with design changes. While waiver by the government may be appropriate when its inaction lulled the contractor into a sense of entitlement to relief (notwithstanding the contractor's failure to provide notice), using waiver here would simply have been unfair to the owner.

19.7 AIA Approach: Concealed Conditions

AIA Document A201-2007, § 3.7.4, uses language similar to the federal DSC clause with its Type I and II substantive bases for additional compensation. The observing party must give a notice "promptly before conditions are disturbed and in no event later than 21 days after first observance of the conditions." Although many notices required by AIA documents can be given to the architect, this notice must go to both the owner and the architect if the contractor is the observing party. Although only "notice" (and not "written" notice) is required, certainly it is safer for a contractor discovering such conditions to provide notice in writing (including to confirm any oral notice already given).

The Type I claim under § 3.7.4 requires that conditions be encountered that "differ materially from those indicated in the Contract Documents." Note that AIA Document A201-2007, § 1.1.1, excludes bidding requirements from the contract documents. (A101-2007, § 9.1.7, recognizes that some may wish to include them.)

In the same vein, suppose subsurface information is not included within the contracts documents (as defined by the AIA) but is simply made available at the geotechnical engineer's office. A literal interpretation of §§ 1.1.1 and 3.7.4 could bar the contractor's Type I claim.

However, it can be contended that including a DSC, yet giving the contractor relevant information in something *other* than a contract document, would be at least a violation of the covenant of good faith and fair dealing and possibly a fraud. Put another way, assuring the contractor it will receive an equitable adjustment if it encounters something unexpected and then refusing to grant an equitable adjustment in a Type I claim—the most common basis for equitable adjustments—would violate the obligation of good faith and fair dealing discussed in Section 16.2B.

With regard to a Type II claim, in AIA Document A201-2007, § 3.2.1, the contractor represents that it has visited the site and familiarized itself with local conditions under which the work is to be performed and "correlated personal observations with the requirements of the Contract Documents." Conditions that could have been observed at a normal site visit cannot justify additional compensation.

A201-2007, § 3.7.4, creates a complicated system to determine the existence of a DSC and the extent of any equitable adjustment. Either party observing an apparent DSC must provide prompt notice to the architect. The architect investigates.

If the architect determines that there is a DSC, he recommends an equitable adjustment. If the owner and contractor cannot agree on the adjustment, this question is referred to the Initial Decision Maker (who could be the architect) under Article 15. If the architect determines that there is no DSC and that no adjustment is warranted, he notifies the owner and contractor in writing, giving his reasons. If either party objects, then the claims resolution procedure under Article 15 is again invoked.

A201-2007, § 3.7.4 provides no review of the architect's decision that a DSC exists. This decision, favorable to the contractor, is apparently binding on the owner. In addition, A201 gives no guidance as to the amount of the equitable adjustment, but presumably the vast federal jurisprudence can be looked to, especially in light of the identical language used to define a DSC.

Whether the occasional owner who builds once in a lifetime is better off with such a provision is debatable. It is also questionable whether lenders will want a provision that can substantially expand the ultimate cost of the project.

In this chapter's scenario, MMI and Prime Builder, Inc. used AIA A201-2007, meaning that the contractor could bring a DSC claim. Here, Prime Builder encountered an unusual number of boulders when it began excavation. Assuming the contractor provided prompt notice, may it bring a Type I DSC claim against MMI?

Recall that Prime Builder did not examine the actual geotechnical report prior to bidding. Suppose that report warned of boulders in the area. MMI would argue that the contractor's failure to examine the report meant that it did not rely on the contract documents when creating its bid, and so could not claim the actual conditions differed from these contract "indications." Prime Builder might counter that there was no reason to examine the report because it was provided "for information only," meaning MMI was not standing behind the accuracy of the report.

A court would likely find the Type I claim would be defeated by the contractor's failure to examine all relevant contract documents. MMI's disclaimer of responsibility does not mean the information in the report is irrelevant.

19.8 EJCDC Approach

The Engineers Joint Contracts Documents Committee (EJCDC) employs a more complex methodology in its 2007 "Standard General Conditions of the Construction Contract," C-700. It divides coverage into subsurface (¶ 4.3) and underground facilities physical conditions (¶ 4.4). Paragraph 4.2(A) states that information will be given in the Supplementary Conditions. Under ¶ 4.2(B), the contractor may rely on "the accuracy of the 'technical data'" even though the data are not contract documents. But ¶ 4.2(B)(1) states that the contractor cannot rely on the completeness of the information for purposes of execution or safety nor under ¶ 4.2(B)(2) on "other data, interpretations, opinions and information."

Paragraph 4.2(B)(3) precludes a contractor's claim against the owner or engineer with respect to the contractor's interpretation or conclusions drawn from the technical data. This is a sensible line of demarcation and fits the DSC system.

To claim an equitable adjustment, under ¶ 4.3(A), a written notice must be given describing the technical data the contractor claims it intends to rely on and why the data are materially inaccurate or stating that the condition differs materially from that shown or indicated in the contract documents. The Type II language of ¶ 4.3(A)(4) is similar to that in the federal procurement system and the AIA.

After a claim is made, under ¶ 4.3(B), the engineer can review the condition and determine whether the owner should obtain "additional explorations and tests." Under that paragraph, he advises the owner of his findings and conclusions. If he concludes a change is required, he issues a work change directive. But this does not guarantee price or time changes in the contract. Under ¶ 4.3(C), an equitable adjustment is granted only to the extent that the condition causes an increase or decrease in the cost or time. The inability of the parties to agree on the amount of adjustment in price or time under ¶ 4.3(C)(3) is a dispute that falls into the disputes clause, ¶ 10.5, under which the engineer plays roughly the same role as does the architect under A201-1997.

Underground facilities are treated differently. Information is given by the owners as to these facilities. According to ¶ 4.4(A)(1), neither the owner nor the engineer is responsible for its "accuracy or completeness." The contractor must review and check such information. But if uncovering reveals an underground facility "not shown or indicated or indicated with reasonable accuracy," ¶ 4.4(B)(2) allows an equitable adjustment "to the extent that they are attributable to the existence or location of any Underground Facility that was not shown or indicated or not shown or indicated with reasonable accuracy . . . and the Contractor did not know of and could not have reasonably expected to be aware of or to have anticipated." Again, claims are handled under ¶ 10.5.

EJCDC documents are more detailed than those of the AIA, largely because of the type of engineering projects for which they are designed. But their classifications and detail would be useful in many building contracts.

19.9 The FIDIC Approach

Although this book emphasizes domestic construction contracts, rather than those made in an international context, occasionally it is useful to look at a particular problem from the vantage point of a standardized contract used in international transactions. The most commonly used standard contract is the one generally referred to as the FIDIC contract, published by the Federation Internationale des Ingenieurs-Conseil (International Federation of Consulting Engineers).

The FIDIC issued its Conditions of Contract for Construction for Building and Engineering Works Designed by the Employer in 1999, generally known as the Redbook. It demonstrates that there are many ways to deal with unforeseen subsurface conditions encountered during the work.

Section 19.4 sets forth different risk allocation plans to deal with unforeseen subsurface conditions. One is to furnish information to the contractor but disclaim responsibility for its accuracy. Another is to use differing site conditions (DSC) clauses, allowing the contractor additional recovery under either a Type I or Type II claim.

The FIDIC blends these approaches. It uses a modified-disclaimer method and a Type II DSC.

Subclause 4.10 is a modified-disclaimer provision. It requires the employer (owner) to make available to the contractor all relevant data in its possession "on sub-surface and hydrological conditions at the Site, including environmental aspects" 28 days prior to tender (bid) submissions as well as any data that comes into the employer's possession after that date. The contractor must interpret this data.[26]

This places all of the risks on the contractor. It includes an assumption that it has inspected and examined the site. The availability of this data means that the contractor is satisfied as to "the form and nature of the Site, including subsurface conditions," as well as hydrological and climatic conditions.

But Subclause 4.10 limits the disclaimer to the "extent which was practicable (taking into account the cost and time)." This exception to the disclaimer takes into account attacks on such disclaimers by contractors—that their effect is to force the contractor to conduct its own tests in order to verify the information made available to it. The limitation in Subclause 4.10 takes into account the practicalities of prebid activities.

FIDIC contracts are likely to be used in large infrastructure and engineering projects. The contractors are likely to be well financed. But these are often mega-projects in strange, inhospitable, and unexplored places where it is difficult to gather reliable information.

Subclause 4.11 hammers away at the disclaimer, stating the contractor is satisfied with the contract price and that it has based its price on the site data available.

Subclause 4.12 enacts a Type II DSC. If "the contractor encounters adverse physical conditions which he considers to have been unforeseeable," he must notify the engineer. Physical conditions include subsurface and hydrological conditions but exclude climatic conditions. The engineer inspects, investigates, and determines whether the conditions were unforeseeable and the adjustments to be made. If unforeseen conditions cause the contractor to suffer delay or increased cost, the contract price and time are adjusted.

There are two other interesting aspects to the FIDIC approach. First, under Subclause 4.12, if the engineer finds that there were more favorable conditions than could have been expected, this can offset any cost increase. The net effect of any adjustment may not reduce the contract price. Second, the contract defines cost to include overhead but not profit.[27]

REVIEW QUESTIONS

1. What have the courts stated, in cases such as *Stees v. Leonard*, about the liability for subsurface conditions for a contractor that both designed and built the home for the owner?

2. What are the three options an owner has if he chooses to do risk distribution in the contract for construction in regard to subsurface problems?

3. What are the advantages and the disadvantages of the differing site conditions method?

4. What is the difference between a Type I and Type II differing site conditions clause?

5. What are some examples of a differing site condition that are not subsurface conditions?

6. What is required in the AIA Document A201-2007 for a contractor to give notice for a differing site conditions claim?

7. What are the two categories of physical conditions in the Engineers Joint Contracts Documents Committee (EJCDC) 2007 "Standard General Conditions of the Construction Contract," C-700?

8. What is the process that is followed by the engineer after a claim has been made for differing site conditions under the EJCDC Standard General Conditions of the Construction Contract?

9. What type of approach does the FIDIC "Redbook" use in regard to unforeseen subsurface conditions?

10. On which types of projects are FIDIC contracts likely to be used and where are these projects often located?

ENDNOTES

[1] 75 Fed.Cl. 696, 700 n. 6 (2007).
[2] 20 Minn. 494 (1874).
[3] 211 W.Va. 578, 567 S.E.2d 294 (2002).
[4] *Harold A. Newman Co. v. Nero*, 31 Cal.App.3d 490, 107 Cal.Rptr. 464 (1973); *Brooks v. Hayes*, 133 Wis.2d 228, 395 N.W.2d 167 (1986).
[5] See Section 16.3B.
[6] See Stein & Popovsky, *Design Professional Liability for Differing Site Conditions and the Risk-Sharing Philosophy*, 20 Constr. Lawyer, No. 2, April 2000, p. 13.
[7] 193 Ct.Cl. 587, 435 F.2d 873, 887 (1970).
[8] Compare *City of Columbia, Mo. v. Paul N. Howard Co.*, 707 F.2d 338 (8th Cir.), cert. denied, 464 U.S. 893 (1983) (disclaimer not enforced) with *R. Zoppo Co. v. City of Dover*, 124 N.H. 666, 475 A.2d 12 (1984) (disclaimer enforced).
[9] 136 Ohio App.3d 166, 736 N.E.2d 69 (2000).
[10] *Empire Paving, Inc. v. City of Milford*, 57 Conn.App. 261, 747 A.2d 1063 (2000); *Green Constr. Co. v. Kansas Power & Light Co.*, 1 F.3d 1005 (10th Cir.1993) (Kansas law).
[11] For an in-depth discussion of the federal DSC clause, see J. CIBINIC, JR., R. NASH, JR. & J. NAGLE, ADMINISTRATION OF GOVERNMENT CONTRACTS, Ch. 5 (4th ed. 2006) and Chu, *Differing Site Conditions: Whose Risk Are They?*, 20 Constr. Lawyer, No. 2, April 2000, p. 5.
[12] 48 CFR § 52.236-2 (2012).
[13] 70 Fed.Cl. 253, 269–72 (2006), aff'd, 499 F.3d 1357 (Fed.Cir.2007).
[14] See *Randa/Madison Jt. Venture III v. Dahlberg*, 239 F.3d 1264 (Fed.Cir.2001) (geotechnical reports made available for inspection); *T. L. James & Co., Inc. v. Traylor Bros., Inc.*, 294 F.3d 743 (5th Cir.2002) (Louisiana law; dredging contractor is bound by information found in maps in possession of the public owner indicating subsurface obstructions in a river).
[15] In *H. B. Mac, Inc. v. United States*, 153 F.3d 1338 (Fed.Cir.1998), the court rejected the argument that the "reasonably prudent contractor" standard should be relaxed where the contractor was a "small, disadvantaged business."
[16] *Conner Brothers Constr. Co., Inc. v. United States*, 65 Fed.Cl. 657, 679–82 (2005).
[17] *Mergentime Corp. H/T Constr., Inc. (JV)*, ENG BCA No. 5756, 94-3 BCA ¶ 27,119; *Municipality of Anchorage v. Frank Coluccio Constr. Co.*, 826 P.2d 316 (Alaska 1992).

[18] GSBCA No. 16588, 06-2 BCA ¶ 33,403, p. 165,616.
[19] *Comtrol, Inc. v. United States,* 294 F.3d 1357, 1363–64 (Fed.Cir.2002).
[20] *Comtrol, Inc. v. United States,* supra note 19, 294 F.3d at 1362.
[21] GSBCA No. 15401, 02-1 BCA ¶ 31,708. See also *Condon-Johnson & Assocs., Inc. v. Sacramento Mun. Utility Dist.,* 149 Cal.App.4th 1384, 57 Cal.Rptr.3d 849 (2007), review denied July 25, 2007.
[22] *Housatonic Valley Constr. Co, Inc.,* AGBCA No. 1999-181-1, 00-1 BCA ¶ 30,869 (wet soil in the Oregon forest not a Type II condition).
[23] 43 Fed.Cl. 306, motion for recon. denied, 44 Fed.Cl. 298 (1999).
[24] 43 Fed.Cl. at 327 (footnote omitted).
[25] Waiver was upheld in *Brinderson Corp. v. Hampton Roads Sanitation Dist.,* 825 F.2d 41 (4th Cir.1987); *Metropolitan Paving Co. v. City of Aurora, Colo.,* 449 F.2d 177 (10th Cir.1971); and *Kenny Constr. Co. v. Metropolitan Sanitary Dist. of Greater Chicago,* 52 Ill.2d 187, 288 N.E.2d 1 (1972).
[26] FIDIC, "Conditions of Contract for Construction for Building and Engineering Works Designed by the Employer," Subclause 1.1.3.1 (1999).
[27] Id. at 1.1.4.3.

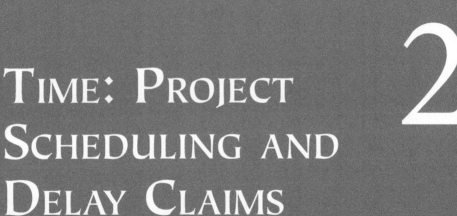

CHAPTER 20

Time: Project Scheduling and Delay Claims

Scenario

Mega Manufacturer, Inc. (MMI) sought to build a new factory. MMI retained Awesome Architects, Inc. (AAI), an architectural firm, to create the plans and specifications.

Based upon this design, the owner entered into a $2 million fixed-price contract with Prime Builder, Inc. The contract was awarded on February 15, and the notice to proceed specified a start date of March 1. Eager to fulfill anticipated customer orders, MMI included in the contract a "liquidated damages" clause of $10,000 per day. Work was to be completed on October 31; however, the contractor would be entitled to a time extension if delayed by "acts of God."

At the February 16 pre-construction conference, MMI warned the contractor that (1) the county had not issued a permit for MMI's drainage plan—but should "any day now,"—and (2) work could not commence until AAI approved project schedule.

Prime Builder's schedule indicated it would complete the project one month early. AAI rejected the plan as too ambitious, noting that it contained virtually no "float" and contemplated several trades working side-by-side in perfect harmony. After several back-and-forths, AAI approved a new schedule on March 15 showing the same method of performance, with more time allotted to each trade and a completion date of October 31 (as the contract had specified).

Meanwhile, the county approved the drainage plan on March 31, so work began on April 1.

Prime Builder complained that it could not perform its work efficiently and smoothly because AAI did not review submittals in a timely manner, wrongly rejected work which had to be redone, and had created a defective design that had to be redone in part. As the October 31 completion date approached, Prime Builder was well short of substantial completion. Then on November 1, a 100-year snowstorm hit the area. All work was stopped for ten days. Shifting to an around-the-clock work schedule, Prime Builder (at great effort and expense) completed the project on November 31.

MMI withheld from final payment $300,000, which was the liquidated damages rate of $10,000 per day times 30 days (the month of November). Prime Builder countered that various wrongful acts of MMI and AAI not only prevented it from completing on time but also prevented it from finishing early as it had originally planned. The contractor also argued it should not have to pay liquidated damages for ten days of weather-related delays.

20.1 The Law's View of Time: Overview

Time permeates the construction process. The timing and duration of a construction project is of keen interest to both owner and contractor. Both parties look primarily to the contract to allocate time-related risks.

Suppose the contract is silent on the question of time; this will not prevent the contract from being enforceable. The law has not looked at time as part of the basic construction contract exchange—that is, money in exchange for the project. The court would likely cure this lacuna by adding an implied term, requiring the contractor to complete the project in a reasonable time.

The most complicated time-related issue is how to deal with delayed performance, an unfortunately common occurrence. Timely completion depends on proper performance by the many participants as well as on optimal conditions for performance, such as weather and anticipated subsurface conditions. Ascertaining the cause (and hence the responsibility) for delays is rarely straightforward. For this reason, an owner who terminates the contractor because the project was not timely completed does so at the risk that the cause of the delay will be traced to a source other than the contractor. In such a case, the termination would be found wrongful and the owner, not the contractor, would be in breach of contract.

Another legal complication is the computation of damages. The owner's basic measure of recovery for unexcused contractor delay is lost use of the project. The contractor's basic measure of recovery for owner-caused delay is added expense. Lost use is difficult to establish in noncommercial projects. Added expense is even more difficult to measure.

Because of measurement problems, each contracting party—whether it pictures itself the potential claimant or the party against whom a claim will be made—would like a contractual method to deal with delay claims—either to limit them or to agree in advance on amount. This adds another legal layer to the analysis: not only the duration and cause of the delay, but also the enforceability of any contract remedy or defense.

20.2 Commencement

The very nature of construction sometimes makes complicated what in other contracts is simple. In an ordinary contract, such as an employment contract or a long-term contract for the sale of goods, the commencement dates usually are simple to establish.

In construction, performance begins when site access is given. This cannot always be precisely forecast. The owner may need to obtain permits, easements, and financing before the contractor can be given site access.

To avoid responsibility for site access delay and to measure the contractor's time obligation fairly, the commencement of the time commitment period, when measured in days, is often triggered by the contractor's being given access to the site, usually by a notice to proceed (NTP). The NTP should specify the commencement date; if not, the date should begin when the NTP is received.

If an NTP is not used, the contractor could conceivably commence performance before the commencement date. This could disadvantage the lender who wishes to perfect its security interest before work begins in order to have priority over any mechanics' liens. (See Section 21.7.)

Similarly, early commencement before insurance is in place can create a coverage gap. The American Institute of Architects' general conditions, A201-2007, § 8.2.2, states that the contractor should not begin work before the effective date of insurance unless with the owner's written consent. Under the owner/contractor agreement, A101-2007, § 3.1, the commencement date is the date of the contract unless a different date is specified.

The Engineers Joint Contract Documents Committee uses a neater and sharper method of determining commencement of contract time. EJCDC C-700, ¶ 2.3 (2007), specifies that the project time starts thirty days after the effective date of the contract or as specified in the NTP.

> In this chapter's scenario, the NTP specified a commencement date of March 1. Yet at the pre-construction conference, MMI warned that the county had not yet approved the drainage plan. The project's commencement date will be a reasonable time after all preconditions to site access (which are the owner's responsibility) have been satisfied, which in this case is on April 1.

20.3 Completion

Construction contracts generally define project completion. For example, both the AIA and EJCDC specify substantial completion as the date of project completion, although not project acceptance (which measures the right to final payment). See AIA A201-2007, § 9.8 and EJCDC C-700, ¶ 14.04 (2007). Some construction contracts use final (not substantial) completion as the date of project completion.

If a contract is silent on a completion date, time is a "soft" concept. If the contract specifies a completion date, the courts view that date as an essential contract term. See Section 20.5 ("time of the essence" clauses).

20.4 Categorizing Causes of and Remedies for Delay

Causes of Delay. Allocating *responsibility* for project delay as an initial matter requires categorization of the delay. Delays are categorized in three ways:

1. the cause of the delay
2. whether the delay is excusable or nonexcusable; and
3. whether the delay is compensable or noncompensable.

Every delay analysis involves viewing the delay through the lens of all three categories. (Sometimes this view is implicit or unarticulated.)

In *England v. Sherman R. Smoot Corp.*, the U.S. Court of Appeals for the Federal Circuit examines the first category—the cause of the delay—and how the cause relates to the other two categories. In that case, the contractor argued that the government's grant of a time extension created a legal presumption (or inference) that the government was admitting responsibility for the project delays:

> There are three potential causes of delay in performance of a contract: the contractor's actions, the government's actions, and forces outside the control of both parties. A delay in a construction contract is excusable if it arises from either the government's action or external forces. Thus, the mere grant by the government of a contract extension does not indicate that the government is at fault; rather, one of a number of other events external to the government could be responsible. In such a situation, a presumption that the government is responsible for the delay is unwarranted, and nothing in the Federal Acquisition Regulations supports such a presumption.[1]

Similarly, in *Redland Co., Inc. v. United States*,[2] the U.S. Court of Federal Claims explained that the government's *failure to assess* liquidated damages (a form of owner damages for delay discussed in Section 20.12) does not create a presumption that the government was at fault. An owner's decision that it is not entitled to delay damages does not necessarily mean the owner thought it was at fault in causing the project to be completed late. Late completion may have been caused by events not attributed to either the owner or contractor.

Remedies for Delay. It is important to divide *remedies* for delay into two categories.

First, the party whose performance has been delayed may seek relief from its obligation to perform by a particular time. For example, a contractor who has agreed to complete the project by a given date may justify its failure to do so by pointing to causes that it claims excuse its obligation to complete by the time specified. Here, the contractor is using project delay as a shield to its own liability.

Second, this contractor may claim additional compensation based on an increase in its anticipated spending because events occurred that delayed its performance. Here, the contractor is using project delay as a sword by which to obtain a damages award.

As a general rule, the law has been more willing to excuse performance than to grant additional compensation. Because of this differentiation, speaking simply of risk allocation of events that impede performance can be misleading.

In addition, the law can take into account the blameworthiness of the party causing the delay. For example, in *Broome Construction, Inc. v. United States*,[3] the U.S. Court of Claims held the government not liable for delay in making a work site available where it sought to do so in good faith. The contractor assumed the risk of this delay but would not have assumed the risk of negligently caused delay.

20.5 Common Law Allocation of Delay Risks

Under the common law, a party who has agreed to perform by a specific time generally assumes the risks of most events that may delay its performance. In the absence of any common law defenses, such as impossibility or mutual mistake, a contractor will not be relieved of its obligation to perform as promised, let alone receive any additional compensation. Relief, if any, must come from the contract.

The contractor normally does not assume the risk that it will be unreasonably delayed by the owner or someone for whom the owner is responsible. Suppose, however, an owner constructs an addition to a functioning plant. The contract provides that the owner can order the contractor to suspend work if the owner's manufacturing operation requires a suspension. The contractor has assumed the risk of these delays. However, that assumption of risk arose from the parties' contract—not from the common law. Nor would this assumption of risk extend to delays caused by the owner's bad faith.

20.6 Contract Allocation of Fault

Force Majeure Clause. Many events can occur that cause delay in construction. The common law placing almost all of these risks on the contractor has led to the frequent use of *force majeure clauses*, which single out specific events and general causes as justifying relief to the contractor.[4]

The purpose of a *force majeure* clause is to relieve the contractor from the risk of rare events over which it has no control. This list may include both man-made and natural events, such as (respectively) labor strikes and floods or "acts of God." Often, a *force majeure* clause will include a "catch-all" phrase, such as "any other event beyond the control of the contractor." Such a phrase would grant relief if events, though unspecified, occur, delay the contractor's performance, and are beyond the control of the contractor.

The purpose of a *force majeure* clause is *not* to displace the essential allocation of responsibility inherent in any prime contract. For this reason, in *Hutton Contracting Co., Inc. v. City of Coffeyville*, a federal court of appeals (applying Kansas law) refused to invoke the clause to relieve a contractor (Hutton) from responsibility for the delays of its equipment supplier in supplying poles. The clause in the parties' contract provided: "The time for Completion of Construction shall be extended for the period of any reasonable delay which is due exclusively to causes beyond the control and *without the fault of [Hutton]*, including Acts of God, fires, floods, and acts or omissions of [the City] with respect to matters for which [the City] is solely responsible."[5] In ruling that the *force majeure* clause does not extend to the acts of Hutton's supplier, the court reasoned,

> The most reasonable interpretation of "fault of [Hutton]" in the *force-majeure* clause is "fault of Hutton and those to whom it delegates its responsibilities under the contract." The contract did not specify a source of the poles. The City was concerned only with the ultimate performance—not whom Hutton employed to reach the result.... [¶] We do not suggest that Hutton was at fault in its choice of supplier, only that it is responsible to the City for its supplier's delays when those delays are not themselves excused by a force majeure. Hutton does not suggest that the supplier's failure was caused by a natural catastrophe or the like. In short, a delay by a subcontractor or supplier is not itself a force majeure.[6]

The AIA's *force majeure* clause (although not titled as such) is A201-2007, § 8.3.1. Its list of excusable delays for the contractor includes "labor disputes, fire, unusual delay in deliveries, unavoidable casualties or other causes beyond the Contractor's control." For Hutton to come within the protection of § 8.3.1, it would have to prove that the late delivery of poles was an "unusual delay in deliveries." Finally, § 8.3.1 gives the architect the right to grant a time extension for any cause that justifies the delay—a flexible (and imprecise) contract standard.

The EJCDC's *force majeure* clause (also not titled as such) is C-700, ¶ 12.03(A) (2007). A contractor is entitled to a time extension for any delay "beyond the control of the Contractor," defined to include acts or neglect of the owner, utilities, or other prime contractors, as well as "fires, floods, epidemics, abnormal weather conditions, or acts of God."

Weather. In A201–2007, § 15.1.5.2, the AIA allows the contractor to seek additional time in the event of "adverse weather conditions," but only if the contractor documents "data substantiating that weather conditions were abnormal for the period of time, could not be reasonably anticipated, and had an adverse effect on the scheduled construction." These requirements are difficult to meet. They reflect a belief by the AIA that weather generally is a risk assumed by the contractor and that only in extraordinary circumstances should weather be the basis for a time extension.

The EJCDC standard is less rigorous. As noted immediately above, C-700, ¶ 12.03(A) (2007) includes "abnormal weather conditions" within the *force majeure* clause.

> In this chapter's scenario, the project was struck by a 100-year snowstorm. The contract allowed for a time extension for acts of God, and this storm should qualify.

"Time Is of the Essence" Clause. As noted in Section 20.1, time is not ordinarily regarded as the essence of a prime contract. A delay may subject a party to a claim for damages, but it is not in itself grounds for rescission of the contract on par with owner nonpayment, for example. With regard to project completion, if the contract is silent, time is a "soft" concept. By contrast, the U.S. Court of Claims in *DeVito v. United States* noted that "[t]ime is of the essence in any contract containing fixed dates for performance."[7]

A "time is of the essence" clause does not *allocate* delay risks. Rather, courts view such a clause as evidencing the owner's intention that timely performance is a fundamental obligation of the contractor. One court held that, in the absence of a "time of the essence" clause, an owner was precluded from seeking damages for missed milestone dates on the project schedule.[8]

But the Texas Supreme Court in *Mustang Pipeline Co., Inc. v. Driver Pipeline Co., Inc.* ruled that the inclusion of such a clause meant that the contractor's failure to complete the project on time was a material breach as a matter of law.[9]

> In this chapter's scenario, the contract contained a completion date of October 31. Absent an excuse, the contractor's failure to complete by that date will be a material breach of contract.

20.7 Measuring the Impact of Delay: Project Schedules

A project schedule is a formal summary of the planned activities, their sequence, the time required, and the conditions necessary for their performance. A schedule alerts the major participants of the tasks they must accomplish to keep the project on schedule. It can reduce project cost by increasing productivity and efficiency, facilitate monitoring of the project, and be used to support or disprove delay claims.

The schedule for a very simple project, such as the construction of a garage, simply may be starting and completion dates. A somewhat more complex project, such as a residence, may add designated stages of completion, mainly as benchmarks for progress payments. When construction moves upscale, for example, from a simple commercial structure to a nuclear energy plant, the schedule will take on more complex characteristics.

The most commonly used schedule is the bar chart, which sometimes is referred to as a Gantt chart after its inventor. One such bar chart[10] is shown in Figure 20.1. Bar charts continue to be used where feasible because they are easy and inexpensive to prepare and simple to understand.

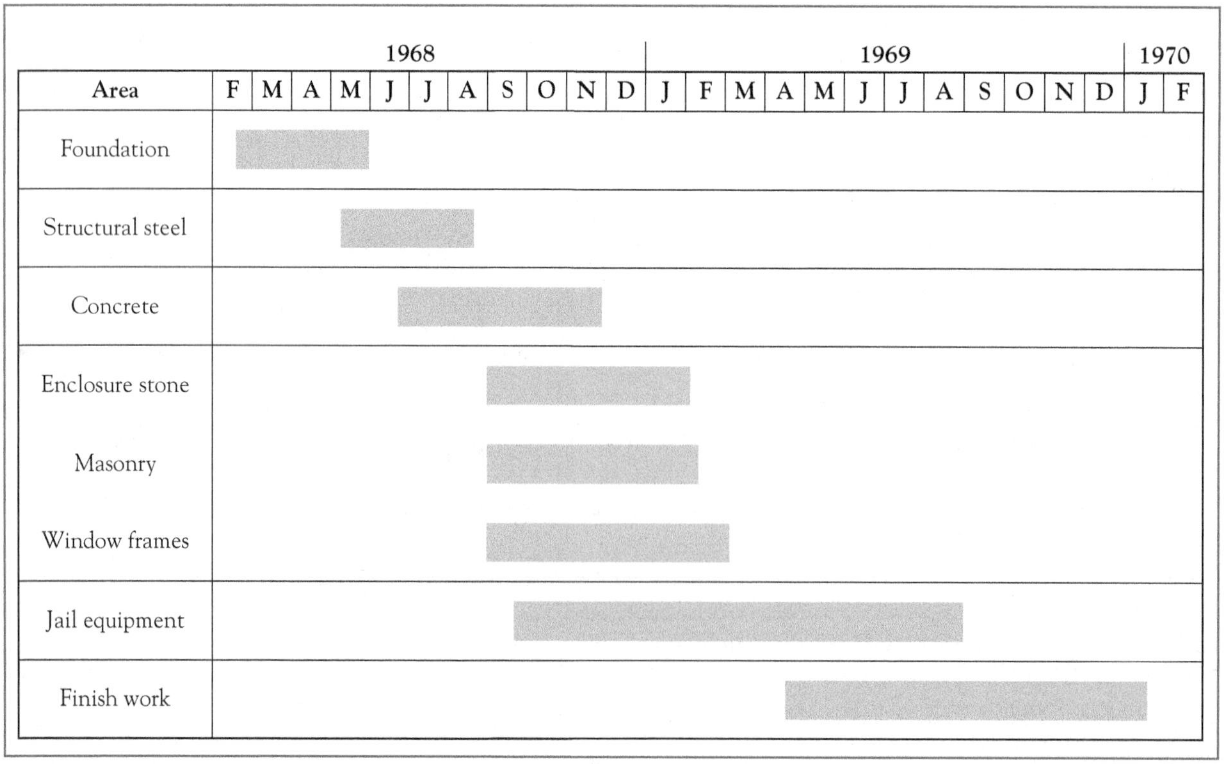

FIGURE 20.1 Bar chart schedule.

Bar charts have deficiencies. They provide no logical relationship between work packages. There are limits to the number of work packages that can be represented in a bar chart—perhaps thirty to fifty—until the level of detail becomes unwieldy. Rates of progress within a package may not be uniform. The different activities are represented equally. In case of a delay, management cannot determine which activities have priority over others, and the courts cannot assess the significance of the delay.

On complex projects, a critical path method (CPM) schedule is used. A CPM schedule lacks the intuitive simplicity of a bar chart but is capable of showing many more activities and, even more importantly, the logical relationship between the different activities: how a delay on one activity affects other activities.

Contracts prepared by experienced public or private owners, particularly private owners under the influence of their lenders, usually prescribe great detail in (and take careful control over) the contractor's schedule. This can manifest itself in language requiring that the schedule be on a form approved by the owner or the owner's lender; that each monthly schedule specify whether the project is on schedule (and if not, the reasons therefore); that monthly schedule reports include a complete list of suppliers and fabricators, the items that they will furnish, the time required for fabrication, and scheduled delivery dates for all suppliers; and that the contractor hold weekly progress meetings and report in detail as to schedule compliance.

20.8 AIA and EJCDC Approaches to Scheduling

The approach taken by the AIA is reflected in AIA Document A201-2007, § 3.10.1. It requires the contractor "promptly after being awarded the Contract" to submit its construction schedule to the owner and architect. The schedule must provide "for expeditious and practicable execution of the Work." It must be revised during performance "at appropriate intervals as required by the conditions of the Work."

Neither details nor schedule type are specified. Indeed, while payments are keyed to work progress (see Section 17.2), and failure to supply enough skilled workers may be grounds for termination (see A201-2007, § 14.2.1.1), A201 contains few levers to obtain timely completion.

The EJCDC takes progress much more seriously than does the AIA. For example, the EJCDC's C-700 (2007), ¶ 2.5(A), requires the contractor to submit within ten days after the effective date of the contract a *preliminary* progress schedule, a *preliminary* schedule of values, and a *preliminary* schedule of submittals.

Paragraph 2.6 requires a preconstruction conference before any work is started. At that conference, among other topics, the preliminary schedule required by Paragraph 2.5A is discussed.

Paragraph 2.7 requires another conference to review "for acceptability to Engineer" the preliminary schedules. The contractor has ten days to make corrections and adjustments and "to complete and resubmit the schedules." No progress payments are made "until acceptable schedules are submitted to Engineer."

20.9 The Critical Path Method (CPM) and Float

Description of CPM. This description of the **critical path method** must be simple, its goal mainly to point out the essential characteristics of the process and note the effect of float or slack time.

To show how a CPM schedule operates, a very simple construction project will be used as illustration, without all of the complexities of arrow diagrams, precedence diagrams, and nodes.

First, the contractor divides the total project into different activities or work packages. A major project may have thousands of activities with each subcontractor generally performing a different activity.

Next, the contractor determines the activities that must be completed before other activities can be started. These constraints are the key to the CPM schedule. For example, usually excavation must be completed before foundation work can be begun. Conversely, plumbing and electrical work usually can be performed at the same time, as neither depends on the other. Subcontractors performing this work can work side by side.

Finally, the contractor estimates how long it will take subcontractors to complete their activities. This estimate is made after the contractor consults with its subcontractors and analyzes the design drawings. These data influence the number of days allocated to each activity.

In the sample project, the following are activities and their respective durations and constraints:

Activity	Duration (days)	Constraint
1. Excavation	7	None
2. Formwork	5	Excavation
3. Plumbing	4	Excavation
4. Electrical	2	Excavation
5. Concrete pour	5	Formwork and plumbing
6. Roof	4	All

Constraints dictate the form of the CPM schedule. Because excavation has no constraints, it can be performed first. Once it is completed, the formwork, plumbing, and electrical activities can be performed. The concrete pour activity cannot be performed until the formwork and plumbing are completed. The roof cannot be installed until the concrete pour and electrical have been completed. Figure 20.2 illustrates the CPM schedule for this project. The total project under this schedule should be completed in twenty-one days.

Critical Path and Float. The critical path, which is the longest path on this simple schedule, consists of activities that will delay the total project if they are held up. In the preceding example, excavation, formwork, concrete pour, and roof work are on the critical path. A delay to any of these activities will hold up the entire project.

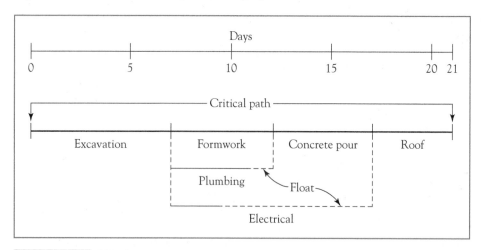

FIGURE 20.2 CPM chart schedule.

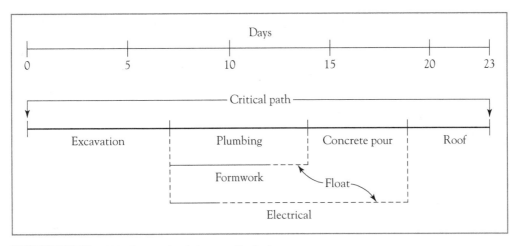

FIGURE 20.3 CPM chart schedule as adjusted.

In contrast, plumbing and electrical activities are not on the critical path. Their delay, up to a point, will not hold up the total project. If electrical work is delayed seven days, the total project will not be held up. The number of days each noncritical path activity can be delayed before the total project is affected is called **float** or slack time. In this illustration, plumbing and electrical work have one day and eight days of float, respectively.

If a noncritical path activity is delayed beyond its float period, it becomes part of the critical path. Moreover, some activities that were previously on the critical path will no longer be there. Suppose there is a three-day delay to plumbing. Originally, plumbing had one day of float. Now the CPM must be adjusted as shown in Figure 20.3.

The total project has now been delayed two days. Plumbing has become part of the critical path, and formwork has moved off the path.

This method can be illustrated by *Morris Mechanical Enterprises, Inc. v. United States*.[11] The contractor was to deliver and install a chiller within 120 days. The contractor delivered the chiller 231 days late. As a result, the government withheld $23,100 as liquidated damages from the final payment.

In ruling for the contractor, the court pointed to the CPM. The schedule showed that delivery and installation of the chiller were originally on the critical path. However, they were later taken off because of delays to other activities for which the contractor was not responsible. The chiller was to be installed in an equipment room. Another contractor who was responsible for completing the equipment room had difficulty procuring materials. When the chiller was actually delivered, the room had not yet been completed. Even though the chiller was delivered 231 days late, the contractor did not delay the total project, inasmuch as its performance was no longer on the critical path. The court relieved the contractor by concluding it should have been given a time extension, precluding the agency from deducting from the unpaid contract balance.

Requiring the contractor to construct and maintain a CPM schedule has at least three advantages. First, it should require the contractor to work more efficiently. Second, it gives the owner notice of the actual progress of the work. Third, from a litigation standpoint, requiring the contractor to maintain a CPM schedule helps prove or disprove the impact of an owner-caused delay. (Schedules are more persuasive evidence if they are actually used during the construction of the project. By contrast, a schedule compiled after completion and with an eye toward litigation will be given little probative value by the courts.)

A CPM schedule has disadvantages. First, it will increase the total contract price. Such schedules are expensive to create and maintain. Second, the contractor may believe such a requirement to be an unnecessary intrusion into its work. In such a case, the contractor's creation and maintenance of the schedule during construction may be haphazard.

CPM Use in Delay Claims. Certain commentators have stated that the following questions must be asked in evaluating a delay claim in which CPM is going to be employed.

1. How was the project actually constructed?
2. What are the differences between the project as planned and as constructed with respect to activities, sequences, durations, manpower, and other resources?
3. What are the causes of the differences or variations between the project as planned and the actual performance?
4. What are the effects of the variances in activities, sequence, duration, manpower, and other resources as they relate to the costs experienced both by the contractor and by the owner.[12]

This type of evidence to present a delay claim (or, as in *Morris*, as a defense to an owner's delay claim) invariably requires the assistance of expert testimony: **forensic scheduling**.[13]

20.10 Causation: Concurrent Causes

A variety of participants and events can cause delays. Delay can be caused by acts of the owner or someone for whose acts the owner is responsible, such as the design professional. Sometimes delays are caused by the contractor or someone for whose acts the contractor is responsible, such as a subcontractor. Sometimes delays are caused by events not chargeable to either owner or contractor, such as non-negligent fires, unpreventable labor difficulties, or unforeseeably extreme weather. Delays can be caused by third parties, such as a union shutting down the project over a labor dispute.

The multiple causes for delay create a number of legal problems. For example, delays caused by events not chargeable to either party will not justify additional compensation unless there is a warranty by one party that those events will not occur. If a delay is caused by both owner and contractor, it may be extremely difficult to determine which portion of the delay is caused by either party. This can mean that neither party will be able to recover losses it suffers because of the delay.

For example, in *Singleton Contracting Corp. v. Harvey*,[14] the federal government's failure to correct the design and the contractor's failure to provide a certificate of insurance were concurrent delays; thus, neither was entitled to compensation for that delay period. For the government, this meant no recovery of liquidated damages (a per diem charge for each day of delay). For the contractor it meant its remedy was limited to a time extension.

Where a clear apportionment (division) can be made between owner-caused delay and contractor-caused delay—as where those delays are sequential rather than concurrent—the delay damages caused by each party may be divided accordingly. The apportionment also can be used to determine the amount of any time extension (see Section 20.10). For example, in *R.P. Wallace, Inc. v. United States*,[15] sequential delays by the federal government and contractor were apportioned to reduce liquidated damages.[16]

> In this chapter's scenario, work was to begin on March 1. This commencement date was delayed by two events: (1) a 15-day delay in Prime Builder submitting an acceptable schedule and (2) a 30-day delay in MMI obtaining a county permit. Neither party will be entitled to delay damages for the first 15 days of delay, since they concurrently caused the delay. Prime Builder is entitled to 15 days of delay because the second half of the month's delay was solely MMI's fault.
>
> What is the practical effect of such a conclusion? In a contract such as this one that contains a liquidated damages clause (see Section 20.12), MMI's entitlement to liquidated damages will be reduced by 15 days.

20.11 Time Extensions

Suppose the contractor is not responsible for the project's delay and it will need additional time (and perhaps additional money) to complete the work. If the owner does not give the contractor additional time, the contractor will need to increase its labor commitment, which will expose the owner to a claim of acceleration (see Section 20.14). If the owner is able to delay the time of project acceptance, her cheapest course of action is to grant the contractor a time extension.

Role of Design Professional. In Section 18.3, it was noted that a change order may specify not only a modification of the contract price but also of the time for performance. This section addresses time extensions to which the contractor is entitled because of project delays that are not the contractor's fault.

Ideally, both owner and contractor should agree on the issuance and extent of a time extension. In the absence of such an agreement (in many construction contracts) the resolution of these issues—at least in the first instance—is given to the design professional, Initial Decision Maker (IDM), or construction manager. Whether a time extension should be granted usually requires that the *force majeure* clause be applied to the facts that are asserted to justify a time extension.

This is the AIA approach: A201-2007, § 8.3.1, entitles the contractor to a time extension for delay arising out of any act or neglect of the owner or architect, changes to the work, specified reasons not the fault of the contractor (such as labor disputes or fire), or "by other causes that the Architect determines may justify delay." (Significantly, § 8.3.3 specifically states that it does not bar the contractor from recovering delay damages. In this regard, the AIA provision is directly at odds with the use of no-damages-for-delay clauses addressed in Section 20.13B.)

Duration of Extension. Suppose a time extension is justified. How is the amount of time extension determined? The contractor's project schedule, coupled with an analysis of any float, may be used to determine entitlement and the duration of any time extension.

In *E. C. Ernst, Inc. v. Manhattan Construction Co. of Texas*, a federal district court faced the imprecision of measuring the exact impact of delay. That court arbitrarily granted a time extension of 65 days for a 131-day delay because this amount was "as accurate an estimate as can be made from the actual resulting delay."[17] Whoever determines the amount of delay will be given considerable latitude. However, it is vital for both owner and contractor to keep careful and detailed contemporaneous records (see Section 20.13D).

Notices. Usually, time-extension mechanisms provide that the contractor must give notice of the occurrence of an event that is to be the basis for a time-extension claim and its probable effect.[18] The requirement of a notice often raises the question of its waiver by the owner, in particular where the owner or the design professional was aware of the delay-causing event and that a claim would be made and where the owner was not harmed by failure to give the notice.

In the context of a time-extension mechanism, the notice of an intention to claim a time extension serves a number of useful functions. The notice informs the design professional or owner that people for whom it is responsible, such as other separate contractors, consultants, or the design professional, are delaying the contractor. This can enable the owner or design professional to eliminate the cause of the delay and minimize future delays or damage claims. A timely notice should permit the design professional to determine what has occurred while the evidence is still fresh and witnesses remember what actually transpired. A notice requirement also prevents a delay claim from unexpectedly being asserted months and even years later—often after project completion. It can prevent false claims.

20.12 Contractor-Caused Delay: Owner Remedies

Delay as justification for termination is discussed in Section 22.4. This section deals with the recovery of damages for the contractor's delayed performance. Although Chapter 4 discusses in greater detail the measurement of claims that owners and contractors have against each other, and Section 4.7C deals with the topic of owner damages for delay, this section primarily addresses damage liquidation—the contractual method used most frequently to deal with contractor delay.

Actual Damages. The damage formula applied most frequently by the common law to delayed contractor performance is the value of the lost use of the project caused by the delay. For example, delayed completion of a residence to be occupied by the owner will be measured by the lost rental value.[19] Although the homeowner may have suffered other losses, such as the inconvenience of living in a motel or with relatives or having to transport a child to a more distant school, such consequential damages are generally difficult to recover.

In commercial construction, the owner very likely will be able to recover the lost use value in the event of unexcused delay by the contractor.[20] Even in projects that have readily ascertainable commercial value, losses are often suffered that may be difficult to recover. This problem is even greater when the project is built for a public entity that intends to use it as a school, an office building, or a freeway. Although some public projects have a readily ascertainable use value, most do not.

Yet actual damages may go beyond lost rental value. Delayed completion may also cause a private owner to suffer a variety of losses (both direct and indirect), including higher financing costs, higher administrative expenses, the cost of an idled workforce, and other damages.

An owner faces a daunting task of proving all of these damages with reasonable certainty. In addition, the cascading consequences of delayed completion means that the owner would have to establish that his claimed damages were direct rather than consequential.[21]

> In this chapter's scenario, MMI wanted the project completed on time because it anticipated several customer orders. MMI's proof of actual damages would include the difficult task of proving it would have obtained the orders, the size of those orders, and the profits MMI would have earned. It is unlikely MMI would be awarded these damages.

Liquidated Damages Clauses. Because proving actual delay damages is very difficult, construction contracts commonly include provisions under which the parties agree that certain types of unexcused delay will result in damages of a specific amount. These are known as *liquidated damages clauses*. (The term "liquidated" means a readily defined or set amount.) The purpose of the liquidated damages clause is to compensate an owner for the contractor's unexcused delay without the owner having to go through the expense, time, and difficulty of proving actual damages. The liquidated damages normally consist of a specified sum of money for each day of delay—the *per diem* rate—after the contract's completion date.

Most modern courts generally recognize the difficulty of proving damages, particularly those relating to delay, and the certainty liquidated damages clauses can provide both parties. In addition, the law is willing to be relieved of the burden of measuring damages. Although the results are by no means unanimous, clauses liquidating damages for construction delay are generally enforced if they are reasonable, as judged by the circumstances existing at the time the contract was made.[22] Often a comparison is made between the amount stipulated and the contract price. Another factor the courts examine is whether the *per diem* rate approximates the owner's expected increased costs caused by the delay.

Public owners also seek the advantages of liquidated damages clauses. This relieves the government of the burden of proving actual damages, and the injury of public inconvenience is inherently difficult to quantify. For example, in *Bethlehem Steel Corp. v. City of Chicago*, the liquidated damages clause expressly referred to "great inconvenience to the public" as a reason to impose *per diem* damages of $1000 per day on a highway painting project.[23] (The very low *per diem* rate is explained by the fact that the contract was from the early 1960s.) The court enforced the clause. Some states require such clauses in public construction contracts.[24]

For historical reasons, an unreasonably high *per diem* rate is called a *penalty*. Its purpose is to punish the contractor rather than compensate the owner, and the clause will not be enforced. To counter the defense that the *per diem* sum is a penalty, the owner may be required to explain the basis for the liquidated damages. In *Carrothers Construction Co., L.L.C. v. City of South Hutchinson*,[25] the city asked its consultant, an engineering firm (MKEC), to assist in drafting the contract, including specifying the liquidated damages rate. In upholding the clause, the Kansas Supreme Court summarized the testimony of the MKEC employee in how he derived the *per diem* rate of $850:

> MKEC considered several factors in determining prospective liquidated damages caused by delay in the work completion including: (1) the City's cost to monitor the project; (2) additional labor costs for city employees, environmental department staff, structural and electrical staff, and controls department staff; (3) additional utilities use; (4) costs to engage another consultant; (5) legal expenses; (6) equipment rental to address flow situations; (7) action by the Kansas Department of Health and Environment if the treatment plant failed to operate within permit limits at the time construction should be finished; and (8) other unknowns if the project was not timely completed. In addition, MKEC considered the project length, the estimated contract value, the general practices by MKEC, and the amounts agreed to between contractors and owners on other comparable projects.[26]

A liquidated damages clause is a contractual remedy for delay that displaces recovery of actual damages. The consequences may be surprising to both the owner and contractor. Owners must understand that these clauses preclude them from recovering actual damages—even if those damages are apparently well in excess of the liquidated sum.[27] On the other hand, contractors are precluded from arguing that the owner's actual damages were less than the liquidated amount or even nonexistent.[28] As explained by two commentators, "liquidated damages become a ceiling as well as a floor for establishing an owner's recovery for contractor-caused delay."[29]

Liquidated damages clauses must be drafted carefully. First, the applicable law must be determined. Second, the clause must be tailored to the particular type of delay to which it is expected to apply. Third, the amount selected should take into account the importance of timely completion, the likely lost use value, and the likelihood that the amount selected will actually achieve the objective.

> In this chapter's scenario, the contract contained a liquidated damages rate of $10,000 on a $2 million contract. This is a daily rate of .5 percent of the entire contract price. MMI will contend this is a reasonable approximation of its actual damages, especially in light of lost sales. Prime Builder will counter that this was a penalty, making the clause unenforceable.

20.13 Owner-Caused Delay

20.13A Sources of Owner Delays and Some Contract Defenses

Increasingly, contractors make large claims for what are sometimes called "delay and disruption", "inefficiency", or "lost productivity" claims. These claims are premised on the owner's implied obligation not to prevent the contractor from performing its obligations in a logical, orderly, and efficient manner.

An owner who interferes in or disrupts the contractor's orderly and efficient manner of performance causes the contractor to suffer two types of injury: increased performance costs and delay damages. A contractor's damages for increased performance costs is addressed in Section 4.7C. This section deals with the contractor's delay claim and the owner's response to such a claim.

Before looking at these responses, it is important to examine the nature of the owner misconduct. Some are based on the owner, or those for whom the owner is responsible, not doing a proper job of communicating information (*Spearin* claims),[30] describing site conditions (Type I DSC claims),[31] allowing timely access so that performance was delayed and occurred during winter conditions,[32] designing (excessive changes or changes beyond the power granted the owner by the changes clause),[33] or administering the project.[34]

As illustrations, cases have pointed to the contractor's failure to coordinate the project (where that was the owner's responsibility);[35] failure to deliver owner-supplied material on time;[36] acceleration (see Section 20.14), which generates excessive working hours and overcrowded conditions,[37] and constant revision of drawings causing confusion and interruption of the orderly progress of the work.[38] These owner actions undermine the conditions necessary for good productivity.

Before examining the troublesome question of the validity of **no-damages-for-delay clauses**, it is important to note other methods that can be used to eliminate or reduce the likelihood of such claims. First, the contract may specify that the owner has the right to delay the contractor and that the interference is not a contract breach. Second, the surrounding facts and circumstances may indicate that the contractor or subcontractor could expect to have its performance interrupted. For example, a subcontractor who knew that it was working on an existing, functioning hospital could not expect to hold the prime contractor responsible for the inevitable delays.[39]

> In this chapter's scenario, Prime Builder alleged that AAI did not review submittals in a timely manner, wrongly rejected work which had to be redone, and had created a defective design that had to be redone in part. The contractor made a lost productivity claim, entitling it to both damages and time extensions. Any time extension would defeat application of the liquidated damages clause for that period.

20.13B No-Damages-for-Delay Clauses

Many public and some private construction contracts meet the delay damage problem head on. These contracts contain no-damages or no-pay-for-delay clauses. Such clauses attempt to place the entire risk for delay damages on the contractor and to limit the contractor to time extensions.

Such clauses have become very controversial. Owners—usually public entities, but increasingly private owners—justify these clauses as a means of fiscal control. Public owners are often limited as to what they can spend by appropriation bills or bond issues. Also, cost overruns can be politically devastating. Public owners want *all* costs, including delay and disruption, to be put "up front" in the contract price. They do not want potentially open-ended delay or disruption claims at the end of the project.

Contractors oppose such clauses. First, they claim that in public contracts they have no choice but to take these clauses. They also say that such costs cannot be rationally priced and included in the contract price. Most such claims result from poor administrative practices, such as excessive change orders, delay in furnishing necessary information, and dilatory submittal approvals. How can a contractor know, at the time it bids, the likelihood of such delay and disruption and what they will cost? Also, contractors say if they do price these risks and put them in their bid prices, in the cutthroat competition of the construction world, they will lose the awards to contractors who will not. The latter will take their chances that they will not suffer such losses or that they can avoid these clauses in court.

Finally, contractors point to the moral hazard such clauses create. An owner and her design professional insulated by such clauses will not do their best to administer the project efficiently.

These objections have not gone unnoticed by courts and legislatures. Although a few courts will interpret the clauses "as written," most will find exceptions that will justify not applying such a clause literally.[40] In *U.S. for Use and Benefit of Williams Electric Co., Inc. v. Metric Constructors, Inc.*,[41] the court, after stating that generally such clauses are enforced, then added:

> A majority of jurisdictions, however, recognize certain exceptions to such clauses. . . . Among the recognized exceptions are (a) delay caused by fraud, misrepresentation, or other bad faith; (b) delay caused by active interference; (c) delay which has extended such an unreasonable length of time that the party delayed would have been justified in abandoning the contract; (d) delay that was not contemplated by the parties and (e) delay caused by gross negligence.[42]

It also noted,

> The most contested of the exceptions is that for "delay not contemplated by the parties." Under this exception, a number of courts find that a "no damage" provision will not bar claims resulting from delays caused by the contractee if the delays "were not within the contemplation of the parties at the time they entered into the contract." *Corinno Civetta Constr. v. City of New York*, 67 N.Y.2d 297, 502 N.Y.S.2d 681, 686, 493 N.E.2d 905, 910 (1986). The rationale for this exception, as stated by the *Corinno Civetta* court, is that "[i]t can hardly be presumed . . . that the contractor bargained away his right to bring a claim for damages resulting from delays which the parties did not contemplate at the time."[43]

Another legal question is whether a *Spearin* claim—delays caused by design flaws, attributable to the owner (see Section 16.3B)—come within the exceptions (a) or (b) for misrepresentations or bad faith *or* active interference. In *Triple R Paving, Inc. v. Broward County*,[44] a road construction project included widening a bridge. During construction, delays resulted from a horizontal sight distance design flaw. There was evidence that an employee of the engineering firm had promised to personally check the sight distance but failed to do so. The court found that both of these exceptions to the contract's no-damages clause applied.

On the other hand, some courts have questioned whether such judicially devised exceptions to a no-damages clause should be created. These courts hold that a clear, unambiguous clause which precludes a contractor's recovery of damages for "any delays" extends even to uncontemplated delays, absent some allegation of intentional wrongdoing, gross negligence, fraud, or misrepresentation.[45]

The ceaseless struggle between private autonomy (freedom of contract) and judicial control of what can be overreaching and abuse of contract power are demonstrated in the varying attitudes of the courts toward "no-damage" clauses. The popularity of delay or disruption claims by contractors will force courts to pass on the validity of such exculpatory clauses. Undoubtedly, case decisions in different jurisdictions will apply different tests and will come to different outcomes.

Legislatures have entered into the fray of using no-damages-for-delay clauses (particularly on public works contracts) and whether such clauses should include limitations. A California statute permits public agencies to allow a time extension in lieu of delay damages for owner-caused delays, but only if those delays are reasonable and within the contemplation of the parties; otherwise, delay damages are recoverable. The statute further provides that no public contract can waive this legislative protection.[46] Oregon and Virginia provide that a contractor cannot waive claims for unreasonable delay in public contracts.[47] Colorado bars no-damage clauses in public contracts.[48] Some states regulate private contracts as well: Ohio and Washington ban all no-damages clauses.[49]

20.13C Subcontractor Claims

Delays caused by the owner not only harm the prime contractor but also may harm subcontractors. The absence of a contract between subcontractors and the owner generally precludes direct legal action and, as a rule, precludes direct negotiations between subcontractors and owners over delay claims. Often the prime contractor processes the subcontractor's claim. This processing is dealt with in Section 21.14.

20.13D Records

In delay disputes, the party with the best records has a great advantage. Each party should keep job records, such as the site representative's daily field reports, correspondence, memoranda, photographs or video recordings, and change orders. These records should include data on labor, equipment, and materials used for each activity and should document the cause and impact of every delay.

20.14 Owner Acceleration of Contractor Performance

20.14A The Changes Clause

One way to accelerate the completion date is to use a specific directive by the owner that the contractor must complete in a time shorter than originally agreed. Power to accelerate is usually determined by the changes clause (see Section 18.4B).

20.14B Constructive Acceleration

Constructive acceleration originated in federal procurement law but has been adopted by some states.

Constructive acceleration is based on the owner's unjustified refusal to grant a time extension. To establish constructive acceleration, a contractor must prove five elements:

1. the contractor experienced an excusable delay and is entitled to an extension;
2. the contractor properly requests a time extension;
3. the owner denies the time extension;
4. the owner demands completion by the original completion date; and
5. the contractor incurs reasonable increased costs caused by its actual acceleration.[50]

The justification for constructive acceleration is that denying a deserved time extension can force additional expenses when work is not performed in the order planned. The contractor's hurdle is to establish that it was entitled to a time extension. In one federal board of contract appeals decision, the government's "cure notice" (a preliminary declaration that the contractor was in breach and must explain how it will cure that breach) issued to the contractor constituted an order to accelerate, where the contractor was at the same time stymied by a design defect.[51]

How is a contractor's acceleration claim different from a claim it had been delayed by the owner?

Suppose that the contractor continues to perform as it would have performed had an extension been granted. This will very likely lead to untimely completion. If the time extension should have been granted and it is granted later (by agreement, by an arbitrator, or by a court), any attempt by the owner to recover actual or liquidated damages would not succeed. This is the same result as a successful delay claim by the contractor. However, the constructive acceleration doctrine allows the contractor to speed up its performance and recover any additional expenses it can establish, as well as to use the wrongful denial of the time extension as a defense against any claim that the owner might bring against the contractor for late completion.[52]

> In this chapter's scenario, after the project suffered a 100-year snowstorm, Prime Builder worked around the clock to complete the work as soon as possible, as the contract's completion date already had passed. If the contractor was not at fault for the project's delays, its efforts would constitute an acceleration.

20.14C Voluntary Acceleration: Early Completion

Delays are so common in construction that attention is rarely paid to the legal effect of the contractor's completing early or claiming it would have completed early had it not been delayed by the owner.

Some owners may find early completion desirable. This can be evidenced by a penalty/bonus clause, as discussed in Section 20.15. Yet early completion may, if unexpected, also frustrate owner plans. For example, suppose a contractor building a factory finishes substantially earlier than planned. The owner may have to take possession before he can install his machinery. Early completion can require payments in advance of resource capabilities. This can be as disruptive as late completion.

AIA Document A101-2007, § 3.3, requires the contractor to substantially complete the project "not later than" a specified date. This appears to give the contractor the freedom to complete early even if it disrupts owner plans.

Construction contracts of any magnitude usually have schedules. It is unlikely the owner will be greatly surprised by early completion. Yet even awareness during construction that performance will be completed earlier than required may not enable the owner to make the adjustments needed to avoid economic losses.

The obligations of good faith and fair dealing require that a contractor notify an owner if it intends to finish much earlier than expected or when it appears this is likely. If this notification is made or the owner is aware of that prospect, the contractor should receive additional compensation if the owner interferes with any realistic schedule under which the contractor would have completed earlier than required by the contract.[53] To establish an early completion claim, a contractor must prove three elements:

1. from the outset of the contract, the contractor intended to complete early;
2. the contractor had the capacity to complete early; and
3. the contractor would have completed early but for the delay caused by the owner.[54]

> In this chapter's scenario, MMI's original schedule specified an early completion date. The architect rejected the schedule as unrealistic. In effect, the validity of AAI's argument goes to the second element of an early completion claim: that the contractor had the capacity to complete early.

20.15 Bonus/Penalty Clauses: An Owner Carrot

Many of a large project owner's contract strategies might be described as threatening the contractor into timely performance. These strategies include liquidated damages clauses, time-is-of-the-essence clauses, the right to accelerate performance, a requirement that the contractor prepare and then periodically update a project schedule, and the right to terminate the contract if the contractor falls behind in performance. By contrast, a bonus/penalty clause is intended to spur the contractor to early completion in return for defined financial incentives.

One form of a bonus/penalty provision is to couple a fixed financial bonus for each day of early completion with a liquidated damages clause that forfeits the same amount for each day of delay. For example, an Ohio statute provides: "When a bonus is offered for completion of a contract prior to a specified date, [a municipality] may exact a prorated penalty in like sum for each day of delay beyond the specified date."[55] Although a bonus component is certainly not necessary to make a liquidated damages clause enforceable, the "mutuality" of the bonus makes enforcement more likely. A bonus clause may also make it appear that the amount actually had been bargained by the contractor and owner.

A bonus clause should not be used unless it is very important to obtain early completion and if the owner has the resources to monitor the work to ensure that speed of performance is not done at the expense of quality. For these reasons, bonus/penalty clauses are most commonly found in public works contracts that are of an urgent nature. The anecdotal evidence is that incentive or bonus clauses are effective. A newspaper article covering replacement of the collapsed Interstate 35W bridge in Minneapolis reports:

> As construction crews worked 24 hours a day, seven days a week, pushing ahead, even through the ferocious Minnesota winter, the job was completed in only 11 months, and for less than the $250 million earmarked by Congress. The builders are eligible for a bonus of up to $27 million for finishing over three months early.[56]

Similarly, use of a bonus clause on highway repair work in Southern California following the January 17, 1994 Northridge earthquake reportedly resulted in "unbelievable" and "remarkable" success.[57]

Of course, not all incentive contracts results in a successful project. In *Koppers Co. v. Inland Steel Co.*, the final construction cost was much higher than the owner had anticipated. The court ruled that the owner's damages claim was not limited by the bonus/penalty agreement.[58]

REVIEW QUESTIONS

1. What is the most complicated time-related issue in a construction project?

2. What are the three categories of delays?

3. What are the two categories for remedies for delay?

4. What are the primary purposes of a force majeure clause and what are some examples of force majeure?

5. What are four questions that must be asked in evaluating a delay claim in which CPM is going to be employed?

6. If a contractor has a time extension that is justified, how is the amount of time determined?

7. What is the damage formula that is most frequently applied by the common law to delayed contractor performance?

8. What is the purpose of a liquidated damages clause?

9. What are the five elements a contractor must prove to establish construction acceleration?

10. What are the three elements a contractor must prove in order to establish an early completion claim?

ENDNOTES

[1]   388 F.3d 844, 857 (Fed.Cir.2004) (footnote omitted).
[2]   97 Fed.Cl. 736, 751 (2011).
[3]   203 Ct.Cl. 521, 492 F.2d 829 (1974).
[4]   Wright, *Force Majeure Delays,* 26 Constr. Lawyer, No. 4, Fall 2006, p. 33.
[5]   487 F.3d 772, 778–79 (10th Cir.2007) (brackets are those of the court; emphasis added).
[6]   Id. at 778–79.
[7]   188 Ct.Cl. 979, 413 F.2d 1147, 1154 (1969).
[8]   *Peter Kiewit Sons' Co. v. Iowa Southern Utilities Co.,* 355 F.Supp. 376, 391 (S.D.Iowa 1973).
[9]   134 S.W.3d 195, 199–200 (Tex.2004).
[10] B. BRAMBLE & M. CALLAHAN, CONSTRUCTION DELAY CLAIMS 11-5 (3d ed. 2000).
[11] 1 Cl.Ct. 50, 554 F.Supp. 433 (1982), aff'd, 728 F.2d 497 (Fed.Cir.), cert. denied, 469 U.S. 1033 (1984).
[12] Wickwire, Hurlbut & Lerman, *Critical Path Method Techniques in Contract Claims: Issues and Developments, 1974–1988,* 18 Pub.Cont.L.J. 338, 341 (1989), analyze these issues in detail.
[13] See Fletcher & Stipanowich, *Successful Forensic Schedule Analysis,* 1 J ACCL, No. 1, Winter 2007, p. 203.
[14] 395 F.3d 1353 (Fed.Cir.2005).
[15] 63 Fed.Cl. 402 (2004).
[16] For a general discussion, see Kutil & Ness, *Concurrent Delay: The Challenge to Unravel Competing Causes of Delay,* 17 Constr. Lawyer, No. 4, Oct. 1997, p. 18.
[17] 387 F.Supp. 1001, 1012-1013 (S.D.Ala.1974), aff'd 551 F.2d 1026, rehearing denied in part, granted in part, 559 F.2d 268 (5th Cir.1977), cert. denied sub. nom. *Providence Hosp. v. Manhattan Constr. Co. of Tex.,* 434 U.S. 1067 (1978).
[18] E.g., AIA Doc. A201-2007, §§ 15.1.2, 15.1.5.1.
[19] *Miami Heart Institute v. Heery Archs. & Eng'rs,* 765 F.Supp. 1083 (S.D.Fla.1991), aff'd, 44 F.3d 1007 (11th Cir.1994) (delay claim by owner against architect).
[20] *Ryan v. Thurmond,* 481 S.W.2d 199 (Tex.Ct.App.1972).
[21] Heckman & Edwards, *Time Is Money: Recovery of Liquidated Damages by the Owner,* 24 Constr. Lawyer, No. 4, Fall 2004, p. 28.
[22] *Carrothers Constr. Co., L.L.C. v. City of South Hutchinson,* 288 Kan. 743, 207 P.3d 231 (2009). See Annot., 12 A.L.R.4th 891 (1982).
[23] 350 F.2d 649, 650 (7th Cir.1965).
[24] Cal.Pub.Cont.Code § 10226.
[25] Supra note 22.
[26] 207 P.3d at 237. However, in *Rohlin Construction Co., Inc. v. City of Hinton,* 476 N.W.2d 78 (Iowa 1991), a *per diem* rate of $400 on a road-building contract was rejected for lack of proof of how this figure was calculated.
[27] *Worthington Corp. v. Consolidated Aluminum Corp.,* 544 F.2d 227 (5th Cir.1976).
[28] *Southwest Eng'g Co. v. United States,* 341 F.2d 998 (8th Cir.), cert. denied, 382 U.S. 819 (1965).
[29] Heckman & Edwards, supra note 21, at 31.
[30] See Section 16.3B.
[31] See Section 19.6.
[32] *Jody Builders Corp.*, PSBCA No. 5047, 08-2 BCA ¶ 33,959.
[33] See Section 18.4B.

[34] See Section 10.4 (review of submittals).
[35] *Howard P. Foley Co. v. J. L. Williams & Co. Inc.*, 622 F.2d 402 (8th Cir.1980) (Arkansas law).
[36] Ibid.
[37] *S. Leo Harmonay, Inc. v. Binks Mfg. Co.*, 597 F.Supp. 1014 (S.D.N.Y.1984), aff'd without opinion, 762 F.2d 990 (2d Cir.1985) (New York law).
[38] Ibid; Jones, *Lost Productivity: Claims for the Cumulative Impact of Multiple Change Orders,* 31 Pub. Cont.L.J., No. 1, Fall 2001, p. 1.
[39] *Port Chester Elec. Constr. Corp. v. HBE Corp.*, 978 F.2d 820, 821 (2d Cir.1992) (New York law).
[40] *J.A. Jones Constr. Co. v. Lehrer McGovern Bovis, Inc.*, 120 Nev. 277, 89 P.3d 1009 (2004). See Lesser & Wallach, *Risky Business: The "Active Interference" Exception to No-Damage-for-Delay Clauses,* 23 Constr. Lawyer, No. 1, Winter 2003, p. 26.
[41] 325 S.C. 129, 480 S.E.2d 447 (1997).
[42] 480 S.E.2d at 448.
[43] Id. at 450. See also *Markwed Excavating, Inc. v. City of Mandan,* 791 N.W.2d 22 (N.D.2010), which applied a no-damages clause in a public contract to uncontemplated delays.
[44] 774 So.2d 50 (Fla.Dist.Ct.App.2000), reh'g denied, Jan. 21, 2001.
[45] *Gregory and Son, Inc. v. Guenther and Sons,* 147 Wis.2d 298, 432 N.W.2d 584, 587 (1988).
[46] Cal.Pub.Cont. Code § 7102.
[47] See Or.Rev.Stat. § 279C.315 and Va. Code Ann. § 2.2-4335.
[48] Colo.Rev.Stat. § 24-91-103.5.
[49] See Ohio Rev. Code Ann. § 4113.62(C)(1) and Wash. Rev. Code Ann. § 4.24.360. The statutes are canvassed in Dunne, *Legislative Update: No Damage for Delay Clauses,* 19 Constr. Lawyer, No. 2, Apr. 1999, p. 38.
[50] *Fraser Constr. Co. v. United States,* 384 F.3d 1354, 1361 (Fed.Cir.2004) (claim denied); *Ace Constructors, Inc. v. United States,* 70 Fed.Cl. 253, 280–81 (2006), aff'd, 499 F.3d 1357 (Fed.Cir.2007) (claim granted).
[51] *Clark Constr. Group, Inc.,* JCL BCA No. 2003-1, 05-1 BCA ¶ 32,843 at pp. 162,559–562.
[52] See generally Dale & D'Onofrio, *Reconciling Concurrency in Schedule Delay and Constructive Acceleration,* 39 Pub. Cont. L.J., No. 2, Winter 2010, p. 161; Gourlay Jr., *Constructive Acceleration and Concurrent Delay: Is There a "Middle Ground"?* 39 Pub. Cont. L.J., No. 2, Winter 2010, p. 231.
[53] *BECO Corp.,* ASBCA No. 27090, 82-2 BCA ¶ 16,124.
[54] *Interstate General Government Contractors, Inc. v. West,* 12 F.3d 1053, 1059 (Fed.Cir.1993) (claim denied); *Fru-Con Constr. Corp.,* ASBCA No. 53544, 05-1 BCA ¶ 32,936 at pp. 163, 160–64 (claim denied).
[55] Ohio Rev. Code Ann. § 731.15. See also Cal.Pub.Cont. Code § 10226.
[56] Nizza, *Minnesota Bridge Reopens a Year After Collapse,* N.Y.Times (Sept. 18, 2008), found at http://thelede.blogs.nytimes.com/2008/09/18/minnesota-bridge-reopens-a-year-after-collapse/?scp=1&sq=I-35W+%26+bonus&st=nyt.
[57] See Tyler, Note, *No (Easy) Way Out: "Liquidating" Stipulated Damages for Contractor Delay in Public Construction Contracts,* 44 Duke L.J. 357, 365–366 (1994).
[58] 498 N.E.2d 1247 (Ind.App.1986), transfer denied Nov. 25, 1987.

CHAPTER 21

The Subcontracting Process: An "Achilles Heel"

Scenario

Mega Manufacturer, Inc. (MMI) sought to build a new factory. MMI retained Awesome Architects, Inc. (AAI), an architectural firm, to create the plans and specifications and to administer the project.

Based upon this design, the owner entered into a $2 million fixed-price contract with Prime Builder, Inc. The prime contract mandated arbitration of any claims between MMI and Prime Builder.

Prime Builder hired several subcontractors. Each subcontract provided that the subcontractor would be paid "if and only after" Prime Builder was paid for the subcontract work by MMI. No subcontract required arbitration of disputes.

Midway through performance, MMI suffered financial reverses and was slow in paying Prime Builder. The contractor in turn stopped paying the subcontractors. Eventually, Prime Builder terminated the prime contract because it had not been paid.

AAI filed a mechanics' lien against the improved property because it had not been fully paid for its design work.

Subcontractor #1 sued Prime Builder for nonpayment.

Subcontractor #1's window supplier filed a mechanics' lien against the improved property.

Subcontractor #2 sued MMI for unjust enrichment.

Besides suing MMI for nonpayment, Prime Builder sued all three subcontractors for defective performance, seeking the cost of repairing their defective work.

21.1 Subcontracting: An Introduction

While the owner under the design–bid–build delivery method looks to the prime contractor to build the project, the contractor often looks to subcontractors to perform different parts of that work. A subcontractor, in turn, may hire a sub-subcontractor to perform portions of the subcontract work. A subcontractor is sometimes called a first-tier subcontractor, and a sub-subcontractor is a second-tier subcontractor.

Contractor, subcontractor, or sub-subcontractor—all are builders. This means that many of the legal issues already discussed in this book with regard to the prime contractor (in particular Section 7.4) to a great degree also apply to subcontractors of any tier. These issues will not be repeated here. Instead, this chapter will focus on legal issues peculiar to subcontractors.

21.2 The Subcontract: Source of Rights and Duties

Basic Structure. The subcontract is the principal source of contract rights and duties between prime contractor and first-tier subcontractor. Similarly, the principal source of contract rights and duties between first and second-tier subcontractors is the sub-subcontract, and so on down the subcontract chain. However, this source is not exclusive. As with any contract, the express terms will be supplemented by terms implied judicially into the subcontract relationship as well as an increasing number of state statutes and regulations.

In addition, each contract on the subcontract chain can be and frequently is affected by contracts higher up the chain. The subcontract relationship is usually affected and may be controlled by terms in the prime contract. Correspondingly, second-tier subcontract relationships are affected and may be controlled by both first-tier subcontract and prime contract provisions. For convenience, discussion focuses on the relationship between prime contractor and first-tier subcontractor.

The current American Institute of Architects (AIA) standard form subcontract is AIA A401-2007. This document is reproduced in Appendix H, which is found on this book's website.

Flow-Through or Conduit Clauses. To tie the various tiers of builders into the same legal framework, the prime contract will usually contain a **flow-through** or conduit clause. Such a clause requires the prime contractor to tie the subcontractors to provisions of the prime contract that affect their work. The legal terminology by which the subcontractor is bound by the terms of the prime contract is called **incorporation by reference**. The referenced parts of the prime contract are incorporated into the subcontract, becoming part of the subcontract itself.

The principal reason for a flow-through or conduit clause is to ensure that subcontractors commit themselves to the *performance requirements* of the prime contract. This way, the use of many contracts does not inadvertently create a gap in the performance obligations of the prime contractor to the owner.

Yet a conduit clause may go beyond this narrow purpose so that it conveys upon the subcontractor *benefits* as well as *obligations*. As an illustration, AIA Document A201-2007, § 5.3—using numbered sentences and italics for sake of clarity—provides:

> [1] By appropriate agreement . . . the Contractor shall require each Subcontractor, to the extent of the Work to be performed by the Subcontractor, to be bound to the Contractor by the terms of the Contract Documents, and to *assume toward the Contractor all the obligations and responsibilities . . . which the Contractor . . . assumes toward the Owner* and Architect. [2] Each subcontract agreement shall preserve and protect the rights of Owner and Architect under the Contract Documents with respect to the Work to be performed by the Subcontractor . . . and shall allow the Subcontractor,

unless specifically provided otherwise in the subcontract agreement, the *benefit of all rights, remedies and redress against the Contractor that the Contractor . . . has against the Owner.*

As indicated in the italicized language, AIA A201-2007, § 5.3, confers both obligations and benefits on the subcontractor. Sentence 1 ties the subcontractor to the *obligations* of the prime contract. By contrast, sentence 2 gives to the subcontractor the same benefits that the contractor has against the owner (unless the subcontract specifically provides otherwise). For example, prime contracts usually contain *force majeure* provisions (see Section 20.6) that excuse delayed performance if certain events occur. Such a provision for the flow-through of benefits gives identical rights to the subcontractor if a claim is made against it by the prime contractor for delay.

Provisions for the flow-through of benefits raise problems of contract interpretation. One of the most common problems involves dispute resolution. In the typical case, the prime contract contains an arbitration clause (requiring disputes between the owner and contractor to be arbitrated), but the subcontract does not. If the contractor brings a lawsuit against the subcontractor, may the subcontractor use a flow-through provision to force the contractor to arbitrate its claim? In *Maxum Foundations, Inc. v. Salus Corp.*,[1] involving AIA contracts, the subcontract incorporated the general conditions into the subcontract and also contained a flow-through provision. The court held that this was sufficient to incorporate the arbitration clause of A201 into the subcontract. The court was influenced by the policy of federal courts favoring arbitration.

The use of flow-through provisions makes understanding the content of subcontracts particularly complex. For example, AIA Document A201-2007, § 2.2.1, gives the prime contractor a right to inquire into the owner's financial arrangements. Does a "benefits" flow-through provision give a similar right to the subcontractor to inquire into the financial sources of the prime contractor?

This "benefits" flow-through provision demonstrates some of the difficulties in construction contract drafting. It seems attractive to include a clause that helps the subcontractor in its often difficult negotiations (if there are any at all) with the prime contractor. However, often drafters do not think about the possible applications of a general clause—both as to substance and procedures. This well-meaning attempt to aid the subcontractors may only create more problems in construction contract administration and in dispute resolution.

> In this chapter's scenario, the prime contract obligated arbitration of any claims between MMI and Prime Builder. If the subcontracts have a flow-through clause similar to A201-2007, § 5.3, the subcontractors will be able to argue that the prime's defective workmanship claim against them must be arbitrated—even if the subcontracts themselves contain no arbitration provision.

21.3 The Subcontractor Bidding Process

21.3A Statement of the Problem

In technical projects requiring specialized skills, the prime contractor may do very little work itself. In such projects, the cost to the prime contractor of work to be done by subcontractors is not likely to be known until the subcontractors submit sub-bids. The prime contractor cannot bid until hearing from all of the prospective subcontractors—something that does not usually occur until close to the time to submit the bid to the owner. In submitting its own bid, the prime contractor relies on the subcontractors' bids.

A prime contractor uses the bids given by subcontractors to compute its bid to the owner. Suppose after the owner awards the prime contractor the contract, a subcontractor withdraws its bid, usually because it contends that its bid had been inaccurately computed or communicated. Can the prime contractor hold the subcontractor to its bid by contending that the prime had relied on the sub-bid in making its own bid?

21.3B Irrevocable Sub-Bids: Promissory Estoppel

The early cases dealing with this problem used traditional contract analysis to allow the subcontractor to revoke its bid. Courts reasoned that, since the contractor and subcontractor had no contract before the prime contractor's bid was accepted by the owner, the contractor and subcontractor still had no contract until one was signed. If the subcontractor withdrew its bid beforehand, it could not be liable to the contractor for breach of contract.

Prime contractors remained without relief until the California Supreme Court in *Drennan v. Star Paving Co.*[2] extended the concept of "promissory estoppel" to the sub-bid process. In that case, the subcontractor had submitted the lowest sub-bid for the paving portion of the work, which was a sub-bid the prime contractor used in computing its overall bid. The prime contractor listed the defendant on the owner's bid form as required by statute. (Subcontractor listing laws on public contracts are discussed in Section 21.16.) The prime contractor was awarded the contract and stopped by the subcontractor's office the next day to firm up the "subcontract." On arrival, the subcontractor immediately informed the prime contractor that it had made a mistake in preparing its bid and would not honor it. The prime contractor sued and was awarded the difference in cost between the subcontractor's sub-bid and the cost of a replacement subcontract.

The *Drennan* court acknowledged that using the sub-bid did not create a bilateral (or two-sided) contract between the plaintiff and the defendant. But the court held the subcontractor to its bid because the bid was a promise relied on reasonably by the prime contractor when it submitted its own bid.

The *Drennan* decision is followed in the Restatement (Second) of Contracts, § 90. To satisfy the doctrine of promissory estoppel, there must be: a clear and definite offer; a reasonable expectation that the offer will induce reliance; actual and reasonable reliance by the offeree; and an "injustice" that can be avoided only by enforcement of the offer. To be clear and definite, the offer must be more than a mere estimate or price quote. Courts rely upon industry custom to conclude that a prime contractor's reliance is reasonable. The final element of injustice is met where the prime contractor will suffer a detriment (financial injury) if the subcontractor's offer is not enforced.

Promissory estoppel is an equitable doctrine, meaning the court seeks to promote a fair outcome. Misconduct by the contractor—such as bid shopping, snatching up an unreasonably low bid, or unreasonable terms of acceptance—will release the subcontractor from being held to its offer (see Sections 21.3C and D). The prime contractor's damages are the difference in price between the withdrawn bid and the replacement subcontractor's cost.

A recent article concludes that a majority of states follow promissory estoppel, as very few reject it outright, and in several states the issue remains undecided.[3]

21.3C Bargaining Situation: Bid Shopping and Peddling

The *Drennan* rule improves the prime contractor's already powerful bargaining position. Although the prime contractor under the *Drennan* rule is not free to delay its acceptance or to reopen bargaining with the subcontractor and still claim a right to accept the original bid, it can at least (for a short period) seek or receive lower bid proposals from other subcontractors.

Under the *Drennan* rule, until the prime contractor is ready to sign a contract with the subcontractor whose bid is used, the subcontractor is not assured of getting the job. Before the award of the prime contract, the plurality of competing prime contractors' bidding on a project tends to diffuse their bargaining power over subcontractors. This competition before the award of the contract should result in lower sub-bids and, consequently, lower overall bids by the prime contractors, which is a definite benefit to the owner. Although subcontractors often wait until the last minute to submit their sub-bids in an effort to minimize the prime contractor's superior bargaining position, a substantial amount of competition still exists among the subcontractors themselves.

After award of the contract, the relative bargaining strengths of the successful prime contractor and the competing subcontractors change drastically. The prime contractor now has a "monopoly" and a substantially superior bargaining position over the subcontractors under the *Drennan* rule. The sub-bids used provide the prime contractor with a protective ceiling on the cost of the work with no obligation to use the subcontractors. The prime contractor is therefore free to look elsewhere for yet a better price and is able to increase its profits by engaging in postaward negotiations.

Postaward negotiations have become controversial. Sometimes they are called "bid shopping"; the prime contractor uses the lowest sub-bid to "shop around" with the hope of getting still lower sub-bids. "Bid peddling" is the converse with other subcontractors attempting to undercut the sub-bid to the prime—in essence engaging in a second round of bidding. Subcontractors often refer to these postaward negotiations as "bid chopping" and "bid chiseling." (These tactics also can be used before prime bids are submitted.)

Whether bid shopping (initiated by the prime contractor) or bid peddling (initiated by subcontractors), the end result is that postaward competition hurts both owners and subcontractors and benefits only the prime contractor. Both subcontractors and owners condemn postaward competition. A Connecticut appellate court, in *Johnson Electric Co., Inc. v. Salce Contracting Assocs., Inc.*,[4] held that bid shopping by a prime contractor constituted an "unfair trade practice," which was prohibited by statute.

Owners too are hurt by bid shopping that occurs postaward, since the prime contractor keeps the difference between the original sub-bid and the "shopped" subcontract price rather than pass these savings on to the owner. For this reason, bid shopping may violate the Constructor Code of Ethics Rule II, which prohibits a constructor from engaging in a "deceptive practice" (see Section 13.4).

The mere threat of bid shopping distorts the bidding process. Subcontractors who fear bid shopping often wait until the last minute to submit their sub-bids to the prime contractor to give the prime contractor as little time as possible to bid shop. This last-minute rush is the cause for many mistakes by both subcontractors and prime contractors. Some subcontractors simply refrain from bidding on jobs where bid shopping is anticipated to save the expense of preparing a bid. To that extent, competition among subcontractors is diminished, and higher prices can result.

In addition, subcontractors feel they must pad their bids to make allowance for the eventual postaward negotiations. This "puffing" raises the cost to the owner, as the inflated bid is the bid the prime contractor uses to compute its overall bid. Any subsequent negotiations that result in reducing the price benefit only the prime contractor.

Prime contractors respond by noting that sub-bids are often unresponsive to the specifications and require further clarification and negotiation. This may be especially true when prime contractors are dealing with subcontractors with whom they have never dealt. They must investigate the subcontractor's reputation and work experience before making a firm contract. A prime contractor who goes through the effort of deciding upon a specific subcontractor is, at the very least, inconvenienced by a withdrawal of a sub-bid on the eve of project commencement. In *Arkansas Contractors Licensing Board v. Pegasus Renovation Co.*,[5] the court affirmed a licensing board ruling that a subcontractor who repeatedly backed out of bids engaged in "misconduct in the conduct of the contractor's business," and these actions justified revocation of the subcontractor's license.

As discussed in Section 21.16, some states have passed listing laws that require a public works prime contractor to list their subcontractors at the time they submit their bids and then restrict the grounds upon which the listed subcontractors may be substituted. One purpose of these statutes is to prevent bid shopping.[6]

21.3D Avoiding *Drennan*

Subcontractors can employ different strategies to avoid their sub-bids becoming firm offers that bind them and not the contractor. They can call their bids "requests for the prime to make offers" to them or "quotations," which are given only for the prime contractor's convenience. They can state in their bids that the bid is provided for information only and is not a firm offer.[7] They may also try to include language stating that using the bid constitutes an acceptance that ties the prime contractor to them. They may refuse to submit bids unless they receive a promise by the prime contractor to accept the bid if it is low and the prime contractor is awarded the contract. However, the difficulty with most methods of avoiding the *Drennan* rule is that either the subcontractors do not have the bargaining power to implement them or the process does not make it convenient to use them.

Courts will not apply promissory estoppel if the conditions for finding a contract are not present. The *Drennan* rule can be avoided if many crucial areas have been left for further negotiation or if there is no "meeting of the minds" as to the scope of work. Also, promissory estoppel is inappropriate if the prime contractor does not accept the bid within a reasonable time or proposes a subcontract that contains new unreasonable or onerous terms. As an example of the latter situation, in *Hawkins Construction Co. v. Reiman Corp.*,[8] there was no binding subcontract when the prime contractor returned a document containing new terms, including a no-damages-for-delay clause (see Section 20.13B).

21.4 Subcontractor Selection and Approval: The Private Owner's Perspective

Owners wish to have competent contractors building their projects. They want subcontractors treated fairly and given an incentive to perform the work expeditiously. Some owners contract directly with the specialty trades to avoid the prime contractor as an "intermediary" between owner and specialized trades. Some contract directly with the specialized trades and assign those contracts to a main or prime contractor. Some owners dictate to the prime contractor which subcontractors will be used; this is known in England as the *nominated subcontractor* system.

These methods of direct intervention are not common in the traditional contracting system. Some owners leave subcontracting exclusively to the prime contractor. Others take a role that gives them *some* control but does not involve the owner *directly* with the subcontractor. Using the prime contractor as a buffer is done for administrative and legal reasons. The subcontracting system requires a well-defined organizational and communication structure under which each participant knows what it must do and with whom it must deal. From a legal standpoint, the owner does not want to be responsible for subcontract work, does want the prime contractor to be responsible for defective subcontract work, and does not want to be responsible if subcontractors are not paid.

One "part-way" control is to require that prime contractors list their subcontractors at the time they make their bids. As noted Section 21.16, statutes in some states, called *listing laws*, impose this on prime contractors in public projects. Listing laws usually regulate substitution of listed subcontractors by providing specific justifications for substitutions and a procedure for determining the grounds for replacing one subcontractor with another.

Listing is used by the American Institute of Architects (AIA) in a more limited way. AIA Document A201-2007, § 5.2.1, requires the contractor "as soon as practicable after award of the contract" to furnish the owner and architect the names of subcontractors and those who will furnish materials and equipment fabricated to a special design. The owner and architect have 14 days to raise any "reasonable objection" to a proposed subcontractor. A price adjustment is made if the contractor must use a replacement.

The Engineers Joint Contracts Documents Committee (EJCDC), in its C-700, ¶ 6.6B (2007), does not require that the identity of subcontractors be furnished—only if the Supplementary Conditions require identification of subcontractors and suppliers are their names submitted to the owner for acceptance. In that event, the owner can revoke any acceptance "on the basis of reasonable objection after due investigation." Here too, a price adjustment is made if the contractor must use a replacement.

But ¶ 6.6A states that the prime will not use a subcontractor "against whom Owner may have reasonable objection." That paragraph also provides, as does A201-2007, § 5.2.2, that the prime cannot be forced to use a subcontractor against whom the prime has "reasonable objection."

Owner intervention into the relationship between prime contractor and subcontractor may be essential, but carries risks. If the risks are so great but the need for intervention so strong, the owner should consider a method other than the traditional contracting system, such as separate contractors (multiple primes).

21.5 Subcontractor Payment Claims Against Prime Contractor: "Pay When Paid" Clause

One particularly sensitive area in the subcontract relationship relates to money flow with subcontractors frequently contending that they invest substantial funds in their performance and are entitled to be paid as they work and to be completely paid when their work is completed. This issue surfaces around two concepts: line item retention and payment conditions. The subcontractors wish to divorce their work and payment for it from the prime contract. They argue *for* line item retention and *against* payment conditions.

Line item retention gives the subcontractor the right to be paid after it has fully performed. Delay usually occurs because the owner holds back retainage, which is a designated amount of the contract price for the entire performance of the prime contractor and all subcontractors. When all the work is completed and the project accepted, the owner will pay the retainage. However, early finishing subcontractors may have to wait a substantial period of time after they have completed their work for the retention allocated to their contract because the entire project is not yet completed. They would like to disassociate their contract from the rest of the subcontracts and the prime contract.

A payment condition (**pay when paid** clause or more realistically, from the position of the prime, "pay *only* if paid") makes payment to the prime contractor a condition of the prime contractor's obligation to pay the subcontractor. Prime contractors seek to create such a condition by including language in the subcontract stating that the prime contractor will pay "if paid by the owner," "when paid by the owner," or "as paid by the owner." An endless number of cases have interpreted this language. Does the language create a *condition* to payment—in which case the prime contractor does not have an obligation to pay the subcontractor until the owner has paid the prime contractor for the subcontract work—or does the language simply indicate that the payment flow contemplated some delay with payment in any event being required after a reasonable time expires?

All courts agree (or at least so they state) that the parties can make a payment condition under which the subcontractor assumes the risk that it will not be paid for its work. The legal issue has centered around the requisite degree of specificity needed to create such a condition.[9]

> In this chapter's scenario, each subcontract provided that the subcontractor would be paid "if and only after" Prime Builder was paid for the subcontract work by MMI. MMI failed to pay the prime, who then stopped paying the subcontractors. If the subcontract language is viewed as creating a condition to payment, Prime Builder will have a defense to Subcontractor #1's claim for breach of contract.

The AIA's standard form subcontract does not include a payment condition. Document A401-2007, §§ 11.3 and 12.1 both state that if the contractor does not receive payment "for any cause which is not the fault of the Subcontractor," then the contractor shall pay the subcontractor "on demand." (Under § 11.3, the subcontractor would receive a progress payment; under § 12.1 it would receive final payment.) In addition, § 4.7 allows the subcontractor to stop work if payment is over fourteen days late through no fault of the subcontractor. Most prime contractors would rather advance payment to a fault-free subcontractor than risk that subcontractor's stopping work.

The struggle between primes and subcontractors in some states has moved to the legislative arena. The statutes take many forms. North Carolina bans payment conditions.[10] Missouri and Maryland prohibit payment conditions from being a defense to a subcontractor's mechanics' lien rights.[11]

Finally, suppose the prime contractor delays payment, rather than denies it has a duty to pay. In addition to common-law rights, prompt payment statutes may give subcontractors additional remedies against the prime contractor (see Section 17.8).

21.6 Subcontractor Payment Claims Against Property, Funds, or Entities Other Than Prime Contractor

Lack of a contractual relationship between the owner and subcontractors makes subcontracting the Achilles heel of the construction process. In a design–bid–build project delivery system, the common law gives subcontractors very little leverage by which to obtain payment in the face of a recalcitrant prime contractor.

Yet no major construction project can go forward without extensive use of subcontractors. Increasing the probability of subcontractor payment is essential to a thriving construction industry. For this reason, both legislatures and the industry have devised payment mechanisms for subcontractors in addition to a breach of contract action against the prime contractor. Several of the following sections address these additional mechanisms.

21.7 Mechanics' Liens

Legal Complexity. Participants in the construction process who can trace their labor and materials in various ways into real property improvements of another are given lien rights against the property in the event they are not paid by the party who has promised to pay them. A lien is a security interest (or right) in the improved property. The remedy accorded a lien holder is the right to demand a judicial foreclosure or sale of the property and be paid out of the proceeds.

Mechanics' lien laws are unique to the United States and Canada. Within the United States, they are complicated and vary considerably from state to state. For that reason, it would be inadvisable to attempt a summary of all aspects of these statutory protections accorded certain participants in the construction process. Instead, this discussion focuses on rationales for such protection and salient features and current criticisms of lien laws.

Mechanics' lien statutes create many requirements for the creation of lien rights. Who will be accorded statutory protection, how such protection is achieved, and the nature of the protection are often resolved by reference to the complicated, almost unreadable statutes. Frequently, such interpretation questions are resolved by holding that the statutes are designed to prevent unjust enrichment by protecting unpaid subcontractors, who are the intended beneficiaries of the legislative protection.

Matters can become even more complicated in jurisdictions that state that the standards for perfecting a mechanics' lien will be *strictly* required, but once the lien has been perfected, the remedy will be administered *liberally*. Generally, any failure to comply with lien perfection requirements will invalidate the lien. Substantial compliance is insufficient. Strained interpretations and language distortion often result, yet construing these statutes to protect subcontractors does not invariably result in lien protection. The principal legacy of this approach is legal uncertainty and unpredictability.

Overview. The usual justification given for granting these liens is unjust enrichment. Those whose labor or materials have gone into the property of another should have lien rights in the property when they are not paid as promised.

The unjust enrichment rationale loses some of its attractiveness when "double payment" is considered. An owner who pays the prime contractor may have to pay again to an unpaid subcontractor if it wants to remove the lien. States increasingly protect residential owners from double payment. Michigan prevents a lien from attaching to a residence to the extent of owner payments made to the prime contractor, while Maryland caps the lien at the amount by which the owner is indebted under the prime contract at the time of the lien notice.[12] However, the Texas "trapping statute" allows subcontractors to file a pre-lien notice with the owner. The owner is personally liable for payments made to the prime contractor after receipt of the notice.[13]

Mechanics' lien laws provide a quick and effective remedy for unpaid workers who cannot wait until a full trial to collect their wages. Quick and certain remedies can induce workers to work on construction by assuring them they will be paid.

Lien laws were first enacted to spur development of Washington D.C. more than 200 years ago. Over time, the list of lien beneficiaries was expanded to include not only laborers but also all those who participate directly in the construction process. The state gives credit to prime contractors by granting subcontractors lien rights, which encourages people to furnish labor and materials for construction. This state credit was especially needed to bolster an unstable construction industry composed of many contractors unwilling or unable to pay subcontractors and suppliers. This credit extension is probably the principal reason for giving lien rights today.

Expansion of lien laws is undoubtedly also traceable to the realities of the political process. Once some participants in the construction process have received lien rights on a frequently asserted unjust enrichment theory, it was relatively easy to expand the list of lien beneficiaries. Those who might oppose lien expansion, such as owners, are often unrepresented as an organized group in the legislatures. This too may have accounted for expansion of lien beneficiaries and lien rights.

The desire by participants to expand mechanics' lien rights is understandable, because the mechanics' lien is a more effective remedy than a money award. However, legislatures that respond to such pressures often ignore the fact that these liens come at the expense of others: In the case of subcontractor liens, they come at the expense of the unsecured creditors of prime contractors; in the case of prime contractor liens, they come at the expense of the unsecured creditors of the owner.

As noted, mechanics' liens are security interests on the improved property. Security interests cannot be filed against public property; therefore, mechanics' liens are not available on public works projects. Instead, those subcontractors and suppliers either have stop notice rights (see Section 21.9) or rights against a payment bond (see Section 21.8).

Finally, because liens encumber the property (that is, interfere with the owner's ownership interest) before entitlement to the lien is established, some owners have argued that mechanics' liens are unconstitutional. This position has been rejected.[14]

Claimants and Lienable Work. Those entitled to liens are usually set forth in the statute, and the list typically is lengthy and expanding. Among the many lien recipients are prime contractors, subcontractors, suppliers, laborers, and design professionals. One unique problem for design professionals is that mechanics' liens become effective only upon the first visible improvement to the property. (Visible improvement also defines lien priority, a topic discussed later in this section.) If a design professional performs work in preparation for the improvement to begin—such as the design or planning activities—but the project is cancelled before physical work starts, the design professional lien right is not created and her services are not lienable.[15]

Lien claimants are divided into two principal categories: those who have direct contract relations with the owner and those who do not. Typical illustrations of the first are design professionals and prime contractors. Illustrations of the second are subcontractors, laborers, and suppliers to prime contractors and subcontractors. Owners can avoid liens by paying their design professionals and prime contractors. But because the majority of the difficult problems are generated by lien claimants not connected by contract with the owner and because this chapter focuses on subcontractor problems, the discussion centers around the second class of lien claimants.

> In this chapter's scenario, AAI filed a mechanics' lien against the improved property because it had not been fully paid for its design work. Whether AAI will be able to file a lien for work done before the first physical improvement of the property will depend on the wording of the mechanics' lien statute. The fact that the plans were used in the construction—being integral to the actual physical improvement—will weigh in favor of lienability.

Because of the lien statutes, whether particular work qualifies for a lien is often unclear. Some statutes use generic terms, such as *improvement, building,* or *structure*. Others that attempt to be detailed do not always keep up with changes in the construction process. For example, liens have been denied where the lien claimant had placed engineering stakes and markers;[16] where a claimant had graded and installed storm and sanitary sewers, paving, and curbing,[17] and where a claimant had performed demolition work.[18] Most states deny liens to lessors of equipment used in construction unless the items are consumed in the process of use.[19]

> In this chapter's scenario, Subcontractor #1's window supplier filed a mechanics' lien against the improved property. As the supplier of a first-tier subcontractor, the supplier will probably be within the class of claimants entitled to file a lien. It may have to prove the windows were actually incorporated into the building for its claim to be lienable. If so, it will hope the building (when sold) generates enough income to pay off other liens with higher priority.

In these examples, the claimants contributed directly to the physical improvement. However, a variety of businesses surrounding the construction project are essential for its existence, yet do not contribute labor or materials to the project and so are not "claimants" for purposes of the lien laws. Thus, mechanics' liens were denied for: an attorney who provided legal services to a developer;[20] an insurer seeking recovery of unpaid premiums by a subcontractor;[21] a workers' compensation insurer seeking unpaid premiums by a contractor;[22] a creditor who paid subcontractors directly on behalf of prime contractor;[23] and a business entity that had helped subcontractors assemble a workforce, dealt with payroll, and advanced funds.[24]

In short, both the identity of the claimant and the type of service or item that may be liened can be subject to dispute.[25]

Lien Priority. If the lien amounts exceed the value of the property after parties with security interests that take priority are paid, all claimants are treated equally. In some states, liens of prime contractors are subordinated to other lien claimants, whereas in other states, laborers are sometimes given preference.

As between lien claimants and others with security interests in the property (such as the seller of the property who retains a security interest or a construction lender), the party who perfects its interest first takes priority. For this reason, lenders will not make construction loans if work has begun on the project for fear that their security interest will not take priority over those who have already begun work. In *Imperial Developers, Inc. v. Calhoun Development,*

LLC,[26] the court described an industry practice of lenders photographing the worksite the day the loan mortgage is filed to prove there was no visible improvement at the time so as to retain their lien priority.

Accordingly, lenders perfect their security interest before work begins. As a result, subcontractor lien claims can become valueless if trouble develops and prior security holders foreclose on the property. This occurs frequently because of market imperfections. The lender is typically able to buy in at less than the amount owing on the construction loan because the liens of other claimants are not extensive enough to justify bidding in or they may not have sufficient funds to be able to compete with the lender.

Claimants' Entitlement to Compensation. Lien claimants must establish that they have performed work under the terms of a valid contract. As has been emphasized throughout this book, there has been a proliferation of legal controls on the construction process. (Licensing laws, land use controls, building and housing codes, and the controls imposed on projects built in part with public funds are illustrations.) As a result, a lien may be denied because of a technical violation of a law or regulation. For example, in *Stokes v. Millen Roofing Co.*,[27] the court denied a mechanics' lien to an unlicensed contractor. The case produced four opinions—the majority, two concurrences and a dissent—all addressing the same issue: the equity of preventing payment based solely on the contractor's unlicensed status where the contractor produced quality workmanship and the owners hired it knowing it was unlicensed.

Subcontractors in Massachusetts must establish that their claim is for money "due or to become due under the original contract." This means that a breach by the prime contractor—without the fault of the subcontractor—deprives the subcontractor of a lien right.[28]

No-Lien Contracts. As noted earlier, one criticism of mechanics' lien laws is that they can compel an inexperienced owner to pay twice for the same work. The owner may pay the prime contractor and then have to pay an unpaid subcontractor to remove a lien.

One way owners seek to avoid this situation is through the use of no-lien contracts imposed upon the subcontractors. The enforceability of such no-lien contracts depends on whether the subcontractor must have consented to the waiver.[29] Legislatures have started to regulate these waivers.[30]

What is the effect of a prime contractor's failure to deliver the owner a lien-free project on a no-lien contract? In *Solar Applications Engineering, Inc. v. T.A. Operating Corp.*,[31] the owner argued that the subcontractors' filing of liens defeated the contractor's entitlement to final payment. The court instead interpreted the no-lien requirement as a covenant rather than a condition. The owner may reduce the retainage by the amount of subcontractor liens, but the owner may not use the liens as grounds for forfeiture of the retainage.

Criticism. The brief examination of some of the salient characteristics of mechanics' lien protection has revealed the weaknesses of the remedy. Most important, the statutes are complex and change frequently. Carelessness in compliance can result in the lien being lost. On the other hand, strict compliance is costly—in smaller jobs, perhaps more than the value of the lien. The lien is most important in construction projects that fail. It is in such situations that it is most likely that lien claimants will find their claims wiped out because prior security holders have foreclosed and the funds left over for lien claimants are nonexistent. Because mechanics' liens differ from state to state, a large contractor with operations in several states must deal with different lien creation and enforcement requirements.

These deficiencies have led some critics to contend that the mechanics' lien system gives subcontractors a false belief that they will receive compensation even in the case of default by the owner or contractor. In the view of these critics, abolition of mechanics' lien acts would not imperil subcontractors' payment options and might well compel them to demand contractual protections, such as bonding.[32] In any event, subcontractors have sought other forms of legislative and judicial relief, as discussed in the remainder of this chapter.

21.8 Payment Bonds

As explained in Section 25.10, a payment bond is a guarantee of payment backed up by a third party (a surety). A payment bond can be a substitute for a mechanics' lien right. Mechanics' liens are not available on public contracts, because foreclosure by a lien claimant is not allowed against public buildings. As a result, the federal government enacted the Miller Act, which is a compulsory bonding system. Many states have enacted comparable legislation for state construction projects. A private owner may also require the prime contractor to acquire a payment bond to prevent (or at least reduce the likelihood of) a mechanics' lien being filed against the improved property.

In a related manner, a private owner may obtain a lien-release bond. This extinguishes the lien and redirects the claimant's remedy exclusively to that bond.

21.9 Stop Notices

The ineffectiveness of mechanics' liens in private work and the unavailability in public work has led some states to supplement mechanics' lien protection by enacting stop notice laws. A stop notice imposes a lien not on the improved property but on the unpaid contract funds. Like mechanics' lien laws, these statutes vary from state to state.[33] Essentially, compliance stops the payment flow. For this reason, a stop notice is more effective than a mechanics' lien in obtaining payment.

21.10 Trust Fund Legislation: Criminal and Civil Penalties

A **trust** is a financial arrangement under which one person holds money or property for the benefit of another person. The person holding the item is the *trustee*, and the person for whom the item is held is the *beneficiary* of the trust. A trustee is subject to a fiduciary duty (see Section 2.3) in favor of the beneficiary.

A number of states have enacted legislation designed to prevent prime contractors from diverting funds received from the owner and meant for payment of subcontractors and suppliers. These trust fund statutes make the prime contractor a trustee with a statutory duty to use the trust funds (the owner's payment) to pay the trust fund beneficiaries (the subcontractors and suppliers). Some trust fund statutes apply to public works, some to private works, and some to both.

To enforce the trust, the court can order an accounting (determine what has been paid and to whom), set aside any unauthorized payments, award damages for breach of trust, or terminate the contractor's authority to apply trust assets. However, because funds often disappear, the main effect of trust fund statutes is to provide harsh penalties for those who violate the trust. Sanctions for violation of the trust vary. Frequently, the breach of trust caused by diversion is a crime. For example, in New York, a diverting prime contractor is guilty of larceny.[34] In New Jersey the contractor can be convicted of theft.[35] In Oklahoma, the managing officers of a corporate contractor are guilty of embezzlement.[36] California does not have a trust fund statute, but it does make it a crime to divert funds intended for subcontractors and suppliers and to submit a false voucher to receive payment from a construction lender.[37] The purpose of penal (criminal) sanctions is to deter diversion and can result in payments to subcontractors and suppliers.[38]

21.11 Nonstatutory Claims Against Third Parties

Suppose an unpaid subcontractor has no remedy against the prime contractor (perhaps because the subcontract contained a pay-if-paid clause, or the contractor had filed for bankruptcy), failed to file a timely mechanics' lien, and the project did not include a payment bond. In this dire situation, a subcontractor may seek relief against third parties.

Owners. Sometimes, especially on a troubled project in which the prime contractor was financially strapped, an owner may reach out to the subcontractors and urge them to stay on the job. The owner may also make statements that he will "take care of" the subcontractor or words to that effect. In this situation, the subcontractor may attempt a claim against the owner, if not based on breach of contract then perhaps promissory estoppel.

More commonly, owners are sued under a theory of unjust enrichment. Claimants assert that their work or materials have benefited the owner and that it would be unjust for the owner to retain this benefit without paying the claimants.

As a rule, such claims have not been successful because the owner can show it has paid someone (usually the prime contractor) or that the retention of benefit was not unjust because the claimant could have protected itself by using the statutory remedies.[39] One case granted recovery where the owner had not paid anyone, and the prime contractor left town before receiving payment.[40]

> In this chapter's scenario, Subcontractor #2 sued MMI for unjust enrichment. Since MMI did not pay Prime Builder, the owner was clearly enriched by Subcontractor #2's work. However, MMI may contend that the Mechanics' Lien Act gives Subcontractor #2 a remedy that displaces its common law rights. Courts generally find that the statutory and common law rights are cumulative and would reject the owner's defense.

Design Professionals. Unpaid subcontractors or suppliers may attempt to recover from the design professional if the latter did not perform in accordance with the professional standards established by the law when approving payments to the prime contractor. An Oklahoma court found an architect on a public works project liable to subcontractors for failing to verify that the prime contractor had obtained a payment bond.[41] A South Carolina court ruled that an architect had a duty of care to ensure a prime contractor paid its subcontractors.[42]

21.12 Joint Checks

One way owners and prime contractors seek to avoid liens and other claims is to issue joint checks. For example, the owner may issue a joint check to the prime and to the subcontractors, and the prime contractor may issue joint checks to subcontractors and their sub-subcontractors and suppliers. This is discussed in Section 17.4.

21.13 Performance-Related Claims Against Prime Contractor

Suppose a subcontractor asserts that its cost of performance was wrongfully increased because of acts or omissions by the prime contractor or someone for whom the prime contractor was responsible. This type of claim broadly mirrors disputes between an owner and a prime contractor with this distinction: subcontractors as a general rule have less economic power than do prime contractors. As phrased by one commentator, "[i]t is always a battle of the forms between the subcontractor and the general contractor with the subcontractor most often coming out on the losing end."[43]

Generally, the law implies an obligation on the part of the prime contractor to take reasonable measures to ensure that the subcontractor can perform expeditiously and is not unreasonably delayed. However, terms will not be implied if express provisions in the subcontract directly deal with this matter. Commonly, subcontract provisions place affirmative responsibilities on the prime contractor, such as requiring that the site be ready by a particular date or the work be in a sufficient state of readiness by a designated time to enable the subcontractor to perform specific work.

Conversely, contract provisions may indicate that the subcontractor has assumed certain risks regarding the sequence of performance. For example, the subcontract might require that the work be performed "as directed by the prime contractor," and such a provision would give the prime contractor wide latitude to determine when the subcontractor will be permitted to work. For example, in *Keeney Construction v. James Talcott Construction Co., Inc.*,[44] the court ruled that an "as directed" clause precluded the subcontractor's delay claim.

Additional clauses in the prime contract to which the subcontract refers or that are incorporated into the subcontract are also relevant. The contract can specify that certain delay-causing risks were contractually assumed risks or grant only a time extension.

In addition, contract clauses may control by denying recoverability of delay damages. For example, a no-damages-for-delay clause that limited the subcontractor to a time extension in the subcontract, especially if tracked with a no-damage clause in the prime contract, would very likely preclude recovery of delay damages by the subcontractor against the prime contractor for delays caused by owner or prime contractor.[45]

The problems of tracking or parallelism in prime contracts or subcontracts complicate these claims. As a general rule, prime contracts and subcontracts are parallel in terms of rights and responsibilities. For example, should a prime contractor be held liable for delay to the subcontractor for acts caused by the owner when the prime contractor is precluded from recovering from the owner because of a no-damage clause?

The bargaining situation sometimes permits the prime contractor to better its position through the subcontract. For example, the prime contractor might be able to include a no-damage clause in the subcontract when the prime contract allows the prime contractor its delay damages against the owner. Although a careful subcontractor might be able to preclude this possibility, time or realities of the process often frustrate parallel rights. An abuse of prime contractor bargaining power can result in discontented subcontractors and poor performance. It is becoming increasingly common for owners to insist that prime contracts contain provisions requiring the prime contractor to give the subcontractor benefits parallel to those given the prime contractor in its contract with the owner (see Section 21.2).

21.14 Pass-Through Claims Against Owner: Liquidating Agreements

When a project is disrupted or delayed by the owner or another responsible entity, all participants incur expenses. Frequently, they seek to transfer these expenses to the owner. Because of the high costs of pursuing claims, subcontractors often pool their claims with those of the prime, and the prime will present them against the owner.

One mechanism for accomplishing this is a liquidating agreement. Under such an agreement, the prime confesses liability to the subcontractor for owner-caused delay (for example), the subcontractor releases the prime from all other liability, and the subcontractor is relegated to whatever delay damages the prime can recover from the owner. In addition to being more efficient than multiple, individual subcontractors claims, such liquidating agreements avoid the problem of the subcontractor not being in privity of contract with the owner.[46] These arrangements are common on public contracts, as discussed in Section 21.16.

21.15 Owner Claims Against Subcontractors

Much of the prior discussion in this chapter has focused on a variety of subcontractor claims—both common law and statutory—against different project actors. Of course, a subcontractor may be exposed to a claim of defective construction brought against it by the prime. As noted in Section 21.1, this analysis parallels an owner claim against the prime and so is not discussed here. However, this does not exhaust the topic of subcontractor liability for defective construction.

Suppose the owner believes subcontract work is defective, but it has no remedy against the prime contractor. May the owner sue the subcontractor directly for economic damages?

Owners employ three theories to recover losses they have suffered that were caused by a subcontractor breach. First, an owner claims it is a third-party beneficiary to the subcontract. Here, the owner is contending that the prime contractor and subcontractor contracted for the owner's benefit, and as a result, the owner may sue the subcontractor for breach of contract.

Second, the owner may assert that the subcontractor's breach was negligent in that it failed to live up to the legal standard of care. The owner would contend that the subcontractor owed the owner a duty of care and had breached that duty, causing the owner's injury. An Illinois appellate court in *Minton v. The Richards Group of Chicago*[47] ruled that, while the prime contractor was judgment proof (unable to pay a judgment, perhaps because of bankruptcy), the economic loss rule did not apply to bar the owner's negligence claim against the subcontractor.

Third, the owner may contend that the prime contractor was merely a conduit between owner and subcontractor—essentially a contention based on the prime contractor's contracting as an agent of the owner.

21.16 Public Contracts

Special rules apply to subcontractors on public works projects. This section addresses two of these rules: the competitive bidding process and owner liability.

Some states have passed subcontractor *listing laws* which require prime contractors to list their subcontractors at the time they make their bids.[48] Although undoubtedly some impetus for such laws came from subcontractor trade associations, as expressed by one court, listing laws also were enacted "because of the indirect effects of ruthless bid-shopping, which include poor workmanship provided by subcontractors who, desperate to retain their subcontracts, shave their profits and expenses below a level which guarantees quality work, subcontractors' insolvencies, and construction workers' lost wages."[49]

A prime contractor who listed a subcontractor is not stuck with that subcontractor regardless of subsequent events. Listing laws usually regulate substitution of listed subcontractors by providing specific justifications for substitutions and a procedure for determining the grounds for replacing one subcontractor with another.[50]

Subcontractors who wish to sue a public owner are, in the absence of privity, confronted with the defense of sovereign immunity. The federal government is immune from contract liability to parties with whom it does not have an express or implied contract.[51] For this reason, any subcontractor who wishes to bring a claim against the government must enter into a "pass-through agreement" with the prime contractor. In such an agreement, the contractor admits liability to the subcontractor and the contractor then brings the subcontractor claims against the government. In return, the subcontractor agrees to be satisfied by any recovery the contractor receives and passes through to the subcontractor. However, under the *Severin* doctrine,[52] the prime cannot prosecute the claim unless it is at least potentially liable to the subcontractor. The Federal Circuit Court of Appeals has placed upon the government the burden of asserting and proving that the prime contractor is not responsible for the subcontractor's costs.[53] State courts have adopted the *Severin* doctrine.[54]

REVIEW QUESTIONS

1. What is the principal reason for a "flow-through" or conduit clause?

2. What elements are needed to satisfy the doctrine of promissory estoppel?

3. What is the difference between bid shopping and bid peddling?

4. What is the difference between line-item retention and payment conditions?

5. What is a mechanics' lien? What remedy is accorded a lien holder?

6. When do mechanics' liens become effective for design professionals?

7. If there are several lien claimants with security interests in a property, which claimant has priority?

8. What is the difference between a payment bond and a stop notice?

9. What is a liquidating agreement and how does it function?

10. What are the three theories owners employ to recover losses they have suffered that were caused by subcontractor breach?

ENDNOTES

[1] 779 F.2d 974 (4th Cir.1985), appeal after remand, 817 F.2d 1086 (4th Cir.1987).
[2] 51 Cal.2d 409, 333 P.2d 757 (1958).
[3] Kovars & Schollaert, *Truth and Consequences: Withdrawn Bids and Legal Remedies,* 26 Constr. Lawyer, No. 3, Summer 2006, p. 5.
[4] 72 Conn.App. 342, 805 A.2d 735, certification denied, 262 Conn. 922, 812 A.2d 864 (2002).
[5] 347 Ark. 320, 64 S.W.3d 241 (2001).
[6] For a discussion of this and other strategies for countering bid shopping and bid peddling, see Gregory & Travers, *Ethical Challenges of Bid Shopping,* 30 Constr. Lawyer, No. 3, Summer 2010, p. 29.
[7] *Fletcher-Harlee Corp. v. Pote Concrete Contractors, Inc.,* 482 F.3d 247 (3d Cir.2007).
[8] 245 Neb. 131, 511 N.W.2d 113 (1994).
[9] For further discussion, see Alsbrook, *Contracting Away an Honest Day's Pay: An Examination of Conditional Payment Clauses in Construction Contracts,* 58 Ark.L.Rev. 353 (2005); Hill & McCormack, *Pay-If-Paid Clauses: Freedom of Contract or Protecting the Subcontractor from Itself?* 31 Constr. Lawyer, No. 1, Winter 2011, p. 26.
[10] N.C.Gen.Stat. § 22C-2.
[11] Mo.Rev.Stat. § 431.183; Md.Real Prop. Ann. Code § 9-113(b).

[12] Mich.Comp.Laws Ann. § 570.1118a. Md.Real Prop. Code § 9-104(f)(3), expansively interpreted to apply to homeowner renovations in *Ridge Heating, Air Conditioning & Plumbing, Inc. v. Brennen,* 366 Md. 336, 783 A.2d 691 (2001).

[13] Tex.Prop.Code Ann. §§ 53.081–53.084, applied in *Don Hill Constr. Co. v. Dealers Elec. Supply Co.,* 790 S.W.2d 805 (Tex.Ct.App.1990).

[14] *Vernon Hills III Ltd. Partnership v. St. Paul Fire and Marine Ins. Co.,* 287 Ill.App.3d 303, 678 N.E.2d 374 (1997) and *Gem Plumbing & Heating Co., Inc. v. Rossi,* 867 A.2d 796 (R.I.2005).

[15] See *New England Sav. Bank v. Meadow Lakes Realty Co.,* 243 Conn. 601, 706 A.2d 465 (1998). For a review of the lien rights of design professionals, see Annot., 31 A.L.R.5th 664 (1995).

[16] *South Bay Eng'g Corp. v. Citizens Sav. & Loan Ass'n,* 51 Cal.App.3d 453, 124 Cal.Rptr. 221 (1975).

[17] *Sampson-Miller Assoc. Companies v. Landmark Realty Co.,* 224 Pa.Super. 25, 303 A.2d 43 (1973). The court noted a number of states where liens are available for preliminary work, such as California, Hawaii, Texas, and Illinois.

[18] *John F. Bushelman Co. v. Troxell,* 44 Ohio App.2d 365, 338 N.E.2d 780 (1975).

[19] Annot., 3 A.L.R.3d 573 (1965).

[20] *Nickel Mine Brook Assocs. v. Joseph E. Sakal, P.C.,* 217 Conn. 361, 585 A.2d 1210 (1991).

[21] *Thompson & Peck, Inc. v. Division Dry-wall, Inc.,* 241 Conn. 370, 696 A.2d 326 (1997).

[22] *CIT Group/Equipment Financing, Inc. v. Horizon Potash Corp.,* 118 N.M. 665, 884 P.2d 821 (App.1994).

[23] *Integon Indemnity Corp. v. Bull,* 311 Ark. 61, 842 S.W.2d 1 (1992).

[24] *Primo Team v. Blake Constr. Co.,* 3 Cal.App.4th 801, 4 Cal.Rptr.2d 701 (1992); *Onsite Eng'g & Management, Inc. v. Illinois Tool Works, Inc.,* 319 Ill.App.3d 362, 744 N.E.2d 928 (2001).

[25] Zimmerman & Orien, *Can I Lien That?* 27 Constr. Lawyer, No. 4, Fall 2007, p. 28; Zimmerman & Orien, *Can I Lien That, Too?* 28 Constr. Lawyer, No. 4, Fall 2008, p. 35.

[26] 790 N.W.2d 146, 150 (Minn.2010).

[27] 466 Mich. 660, 649 N.W.2d 371 (2002).

[28] Mass.Gen.Laws chapter 254, § 4; *BloomSouth Flooring Corp. v. Boys' and Girls' Club of Taunton Inc.,* 440 Mass. 618, 800 N.E.2d 1038 (2003).

[29] *Pero Bldg. Co., Inc. v. Donald H. Smith,* 6 Conn.App. 180, 504 A.2d 524 (1986) (contractor cannot bind subcontractors and suppliers). But lien waivers were upheld in *First American Bank of Va. v. J.S.C. Concrete Constr., Inc.,* 259 Va. 60, 523 S.E.2d 496 (2000).

[30] Cal.Civ.Code § 3262 (bars owner or prime from waiving liens of others, except with their written consent); 49 Pa.Stat. § 1401 (lien waivers are void except for residential work); and Wis. Ann.Stat. § 779.135(1) (lien waivers are void).

[31] 327 S.W.3d 104 (Tex.2010).

[32] See Sweet, *A View From the Tower,* 18 Constr. Lawyer, No. 1, Jan. 1998, p. 47.

[33] Annot., 4 A.L.R.5th 772 (1992).

[34] N.Y.—McKinney's Lien Law § 79-a.

[35] N.J.Stat.Ann. 2C:20-9, narrowly interpreted in *Houdaille Constr. Materials, Inc. v. American Tel. & Tel. Co.,* 166 N.J.Super. 172, 399 A.2d 324 (Law Div.1979).

[36] Okla.Stat.Ann. tit. 42, § 153.

[37] Cal.Penal Code §§ 484b, 484c.

[38] Criminal convictions were successfully obtained in *State v. Cohn,* 783 So.2d 1269 (La.2001); *State v. Spears,* 929 So.2d 1219 (La.2006); *People v. Brickley,* 306 A.D.2d 551, 760 N.Y.S.2d 266 (2003).

[39] *DJ Painting, Inc. v. Baraw Enterprises, Inc.,* 172 Vt. 239, 776 A.2d 413 (2001) (no unjust enrichment of owner by subcontractor, where owner had paid prime).

[40] *Costanzo v. Stewart,* 9 Ariz.App. 430, 453 P.2d 526 (1969).

[41] *Boren v. Thompson & Assocs.,* 999 P.2d 438 (Okla.2000).

[42] *Cullum Mechanical Constr., Inc. v. South Carolina Baptist Hosp.,* 344 S.C. 426, 544 S.E.2d 838 (2001).

[43] Sklar, *A Subcontractor's View of Construction Contracts,* 8 Constr. Lawyer, No. 1, Jan. 1988, pp. 1, 20.

[44] 309 Mont. 226, 45 P.3d 19 (2002).

[45] *McDaniel v. Ashton-Mardian Co.,* 357 F.2d 511 (9th Cir.1966) (subcontractor assumed the risk of government-caused delays).

[46] See Calvert & Ingwalson, *Pass Through Claims and Liquidation Agreements*. 18 Constr. Lawyer, No. 4, Oct. 1998, p. 29.
[47] 116 Ill.App.3d 852, 72 Ill.Dec. 582, 452 N.E.2d 835 (1983).
[48] Cal.Pub.Cont.Code §§ 4100–4114.
[49] *Clark Pacific v. Krump Constr., Inc.,* 942 F.Supp. 1324, 1338 (D.Nev.1996).
[50] *Golden State Boring & Pipe Jacking, Inc. v. Orange County Water Dist.,* 143 Cal.App.4th 718, 49 Cal.Rptr.3d 447 (2006).
[51] 28 U.S.C. § 1491.
[52] *Severin v. United States,* 99 Ct.Cl. 435 (1943), cert. denied, 322 U.S. 733 (1944).
[53] *E. R. Mitchell Constr. Co. v. Danzig,* 175 F.3d 1369, 1371 (Fed.Cir.1999), rehearing denied, *en banc* suggestion declined, Aug. 25, 1999.
[54] *Howard Contracting, Inc. v. G. A. MacDonald Constr. Co., Inc.,* 71 Cal.App.4th 38, 83 Cal.Rptr.2d 590 (1998); *Frank Coluccio Constr. Co. v. City of Springfield,* 779 S.W.2d 550 (Mo.1989).

CHAPTER 22

Terminating a Construction Contract: Sometimes Necessary but Always Costly

Scenario

Oliver and Olivia Owners decided to build their dream home. The Owners retained Awesome Architects, Inc. (AAI), an architectural firm, to create the plans and specifications and to administer the project.

The Owners hired Prime Builder, Inc. as prime contractor, using an American Institute of Architects General Conditions of the Contract for Construction, AIA A201-2007. Prime Builder hired several subcontractors.

Prime Builder began performance. The city's building inspector happened to be driving by and saw the new construction. Curious, he checked and found that no building permit had been issued. He ordered the work stopped until the Owners put in an application for a permit, which was then approved.

AAI, meanwhile, was unhappy with the quality of the contractor's work. Within a week after the foundation was poured, it began to crack. AAI determined the soil preparation work was not properly done. It ordered Prime Builder to remove the foundation and redo the soil preparation, which the contractor did.

Prime Builder began to suffer reduced cash flow, which caused it to delay paying subcontractors. As a result, subcontractors began not to show up, and the entire project slowed down. Just a month before the completion date, the work was only half done.

The Owners despaired that the project would never be completed. They asked AAI to determine whether the prime may be terminated for cause. AAI thought termination would be justified, as it seemed impossible for Prime Builder to finish on time. The next day the Owners mailed a letter to Prime Builder, informing the contractor that the contract was immediately terminated for cause.

Claiming the termination was wrongful, Prime Builder sued the Owners for breach of contract.

22.1 Termination: A Drastic Step

Termination does not occur frequently in construction contracts. One reason for this is the difficulty of determining whether a legal right to terminate exists. Often each party can correctly claim the other has breached. It may be difficult to determine whether a party wishing to terminate is sufficiently free from fault and can find a serious deviation on the part of the other. Another reason is the often troublesome question of whether the right to terminate has been lost. A third reason, the serious consequences of terminating without proper cause, is treated in this section.

The case of *Indiana & Michigan Electric Co. v. Terre Haute Industries*[1] illustrates the danger of an improper termination. The jury trial lasted 81 days and generated a 36,000-page transcript, which was bound into 138 volumes. The reported appellate court decision filled 33 double-columned pages in the regional reports. The contract was for the installation of pollution control equipment for an energy plant using coal-fired generators. The state pollution control board had been exerting pressure on the utility to install an electrostatic precipitator and to meet strict deadlines for its installation.

The contractor submitted the low bid of approximately $7 million. At a point at which approximately 80 percent of the by now contract price of $8 million had been paid, the utility and the contractor disagreed as to whether the contractor was on schedule. Deciding that the contract would not be completed on time, the utility terminated the contractor because of its alleged failure to maintain the schedule. Evidence at the trial indicated that at the time of termination the project was 60 percent complete. The utility spent an additional $5 million to complete the work. At the time of termination, the utility took possession of construction equipment and used it for approximately four months based on a power given to it in the termination clause of the construction contract.

The contractor's claim and the judgment it obtained in the trial court indicate the high degree of risk that termination can entail. The judgment was for $17 million (all figures approximations), which was broken down as

- retainage and interest, lost profits on the project, extras, and additional expenses—$2 million
- loss of future business—$3 million
- punitive damages—$12 million

In addition, the trial court did not grant the utility's counter-claim, because it concluded the termination by the utility had not been proper.

On appeal, the court reduced the award drastically, concluding that the contractor was not entitled to one small expense item, future profits, and punitive damages, leaving an award of $2 million. (The court's ruling on punitive damages is analyzed in Section 22.4.)

This case demonstrates the difficult position in which the utility found itself when it decided to terminate the contractor's performance. It was difficult to determine who was responsible for the delay. Acrimonious disputes had arisen throughout the entire performance. Although the court ultimately concluded that punitive damages were not appropriate for this wrongful termination, at the time of the termination an award of open-ended punitive damages certainly could not be excluded as a possibility. To be sure, not terminating carried risks as well. Undoubtedly, the economic dislocation and legal exposure make termination in construction disputes relatively rare.

The *Indiana Power* case shows the risks of termination. Were there adequate grounds to terminate? What is the cost of being wrong? Other factors must be taken into account—even when there are grounds to terminate. What happens after termination? What will be the effect of termination on the contractor?

It may sometimes be necessary to terminate a construction contract. Performance may be going so badly and relations may be so strained that continued performance would be a disaster. But because of the reasons mentioned, the drastic step of termination should not be taken precipitously.

22.2 Termination by Agreement of the Parties

Just as parties have the power to make a contract, they can "unmake" it. In legal parlance, exercising this power may be described as *rescission, cancellation, mutual termination,* or some other synonym. However described, the parties have agreed that each is to be relieved from any further performance obligations. In lay terms, they have "called the deal off."

The legal requirements for such an arrangement are generally the same as those for making a contract: manifestations of mutual assent, consideration, a lawful purpose, and compliance with any formal requirements.

Just as most construction contracts need not be expressed by a written memorandum, formal requirements rarely impede enforcement of contracts of mutual termination. However, because proof is desirable, such agreements usually will be written.

22.3 Contractual Power to Terminate: Introduction

Construction contracts frequently contain provisions giving one or both parties the power to terminate the contract. These provisions are a backdrop for material to be discussed in Section 22.4—the common law right to terminate a contract. Although contracts are not always clear on this point, as a general rule, common law termination rights have not been eliminated by express contract termination provisions; instead, specific provisions are considered to be illustrations or amplifications of common law doctrines. These doctrines still are applicable.

Construction contracts drafted by an owner generally give explicit termination rights only to the owner. In contrast, construction contracts published by professional associations provide that either owner or contractor can terminate for certain designated defaults by the other.

Although variations exist, it is useful to briefly compare the federal procurement, American Institute of Architects (AIA), and Engineers Joint Contract Documents Committee (EJCDC) approaches to the contractual power to terminate. All three provide the *owner* with the option to terminate the contractor for cause (also called a "default" termination) or at the convenience of the owner.

The AIA and EJCDC documents give the *contractor* the right to terminate, but the federal government does not.

22.4 Default Termination

Termination by Owner. The Federal Acquisition Regulations (FAR) control federal procurement. They provide that the government may "terminate the contract completely or partially" if the contractor does not perform the required services within the specified time, does not perform any other contract provision, or does not make progress "and that failure endangers performance of the contract."[2] This resembles the common law requirement that discharge of the contract requires a material breach.

Article 14 of AIA A201-2007 addresses termination or suspension of the contract. The owner's rights are covered in §§ 14.2 through 14.4.

The EJCDC General Conditions, C-700 (2007), Article 15, addresses suspension and termination of the work. Paragraphs 15.1 through 51.3 address termination or suspension by the owner.

The focus in this subsection is on the owner's right to terminate for cause, which is also called a default termination.

Both the AIA and EJCDC list specific events that justify an owner's termination of the prime contract, including that the contractor: failed to supply adequate workers or materials;[3] disregarded laws;[4] or committed a substantial breach of the contract documents.[5] This last ground for termination corresponds to the common law standard of a material breach.[6]

The AIA and EJCDC diverge at points. The AIA adds as a ground for a default termination the contractor's failure to pay subcontractors,[7] and the EJCDC includes "repeated disregard of the authority of Engineer."[8]

Both associations require the owner to give the contractor and its surety seven days by written notice before the termination order becomes effective.[9] Under the AIA, the owner may not terminate for cause unless he first obtained "certification by the Initial Decision Maker" (or architect if there is no IDM).[10] It is this requirement—that the IDM or architect certify that termination for cause is warranted—that exposes the design professional to a claim of interference with contract by a terminated contractor.[11] The role of a design professional in a termination decision is further discussed in Section 22.6.

Unlike the AIA, the EJCDC gives the contractor an opportunity to use the seven-day notice period to attempt to cure its nonperformance; termination is cancelled if the contractor cures within 30 days from receipt of the notice.[12] Federal Procurement requires a ten-day cure period before there can be a default termination.[13] Even without a specific cure provision, South Carolina held that, before there can be a termination, good faith requires that the contractor be given an opportunity to cure.[14] Of course, if the defect is not curable within the cure period, the termination can take place in accordance with the termination clause without waiting for any cure period to elapse.

Both the AIA and EJCDC empower the owner to exclude the contractor from the site and take possession of its work, tools, materials, and equipment; accept assignment of the subcontracts; and complete the work as "the Owner may deem expedient."[15] The contractor is not entitled to further payment until the work is complete,[16] and the owner may use the contract balance to complete the work.[17] The EJCDC adds that, in doing so, the owner "shall not be required to obtain the lowest price for the Work performed."[18]

> In this chapter's scenario, the Owners sought to terminate the prime contract for cause, because they feared Prime Builder could not complete the job on time. AIA A201-2007, § 14.2.1 lists several reasons to terminate for cause, but failure to complete on time is not one of them
>
> That said, § 14.2.1 does permit a default termination if the contractor is "guilty of substantial breach" of a contract provision. Failure to meet the completion date would be a material breach (see Section 20.6). However, is the contractor entirely to blame for this delay? We know the Owners failed to obtain a building permit before construction began. If this was a substantial cause in the contractor's delay, then the termination might have been wrongful.

Termination by Contractor. Both the AIA and EJCDC give the contractor the contractual right to terminate the contract.[19]

The AIA approach is discussed first. AIA A201-2007, § 14.1, giving this right to the contractor, must be looked at carefully. Section 14.1.1 lists four reasons justifying contractor suspension or termination, including nonpayment or failure to provide evidence of owner solvency. However, occurrence of these events does not automatically give the contractor a power to terminate. Instead, the work must have stopped for a period of 30 consecutive days, and the contractor must have given the owner and architect seven days' written notice of the intent to suspend or terminate.

Another ground exists. Under § 14.1.4, if the work is stopped for 60 consecutive days through no fault of the contractor or subcontractors or if the owner has repeatedly failed to meet its more important obligations (such as payment), the contractor may terminate—but only after giving seven days' written notice.

Under § 14.1.3, a contractor who properly terminated the contract is entitled to payment for work performed, including reasonable overhead and profit, the costs incurred by reason of the termination, and any remaining damages.

Paralleling the discussion of the AIA, EJCDC C-700, ¶ 15.4 gives the contractor the right to stop or terminate the work. Under ¶ 15.4(A), the contractor may terminate if the work is suspended for more than 90 consecutive days or payment is not made within 30 days of when due. As with the AIA, the contractor must give the owner and engineer seven days' written notice. Unlike the AIA, the EJCDC explicitly allows the contractor to remedy the reason for the threatened termination. Unlike the AIA, the EJCDC does not have a provision that expressly spells out the contractor's remedies.

Wrongful Termination for Default. Under the common law, a wrongful termination for default subjects the terminating party to liability for breach of contract. The termination may be wrongful either because the grounds were not supported by the facts (for example, the contractor was accused of being behind schedule, yet trial reveals the fault for delayed performance lay with the owner) or the contractual procedures to terminate were not followed.

While principles of good faith and fair dealing may play a role in the *manner* of termination—for example, requiring an opportunity to cure—courts are reluctant to characterize the actual *decision* to terminate as done in bad faith. This reluctance is due in part to the possibility that a "bad faith" termination would be sufficiently wrongful to be considered a tort and sufficiently intentional to be considered the basis for punitive damages. This position is illustrated by the unwillingness of the court in *Indiana & Michigan Elec. Co. v. Terre Haute Industries*[20] to impose punitive damages after it concluded that the owner did not have grounds for terminating the contract. The award of punitive damages had been based on a finding that termination had been malicious and oppressive.

The appellate court found that, at worst, the conduct was "substandard business practice, and arrogance."[21] It noted that the law did not impose punitive damages simply because "the contracting party or his agents are disagreeable people."[22] Awarding punitive damages, according to the court, would "let all disputes and quarrels over broken contracts and disappointed business ventures become the subject of acrimonious litigation over punitive damages."[23]

As noted in Section 22.5, a wrongful default termination may be converted into a termination for convenience under federal procurement law.

22.5 Termination or Suspension for Convenience

Owner Suspension. Both the AIA and EJCDC give the owner the right to suspend all or a portion of the work "without cause."[24] The EJCDC allows the owner to suspend for up to 90 consecutive days. The AIA has no explicit time limit; however, a work stoppage (not the fault of the contractor) for more than 30 consecutive days entitles the contractor to terminate the contract.[25] Under both associations' standard documents, the contractor is entitled to an adjustment in the contract price and time.[26]

Contractor Suspension. The AIA A201-2007, § 9.7, gives the contractor the right to suspend the work for nonpayment. With this exception, the contractor has few options in the face of an uncooperative owner.

By contrast, EJCDC C-700, ¶ 15.4(B) gives the contractor the right to stop work until it has been fully paid. This intermediate step of stopping work gives the parties room to resolve payment disputes and preserve project viability

Termination for Convenience. Only the owner may terminate for convenience.

Termination for convenience was pioneered by federal procurement regulations.[27] The only limit on the contracting officer's (the government representative's) decision is that the convenience termination "is in the Government's interest." The contractor must stop work, place no further orders, cancel orders that have been placed, and perform other acts designed to terminate performance and protect the interests of the government. The contractor is reimbursed for work performed, unavoidable losses suffered, and expenditures incurred to preserve and protect government property. The contractor is also paid a designated profit for *work performed* but is denied its lost, anticipated profit (as would be available under the common law).

Private contracts have increasingly embraced the right to terminate for convenience. This right is also found in subcontracts, particularly those tied to prime contracts where the owner has this power. An example is EJCDC C-700, Standard General Conditions of the Construction Contract, ¶ 15.03 and AIA Document A201-2007, § 14.4. Differences from the federal regulations may exist. For example, under A201-2007, § 14.4.3, the contractor receives payment for work performed, the costs incurred by reason of the termination, and reasonable overhead and profit on work *not* performed (whereas federal regulations permit profit only on work performed).

In addition, federal procurement law has developed the concept of *constructive* convenience termination. Under this, a contract with a convenience termination clause converts a *wrongful* government default termination into a *convenience* termination.

A contract giving the owner the power to terminate without cause appears unlimited. In fact, this power is not completely unrestricted. Some courts hold the public owner may terminate if the project is no longer needed or has become outmoded or uneconomical.[28] Others use a "good faith" or "abuse of discretion" (by the contracting officer) standard.[29]

In a recent decision, Maryland's highest court analyzed good faith limitations on a termination for convenience clause on a private works project. In *Questar Builders, Inc. v. CB Flooring*, LLC,[30] a subcontract contained a provision allowing the prime contractor to terminate for its convenience. The court rejected the contractor's contention that this meant it could terminate the subcontract "for any reason whatsoever, including a bad reason or no reason,"[31] without thereby incurring any liability. The court instead ruled that termination for convenience clauses are valid—they do not make the contract entirely one-sided—because the contractor's termination right must be exercised in good faith. Viewing a termination for convenience clause as a risk-allocation tool, the court stated that the contractor may terminate for convenience only if it reasonably believed that doing so was necessary to avoid significant financial harm. The contractor may not terminate for convenience in order to avoid issuing a change order giving the subcontractor additional compensation and/or time.

22.6 Role of Design Professional

Despite the AIA's movement toward reducing the activities and responsibilities of the architect, the owner's power to terminate for cause requires the architect—or any Initial Decision Maker (IDM), but this analysis assumes no IDM is present—to certify "that sufficient cause exists to justify such action." See A201-2007, § 14.2.2. No similar requirement exists for a termination by the contractor.

Termination is a drastic step. Why require the architect to certify that there are adequate grounds? Such a decision, although involving some issues for which the architect may be trained, requires legal expertise rather than design skill. The owner may want the architect's advice. That need not require giving power to the architect to decide a sensitive and liability-exposing issue.

The only possible justification is that such a preliminary step can act as a brake on any hasty, ill-conceived decision by the owner to terminate. It would be better to have the owner's attorney perform this function, especially if the architect's conduct is itself an issue in the termination.

While certification by the architect affords advantages to the owner, this exposes the design professional to a contractor claim of tortious interference with contract. As explained in Section 11.6E, the "advisor's privilege" defense will most likely entitle the architect to a pretrial ruling in its favor.

> In this chapter's scenario, the Owners asked AAI if termination was proper. Absent bad faith by AAI, its advice to the Owners—even if factually wrong because the Owners were substantially responsible for the contractor's delayed performance—would still come within the "advisor's privilege" defense.

22.7 Waiver of Termination and Reinstatement of Completion Date

Suppose one party has the power to terminate but does not exercise it. Has the power to terminate been lost? Under what conditions can it be revived?

Cases that have dealt with this problem have usually involved a performing party—usually the contractor—who has not met the completion date, but for various reasons, the owner decides not to terminate. The contractor continues to perform, believing the power to terminate will not be exercised. At some point during continued performance, the owner wishes either to terminate immediately or to set a firm date for completion. If not met, this will be grounds for termination.

Clearly, a termination after the contractor has been led to believe there will be no termination would be improper.[32] The more difficult question relates to the requirements for reinstating a firm deadline that will allow termination if performance is not met by that time. The parties can agree to extend the time for completion with the clear understanding that failure to comply would entitle the owner to terminate.

The most difficult question involves unilateral attempts by the owner to reinstate a firm completion date and revive the right to terminate. The U.S. Court of Claims in *DeVito v. United States* outlined this path for the government to follow to reinstate a completion date:

> When a due date has passed and the contract has not been terminated for default within a reasonable time, the inference is created that time is no longer of the essence so long as the constructive election not to terminate continues and the contractor proceeds with performance. The proper way thereafter for time to again become of the essence is for the Government to issue a notice under the Default clause setting a reasonable but specific time for performance on pain of default termination. The election to waive performance remains in force until the time specified in the notice, and thereupon time is reinstated as being of the essence. The notice must set a new time for performance that is both reasonable and specific from the standpoint of the performance capabilities of the contractor at the time the notice is given.[33]

22.8 Notice of Termination

Termination clauses often require that a notice of termination be sent by the terminating party to the party whose performance is being terminated and (in many contracts) to the lender or surety. The notice usually states that termination will become effective a designated number of days after dispatch or receipt of the notice.

What are the parties allowed to do and what are they required to do during this notice period? There are three possible uses of a notice period. It may allow the terminating party to "cool off." Construction performance problems often generate animosity, and before the important step of termination is effective, the terminating party may wish to rethink his position.

The notice period can permit a defaulting party to cure defaults in order to keep the contract in effect for the benefit of both parties. Whether the notice period is designed to "cure" may depend on the facts that give rise to termination and on the notice period. If the intent is to allow the party to cure, then the notice period must be sufficiently long for the contractor to have a realistic opportunity to do so.

Finally, the notice period can be used to wind down operations and to protect the site. This allows each party to cut losses and make new arrangements. Such a position does not allow cure, although the owner who terminates should have the option of ordering that the work to be done can be completed by the effective date of termination.

In sum, the termination clause should clearly indicate whether it permits *cure,* provides a *cooling-off period,* or sets into motion a *winding down* of the project.

> In this chapter's scenario, the Owners terminated Prime Builder effective immediately. This notice violated A201-2007, § 14.2.2, which requires giving the terminated contractor seven days' written notice. This may be ground for the contractor to argue that the termination was wrongful and should be converted to a termination for convenience.

22.9 Termination Under Common Law

Material Breach. Regardless of whether the contract expressly creates a power to terminate, termination is allowed under the common law in the event of a material breach. Even with a contract clause dealing with termination, many of the factors that determine materiality can be influential when such clauses are interpreted. Also, as indicated, the termination clause may not be the exclusive source for determining when a party has the power to terminate.

Rather than establishing fixed rules, such as the importance of the clause breached, the law examines all of the facts and circumstances surrounding the breach to determine whether it would be fair to permit termination. The Restatement (Second) of Contracts, § 241, lists factors that are significant in determining whether a particular breach is material. They include:

(a) the extent of injury caused by the breach;
(b) whether the injured party can be adequately compensated;
(c) whether the nonperforming party will suffer forfeiture;
(d) the likelihood of cure by the nonperforming party; and
(e) whether the nonperforming party acted in conformity with the standards good faith and fair dealing.

Perhaps the most relevant factors are the particular nature of the nonperformance, the likelihood of future breaches, and the possibility of forfeiture. Nonpayment—probably the most frequently asserted justification for termination—was discussed in Section 17.8.

A canvass of the case law reveals other examples of material breach. The owner's failure to make an equitable adjustment when unforeseen underground conditions were discovered and the contractor was in financial trouble allowed the prime contractor to terminate.[34] A prime contractor's failure to have the site ready for the flooring subcontractor was a material breach.[35] Failure to make satisfactory progress justified termination.[36] Hindering the subcontractor's operation was a material breach when coupled with the prime contractor's having stopped payment on a check that the subcontractor was about to negotiate.[37] These obviously incomplete illustrations are not designed to indicate that breaches of the type described will always be considered material. As emphasized in this subsection, determining whether a breach is material requires a careful evaluation of the facts and circumstances surrounding the breach as well as the effect of termination.

Anticipatory Repudiation. The previous discussion deals with breaches that have occurred. This subsection deals with breaches that may occur in the future.

It may appear that one party may not be able to perform when the time for performance arrives, or one party may state it will not perform when the time for performance arrives. The contractor may discharge some of its employees, or a number of employees may quit. The contractor may cancel orders for supplies, or its suppliers may indicate they will not perform at the time for performance. In such cases, the owner may realize the contractor will be unable to perform.

A breach by "anticipatory repudiation" occurs when one party indicates to the other that it cannot or will not perform. Each party is entitled to reasonable assurance that the other will perform in accordance with the contract.[38] If one party indicates that it will not or cannot perform, the other party loses this assurance. Should the latter be required to wait and see whether the threat or the indication of inability to perform will come to fruition?

Prospective inability deals with probabilities. An owner need not wait to see whether actual nonperformance or defective performance will occur. He can demand assurance and, in the absence of this assurance, legally terminate any obligation to use the contractor.

22.10 Keeping Subcontractors After Termination

An owner who exercises his power to terminate the prime contract may want to employ a successor to continue the project. Alternatively, the surety on a performance bond may exercise its option to complete the project with a successor.

Default by the prime contractor is likely to result in failure to pay the subcontractors. This would give subcontractors the power to terminate their obligations under their subcontracts. The owner or surety may wish to take an "assignment"[39] of some subcontracts (those of subcontractors who are performing well and are not owed an excessive amount of money) without having to renegotiate with them. (An assignment transfers the subcontract from the prime contractor to the owner, without thereby altering the subcontractor's rights and obligations.) Renegotiation always causes delay and may result in a higher contract price. Yet most subcontracts contain nonassignment clauses that preclude assignment of any rights under the contract without the consent of the party whose performance is being assigned—in this case, the subcontractors.

To avoid this tangle, it has become common for the prime contract to include an assignment conditioned on default by the prime to the owner of those subcontracts that the owner wishes to take over with a promise by the prime contractor to obtain the subcontractors' consent to these assignments.[39]

REVIEW QUESTIONS

1. What are the legal requirements to "unmake" or rescind a contract?

2. Under the Federal Acquisition Regulations, when may the government terminate the contract completely or partially?

3. What are the similarities and differences between the provisions of the AIA and the EJCDC in regard to owner's termination of the prime contract?

4. Which parties in a construction contract can terminate for convenience?

5. What are some grounds for the contractor to terminate the contract under the AIA A201-2007 Document?

6. What is the concept of constructive convenience termination under federal procurement law?

7. What did the court state in *DeVito v. United States* was necessary for the government to reinstate a firm completion date and revive the right to terminate after the project had already exceeded the original completion date and the owner had not yet exercised their right to terminate?

8. What are the three possible uses of the notice period when a party has been notified of a termination?

9. According to the Restatement (Second) of Contracts, § 241, what are the significant factors in determining whether a particular breach is material?

10. What is a breach by anticipatory repudiation?

ENDNOTES

[1] 507 N.E.2d 588 (Ind.Ct.App.1987).
[2] 48 CFR § 49.402-1 (2012).
[3] AIA A201-2007, § 14.2.1.1; EJCDC C-700, ¶ 15.2(A)(1).
[4] A201-2007, § 14.2.1.3; EJCDC C-700, ¶ 15.2(A)(2).
[5] A201-2007, § 14.2.1.4; EJCDC C-700, ¶ 15.2(A)(4).
[6] See Sections 3.7B and 22.9.
[7] A201-2007, § 14.2.1.2.
[8] EJCDC C-700, ¶ 15.2(A)(3).
[9] A201-2007, § 14.2.2; EJCDC C-700, ¶ 15.2(B).
[10] A201-2007, § 14.2.2.
[11] See Section 11.6E.
[12] EJCDC C-700, ¶ 15.2(D).
[13] 48 CFR § 52.249-8 (2012). Curing in federal procurement is canvassed in *Empire Energy Management Systems, Inc. v. Roche*, 362 F.3d 1343 (Fed.Cir.2004).
[14] *Bensch v. Davidson*, 354 S.C. 173, 580 S.E.2d 128 (2003).
[15] A201-2007, § 14.2.1 - .3; EJCDC C-700, ¶ 15.2(B)(1)–(3).
[16] A201-2007, § 14.2.3; EJCDC C-700, ¶ 15.2(C).
[17] A201-2007, § 14.2.4; EJCDC C-700, ¶ 15.2(C).
[18] EJCDC C-700, ¶ 15.2(C).
[19] AIA A201-2007, § 14.1; EJCDC C-700, ¶ 15.4(A).
[20] Supra note 1.
[21] 507 N.E.2d at 617.
[22] Ibid.
[23] Ibid.
[24] A201-2007, § 14.3.1; EJCDC C-700, ¶ 15.1.
[25] A201-2007, § 14.1.1.
[26] A201-2007, § 14.3.2; EJCDC C-700, ¶ 15.1.

[27] 48 CFR § 52.249-1,-2, et. seq. (2012).
[28] *Torncello v. United States*, 231 Ct.Cl. 20, 681 F.2d 756 (1982).
[29] *Krygoski Constr. Co. v. United States*, 94 F.3d 1537 (Fed.Cir.), cert. denied, 520 U.S. 1210 (1997). This case was followed in a state public works project, *Capital Safety, Inc. v. State Division of Bldgs. & Constr.*, 369 N.J.Super 295, 848 A.2d 863 (App.Div.2004).
[30] 410 Md. 241, 978 A.2d 651 (2009).
[31] 978 A.2d at 671.
[32] *DeVito v. United States,* 188 Ct.Cl. 979, 413 F.2d 1147 (1969).
[33] Id., 413 F.2d at 1154.
[34] *Metropolitan Sewerage Comm'n v. R. W. Constr., Inc.*, 72 Wis.2d 365, 241 N.W.2d 371 (1976), appeal after remand, 78 Wis.2d 451, 255 N.W.2d 293 (1977) (jury verdict approved).
[35] *Great Lakes Constr. Co. v. Republic Creosoting Co.,* 139 F.2d 456 (8th Cir.1943).
[36] *Aptus Co. v. United States,* 61 Fed.Cl. 638 (2004); *Mustang Pipeline Co., Inc. v. Driver Pipeline Co., Inc.,* 134 S.W.3d 195 (Tex.2004).
[37] *Citizens Nat'l Bank of Orlando v. Vitt,* 367 F.2d 541 (5th Cir.1966).
[38] Restatement (Second) of Contracts §§ 251, 252 (1981) followed in *Danzig v. AEC Corp.,* 224 F.3d 1333 (Fed.Cir.2000), cert. denied sub nom. *ABC Corp. v. Pirie,* 532 U.S. 995 (2001).
[39] AIA Doc. A201-2007, §§ 5.4 and 14.2.2.2.

PART

RISK MANAGEMENT AND DISPUTE RESOLUTION

CHAPTER 23

Apportioning or Shifting Losses: Contribution and Indemnity

CHAPTER 24

Insurance

CHAPTER 25

Surety Bonds: Backstopping Contractors

CHAPTER 26

Claims and Disputes: Emphasis on Arbitration

CHAPTER 23

Apportioning or Shifting Losses: Contribution and Indemnity

Scenario

In this scenario, words in quotation marks (such as "contribution") will be defined later in the chapter.

Mega Manufacturer, Inc. (MMI) sought to build a new factory. MMI retained Awesome Architects, Inc. (AAI), an architectural firm, to create the plans and specifications.

Based upon this design, the owner entered into a $2 million fixed-price contract with Prime Builder, Inc. Prime Builder hired several subcontractors, including a foundation subcontractor (Foundation, Inc.), an electrical subcontractor (Electrical, Inc.), and a roofing subcontractor (Roofer, Inc.).

The foundation subcontract required Foundation to "indemnify" MMI and Prime Builder against any damages, loss or claim for personal injury, or property damage arising out of Foundation's performance of its subcontract—"however caused."

The Electrical subcontract required Electrical to indemnify Prime Builder against any damages, loss or claim for personal injury, property damage, or loss of use—but "only to the extent caused by" Electrical's negligence or breach of contract.

The Roofer subcontract contained no indemnity provision.

Things did not go smoothly on the project.

The newly built foundation failed a compression test. The problem was traced to a defective soil report relied upon by AAI in creating the foundation's design. Foundation, Inc. was required to redo the foundation. This delayed the start of work by other subcontractors.

An employee of Electrical, Jane Sparks, suffered an electrical shock while working on a line she thought was turned off. It turned out that Electrical's foreman forgot to tell Prime Builder to turn off electricity to that part of the building. Sparks had tested the line before working on it, and the volt meter had shown the line was dead. The volt meter was manufactured by Atoms Manufacturing, Inc.

Prime Builder's foreman, while inspecting the half-completed roof first thing in the morning (and before Electrical's employees had arrived), slipped on ice and fell off the roof. He had not been wearing a harness, in violation of both OSHA regulations and company policy.

Litigation ensued.

The foundation's failure resulted in MMI imposing liquidated damages (see Section 20.12) on Prime Builder. The contractor brought a "contractual indemnity" claim against Foundation to recover those damages.

Sparks sued Prime Builder for negligence in failing to make sure the electricity was cut off to an area of the building where Electrical was working. Sparks separately sued Atoms Mfg. for products liability.

Prime Builder brought (1) a "contribution" claim against Atoms Mfg. and (2) a "contractual indemnity" claim against Electrical.

Prime Builder's foreman sued Roofer for premises liability, on the ground that his access to the roof had not been blocked until the safety of accessing the roof could be assessed. Roofer brought a "contribution" claim against the prime contractor.

23.1 Loss Shifting, Responsibility Apportionment, and Risk Management: An Overview

Construction is risky business. Some risks are of a physical nature, such as a worker falling off a scaffold (personal injury) or a welder's spark starting a fire which guts a half-completed project (property damage). Other risks are economic or financial: an owner is left with an uncompleted project if the prime contractor absconds with the most recent progress payment, resulting in unpaid subcontractors who file mechanics' liens.

Focusing now on physical harm, there are two fundamental approaches to managing liability associated with these risks. One approach is to *shift the loss* from the person sued by the plaintiff to another. This is done through indemnification (whether contractual or noncontractual).

When the loss is caused by multiple defendants, the issue of *liability apportionment* among the various defendants arises. The concern is that one defendant does not pay the entire judgment, while the others pay nothing. Apportionment is done through either comparative negligence or contribution.

Loss shifting and risk management are complex technical areas of the law. Both in this chapter and the next two on insurance and suretyship, technical terminology predominates. Use of technical terminology will be minimized, but it cannot be avoided.

Chapter 24 discusses insurance, which also largely covers risk management of physical harm. Chapter 25 covers risk management of *economic loss* through the use of suretyship.

23.2 First Instance and Ultimate Responsibility Compared

Today there is increasing likelihood that those who suffer losses incident to the construction process will be compensated, especially where those losses involve personal harm. Although claimants do not always receive judicial awards, the tendency has been to expand liability to ensure that those who suffer losses are compensated.

Lawsuits today generally begin with claims against a number of defendants, which is both legally permissible and relatively inexpensive. Typically, these defendants make claims against each other as well as claims against parties who have not been sued by the original claimant. The result is a multiparty lawsuit that generally involves as many as a half-dozen interested participants as well as an almost equal number of sureties and insurers. For example, the injured employee of a prime contractor is likely to sue all contractors who are not immune, such as other separate contractors (multiple primes) and subcontractors, the owner, the design professional, any construction manager, and (depending on the facts) those who have supplied equipment or materials. Each of the defendants will very likely bring claims against the others. If there is a building defect, such as the failure of an air conditioning system, the owner is likely to assert claims against the design professional, the contractor, and the manufacturers and sellers of the system. Again, those against whom claims have been asserted are likely to assert claims against each other.

These lawsuits show two levels of responsibility. There is *first-instance* responsibility to the original claimant, such as the injured worker or the owner. (This lawsuit is also referred to as the *underlying claim*.) Resolving this issue depends on whether any of the defendants or someone for whom they are responsible had a duty to the claimant, failed to live up to the standard required by the contract or by the law, and was the legal and proximate cause of the claimant's injury.

After this determination is made, the next level is *ultimate* responsibility, which must be addressed. Who among those responsible will ultimately bear the loss? This inquiry—the focal point of this chapter—has developed unbelievably complex and costly legal controversies.

23.3 Responsibility Apportionment Among Multiple Wrongdoers: Stating the Problem

Introduction. This book at this point has almost exclusively discussed claims by a plaintiff against one or more defendants. This section addresses special issues that arise when *multiple* defendants have contributed to the plaintiff's injury.

As is not uncommon in construction disputes, the injured party may feel that more than one person or business entity caused the harm. An owner may lay the blame for defects on both the architect and contractor. A subcontractor may attribute damage to its work to both another subcontractor and a manufacturer's defective product. In this situation, the plaintiff sues multiple defendants in the same legal action.

Suppose the injured party seeks a remedy from only one defendant, and that defendant believes others were the cause of the plaintiff's injury. For example, an owner sues the prime contractor for an injury to the owner's employee while visiting the site under construction. The prime believes a subcontractor's negligence was the cause of the employee's injury. The prime then brings its own lawsuit against this subcontractor. As another example, the owner may blame settlement of the building on the architect's design of the foundation. The architect, in turn, may sue the geotechnical engineer, claiming the engineer's negligence was the source of any defects in the foundation design. Under either example, the result is that multiple defendants (the prime and the subcontractor, or the architect and engineer) are in the same lawsuit.

A number of defendants may be sued in the same legal action even though they did not act together. They are co-defendants because each may have played a substantial role in causing the injury, and for procedural convenience, all of the claims are decided in one lawsuit. The defendants are concurrent wrongdoers—not in the sense that their wrongdoing occurred at the same time but in the sense that each played a substantial role in causing an indivisible loss to the claimant.

Suppose three defendants are held liable to the plaintiff and one defendant pays the entire award. It would be unfair for the one defendant to pay the entire award while the two co-defendants pay nothing. This section addresses how that loss transfer (or apportionment) among multiple defendants occurs.

Direct Versus Third-Party Action. In the previous examples, the introduction of the additional defendants happened in two different ways. In the first, the plaintiff sues multiple defendants (A and B). In the latter (of the owner's injured employee and the defective foundation), the plaintiff sues one defendant (A), but A believes another party (B) also contributed to (or even was the sole cause of) the plaintiff's injury, so A sues B to make sure B contributes to paying for the plaintiff's damages.

This distinction as to how B was brought into the litigation is procedural in nature. The end result is that (in both situations) both A and B are potentially liable to the plaintiff. Nonetheless, this procedural difference affects the apportionment devices available to the defendants.

In the first situation, B was sued directly by the plaintiff. B is in the litigation by virtue of a *direct action* (or lawsuit) against him by the plaintiff.

In the second situation, B was brought into the litigation when sued by A. B is in the litigation by virtue of a *third-party complaint* (or a third-party action) brought against him by A. In this situation, A is the third-party plaintiff (to distinguish him from the underlying or original plaintiff), and B is the third-party defendant. See Figure 23.1.

> In this chapter's scenario, Sparks sued Prime Builder for negligence, and the prime brought a contribution claim against Atoms Mfg. and a contractual indemnity claim against Electrical. Sparks is bringing a direct action against Prime Builder. Prime Builder is bringing third-party complaints against Atoms Mfg. and Electrical.

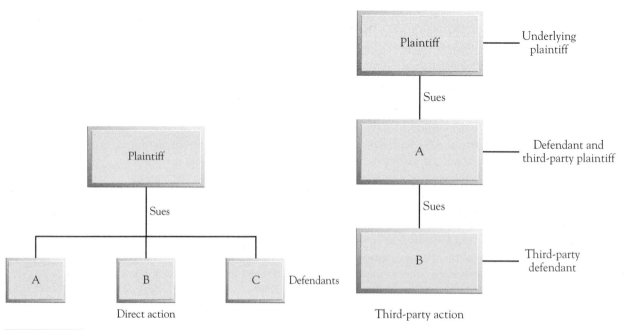

FIGURE 23.1 Direct versus third-party action.

23.4 Loss Shifting and Liability Apportionment by Operation of Law: Three Devices

As explained in Section 23.1, this chapter addresses two distinct yet related concepts: loss shifting (from one person to another) and liability apportionment (among multiple defendants).

Loss shifting is done by **indemnity**. Indemnity may be by operation of law (not based on an agreement) or by contract.

Liability apportionment (or sharing) is done by **contribution** or **comparative negligence**. Both contribution and comparative negligence are legal devices that do not arise from agreement.

The remainder of this section will discuss the three devices that exist by operation of law: noncontractual indemnity, contribution, and comparative negligence.

Contractual indemnity is the topic of the remainder of this chapter. It is the most prevalent loss-transfer device and involves its own, many complexities.

Noncontractual Indemnity. Noncontractual indemnity is a device for shifting the *entire liability* to the plaintiff from the person sued by the plaintiff (defendant A) to another person (defendant B). Plaintiff sues A in a direct action; A then brings a third-party complaint for noncontractual indemnity against B. A's loss (primarily consisting of the damages owed to the plaintiff) is shifted from the **indemnitor** (A) to the **indemnitee** (B). An indemnitor indemnifies the indemnitee; the indemnitee is the beneficiary of an indemnification.

As noted, noncontractual indemnity is a legal device permitted by the courts. While this book uses the term "noncontractual indemnity," other terms invoked by the courts are equitable indemnity (used in California), implied indemnity, quasi-contractual indemnity, and common law indemnity.

The fundamental principle underlying noncontractual indemnity is unjust enrichment. If A's liability to the plaintiff cannot be transferred to B, B will be unjustly enriched. The law seeks to avoid this inequitable result.

Suppose a worker employed by a prime contractor is injured because of deliberate safety violations by his employer. Under certain circumstances, the worker can recover from the owner. Liability (in such a case) may be based on the owner's: statutory duty to furnish a safe workplace, failure to determine whether safe practices were being followed, or failure to discharge the contractor after becoming aware of the violations.

In this example, the *degree* of wrongdoing between the owner and contractor is qualitatively different. The owner's liability to the employee is *passive,* secondary, or vicarious, and that of the contractor is *active,* primary, or direct. The contractor would be unjustly enriched and the owner unjustly impoverished if financial responsibility for the employee's injury stayed with the owner. To rectify this situation, the law allows the owner to receive indemnification from the contractor. It would be unfair for the owner to bear the cost, as the contractor's negligence was the greater contributor to the injury.

Similarly, suppose a member of the public is injured when entering property where construction work is taking place. The contractor's negligence in failing to put up a fence allowed the person to enter the land. However, the owner may be liable to entrants upon his property simply by virtue of his status as landowner. (Premises liability is discussed in Section 5.6.) In such a situation the owner is *vicariously* liable for the negligence of the contractor (see Figure 23.2). If the owner pays the entrant for his physical injuries, the owner may sue the contractor in noncontractual indemnity to recoup (recover) that payment to the entrant.

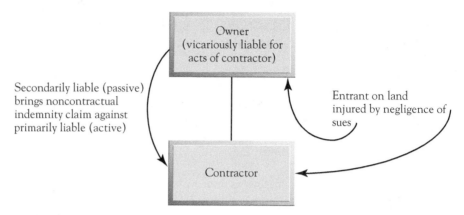

FIGURE 23.2 Noncontractual (equitable) indemnity illustrated.

In this chapter's scenario, Sparks sued Prime Builder, and the prime brought a noncontractual claim against Atoms Mfg. Prime Builder would argue that is own negligence (if any) was passive, while Atoms Mfg.'s negligence or strict liability was egregious given the danger of selling a defective volt meter.

Contribution. Contribution apportions (or divides) damages owed to the plaintiff among the defendants who caused the plaintiff a physical injury. Contribution arises only if a plaintiff suffered an *indivisible physical injury* tortiously caused by multiple defendants. If the plaintiff recovers for the entire loss from one defendant, that defendant will use contribution to shift that loss to the other co-defendant(s).

An injury is indivisible if it cannot be traced to a single defendant. If one defendant injured the plaintiff's arm and another defendant injured the plaintiff's leg, then plaintiff's injuries are not indivisible. Plaintiff may recover for the injured arm only from the first defendant. However, if two defendants concurrently and tortiously injured plaintiff's arm, the plaintiff may sue both defendants for the one injury. Moreover, the law permits the plaintiff to recover his entire damages from just one of the defendants.

Assume the plaintiff sues three defendants (A, B, and C) for an indivisible injury that the plaintiff claims was tortiously caused by the defendants. If the plaintiff recovers for the entire loss from one defendant (A), that defendant will use contribution to shift that loss to the other co-defendants (B and C). A brings a contribution claim against the two nonpaying co-defendants so that each pays a portion of the judgment owed by the underlying (or original) plaintiff. How is this apportionment of liability among wrongdoers calculated?

The majority rule is that apportionment is based on the percentage of fault. Suppose A is 50 percent at fault, B is 30 percent at fault, and C is 20 percent at fault in causing the plaintiff's injury. If A pays the plaintiff the entire judgment, then A may recover 30 percent of that amount from B and 20 percent of that amount from C.[1]

The minority rule is that apportionment is done pro rata, with A, B, and C each responsible for one-third of the award. That there are different degrees of fault between the different co-defendants is irrelevant.

Of particular significance to the construction industry, contribution applies where the indivisible injury caused by the multiple defendants involved *physical* harm—whether death, personal injury, or property damage. In addition, the defendants, in inflicting the injury upon the plaintiff, must have each committed a *tort*. This means that no right of contribution exists if an owner sues the architect and prime contractor for the same *financial injury* (defective construction) caused by each defendant's *breach of contract*.[2]

In this chapter's scenario, Prime Builder's foreman (who fell off the roof) sued Roofer for negligence, and Roofer brought a contribution claim against the prime contractor. Roofer would argue that any negligence by the foreman (not wearing a safety harness) should be imputed (attributed) to Prime Builder, so that both Roofer and the prime tortiously caused an indivisible injury to the foreman. Contribution should be allowed.

If apportionment is based on the percentage of fault, certainly Prime Builder's percentage of fault will be greater. The foreman went on the roof without a mandatory safety device, while Roofer (at most) failed to erect a safety obstacle.

On the merits, since the construction site is off-grounds to the general public and the danger of an icy roof is open and obvious to a sophisticated construction worker, Roofer should not be liable under a theory of premises liability (see Section 5.6).

Comparative Negligence. Currently, the great majority of states, through legislative or judicial initiative, have embraced comparative negligence. Under comparative negligence, when a plaintiff obtains a damages award against multiple tortfeasors, each defendant is liable to the plaintiff only for that defendant's percentage liability of fault. If there are two defendants (A and B), and the jury finds A was 80 percent at fault and B 20 percent at fault, B must pay the plaintiff 20 percent of the award amount—regardless of whether plaintiff obtains any recovery from A.

New Regime of Liability Apportionment. Comparative negligence (in which liability is proportionate to the degree of the defendant's fault) is fundamentally inconsistent with the traditional form of noncontractual indemnity, which sought to transfer the entirety of the loss from one tortfeasor to another. Because the great majority of states have adopted comparative negligence, one commentator has concluded that the "active-passive kind of indemnity is now in disrepute and probably has little or no future in apportioning responsibility among tortfeasors."[3]

Indeed, the new regime of loss apportionment among multiple tortfeasors who have caused an indivisible injury in actuality is fault-based contribution. Otherwise stated—regardless of the theory of liability alleged (whether contribution or noncontractual indemnity)—liability is based on comparative fault.

23.5 Contractual Indemnity Compared to Exculpation, Liability Limitation, and Liquidated Damages

Before examining contractual indemnity, that device for loss transfer must be distinguished from other clauses that also seek to distribute losses. Distinct differences exist between the various loss-distribution clauses, but there is also much overlap. Indemnification—the principal topic in this chapter—has become controversial and highly regulated because it can be (in effect) *exculpatory*; that is, it can relieve one party of the cost of liability that he would otherwise have to bear. For that reason, it is instructive to look carefully at both exculpation and indemnification.

One of the most simple (yet effective) ways of relieving a party from responsibility is by a contractual exculpation where a party that may suffer losses agrees he will not pursue the party that is legally responsible for such losses. For example, suppose a patient, on being admitted to a hospital, signs a form in which he agrees the hospital will not be responsible if he is harmed as a result of the hospital's negligence. The hospital would be seeking to relieve itself from liability. Although the law accords considerable freedom to contracting parties to make their own rules and distribute losses as they wish, the hospital admissions room is hardly the place for the contract process. For that reason, such an agreement would be held invalid.

In addition to the potential for abuse in the example just given, an exculpatory clause may and often does eliminate the exculpated party's incentive to act with due care, because the clause deprives the plaintiff of his right to use the legal process when he would otherwise have had the opportunity to do so. For these reasons, clauses that exculpate one party from the consequences of his future negligence are often found to violate public policy and will not be enforced. (Yet even with the possibility of exculpation, a defendant may have an incentive to act with due care—whether out of fear for increased workers' compensation insurance rates or simply concern for the safety of others.)

Indemnification, in contrast, does not preclude first-instance liability. It deals with ultimate responsibility by shifting the loss from one party to another. For example, if an injured worker recovers against the owner, permitting the owner to shift that expense to the contractor by indemnification in no way precludes the injured worker from receiving compensation. A shift of ultimate responsibility from owner to contractor occurs. If, however, the owner is not able to recover against the contractor, for example because the contractor has filed for bankruptcy, the owner remains liable to the worker. The contingent nature of the owner's contractual right to be indemnified means that he continues to have an incentive to act with due care toward the worker. These twin facts—that the worker receives a remedy and the owner has an incentive to act with due care—show that indemnity agreements, unlike exculpatory clauses, do not violate public policy. (The owner may require the contractor to back up its indemnity promise with insurance, which is a topic discussed in Section 23.10.)

If the owner was in any way at fault in causing the worker's injury, then risk-shifting through indemnification resembles exculpation. Not only is the owner relieved of the consequences of his negligence, but indemnity agreements are often written in favor of the stronger party and against the weaker. Concern with the fairness of indemnity clauses, in particular when used in the construction industry, has led about half the states to regulate such clauses (see Section 23.8). In addition, courts often interpret such clauses narrowly where the indemnified party was itself negligent (see Section 23.9).

Another contract device that can be compared with an indemnity clause is a *limitation of liability* clause, which seeks to limit (but not eliminate) the legal remedy. For example, sellers of machinery sometimes seek to limit their liability to repair and replace defective parts. Similarly, some design professionals (in particular, geotechnical engineers) seek to limit their liability to their client to a designated amount of money or their fee (see Section 12.7D). If the actual damages are less than the specified amount, only actual damages can be recovered. But a liability limitation sets a ceiling on the damages paid by the machinery seller or design professional (in those two examples).

Many of the same considerations that have been discussed with regard to exculpatory clauses apply to liability limitations. When they are determined in a proper setting by parties of relatively equal bargaining power and when the language clearly expresses an intention to limit the liability, they are given effect.[4]

Contract clauses sometimes stipulate the amount of damages in advance. For example, liquidated damages clauses (see Section 20.12) are frequently used for unexcused time delay. They are generally given effect as long as they are reasonable. However, some of the same considerations that relate to bargaining power and appropriateness of advance agreement will be taken into account. Liquidated damages clauses (unlike indemnity clauses) do not shift losses.

If the surrounding circumstances justify enforcement of an exculpatory clause or one that limits liability, those same circumstances may justify enforcement of an indemnification clause—even if it has an exculpatory element (i.e., it calls for indemnification of a negligent indemnitee).

23.6 Parsing Indemnity Clauses

Terminology. This section uses the prime contract as an illustration. See Figure 23.3.

The prime contractor is the *indemnitor*: that is, the party promising to indemnify. The owner is the *indemnitee*: that is, the party to whom indemnification has been promised. The analysis in this section equally applies to a subcontract where the prime contractor is the indemnitee and the subcontractor the indemnitor.

Components. Each indemnity clause consists of the following components:

1. identifying the indemnitor and indemnitee(s)
2. a triggering event or events
3. the covered loss or losses
4. a causal connection
5. covered damages or expenses

Looking at a truncated version of the AIA indemnity clause, A201-2007, § 3.18.1, and numbering the different subclauses for ease of reading provides:

> [1] Contractor shall indemnify and hold harmless Owner [and] Architect from and against [2] all claims, damages, losses and expenses, including but not limited to attorneys' fees, [3] arising out of or resulting from performance of the Work, provided that such claim, damage, loss or expense is attributable to [4] bodily injury, sickness, disease or death, or to injury to or destruction of tangible property (other than the Work itself), [5] but only to the extent caused by the negligent acts or omissions of the Contractor [or] a Subcontractor, [6] regardless of whether or not such claim, damage, loss or expense is caused in part by a party indemnified hereunder.

Parsing this indemnity provision, its components consist of the following:

- The contractor is the indemnitor and the owner and architect the indemnitees.
- The triggering event is the negligence of the contractor or subcontractor (even if the indemnitee was also negligent).
- The covered losses are "all claims, damages, losses and expenses," but only for bodily injury, death or tangible property damage.
- The causal connection is "work related," that is, the covered losses arose out of or resulted from "performance of the Work."
- The covered expenses are "claims, damages, losses and expenses, including but not limited to attorneys' fees," but "only to the extent caused by" the contractor's or subcontractor's negligence.

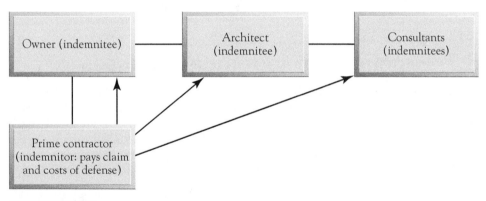

FIGURE 23.3 AIA Document A201-2007, § 3.18: Comparative indemnity clause.

Suppose a subcontractor's employee is injured on the job and sues the owner, architect, and prime contractor for negligence. The personal injury action is the *underlying claim*. Would the owner and architect be entitled to a contractual indemnity action against the contractor to transfer the risk of the underlying claim to the contractor? The answer is yes: the underlying plaintiff is suing the owner and architect for personal injury allegedly caused by a negligent contractor during performance of the work. However, the duty to indemnify is limited to the extent of damages caused by the contractor's negligence; the contractor is not obligated to indemnify for damages caused by the owner's or architect's own negligence.

Now suppose the contractor negligently drops solar panels while carrying them up to the roof. The panels fall on and destroy a parked car. The owner of the car sues the owner of the project to recover the cost of a new car. The project owner pays the driver, then brings a contractual indemnity claim against the contractor to recover that amount.

Here, the injury is property damage, not personal injury. Subclause [4] of the AIA provision extends the duty to indemnify to "injury to or destruction of tangible property (other than the Work itself)." Under the AIA, the contractor must indemnify the owner for damage to the car (which is not part of the Work). However, it has no duty to indemnify the owner for the damage to the solar panels, as that is part of the Work itself. The reason for this bifurcation in types of property damage is that tangible injury to the Work may be covered by property insurance procured by the owner under the AIA's General Conditions, A201-2007, § 11.3 (see also Section 24.4).

It is important to understand that A201-2007, § 3.18.1, is just one example of an indemnity clause. Non-AIA contracts may contain differently worded indemnity provisions. However, the same five components would be present, and the same analysis—matching those components to the facts of the case—would apply.

> In this chapter's scenario, Prime Builder brought a contractual indemnity claim against Foundation to recover liquidated damages assessed by MMI. The Foundation subcontract required Foundation to indemnify MMI and Prime Builder against any damages, loss, or claim for personal injury or property damage arising out of Foundation's performance of its subcontract "however caused."
>
> Looking first at the causal connection element, Foundation was obligated to indemnify Prime Builder for damages "however caused." This could be interpreted to cover losses incurred by the indemnitee (the prime); after all, neither party to the indemnity agreement was the cause of the failure (which was traced to a design defect). A court would enforce this allocation of risk among two innocent parties.
>
> However, the duty to indemnify extended only to losses or claims for personal injury or property damage. Since the assessment of liquidated damages did not arise out of physical harm, the indemnity provision would not apply.

23.7 Functions of Indemnity Clauses

Indemnification plays an important function in the construction industry. As an illustration, suppose an owner plans to make a construction contract with a prime contractor. The owner recognizes the increased likelihood that claims will be made against it that are based on the contractor's performance (that are sometimes, but not always, negligent). Recognizing that the law increasingly makes the owner responsible to third parties or at least that the law has made it more likely for third parties may make claims against the owner, the owner may say to the contractor,

> I have turned over the site to you. It is your responsibility to see to it that the building is constructed properly. You must not expose others to unreasonable risk of harm. The increasing likelihood that

I will be sued for what you do makes it fair that you relieve me of ultimate responsibility for these claims by your agreeing to hold me harmless or to indemnify me.

Alternatively, a reassuring proposal may come from the contractor, who may say to the owner,

I know you may be concerned about the possibility that a claim will be made against you by a third party during the course of my performance and that you will have to defend against that claim and either negotiate a settlement or even pay a court judgment. I always conduct my work in accordance with the best construction practices, and I have promised in my contract to do the work in a proper manner. I am so confident that I will do this that I am willing to relieve your anxiety by holding you harmless or by indemnifying you if any claim is made against you by third parties relating to my work. You will have nothing to worry about, as I will stand behind my work. If you are concerned about my ability to pay you, I will agree to back it up with liability insurance coverage.

In this context, indemnification acts to seal a deal when one party is anxious. The same scenario can be played in a slightly different way. Suppose the architect asks the owner to obtain indemnification for him from the prime through the prime contract in the manner accomplished by AIA Document A201-2007, § 3.18. The architect may be saying to the contractor through the owner,

The law may hold me accountable for injury to your workers or to employees of your subcontractors because they may connect their injury with something they claim I did or should have done. You are being paid for your expertise in construction methods and your knowledge of safety rules. These are not activities in which I have been trained or in which I claim to have great skill or experience. For that reason, if a claim is made against me for conduct that is your responsibility, I want you to hold me harmless and indemnify me.

In addition, either owner or architect may back up its request for indemnification from the prime contractor by noting that it is exposed to potentially open-ended tort liability if claims are made by employees of the contractor or other subcontractors, whereas the actual employer—either the prime contractor or subcontractor—who is most directly responsible need only pay the more limited liability that workers' compensation law imposes on it.

Indemnification plays another indirect function: It can shift the cost of construction site accidents to the party in the best position to prevent the accident, thereby incentivizing safety. In a design–bid–build project, the prime contractor typically promises to indemnify the owner and architect against claims for physical harm to person or property. Between these three parties—the owner, architect, and prime—the prime is the one best situated to prevent the accident from first occurring. Moreover, to the extent that indemnity promises are often required to be backed up by insurance covering the indemnity obligation (see Section 23.10), having the insured (the person obtaining insurance) also be the person best capable of preventing the loss should mean the insurance is at the cheapest possible cost.[5]

23.8 Statutory Regulation

Notwithstanding the benefits contractual indemnification brings to the construction process, some legislatures and some courts have not permitted full contractual freedom to the parties to decide how risks should be allocated by indemnification. Undoubtedly, much of this unwillingness is influenced by the beliefs that, if indemnification is freely permitted, it will encourage carelessness and is forced on the weaker party on a take-it-or-leave-it basis.

Focusing first on the legislative arena, the concern over indemnification has led to frequent statutory regulation of indemnification clauses, particularly in the construction industry.

These **anti-indemnity statutes** do not view indemnification as a method of reassuring a nervous contract maker, obtaining insurance at the best possible cost, or placing the risk on the party that can avoid the harm most cheaply. Those who regulate indemnification look principally at the exculpatory aspects of all-or-nothing indemnification and the means by which the stronger party obtains indemnification from the weaker party.

Although a detailed examination of anti-indemnity statutes is beyond the scope of this book,[6] these statutes generally are of two types: "sole negligence" and "own negligence."

A *sole negligence* statute prohibits indemnification of losses caused by the indemnitee's sole negligence.[7] Under these statutes, an indemnitor who is fault-free is not required to indemnify, regardless of the wording of the indemnity clause. These statutes are motivated by the view that the stronger party who also caused the loss should not be allowed to transfer its liability to a weaker party who was without fault. Still, an indemnitor who was only 1 percent at fault may (if the indemnity clause so provides) be liable for the entire loss. While most anti-indemnity legislation consists of "sole negligence" statutes, they are rarely applied to preclude enforcement of the indemnity clause. Most claims for which indemnification is sought do not result exclusively from the negligence of the indemnitee. To the contrary, the interdependent nature of elements in the construction project means that claims are usually attributable to both the indemnitee and the indemnitor.

Own negligence statutes prevent clauses under which the indemnitor would have to pay for part of the loss caused by the indemnitee's negligence. Each party must bear the cost of the damage it had caused. The parties are not allowed by agreement to circumvent comparative negligence.[8]

Regardless of the type of anti-indemnity statute involved, there is almost universal agreement that they should not extend to agreements to procure insurance (whether general liability or workers' compensation) or bonds. After all, the motivation behind these statutes is to increase the likelihood of compensation for the injured party by helping ensure the party with the strongest bargaining power does not exempt itself from any liability. An agreement to procure insurance or bonding brings another source of funding to the project, thereby increasing the likelihood the victim will be compensated.[9]

As already noted, one motivation for anti-indemnity statutes is to prevent prime contractors from imposing expensive insurance requirements on subcontractors—a cost that ultimately is transferred to the owner. Perhaps to bring down the cost of home construction, California bars residential builders from demanding indemnification from their subcontractors for losses that did not arise out of the subcontractor's work.[10]

Some statutes apply to public works. California prohibits public agencies from seeking indemnification against design professionals who were fault-free,[11] and Colorado prohibits public agencies from being indemnified for the losses caused by their own negligence.[12] Indiana prohibits indemnity agreements for the indemnitee's sole negligence or willful misconduct on highway construction contracts.[13]

> In this chapter's scenario, Electrical agreed to indemnify Prime Builder against any damages or claim—but "only to the extent caused by" Electrical's negligence or breach of contract. This agreement is enforceable under either a "sole negligence" or "own negligence" anti-indemnity statute.

23.9 Common Law Regulation: Specificity Requirements

Long before the enactment of the anti-indemnification legislation just described, the common law looked with hostility at such clauses to the extent that they purported to require indemnification of a negligent indemnitee. Such clauses were enforced only if the language clearly stated that indemnification would be made even if the loss had been caused in whole or in part by the indemnitee's negligence.[14] Some even required that negligence be mentioned specifically.[15] Regulating such clauses reflected the belief that indemnification often exculpated the indemnitee and could be a disincentive for good safety practices as well as the belief that such clauses were forced on weaker parties. In operation, the effectiveness of such an approach depended on the skill of the person drafting the indemnification clause.

In recent years, there has been a tendency to look on indemnification clauses neutrally as representing a rational attempt to distribute losses efficiently by facilitating insurance coverage. Courts espousing a more favorable attitude toward such clauses seek to determine the intention of the parties as they would when interpreting any other clause.[16]

23.10 Indemnitor Required to Procure Insurance

Owners frequently require that contractors procure insurance covering the risks specified in the indemnification clause. Even without this requirement, a prudent contractor will be certain that its insurance will cover this risk. Generally, liability policies cover only liability imposed by law and not that imposed or assumed by contract. However, modern standard form commercial liability policies provide coverage for liability assumed under an "insured contract," which has been interpreted to mean an indemnity agreement.[17] Contractors should be certain—as should owners—that the liability insurance policy covers this risk.

As noted, agreements to procure insurance to back up an indemnity provision are generally not invalidated by anti-indemnity legislation. Sometimes clarification that insurance does not run afoul of anti-indemnity legislation is done expressly by statute. When courts have faced this issue without the benefit of a statute, they have recognized the difference between an indemnity clause with exculpatory features and a liability insurance policy.[18] The former is often thought to induce carelessness on the site, whereas the latter is looked on as a proper method of distributing risks.

Another difficulty that arises when an indemnity obligation is backed up by insurance is whether the insurance obligation remains in force if the indemnity clause is found invalid—either under an anti-indemnity statute or a common law analysis. Some courts find that the two obligations rise and fall together, so a party not liable for indemnification also has no duty to provide insurance to cover the loss.[19]

Other courts find the insurer's duty to provide coverage remains—even if its insured's liability under the indemnity agreement has been negated. These courts point out that the owner could hardly have intended the requested insurance coverage to fail precisely when its indemnity rights against the contractor are lost. They also note that the insurance coverage trigger (usually under an additional insured endorsement) may be broader than the duty to indemnify.[20] Courts also reason that the insurer, having accepted a premium, should not be allowed to escape coverage for reasons having nothing to do with the policy itself.[21]

REVIEW QUESTIONS

1. What are two fundamental approaches to managing liability associated with the risks of physical harm?

2. What is the difference between a direct action and a third-party complaint?

3. In which two ways can indemnity be conferred?

4. What is the fundamental principle underlying noncontractual indemnity?

5. What is the difference between the majority rule and minority rule of apportionment?

6. What is contractual exculpation?

7. What is the difference between an indemnitor and an indemnitee?

8. What are the five components of an indemnity clause?

9. What is an indirect function of indemnification?

10. What are the two types of anti-indemnity statutes and how do they work?

ENDNOTES

[1] *Pacesetter Pools, Inc. v. Pierce Homes, Inc.*, 86 S.W.3d 827 (Tex.App.Ct.2002); Restatement (Third) of Torts: Apportionment of Liability § 23(b) (2000).

[2] *Board of Educ. of Hudson City School Dist. v. Sargent, Webster, Crenshaw & Folley*, 71 N.Y.2d 21, 517 N.E.2d 1360, 523 N.Y.S.2d 475 (1987).

[3] D. DOBBS, TORTS, 1079 (2000) (footnote omitted).

[4] *RSN Properties, Inc. v. Engineering Consulting Services, Ltd.*, 301 Ga.App. 52, 686 S.E.2d 853 (2009).

[5] *Leitao v. Damon G. Douglas Co.*, 301 N.J.Super. 187, 693 A.2d 1209, 1211-12 (App. Div.), certification denied, 151 N.J. 466, 700 A.2d 879 (1997); *Di Lonardo v. Gilbane Bldg. Co.*, 114 R.I. 469, 334 A.2d 422 (1975).

[6] The statutes are categorized and discussed in M. SCHNEIER, CONSTRUCTION ACCIDENT LAW: A COMPREHENSIVE GUIDE TO LEGAL LIABILITY AND INSURANCE CLAIMS, 462-82 (Am.Bar Assn.1999) and Gwyn & Davis, *Fifty-State Survey of Anti-Indemnity Statutes and Related Case Law*, 23 Constr. Lawyer, No. 3, Summer 2003, p. 26.

[7] E.g., Cal.Civ.Code § 2782(a) (bars indemnification for both sole negligence and willful misconduct of the indemnitee); Cal.Civ.Code § 2782(c)(1) (bars indemnification for private real estate owner's active negligence, but subsection (c)(3) exempts a homeowner performing a home improvement on a single family dwelling); N.J.Stat.Ann. § 2A:40A-1; and Mich.Comp.Laws Ann. § 691.991.

[8] E.g., Minn.Stat.Ann. § 337.2; N.Y.McKinney's Gen.Oblig.Law § 5-322.1; and Wash.Rev.Code § 4.24.115.

[9] 740 Ill.Comp.Stat.Ann. § 35/3, interpreted in *Bosio v. Branigar Org., Inc.*, 154 Ill.App.3d 611, 506 N.E.2d 996 (1987).

[10] Cal.Civ.Code § 2782(d).

[11] Id., § 2782.8. Also, Cal.Civ.Code § 2782(b) shields a contractor from liability for a public agency's active negligence.

[12] Colo.Rev.Stat. § 13-50.5-102(8).

[13] Ind.Code § 26-2-5-1.

[14] For a few of the many cases, see *Becker v. Black & Veatch Consulting Eng'rs*, 509 F.2d 42 (8th Cir.1974) and *Greer v. City of Philadelphia*, 568 Pa. 244, 795 A.2d 376 (2002), reaffirming *Ruzzi v. Butler Petroleum Co.*, 527 Pa. 1, 588 A.2d 1 (1991).

[15] *Ethyl Corp. v. Daniel Constr. Co.*, 725 S.W.2d 705 (Tex.1987).

[16] *Washington Elementary School Dist. No. 6 v. Baglino Corp.*, 169 Ariz. 58, 817 P.2d 3, 6 (1991) (interpreting AIA contract); *Morton Thiokol, Inc., v. Metal Bldg. Alteration Co.*, 193 Cal.App.3d 1025, 238 Cal.Rptr. 722 (1987), review denied Oct. 17, 1987 (focuses on intention of the parties).

[17] *Mid-Continent Cas. Co. v. Global Enercom Management, Inc.*, 323 S.W.3d 151 (Tex.2010).

[18] *Kinney v. G.W. Lisk Co.*, 76 N.Y.2d 215, 556 N.E.2d 1090, 557 N.Y.S.2d 283 (1990).
[19] *Hurlburt v. Northern States Power Co.*, 549 N.W.2d 919 (Minn.1996).
[20] *Acceptance Ins. Co. v. Syufy Enterprises*, 69 Cal.App.4th 321, 81 Cal.Rptr.2d 557 (1999), review denied Apr. 14, 1999 (insurance coverage for owner's own negligence involving a premises defect not part of the contract work).
[21] *Heat & Power Corp. v. Air Products & Chemicals, Inc.*, 320 Md. 584, 578 A.2d 1202, 1208 (1990).

WORK INJURY
CLAIM FORM

LS

What area of the work... ...ing in when you were injured?

What is the street address where the incident oc...

Maiden name

Suburb

mangostock/Shutterstock

CHAPTER 24

INSURANCE

SCENARIO

Mega Manufacturer, Inc. (MMI) sought to build a new factory. MMI retained Awesome Architects, Inc. (AAI), an architectural firm, to create the plans and specifications.

Based upon this design, MMI entered into a $2 million fixed-price contract with Prime Builder, Inc. Prime Builder hired several subcontractors, including a excavation subcontractor (Excavator) and a window installer (Window Wizard, Inc.). Under the design contract, AAI was to inspect the work to determine the contractor's entitlement to progress payments. The contract excluded AAI's responsibility for the contractor's means and methods of performance.

The project suffered from heavy rains. During one inspection, the AAI employee (a licensed architect) measured the amount of excavation work completed. He noticed the trench where Excavator's employees were working was water logged, but he said nothing. Shortly after he left, the trench caved in, injuring an Excavator employee.

Late on a Friday, Window Wizard installed several windows with wooden frames. The employee forgot to caulk the windows. That weekend, a hurricane struck and the project was not resumed until the following week. When Window Wizard returned, it discovered the window frames had begun to rot. In addition, the drywall surrounding the windows were also molding. As a result (1) Prime Builder replaced the drywall and (2) Window Wizard replaced the windows.

Assuming all parties carried the customary insurance, what are the insurance consequences of these various events?

24.1 Insurance: Risk Spreading

While contractual indemnity is increasingly common in construction contracts, insurance is the primary mechanism for risk management. As discussed in Sections 23.8 and 23.9, indemnity clauses are regulated by statute and are narrowly construed by the courts, especially where the indemnitee was negligent. Even if such a clause is enforceable, the right to contractual indemnity is only as good as the solvency of the indemnitor. These are the reasons (as noted in Section 23.10) that a party's duty to indemnify is often coupled with a requirement to back up that duty with insurance coverage.

Insurance spreads the risk of economic calamity from accidental events. Activities are subject to predictable—although unexpected (or fortuitous)—risks. The activity of driving a car is subject to the risk of a car accident or that the car will be stolen. The activity of owning a house is subject to the risk that the house will burn down or be burgled. Other risks are not viewed as predictable. Although either the car or the house may be hit by a meteorite, the possibility of such an event is extremely small.

As a practical matter, it is impossible for an individual to predict the economic cost of even predictable risks—let alone have the financial reserves with which to safeguard against them. Moreover, the unexpected nature of a loss from these risks actually occurring means any financial reserve must be tied up indefinitely. This is both economically inefficient and counterproductive.

The function of insurance is to transfer, in exchange for the payment of a premium (the price of the insurance policy or contract), the risk of a financial loss from predictable, unexpected (or fortuitous), and unintentional perils. The insurance company pools (collects) the premiums from all of those insured, so it has a large reserve from which to pay out covered losses.

24.2 Construction Insurance: An Overview

As just noted, insurance covers losses that are predictable—but also fortuitous (unexpected) and unintentional. Not all predictable risks are insurable (covered by an insurance policy). Normal business risks, such as the possibility that a new apartment building will not attract tenants, are not insurable.

> In this chapter's scenario, suppose the project's completion was delayed by the unusually bad weather. This caused MMI to incur delay costs in terms of lost use of the new facility. This is not an insurable loss under MMI's property insurance.

Construction projects as a general rule engender two types of insurable risks: bodily injury (including death) and property damage. Insurance generally protects against the peril of *physical harm* to person or property.

For purposes of this chapter, three forms of insurance are most relevant.

1. *Property insurance*: Covers damage to tangible property; usually maintained by the owner.
2. *Liability insurance*: Covers liability claims against the insured; primarily maintained by the contractor (including subcontractor) and design professional.
3. *Workers' compensation and employer's liability insurance*: Covers employee personal injury claims for job-related accidents; maintained by all employers.

Liability insurance (the second category) itself divides into two major groups.

1. *Commercial general liability (CGL) insurance*: Protects the insured from claims of general negligence.
2. *Professional liability insurance*: Protects the insured from claims of professional negligence (or malpractice).

This basic division of insurance procurement is represented in Figure 24.1.

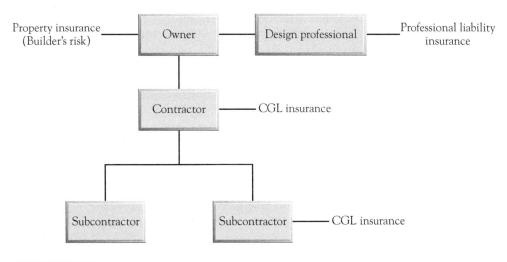

FIGURE 24.1 Simplified illustration of insurance coverage

24.3 Introduction to Insurance Industry and Policy

Standardized Insurance Policies. Standardized policies are the norm in the insurance industry. The most widely used provider of standard forms is the Insurance Services Office, Inc. (ISO), which is an association founded by insurance companies for the purpose of writing standardized liability policies. In *Hartford Fire Insurance Co. v. California*, the U.S. Supreme Court, writing in 1993, described the importance of the ISO:

> [M]ost primary insurers rely on certain outside support services for the type of insurance coverage they wish to sell. Defendant Insurance Services Office, Inc. (ISO), an association of approximately 1,400 domestic property and casualty insurers . . . is the almost exclusive source of support services in this country for CGL insurance. ISO develops standard policy forms and files or lodges them with each State's insurance regulators; most CGL insurance written in the United States is written on these forms. All of the "traditional" features of CGL insurance relevant to this litigation were embodied in the ISO standard CGL insurance form that had been in use since 1973 (1973 ISO CGL form).[1]

"Manuscript" policies contain provisions (also called special endorsements) that are tailored to particular insureds.

Regulation. All states regulate insurance to varying extents. State laws usually provide requirements of capital and financial capacity to ensure the solvency of insurers, and they increasingly regulate claims settlement practices. State insurance regulatory agencies determine which insurers will be permitted to do business in the state. In some states, regulators can determine coverage, exclusions, and premiums.

Courts have also regulated insurance, mainly through interpreting insurance policies when disputes arise. The law interprets ambiguities against the insurer unless the insurance policy is negotiated between a strong insured, such as a group of hospitals, and an insurer. Some courts seek to determine the reasonable expectations of the insured or some average insured. They will protect that expectation, with some courts even disregarding insurance policy language in order to protect that expectation.

Finally, there are many technical contractual requirements before the insurer's obligation to defend and indemnify matures, such as complying with warranties or representations of activities, giving notice of claims, furnishing proof of loss, or cooperating with the insurer in defending claims. Technically, a failure to comply with any of these requirements can result in loss of coverage. But the law often protects the insured by requiring the insurer to show that the failure to comply prejudiced (disadvantaged) the insurer.

Premiums. Premiums have become an increasingly important overhead cost for design professionals and contractors. The amount of premiums for any individual insured is determined by a number of factors, such as type of services performed, experience of the insured, locality in which work or projects are located, gross billings of the insured, contracts under which services are performed, and claims record of the insured.

Premiums can rise because of underwriting predictions (selection of risks) by insurers, a decline in the value of the insurer's portfolio (investments made by the insurer using the premiums it has received), a downturn in the economy that can induce insureds to cancel coverage, high claims pay-outs, and increased cost of defending claims.

Deductible Policies. Insurance companies increasingly seek to reduce their risk by excluding from coverage claims, settlements, or court awards below a specified amount. Policies that exclude smaller claims are called *deductible policies*. Generally, the higher the deductible, the lower the premium cost. A high deductible means that the insured bears substantial risks. The recent tendency to raise deductible amounts makes the insured increasingly a self-insurer for small claims. Some policies include in the deductible the cost of defense. For example, if the deductible is $5,000 and a claim of $3,000 is paid to a claimant, the insured bears any cost of defense up to the deductible amount.

Deductible policies can create a conflict of interest between insurer and insured. When small claims are made, the insurer may prefer to settle the claim rather than incur the cost of litigation. The insured may oppose such a settlement, believing that it is an admission of negligence and that payment to the claimant will come out of the insured's pocket.

Policy Limits. American policies generally limit insurance liability. Suppose a claim for $500,000 is made, and the policy limit is one million dollars. The claimant offers to settle for $400,000. The insurance company exercises its right to veto settlement and refuses to settle. The claim is litigated, and the claimant recovers $1,125,000. The insured may contend that had the insurer settled, he would not have had to pay amounts in excess of the policy limit ($125,000). Courts are sympathetic to this argument. Insurance companies are generally liable for amounts over the policy limit if their refusal to settle was unreasonable in light of all the circumstances.[2]

Policies also contain an aggregate limit that applies to all claims made during the policy period. The maximum amount available to settle all claims arising out of one negligent act is usually the limit of liability per claim, but the maximum amount for all claims is the policy's aggregate limit of liability.

Notice of Claim: Cooperation. Insurance policies usually state that the insured must notify the insurance company when an accident has occurred or when a claim has been made. The notice is to enable the insurer to evaluate the claim and gather evidence for a possible lawsuit.

In this chapter's scenario, AAI should notify its insurer immediately of the trench collapse. AAI should not wait to see whether a claim arising out of the collapse is actually brought against it.

In addition to requiring that the insured notify the insurer, policies usually require the insured to cooperate with the insurer in handling the claim. The insured must give honest statements (a false swearing clause may make any dishonest statement the basis for denying coverage). He must also make reasonable efforts to identify and locate witnesses and supply the insurer with material that may be important in defending the claim. The insured must also comply with any notice to attend hearings that have to do with the claim, such as depositions and trials. Although insurance policies often state that coverage will be denied if there is a failure to cooperate, courts are not always willing to make failure of cooperation a sufficient basis to deny coverage unless the insurer can prove it was prejudiced by the lack of cooperation.

Duty to Defend. In most liability insurance policies, the insurer promises to defend and indemnify. *Defend* means providing the insured with a lawyer to handle the litigation against the insured by the claimant and also to cover litigation costs. *Indemnify* means to pay for a judgment rendered against the insured (or a settlement agreed to by the insured) in that litigation. The duty to indemnify is capped by the policy's limits.

One of the most difficult areas of law surrounds the question of when the insurer has a duty to defend. The duty to defend is broader than the duty to indemnify, as policies usually state that the insurer will defend claims that are groundless, false, or fraudulent. Whether or not there is a duty to defend first depends on the complaint made against the insured. If coverage appears likely from the complaint, the insurer must defend. However, the complaint does not limit the duty to defend. The insurer will have to defend if facts brought to its attention indicate a possibility of coverage.

Suppose the insurer decides it has a duty to defend. During trial it learns the insured's liability is based on some acts that are covered by the policy but also other acts that are excluded from coverage. For example, suppose the claimant's loss could be attributable either to defective plans and specifications, which would be covered, or to an express warranty of a successful outcome, which would not. As the defense is within the control of the insurer, the insurer's attorney could use that control to obtain a judicial conclusion that placed liability on conduct not covered. (Also, the attorney is ordinarily more concerned with preserving his relationship with the insurer than with the interests of the insured.)

These considerations led a California court, in the case of *San Diego Federal Credit Union v. Cumis Insurance Society, Inc.*,[3] to conclude that if a conflict of interest arises between the insurer and insured, the insurer must pay for an independent attorney for the insured. Shortly after this decision, the California legislature enacted a statute that both codifies the right to independent counsel where there are certain conflicts of interest and regulates who can be appointed as independent counsel and his fees.[4]

Even if the insurer need not pay for the cost of an independent attorney for the insured, it may be advisable for the insured to retain independent counsel at his own expense.

Settlement. The typical policy, although granting the power to settle to the insured, places sharp restraints on that power by making the insured take certain risks if he refuses to settle when the insurer suggests he do so.

Settlement provisions typically state that the insurer will not settle without consent of the insured. If the insured fails to consent to a recommended settlement and elects to contest the claim and continue legal proceedings, the insurer's liability for the claim does not exceed the amount for which the claim would have been settled, plus claims expense incurred up to the date of such refusal. In other words, the insured will take the risk of the settlement having been a good one.

Multiparty Policies. In a transaction as complex as construction and with its host of participants, inefficiencies can develop if each party carries its own liability insurance. This situation has led some owners to require "wrap-up" policies for those engaged in construction and project-wide professional liability insurance for all of the professionals involved in the project. Such insurance use is increasing.[5]

24.4 Property Insurance

24.4A Introduction

Property insurance provides coverage to a building against enumerated perils, such as fire or wind damage. The insured is the owner of the building.

The peril of fire, for example, is different for a building under construction as compared to a completed building. For this reason, an owner's property insurance policy for a completed building typically excludes coverage for properties under construction.

The "builders risk" coverage form is the property insurance policy most often secured to insure against the risk of loss to construction projects.[6] Owners purchase builder's risk insurance to protect themselves against losses they would otherwise suffer directly. Accordingly, standard form documents place upon the owner the obligation to acquire property insurance covering the project.[7]

24.4B Coverage for Project Destruction During Contract Performance

During project construction, the project may be partially or totally destroyed either because of contractor (or subcontractor) negligence or by circumstances for which neither party can be charged. The work may be destroyed by fire (started by a subcontractor while welding), unstable subsurface conditions, or violent natural acts such as earthquakes or hurricanes.

When the project is destroyed during construction, the owner receives proceeds (money) from its property insurer to repair the damage. Both by contract agreement and judicial interpretation, insurance coverage preempts (bars) subrogation claims: litigation of damages against the negligent contractor or subcontractor brought by the property insurer in the name of the owner. This topic is discussed in Section 16.9C.

24.5 CGL Insurance and Defective Construction Claims

As noted in Section 24.2, the prime contractor maintains *commercial* (formerly called "comprehensive") *general liability* (*CGL*) insurance. The primary purposes of CGL insurance are to provide the insured contractor with a legal defense against claims of personal harm or property damage brought by third parties and to "indemnify" the contractor if the claims are successful (by paying the judgment). As an example, if a worker's tool fell and struck a parked car, the contractor's liability insurer would defend the contractor and pay the car owner's claim. As another example, the claimants may be injured members of the public (whether passers-by or those visiting the construction site) or injured workers on the job who are not employees of the contractor (as the latter are usually covered by workers' compensation).

To be more specific: CGL policies provide the insured with a legal defense and indemnification (up to the policy limits) against any claims of liability brought against the insured that assert "bodily injury" or "property damage" caused by an "occurrence." An "occurrence" is defined in the insurance policy as an "accident." Hence, the claim must have been in some way unexpected and unintended by the insured. Coverage for claims alleging "bodily injury" is relatively straightforward.

A more difficult question is whether an owner's claim of defective workmanship constitutes "property damage" caused by an "occurrence." Suppose the insured is the prime contractor, and the roofing subcontractor negligently installs the roof, causing the completed building to suffer water damage. Should the contractor's liability insurance provide coverage?

There is little doubt that the water damage constitutes "property damage" for purposes of the insurance policy. But was negligence by the subcontractor an "occurrence"—an "accident" from the perspective of the insured? More broadly, are these types of claims—arising out of a breach of contract—the proper subject of liability insurance (in contrast to the damaged car caused by tortious conduct)?

Liability insurance is intended primarily to cover unusual, unexpected losses, such as a tool falling from a height on a car parked nearby. Conversely, CGL policies are not intended to cover ordinary business losses (defective work being an example). However, sometimes courts confuse the *purpose* of CGL insurance with the *scope of coverage*. Rather than examine whether the claim against the insured involves "property damage" caused by an "occurrence" as the policy requires, they broadly exclude from coverage all breach of contract claims alleging economic losses, including owner claims for defective workmanship.[8]

However, a more sophisticated analysis is to determine coverage based on the policy language—not on the theory of liability or the broad purpose of commercial liability insurance. Courts that employ this analysis hold that contract liability claims are not automatically excluded from CGL coverage.[9] These courts also reject the argument that a contractor's defective workmanship cannot be an "accident" (and hence not an "occurrence") because that work was intentionally (although negligently) done.[10] Otherwise stated, the contractor's CGL insurance covers the risk of property damage caused by negligent conduct—even if that negligent conduct is faulty workmanship.

Even if the underlying claim involves "property damage" or "bodily injury" caused by an "occurrence," coverage may be defeated by the application of policy exclusions. Under these standard provisions, CGL policies exclude work products, property in control of the insured (this should be handled by property insurance), or the contractor's own work that needs to be repaired or replaced (but not if the insured is the prime contractor and the work was done by a subcontractor). In sum, unless the defective workmanship caused collateral damage to nondefective work (such as water damage to previously installed drywall) or to property entirely outside the scope of the construction contract (such as the owner's personal property), defective workmanship claims generally fall within the scope of the exclusions.[11]

> In this chapter's scenario, Window Wizard was negligent in not caulking the newly installed window. Both the window frames (the insured's work) and the surrounding drywall (not part of the insured's scope of work) suffered water damage and rot.
>
> Assuming negligent workmanship is considered an "occurrence," Window Wizard's CGL insurance would cover the cost of replacing the drywall. However, the cost of replacing the windows would be excluded under the exclusion for property damage to a contractor's own work that needed to be repaired or replaced because of the contractor's operations performed on that work.

24.6 Professional Liability Insurance

24.6A Requirements of Professional Liability Insurance

The law does not require design professionals to carry professional liability (also called "errors and omission" or E&O) insurance—nor did the AIA prior to 2007. However, clients increasingly require that design professionals have and maintain professional liability insurance. Recognizing this and the benefits of proper insurance coverage, AIA B101-2007, § 2.5, requires that the architect maintain general liability, automobile, workers' compensation, and professional liability insurance. (Design professionals should carry commercial general liability (CGL) insurance to protect against claims of ordinary negligence, such as a fire caused by leaving a coffee pot plugged in.)

Clients may have a claim against the design professional for losses relating to the project or because they have satisfied claims of third parties that are directly traceable to the design professional's failure to perform in accordance with the legal standards. To make any claim collectible, they may require the design professional to carry professional liability insurance. If the design professional has adequate professional liability insurance, third parties injured as a result of his conduct may choose to bring legal action against the design professional directly rather than against the owner.

Even if not required to, many design professionals carry such insurance. One reason is to protect their nonexempt assets from being seized if a judgment is obtained against them. Another is that many design professionals do not wish to see people go uncompensated who suffer losses because of the design professional's failure to live up to the legal standard.

24.6B Policy Types: Occurrence or Claims-Made

In the construction process, a long time lag can exist between the act or omission claimed to be the basis for liability and the making of the claim. Coverage in such cases depends on whether the policy is a "claims-made" or an "occurrence" policy. Claims-made policies cover only claims made during the policy period, regardless of when the act giving rise to the claim occurred. An occurrence policy, in contrast, gives coverage if the act or omission occurs during the policy period.

Professional liability insurance is primarily written on a claims-made basis. This avoids the "tail" of potential liability at the end of any occurrence policy. Proper insurance underwriting and rate making require the insurer to predict payouts in a designated period. Particularly in states that do not begin the statutory period for bringing claims until discovery that there is a claim and against whom it can be made, occurrence policies can result in coverage many years after the premium has been fixed. This led the professional liability insurers to base policies on claims made. It is easier to predict the value of claims that will be made during the period of the policy than to look far into the future where a prediction is needed for setting premiums under an occurrence policy.

At any given time, a variety of claims-made policies can be available. Some are hybrid. Such policies specifically protect against claims made during the policy period and require that the act giving rise to the claim occur during the policy period. Most courts have sustained claims-made policies and "hybrid" policies consisting of claims-made *and* occurrence policies where both the occurrence and the claim must occur during the policy period for there to be coverage.

Insurers usually offer retroactive or "prior acts" coverage, giving coverage regardless of when the act occurs. Retroactive coverage usually requires the insured to represent that he is unaware of any facts that could give rise to a claim or that he was insured when the act occurred and that he has carried insurance throughout his career.

> In this chapter's scenario, suppose Excavator's injured employee brings a claim against AAI. If AAI had a claims-made policy and both the bodily injury occurred and the claim were brought during the policy period, AAI would have coverage.

24.6C Coverage and Exclusions: Professional Services

In the absence of a special endorsement, professional liability insurance policies generally cover liability for performing normal professional services.[12] Conversely, coverage under commercial general liability (CGL) policies carried by contractors usually excludes professional services.

Often, claims against the professional liability insurer are denied because the insurer asserts that the services in question were not professional. Similarly, claims against a CGL insurer are denied because the insurer asserts that the services were professional and not covered. This situation leaves the insured—the design professional—with the expensive (and frankly frightful) prospect of facing a claim without the protection of insurance. (Purposefully not carrying insurance as a strategy to avoid claims is not recommended.)

However, where it is unclear whether a claim is covered by professional liability or CGL coverage and the insured maintains both policies, one insurer will volunteer to assume the insured's defense. If that insurer pays the claim, it often brings a claim against the other insurer based upon its having paid a claim that should have been paid by the other.

When is a claim subject to professional liability insurance, and when is it subject to general liability insurance? Professional services are those that are predominantly mental—not physical or manual in nature. A federal court held that the draftsman who created the design was engaged in specialized intellectual labor and that the exclusion for professional services applied to the insured's CGL policy.[13] Professional liability coverage can also encompass site services, such as monitoring the work as it proceeds for issuing payments and completion certificates, as all involve intellectual judgment.

American Motorists Insurance Co. v. Republic Insurance Co.[14] involved the question of whether the architect's preparation and submission of a competitive design-build bid is a professional service and is covered under the architect's professional liability policy. The case involved a competitive design system with the insured being the successful bidder. After the award of the bid, an unsuccessful bidder sued the insured for misrepresentations and other tortious conduct. The insured's professional liability insurer refused to defend, but the action was defended successfully by the insured's CGL insurer. The latter then sought to recover a *pro rata* (proportionate) share of its defense costs from the professional liability insurer.

The professional liability insurer argued that bid preparation and submission are merely preparation to render professional services and not the actual rendering of such services. However, the Alaska Supreme Court did not accept this contention, noting that only an architect using his specialized knowledge, labor, or skills could have prepared the bid. The bid consisted of two booklets approximately 160 pages in length and providing considerable detail as to the design. The court noted that the policy did not define professional services and that (in such cases) the term was considered ambiguous, requiring an interpretation of the policy that favored coverage.

The same type of issue came before the court in *Camp Dresser & McKee, Inc. v. Home Insurance Co.*[15] The original claim was made by an injured worker at a plant who attempted to throw ash onto a head pulley of a conveyor that was missing a safety guard, thus exposing the worker's hand and arm to injury. Camp Dresser, the insured, was an engineering firm that acted as a "project manager" to coordinate the work and perform services similar to that of a construction manager. Camp Dresser's professional liability insurer defended the case and paid for the settlement in excess of the $150,000 deductible under the professional liability policy. But Camp Dresser also notified its CGL insurer because that policy had no deductible.

The CGL insurer denied coverage based on a "professional services" exclusion in the policy. The exclusion defined professional services to include "supervisory, inspection or engineering services."

The court held in favor of Camp Dresser, noting that the allegations in the injured worker's complaint of inadequate control and failure to warn may constitute general negligence. The court explained that professional services require specialized knowledge and mental rather than physical skills. While the endorsement included the word "supervisory," that word was ambiguous as to whether it described supervision of purely professional activities or, more broadly, management or control of the aspects of a project involving professional and nonprofessional activities. Because the policy did not make clear in what sense the term was used, Camp Dresser received coverage.

These cases demonstrate that the borderline between professional and nonprofessional services can be unclear. This may mean there is coverage under both types of policies or, instead, that neither policy will cover these services. An insured who can anticipate this event and can be concerned that it will have to pay the formidable costs of defense might consider procuring business legal expense insurance.

A design professional should examine the policy for exclusions. Exclusions may extend to activities relating to boundary surveys, failure to advise on required insurance or surety bonds, failure to complete construction documents, and failure to respond to submittals in the time promised—unless those losses were due to improper design. Other exclusions may be for estimates of probable construction costs or for claims of copyright, trademark, and patent infringements. Another possible area of exclusion is related to the handling of hazardous waste. Consultation with a competent insurance counselor is vital.

> In this chapter's scenario, suppose Excavator's injured employee brings a claim against AAI. Would AAI be entitled to coverage under its professional liability or general liability insurance?
>
> Since AAI's employee was on the site only to measure the length of the trench (assume payment was made on a unit price basis), the professional liability insurer would argue that the employee's acts did not require professional skills and that coverage should be under the CGL policy.
>
> The CGL insurer would counter that liability is being asserted against AAI based upon its knowledge as a design professional of the hazardous nature of the trench. It would try to shift coverage on that basis to the professional liability insurer.

24.6D Preparing to Face Claims

Design professionals should anticipate the likelihood that claims will be made against them. With this in mind, the design professional must be able to document in the clearest and most objective way that a proper job was done. For example, expanded liability should not deter design professionals from using new designs, materials, or products. Design professionals must, however, prepare for the possibility that (if things go wrong) they will be asked to explain their choice.

Using new materials as an example, the design professional should accumulate information allowing him to predict the performance of any contemplated new materials. This information should be obtained from unbiased people who have used the materials on comparable projects. A list of such people can be requested from manufacturers, whose representatives should be questioned about any bad results. The manufacturer can be notified as to the intended use in the project and asked for technical data that include limitations of the materials. Sometimes it is possible to have a manufacturer's representative present when a new material is being installed to verify installation procedures. Any representations or warranties obtained should be accessible.

Design professionals should be able to reconstruct the past quickly and efficiently. A system for efficient making, storing, and retrieving of memoranda, letters, e-mails, and contracts is essential. Legal advice should determine the proper time to preserve records. If major design decisions have to be made, the design professional should indicate the advantages and disadvantages and obtain a final written approval from the client. Records should show when all communications are received and responses made. If work is to be rejected, the design professional should support his decision in communications to client and contractor. Similarly, if any previous approvals are to be withdrawn, written notice should be given to all interested parties. Records should be kept of all conferences, telephone calls, and discussions that may later need to be reconstructed in the event of a dispute.

In regard to recordkeeping, the instability of many design professional relationships can be troublesome. Design professionals dissolve partnerships frequently. Where dissolution occurs, records that should be kept are often lost or destroyed. When rearrangements occur, those involved should separate records and see that those who may need them have them.

At all stages of their practice, design professionals need competent legal services at prices they can afford. Certainly, legal advice obtained only after disputes have arisen is insufficient. Younger groups of design professionals should consider negotiating with those who provide legal services for prepaid service plans.

Finally, the operations of the design professional, including contracts used, records kept, and compliance with laws regulating employers, should be evaluated periodically.

24.7 Insurance and Alternative Project Delivery Methods

The basic distinction between CGL and professional liability insurance traces back to the division of labor between constructor and designer inherent in the design–bid–build traditional project delivery system. As has been repeatedly emphasized in this book, that fundamental division of labor has largely broken down. Not only are constructors increasingly brought into the design process, but the industry has seen the introduction of new actors (such as construction management, project management, and design/build) that bridge (or at least blur) that division of labor. What are the insurance consequences of such developments?

As a preliminary matter, all project participants—regardless of their responsibilities or roles—should carry general liability insurance. Any participant may be accused of general negligence. Similarly, any employer should carry workers' compensation and employer's liability insurance in the event an employee is injured or killed on the job site. Standard policies such as automobile insurance also must be maintained.

Second, participants should communicate effectively with their insurance agents as to the type of activities they will engage in on the project. Coverage is a matter of conduct, not job title. Only an insurance agent can correlate the activity with the type of policy required.

Increasingly, the contractor and specialty subcontractors participate in creation of the design. Those contractors should insure against claims based upon design; yet this "professional" service is excluded from CGL coverage. These contractors should supplement their CGL insurance with professional liability insurance.[16]

The most difficult issue is coverage for new actors in the construction process. Reliance upon CGL coverage alone is clearly inadequate; current policies not only exclude professional services but also services performed as a construction manager.

Design/build clearly bridges the distinction between construction and design. As made clear from the *Camp Dresser* decision discussed earlier, the mere creation of a bid on a design/build project involved conduct subject to professional liability coverage—not CGL coverage.[17]

In a recent Louisiana case, *North American Treatment Systems, Inc. v. Scottsdale Insurance Company*,[18] a project manager (PM) was hired to act as a liaison between the owner and the prime contractor. He hired an engineer to perform engineering services because the PM was not a licensed engineer. The PM was to provide professional services required for design, engineering, construction management, and operations management for construction and operation of the plant being built. A subcontractor was negligent, and damages occurred.

The owner's property damage insurer compensated the owner, and that insurer sued the PM to recover the insurer's payment to the owner. The PM sought a defense from its CGL insurer. The issue was whether the PM's services were professional and not covered because it was excluded by the CGL policy. The court on appeal held that the PM was performing professional services because its work took particular skill and training. Clearly, the PM should have obtained professional liability insurance.

REVIEW QUESTIONS

1. As a general rule, construction projects engender which types of insurable risks?

2. Which three forms of insurance are most relevant to construction projects?

3. Which are the primary types of insurance obtained by the owner, design professional, and contractor or subcontractor?

4. What are some factors that can determine the amount of premiums required for any individual insured?

5. What are some examples of how an insured must cooperate with the insurer in handling a claim? What is the possible consequence to the insured of its failure to cooperate?

6. Which type of insurance is the property insurance policy most often secured to insure against the risk of loss to construction projects?

7. What is a primary purpose of commercial general liability insurance?

8. What are two reasons why a design professional would carry professional liability insurance—even if it was not required by his contract with an owner?

9. What are some examples of activities that may be exclusions in a professional liability insurance policy?

10. What strategies should design professionals incorporate in regard to their documentation in anticipation for facing claims?

ENDNOTES

[1] 509 U.S. 764, 772 (1993) (citations to the record omitted).

[2] *Comunale v. Traders & General Inv. Co.,* 50 Cal.2d. 654, 328 P.2d 198 (1958); *Crisci v. Security Ins. Co.,* 66 Cal.2d 425, 426 P.2d 173, 58 Cal.Rptr. 13 (1967) (insured also recovered for her emotional distress).

[3] 162 Cal.App.3d 358, 208 Cal.Rptr. 494 (1984).

[4] Cal.Civil Code § 2860.

[5] Kaplan, Bunting & Iannone, *OCIPs, CCIPs, and Project Policies,* 29 Constr. Lawyer, No. 3, Summer 2009, p. 11. To explain the title: the authors refer to wrap-up insurance as a controlled insurance program (CIP). Where the one who purchases the insurance (the sponsor) is the owner, the project has an OCIP ("O" stands for owner). When the sponsor is the prime contractor (or construction manager or design-builder), the project has a CCIP ("C" stands for contractor).

[6] Bell, Dunn & Costner, *Confronting Conventional Wisdom on Builders Risk: From Named-Insured to Concurrent Causation,* 31 Constr. Lawyer, No. 4, Fall 2011, p. 15.

[7] AIA A201-2007, § 11.3; EJCDC C-700, ¶¶ 5.06–5.10 (2007).

[8] *Heile v. Herrmann,* 136 Ohio App.3d 351, 736 N.E.2d 566 (1999); *Erie Ins. Property & Cas. Co. v. Pioneer Home Improvement, Inc.,* 206 W.Va. 506, 526 S.E.2d 28 (1999).

[9] *Fejes v. Alaska Ins. Co., Inc.* 984 P.2d 519, 523–24 (Alaska 1999); and *Vandenberg v. Superior Court,* 21 Cal.4th 815, 982 P.2d 229, 243–46, 88 Cal.Rptr.2d 366 (1999).

[10] *American Empire Surplus Lines Ins. Co. v. Hathaway Development Co., Inc.*, 288 Ga. 749, 707 S.E.2d 369 (2011); *Lamar Homes, Inc. v. Mid-Continent Casualty Co.*, 242 S.W.3d 1, 16 (Tex.2007).

[11] *American Equity Ins. Co. v. Van Ginhoven*, 788 So.2d 388 (Fla.Dist.Ct.App. 2001); *Supreme Services and Specialty Co., Inc. v. Sonny Greer, Inc.*, 958 So.2d 634 (La.2007); and *Alverson v. Northwestern Nat. Cas. Co.*, 559 N.W.2d 234 (S.D.1997).

[12] See Annot., 83 A.L.R.3d 539 (1978).

[13] *Stone v. Hartford Cas. Co.*, 470 F.Supp.2d 1088 (C.D.Cal.2006) (California law).

[14] 830 P.2d 785 (Alaska 1992).

[15] 30 Mass.App.Ct. 318, 568 N.E.2d 631 (1991).

[16] *Harbor Ins. Co. v. OMNI Constr. Inc.*, 912 F.2d 1250 (D.C.Cir.1990). See also Jones & Myers, *Risk by Design: Why Contractors Need Professional Liability Insurance*, 24 Constr.Litg.Rep., No. 10, Oct. 2003, p. 399.

[17] See generally HICKMAN, DESIGN-BUILD RISK AND INSURANCE (IRMI 2002); Abramowitz, *Professional Liability Insurance in the Design/Build Setting*, 15 Constr. Lawyer, No. 3, Aug. 1995, p. 3.

[18] 943 So.2d 429 (La.App.2006), writ denied, 949 So.2d 423, 424 (La.2007).

CHAPTER 25

Surety Bonds: Backstopping Contractors

Scenario

Mega Manufacturer, Inc. (MMI) sought to build a new factory. MMI retained Awesome Architects, Inc. (AAI), an architectural firm, to create the plans and specifications and administer the project.

Based upon this design, the owner entered into a $2 million fixed-price contract with Prime Builder, Inc. Under the prime contract, Prime Builder was required to obtain payment and performance bonds, which it acquired from Surety, Inc. To obtain the bonds, Prime Builder and its individual owners entered into an indemnity agreement broadly promising to reimburse Surety for its expenses incurred "by or under the bond."

Prime Builder hired several subcontractors, including a framing subcontractor (Framer) and a roofing subcontractor (Roofer). Framer built the building's framework with lumber purchased from a supplier. Roofer assembled materials on the site in preparation for building the roof.

AAI was unhappy with the quality of the work and consistently approved only partial payments for Prime Builder. The prime contractor began cutting back on payments to subcontractors and suppliers. When these parties complained to Surety, Surety investigated and discovered that Prime Builder's finances were in disarray. On February 25, Surety informed MMI to forward all future progress payments to it.

Instead, MMI sent a $100,000 progress payment on March 1 to its contractor.

On March 20, MMI declared Prime Builder in default and terminated the prime contract. MMI notified Surety of its actions.

On March 21, a fire destroyed Roofer's materials (which were still in storage on the ground) before they could be removed.

On March 25, Prime Builder wrote a long letter to Surety, explaining that its cash flow problems were caused by AAI's failure to certify payments and as a result that its termination by MMI was wrongful. The prime asked Surety not to assume its bond obligations.

Instead, Surety hired Prime Builder to complete the project. MMI objected, stating that it would not allow Prime Builder back unto the site.

25.1 Introduction

Overview. Chapter 25 continues the discussion of risk management begun in the previous chapters on contribution, indemnity, and insurance. Surety bonds primarily backstop (guarantee) the contractor's performance and payment obligations.

Repeated reference will be made to the American Institute of Architects performance and payment bonds, AIA Document A312-2010. Somewhat confusingly, both bonds are numbered A312, and each begins at Section 1. Thus, a citation to A312-2010, § 3, does not indicate whether the reference is to the performance bond or the payment bond. However, this should not create confusion, as these bonds to the most part are analyzed in separate sections of this chapter.

The Engineers Joint Contract Documents Committee (EJCDC) performance bond is EJCDC C-610 (2010), and the payment bond is EJCDC C-615 (2010). They are identical to the AIA bonds.

AIA Doc. A312-2010 is reproduced in Appendix I.

Terminology. The surety bond transaction is a peculiar arrangement and differs procedurally from most contracts. The typical surety arrangement is essentially triangular. The "surety" obligates itself to perform or to pay a specified amount of money if the **principal debtor** (usually called the **principal**) does not perform. The person to whom this performance is promised is usually called the **obligee** (sometimes called the "creditor"). In the building contract context, the surety is usually a professional bonding company. The principal is the prime contractor or, in the case of subcontractor bonds, a subcontractor. The obligee is the owner or, in the case of a subcontractor bond, the prime contractor.

The Restatement (Third) of Suretyship and Guaranty (1996) uses somewhat different terminology: the principal is the **principal obligor**, the surety is the **secondary obligor**, and the term "obligee" is unchanged. This language reflects the fact that the principal has the primary obligation to perform, and the surety has only a back-up or "secondary" obligation to perform. This text will largely refrain from using the Restatement's terminology.

Mechanics of Suretyship. The owner specifies in the contract or bidding documents whether it wants the successful prime contractor to obtain specific bonds. The owner indirectly pays the cost of the bond, because the bidder adds the bond premium to its costs when computing its bid. The bond is issued to the owner as obligee. The bond "runs to" the owner in that performance by the surety has been promised to the owner even though the prime contractor (the principal) applied for the bond.

Although the owner pays the bond premium (as part of the contract price), the surety will not issue the bond unless the contractor enters into an indemnity agreement with the surety promising to reimburse it for losses incurred under the bond. This indemnity arrangement makes sense, since it is the contractor's default that triggers the surety's liability. The surety's right of reimbursement from the principal also means that bond premiums are much lower than a contractor's liability insurance premiums (where reimbursement from the insured is prohibited). However, tensions arise if the contractor views the surety as incurring costs in an excessive manner—literally at the contractor's expense.

For historical reasons, the amount of the bond is called the **penal sum**. In the United States, this usually corresponds to 100 percent of the contract price, although sometimes it will be 50 percent. (In international contracts, the penal sum is closer to 10 percent.) Just as an insurer's liability is capped by the policy limits, so too the surety's liability should not exceed the penal sum. Section 25.8 notes that the surety's liability may exceed the penal sum if the surety engaged in bad faith conduct or breached regulations.

Another distinction of terms is between the conditional nature of bonds used in the American construction industry and "on demand" or unconditional bonds commonly used in British and international construction projects. These will be discussed in Section 25.10.

25.2 Need for Bonds in Construction Industry

Surety bonds play a vital part in the construction process. The contracting industry is volatile: Bankruptcies are not uncommon, and a few unsuccessful projects can cause financial catastrophe. Estimating costs is difficult and requires much skill. Fixed-price contracts place many risks on the contractor, such as price increases, labor difficulties, subsurface conditions, and changing governmental policy.

Some construction companies are poorly managed and supervised. Often they are undercapitalized and rely heavily on the technological skill of a few individuals. If these people become unavailable, difficulties may arise. For many construction companies, credit may be difficult to obtain. Some contractors do not insure against the risks and calamities that can be covered by insurance. Finally, economic downturns in the larger economy almost always hit the building industry.

In most construction projects, bonds are needed to protect the owner. The owner in most projects would like to have a financially solvent surety if the successful bidder does not enter into the construction contract (bid bond), the prime contractor does not perform its work properly (performance bond), or the prime contractor does not pay its subcontractors or suppliers (payment bond). Bond requirements can also act as preliminary screen for contractor selection.

If these events do not occur, the amount paid for a surety bond may seem wasted. Some institutional owners believe there is no need for a surety bond system if the prime contractor is chosen carefully and if a well-administered payment system eliminates the risk of unpaid subcontractors and suppliers. Such owners may choose to be self-insurers and not obtain bonds. They realize there may be losses, but they believe the losses over a long period will be less than the cost of bond premiums.

Even where a bond is not required at the outset, it is best to include a provision in the prime contract that will require the prime contractor to obtain a bond before or during performance if the owner so requests. Usually the owner pays the cost of a bond issued after the price of the project is agreed on.

Public construction frequently requires performance and payment bonds. The latter are required to protect subcontractors and suppliers who have no lien rights on public work. The Miller Act[1] requires federal prime contractors to obtain performance bonds and payment bonds based on the contract price, and similar requirements exist in state public contracting under "Little Miller Acts." In addition, often local housing development legislation requires that the developer furnish bonds to protect the local government if improvements the developer has promised are not made. Legal advice should be obtained to determine whether the project requires bonds.

The invitation to bid usually states whether the contractor is required to obtain a surety bond, the type or types of bonds, and the amount of the bonds. In some cases, the owner wishes to approve the form of bond and surety used. Also, the owner may want to have the right to refuse any substitution of the surety without its express written consent given before substitution.

25.3 Function of Surety: Insurer Compared

A surety's function is to assure one party that the entity with whom he is dealing will be backed up by someone who is financially responsible. Sureties are used in transactions where people deal with individuals or organizations of doubtful financial capacity. They provide credit. Sureties must be distinguished from insurers, although each provides financial security.

An insured is concerned that unusual, unexpected events will cause him to suffer losses or expose him to liability. Although he can self-insure—that is, bear the risk itself—he usually chooses to indemnify himself against this risk by buying a promise from an insurance company in exchange for paying a premium. The insurer distributes this risk among its policyholders.

In liability insurance, the insured himself may be at fault and cause a loss to the insurer. But the insurer cannot recover its loss from its own insured. Although it may seek to recover its losses from third parties through subrogation (stepping into the position of the person it has paid—its insured—and thereby acquiring any claims of its insured against those who caused the loss), it cannot recover from those named as insureds in the policy.[2]

By contrast, a surety, in addition to dealing with ordinary construction contract performance problems (that can be considered business losses, not the "accidents" central to insurance), seeks through an indemnity agreement to recover any losses it has suffered from the principal on whom it has written a bond as well as individual shareholders of a corporate principal. Contractors who are liable because of defective workmanship may seek to recover from their commercial general liability (CGL) insurers, preferring this arrangement to having their sureties pay the loss. (See Section 24.5.) Their sureties would seek to recover from them in an indemnity action; their CGL insurers could not.

25.4 Ancillary Bonds

Surety bonds contribute stability to the construction industry. Bonds back up business expectations—mostly from contractors to others, especially the owner and subcontractors. (Where the subcontractor is the principal, the payment bond benefits sub-subcontractors, and the performance bond benefits the contractor.) This chapter primarily discusses payment and performance bonds. However, the industry also utilizes other bonds, some of which are briefly listed here.

Bid Bonds. The function of a bid bond is to provide the owner with a financially responsible party who will pay all or a portion of the damages caused if the bidder to whom a contract is awarded refuses to enter into it.

License Bonds. As noted in Chapter 8, virtually all states license contractors and subcontractors. Some require the builder to purchase a license bond as a condition to receive a license. Such a bond promises to indemnify the obligee against financial loss caused by the licensee's noncompliance with specified laws or regulations. In California, a license bond is for the benefit of: a homeowner contracting for home improvements that may be damaged by a violation of the license law; any person that is injured as a result of the licensee's willful and deliberate violation of the license law or fraud in the performance of the construction contract; any employee of the licensee who was not paid wages; and any fringe benefit trust not paid fringe benefits.[3]

Lien Release Bonds. In some states, owners can post a bond that can preclude a lien from being filed or dissolve a lien that has been filed. The party filing the mechanic's lien must pursue its rights against the bond. The owner's title (or ownership rights) for the property remains free and clear.[4]

Subdivision Bonds. As a condition for approving a subdivision development, the local government may require the developer to build streets, sewers, and drains, so that these public improvements are not the responsibility of the government. A subdivision bond is (in effect) a performance bond for these public improvements.[5]

25.5 Performance Bonds: Surety's Promise to Owner

This discussion will assume that the principal is the contractor and the obligee is the owner. See Figure 25.1. As noted, it is also possible for a subcontractor to be the principal, in which case the prime contractor is the obligee. The protections of the obligee and duties of the surety do not vary based on the obligee's identity.

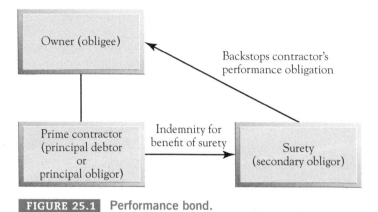

FIGURE 25.1 Performance bond.

A performance bond provides protection for the project owner in the event the bonded contractor defaults. The AIA performance bond, A312-2010, § 1, states that the contractor and surety "jointly and severally[ ] bind themselves . . . to the Owner for performance of the Construction Contract, which is incorporated herein by reference."

"Joint and several liability" of two or more persons means that each person is individually liable for the entire debt. The joint and several liability of the contractor and surety means that the surety backstops the contractor to ensure the project is built. By the same token, the surety often argues that its obligations to the owner are coextensive with those of the contractor—the owner may not impose upon the surety greater obligations than are imposed on the contractor, and the contractor's defenses to the owner's claims are equally available to the surety.

Consistent with industry custom, AIA A312-2010, § 2, phrases the surety's obligation using archaic, *conditional* terminology: "If the Contractor performs the Construction Contract, the Surety . . . shall have no obligation under this Bond . . ." Of course, what constitutes "perform[ing] the Construction Contract" can be difficult to determine.

25.6 Triggering the Performance Bond Obligation

AIA A312-2010, § 3, lists three steps the owner should take in order to trigger the surety's bond obligations:

1. Hold a conference with the surety and contractor.
2. Declare the contractor in default and terminate the contract.
3. Tender (offer) the balance of the construction funds to the surety.

Conference. As noted, so long as the contractor is performing the construction contract, the surety has no obligation under the bond. Just as minor deficiencies by the contractor would not justify an owner's default termination of the contract,[6] so too not every contract breach by the contractor would justify the owner's invocation of the bond. The AIA attempts to temper the owner's actions by imposing upon him an obligation to meet with the contractor and surety if the owner is "considering" declaring the contractor in default. Without prejudice to the owner's rights, the owner, surety, and contractor may agree to give the contractor a reasonable time to perform the contract. However, the owner need not hold a conference if such a meeting is likely to be ineffectual.

Declaration of Default. The key act by the owner to alert the surety that its bond obligations have been triggered is the owner's declaration that the contractor is in default and that the construction contract has been terminated. That declaration is the dividing line from the surety's perspective between merely a troubled project and one under which it may have to assume its bond obligations.

By coupling termination of the contract with the declaration of default, the AIA makes clear that a default involves more than a breach of contract; the breach must be material with the principal unable to properly complete the project within a reasonable time. Termination of the contractor also frees up the surety to decide the next step, including whether it will (or through the hiring of an independent contractor) complete the contract itself. According to a federal court of appeals in *L & A Contracting Co. v. Southern Concrete Services, Inc.*:

> Serious legal consequences attend a "declaration of default", particularly in cases such as this case involving multi-million-dollar construction projects. Before a declaration of default, sureties face possible tort liability for meddling in the affairs of their principals. After a declaration of default, the relationship changes dramatically, and the surety owes immediate duties to the obligee. Given the consequences that follow a declaration of default, it is vital that the declaration be made in terms sufficiently clear, direct, and unequivocal to inform the surety that the principal has defaulted on its obligations and the surety must immediately commence performing under the terms of its bond. Sureties deprived of a clear rule for notices of default would be reluctant to enter into otherwise profitable contracts.[7]

Tender of Contract Balance. By tendering (offering) the contract balance to the surety, the owner deprives himself of the means by which to assume control of contract performance. The task of completing the project—or at least deciding what is to be done with the contract balance—has shifted to the surety.

> In this chapter's scenario, MMI did not hold a conference with Prime Builder and Surety, but it did declare the prime in default and terminated the prime contract. MMI presumably also tendered the contract balance to Surety. Under these conditions, it is unlikely Surety will be able to contest that its responsibility under the performance bond have been triggered. The lack of a conference is probably irrelevant, since Surety had already concluded its principal (Prime Builder) was in financial difficulty.

25.7 Performance Bond Surety's Options

Under AIA A312-2010, § 5, the performance bond surety has the option of fulfilling its bond obligations by choosing one of four courses of action:

1. Arrange for the original contractor, with the owner's consent, to complete the contract.
2. Complete the project itself, including through use of independent contractors.
3. Obtain bids or negotiated proposals from qualified contractors acceptable to the owner and arrange for the owner to continue work with that new contractor, backed up by new performance and payment bonds.
4. Waive its right to perform and complete, and instead (i) pay the owner its costs of completion or (ii) deny liability in whole or in part.

These options may come as a surprise to owners. Many owners will have the expectation that, once the contractor has been declared in default, the performance bond surety will mobilize manpower and equipment to the project site and immediately go about completing the project.

Instead, the surety may simply buy back the bond by tendering the contract balance (or just a part of the balance) to the owner, leaving it up to the owner to then undertake completion of the project. Whichever option the surety chooses must be made with "reasonable promptness" and not later than seven days after receipt of written notice from the owner demanding that the surety perform (see A312-2010, § 6). One option *not* available to the surety is to do nothing, force the owner to complete the performance himself, and then offer to reimburse the owner for his reasonable costs.[8]

The contractor too may be unhappy with the surety's decisions. Recall that the principal, as a condition to receiving the bond, had agreed to indemnify the surety for losses incurred under the bond. (This indemnity obligation typically extends to a corporate contractor's owners.) The contractor may fear that the surety will not complete the project in the most economical manner possible, yet it will then demand reimbursement of its costs from the contractor. In addition, the contractor may strongly believe the project's failure was the fault of the owner.

In sum, the performance bond surety may be placed in the dilemma of having to choose between the owner and the contractor. If after an investigation the surety sides with the contractor (believing the owner's declaration of default was wrong), the owner will contend that the surety breached its bond obligations and may seek extracontractual damages (in excess of the bond's penal sum). For example, in *Republic Insurance Co. v Prince George's County*,[9] the surety was liable for higher costs of completion caused by the surety's delay in undertaking completion.

Yet, if the surety sides with the obligee and completes the project, the contractor may sue the surety for breach of the covenant of good faith and fair dealing. Such a claim is difficult to prove, as the courts protect the business judgment of the surety as long as that judgment was exercised in good faith. For example, one federal district court, in *Liberty Mutual Insurance Co. v. Aventura Engineering & Construction Corp.*, rejected the contractor's argument that the surety may not settle the owner's claims until the contractor had litigated the propriety of the owner's default termination. The court reasoned:

> The situation [advocated by the contractor] places the surety in an unacceptable dilemma: If it does not honor the claim on the bond, the owner sues the surety and the surety incurs the cost of litigation and of a potential judgment. On the other hand, if the surety honors or settles the claim on the bond, the principal sues under a bad faith theory in an attempt to avoid its obligation to indemnify the surety and the surety once again has to absorb, or at least advance, the cost of litigation and a potential judgment. This lose-lose situation is not a reasonable interpretation of the Indemnity Agreement or the Bond.[10]

One commentator recommends that it is in the best interests of a terminated contractor for the surety to complete the contract under a reservation of rights—rather than immediately contest the propriety of the termination. This way, the surety can best control the costs of completion and recover these costs by proving the termination of the contractor was wrongful. In sum, the best approach in a default termination would be for the surety and contractor together to complete the project (thereby minimizing the costs) and then recover the costs in a wrongful termination action against the owner.[11]

In this chapter's scenario, MMI declared Prime Builder in default and terminated the contract, but the contractor believes the termination was wrongful. Surety is faced with competing demands from its obligee and principal. Surety can argue that it already investigated Prime Builder's finances and discovered they were unstable.

Surety can also tell Prime Builder that hiring it to complete the project will be the best way for this conflict to be resolved. Under the indemnity agreement, Prime Builder will have to reimburse Surety for its out-of-pocket costs. This is an incentive for Prime Builder to complete the work as cheaply as possible.

25.8 Surety's Defenses

As noted in Section 25.5, the surety's duties are coextensive with those of the contractor. This generally means a surety may raise any defense to performance that would be available to the contractor, including that the default termination of the contractor was wrongful or the owner's claim was untimely. (Time requirements on asserting claims are discussed in Section 25.13.)

An owner's actions that increase the surety's risk or prejudice the surety's options may result in *discharge* (or release) of the surety. For example, an owner who materially increases the contractor's scope of work without consent of the surety may thereby discharge the surety. A New York court held that the owner's issuance of change orders without the surety's consent discharged the surety.[12] Incidentally, this result would not be permissible under the AIA performance bond, A312-2010, § 10, which waives the surety's right to notice of any change to the underlying construction contract.

In addition, an owner's failure to comply with his obligations under the bond may result in discharge of the surety. A surety was discharged when the owner failed to give the surety notice of its termination of the contractor, and the owner then hired his own contractor to complete the contract.[13] Similarly, an owner discharged the surety by refusing to allow the surety to use the original contractor as the completion contractor.[14]

> In this chapter's scenario, if MMI refuses to let Surety employ Prime Builder to complete the project, Surety can argue its bond obligations have been discharged and it has no further responsibilities to the owner.

The surety may also argue that the owner's attempts to help the contractor financially or delay in declaring the contractor in default discharged the surety from its bond obligations. For example, to keep the project moving forward, the owner may advance monies to the contractor for work the contractor has not yet performed. However, any diminishment in the contract balance reduces the funds from which the surety may obtain payment if it is later called upon to complete the contractor's work. The surety may argue that the owner's prepayment increased the surety's risk and discharged the surety from its obligations under the bond. However, so long as the payments are used to build the project (thereby reducing the performance bond surety's ultimate liability), the fact that the payments were made ahead of schedule should not discharge the surety. The owner must have reasonable latitude to deal with a financially distressed contractor without fear of losing its protections under the bond.[15]

25.9 Performance Bond Surety's Liabilities

A performance bond surety's promise to the owner is to perform the construction contract. Clearly, bond liability includes correction of known defective work and completion of the project.

But an owner may suffer other injuries from defective performance. If the principal completes late, is the surety liable for delay damages? To determine whether delay comes within the bond commitment, the language of the bond is crucial. The Florida Supreme Court in *American Home Assurance v. Larkin General Hospital*,[16] pointing to the bond language and emphasizing that the most important responsibility of the prime contractor is to build the project, did not allow the obligee to recover delay damages from the surety. Its decision may have been traceable to a belief that delays are inherent in all construction projects.

But other courts have allowed recovery against the surety for the contractor's delayed performance.[17] AIA Document A312-2010, § 7.3, requires the surety to pay any liquidated or actual damages caused by the contractor's delay. The recovery against the surety should be the same as the amount that would be recovered from the prime contractor.

Generally, the surety will not be responsible for punitive damages that might be awarded against the prime contractor.[18] But the surety, as noted in Section 25.15, might be liable for punitive damages for its bad faith refusal to settle the claim made on the bond.

What is the performance bond surety's liability if it delays paying the owner? If the surety's failure to pay promptly delays completion of the project, resulting in higher performance costs, the surety will be liable for these increased costs—again without regard to the bond limit.[19]

25.10 Payment Bonds: Functions

The payment bond is an undertaking by the surety to pay unpaid subcontractors and suppliers (see Figure 25.2). An understanding of the function of a payment bond requires a differentiation between private and public construction work. All states give unpaid subcontractors and suppliers mechanics' lien rights if they improve private construction projects. This process was described in Section 21.7.

Although there are various ways to avoid liens, one method has been to require prime contractors to obtain payment bonds. A payment bond obligates a surety to pay subcontractors and suppliers if the prime contractor does not pay them. The owner would prefer to direct subcontractor and supplier payment claims to the bond rather than have his property encumbered by a lien.

Also, subcontractors are more likely to make bids when they can be assured of a surety if they are unpaid. The competent subcontractor who deals with a prime contractor of uncertain financial responsibility should add a contingency to its bid to cover possible collection costs and the risk of not collecting. Having a payment bond should eliminate the need for this cost factor. In addition, subcontractors and suppliers should be more willing to perform properly and deliver materials as quickly as possible when they are assured they will be paid. Although they have a right to a mechanics' lien, the procedures for perfecting the lien and satisfying the unpaid obligation out of foreclosure proceeds are cumbersome and often ineffective. Payment bonds are preferable to mechanics' liens.

Generally, subcontractors and suppliers cannot impose liens on public work. In some states, they can file a stop notice that informs the public agency that a subcontractor has not been paid and requires the owner to hold up payments to the prime contractor (see Section 21.9). However, the stop notice applies only to the unpaid balance still held by the owner when the notice was filed. A payment bond for the full contract price provides subcontractors with greater fiscal protection.

Competent subcontractors and willing suppliers are essential to the construction industry. These important components should have a mechanism that lets them collect for their work. Payment bonds do this. Without a reliable payment mechanism, a substantial number of subcontractors might go out of business. Eliminating competent subcontractors can have the unfortunate effect of reducing competition and the quality of construction work.

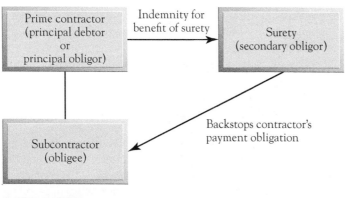

FIGURE 25.2 Payment bond.

25.11 Who Can Sue on the Payment Bond?

In a performance bond, the owner is both the obligee and the claimant. By contrast, in a payment bond, the obligee is the owner, but the claimant is an unpaid subcontractor or supplier. This means that, absent unusual bond language, unpaid subcontractors and suppliers may sue directly on the payment bonds as third-party beneficiaries. The beneficiaries of public works bonds are established by statute and usually mirror the claimants entitled to a mechanics' lien on a private works project.

Coverage (or lack there of) under payment bonds sometimes depends on where the claimant is located in the chain of contracts ultimately leading to the prime contractor. The farther down a claimant is on the chain, the less likely it will be covered by the bond. A further complication is whether a remote claimant's contract is with a subcontractor or a supplier (also called a materialman). Because subcontractors provide on-site services (whereas suppliers do not), those who contract with a subcontractor are more likely to come within the scope of a payment bond than those who contract with a supplier.

Planning in the context of a traditional construction organization should include ensuring that unpaid subcontractors and suppliers can bring action on these bonds. This can be accomplished by including language similar to that in AIA bonds. Under AIA A312-2010, § 2, payment bond coverage extends to "any person or entity seeking payment for labor, materials or equipment furnished for use in the performance of the Construction Contract."

When a nontraditional system is used, such as separate contracts (multiple primes) or the party most interested in performance is not a nominal owner but an ultimate user, planning must consider who can bring legal action on any bonds that are supplied by contractors. That choice must be expressed in the contract and the bond.

> In this chapter's scenario, Framer is clearly a third-party beneficiary to the payment bond, as its labor was used to build the improvement. As the supplier of a first-tier subcontractor, the lumber supplier should also qualify as a claimant.
>
> Roofer's materials were destroyed before they were incorporated into the project, and Surety may be able to argue that, although a first-tier subcontractor, Roofer has no claim against the bond.

25.12 Payment Bond Liability

Once it establishes its status as a claimant, an unpaid subcontractor or supplier can recover on the payment bond for the reasonable value of the work it has performed or the materials it has supplied. Unless profit can be encompassed in the preceding formula, however, some courts have difficulty awarding profits against the surety—probably a reflection of the view that a payment bond is a substitute for lien rights.[20] A subcontractor was allowed a claim against the surety for increased costs of performance.[21]

A Miller Act payment bond covers all "sums justly due" the claimant. A federal court of appeals, in *United States for Use and Benefit of Perlun Construction Co. v. Harvesters Group, Inc.*,[22] interpreted this to include delay damages. The court reasoned that only by allowing for full recovery of costs, including those caused by delay, would subcontractors receive the financial protection intended by the statute.

Suppose the subcontractor claimant was not paid because the prime contractor had not been paid by the owner, and the subcontract contained a "pay when paid" clause, conditioning the contractor's obligation to pay the subcontractor on the contractor's payment by the owner (see Section 21.5). Even when such clauses are enforced, most courts—particularly when interpreting public works payment bonds—will not allow the surety to assert the contractor's defense.

25.13 Asserting Claims: Time Requirements

The bond usually requires that claimants give the surety certain notices. Likewise, statutes requiring public work to be bonded, such as the federal Miller Act and state "Little Miller Acts," often specify that notices must be given within certain periods of time to designated persons. A court held that a town's failure to give the performance bond surety a bond-mandated fifteen-day cure notice, before undertaking performance itself, discharged the surety.[23]

On the other hand, the bond may also impose time limits on a surety's response to an obligee's claim. AIA Doc. A312-1984, ¶ 6.1, required the payment bond surety to respond to a claim "within 45 days after receipt of the claim, stating the amounts that are undisputed and the basis for challenging any amounts that are disputed." Where a surety failed to make any response within the forty-five-day period, Maryland's high court in *National Union Fire Ins. Co. of Pittsburgh v. David A. Bramble, Inc.*[24] ruled the surety was precluded from contesting the claim.

AIA Doc. A312-2010 (payment bond) seeks to change this result. AIA 312-2010, § 7.1, increases the period of time in which the surety must answer a claimant's claim from forty-five days to sixty days. Also, § 7.3 adds that the surety's failure to discharge its obligations under § 7.1 "shall not be deemed to constitute a waiver of defenses" available to the surety or contractor. However, "the surety shall indemnify the Claimant for the reasonable attorney's fees the Claimant incurs thereafter to recover any sums found to be due and owing the Claimant." In other words, if the surety fails to answer the claim within sixty days, the surety will finance the claimant's attorney's efforts to enforce the claim.

25.14 Reimbursement of Surety

Sureties do not expect to take a loss. They do not see themselves as insurers. As a result, they assert defenses that the principal debtor could have asserted. They seek bond language protection. Finally, they seek to recoup payments made under the bond—first from the principal, and if unsuccessful, from third parties, including the obligee.

As a condition to receiving bonding, the prime contractor and often also the individuals who control it agree to indemnify the surety for any payments made in good faith under the bond. The agreement to indemnify extends also to the surety's settlement of any claims brought against the bond—whether or not the prime contractor turns out to have been liable to the bond claimants paid by the surety.[25] While courts will enforce such clauses, a surety who acts in bad faith will either lose its right of indemnification or have that right reduced to the extent of losses caused by improper settlement practices.[26]

The surety's indemnity right will be of little use if, as often happens, a defaulting principal is not in a position to reimburse the surety. The surety may turn next to the obligee as a source of recovery. When the prime contractor defaults and the surety takes over, the surety usually notifies the owner that it should be paid all payments that would have gone to the prime contractor. In addition, the surety usually demands at the end of the job any retainage that the owner has withheld to secure the owner against claims.

In seeking the retainage, the surety usually competes with other creditors of the prime contractor, the taxing authorities, and the trustee in bankruptcy if the prime has been declared bankrupt. As a rule, many more claims than can be satisfied exist, and the result is a complicated lawsuit. Typically, the owner pays the retainage into court and notifies all claimants, and the court determines how the fund is to be distributed.

In addition to pursuing indemnity rights against the principal and demanding the contract balance from the owner obligee, the surety who takes over after a default succeeds to any claims the prime contractor may have had against the owner or third parties. It is common for the surety to ascribe its difficulties to the owner (perhaps that the owner made progress payments despite the surety's warning that the contractor may default), the design professional (usually a claim for over-certification of progress payments), subcontractors, or other third parties.

Sureties make strong efforts to be reimbursed and often succeed in salvaging a substantial amount of their loss when they are called on to respond for their principals' default.

> In this chapter's scenario, Surety may bring a claim against MMI for paying Prime Builder the March 1 progress payment after Surety had requested that all future payments be made to Surety.
>
> Surety might also sue AAI for wrongful undercertification of progress payments to Prime Builder. Surety's damages claim would be a complicated one. If the contractor's performance was proper and it should have been paid in full, then the undercertification simply meant the contract balance—tendered to Surety—was larger than it should have been. AAI could argue the Surety was not damaged and had no claim to pursue against the architect.

25.15 Regulation: Bad Faith Claims

Surety companies are regulated by the states in which they operate. In addition, sureties who wish to write bonds for federal projects must qualify under regulations of the U.S. Treasury Department. The financial capability of a surety limits the size of the projects a surety can bond. Bond dollar limits place a ceiling on exposure. In larger projects, there may be co-sureties or the surety may be required to reinsure a portion with another surety. Surety rates are usually regulated and are based on a specified percentage of the limit of the surety bond.

As noted, surety bonds contain what is called a *penal sum*, which is the amount of the bond obligation (usually the same as the contract price, although sometimes less). The surety seeks to limit its obligation to the amount of the penal sum, and this strategy generally is effective.

However, the law has begun to inquire into the settlement practices of sureties just as it has examined the settlement practices of insurance companies (see Section 24.3). As a result, some exceptions have emerged to the general rule that the penal sum limits the surety's obligation. An important trial court opinion, *Continental Realty Corp. v. Andrew J. Crevolin Co.*,[27] held that the surety's liability would not be limited by the bond limit, mainly because of improper settlement tactics of the surety. Similarly, in another federal case, *Insurance Co. of North America v. United States*,[28] where the surety was not cooperative in settlement discussions concerning an award of prejudgment interest, the bond limit did not place a ceiling on the liability of the surety.

Bad faith settlement practices by the surety includes its duty to investigate once the obligee declares the principal debtor (the prime contractor) in default. In the federal court of appeals case of *United States Fidelity & Guarantee Co. v. Braspetro Oil Services, Co.*,[29] the owner properly declared the contractor in default. In bad faith, the performance bond surety threatened the owner with a lengthy investigation before performing its bond obligation. This threat forced the owner to complete performance with its own funds. The court held the owner was entitled to prejudgment interest even if this exceeded the penal sum of the bond.

The law has taken steps to deter *insurers* from conduct unfair to their policyholders. Most states allow claims based on tortious bad faith conduct by the insurer to policyholders. This technique has enabled claimants to obtain awards that exceed the policy limits, usually through awards for emotional distress or punitive damages.

Building on increased judicial regulation of insurers, some states have allowed similar claims against *sureties*. When bad faith claims are allowed and a surety's conduct falls below that required by law, the surety's obligations can exceed the stated limit of the bond. This is accomplished by finding that the surety has committed a tort that (in some instances) may justify the award of punitive damages as well as normal tort damages.[30]

While a claim of bad faith settlement practices would be brought by a beneficiary on a payment bond, most courts deny a principal from suing the performance bond surety for bad faith or breach of the implied covenant of good faith and fair dealing. A principal is not analogous to an insured for purposes of the unfair insurance claims practice statute.[31] Obligees too may not sue the performance bond surety for breach of the covenant of good faith and fair dealing.[32]

25.16 Bankruptcy of Contractor

If during the course of the contractor's performance the contractor is adjudicated a bankrupt, the trustee in bankruptcy (the person who takes over the affairs of the bankrupt contractor) can determine whether to continue the contract. Usually the trustee does not (because of the contractor's lack of resources). If the contract is not continued, the bankrupt contractor has no further obligation to perform under the contract. The owner has a claim against the bankrupt contractor but is not likely to recover much. Ending the contractor's obligation, however, should not release the surety. This is the risk contemplated when the surety bond is purchased.[33]

25.17 International Contracts

As explained in Section 25.5, the construction contract and the bond are linked together. The contract is incorporated by reference into the bond. The surety's obligations are coextensive with those of the contractor. The surety's obligations to the owner are not greater than those of the contractor under the construction contract, and the obligations of the owner to the surety are not greater than the owner's obligations under the construction contract.

The dependent nature of the bond in American construction practices may be contrasted to the security instruments that the contractor usually is required to furnish the employer (owner) in construction projects in other countries and in international engineering projects. In those transactions, the contractor usually is expected to furnish an unconditional bank guarantee, a standby letter of credit, or an "on demand" bond. These are unconditional promises by the issuing bank or surety that are not tied to default by the contractor. These are powerful securities that can be used in an abusive way. They can be unfair to the contractor who will be expected to indemnify the bank or surety. As a result, they have spawned considerable litigation in Great Britain and in international transactions.[34]

REVIEW QUESTIONS

1. What are the requirements for a prime contractor under the Miller Act?

2. What is the function of a surety?

3. What is the function of a bid bond?

4. According to AIA A312-2010 § 3, what are the three steps the owner should take in order to trigger the surety's bond obligations?

5. What are the four courses of action the performance bond surety has the option of using in fulfilling its bond obligations, according to the AIA A312-2010?

6. What are some actions that an owner may take that could discharge the surety (release the surety from its obligations under the bond)?

7. What is the purpose of a payment bond?

8. According to the AIA A312-2010 document, who may sue on a payment bond?

9. What did the federal court of appeals state was included in "sums justly due" in a Miller Act payment bond in the case, *United States for Use and Benefit of Perlun Construction Co. v. Harvesters Group, Inc*? What was the basis for the court's reasoning?

10. What is a consequence to the surety if a court finds it acted in bad faith in the (non) performance of its bond obligations?

ENDNOTES

[1] 40 U.S.C. §§ 3131-3134 (formerly 40 U.S.C. §§ 270a–270f).

[2] Subrogation is discussed in Section 16.9C.

[3] Cal.Bus. & Prof.Code § 7071.5. Alaska's license bond statute, Alas.Stat. § 08.18.071, is interpreted in *Alaska Nat'l Ins. Co. v. Northwest Cedar Structures, Inc.*, 153 P.3d 336 (Alas.2007).

[4] Cal.Civ.Code § 3144.5, interpreted in *Dennis Elec., Inc. v. United States Fid. & Guar. Co.*, 219 Cal.App.3d 1228, 269 Cal.Rptr. 26 (1990).

[5] Conn.Gen.Stat.Ann. § 8-25(a); Wash.Rev.Code Ann. § 58.17.130.

[6] See Section 22.9.

[7] 17 F.3d 106, 111, reh'g denied, 22 F.3d 1096 (5th Cir.1994). See also *Bank of Brewton, Inc. v. International Fid. Ins. Co.*, 827 So.2d 747 (Ala.2002); *Insurance Co. of North America v. United States*, 951 F.2d 1244 (Fed. Cir.1991) (Miller Act).

[8] *National Fire Ins. Co. of Hartford v. Fortune Constr. Co.*, 320 F.3d 1260 (11th Cir.), cert. denied, 540 U.S. 873 (2003).

[9] 92 Md.App. 528, 608 A.2d 1301 (1992), cert. dismissed as improvidently granted, 329 Md. 349, 619A.2d 553 (1993).

[10] 534 F.Supp.2d 1290, 1312 (S.D.Fla.2008).

[11] Worley, *Recovery of the Surety's Costs Following Wrongful Termination,* 30 Constr. Lawyer, No. 2, Spring 2010, p. 40.

[12] *In re Liquidation of Union Indem. Ins. Co. of New York*, 220 A.D.2d 339, 632 N.Y.S.2d 788 (1995).

[13] *Seaboard Sur. Co. v. Town of Greenfield*, 370 F.3d 215 (1st Cir.2004); *Elm Haven Constr. Ltd. Partnership v. Neri Constr. LLC*, 376 F.3d 96 (2d Cir.2004).

[14] *St. Paul Fire & Marine Ins. Co. v. VDE Corp.*, 603 F.3d 119 (1st Cir.2010).

[15] *United States Fid. & Guar. Co. v. Braspetro Oil Services, Co.*, 369 F.3d 34, 61-66 (2d Cir.2004) (efforts by owner to keep project afloat did not discharge surety); *John T. Callahan & Sons, Inc. v. Dykeman Elec. Co., Inc.*, 266 F.Supp.2d 208 (D.Mass.2003) (surety is not discharged by prime's delay in declaring a struggling subcontractor in default).

[16] 593 So.2d 195 (Fla.1992). See also *Downington Area School Dist. v. International Fidelity Ins. Co.*, 769 A.2d 560 (Pa.Cmwlth.), appeal denied, 567 Pa. 731, 786 A.2d 991 (2001).

[17] *MAI Steel Service, Inc. v. Blake Constr. Co.*, 981 F.2d 414 (9th Cir.1992); *Cates Constr., Inc. v. Talbot Partners*, 21 Cal.4th 28, 980 P.2d 407, 86 Cal.Rptr.2d 855 (1999). See Douglas, *Delay Claims Against the Surety,* 17 Constr. Lawyer, No. 3, July 1997, p. 4.

[18] Annot., 2 A.L.R.4th 1254 (1980).

[19] *Republic Ins. Co. v Prince George's County,* supra note 9.

[20] *MAI Steel Service, Inc. v. Blake Constr. Co.*, 981 F.2d 414 (9th Cir.1992).

[21] Ibid.

[22] 918 F.2d 915 (11th Cir.1990).

[23] *Seaboard Sur. Co. v. Town of Greenfield,* supra note 13.

[24] 388 Md. 195, 879 A.2d 101 (2005).
[25] *Gulf Ins. Co. v. AMSCO, Inc.*, 153 N.H. 28, 889 A.2d 1040 (2005).
[26] *Auto-Owners Ins. Co. v. Southeast Floating Docks, Inc.*, 571 F.3d 1143 (11th Cir.2009); *PSE Consulting, Inc. v. Frank Mercede and Sons, Inc.*, 267 Conn. 279, 838 A.2d 135 (2004).
[27] 380 F.Supp. 246 (S.D.W.Va.1974). As the case was settled, there was no appeal.
[28] Supra note 7.
[29] Supra note 15, 369 F.3d at 78-81.
[30] *Riva Ridge Apartments v. Robert G. Fisher Co., Inc.*, 745 P.2d 1034 (Colo.App.1987).
[31] *Insurance Co. of the West v. Gibson Tile Co., Inc.*, 122 Nev. 455, 134 P.3d 698 (2006); *Masterclean, Inc. v. Star Ins. Co.*, 347 S.C. 405, 556 S.E.2d 371 (2001).
[32] *Cates Constr., Inc. v. Talbot Partners,* supra note 17.
[33] Restatement (Third) of Suretyship and Guaranty § 34(1)(a).
[34] Bailey, *Unconditional Bank Guarantees* [2003] Int'l Constr.L.Rev., Part 2, p. 240.

CHAPTER 26

Claims and Disputes: Emphasis on Arbitration

Scenario

Mega Manufacturer, Inc. (MMI) sought to build a new factory. MMI retained Awesome Architects, Inc. (AAI), an architectural firm, to create the plans and specifications and to administer the project. Based upon this design, the owner entered into a $2 million fixed-price contract with Prime Builder, Inc. MMI used the 2007 American Institute of Architects (AIA) standard documents dealing with construction services: A101/201-2007.

During performance, AAI refused to authorize payment for certain work because it did not conform to the design. Prime Builder submitted a claim, contending the design was ambiguous. As Prime Builder read the requirements, the work complied with the contract documents. Privately, the AAI's architect believed the contractor's reading of the design made sense, although it was not what he intended. Desiring the project to comply with his intentions, the architect denied the claim.

The contractor requested mediation with the owner, and the owner agreed. The mediator, a retired engineer, tried to get each side to see the other's understanding of the design requirements. The owner also saw the contractor's perspective, and told the mediator he would settle by paying half the cost of the increased work. The contractor rejected that settlement and demanded arbitration.

During arbitration, the contractor discovered that the arbitrator (a lawyer) had been college classmates with AAI's architect and that (after a 20-year hiatus) they had rediscovered each other on Facebook. The lawyer–arbitrator had not revealed this relationship before agreeing to arbitrate, and none of the parties were aware of it.

The arbitrator entered an award in favor of the owner. When the owner sought to have the award confirmed by a trial court, the contractor (having learned of the relationship) instead asked the court to vacate the award because the arbitrator was biased.

26.1 Claims Resolution: Two- or Three-Step Process

Methods of Dispute Resolution. Most major (even minor) construction projects generate disputes or at least disagreements. Disputes or disagreements may be resolved in one of two ways (or by a combination of the two).

A *voluntary* resolution is negotiated by the parties themselves. Sometimes, resolution is *imposed* by third parties, such as a judge or arbitrator.

The law favors voluntary resolution of disputes by the parties. The parties know best what are the issues, facts, and considerations to be weighed in resolving the dispute. A negotiated resolution will be upheld by the courts, absent evidence of a contract defense, such as ambiguity or economic duress (unfair bargaining).

Sometimes a mediator—an outside person not an agent of either party—is brought in to facilitate negotiations. The purpose of a mediator is not to decide issues but to help the parties settle their disputes.[1] The mediator listens to both sides to facilitate settlement but does not impose a solution on the parties. A mediated settlement is viewed by the courts as a voluntary negotiation. To promote an open and honest relationship with the mediator, communications (including proposed settlement offers) made during the mediation process remain confidential and cannot be used in a later dispute resolution or litigation.

> In this chapter's scenario, MMI during mediation offered to settle with Prime Builder by paying for half the cost of the work required by AAI. This settlement offer may not be introduced into evidence during the subsequent arbitration or litigation.

A solution imposed by a judge or arbitrator can be reviewed by an appellate court. But review of a judicial decision is limited as a rule to issues of law. Judicial review of an arbitration award rarely results in an award being set aside (see Section 26.6). However, failure to appeal within an allowed period of time (so that no judicial review is possible) will make the decision "final" or binding on the parties. (Failure to timely appeal a design professional's initial decision—discussed immediately next—also makes that decision binding on the parties.)

Initial Administrative Review. Efficient dispute resolution is crucial to the successful prosecution of the project. It must be emphasized that the vast majority of disputes in the construction process are resolved through negotiation. The parties must be willing to gather information and look at the dispute from the other party's position as well as their own. They should recognize the horrendous costs that can be incurred in terms of time, money, and outcomes not from negotiation but because of the intervention of third parties—in the form of either litigation or alternative dispute resolution (ADR), such as arbitration.

One method to keep third-party intervention (either litigation or ADR) at bay is to have an internal or *administrative* dispute-resolution mechanism in place. If disputes (as a preliminary matter) must be reviewed by the owner's representative acting as an initial arbiter or judge, this may stop minor disputes from spiraling into much larger claims. Of course, further review of this initial decision is necessary to preserve fairness to the contractor.

This two-step process—an administrative decision with the option of review by a third party arbiter or judge—is the template used by many sophisticated owners. For example, in federal procurement, a contractor who was awarded a contract by a government agency must submit its claims to the agency's contracting officer for initial decision. The contractor may then appeal that decision either to the board of contract appeals or to the U.S. Court of Federal Claims.[2]

EJCDC. The Engineers Joint Contract Documents Committee's "Standard General Conditions of the Contract for Construction" uses a three-step process with mediation as a new, in-between step.[3] The EJCDC requires all claims to be referred initially to the engineer for decision. Either party dissatisfied with the engineer's decision may submit the claim to mediation.

Under the EJCDC, if mediation fails, the claim may be submitted for resolution using either the ADR method chosen by the parties or litigation. Here, the three-step (including mediation) claims resolution process begins with an initial decision by the engineer (administrative review) followed by the introduction of third parties.

AIA. In its first Uniform Contract issued in 1888—endorsed by the National Association of Builders (the predecessor to the Associated General Contractors of America)—the AIA utilized a two-step dispute resolution template: the owner or contractor submits a claim to the project architect for *initial decision*, which is subject to review by arbitration.[4] This *two-step* process continued unchanged for over a century. For example, in the 1987 documents, AIA A201-1987, ¶ 4.5.1, the architect's initial decision could be followed by arbitration or litigation.

In 1997, mediation was added as a third step in the claims process. AIA A201-1997, ¶¶ 4.5 and 4.6, sandwiched mediation between the architect's initial decision and either arbitration or litigation.

Even more significant changes to the AIA claims process came a decade later. Contractors expressed concern that architects could not be objective in evaluating their claims, especially since the architect is paid by the owner and the contractor's claim may require the architect to find his own design to be defective.

In response, AIA in A201-2007, § 1.1.8, created a new project position: the Initial Decision Maker (IDM). If the parties do not designate an IDM, the architect is the IDM. Regardless of who occupies the position, the IDM's primary roles are to issue an initial decision when presented with a contractor or owner claim (see A201-2007, § 15.2) and also to certify that grounds exist for the owner to terminate the contractor for cause (see A201-2007, § 4.2.2). Further discussion of the IDM is found in Section 26.5 of this chapter.

> In this chapter's scenario, the architect made the initial decision concerning the contractor's claim. Thus, either the parties did not designate an IDM, or they specified the architect as the IDM.

A second significant change to the AIA claims process made in 2007 is that mediation is a condition precedent to "binding dispute resolution," defined as arbitration *or* litigation. A201-2007, §§ 15.3.1 and 15.4.1. For the first time in nearly a century and a quarter, the AIA now requires an election to submit a dispute to arbitration. See the check-off mechanism in A101-2007, § 6.2, used by the parties to choose between arbitration, litigation, or another dispute resolution method once mediation has failed.

No Initial Review. Of course, there is no legal requirement that the design professional make an initial decision of a claim. In that situation, any disputes between owner and contractor would have to be negotiated by the parties, submitted to arbitrators if the parties have agreed to use this process, or decided by a court.

26.2 Reasons for Initial Design Professional Decision

As noted, both the EJCDC and the AIA (with the exception of designating an IDM) require the design professional—a party hired by the owner—to make an initial decision when presented with a claim by either the owner or contractor. What are the reasons for practice of design professional as (initial) judge?

To a continental European, the design professional as judge seems incongruous. With the exception of Great Britain, European construction administration might include the design professional giving his interpretation of design documents he drafted, but he would not be given a "judging role." The close association between owner and design professional based on the former's selection and payment of the latter precludes this role being given to her.

Despite the close association between owner and design professional, most American construction contracts—public and private—give the design professional broad decision-making powers. A number of reasons can be given for the development of this system.

First, the stature and integrity of the design professions may give both parties to the construction contract confidence that the decisions will reflect technical skill and basic elements of fairness.

Second, the design professional's role in design before construction equips the design professional with the skill to make decisions that will successfully implement the project objectives of the owner. In a sense, the role as interpreter and judge is a continuation of design.

Third, owners are often unsophisticated in matters of construction and need the protection of a design professional to obtain what they have been promised in the construction documents. Implicit is the assumption that, without a champion and protector, the owner might be taken advantage of by the contractor. Couple this with the owner's bargaining strength, and the present system results.

Fourth, even assuming that complete objectivity is lacking and that the contractor rarely has much choice, the alternative would be worse.

Suppose the design professional did not act as interpreter and judge. Such matters would have to be resolved by owner and contractor, which for many owners would require professional advice. If owner and contractor cannot agree, the complexity of construction documents and performance will necessitate many costly delays, because the alternative forums for owners and contractors who cannot agree—litigation and arbitration—still involve time and expense.

Fifth, despite the dangers of partiality and conflict of interest (the design professional may overlook defective workmanship to induce the contractor not to press a delay claim or a claim for extras based on the design professional's negligence, or the design professional may find a defect due to poor workmanship rather than expose himself to a claim for defective design), the system seems to have kept the project going notwithstanding inevitable disputes. Perhaps a quick decision that may at times be unfair was better than a more costly, cumbersome system that might give better and more impartial decisions.

The potential for bias toward the owner and decision making often involving the design professional's own prior design work are bound to reflect themselves in the judicial treatment of the design professional's decisions and whether the design professional will be held responsible for decisions (discussed in Section 26.4).

26.3 Procedural Matters Concerning the Initial Decision

Requirements of Elemental Fairness. The architect's or engineer's role as interpreter or judge invites comparison with arbitration and litigation. Should the design professional conduct a hearing similar to that used in arbitration or litigation? Clearly, the formalities of the courtroom would be inappropriate and unnecessary. Even the informal hearings conducted by an arbitrator would not be required. No hearing at all is necessary unless explicitly required by the contract.

Cogent reasons exist for some semblance of elemental fairness to both parties. First, continued good relations on the project necessitate a feeling on the part of the participants that they have been treated fairly. Each party should feel it has been given a fair chance to state its case and be informed of the other party's position. A fair chance need not necessarily include even an informal hearing. The architect should listen to the positions of each party where feasible before making a decision.

The exact nature of what is fair will depend on the facts and circumstances existing at the time the matter is submitted to the design professional. Small matters and those that require quick decisions may not justify the procedural caution that would be necessary where large amounts of money are at stake or where an urgent decision is less important.

The second reason for elemental fairness is the likelihood that a decision made without it will not be accorded much *finality*—in other words, it will not be accepted by the courts. For example, *John W. Johnson, Inc. v. Basic Construction Co.*,[5] involved a dispute in which the architect had ordered a prime contractor to terminate a particular subcontractor. The trial judge stated,

> This amazing directive was issued by the architect's office without notice to the plaintiff and without giving the plaintiff any opportunity to be heard, orally or in writing, formally or informally. This action on the part of the architect's office was contrary to the fundamental ideas of justice and fair play. The suggestion belatedly made at the trial that it was not appropriate for the architect to maintain any contacts with subcontractors is fallacious in this connection. Any such principle as that did not bar the architect's representative from according a hearing to the subcontractor before directing that his subcontract be cancelled.[6]

Termination is a serious matter, and more process fairness can be expected in such matters. But this case also reflects judicial unwillingness to uphold rash, impetuous decisions.

The third reason relates to the immunity sometimes given to design professionals when they act in a quasi-judicial role (explored in Section 12.7E). The more design professionals act like judges or arbitrators, the more likely they will be given judicial protection from a lawsuit by someone dissatisfied with the decision.

Standard of Interpretation. When presented with a dispute, how should the design professional interpret the relevant contract language? Contracts that give interpreting and judging powers to the design professional often specify general standards of interpretation for the design professional to use. For example, AIA Document A201-2007, § 4.2.12, requires that all "interpretations and decisions of the Architect . . . be consistent with the intent of and reasonably inferable from the Contract Documents." Despite the limits and guides, the design professional often faces the formidable task of deciding disputes over the meaning of contract document language. Their prior participation in developing contract documents makes this task even more difficult.

The design professional should judge the contract documents from the perspective of an honest contractor examining them before bid or negotiation. The owner's preparation of the contract documents through the design professional takes a long time—much longer than the contractor has to examine and bid. Any ambiguities the contractor should not have been expected to notice, and to which attention should not have been directed before bid submission or negotiation, should be resolved in favor of the contractor. Conversely, unclear language to which the contractor should have directed attention should be resolved against it.

Admittedly, a standard that looks at the honest contractor can place design professionals in a difficult position where the language in question was derived from the drawings and specifications they prepared. (The discussion assumes that the language in question was not part of the basic contract or general conditions.) Openly acknowledging that the specifications were unclear can be a confession of professional failure. The standard of interpretation suggested—that of favoring the contractor under certain circumstances—can inhibit the design professional from ever finding language unclear for fear that it would reflect on his work. An honest design professional should be fair to both owner and contractor despite this possibility.

Perhaps it will be expecting too much for the design professional to step back from his own work and judge it objectively. This possibility may be a reason to accord less finality to the decision. It is certainly a reason the A201-2007 uses an IDM to review contractor claims.

> In this chapter's scenario, the contractor's claim was that the design was ambiguous and its interpretation of the design requirements was reasonable and should be given effect. The architect realized the ambiguity only when confronted with the contractor's claim. However, the architect wanted the project to be built as the architect envisioned it and so denied the claim.
>
> Here, the architect confuses the project conception with the question of who pays for that end result—a distinction perhaps more familiar to lawyers. The purpose of the claim is not to decide how the completed project will look, but who pays for the added cost of complying with the architect's design intent, when the contractor's differing interpretation was also reasonable (and cheaper to implement).
>
> For these reasons, the design professional's decision is initial only, subject to further review.

Form of Decision. The contract clause giving the design professional the power to interpret documents can require that a particular form be followed when a decision is made. AIA Document A201-2007, § 4.2.12, requires that interpretations be in writing or in the form of drawings. A201-2007, § 15.2.5, requires that decisions resolving claims must be in writing with the reasons for the decision stated.

For this and other reasons, it is generally advisable for decisions to be made in writing and communicated as soon as possible to each party. When it is not feasible to make a decision in writing on the spot, any oral decision should be confirmed in writing and sent by a reliable means of communication to each party. A written communication giving the design professional's interpretation of his decision need not give reasons to support the decision. The essential requirement is that the parties know that a decision has been made and that they know the nature of the decision. However, the process will work more smoothly if the participants are given a reason for the decision. The decision need not be elaborate or detailed but should specify the relevant contract language and facts and the process by which the decision has been made.

Costs. The architect's costs during the construction phase—including interpreting the contract, reviewing submittals, inspecting the contractor's work, and responding to the contractor's requests for information—are included within his basic fee. If the architect is not the Initial Decision Maker, the architect's cost in assisting the IDM is an additional service justifying a cost beyond the basic fee.[7]

The AIA documents do not specify how the IDM is to be paid. One reason for creation of the IDM position was the concern of contractors that the architect, because he is paid by the owner, would be biased in favor of the owner. However, if the cost of the IDM is borne by the owner, then the possibility of bias in favor of the owner remains. If instead the cost of the IDM is borne equally by the owner and contractor, this expense must be budgeted in advance by the contractor when entering into the contract. In either case, the payment mechanism of the IDM should be addressed by a separate agreement between the owner and contractor. This may be best accomplished through a separate contract signed by the owner, contractor, and IDM.

The EJCDC provides in its E-500, ¶ A2.2(6) (2008), that evaluating "an unreasonable claim or an excessive number of claims" is a *required* additional service justifying additional compensation above the basic fee.

The costs incurred by the parties, such as transporting witnesses to any informal hearing, obtaining any expert testimony, or attending any informal hearings, will be borne by the parties who incur them unless the contract provides otherwise. As most dispute resolution done by the design professional is informal, costs of this sort are not likely to be comparable to those incurred in an arbitration or litigation. Under A201-2007, § 15.1.3, if the IDM elects to consult with experts or persons with specialized knowledge, the owner must cover any cost involved.

Suppose the IDM orders that work be uncovered. Uncovering work is costly. Often contracts specify who will pay for the cost of uncovering and recovering. AIA Document A201-2007, § 12.1.1, states that if work had been *improperly* covered by the contractor, such as covering work despite a request by the architect that it not be covered, the cost must be borne by the contractor even if the work had been properly performed.

A201-2007, § 12.1.2, deals with work properly covered. If the work is found "in accordance with the Contract Documents," costs are borne by the owner. If the work did not comply, the contractor must pay the cost of uncovering and recovering unless the deviation was caused by the owner or a separate contractor (on a project involving multiple primes). Similarly, § 13.5.3 places the entire cost of "testing, inspection or approval" on the contractor if these processes reveal that the contractor has not complied with the contract documents.

Should the contractor be required to pay the entire cost if *any* deviation is discovered? One purpose of having the architect visit the site periodically is to observe work before it is covered. The assumption under § 12.1.2 is that the architect did not see the defective work before it was covered or request the contractor not to cover it until he had a chance to inspect it. Under such circumstances, the architect may be fearful of a claim being made against him unless some defective work is found.

The all-or-nothing solution is justified only if there were major deviations or if the work was covered to hide defective work. If the defect is slight or inadvertent, it is more fair to share the cost of covering and uncovering.

26.4 Finality of Initial Decision

Range of Finality. Parties to a contract can create a mechanism by which disputed matters or other matters that require judgment can be submitted to a third party for a decision. The finality of that decision—that is, whether it can be challenged and the extent of the challenge—can range from 0 to 100 percent. It can be absolutely unchallengeable. At the other extreme, a decision by a third party can be simply advisory.

Issuance of an initial decision by the Initial Decision Maker is a condition precedent (a precondition) to mediation in AIA A201-2007, § 15.2.1. The AIA does not specify whether the mediator should give deference to the IDM's decision. However, mediation should be viewed as a fresh start to resolution of the dispute, and the mediator and parties are not in any way bound by the IDM's decision.

Most impartial, third-party decisions have some finality—but not the finality of an arbitral award, as noted in Section 26.7, or a court ruling. But such decisions are clearly more than advisory. Unless they are shown either to have been dishonestly made or to be clearly wrong, they are likely to be upheld if challenged in court.[8] One court held the decision to be binding unless there was fraud or bad faith.[9] Another court stated that the decision is conclusive if honestly made unless it is clear that the design professional made a serious mistake.[10] The Restatement (Second) of Contracts makes the decision binding as long as it is made honestly and not on the basis of gross mistake as to the facts.[11]

Other courts, when addressing the issue of finality, express concern that an architect or engineer cannot be said to be a disinterested third party. In discussing the reasons for limiting the effect of the refusal by an architect to issue a certificate for payment, a New York judge in *Arc Elec. Construction Co. v. George A. Fuller Co.* stated, "The rule is based upon the fact that the architect, in contracts of this sort, rarely a disinterested arbiter, is usually the representative of the party, often the owner, who must ultimately bear the cost of the work."[12]

Another factor that may bear on the degree of finality to be given design professional decisions is the process by which the decision is made. If it appears to have been made precipitously and without elemental notions of fairness, less finality, if any, will be accorded.[13]

Subject Matter of Dispute. The subject matter of the dispute may also bear upon the degree of finality. Disputes submitted to the design professional can range from purely factual (Did the work meet certain specific standards?) to matters that though sometimes called legal are really factual (How should this clause be interpreted?) to legal questions (Is the substantial performance doctrine applicable where the design professional is to judge performance?). The more factual or technical the dispute, the greater the court's deference to the decision. The more legal the dispute, the less finality will be accorded.[14]

AIA Document A201-2007, § 4.2.13, states that architect decisions as to aesthetic effect "will be final if consistent with the intent expressed in the Contract Documents." The very subjectivity of aesthetic effect and the power for abuse that such a clause can create will inevitably lead to a difference of opinion among courts as to the enforceability of such a clause.[15]

Review Process. Either party has a specified amount of time to appeal an initial decision either to mediation, arbitration, or litigation. Failure to seek timely review will make the initial decision binding.

For example, under A201-2007, § 15.2.6.1, either party within 30 days from the date of the initial decision may demand that the *other party* agree to mediate within 60 days of the initial decision. If that demand is made and the other party does *not* ask for mediation within the time allowed, both parties waive their rights to mediate or to pursue "binding dispute resolution proceedings with respect to the initial decision."

Suppose the architect's decision is final and binding unless arbitration is invoked. If the arbitration clause has been deleted or arbitration waived, the court will extend a considerable amount of deference to the architect's decision.[16] (If arbitration is sought, the arbitrators need pay no attention to the architect's decision. This can encourage arbitration by the party that is dissatisfied with the architect's decision.)

Other Considerations. If a payment certificate is issued and the *owner* refuses to pay, it is likely that the certificate will be considered final. Undoubtedly, this recognizes that the owner (for all practical purposes) has selected the design professional and should be given less opportunity to challenge the decision.

If a subcontractor's rights are at stake, less finality will be given.[17] The subcontractor had even less of a role in selecting the design professional. Giving power to the design professional is often accomplished by incorporating prime contract general terms by reference into the subcontract. The subcontractor may not have had much opportunity to present its case to the design professional.

Another problem relates to the interaction between language giving finality to a design professional decision and language that bars acceptance of the project from waiving claims for defective work subsequently discovered.[18] In *City of Midland v. Waller*,[19] the Texas Supreme Court sensibly reconciled these two clauses by concluding that the issuance of a certificate is conclusive where defects are patent or obvious but not where defects are latent (that is, not reasonably discoverable).

26.5 The Initial Decision Maker: Some Observations

Creation of the Initial Decision Maker (IDM) position is one of the most significant changes made in A201-2007. This change was made in response to several contractor concerns. Historically, contractors have viewed the architect as biased, because the architect was both selected and paid by the owner. Many also believed the architect could not be impartial in response to allegations of negligent design or failure to timely respond to requests made during the construction phase. The architect likely would be reluctant to render an initial decision blaming himself—both for psychological reasons and in the event of future legal disputes with the owner.

While good reasons may exist for the creation of an IDM position, A201-2007 suffers from a lack of detail as to how the new system would work, especially if the IDM is someone other than the architect. There is no explanation of how the IDM is selected. While the architect must be licensed, there is no similar requirement for the IDM; indeed, the IDM might be a retired judge, lawyer, contractor, or scientist, rather than a design professional.

Unless the IDM is put "on staff" from the beginning of the project waiting for disputes to arise, there may be a significant delay while the IDM is "brought up to speed" on the facts leading up to the present dispute. In any event, significant costs may be involved with employment of an outside IDM; yet A201-2007 gives no indication as to how the IDM is to be paid. (By comparison, the owner and contractor must share equally in the cost of the mediator; see A201-2007, § 15.3.3.)

There is also no provision shielding the IDM from potential liability and no insurance requirement. If the IDM is a contractor and does not carry professional liability insurance, he may not be covered against a claim. In short, the general lack of guidance in A201 as to the nature of this new arrangement will likely generate disputes.[20]

26.6 Statutory Framework of Arbitration

FAA and UAA/RUAA. Although a creature of contract, arbitration is so well established in modern commerce only because statutory authorization—beginning with the Federal Arbitration Act (FAA)[21] enacted in 1925—make arbitration awards enforceable in court. Prior to passage of the FAA, courts were openly hostile to arbitration and often found agreements to arbitrate future disputes unenforceable.

In addition to the FAA, almost all states currently have arbitration statutes. These arbitration acts regulate confirmation, modification, or vacation of arbitration awards. (Confirmation converts the award into a court judgment.) Many of the modern state statutes were adopted from the Uniform Arbitration Act (UAA). A small but growing number of states have adopted the Revised Uniform Arbitration Act (RUAA), which was published in 2000.[22] Clearly, these statutes have greatly encouraged the use of arbitration. Yet having two layers of arbitration regulation—federal and state—has led to conflicts over federal preemption, which is a topic addressed in Section 26.7.

Enforcement and Limited Judicial Review. Suppose a party receives a favorable arbitration award. While the arbitration process to this point hopefully has been entirely free of judicial involvement, *enforcement* of an award requires judicial assistance. The successful party to the arbitration seeks a court order "confirming" the award and reducing it to a judgment.

At the same time, the party seeking to challenge the award may go to court and ask that the award be modified or (more likely) vacated. (A successful party may also seek to modify or correct an award if, for example, the party believes the arbitrator accidentally left out a component of the damages award.)

State arbitration statutes specify grounds for reviewing an arbitrator's award. Section 12 of the Uniform Arbitration Act (also, RUAA Section 23), which was enacted in whole or with minor variations in a substantial number of states, permits an award to be vacated (set aside) if there has been corruption or fraud, partiality (bias) by the arbitrator, excess jurisdiction taken, procedural misconduct, or lack of a valid agreement to arbitrate.

Section 13 of the UAA (also, RUAA Section 24) allows modification or correction of an award if there has been an "evident miscalculation" or "evident mistake," if the arbitrators decided a matter not submitted to them and the award may be corrected without affecting its merits, or if the award is "imperfect in a matter of form."

Similar language is contained in the Federal Arbitration Act.[23] Grounds for vacating are limited and principally look to serious procedural misconduct on the part of the arbitrators. Courts are reluctant to upset the award because the arbitrator refuses to admit evidence, particularly after the hearing has been closed.

The limited review does not mean arbitration awards are never upset. For all practical purposes, however, an arbitrator's decision is final.[24] For this reason, any legal challenges to arbitration are usually aimed at preventing the process from going forward in the first place. Sadly, for a dispute resolution process which is supposed to operate relatively free of judicial involvement, these legal disputes have become increasingly prevalent and their resolution unusually complex. They are addressed in the following section. However, the legal issues may only be highlighted, as an in-depth discussion is beyond the purpose of this book.

26.7 Abuse of Arbitration, State Regulation, and Federal Preemption

Introduction. Modern law looks favorably upon arbitration. Arbitration relieves courts from having to resolve many disputes. This enables courts to use their facilities and skills to resolve other disputes that cannot be handled by arbitration, such as criminal cases, family disputes, or accident claims. Classically, and in the ideal, a properly structured arbitration can provide skilled dispute resolution. When arbitration was first developed, arbitrators were selected from a pool of persons who were experts in the context of the transaction and possessed knowledge of the customs and risks of the relevant market.

Ideally, arbitration should be quicker and cheaper than the judicial resolution of disputes. Also, arbitration is private. Most disputants dislike public airing of their disputes.

Yet arbitration can be abused, usually by the party with the economic power to require that disputes be resolved by arbitration. As a rule, this is the party that provides goods and services to another. Arbitration can be the method by which the manufacturer or supplier denies the consumer the right to have his claims addressed in a court presided over by a neutral judge and decided by an impartial jury. Often the party that agrees to arbitrate does not realize what it is getting and what it is giving up. Notwithstanding these concerns, the U. S. Supreme Court has repeatedly made clear, such as in the 2011 case of *AT & T Mobility LLC v. Concepcion*,[25] that arbitration agreements in consumer contracts are binding and enforceable.

The law has become concerned that the economically more powerful party could use arbitration to exert undue influence over the disputes *process*. The arbitration contract could control the jurisdiction of the arbitrator, the cost of the arbitrator's fee, the identity of the arbitrator, the place and structure of the hearing, the nature of the award, the remedies allowed, and the finality of the award.

As discussed in the remainder of this section, traditionally states have regulated arbitration—first through the courts and later primarily by statute. Yet the Supreme Court's expanded invocation of the Federal Arbitration Act has resulted in federal preemption (displacement) of much of that regulation.

State Judicial Regulation. During much of the twentieth century, regulation was done by the courts. Parties to a contract that contained an arbitration provision objected on grounds of fairness. Courts refused to allow the mechanism of arbitration to be a vehicle by which the stronger party unduly extended its power over the dispute resolution process. Two California cases from the 1970s are illustrative. In one, the contract was for a home improvement, but arbitration included a $720 filing fee.[26] The other involved a subcontract for a California project, but the prime contractor, whose home office was in New Jersey, required arbitration in New Jersey with one of three arbitrators picked by a New Jersey trial judge.[27] Both courts refused to enforce the arbitration clauses as written.

While modern law looks favorably upon arbitration, judicial scrutiny has not entirely abated. Two California appellate court cases from the twenty-first century came to opposite conclusions on enforceability of an arbitration provision in sale agreements on large subdivision housing projects.[28] Yet another court upheld a clause that required arbitration in California despite the subcontractor being a Mississippi company and the project a federal one in Mississippi.[29]

State Statutes. Concern with the one-sided nature of some arbitration clauses generated an outpouring of state statutes. They were designed to ensure that parties asked to assent to arbitration clauses knew what they were getting and to prevent the system from operating in such a way as to bar a dispute from being resolved in a convenient forum by a neutral person after a fair hearing. While the details varied from state to state, these statutes sought to make certain that all the parties, particularly the ones inexperienced and lacking bargaining power, were aware of what they were getting and what they were giving up.

Some statutes compelled arbitration clauses to be in capital letters, be of a different print and a different color than the rest of the contract, be signed separately, and be signed by both the parties and their attorneys, so that the signer knew he was giving up its "day in court." Some barred agreements to arbitrate future disputes in consumer transactions. Others barred clauses that required that the arbitration be held in an inconvenient place.

Virtually all (if not all) of these statutes are no longer in effect—they have been preempted by federal law, which is addressed next.

Federal Preemption. The outpouring of state statutes just mentioned led to a new problem: federal preemption. This doctrine bars the application of state law that frustrates the policy of encouraging arbitration expressed in the FAA. The U. S. Supreme Court has interpreted the FAA broadly as applying to all transactions that affect interstate commerce (commerce between different states). Under this test, virtually any construction project of a significant size involves interstate commerce and hence is subject to the FAA. For example, in *Citizens Bank v. Alafabco, Inc.*,[30] the Supreme Court ruled that it is sufficient if materials used on the project were delivered from a different state from where the project is located for the transaction to involve interstate commerce.

Once a court finds that the underlying transaction affects interstate commerce and that the FAA applies, it must then determine whether the state law is preempted. The FAA § 2 requires arbitration agreements to be treated as equally valid and enforceable as is any contract under state law.

26.8 Common Law Contract-Based Defenses to Arbitration

As noted previously, the FAA § 2 requires arbitration agreements to be treated as equally valid and enforceable as is any contract under state law. Specifically, § 2 declares arbitration agreements "shall be valid, irrevocable, and enforceable, save upon such grounds as exist at law or in equity for the revocation of any contract." Any common law defense to enforcement of a contract is equally applicable to an arbitration agreement.[31]

In a typical situation, the prime contract contains an arbitration provision, but the owner sues the contractor for construction defects. The contractor asks the court to stay (stop) the litigation and to direct the owner to arbitrate his claim. The owner counters that the arbitration agreement is unenforceable for one of the reasons discussed in this section.

Unconscionability. Courts may refuse to enforce contract terms that are unconscionable. For equity courts (in "old England"), these were terms that shocked the conscience of the judge.

An unconscionability analysis contains two components: procedural unconscionability and substantive unconscionability. *Procedural* unconscionability first examines whether the contract was one of adhesion—one not freely bargained but adhered to on a "take it or leave it" basis. If so, then the analysis focuses on whether the weaker party was oppressed or surprised. Oppression occurs where a contract involves a lack of negotiation and meaningful choice; surprise is where the allegedly unconscionable provision is hidden within a long document containing small print. *Substantive* unconscionability is concerned with contractual terms that produce unfair or one-sided results.

Procedural and substantive unconscionability are viewed in totality and need not be of equal weight. A close case of procedural unconscionability can be made up for by a strong case of substantive unconscionability.

Note that one element of procedural unconscionability is that the contract is adhesive. For this reason, in the construction industry claims of unconscionability often involve large housing subdivisions. The developer's pre-printed (and often unreadable) sales contract is the same for each purchaser and requires arbitration of construction defect disputes. In this setting, there may be little room for negotiation. Yet courts are divided on whether a developer's use of pre-printed sales agreements for the entire subdivision establishes that these contracts are adhesive and that these buyers therefore may invoke the unconscionability doctrine.[32]

In one such case, the subdivision homeowners filed suit against the developer, contending that its defective construction exposed them to personal injury from radon gas. The developer (Richmond) sought to enforce an arbitration provision in the sales agreements. The owners countered that the arbitration provision was unconscionable. After deciding that the sales agreements were adhesive and procedurally unconscionable, the West Virginia Supreme Court in *State ex rel. Richmond American Homes of West Virginia, Inc. v. Sanders* cited these facts in support of a determination of substantive unconscionability:

> The circuit [trial] court found substantive unconscionability because the arbitration process established by the Purchase Agreement was unduly oppressive in that it exculpated Richmond from its misconduct, and substantially impaired the plaintiffs' right to pursue remedies for their losses. Richmond's Purchase Agreement collectively created an arbitration process that either implicitly or explicitly limited the ability of the plaintiffs to fairly pursue remedies for their injuries and damages caused by radon gas in their new homes. For instance, the Agreement explicitly states that Richmond "disclaims liability, and [Plaintiff] expressly waives any and all claims for property and/or personal injury or other economic loss resulting from . . . radon gas[.]" The Agreement also says that Richmond is not "liable for any special, indirect or consequential damages," which the circuit court found would exculpate Richmond from damages for pain and suffering, intentional and negligent infliction of emotional distress, and punitive damages. The Agreement also says Richmond is not liable for diminution in the value of the home as a result of its misconduct. Further, the Agreement categorically attempts to eliminate various statutory and common law rights, including the plaintiffs' warranties under the Magnuson–Moss Act and "all warranties of fitness, merchantability and habitability."[33]

Here, the arbitration provision was part-and-parcel of a broad exculpatory contract that ran solely to the benefit of the developer. Such a contract would not be enforceable in a court. The *Richmond* court essentially refused to allow the arbitration process to be used to achieve the result the developer sought.

Other Defenses: Mutuality, Termination of the Contract, and Conditions Precedent. After unconscionability, the main attacks to the validity of an arbitration provision in a construction contract are (1) "mutuality," (2) the effect of termination of the underlying construction contract, or (3) frustration of the disputes process due to nonoccurrence of a condition precedent to arbitration.

An attack on an arbitration clause that it lacks mutuality may arise when one party is required to arbitrate but the other is not. Courts generally hold that the fact that one party can choose whether to arbitrate or litigate (while the other party must arbitrate) as a rule will not automatically invalidate the arbitration—even in consumer transactions.[34]

Discussion in this section has centered on the validity of agreements to arbitrate. Suppose a valid contract with an arbitration clause is terminated? What effect does termination have on the agreement to arbitrate?

Termination not based on a formation defect (such as fraud, mistake, or duress) should not and does not abrogate any broad arbitration clause.[35] This can become cloudy, because some dispute resolution systems include steps that must be taken prior to arbitration; for example, the AIA system requires an initial decision by a design professional or IDM. These steps are usually a condition precedent to arbitration. But the power of the architect or IDM under A201-2007, § 15.2.1, terminates 60 days after final payment is due.

Suppose the IDM or architect is no longer "on board" because her power expired. This should mean that the dispute can proceed directly to arbitration because the condition precedent had been excused.[36] However, a New York intermediate court, in *Lopez v. 14th Street Development LLC*,[37] held that the failure to have an interim decision eliminates the arbitration requirement. This frustrates the intention of the parties, which was to use arbitration as a dispute resolution method.

Waiver of Arbitration. The law generally favors arbitration. Yet a party may indicate by its acts that it chooses to litigate even though it had agreed to arbitrate. If it does so, the other party has a choice. It can compel arbitration by filing a motion in court to hold off litigation until the dispute is submitted to arbitration. Alternatively, it can decide to have the dispute handled in court.

The issue of wavier arises when the party who took steps inconsistent with the desire to arbitrate changes its mind and seeks to arbitrate. The court or (in some cases the arbitrator) must decide whether that party has waived (or lost) its right to arbitrate.

Two factors are relevant in deciding whether the right to arbitrate has been waived. The first is whether the acts by one party—actively pursuing litigation or just "blowing hot and cold" on arbitration by first seeking another way of resolving the dispute and then seeking to go back to arbitration—were so inconsistent with the desire to arbitrate that it is clear that party no longer wishes to arbitrate. The second is whether this conduct prejudiced (disadvantaged) the other party.[38]

Special attention should be paid to the filing of mechanics' liens. A contractor considering filing an arbitration demand must also be concerned with the short time limits for filing a lien. The majority rule is that filing a lien, which is simply a way of ensuring payment, is not grounds for a finding of waiver. The propriety of the lien can be decided by the arbitrator. The AIA expressly states that the filing of a lien does not constitute waiver of the right to arbitrate.[39]

26.9 The Arbitration Process: Introduction

The remainder of this chapter will assume the parties agreed to conduct arbitration under the auspices of the American Arbitration Association (AAA) Construction Industry Arbitration Rules (CIA Rules) (2009). (An alternative arbitration association, called JAMS, uses only retired judges and experienced retired attorneys).[40] There are three sets of CIA Rules: Regular Track, Fast-Track, and Large/Complex Rules. Regular Track rules are preceded by R; Fast-Track by F; and Large/Complex by L. The AIA incorporates the AAA CIA Rules into its documents.[41]

The party initiating arbitration usually files a notice of an intention to arbitrate with the AAA and pays a fee based on the amount of the claim. The notice usually contains a statement setting forth the nature of the dispute, the amount involved, and any remedies sought.[42]

26.10 Prehearing Activities: Discovery

A much-debated issue is whether the parties to an arbitration should have the right to examine people who might be witnesses for the other party and examine documents in the other's possession to evaluate the other party's case, to prepare for trial, and possibly to settle. In judicial proceedings, these activities are classified as *discovery*.[43]

Discovery has become controversial. Originally created to preclude the trial from taking unexpected twists and turns because of surprises, it has become largely uncontrolled—often with excessive demands to produce documents and lengthy and often irrelevant questioning of many potential witnesses. To a large degree, it has been run by the lawyers, although increasingly courts are beginning to take control to prevent costly and time-consuming excesses.

Yet despite the increasing concern over full-blown, lawyer-controlled discovery, it seems clear that parties cannot begin to negotiate to settle disputes or even seek to mediate them without having some idea of the merits of the case. Obtaining basic information need not require full-blown discovery. In deciding whether to allow discovery, the AAA takes into consideration the size and complexity of the dispute with the desirability of discovery.

26.11 Selecting Arbitrators and Arbitrator Neutrality

A leading treatise states, "The most important tactical step in an arbitration proceeding is the selection of the arbitrator."[44]

Usually the method for selecting arbitrators is specified in the arbitration clause. Sometimes a particular arbitrator or specific panel of arbitrators is designated by the parties in advance and is incorporated in the arbitration clause. Advance agreement on the arbitrators or a panel of arbitrators should build confidence in the arbitration process. However, in construction contracts such advance agreement is uncommon.

Generally, a procedure to select arbitrators rather than a designation of particular individuals is used. For example, the procedure can require each party to name an arbitrator, with the two-party–appointed arbitrators designating a third or neutral arbitrator. Some arbitration clauses provide that each party will appoint an arbitrator, and only if they cannot agree on the disposition of the dispute do they appoint a third arbitrator who makes the decision. The AAA Rule R-3 states that the AAA will maintain a National Roster of Construction Arbitrators. If the parties have not chosen their own arbitrators, the AAA sends them a list of names from the national roster.

The issues of arbitrator disclosure and qualification to serve as arbitrator reveals an important potential conflict between the goals of arbitration—rapid resolution of technical disputes by *neutral* experts—and reality: that these persons become expert by virtue of their participation in the industry and may have had a prior relationship with one of the parties. An early U.S. Supreme Court decision, *Commonwealth Coatings Corp. v. Continental Casualty Co.*,[45] refused to confirm the arbitration award where the arbitrator (an engineering consultant) had not revealed economic dealings with one of the parties.

Some state arbitration statutes have dealt with this issue by mandating disclosures by potential arbitrators to a dispute.[46] The AAA's CIA Rule R-20 governs disqualification of an arbitrator, including because of partiality or lack of independence.

Finally, the Revised Uniform Arbitration Act (RUAA) was published in 2000. It is the successor to the Uniform Arbitration Act that was adopted in whole or in part in many states. The RUAA is being presented to state legislatures for adoption. Rule 12 specifies disclosure requirements for arbitrators.

The conflicting policies must be accommodated. A pool of skilled arbitrators is clearly needed. But the parties must believe that they were given a fair hearing if they are expected not to challenge the award. Yet an award should not be invalidated based upon an attenuated prior relationship that does not raise a realistic appearance of bias.[47]

> In this chapter's scenario, the arbitrator did not reveal that he knew the architect and that the two had at least Facebook contact. There is no indication of an economic interest the arbitrator may have in the dispute, and the architect is not one of the parties.
>
> Nonetheless, the arbitrator knew that the other parties were ignorant of his relationship with the architect and also that his award would be an implicit agreement with or rebuke of the architect. RUAA Rule 12 obligates the arbitrator to disclose any "relationship" with a party or witness, independent of the possible existence of an economic interest.
>
> The court should vacate the award based on arbitrator bias.

26.12 Multiple-Party Arbitrations: Joinder and Consolidation

Frequently, a linked set of construction contracts contains identical arbitration provisions. For example, there are identical arbitration clauses in AIA standard contracts between owner and contractor and between owner and architect. Likewise, identical arbitration clauses are contained in the AIA standard contracts between architect and consulting engineer and between prime contractor and subcontractor.

Suppose a building collapses and the owner wants to assert a claim. It may not be certain whether the collapse resulted from poor design or poor workmanship. The first would be chargeable primarily to the design professional and the second to the contractor. The owner may wish to arbitrate. Suppose there are identical arbitration clauses in the owner's contracts with the design professional and contractor. The owner can arbitrate separately with each, but there are disadvantages. Two arbitrations will very likely take longer and cost more than one. The owner may lose both arbitrations. (Although this loss may be unpalatable to the owner, the result may be correct if the design professional performed in accordance with the standards required of design professionals and the contractor executed the design properly.)

Inconsistent findings may result. For example, one arbitrator may conclude that the design professional performed in accordance with the professional standard, whereas the other may not. One may conclude that the contractor followed the design, and the other may not. To avoid this inconsistency and save time and costs, the owner may wish to consolidate the two arbitrations.

Suppose the owner demands arbitration only with the contractor. The contractor may believe that the principal responsibility for the collapse was defective design. In such a case, the contractor might wish to add the design professional as a party to the arbitration or, in legal terminology, "join" the design professional in the arbitration proceedings.

A contractor can be caught in a similar dilemma if it felt the responsibility for the owner's claim against it was work by a subcontractor. In such a case, the prime contractor may wish to seek arbitration with the subcontractor and consolidate the two arbitrations. Similarly, the architect may find it advisable to consolidate any arbitration she might have with her consulting engineer and any arbitration proceedings with the owner.

There is general agreement that the courts will respect any contract language that deals with consolidating existing arbitrations or adding a new party (joinder) to an existing arbitration. But when the parties have not spoken on these issues, there is great variation in court decisions in state and federal courts where one party seeks to consolidate or join while others object.

Some states preclude consolidation unless there is express statutory authority allowing it. Other states hold that, even in the absence of statutory authority, a court has the judicial authority to do so. Some states have enacted statutes granting the court authority to consolidate, while others give the arbitrator this authority.[48] The AAA has always been in favor of consolidation or joinder, and its CIA Rule R-7 establishes a mechanism for doing so.

The AIA for many years did not permit the architect to be a party to a multiparty arbitration without her written consent. After years of criticism of this policy, in 2007 the AIA permitted joinder and consolidation (see AIA Documents A201-2007, § 15.4.4 and B101-2007, § 8.3.4).

For reasons mentioned, consolidation and joinder are generally desirable. The possibility of confusion because of the potentially large number of parties in a consolidated arbitration can be handled by according the arbitrator the power to decide the number of parties and issues that would make consolidation or joinder too confusing. If so, a request to do so could be denied.

26.13 Award

Before making the award, the arbitrators review any documents submitted and listen to or read any transcription of the hearings. They consider any briefs that may have been submitted by the parties or their attorneys. Submission of briefs is uncommon except for disputes involving large amounts of money. The decision need not be unanimous unless the arbitration clause or the rules under which the arbitration is being held so require.

The form of the award can be simple. The arbitrator need not give reasons for the award. There are arguments for and against a reasoned explanation of the decision accompanying the award. An explanation may persuade the parties that the arbitrators have considered the case carefully. This may lead to voluntary compliance, which is obviously better than costly court confirmation.

A disadvantage of giving an explanation is the possibility that the dissatisfied party or parties may refuse to comply and seek to reopen the matter by objecting to the reasons given.

A more persuasive argument against reasons accompanying the award is the additional time and expense entailed. Making the arbitration too much like a court trial can lose some of the advantages of arbitration.

Taking all this into account, a short, reasoned explanation accompanying the award is advisable even if not required.

The rules under which an arbitration is conducted often specify when the award must be made. For example, R-43 of the CIA Rules requires an award be promptly made and, unless otherwise agreed or required by law, "no later than thirty (30) calendar days from the date of closing the hearing, or, if oral hearings have been waived, from the date of the AAA's transmittal of the final statements and proofs to the arbitrator." Failure to make the award by the designated time may terminate the jurisdiction of the arbitrators, although more commonly the parties agree to waive any time deadlines.

If there are no specific time requirements, the award must be made within a reasonable time.

26.14 Other Dispute Resolution Mechanisms

The cost, the time consumed, and the quality of litigation outcome have discouraged taking disputes to court. As a result, new nonjudicial methods have mushroomed. These other methods are collectively referred to as alternative dispute resolution (ADR). The primary ADR method is arbitration.

But as arbitration has become more like litigation as to time, expense, and formalities, emphasis has shifted to new techniques to promote a successful and conflict-reduced project. One strategy, exemplified by such project delivery methods as Partnering (Section 14.15) and

Project Alliance (Section 14.18), seeks to prevent disputes by altering the fundamental relations between the major participants.

Alternative ADR methods to arbitration have been developed. As ADR consists of fluid methodologies, it is beyond the scope of this book to do more than list these methodologies.[49] They include:

1. Mediation-arbitration
2. Mini-trials
3. Dispute review boards
4. Project neutral
5. Project counsel
6. Conflict manager
7. Court-appointed arbitrators: adjudicator (Great Britain)

26.15 Adjuncts of Judicial System

Just as outsiders saw problems in the judicial system, people operating the system were well aware of the system's flaws. Although wholesale restructuring was not considered possible, some judicial systems at federal and state court levels developed techniques to reduce the time and expense of providing a method of resolving private disputes.

One approach, particularly appropriate for construction disputes, is for the court to appoint individuals—called masters or referees—with special powers to expedite construction litigation. Sometimes these individuals are given the authority to set rules for the deposition process, which has become a costly method of obtaining information. Sometimes masters or referees informally act as mediators with a view toward persuading the litigants to settle the dispute. Finally, such individuals are sometimes authorized by the parties to resolve the dispute and are sometimes authorized by the judge to make findings of fact and conclusions of law, which can then be passed on and adopted by the judge if the parties have given up their rights to a jury trial.

Another approach is the use of summary jury trials. A jury is selected as would be a real jury. The judge informs the jury that the parties have agreed in advance to an abbreviated procedure to save time and money. Not until after the proceedings is the jury told that its determinations are not binding.

Each attorney makes what in a real trial would be a combined opening and closing statement. Attorneys may use charts, graphs, or other visual aids that would be used in a normal closing argument. However, no witnesses offer testimony. After the attorneys make the presentation, the judge instructs the jury as at the conclusion of a normal jury trial. After instructions, the jury retires and then presents its verdict.

Because the verdict is not binding, either party can demand a regular trial. But if the parties believe that the verdict is very likely what a real jury would determine, the mock verdict should encourage settlement. The obvious advantages to such a method are avoiding the expensive marshalling of documents, preparing a lengthy pretrial order, participating in a pretrial conference, and preparing and presenting the witnesses. If the parties take their obligation to present their cases in good faith and then negotiate in good faith after the jury verdict, this method should save time and money.

26.16 Public Contracts

Public construction contracts involve considerations not found in private contracts. Many statutes, rules, and regulations govern the award of contracts and the resolution of disputes under those contracts. This book cannot examine the details of these legal restraints because of their complexity and variations. However, the increased and often intense spotlighting of disputes in construction work necessitates some observations regarding dispute resolution in the context of public construction contracts.

26.16A Federal Procurement Contracts

Modern federal procurement began in 1978 when Congress enacted the Contract Disputes Act (CDA).[50] The CDA utilizes a two-step dispute resolution process. The contractor submits claims to the contracting officer for an initial decision. The contractor must certify to the contracting officer that the claim is made in good faith, that the supporting data are accurate and complete to the best of the contractor's knowledge and belief, and the government is in fact liable for the requested amounts.

Once the contracting officer issues a final decision, the contractor may either appeal to the board of contract appeals or bring a lawsuit in the U. S. Court of Federal Claims. The contractor may appeal from either forum to the U.S. Court of Appeals for the Federal Circuit and from there to the U.S. Supreme Court.

The Contract Disputes Act encourages the use of ADR. The CDA § 7103(h) provides that "a contractor and a contracting officer may use any alternative means of dispute resolution" found in the Administrative Dispute Resolution Act (ADRA).[51] Both the contracting officer and the contractor must provide a written explanation for declining a request for ADR.[52] The ADR methods listed in the ADRA are project neutrals[53] and arbitration.[54]

26.16B State and Local Contracts

A detailed treatment of the many state and local laws and regulations cannot be given in this book. It is important to recognize the importance of complying with specialized requirements for disputes under such public contracts of this type. Special rules exist for making claims. Some states have created special courts (often called a court of claims) to deal with these claims.

Also, there is a great variety of contractual methods for resolving claims. Some make a decision of an employee of the public entity a condition precedent to litigation. Some give broad dispute resolution power to an official of the public entity, with her decision being final unless the decision was arbitrary or capricious. Some states require arbitration of contractor claims.

26.17 International Arbitration

Section 1.10 mentioned the difference between domestic construction contracts and those that involve nationals of different countries, particularly contracts made by American construction companies requiring them to build projects in a foreign country. (Increasingly, foreign contractors build in the United States.) Transactions of this type generate some issues of minimal or no importance in domestic contracts.

The contracts themselves may be expressed in more than one language and often generate problems that result from imprecise translation. They may also involve payment in currency that varies greatly in value. Contractors in such transactions may often have to deal with tight and changing laws relating to repatriation of profits and import of personnel and materials. Perhaps most important, neither party may trust the other's legal system, and the contractor may believe it will not obtain an impartial hearing if forced to bring disputes to courts in the foreign country, particularly if the owner is (as so common in lesser developed countries) an instrumentality of the government.

Contractors making these contracts commonly insist on international arbitration to resolve disputes. Such arbitrations are usually held in neutral countries or in centers of respected commercial arbitration. Such awards are enforceable worldwide because of the Convention of the Recognition and Enforcement of Foreign Arbitration Awards, which is known as the New York Convention. Because of the complexities generated by international arbitration, this book does not discuss the subject in detail. Yet the emphasis on dispute resolution worldwide is a justification for some information regarding systems used for international transactions.[55]

The most commonly used contract for international engineering is the one published by the Federation Internationale des Ingenieurs-Conseils (International Federation of Consulting Engineers). The federation's contract for civil engineering construction (generally known as the FIDIC contract) provides for international arbitration.[56]

REVIEW QUESTIONS

1. What are some of the reasons that most American construction contracts, both public and private, give the design professional broad decision making powers?

2. What are the three reasons for elemental fairness?

3. Who has the right to appeal an initial decision? What will happen if a party fails to make a timely appeal of an initial decision?

4. What are some negative issues with the creation of the Initial Decision Maker (IDM) in the AIA A201-2007 document?

5. What are the reasons that an arbitration award may be vacated (set aside) according to the Uniform Arbitration Act?

6. How could an economically more powerful party use arbitration to exert undue influence over the disputes process?

7. What are the two components of unconscionability analysis and how does each function?

8. After unconscionability, what are the main attacks on the validity of an arbitration provision in a construction contract?

9. What two factors are relevant in deciding whether the right to arbitrate has been waived?

10. What are alternative dispute resolution methods to arbitration?

ENDNOTES

[1] J. BEER & C. PACKARD, THE MEDIATOR'S HANDBOOK 6 (4th ed.2012).
[2] Contract Disputes Act, 41 U.S.C. §§ 7103, 7104.
[3] EJCDC Doc. C-700, ¶¶ 10.05, 16 (2007).
[4] Bruner, *Rapid Resolution ADR*, 31 Constr. Lawyer, No. 2, Spring 2011, p. 6.
[5] 292 F.Supp. 300 (D.D.C.1968), aff'd, 429 F.2d 764 (D.C.Cir.1970).
[6] Id. at 304.

[7] AIA B101-2007, § 4.3.1.11.
[8] *Bolton Corp. v. T. A. Loving Co.*, 94 N.C.App. 392, 380 S.E.2d 796, 801, review denied, 325 N.C. 545, 385 S.E.2d (1989).
[9] *Laurel Race Course, Inc. v. Regal Constr. Co.*, 274 Md. 142, 333 A.2d 319 (1975).
[10] *Ingrassia Constr. Inc. v. Vernon Tp. Bd. of Educ.*, 345 N.J. Super 130, 784 A.2d 73, 80 (App.Div.2001).
[11] Restatement (Second) of Contracts § 227, comment c, illustrations 7 and 8 (1981).
[12] 24 N.Y.2d 99, 247 N.E.2d 111, 113, n. 2, 299 N.Y.S.2d 129 (1969). See also *Martel v. Bulotti*, 65 P.3d 192 (Idaho 2003) (applying a common law standard of review to an architect's initial decision and rejecting the narrower standard of review applicable to arbitration awards).
[13] *John W. Johnson, Inc. v. Basic Constr. Co.*, supra note 5.
[14] *Yonkers Contracting Co. v. New York State Thruway Auth.*, 25 N.Y.2d 1, 250 N.E.2d 27, 302 N.Y.S.2d 521 (1969), opinion amended by 26 N.Y.2d 969, 259 N.E.2d 483, 311 N.Y.S.2d 14 (1970).
[15] Compare *Baker v. Keller Constr. Corp.*, 219 So.2d 569 (La.App.1969) (reviewed decision), with *Mississippi Coast Coliseum Comm'n v. Stuart Constr. Co.*, 417 So.2d 541 (Miss.1982) (refused to review).
[16] *Martel v. Bulotti*, supra note 12.
[17] *Walnut Creek Elec. v. Reynolds Constr. Co.*, 263 Cal.App.2d 511, 69 Cal.Rptr. 667 (1968).
[18] See Section 17.9.
[19] 430 S.W.2d 473 (Tex.1968).
[20] Several of these points are made in Lesser & Bacon, *Meet the 2007 A201 "General Conditions of the Contract for Construction"—Part I: Letting Go of Paper and Making Way for the IDM,* 29 Constr.Litg.Rep., No. 4, Apr. 2008, p. 143.
[21] 9 U.S.C. §§ 1-16 (the FAA's general provisions).
[22] Ness, *Legislative Update: The Revised Uniform Arbitration Act of 2000,* 21 Constr. Lawyer, No. 4, Fall 2001, p. 35.
[23] 9 U.S.C. §§ 10, 11.
[24] Toomey & Euteneuer, *The Arbitrators Have Decided the Construction Dispute: What Do I Do Now?* 32 Constr. Lawyer, No. 2, Spring 2012, p. 20.
[25] 131 S.Ct. 1740 (2011).
[26] *Spence v. Omnibus Industries*, 44 Cal.App.3d 970, 119 Cal.Rptr. 171 (1975).
[27] *Player v. George M. Brewster & Son, Inc.*, 18 Cal.App.3d 526, 96 Cal.Rptr. 149 (1971).
[28] *Pardee Constr. Co. v. Superior Court,* 100 Cal.App.4th 1081, 123 Cal.Rptr.2d 288 (2002) and *Woodside Homes of California, Inc. v. Superior Court,* 107 Cal.App.4th 723, 132 Cal.Rptr.2d 35 (2003).
[29] *Ellefson Plumbing Co. v. Holmes & Narver Constructors, Inc.*, 143 F.Supp.2d 652 (N.D.Miss.2000).
[30] 539 U.S. 52 (2003).
[31] See Section 3.4A.
[32] *State ex rel. Vincent v. Schneider*, 194 S.W.3d 853 (Mo.2006) (no adhesion); *State ex rel. Richmond American Homes of West Virginia, Inc. v. Sanders*, 228 W.Va. 125, 717 S.E.2d 909 (2011) (unconscionability claim upheld).
[33] Supra note 32, 717 S.E.2d at 922. The Magnuson-Moss Act is a federal consumer protection statute; see Section 16.7C.
[34] *Albert M. Higley Co. v. N/S Corp.*, 445 F.3d 861 (6th Cir.2006) (enforced even if prime had discretion in prime-subcontractor dispute); Nahmias, *The Enforceability of Contract Clauses Giving One Party the Unilateral Right to Choose Between Arbitration and Litigation,* 21 Constr. Lawyer, No. 3, Summer 2001, p. 36.
[35] *Auchter Co. v. Zagloul*, 949 So.2d 1189 (Fla.Dist.Ct.App.2007).
[36] Ibid.
[37] 40 A.D.3d 313, 835 N.Y.S.2d 186 (2007).
[38] Ness & Peden, *Arbitration Developments: Defects and Solutions,* 22 Constr. Lawyer, No. 3, Summer 2002, pp. 10.
[39] A201-2007, § 15.2.8; B101-2007, § 8.2.1, applied in *Brendsel v. Winchester Constr. Co., Inc.*, 392 Md. 601, 898 A.2d 472 (2006).
[40] See http://www.jamsadr.com.

[41] See A201-2007, § 15.4.1 and B101-2007, § 8.3.1.
[42] CIA Rule R–4(a).
[43] See Moseley, *What Do You Mean I Can't Get That? Discovery in Arbitration Proceedings,* 26 Constr. Lawyer, No. 4, Fall 2006, p. 18.
[44] J. ACRET & A. D. PERROCHET, CONSTRUCTION ARBITRATION HANDBOOK, § 6.1 at 749 (2013).
[45] 393 U.S. 145 (1968).
[46] Cal.Code Civ.Pro. § 1281.9, applied in *La Serena Properties v. Weisbach,* 186 Cal.App.4th 893, 112 Cal. Rptr.3d 597 (2010).
[47] *La Serena Properties v. Weisbach,* supra note 46. See generally, Annot., 67 A.L.R.5th 179 (1999).
[48] McCurnin, *Two-Party Arbitrations in a Multiple Party World,* 26 Constr. Lawyer, No.1, Winter 2006, p. 5.
[49] For a comprehensive survey of ADR, see Stipanowich, *Beyond Arbitration: Innovation and Evolution in the United States Construction Industry,* 31 Wake Forest L. Rev. 65 (1996). For a survey of ADR methods, see Bruner, supra note 4; Stipanowich, *The Multi-Door Contract and Other Possibilities,* 13 Ohio St.J. on Disp. Res. 303 (1998). For an incisive analysis, see Hinchey, *Evolution of ADR Techniques for Major Construction Projects in the Nineties and Beyond: A United States Perspective,* 12 Constr.L.J. 14 (1996).
[50] 41 U.S.C. §§ 7101–7109 (but originally at 41 U.S.C. §§ 601–613).
[51] 5 U.S.C. §§ 571–583 (which was added in 1990 as an amendment to the Administrative Procedure Act).
[52] 41 U.S.C. § 7103.
[53] 5 U.S.C. § 573.
[54] Id. §§ 575–581.
[55] Black, Venoit, & Pierson, *Arbitration of Cross-Border Disputes,* 27 Constr. Lawyer, No. 2, Spring 2006, p. 5 (helpful introduction); Bond, *The Arbitration Clause in International Construction Project Contracts,* 28 Constr.Litg.Rep., No. 4, April 2007, p. 167.
[56] For an overview of the FIDIC standard form contracts, see Reilly & Tweeddale, *FIDIC's New Suite of Contracts,* 16 Constr.L.J. 187 (2000).

APPENDIX A

Standard Form of Agreement Between Owner and Contractor

AIA Document A101–2007

APPENDIX A / STANDARD FORM OF AGREEMENT BETWEEN OWNER AND CONTRACTOR

AIA® Document A101™ – 2007

Standard Form of Agreement Between Owner and Contractor where the basis of payment is a Stipulated Sum

AGREEMENT made as of the   day of
in the year
(In words, indicate day, month and year)

BETWEEN the Owner:
(Name, address and other information)

and the Contractor:
(Name, address and other information)

This document has important legal consequences. Consultation with an attorney is encouraged with respect to its completion or modification.

AIA Document A201™–2007, General Conditions of the Contract for Construction, is adopted in this document by reference. Do not use with other general conditions unless this document is modified.

---

for the following Project:
(Name, location, and detailed description)

The Architect:
(Name, address and other information)

The Owner and Contractor agree as follows.

AIA DOCUMENT A101-2007   A-3

TABLE OF ARTICLES

1 THE CONTRACT DOCUMENTS

2 THE WORK OF THIS CONTRACT

3 DATE OF COMMENCEMENT AND SUBSTANTIAL COMPLETION

4 CONTRACT SUM

5 PAYMENTS

6 DISPUTE RESOLUTION

7 TERMINATION OR SUSPENSION

8 MISCELLANEOUS PROVISIONS

9 ENUMERATION OF CONTRACT DOCUMENTS

10 INSURANCE AND BONDS

ARTICLE 1   THE CONTRACT DOCUMENTS
The Contract Documents consist of this Agreement, Conditions of the Contract (General, Supplementary and other Conditions), Drawings, Specifications, Addenda issued prior to execution of this Agreement, other documents listed in this Agreement and Modifications issued after execution of this Agreement, all of which form the Contract, and are as fully a part of the Contract as if attached to this Agreement or repeated herein. The Contract represents the entire and integrated agreement between the parties hereto and supersedes prior negotiations, representations or agreements, either written or oral. An enumeration of the Contract Documents, other than a Modification, appears in Article 9.

ARTICLE 2   THE WORK OF THIS CONTRACT
The Contractor shall fully execute the Work described in the Contract Documents, except as specifically indicated in the Contract Documents to be the responsibility of others.

ARTICLE 3   DATE OF COMMENCEMENT AND SUBSTANTIAL COMPLETION
§ 3.1 The date of commencement of the Work shall be the date of this Agreement unless a different date is stated below or provision is made for the date to be fixed in a notice to proceed issued by the Owner.
(Insert the date of commencement if it differs from the date of this Agreement or, if applicable, state that the date will be fixed in a notice to proceed.)

If, prior to the commencement of the Work, the Owner requires time to file mortgages and other security interests, the Owner's time requirement shall be as follows:

§ 3.2 The Contract Time shall be measured from the date of commencement.

APPENDIX A / STANDARD FORM OF AGREEMENT BETWEEN OWNER AND CONTRACTOR

§ 3.3 The Contractor shall achieve Substantial Completion of the entire Work not later than
(   ) days from the date of commencement, or as follows:
(Insert number of calendar days. Alternatively, a calendar date may be used when coordinated with the date of commencement. If appropriate, insert requirements for earlier Substantial Completion of certain portions of the Work.)

, subject to adjustments of this Contract Time as provided in the Contract Documents.
(Insert provisions, if any, for liquidated damages relating to failure to achieve Substantial Completion on time or for bonus payments for early completion of the Work.)

ARTICLE 4   CONTRACT SUM
§ 4.1 The Owner shall pay the Contractor the Contract Sum in current funds for the Contractor's performance of the Contract. The Contract Sum shall be
Dollars ($), subject to additions and deductions as provided in the Contract Documents.

§ 4.2 The Contract Sum is based upon the following alternates, if any, which are described in the Contract Documents and are hereby accepted by the Owner:
(State the numbers or other identification of accepted alternates. If the bidding or proposal documents permit the Owner to accept other alternates subsequent to the execution of this Agreement, attach a schedule of such other alternates showing the amount for each and the date when that amount expires.)

§ 4.3 Unit prices, if any:
(Identify and state the unit price; state quantity limitations, if any, to which the unit price will be applicable.)

Item	Units and Limitations	Price Per Unit

§ 4.4 Allowances included in the Contract Sum, if any:
(Identify allowance and state exclusions, if any, from the allowance price.)

Item	Price

ARTICLE 5   PAYMENTS
§ 5.1 PROGRESS PAYMENTS
§ 5.1.1 Based upon Applications for Payment submitted to the Architect by the Contractor and Certificates for Payment issued by the Architect, the Owner shall make progress payments on account of the Contract Sum to the Contractor as provided below and elsewhere in the Contract Documents.

§ 5.1.2 The period covered by each Application for Payment shall be one calendar month ending on the last day of the month, or as follows:

§ 5.1.3 Provided that an Application for Payment is received by the Architect not later than the
(   ) day of a month, the Owner shall make payment of the certified amount to the Contractor not later than the   (   ) day of the   (   ) month. If an Application for Payment is received by the Architect after the application date fixed above, payment shall be made by the Owner not later than   (   ) days after the Architect receives the Application for Payment.
(Federal, state or local laws may require payment within a certain period of time.)

§ 5.1.4 Each Application for Payment shall be based on the most recent schedule of values submitted by the Contractor in accordance with the Contract Documents. The schedule of values shall allocate the entire Contract Sum among the various portions of the Work. The schedule of values shall be prepared in such form and supported by such data to substantiate its accuracy as the Architect may require. This schedule, unless objected to by the Architect, shall be used as a basis for reviewing the Contractor's Applications for Payment.

§ 5.1.5 Applications for Payment shall show the percentage of completion of each portion of the Work as of the end of the period covered by the Application for Payment.

§ 5.1.6 Subject to other provisions of the Contract Documents, the amount of each progress payment shall be computed as follows:
- .1 Take that portion of the Contract Sum properly allocable to completed Work as determined by multiplying the percentage completion of each portion of the Work by the share of the Contract Sum allocated to that portion of the Work in the schedule of values, less retainage of percent (%). Pending final determination of cost to the Owner of changes in the Work, amounts not in dispute shall be included as provided in Section 7.3.9 of AIA Document A201™–2007, General Conditions of the Contract for Construction;
- .2 Add that portion of the Contract Sum properly allocable to materials and equipment delivered and suitably stored at the site for subsequent incorporation in the completed construction (or, if approved in advance by the Owner, suitably stored off the site at a location agreed upon in writing), less retainage of percent (%);
- .3 Subtract the aggregate of previous payments made by the Owner; and
- .4 Subtract amounts, if any, for which the Architect has withheld or nullified a Certificate for Payment as provided in Section 9.5 of AIA Document A201–2007.

§ 5.1.7 The progress payment amount determined in accordance with Section 5.1.6 shall be further modified under the following circumstances:
- .1 Add, upon Substantial Completion of the Work, a sum sufficient to increase the total payments to the full amount of the Contract Sum, less such amounts as the Architect shall determine for incomplete Work, retainage applicable to such work and unsettled claims; and
(Section 9.8.5 of AIA Document A201–2007 requires release of applicable retainage upon Substantial Completion of Work with consent of surety, if any.)
- .2 Add, if final completion of the Work is thereafter materially delayed through no fault of the Contractor, any additional amounts payable in accordance with Section 9.10.3 of AIA Document A201–2007.

§ 5.1.8 Reduction or limitation of retainage, if any, shall be as follows:
(If it is intended, prior to Substantial Completion of the entire Work, to reduce or limit the retainage resulting from the percentages inserted in Sections 5.1.6.1 and 5.1.6.2 above, and this is not explained elsewhere in the Contract Documents, insert here provisions for such reduction or limitation.)

§ 5.1.9 Except with the Owner's prior approval, the Contractor shall not make advance payments to suppliers for materials or equipment which have not been delivered and stored at the site.

§ 5.2 FINAL PAYMENT
§ 5.2.1 Final payment, constituting the entire unpaid balance of the Contract Sum, shall be made by the Owner to the Contractor when

.1 the Contractor has fully performed the Contract except for the Contractor's responsibility to correct Work as provided in Section 12.2.2 of AIA Document A201–2007, and to satisfy other requirements, if any, which extend beyond final payment; and
.2 a final Certificate for Payment has been issued by the Architect.

§ 5.2.2 The Owner's final payment to the Contractor shall be made no later than 30 days after the issuance of the Architect's final Certificate for Payment, or as follows:

ARTICLE 6   DISPUTE RESOLUTION
§ 6.1 INITIAL DECISION MAKER
The Architect will serve as Initial Decision Maker pursuant to Section 15.2 of AIA Document A201–2007, unless the parties appoint below another individual, not a party to this Agreement, to serve as Initial Decision Maker.
(If the parties mutually agree, insert the name, address and other contact information of the Initial Decision Maker, if other than the Architect.)

§ 6.2 BINDING DISPUTE RESOLUTION
For any Claim subject to, but not resolved by, mediation pursuant to Section 15.3 of AIA Document A201–2007, the method of binding dispute resolution shall be as follows:
(Check the appropriate box. If the Owner and Contractor do not select a method of binding dispute resolution below, or do not subsequently agree in writing to a binding dispute resolution method other than litigation, Claims will be resolved by litigation in a court of competent jurisdiction.)

☐ Arbitration pursuant to Section 15.4 of AIA Document A201–2007

☐ Litigation in a court of competent jurisdiction

☐ Other *(Specify)*

ARTICLE 7   TERMINATION OR SUSPENSION
§ 7.1 The Contract may be terminated by the Owner or the Contractor as provided in Article 14 of AIA Document A201–2007.

§ 7.2 The Work may be suspended by the Owner as provided in Article 14 of AIA Document A201–2007.

ARTICLE 8 MISCELLANEOUS PROVISIONS

§ 8.1 Where reference is made in this Agreement to a provision of AIA Document A201–2007 or another Contract Document, the reference refers to that provision as amended or supplemented by other provisions of the Contract Documents.

§ 8.2 Payments due and unpaid under the Contract shall bear interest from the date payment is due at the rate stated below, or in the absence thereof, at the legal rate prevailing from time to time at the place where the Project is located. *(Insert rate of interest agreed upon, if any.)*

§ 8.3 The Owner's representative:
(Name, address and other information)

§ 8.4 The Contractor's representative:
(Name, address and other information)

§ 8.5 Neither the Owner's nor the Contractor's representative shall be changed without ten days written notice to the other party.

§ 8.6 Other provisions:

ARTICLE 9 ENUMERATION OF CONTRACT DOCUMENTS

§ 9.1 The Contract Documents, except for Modifications issued after execution of this Agreement, are enumerated in the sections below.

§ 9.1.1 The Agreement is this executed AIA Document A101–2007, Standard Form of Agreement Between Owner and Contractor.

§ 9.1.2 The General Conditions are AIA Document A201–2007, General Conditions of the Contract for Construction.

§ 9.1.3 The Supplementary and other Conditions of the Contract:

Document	Title	Date	Pages

§ 9.1.4 The Specifications:
(Either list the Specifications here or refer to an exhibit attached to this Agreement.)

Section	Title	Date	Pages

§ 9.1.5 The Drawings:
(Either list the Drawings here or refer to an exhibit attached to this Agreement.)

Number	Title	Date

§ 9.1.6 The Addenda, if any:

Number	Date	Pages

Portions of Addenda relating to bidding requirements are not part of the Contract Documents unless the bidding requirements are also enumerated in this Article 9.

§ 9.1.7 Additional documents, if any, forming part of the Contract Documents:
.1 AIA Document E201™–2007, Digital Data Protocol Exhibit, if completed by the parties, or the following:

.2 Other documents, if any, listed below:
(List here any additional documents that are intended to form part of the Contract Documents. AIA Document A201–2007 provides that bidding requirements such as advertisement or invitation to bid, Instructions to Bidders, sample forms and the Contractor's bid are not part of the Contract Documents unless enumerated in this Agreement. They should be listed here only if intended to be part of the Contract Documents.)

ARTICLE 10 INSURANCE AND BONDS
The Contractor shall purchase and maintain insurance and provide bonds as set forth in Article 11 of AIA Document A201–2007.
(State bonding requirements, if any, and limits of liability for insurance required in Article 11 of AIA Document A201–2007.)

This Agreement entered into as of the day and year first written above.

_____   _____
OWNER *(Signature)*   **CONTRACTOR** *(Signature)*

_____   _____
(Printed name and title)   *(Printed name and title)*

CAUTION: You should sign an original AIA Contract Document, on which this text appears in RED. An original assures that changes will not be obscured.

APPENDIX B

GENERAL CONDITIONS OF THE CONTRACT FOR CONSTRUCTION

AIA Document A201–2007

AIA Document A201™ – 2007

General Conditions of the Contract for Construction

for the following PROJECT:
(Name and location or address)

THE OWNER:
(Name and address)

THE ARCHITECT:
(Name and address)

This document has important legal consequences. Consultation with an attorney is encouraged with respect to its completion or modification.

TABLE OF ARTICLES

1 GENERAL PROVISIONS

2 OWNER

3 CONTRACTOR

4 ARCHITECT

5 SUBCONTRACTORS

6 CONSTRUCTION BY OWNER OR BY SEPARATE CONTRACTORS

7 CHANGES IN THE WORK

8 TIME

9 PAYMENTS AND COMPLETION

10 PROTECTION OF PERSONS AND PROPERTY

11 INSURANCE AND BONDS

12 UNCOVERING AND CORRECTION OF WORK

13 MISCELLANEOUS PROVISIONS

14 TERMINATION OR SUSPENSION OF THE CONTRACT

15 CLAIMS AND DISPUTES

AIA Document A201™ – 2007. Copyright © 1888, 1911, 1915, 1918, 1925, 1937, 1951, 1958, 1961, 1963, 1966, 1970, 1976, 1987, 1997 and 2007 by The American Institute of Architects. **All rights reserved.** WARNING: This AIA® Document is protected by U.S. Copyright Law and International Treaties. Unauthorized reproduction or distribution of this AIA® Document, or any portion of it, may result in severe civil and criminal penalties, and will be prosecuted to the maximum extent possible under the law. Purchasers are not permitted to reproduce this document. To report copyright violations of AIA Contract Documents, e-mail The American Institute of Architects' legal counsel, copyright@aia.org.

Reproduced with permission of The American Institute of Architects, 1735 New York Avenue, NW, Washington, DC 20006.

INDEX
(Numbers and Topics in Bold are Section Headings)

Acceptance of Nonconforming Work
9.6.6, 9.9.3, **12.3**
Acceptance of Work
9.6.6, 9.8.2, 9.9.3, 9.10.1, 9.10.3, 12.3
Access to Work
3.16, 6.2.1, 12.1
Accident Prevention
10
Acts and Omissions
3.2, 3.3.2, 3.12.8, 3.18, 4.2.3, 8.3.1, 9.5.1, 10.2.5, 10.2.8, 13.4.2, 13.7.1, 14.1, 15.2
Addenda
1.1.1, 3.11.1
Additional Costs, Claims for
3.7.4, 3.7.5, 6.1.1, 7.3.7.5, 10.3, 15.1.4
Additional Inspections and Testing
9.4.2, 9.8.3, 12.2.1, **13.5**
Additional Insured
11.1.4
Additional Time, Claims for
3.2.4, 3.7.4, 3.7.5, 3.10.2, 8.3.2, **15.1.5**
Administration of the Contract
3.1.3, **4.2**, 9.4, 9.5
Advertisement or Invitation to Bid
1.1.1
Aesthetic Effect
4.2.13
Allowances
3.8, 7.3.8
All-risk Insurance
11.3.1, 11.3.1.1
Applications for Payment
4.2.5, 7.3.9, 9.2, **9.3**, 9.4, 9.5.1, 9.6.3, 9.7.1, 9.10, 11.1.3
Approvals
2.1.1, 2.2.2, 2.4, 3.1.3, 3.10.2, 3.12.8, 3.12.9, 3.12.10, 4.2.7, 9.3.2, 13.5.1
Arbitration
8.3.1, 11.3.10, 13.1.1, 15.3.2, **15.4**
ARCHITECT
4
Architect, Definition of
4.1.1
Architect, Extent of Authority
2.4.1, 3.12.7, 4.1, 4.2, 5.2, 6.3.1, 7.1.2, 7.3.7, 7.4, 9.2.1, 9.3.1, 9.4, 9.5, 9.6.3, 9.8, 9.10.1, 9.10.3, 12.1, 12.2.1, 13.5.1, 13.5.2, 14.2.2, 14.2.4, 15.1.3, 15.2.1
Architect, Limitations of Authority and Responsibility
2.1.1, 3.12.4, 3.12.8, 3.12.10, 4.1.2, 4.2.1, 4.2.2, 4.2.3, 4.2.6, 4.2.7, 4.2.10, 4.2.12, 4.2.13, 5.2.1, 7.4.1, 9.4.2, 9.5.3, 9.6.4, 15.1.3, 15.2
Architect's Additional Services and Expenses
2.4.1, 11.3.1.1, 12.2.1, 13.5.2, 13.5.3, 14.2.4

Architect's Administration of the Contract
3.1.3, 4.2, 3.7.4, 15.2, 9.4.1, 9.5
Architect's Approvals
2.4.1, 3.1.3, 3.5.1, 3.10.2, 4.2.7
Architect's Authority to Reject Work
3.5.1, 4.2.6, 12.1.2, 12.2.1
Architect's Copyright
1.1.7, 1.5
Architect's Decisions
3.7.4, 4.2.6, 4.2.7, 4.2.11, 4.2.12, 4.2.13, 4.2.14, 6.3.1, 7.3.7, 7.3.9, 8.1.3, 8.3.1, 9.2.1, 9.4.1, 9.5, 9.8.4, 9.9.1, 13.5.2, 15.2, 15.3
Architect's Inspections
3.7.4, 4.2.2, 4.2.9, 9.4.2, 9.8.3, 9.9.2, 9.10.1, 13.5
Architect's Instructions
3.2.4, 3.3.1, 4.2.6, 4.2.7, 13.5.2
Architect's Interpretations
4.2.11, 4.2.12
Architect's Project Representative
4.2.10
Architect's Relationship with Contractor
1.1.2, 1.5, 3.1.3, 3.2.2, 3.2.3, 3.2.4, 3.3.1, 3.4.2, 3.5.1, 3.7.4, 3.7.5, 3.9.2, 3.9.3, 3.10, 3.11, 3.12, 3.16, 3.18, 4.1.2, 4.1.3, 4.2, 5.2, 6.2.2, 7, 8.3.1, 9.2, 9.3, 9.4, 9.5, 9.7, 9.8, 9.9, 10.2.6, 10.3, 11.3.7, 12, 13.4.2, 13.5, 15.2
Architect's Relationship with Subcontractors
1.1.2, 4.2.3, 4.2.4, 4.2.6, 9.6.3, 9.6.4, 11.3.7
Architect's Representations
9.4.2, 9.5.1, 9.10.1
Architect's Site Visits
3.7.4, 4.2.2, 4.2.9, 9.4.2, 9.5.1, 9.9.2, 9.10.1, 13.5
Asbestos
10.3.1
Attorneys' Fees
3.18.1, 9.10.2, 10.3.3
Award of Separate Contracts
6.1.1, 6.1.2
Award of Subcontracts and Other Contracts for Portions of the Work
5.2
Basic Definitions
1.1
Bidding Requirements
1.1.1, 5.2.1, 11.4.1
Binding Dispute Resolution
9.7.1, 11.3.9, 11.3.10, 13.1.1, 15.2.5, 15.2.6.1, 15.3.1, 15.3.2, 15.4.1
Boiler and Machinery Insurance
11.3.2
Bonds, Lien
7.3.7.4, 9.10.2, 9.10.3
Bonds, Performance, and Payment
7.3.7.4, 9.6.7, 9.10.3, 11.3.9, **11.4**
Building Permit
3.7.1
Capitalization
1.3

Certificate of Substantial Completion
9.8.3, 9.8.4, 9.8.5
Certificates for Payment
4.2.1, 4.2.5, 4.2.9, 9.3.3, **9.4**, 9.5, 9.6.1, 9.6.6, 9.7.1,
9.10.1, 9.10.3, 14.1.1.3, 14.2.4, 15.1.3
Certificates of Inspection, Testing or Approval
13.5.4
Certificates of Insurance
9.10.2, 11.1.3
Change Orders
1.1.1, 2.4.1, 3.4.2, 3.7.4, 3.8.2.3, 3.11.1, 3.12.8, 4.2.8,
5.2.3, 7.1.2, 7.1.3, **7.2**, 7.3.2, 7.3.6, 7.3.9, 7.3.10,
8.3.1, 9.3.1.1, 9.10.3, 10.3.2, 11.3.1.2, 11.3.4, 11.3.9,
12.1.2, 15.1.3
Change Orders, Definition of
7.2.1
CHANGES IN THE WORK
2.2.1, 3.11, 4.2.8, **7**, 7.2.1, 7.3.1, 7.4, 7.4.1, 8.3.1,
9.3.1.1, 11.3.9
Claims, Definition of
15.1.1
CLAIMS AND DISPUTES
3.2.4, 6.1.1, 6.3.1, 7.3.9, 9.3.3, 9.10.4, 10.3.3, **15**,
15.4
Claims and Timely Assertion of Claims
15.4.1
Claims for Additional Cost
3.2.4, 3.7.4, 6.1.1, 7.3.9, 10.3.2, **15.1.4**
Claims for Additional Time
3.2.4, 3.7.4.6.1.1, 8.3.2, 10.3.2, **15.1.5**
Concealed or Unknown Conditions, Claims for
3.7.4
Claims for Damages
3.2.4, 3.18, 6.1.1, 8.3.3, 9.5.1, 9.6.7, 10.3.3, 11.1.1,
11.3.5, 11.3.7, 14.1.3, 14.2.4, 15.1.6
Claims Subject to Arbitration
15.3.1, 15.4.1
Cleaning Up
3.15, 6.3
Commencement of the Work, Conditions Relating to
2.2.1, 3.2.2, 3.4.1, 3.7.1, 3.10.1, 3.12.6, 5.2.1, 5.2.3,
6.2.2, 8.1.2, 8.2.2, 8.3.1, 11.1, 11.3.1, 11.3.6, 11.4.1,
15.1.4
Commencement of the Work, Definition of
8.1.2
**Communications Facilitating Contract
Administration**
3.9.1, **4.2.4**
Completion, Conditions Relating to
3.4.1, 3.11, 3.15, 4.2.2, 4.2.9, 8.2, 9.4.2, 9.8, 9.9.1,
9.10, 12.2, 13.7, 14.1.2
COMPLETION, PAYMENTS AND
9
Completion, Substantial
4.2.9, 8.1.1, 8.1.3, 8.2.3, 9.4.2, 9.8, 9.9.1, 9.10.3,
12.2, 13.7

Compliance with Laws
1.6.1, 3.2.3, 3.6, 3.7, 3.12.10, 3.13, 4.1.1, 9.6.4,
10.2.2, 11.1, 11.3, 13.1, 13.4, 13.5.1, 13.5.2, 13.6,
14.1.1, 14.2.1.3, 15.2.8, 15.4.2, 15.4.3
Concealed or Unknown Conditions
3.7.4, 4.2.8, 8.3.1, 10.3
Conditions of the Contract
1.1.1, 6.1.1, 6.1.4
Consent, Written
3.4.2, 3.7.4, 3.12.8, 3.14.2, 4.1.2, 9.3.2, 9.8.5, 9.9.1,
9.10.2, 9.10.3, 11.3.1, 13.2, 13.4.2, 15.4.4.2
Consolidation or Joinder
15.4.4
**CONSTRUCTION BY OWNER OR BY
SEPARATE CONTRACTORS**
1.1.4, **6**
Construction Change Directive, Definition of
7.3.1
Construction Change Directives
1.1.1, 3.4.2, 3.12.8, 4.2.8, 7.1.1, 7.1.2, 7.1.3, **7.3**,
9.3.1.1
Construction Schedules, Contractor's
3.10, 3.12.1, 3.12.2, 6.1.3, 15.1.5.2
Contingent Assignment of Subcontracts
5.4, 14.2.2.2
Continuing Contract Performance
15.1.3
Contract, Definition of
1.1.2
**CONTRACT, TERMINATION OR
SUSPENSION OF THE**
5.4.1.1, 11.3.9, **14**
Contract Administration
3.1.3, 4, 9.4, 9.5
Contract Award and Execution, Conditions Relating to
3.7.1, 3.10, 5.2, 6.1, 11.1.3, 11.3.6, 11.4.1
Contract Documents, The
1.1.1
Contract Documents, Copies Furnished and Use of
1.5.2, 2.2.5, 5.3
Contract Documents, Definition of
1.1.1
Contract Sum
3.7.4, 3.8, 5.2.3, 7.2, 7.3, 7.4, **9.1**, 9.4.2, 9.5.1.4,
9.6.7, 9.7, 10.3.2, 11.3.1, 14.2.4, 14.3.2, 15.1.4,
15.2.5
Contract Sum, Definition of
9.1
Contract Time
3.7.4, 3.7.5, 3.10.2, 5.2.3, 7.2.1.3, 7.3.1, 7.3.5, 7.4,
8.1.1, 8.2.1, 8.3.1, 9.5.1, 9.7.1, 10.3.2, 12.1.1, 14.3.2,
15.1.5.1, 15.2.5
Contract Time, Definition of
8.1.1
CONTRACTOR
3
Contractor, Definition of
3.1, 6.1.2

Contractor's Construction Schedules
3.10, 3.12.1, 3.12.2, 6.1.3, 15.1.5.2
Contractor's Employees
3.3.2, 3.4.3, 3.8.1, 3.9, 3.18.2, 4.2.3, 4.2.6, 10.2, 10.3, 11.1.1, 11.3.7, 14.1, 14.2.1.1
Contractor's Liability Insurance
11.1
Contractor's Relationship with Separate Contractors and Owner's Forces
3.12.5, 3.14.2, 4.2.4, 6, 11.3.7, 12.1.2, 12.2.4
Contractor's Relationship with Subcontractors
1.2.2, 3.3.2, 3.18.1, 3.18.2, 5, 9.6.2, 9.6.7, 9.10.2, 11.3.1.2, 11.3.7, 11.3.8
Contractor's Relationship with the Architect
1.1.2, 1.5, 3.1.3, 3.2.2, 3.2.3, 3.2.4, 3.3.1, 3.4.2, 3.5.1, 3.7.4, 3.10, 3.11, 3.12, 3.16, 3.18, 4.1.3, 4.2, 5.2, 6.2.2, 7, 8.3.1, 9.2, 9.3, 9.4, 9.5, 9.7, 9.8, 9.9, 10.2.6, 10.3, 11.3.7, 12, 13.5, 15.1.2, 15.2.1
Contractor's Representations
3.2.1, 3.2.2, 3.5.1, 3.12.6, 6.2.2, 8.2.1, 9.3.3, 9.8.2
Contractor's Responsibility for Those Performing the Work
3.3.2, 3.18, 5.3.1, 6.1.3, 6.2, 9.5.1, 10.2.8
Contractor's Review of Contract Documents
3.2
Contractor's Right to Stop the Work
9.7
Contractor's Right to Terminate the Contract
14.1, 15.1.6
Contractor's Submittals
3.10, 3.11, 3.12.4, 4.2.7, 5.2.1, 5.2.3, 9.2, 9.3, 9.8.2, 9.8.3, 9.9.1, 9.10.2, 9.10.3, 11.1.3, 11.4.2
Contractor's Superintendent
3.9, 10.2.6
Contractor's Supervision and Construction Procedures
1.2.2, 3.3, 3.4, 3.12.10, 4.2.2, 4.2.7, 6.1.3, 6.2.4, 7.1.3, 7.3.5, 7.3.7, 8.2, 10, 12, 14, 15.1.3
Contractual Liability Insurance
11.1.1.8, 11.2
Coordination and Correlation
1.2, 3.2.1, 3.3.1, 3.10, 3.12.6, 6.1.3, 6.2.1
Copies Furnished of Drawings and Specifications
1.5, 2.2.5, 3.11
Copyrights
1.5, **3.17**
Correction of Work
2.3, 2.4, 3.7.3, 9.4.2, 9.8.2, 9.8.3, 9.9.1, 12.1.2, **12.2**
Correlation and Intent of the Contract Documents
1.2
Cost, Definition of
7.3.7
Costs
2.4.1, 3.2.4, 3.7.3, 3.8.2, 3.15.2, 5.4.2, 6.1.1, 6.2.3, 7.3.3.3, 7.3.7, 7.3.8, 7.3.9, 9.10.2, 10.3.2, 10.3.6, 11.3, 12.1.2, 12.2.1, 12.2.4, 13.5, 14
Cutting and Patching
3.14, 6.2.5

Damage to Construction of Owner or Separate Contractors
3.14.2, 6.2.4, 10.2.1.2, 10.2.5, 10.4, 11.1.1, 11.3, 12.2.4
Damage to the Work
3.14.2, 9.9.1, 10.2.1.2, 10.2.5, 10.4.1, 11.3.1, 12.2.4
Damages, Claims for
3.2.4, 3.18, 6.1.1, 8.3.3, 9.5.1, 9.6.7, 10.3.3, 11.1.1, 11.3.5, 11.3.7, 14.1.3, 14.2.4, 15.1.6
Damages for Delay
6.1.1, 8.3.3, 9.5.1.6, 9.7, 10.3.2
Date of Commencement of the Work, Definition of
8.1.2
Date of Substantial Completion, Definition of
8.1.3
Day, Definition of
8.1.4
Decisions of the Architect
3.7.4, 4.2.6, 4.2.7, 4.2.11, 4.2.12, 4.2.13, 15.2, 6.3, 7.3.7, 7.3.9, 8.1.3, 8.3.1, 9.2.1, 9.4, 9.5.1, 9.8.4, 9.9.1, 13.5.2, 14.2.2, 14.2.4, 15.1, 15.2
Decisions to Withhold Certification
9.4.1, 9.5, 9.7, 14.1.1.3
Defective or Nonconforming Work, Acceptance, Rejection and Correction of
2.3.1, 2.4.1, 3.5.1, 4.2.6, 6.2.5, 9.5.1, 9.5.2, 9.6.6, 9.8.2, 9.9.3, 9.10.4, 12.2.1
Defective Work, Definition of
3.5.1
Definitions
1.1, 2.1.1, 3.1.1, 3.5.1, 3.12.1, 3.12.2, 3.12.3, 4.1.1, 15.1.1, 5.1, 6.1.2, 7.2.1, 7.3.1, 8.1, 9.1, 9.8.1
Delays and Extensions of Time
3.2., 3.7.4, 5.2.3, 7.2.1, 7.3.1, 7.4.1, **8.3**, 9.5.1, 9.7.1, 10.3.2, 10.4.1, 14.3.2, 15.1.5, 15.2.5
Disputes
6.3.1, 7.3.9, 15.1, 15.2
Documents and Samples at the Site
3.11
Drawings, Definition of
1.1.5
Drawings and Specifications, Use and Ownership of
3.11
Effective Date of Insurance
8.2.2, 11.1.2
Emergencies
10.4, 14.1.1.2, 15.1.4
Employees, Contractor's
3.3.2, 3.4.3, 3.8.1, 3.9, 3.18.2, 4.2.3, 4.2.6, 10.2, 10.3.3, 11.1.1, 11.3.7, 14.1, 14.2.1.1
Equipment, Labor, Materials or
1.1.3, 1.1.6, 3.4, 3.5.1, 3.8.2, 3.8.3, 3.12, 3.13.1, 3.15.1, 4.2.6, 4.2.7, 5.2.1, 6.2.1, 7.3.7, 9.3.2, 9.3.3, 9.5.1.3, 9.10.2, 10.2.1, 10.2.4, 14.2.1.1, 14.2.1.2
Execution and Progress of the Work
1.1.3, 1.2.1, 1.2.2, 2.2.3, 2.2.5, 3.1, 3.3.1, 3.4.1, 3.5.1, 3.7.1, 3.10.1, 3.12, 3.14, 4.2, 6.2.2, 7.1.3, 7.3.5, 8.2, 9.5.1, 9.9.1, 10.2, 10.3, 12.2, 14, 14.3.1, 15.1.3

Extensions of Time
3.2.4, 3.7.4, 5.2.3, 7.2.1, 7.3, 7.4.1, 9.5.1, 9.7.1, 10.3.2, 10.4.1, 14.3, 15.1.5, 15.2.5
Failure of Payment
9.5.1.3, **9.7**, 9.10.2, 13.6, 14.1.1.3, 14.2.1.2
Faulty Work
(*See* Defective or Nonconforming Work)
Final Completion and Final Payment
4.2.1, 4.2.9, 9.8.2, **9.10**, 11.1.2, 11.1.3, 11.3.1, 11.3.5, 12.3.1, 14.2.4, 14.4.3
Financial Arrangements, Owner's
2.2.1, 13.2.2, 14.1.1.4
Fire and Extended Coverage Insurance
11.3.1.1
GENERAL PROVISIONS
1
Governing Law
13.1
Guarantees (*See* Warranty)
Hazardous Materials
10.2.4, **10.3**
Identification of Subcontractors and Suppliers
5.2.1
Indemnification
3.17.1, **3.18**, 9.10.2, 10.3.3, 10.3.5, 10.3.6, 11.3.1.2, 11.3.7
Information and Services Required of the Owner
2.1.2, **2.2**, 3.2.2, 3.12.4, 3.12.10, 6.1.3, 6.1.4, 6.2.5, 9.6.1, 9.6.4, 9.9.2, 9.10.3, 10.3.3, 11.2.1, 11.4, 13.5.1, 13.5.2, 14.1.1.4, 14.1.4, 15.1.3
Initial Decision
15.2
Initial Decision Maker, Definition of
1.1.8
Initial Decision Maker, Decisions
14.2.2, 14.2.4, 15.2.1, 15.2.2, 15.2.3, 15.2.4, 15.2.5
Initial Decision Maker, Extent of Authority
14.2.2, 14.2.4, 15.1.3, 15.2.1, 15.2.2, 15.2.3, 15.2.4, 15.2.5
Injury or Damage to Person or Property
10.2.8, 10.4.1
Inspections
3.1.3, 3.3.3, 3.7.1, 4.2.2, 4.2.6, 4.2.9, 9.4.2, 9.8.3, 9.9.2, 9.10.1, 12.2.1, 13.5
Instructions to Bidders
1.1.1
Instructions to the Contractor
3.2.4, 3.3.1, 3.8.1, 5.2.1, 7, 8.2.2, 12, 13.5.2
Instruments of Service, Definition of
1.1.7
Insurance
3.18.1, 6.1.1, 7.3.7, 9.3.2, 9.8.4, 9.9.1, 9.10.2, **11**
Insurance, Boiler and Machinery
11.3.2
Insurance, Contractor's Liability
11.1
Insurance, Effective Date of
8.2.2, 11.1.2

Insurance, Loss of Use
11.3.3
Insurance, Owner's Liability
11.2
Insurance, Property
10.2.5, **11.3**
Insurance, Stored Materials
9.3.2, 11.4.1.4
INSURANCE AND BONDS
11
Insurance Companies, Consent to Partial Occupancy
9.9.1, 11.4.1.5
Insurance Companies, Settlement with
11.4.10
Intent of the Contract Documents
1.2.1, 4.2.7, 4.2.12, 4.2.13, 7.4
Interest
13.6
Interpretation
1.2.3, 1.4, 4.1.1, 5.1, 6.1.2, 15.1.1
Interpretations, Written
4.2.11, 4.2.12, 15.1.4
Judgment on Final Award
15.4.2
Labor and Materials, Equipment
1.1.3, 1.1.6, **3.4**, 3.5.1, 3.8.2, 3.8.3, 3.12, 3.13, 3.15.1, 4.2.6, 4.2.7, 5.2.1, 6.2.1, 7.3.7, 9.3.2, 9.3.3, 9.5.1.3, 9.10.2, 10.2.1, 10.2.4, 14.2.1.1, 14.2.1.2
Labor Disputes
8.3.1
Laws and Regulations
1.5, 3.2.3, 3.6, 3.7, 3.12.10, 3.13.1, 4.1.1, 9.6.4, 9.9.1, 10.2.2, 11.1.1, 11.3, 13.1.1, 13.4, 13.5.1, 13.5.2, 13.6.1, 14, 15.2.8, 15.4
Liens
2.1.2, 9.3.3, 9.10.2, 9.10.4, 15.2.8
Limitations, Statutes of
12.2.5, 13.7, 15.4.1.1
Limitations of Liability
2.3.1, 3.2.2, 3.5.1, 3.12.10, 3.17.1, 3.18.1, 4.2.6, 4.2.7, 4.2.12, 6.2.2, 9.4.2, 9.6.4, 9.6.7, 10.2.5, 10.3.3, 11.1.2, 11.2.1, 11.3.7, 12.2.5, 13.4.2
Limitations of Time
2.1.2, 2.2, 2.4, 3.2.2, 3.10, 3.11, 3.12.5, 3.15.1, 4.2.7, 5.2, 5.3.1, 5.4.1, 6.2.4, 7.3, 7.4, 8.2, 9.2.1, 9.3.1, 9.3.3, 9.4.1, 9.5, 9.6, 9.7.1, 9.8, 9.9, 9.10, 11.1.3, 11.3.1.5, 11.3.6, 11.3.10, 12.2, 13.5, 13.7, 14, 15
Loss of Use Insurance
11.3.3
Material Suppliers
1.5, 3.12.1, 4.2.4, 4.2.6, 5.2.1, 9.3, 9.4.2, 9.6, 9.10.5
Materials, Hazardous
10.2.4, **10.3**
Materials, Labor, Equipment and
1.1.3, 1.1.6, 1.5.1, 3.4.1, 3.5.1, 3.8.2, 3.8.3, 3.12, 3.13.1, 3.15.1, 4.2.6, 4.2.7, 5.2.1, 6.2.1, 7.3.7, 9.3.2, 9.3.3, 9.5.1.3, 9.10.2, 10.2.1.2, 10.2.4, 14.2.1.1, 14.2.1.2

Means, Methods, Techniques, Sequences and
Procedures of Construction
3.3.1, 3.12.10, 4.2.2, 4.2.7, 9.4.2
Mechanic's Lien
2.1.2, 15.2.8
Mediation
8.3.1, 10.3.5, 10.3.6, 15.2.1, 15.2.5, 15.2.6, **15.3**, 15.4.1
Minor Changes in the Work
1.1.1, 3.12.8, 4.2.8, 7.1, **7.4**
MISCELLANEOUS PROVISIONS
13
Modifications, Definition of
1.1.1
Modifications to the Contract
1.1.1, 1.1.2, 3.11, 4.1.2, 4.2.1, 5.2.3, 7, 8.3.1, 9.7.1, 10.3.2, 11.3.1
Mutual Responsibility
6.2
Nonconforming Work, Acceptance of
9.6.6, 9.9.3, **12.3**
Nonconforming Work, Rejection and Correction of
2.3.1, 2.4.1, 3.5.1, 4.2.6, 6.2.4, 9.5.1, 9.8.2, 9.9.3, 9.10.4, 12.2.1
Notice
2.2.1, 2.3.1, 2.4.1, 3.2.4, 3.3.1, 3.7.2, 3.12.9, 5.2.1, 9.7.1, 9.10, 10.2.2, 11.1.3, 11.4.6, 12.2.2.1, 13.3, 13.5.1, 13.5.2, 14.1, 14.2, 15.2.8, 15.4.1
Notice, Written
2.3.1, 2.4.1, 3.3.1, 3.9.2, 3.12.9, 3.12.10, 5.2.1, 9.7.1, 9.10, 10.2.2, 10.3, 11.1.3, 11.3.6, 12.2.2.1, **13.3**, 14, 15.2.8, 15.4.1
Notice of Claims
3.7.4, 4.5, 10.2.8, **15.1.2**, 15.4
Notice of Testing and Inspections
13.5.1, 13.5.2
Observations, Contractor's
3.2, 3.7.4
Occupancy
2.2.2, 9.6.6, 9.8, 11.3.1.5
Orders, Written
1.1.1, 2.3, 3.9.2, 7, 8.2.2, 11.3.9, 12.1, 12.2.2.1, 13.5.2, 14.3.1
OWNER
2
Owner, Definition of
2.1.1
Owner, Information and Services Required of the
2.1.2, **2.2**, 3.2.2, 3.12.10, 6.1.3, 6.1.4, 6.2.5, 9.3.2, 9.6.1, 9.6.4, 9.9.2, 9.10.3, 10.3.3, 11.2.1, 11.3, 13.5.1, 13.5.2, 14.1.1.4, 14.1.4, 15.1.3
Owner's Authority
1.5, 2.1.1, 2.3.1, 2.4.1, 3.4.2, 3.8.1, 3.12.10, 3.14.2, 4.1.2, 4.1.3, 4.2.4, 4.2.9, 5.2.1, 5.2.4, 5.4.1, 6.1, 6.3.1, 7.2.1, 7.3.1, 8.2.2, 8.3.1, 9.3.1, 9.3.2, 9.5.1, 9.6.4, 9.9.1, 9.10.2, 10.3.2, 11.1.3, 11.3.3, 11.3.10, 12.2.2, 12.3.1, 13.2.2, 14.3, 14.4, 15.2.7

Owner's Financial Capability
2.2.1, 13.2.2, 14.1.1.4
Owner's Liability Insurance
11.2
Owner's Loss of Use Insurance
11.3.3
Owner's Relationship with Subcontractors
1.1.2, 5.2, 5.3, 5.4, 9.6.4, 9.10.2, 14.2.2
Owner's Right to Carry Out the Work
2.4, 14.2.2
Owner's Right to Clean Up
6.3
Owner's Right to Perform Construction and to Award Separate Contracts
6.1
Owner's Right to Stop the Work
2.3
Owner's Right to Suspend the Work
14.3
Owner's Right to Terminate the Contract
14.2
Ownership and Use of Drawings, Specifications and Other Instruments of Service
1.1.1, 1.1.6, 1.1.7, **1.5**, 2.2.5, 3.2.2, 3.11.1, 3.17.1, 4.2.12, 5.3.1
Partial Occupancy or Use
9.6.6, **9.9**, 11.3.1.5
Patching, Cutting and
3.14, 6.2.5
Patents
3.17
Payment, Applications for
4.2.5, 7.3.9, 9.2.1, **9.3**, 9.4, 9.5, 9.6.3, 9.7.1, 9.8.5, 9.10.1, 14.2.3, 14.2.4, 14.4.3
Payment, Certificates for
4.2.5, 4.2.9, 9.3.3, **9.4**, 9.5, 9.6.1, 9.6.6, 9.7.1, 9.10.1, 9.10.3, 13.7, 14.1.1.3, 14.2.4
Payment, Failure of
9.5.1.3, **9.7**, 9.10.2, 13.6, 14.1.1.3, 14.2.1.2
Payment, Final
4.2.1, 4.2.9, 9.8.2, 9.10, 11.1.2, 11.1.3, 11.4.1, 11.4.5, 12.3.1, 13.7, 14.2.4, 14.4.3
Payment Bond, Performance Bond and
7.3.7.4, 9.6.7, 9.10.3, 11.4.9, **11.4**
Payments, Progress
9.3, **9.6**, 9.8.5, 9.10.3, 13.6, 14.2.3, 15.1.3
PAYMENTS AND COMPLETION
9
Payments to Subcontractors
5.4.2, 9.5.1.3, 9.6.2, 9.6.3, 9.6.4, 9.6.7, 11.4.8, 14.2.1.2
PCB
10.3.1
Performance Bond and Payment Bond
7.3.7.4, 9.6.7, 9.10.3, 11.4.9, **11.4**
Permits, Fees, Notices and Compliance with Laws
2.2.2, **3.7**, 3.13, 7.3.7.4, 10.2.2

PERSONS AND PROPERTY, PROTECTION OF
10
Polychlorinated Biphenyl
10.3.1
Product Data, Definition of
3.12.2
Product Data and Samples, Shop Drawings
3.11, **3.12**, 4.2.7
Progress and Completion
4.2.2, **8.2**, 9.8, 9.9.1, 14.1.4, 15.1.3
Progress Payments
9.3, **9.6**, 9.8.5, 9.10.3, 13.6, 14.2.3, 15.1.3
Project, Definition of the
1.1.4
Project Representatives
4.2.10
Property Insurance
10.2.5, **11.3**
PROTECTION OF PERSONS AND PROPERTY
10
Regulations and Laws
1.5, 3.2.3, 3.6, 3.7, 3.12.10, 3.13, 4.1.1, 9.6.4, 9.9.1, 10.2.2, 11.1, 11.4, 13.1, 13.4, 13.5.1, 13.5.2, 13.6, 14, 15.2.8, 15.4
Rejection of Work
3.5.1, 4.2.6, 12.2.1
Releases and Waivers of Liens
9.10.2
Representations
3.2.1, 3.5.1, 3.12.6, 6.2.2, 8.2.1, 9.3.3, 9.4.2, 9.5.1, 9.8.2, 9.10.1
Representatives
2.1.1, 3.1.1, 3.9, 4.1.1, 4.2.1, 4.2.2, 4.2.10, 5.1.1, 5.1.2, 13.2.1
Responsibility for Those Performing the Work
3.3.2, 3.18, 4.2.3, 5.3.1, 6.1.3, 6.2, 6.3, 9.5.1, 10
Retainage
9.3.1, 9.6.2, 9.8.5, 9.9.1, 9.10.2, 9.10.3
Review of Contract Documents and Field Conditions by Contractor
3.2, 3.12.7, 6.1.3
Review of Contractor's Submittals by Owner and Architect
3.10.1, 3.10.2, 3.11, 3.12, 4.2, 5.2, 6.1.3, 9.2, 9.8.2
Review of Shop Drawings, Product Data and Samples by Contractor
3.12
Rights and Remedies
1.1.2, 2.3, 2.4, 3.5.1, 3.7.4, 3.15.2, 4.2.6, 4.5, 5.3, 5.4, 6.1, 6.3, 7.3.1, 8.3, 9.5.1, 9.7, 10.2.5, 10.3, 12.2.2, 12.2.4, **13.4**, 14, 15.4
Royalties, Patents and Copyrights
3.17
Rules and Notices for Arbitration
15.4.1
Safety of Persons and Property
10.2, 10.4

Safety Precautions and Programs
3.3.1, 4.2.2, 4.2.7, 5.3.1, **10.1**, 10.2, 10.4
Samples, Definition of
3.12.3
Samples, Shop Drawings, Product Data and
3.11, **3.12**, 4.2.7
Samples at the Site, Documents and
3.11
Schedule of Values
9.2, 9.3.1
Schedules, Construction
1.4.1.2, 3.10, 3.12.1, 3.12.2, 6.1.3, 15.1.5.2
Separate Contracts and Contractors
1.1.4, 3.12.5, 3.14.2, 4.2.4, 4.2.7, 6, 8.3.1, 11.4.7, 12.1.2
Shop Drawings, Definition of
3.12.1
Shop Drawings, Product Data and Samples
3.11, **3.12**, 4.2.7
Site, Use of
3.13, 6.1.1, 6.2.1
Site Inspections
3.2.2, 3.3.3, 3.7.1, 3.7.4, 4.2, 9.4.2, 9.10.1, 13.5
Site Visits, Architect's
3.7.4, 4.2.2, 4.2.9, 9.4.2, 9.5.1, 9.9.2, 9.10.1, 13.5
Special Inspections and Testing
4.2.6, 12.2.1, 13.5
Specifications, Definition of the
1.1.6
Specifications, The
1.1.1, **1.1.6**, 1.2.2, 1.5, 3.11, 3.12.10, 3.17, 4.2.14
Statute of Limitations
13.7, 15.4.1.1
Stopping the Work
2.3, 9.7, 10.3, 14.1
Stored Materials
6.2.1, 9.3.2, 10.2.1.2, 10.2.4, 11.4.1.4
Subcontractor, Definition of
5.1.1
SUBCONTRACTORS
5
Subcontractors, Work by
1.2.2, 3.3.2, 3.12.1, 4.2.3, 5.2.3, 5.3, 5.4, 9.3.1.2, 9.6.7
Subcontractual Relations
5.3, 5.4, 9.3.1.2, 9.6, 9.10, 10.2.1, 11.4.7, 11.4.8, 14.1, 14.2.1
Submittals
3.10, 3.11, 3.12, 4.2.7, 5.2.1, 5.2.3, 7.3.7, 9.2, 9.3, 9.8, 9.9.1, 9.10.2, 9.10.3, 11.1.3
Submittal Schedule
3.10.2, 3.12.5, 4.2.7
Subrogation, Waivers of
6.1.1, 11.4.5, **11.3.7**
Substantial Completion
4.2.9, 8.1.1, 8.1.3, 8.2.3, 9.4.2, **9.8**, 9.9.1, 9.10.3, 12.2, 13.7

Substantial Completion, Definition of
9.8.1
Substitution of Subcontractors
5.2.3, 5.2.4
Substitution of Architect
4.1.3
Substitutions of Materials
3.4.2, 3.5.1, 7.3.8
Sub-subcontractor, Definition of
5.1.2
Subsurface Conditions
3.7.4
Successors and Assigns
13.2
Superintendent
3.9, 10.2.6
Supervision and Construction Procedures
1.2.2, **3.3**, 3.4, 3.12.10, 4.2.2, 4.2.7, 6.1.3, 6.2.4, 7.1.3, 7.3.7, 8.2, 8.3.1, 9.4.2, 10, 12, 14, 15.1.3
Surety
5.4.1.2, 9.8.5, 9.10.2, 9.10.3, 14.2.2, 15.2.7
Surety, Consent of
9.10.2, 9.10.3
Surveys
2.2.3
Suspension by the Owner for Convenience
14.3
Suspension of the Work
5.4.2, 14.3
Suspension or Termination of the Contract
5.4.1.1, 11.4.9, 14
Taxes
3.6, 3.8.2.1, 7.3.7.4
Termination by the Contractor
14.1, 15.1.6
Termination by the Owner for Cause
5.4.1.1, **14.2**, 15.1.6
Termination by the Owner for Convenience
14.4
Termination of the Architect
4.1.3
Termination of the Contractor
14.2.2
TERMINATION OR SUSPENSION OF THE CONTRACT
14
Tests and Inspections
3.1.3, 3.3.3, 4.2.2, 4.2.6, 4.2.9, 9.4.2, 9.8.3, 9.9.2, 9.10.1, 10.3.2, 11.4.1.1, 12.2.1, **13.5**
TIME
8
Time, Delays and Extensions of
3.2.4, 3.7.4, 5.2.3, 7.2.1, 7.3.1, 7.4.1, **8.3**, 9.5.1, 9.7.1, 10.3.2, 10.4.1, 14.3.2, 15.1.5, 15.2.5

Time Limits
2.1.2, 2.2, 2.4, 3.2.2, 3.10, 3.11, 3.12.5, 3.15.1, 4.2, 4.4, 4.5, 5.2, 5.3, 5.4, 6.2.4, 7.3, 7.4, 8.2, 9.2, 9.3.1, 9.3.3, 9.4.1, 9.5, 9.6, 9.7, 9.8, 9.9, 9.10, 11.1.3, 11.4.1.5, 11.4.6, 11.4.10, 12.2, 13.5, 13.7, 14, 15.1.2, 15.4
Time Limits on Claims
3.7.4, 10.2.8, **13.7**, 15.1.2
Title to Work
9.3.2, 9.3.3
Transmission of Data in Digital Form
1.6
UNCOVERING AND CORRECTION OF WORK
12
Uncovering of Work
12.1
Unforeseen Conditions, Concealed or Unknown
3.7.4, 8.3.1, 10.3
Unit Prices
7.3.3.2, 7.3.4
Use of Documents
1.1.1, 1.5, 2.2.5, 3.12.6, 5.3
Use of Site
3.13, 6.1.1, 6.2.1
Values, Schedule of
9.2, 9.3.1
Waiver of Claims by the Architect
13.4.2
Waiver of Claims by the Contractor
9.10.5, 11.4.7, 13.4.2, 15.1.6
Waiver of Claims by the Owner
9.9.3, 9.10.3, 9.10.4, 11.4.3, 11.4.5, 11.4.7, 12.2.2.1, 13.4.2, 14.2.4, 15.1.6
Waiver of Consequential Damages
14.2.4, 15.1.6
Waiver of Liens
9.10.2, 9.10.4
Waivers of Subrogation
6.1.1, 11.4.5, **11.3.7**
Warranty
3.5, 4.2.9, 9.3.3, 9.8.4, 9.9.1, 9.10.4, 12.2.2, 13.7.1
Weather Delays
15.1.5.2
Work, Definition of
1.1.3
Written Consent
1.5.2, 3.4.2, 3.7.4, 3.12.8, 3.14.2, 4.1.2, 9.3.2, 9.8.5, 9.9.1, 9.10.2, 9.10.3, 11.4.1, 13.2, 13.4.2, 15.4.4.2
Written Interpretations
4.2.11, 4.2.12
Written Notice
2.3, 2.4, 3.3.1, 3.9, 3.12.9, 3.12.10, 5.2.1, 8.2.2, 9.7, 9.10, 10.2.2, 10.3, 11.1.3, 11.4.6, 12.2.2, 12.2.4, **13.3**, 14, 15.4.1
Written Orders
1.1.1, 2.3, 3.9, 7, 8.2.2, 11.4.9, 12.1, 12.2, 13.5.2, 14.3.1, 15.1.2

ARTICLE 1 GENERAL PROVISIONS
§ 1.1 BASIC DEFINITIONS
§ 1.1.1 THE CONTRACT DOCUMENTS
The Contract Documents are enumerated in the Agreement between the Owner and Contractor (hereinafter the Agreement) and consist of the Agreement, Conditions of the Contract (General, Supplementary and other Conditions), Drawings, Specifications, Addenda issued prior to execution of the Contract, other documents listed in the Agreement and Modifications issued after execution of the Contract. A Modification is (1) a written amendment to the Contract signed by both parties, (2) a Change Order, (3) a Construction Change Directive or (4) a written order for a minor change in the Work issued by the Architect. Unless specifically enumerated in the Agreement, the Contract Documents do not include the advertisement or invitation to bid, Instructions to Bidders, sample forms, other information furnished by the Owner in anticipation of receiving bids or proposals, the Contractor's bid or proposal, or portions of Addenda relating to bidding requirements.

§ 1.1.2 THE CONTRACT
The Contract Documents form the Contract for Construction. The Contract represents the entire and integrated agreement between the parties hereto and supersedes prior negotiations, representations or agreements, either written or oral. The Contract may be amended or modified only by a Modification. The Contract Documents shall not be construed to create a contractual relationship of any kind (1) between the Contractor and the Architect or the Architect's consultants, (2) between the Owner and a Subcontractor or a Sub-subcontractor, (3) between the Owner and the Architect or the Architect's consultants or (4) between any persons or entities other than the Owner and the Contractor. The Architect shall, however, be entitled to performance and enforcement of obligations under the Contract intended to facilitate performance of the Architect's duties.

§ 1.1.3 THE WORK
The term "Work" means the construction and services required by the Contract Documents, whether completed or partially completed, and includes all other labor, materials, equipment and services provided or to be provided by the Contractor to fulfill the Contractor's obligations. The Work may constitute the whole or a part of the Project.

§ 1.1.4 THE PROJECT
The Project is the total construction of which the Work performed under the Contract Documents may be the whole or a part and which may include construction by the Owner and by separate contractors.

§ 1.1.5 THE DRAWINGS
The Drawings are the graphic and pictorial portions of the Contract Documents showing the design, location and dimensions of the Work, generally including plans, elevations, sections, details, schedules and diagrams.

§ 1.1.6 THE SPECIFICATIONS
The Specifications are that portion of the Contract Documents consisting of the written requirements for materials, equipment, systems, standards and workmanship for the Work, and performance of related services.

§ 1.1.7 INSTRUMENTS OF SERVICE
Instruments of Service are representations, in any medium of expression now known or later developed, of the tangible and intangible creative work performed by the Architect and the Architect's consultants under their respective professional services agreements. Instruments of Service may include, without limitation, studies, surveys, models, sketches, drawings, specifications, and other similar materials.

§ 1.1.8 INITIAL DECISION MAKER
The Initial Decision Maker is the person identified in the Agreement to render initial decisions on Claims in accordance with Section 15.2 and certify termination of the Agreement under Section 14.2.2.

§ 1.2 CORRELATION AND INTENT OF THE CONTRACT DOCUMENTS
§ 1.2.1 The intent of the Contract Documents is to include all items necessary for the proper execution and completion of the Work by the Contractor. The Contract Documents are complementary, and what is required by one shall be as binding as if required by all; performance by the Contractor shall be required only to the extent consistent with the Contract Documents and reasonably inferable from them as being necessary to produce the indicated results.

§ 1.2.2 Organization of the Specifications into divisions, sections and articles, and arrangement of Drawings shall not control the Contractor in dividing the Work among Subcontractors or in establishing the extent of Work to be performed by any trade.

§ 1.2.3 Unless otherwise stated in the Contract Documents, words that have well-known technical or construction industry meanings are used in the Contract Documents in accordance with such recognized meanings.

§ 1.3 CAPITALIZATION
Terms capitalized in these General Conditions include those that are (1) specifically defined, (2) the titles of numbered articles or (3) the titles of other documents published by the American Institute of Architects.

§ 1.4 INTERPRETATION
In the interest of brevity the Contract Documents frequently omit modifying words such as "all" and "any" and articles such as "the" and "an," but the fact that a modifier or an article is absent from one statement and appears in another is not intended to affect the interpretation of either statement.

§ 1.5 OWNERSHIP AND USE OF DRAWINGS, SPECIFICATIONS AND OTHER INSTRUMENTS OF SERVICE
§ 1.5.1 The Architect and the Architect's consultants shall be deemed the authors and owners of their respective Instruments of Service, including the Drawings and Specifications, and will retain all common law, statutory and other reserved rights, including copyrights. The Contractor, Subcontractors, Sub-subcontractors, and material or equipment suppliers shall not own or claim a copyright in the Instruments of Service. Submittal or distribution to meet official regulatory requirements or for other purposes in connection with this Project is not to be construed as publication in derogation of the Architect's or Architect's consultants' reserved rights.

§ 1.5.2 The Contractor, Subcontractors, Sub-subcontractors and material or equipment suppliers are authorized to use and reproduce the Instruments of Service provided to them solely and exclusively for execution of the Work. All copies made under this authorization shall bear the copyright notice, if any, shown on the Instruments of Service. The Contractor, Subcontractors, Sub-subcontractors, and material or equipment suppliers may not use the Instruments of Service on other projects or for additions to this Project outside the scope of the Work without the specific written consent of the Owner, Architect and the Architect's consultants.

§ 1.6 TRANSMISSION OF DATA IN DIGITAL FORM
If the parties intend to transmit Instruments of Service or any other information or documentation in digital form, they shall endeavor to establish necessary protocols governing such transmissions, unless otherwise already provided in the Agreement or the Contract Documents.

ARTICLE 2 OWNER
§ 2.1 GENERAL
§ 2.1.1 The Owner is the person or entity identified as such in the Agreement and is referred to throughout the Contract Documents as if singular in number. The Owner shall designate in writing a representative who shall have express authority to bind the Owner with respect to all matters requiring the Owner's approval or authorization. Except as otherwise provided in Section 4.2.1, the Architect does not have such authority. The term "Owner" means the Owner or the Owner's authorized representative.

§ 2.1.2 The Owner shall furnish to the Contractor within fifteen days after receipt of a written request, information necessary and relevant for the Contractor to evaluate, give notice of or enforce mechanic's lien rights. Such information shall include a correct statement of the record legal title to the property on which the Project is located, usually referred to as the site, and the Owner's interest therein.

§ 2.2 INFORMATION AND SERVICES REQUIRED OF THE OWNER
§ 2.2.1 Prior to commencement of the Work, the Contractor may request in writing that the Owner provide reasonable evidence that the Owner has made financial arrangements to fulfill the Owner's obligations under the Contract. Thereafter, the Contractor may only request such evidence if (1) the Owner fails to make payments to the Contractor as the Contract Documents require; (2) a change in the Work materially changes the Contract Sum; or (3) the Contractor identifies in writing a reasonable concern regarding the Owner's ability to make payment when due. The Owner shall furnish such evidence as a condition precedent to commencement or continuation of the Work or the portion of the Work affected by a material change. After the Owner furnishes the evidence, the Owner shall not materially vary such financial arrangements without prior notice to the Contractor.

§ 2.2.2 Except for permits and fees that are the responsibility of the Contractor under the Contract Documents, including those required under Section 3.7.1, the Owner shall secure and pay for necessary approvals, easements, assessments and charges required for construction, use or occupancy of permanent structures or for permanent changes in existing facilities.

§ 2.2.3 The Owner shall furnish surveys describing physical characteristics, legal limitations and utility locations for the site of the Project, and a legal description of the site. The Contractor shall be entitled to rely on the accuracy of information furnished by the Owner but shall exercise proper precautions relating to the safe performance of the Work.

§ 2.2.4 The Owner shall furnish information or services required of the Owner by the Contract Documents with reasonable promptness. The Owner shall also furnish any other information or services under the Owner's control and relevant to the Contractor's performance of the Work with reasonable promptness after receiving the Contractor's written request for such information or services.

§ 2.2.5 Unless otherwise provided in the Contract Documents, the Owner shall furnish to the Contractor one copy of the Contract Documents for purposes of making reproductions pursuant to Section 1.5.2.

§ 2.3 OWNER'S RIGHT TO STOP THE WORK
If the Contractor fails to correct Work that is not in accordance with the requirements of the Contract Documents as required by Section 12.2 or repeatedly fails to carry out Work in accordance with the Contract Documents, the Owner may issue a written order to the Contractor to stop the Work, or any portion thereof, until the cause for such order has been eliminated; however, the right of the Owner to stop the Work shall not give rise to a duty on the part of the Owner to exercise this right for the benefit of the Contractor or any other person or entity, except to the extent required by Section 6.1.3.

§ 2.4 OWNER'S RIGHT TO CARRY OUT THE WORK
If the Contractor defaults or neglects to carry out the Work in accordance with the Contract Documents and fails within a ten-day period after receipt of written notice from the Owner to commence and continue correction of such default or neglect with diligence and promptness, the Owner may, without prejudice to other remedies the Owner may have, correct such deficiencies. In such case an appropriate Change Order shall be issued deducting from payments then or thereafter due the Contractor the reasonable cost of correcting such deficiencies, including Owner's expenses and compensation for the Architect's additional services made necessary by such default, neglect or failure. Such action by the Owner and amounts charged to the Contractor are both subject to prior approval of the Architect. If payments then or thereafter due the Contractor are not sufficient to cover such amounts, the Contractor shall pay the difference to the Owner.

ARTICLE 3 CONTRACTOR
§ 3.1 GENERAL
§ 3.1.1 The Contractor is the person or entity identified as such in the Agreement and is referred to throughout the Contract Documents as if singular in number. The Contractor shall be lawfully licensed, if required in the jurisdiction where the Project is located. The Contractor shall designate in writing a representative who shall have express authority to bind the Contractor with respect to all matters under this Contract. The term "Contractor" means the Contractor or the Contractor's authorized representative.

§ 3.1.2 The Contractor shall perform the Work in accordance with the Contract Documents.

§ 3.1.3 The Contractor shall not be relieved of obligations to perform the Work in accordance with the Contract Documents either by activities or duties of the Architect in the Architect's administration of the Contract, or by tests, inspections or approvals required or performed by persons or entities other than the Contractor.

§ 3.2 REVIEW OF CONTRACT DOCUMENTS AND FIELD CONDITIONS BY CONTRACTOR
§ 3.2.1 Execution of the Contract by the Contractor is a representation that the Contractor has visited the site, become generally familiar with local conditions under which the Work is to be performed and correlated personal observations with requirements of the Contract Documents.

§ 3.2.2 Because the Contract Documents are complementary, the Contractor shall, before starting each portion of the Work, carefully study and compare the various Contract Documents relative to that portion of the Work, as well as the information furnished by the Owner pursuant to Section 2.2.3, shall take field measurements of any existing conditions related to that portion of the Work, and shall observe any conditions at the site affecting it. These obligations are for the purpose of facilitating coordination and construction by the Contractor and are not for the purpose of discovering errors, omissions, or inconsistencies in the Contract Documents; however, the Contractor shall promptly report to the Architect any errors, inconsistencies or omissions discovered by or made known to the Contractor as a request for information in such form as the Architect may require. It is recognized that the Contractor's review is made in the Contractor's capacity as a contractor and not as a licensed design professional, unless otherwise specifically provided in the Contract Documents.

§ 3.2.3 The Contractor is not required to ascertain that the Contract Documents are in accordance with applicable laws, statutes, ordinances, codes, rules and regulations, or lawful orders of public authorities, but the Contractor shall promptly report to the Architect any nonconformity discovered by or made known to the Contractor as a request for information in such form as the Architect may require.

§ 3.2.4 If the Contractor believes that additional cost or time is involved because of clarifications or instructions the Architect issues in response to the Contractor's notices or requests for information pursuant to Sections 3.2.2 or 3.2.3, the Contractor shall make Claims as provided in Article 15. If the Contractor fails to perform the obligations of Sections 3.2.2 or 3.2.3, the Contractor shall pay such costs and damages to the Owner as would have been avoided if the Contractor had performed such obligations. If the Contractor performs those obligations, the Contractor shall not be liable to the Owner or Architect for damages resulting from errors, inconsistencies or omissions in the Contract Documents, for differences between field measurements or conditions and the Contract Documents, or for nonconformities of the Contract Documents to applicable laws, statutes, ordinances, codes, rules and regulations, and lawful orders of public authorities.

§ 3.3 SUPERVISION AND CONSTRUCTION PROCEDURES
§ 3.3.1 The Contractor shall supervise and direct the Work, using the Contractor's best skill and attention. The Contractor shall be solely responsible for, and have control over, construction means, methods, techniques, sequences and procedures and for coordinating all portions of the Work under the Contract, unless the Contract Documents give other specific instructions concerning these matters. If the Contract Documents give specific instructions concerning construction means, methods, techniques, sequences or procedures, the Contractor shall evaluate the jobsite safety thereof and, except as stated below, shall be fully and solely responsible for the jobsite safety of such means, methods, techniques, sequences or procedures. If the Contractor determines that such means, methods, techniques, sequences or procedures may not be safe, the Contractor shall give timely written notice to the Owner and Architect and shall not proceed with that portion of the Work without further written instructions from the Architect. If the Contractor is then instructed to proceed with the required means, methods, techniques, sequences or procedures without acceptance of changes proposed by the Contractor, the Owner shall be solely responsible for any loss or damage arising solely from those Owner-required means, methods, techniques, sequences or procedures.

§ 3.3.2 The Contractor shall be responsible to the Owner for acts and omissions of the Contractor's employees, Subcontractors and their agents and employees, and other persons or entities performing portions of the Work for, or on behalf of, the Contractor or any of its Subcontractors.

§ 3.3.3 The Contractor shall be responsible for inspection of portions of Work already performed to determine that such portions are in proper condition to receive subsequent Work.

§ 3.4 LABOR AND MATERIALS
§ 3.4.1 Unless otherwise provided in the Contract Documents, the Contractor shall provide and pay for labor, materials, equipment, tools, construction equipment and machinery, water, heat, utilities, transportation, and other facilities and services necessary for proper execution and completion of the Work, whether temporary or permanent and whether or not incorporated or to be incorporated in the Work.

§ 3.4.2 Except in the case of minor changes in the Work authorized by the Architect in accordance with Sections 3.12.8 or 7.4, the Contractor may make substitutions only with the consent of the Owner, after evaluation by the Architect and in accordance with a Change Order or Construction Change Directive.

§ 3.4.3 The Contractor shall enforce strict discipline and good order among the Contractor's employees and other persons carrying out the Work. The Contractor shall not permit employment of unfit persons or persons not properly skilled in tasks assigned to them.

§ 3.5 WARRANTY
The Contractor warrants to the Owner and Architect that materials and equipment furnished under the Contract will be of good quality and new unless the Contract Documents require or permit otherwise. The Contractor further warrants that the Work will conform to the requirements of the Contract Documents and will be free from defects, except for those inherent in the quality of the Work the Contract Documents require or permit. Work, materials, or equipment not conforming to these requirements may be considered defective. The Contractor's warranty excludes remedy for damage or defect caused by abuse, alterations to the Work not executed by the Contractor, improper or insufficient maintenance, improper operation, or normal wear and tear and normal usage. If required by the Architect, the Contractor shall furnish satisfactory evidence as to the kind and quality of materials and equipment.

§ 3.6 TAXES
The Contractor shall pay sales, consumer, use and similar taxes for the Work provided by the Contractor that are legally enacted when bids are received or negotiations concluded, whether or not yet effective or merely scheduled to go into effect.

§ 3.7 PERMITS, FEES, NOTICES, AND COMPLIANCE WITH LAWS
§ 3.7.1 Unless otherwise provided in the Contract Documents, the Contractor shall secure and pay for the building permit as well as for other permits, fees, licenses, and inspections by government agencies necessary for proper execution and completion of the Work that are customarily secured after execution of the Contract and legally required at the time bids are received or negotiations concluded.

§ 3.7.2 The Contractor shall comply with and give notices required by applicable laws, statutes, ordinances, codes, rules and regulations, and lawful orders of public authorities applicable to performance of the Work.

§ 3.7.3 If the Contractor performs Work knowing it to be contrary to applicable laws, statutes, ordinances, codes, rules and regulations, or lawful orders of public authorities, the Contractor shall assume appropriate responsibility for such Work and shall bear the costs attributable to correction.

§ 3.7.4 **Concealed or Unknown Conditions.** If the Contractor encounters conditions at the site that are (1) subsurface or otherwise concealed physical conditions that differ materially from those indicated in the Contract Documents or (2) unknown physical conditions of an unusual nature that differ materially from those ordinarily found to exist and generally recognized as inherent in construction activities of the character provided for in the Contract Documents, the Contractor shall promptly provide notice to the Owner and the Architect before conditions are disturbed and in no event later than 21 days after first observance of the conditions. The Architect will promptly investigate such conditions and, if the Architect determines that they differ materially and cause an increase or decrease in the Contractor's cost of, or time required for, performance of any part of the Work, will recommend an equitable adjustment in the Contract Sum or Contract Time, or both. If the Architect determines that the conditions at the site are not materially different from those indicated in the Contract Documents and that no change in the terms of the Contract is justified, the Architect shall promptly notify the Owner and Contractor in writing, stating the reasons. If either party disputes the Architect's determination or recommendation, that party may proceed as provided in Article 15.

§ 3.7.5 If, in the course of the Work, the Contractor encounters human remains or recognizes the existence of burial markers, archaeological sites or wetlands not indicated in the Contract Documents, the Contractor shall immediately suspend any operations that would affect them and shall notify the Owner and Architect. Upon receipt of such notice, the Owner shall promptly take any action necessary to obtain governmental authorization required to resume the operations. The Contractor shall continue to suspend such operations until otherwise instructed by the Owner but shall continue with all other operations that do not affect those remains or features. Requests for adjustments in the Contract Sum and Contract Time arising from the existence of such remains or features may be made as provided in Article 15.

§ 3.8 ALLOWANCES
§ 3.8.1 The Contractor shall include in the Contract Sum all allowances stated in the Contract Documents. Items covered by allowances shall be supplied for such amounts and by such persons or entities as the Owner may direct,

but the Contractor shall not be required to employ persons or entities to whom the Contractor has reasonable objection.

§ 3.8.2 Unless otherwise provided in the Contract Documents,
.1 allowances shall cover the cost to the Contractor of materials and equipment delivered at the site and all required taxes, less applicable trade discounts;
.2 Contractor's costs for unloading and handling at the site, labor, installation costs, overhead, profit and other expenses contemplated for stated allowance amounts shall be included in the Contract Sum but not in the allowances; and
.3 whenever costs are more than or less than allowances, the Contract Sum shall be adjusted accordingly by Change Order. The amount of the Change Order shall reflect (1) the difference between actual costs and the allowances under Section 3.8.2.1 and (2) changes in Contractor's costs under Section 3.8.2.2.

§ 3.8.3 Materials and equipment under an allowance shall be selected by the Owner with reasonable promptness.

§ 3.9 SUPERINTENDENT
§ 3.9.1 The Contractor shall employ a competent superintendent and necessary assistants who shall be in attendance at the Project site during performance of the Work. The superintendent shall represent the Contractor, and communications given to the superintendent shall be as binding as if given to the Contractor.

§ 3.9.2 The Contractor, as soon as practicable after award of the Contract, shall furnish in writing to the Owner through the Architect the name and qualifications of a proposed superintendent. The Architect may reply within 14 days to the Contractor in writing stating (1) whether the Owner or the Architect has reasonable objection to the proposed superintendent or (2) that the Architect requires additional time to review. Failure of the Architect to reply within the 14 day period shall constitute notice of no reasonable objection.

§ 3.9.3 The Contractor shall not employ a proposed superintendent to whom the Owner or Architect has made reasonable and timely objection. The Contractor shall not change the superintendent without the Owner's consent, which shall not unreasonably be withheld or delayed.

§ 3.10 CONTRACTOR'S CONSTRUCTION SCHEDULES
§ 3.10.1 The Contractor, promptly after being awarded the Contract, shall prepare and submit for the Owner's and Architect's information a Contractor's construction schedule for the Work. The schedule shall not exceed time limits current under the Contract Documents, shall be revised at appropriate intervals as required by the conditions of the Work and Project, shall be related to the entire Project to the extent required by the Contract Documents, and shall provide for expeditious and practicable execution of the Work.

§ 3.10.2 The Contractor shall prepare a submittal schedule, promptly after being awarded the Contract and thereafter as necessary to maintain a current submittal schedule, and shall submit the schedule(s) for the Architect's approval. The Architect's approval shall not unreasonably be delayed or withheld. The submittal schedule shall (1) be coordinated with the Contractor's construction schedule, and (2) allow the Architect reasonable time to review submittals. If the Contractor fails to submit a submittal schedule, the Contractor shall not be entitled to any increase in Contract Sum or extension of Contract Time based on the time required for review of submittals.

§ 3.10.3 The Contractor shall perform the Work in general accordance with the most recent schedules submitted to the Owner and Architect.

§ 3.11 DOCUMENTS AND SAMPLES AT THE SITE
The Contractor shall maintain at the site for the Owner one copy of the Drawings, Specifications, Addenda, Change Orders and other Modifications, in good order and marked currently to indicate field changes and selections made during construction, and one copy of approved Shop Drawings, Product Data, Samples and similar required submittals. These shall be available to the Architect and shall be delivered to the Architect for submittal to the Owner upon completion of the Work as a record of the Work as constructed.

§ 3.12 SHOP DRAWINGS, PRODUCT DATA AND SAMPLES

§ 3.12.1 Shop Drawings are drawings, diagrams, schedules and other data specially prepared for the Work by the Contractor or a Subcontractor, Sub-subcontractor, manufacturer, supplier or distributor to illustrate some portion of the Work.

§ 3.12.2 Product Data are illustrations, standard schedules, performance charts, instructions, brochures, diagrams and other information furnished by the Contractor to illustrate materials or equipment for some portion of the Work.

§ 3.12.3 Samples are physical examples that illustrate materials, equipment or workmanship and establish standards by which the Work will be judged.

§ 3.12.4 Shop Drawings, Product Data, Samples and similar submittals are not Contract Documents. Their purpose is to demonstrate the way by which the Contractor proposes to conform to the information given and the design concept expressed in the Contract Documents for those portions of the Work for which the Contract Documents require submittals. Review by the Architect is subject to the limitations of Section 4.2.7. Informational submittals upon which the Architect is not expected to take responsive action may be so identified in the Contract Documents. Submittals that are not required by the Contract Documents may be returned by the Architect without action.

§ 3.12.5 The Contractor shall review for compliance with the Contract Documents, approve and submit to the Architect Shop Drawings, Product Data, Samples and similar submittals required by the Contract Documents in accordance with the submittal schedule approved by the Architect or, in the absence of an approved submittal schedule, with reasonable promptness and in such sequence as to cause no delay in the Work or in the activities of the Owner or of separate contractors.

§ 3.12.6 By submitting Shop Drawings, Product Data, Samples and similar submittals, the Contractor represents to the Owner and Architect that the Contractor has (1) reviewed and approved them, (2) determined and verified materials, field measurements and field construction criteria related thereto, or will do so and (3) checked and coordinated the information contained within such submittals with the requirements of the Work and of the Contract Documents.

§ 3.12.7 The Contractor shall perform no portion of the Work for which the Contract Documents require submittal and review of Shop Drawings, Product Data, Samples or similar submittals until the respective submittal has been approved by the Architect.

§ 3.12.8 The Work shall be in accordance with approved submittals except that the Contractor shall not be relieved of responsibility for deviations from requirements of the Contract Documents by the Architect's approval of Shop Drawings, Product Data, Samples or similar submittals unless the Contractor has specifically informed the Architect in writing of such deviation at the time of submittal and (1) the Architect has given written approval to the specific deviation as a minor change in the Work, or (2) a Change Order or Construction Change Directive has been issued authorizing the deviation. The Contractor shall not be relieved of responsibility for errors or omissions in Shop Drawings, Product Data, Samples or similar submittals by the Architect's approval thereof.

§ 3.12.9 The Contractor shall direct specific attention, in writing or on resubmitted Shop Drawings, Product Data, Samples or similar submittals, to revisions other than those requested by the Architect on previous submittals. In the absence of such written notice, the Architect's approval of a resubmission shall not apply to such revisions.

§ 3.12.10 The Contractor shall not be required to provide professional services that constitute the practice of architecture or engineering unless such services are specifically required by the Contract Documents for a portion of the Work or unless the Contractor needs to provide such services in order to carry out the Contractor's responsibilities for construction means, methods, techniques, sequences and procedures. The Contractor shall not be required to provide professional services in violation of applicable law. If professional design services or certifications by a design professional related to systems, materials or equipment are specifically required of the Contractor by the Contract Documents, the Owner and the Architect will specify all performance and design criteria that such services must satisfy. The Contractor shall cause such services or certifications to be provided by a properly licensed design professional, whose signature and seal shall appear on all drawings, calculations, specifications, certifications, Shop Drawings and other submittals prepared by such professional. Shop Drawings and other submittals related to the Work designed or certified by such professional, if prepared by others, shall bear such professional's written approval when submitted to the Architect. The Owner and the Architect shall be entitled to rely upon the adequacy, accuracy and completeness of the services, certifications and approvals performed or provided by such design professionals, provided the Owner and Architect have specified to the Contractor all performance and design criteria that such services must satisfy. Pursuant to this Section 3.12.10, the Architect will review, approve or take other appropriate action on submittals only for the limited purpose of checking for conformance with information given and the design concept expressed in the Contract Documents. The Contractor shall not be responsible for the adequacy of the performance and design criteria specified in the Contract Documents.

§ 3.13 USE OF SITE
The Contractor shall confine operations at the site to areas permitted by applicable laws, statutes, ordinances, codes, rules and regulations, and lawful orders of public authorities and the Contract Documents and shall not unreasonably encumber the site with materials or equipment.

§ 3.14 CUTTING AND PATCHING
§ 3.14.1 The Contractor shall be responsible for cutting, fitting or patching required to complete the Work or to make its parts fit together properly. All areas requiring cutting, fitting and patching shall be restored to the condition existing prior to the cutting, fitting and patching, unless otherwise required by the Contract Documents.

§ 3.14.2 The Contractor shall not damage or endanger a portion of the Work or fully or partially completed construction of the Owner or separate contractors by cutting, patching or otherwise altering such construction, or by excavation. The Contractor shall not cut or otherwise alter such construction by the Owner or a separate contractor except with written consent of the Owner and of such separate contractor; such consent shall not be unreasonably withheld. The Contractor shall not unreasonably withhold from the Owner or a separate contractor the Contractor's consent to cutting or otherwise altering the Work.

§ 3.15 CLEANING UP
§ 3.15.1 The Contractor shall keep the premises and surrounding area free from accumulation of waste materials or rubbish caused by operations under the Contract. At completion of the Work, the Contractor shall remove waste materials, rubbish, the Contractor's tools, construction equipment, machinery and surplus materials from and about the Project.

§ 3.15.2 If the Contractor fails to clean up as provided in the Contract Documents, the Owner may do so and Owner shall be entitled to reimbursement from the Contractor.

§ 3.16 ACCESS TO WORK
The Contractor shall provide the Owner and Architect access to the Work in preparation and progress wherever located.

§ 3.17 ROYALTIES, PATENTS AND COPYRIGHTS
The Contractor shall pay all royalties and license fees. The Contractor shall defend suits or claims for infringement of copyrights and patent rights and shall hold the Owner and Architect harmless from loss on account thereof, but shall not be responsible for such defense or loss when a particular design, process or product of a particular manufacturer or manufacturers is required by the Contract Documents, or where the copyright violations are contained in Drawings, Specifications or other documents prepared by the Owner or Architect. However, if the Contractor has reason to believe that the required design, process or product is an infringement of a copyright or a patent, the Contractor shall be responsible for such loss unless such information is promptly furnished to the Architect.

§ 3.18 INDEMNIFICATION
§ 3.18.1 To the fullest extent permitted by law the Contractor shall indemnify and hold harmless the Owner, Architect, Architect's consultants, and agents and employees of any of them from and against claims, damages, losses and expenses, including but not limited to attorneys' fees, arising out of or resulting from performance of the Work, provided that such claim, damage, loss or expense is attributable to bodily injury, sickness, disease or death, or to injury to or destruction of tangible property (other than the Work itself), but only to the extent caused by the negligent acts or omissions of the Contractor, a Subcontractor, anyone directly or indirectly employed by them or anyone for whose acts they may be liable, regardless of whether or not such claim, damage, loss or expense is caused in part by a party indemnified hereunder. Such obligation shall not be construed to negate, abridge, or reduce

other rights or obligations of indemnity that would otherwise exist as to a party or person described in this Section 3.18.

§ 3.18.2 In claims against any person or entity indemnified under this Section 3.18 by an employee of the Contractor, a Subcontractor, anyone directly or indirectly employed by them or anyone for whose acts they may be liable, the indemnification obligation under Section 3.18.1 shall not be limited by a limitation on amount or type of damages, compensation or benefits payable by or for the Contractor or a Subcontractor under workers' compensation acts, disability benefit acts or other employee benefit acts.

ARTICLE 4 ARCHITECT
§ 4.1 GENERAL
§ 4.1.1 The Owner shall retain an architect lawfully licensed to practice architecture or an entity lawfully practicing architecture in the jurisdiction where the Project is located. That person or entity is identified as the Architect in the Agreement and is referred to throughout the Contract Documents as if singular in number.

§ 4.1.2 Duties, responsibilities and limitations of authority of the Architect as set forth in the Contract Documents shall not be restricted, modified or extended without written consent of the Owner, Contractor and Architect. Consent shall not be unreasonably withheld.

§ 4.1.3 If the employment of the Architect is terminated, the Owner shall employ a successor architect as to whom the Contractor has no reasonable objection and whose status under the Contract Documents shall be that of the Architect.

§ 4.2 ADMINISTRATION OF THE CONTRACT
§ 4.2.1 The Architect will provide administration of the Contract as described in the Contract Documents and will be an Owner's representative during construction until the date the Architect issues the final Certificate For Payment. The Architect will have authority to act on behalf of the Owner only to the extent provided in the Contract Documents.

§ 4.2.2 The Architect will visit the site at intervals appropriate to the stage of construction, or as otherwise agreed with the Owner, to become generally familiar with the progress and quality of the portion of the Work completed, and to determine in general if the Work observed is being performed in a manner indicating that the Work, when fully completed, will be in accordance with the Contract Documents. However, the Architect will not be required to make exhaustive or continuous on-site inspections to check the quality or quantity of the Work. The Architect will not have control over, charge of, or responsibility for, the construction means, methods, techniques, sequences or procedures, or for the safety precautions and programs in connection with the Work, since these are solely the Contractor's rights and responsibilities under the Contract Documents, except as provided in Section 3.3.1.

§ 4.2.3 On the basis of the site visits, the Architect will keep the Owner reasonably informed about the progress and quality of the portion of the Work completed, and report to the Owner (1) known deviations from the Contract Documents and from the most recent construction schedule submitted by the Contractor, and (2) defects and deficiencies observed in the Work. The Architect will not be responsible for the Contractor's failure to perform the Work in accordance with the requirements of the Contract Documents. The Architect will not have control over or charge of and will not be responsible for acts or omissions of the Contractor, Subcontractors, or their agents or employees, or any other persons or entities performing portions of the Work.

§ 4.2.4 COMMUNICATIONS FACILITATING CONTRACT ADMINISTRATION
Except as otherwise provided in the Contract Documents or when direct communications have been specially authorized, the Owner and Contractor shall endeavor to communicate with each other through the Architect about matters arising out of or relating to the Contract. Communications by and with the Architect's consultants shall be through the Architect. Communications by and with Subcontractors and material suppliers shall be through the Contractor. Communications by and with separate contractors shall be through the Owner.

§ 4.2.5 Based on the Architect's evaluations of the Contractor's Applications for Payment, the Architect will review and certify the amounts due the Contractor and will issue Certificates for Payment in such amounts.

§ 4.2.6 The Architect has authority to reject Work that does not conform to the Contract Documents. Whenever the Architect considers it necessary or advisable, the Architect will have authority to require inspection or testing of the

Work in accordance with Sections 13.5.2 and 13.5.3, whether or not such Work is fabricated, installed or completed. However, neither this authority of the Architect nor a decision made in good faith either to exercise or not to exercise such authority shall give rise to a duty or responsibility of the Architect to the Contractor, Subcontractors, material and equipment suppliers, their agents or employees, or other persons or entities performing portions of the Work.

§ 4.2.7 The Architect will review and approve, or take other appropriate action upon, the Contractor's submittals such as Shop Drawings, Product Data and Samples, but only for the limited purpose of checking for conformance with information given and the design concept expressed in the Contract Documents. The Architect's action will be taken in accordance with the submittal schedule approved by the Architect or, in the absence of an approved submittal schedule, with reasonable promptness while allowing sufficient time in the Architect's professional judgment to permit adequate review. Review of such submittals is not conducted for the purpose of determining the accuracy and completeness of other details such as dimensions and quantities, or for substantiating instructions for installation or performance of equipment or systems, all of which remain the responsibility of the Contractor as required by the Contract Documents. The Architect's review of the Contractor's submittals shall not relieve the Contractor of the obligations under Sections 3.3, 3.5 and 3.12. The Architect's review shall not constitute approval of safety precautions or, unless otherwise specifically stated by the Architect, of any construction means, methods, techniques, sequences or procedures. The Architect's approval of a specific item shall not indicate approval of an assembly of which the item is a component.

§ 4.2.8 The Architect will prepare Change Orders and Construction Change Directives, and may authorize minor changes in the Work as provided in Section 7.4. The Architect will investigate and make determinations and recommendations regarding concealed and unknown conditions as provided in Section 3.7.4.

§ 4.2.9 The Architect will conduct inspections to determine the date or dates of Substantial Completion and the date of final completion; issue Certificates of Substantial Completion pursuant to Section 9.8; receive and forward to the Owner, for the Owner's review and records, written warranties and related documents required by the Contract and assembled by the Contractor pursuant to Section 9.10; and issue a final Certificate for Payment pursuant to Section 9.10.

§ 4.2.10 If the Owner and Architect agree, the Architect will provide one or more project representatives to assist in carrying out the Architect's responsibilities at the site. The duties, responsibilities and limitations of authority of such project representatives shall be as set forth in an exhibit to be incorporated in the Contract Documents.

§ 4.2.11 The Architect will interpret and decide matters concerning performance under, and requirements of, the Contract Documents on written request of either the Owner or Contractor. The Architect's response to such requests will be made in writing within any time limits agreed upon or otherwise with reasonable promptness.

§ 4.2.12 Interpretations and decisions of the Architect will be consistent with the intent of, and reasonably inferable from, the Contract Documents and will be in writing or in the form of drawings. When making such interpretations and decisions, the Architect will endeavor to secure faithful performance by both Owner and Contractor, will not show partiality to either and will not be liable for results of interpretations or decisions rendered in good faith.

§ 4.2.13 The Architect's decisions on matters relating to aesthetic effect will be final if consistent with the intent expressed in the Contract Documents.

§ 4.2.14 The Architect will review and respond to requests for information about the Contract Documents. The Architect's response to such requests will be made in writing within any time limits agreed upon or otherwise with reasonable promptness. If appropriate, the Architect will prepare and issue supplemental Drawings and Specifications in response to the requests for information.

ARTICLE 5   SUBCONTRACTORS
§ 5.1 DEFINITIONS
§ 5.1.1 A Subcontractor is a person or entity who has a direct contract with the Contractor to perform a portion of the Work at the site. The term "Subcontractor" is referred to throughout the Contract Documents as if singular in number and means a Subcontractor or an authorized representative of the Subcontractor. The term "Subcontractor" does not include a separate contractor or subcontractors of a separate contractor.

§ 5.1.2 A Sub-subcontractor is a person or entity who has a direct or indirect contract with a Subcontractor to perform a portion of the Work at the site. The term "Sub-subcontractor" is referred to throughout the Contract Documents as if singular in number and means a Sub-subcontractor or an authorized representative of the Sub-subcontractor.

§ 5.2 AWARD OF SUBCONTRACTS AND OTHER CONTRACTS FOR PORTIONS OF THE WORK

§ 5.2.1 Unless otherwise stated in the Contract Documents or the bidding requirements, the Contractor, as soon as practicable after award of the Contract, shall furnish in writing to the Owner through the Architect the names of persons or entities (including those who are to furnish materials or equipment fabricated to a special design) proposed for each principal portion of the Work. The Architect may reply within 14 days to the Contractor in writing stating (1) whether the Owner or the Architect has reasonable objection to any such proposed person or entity or (2) that the Architect requires additional time for review. Failure of the Owner or Architect to reply within the 14-day period shall constitute notice of no reasonable objection.

§ 5.2.2 The Contractor shall not contract with a proposed person or entity to whom the Owner or Architect has made reasonable and timely objection. The Contractor shall not be required to contract with anyone to whom the Contractor has made reasonable objection.

§ 5.2.3 If the Owner or Architect has reasonable objection to a person or entity proposed by the Contractor, the Contractor shall propose another to whom the Owner or Architect has no reasonable objection. If the proposed but rejected Subcontractor was reasonably capable of performing the Work, the Contract Sum and Contract Time shall be increased or decreased by the difference, if any, occasioned by such change, and an appropriate Change Order shall be issued before commencement of the substitute Subcontractor's Work. However, no increase in the Contract Sum or Contract Time shall be allowed for such change unless the Contractor has acted promptly and responsively in submitting names as required.

§ 5.2.4 The Contractor shall not substitute a Subcontractor, person or entity previously selected if the Owner or Architect makes reasonable objection to such substitution.

§ 5.3 SUBCONTRACTUAL RELATIONS

By appropriate agreement, written where legally required for validity, the Contractor shall require each Subcontractor, to the extent of the Work to be performed by the Subcontractor, to be bound to the Contractor by terms of the Contract Documents, and to assume toward the Contractor all the obligations and responsibilities, including the responsibility for safety of the Subcontractor's Work, which the Contractor, by these Documents, assumes toward the Owner and Architect. Each subcontract agreement shall preserve and protect the rights of the Owner and Architect under the Contract Documents with respect to the Work to be performed by the Subcontractor so that subcontracting thereof will not prejudice such rights, and shall allow to the Subcontractor, unless specifically provided otherwise in the subcontract agreement, the benefit of all rights, remedies and redress against the Contractor that the Contractor, by the Contract Documents, has against the Owner. Where appropriate, the Contractor shall require each Subcontractor to enter into similar agreements with Sub-subcontractors. The Contractor shall make available to each proposed Subcontractor, prior to the execution of the subcontract agreement, copies of the Contract Documents to which the Subcontractor will be bound, and, upon written request of the Subcontractor, identify to the Subcontractor terms and conditions of the proposed subcontract agreement that may be at variance with the Contract Documents. Subcontractors will similarly make copies of applicable portions of such documents available to their respective proposed Sub-subcontractors.

§ 5.4 CONTINGENT ASSIGNMENT OF SUBCONTRACTS

§ 5.4.1 Each subcontract agreement for a portion of the Work is assigned by the Contractor to the Owner, provided that

.1 assignment is effective only after termination of the Contract by the Owner for cause pursuant to Section 14.2 and only for those subcontract agreements that the Owner accepts by notifying the Subcontractor and Contractor in writing; and

.2 assignment is subject to the prior rights of the surety, if any, obligated under bond relating to the Contract.

When the Owner accepts the assignment of a subcontract agreement, the Owner assumes the Contractor's rights and obligations under the subcontract.

§ 5.4.2 Upon such assignment, if the Work has been suspended for more than 30 days, the Subcontractor's compensation shall be equitably adjusted for increases in cost resulting from the suspension.

§ 5.4.3 Upon such assignment to the Owner under this Section 5.4, the Owner may further assign the subcontract to a successor contractor or other entity. If the Owner assigns the subcontract to a successor contractor or other entity, the Owner shall nevertheless remain legally responsible for all of the successor contractor's obligations under the subcontract.

ARTICLE 6   CONSTRUCTION BY OWNER OR BY SEPARATE CONTRACTORS
§ 6.1 OWNER'S RIGHT TO PERFORM CONSTRUCTION AND TO AWARD SEPARATE CONTRACTS
§ 6.1.1 The Owner reserves the right to perform construction or operations related to the Project with the Owner's own forces, and to award separate contracts in connection with other portions of the Project or other construction or operations on the site under Conditions of the Contract identical or substantially similar to these including those portions related to insurance and waiver of subrogation. If the Contractor claims that delay or additional cost is involved because of such action by the Owner, the Contractor shall make such Claim as provided in Article 15.

§ 6.1.2 When separate contracts are awarded for different portions of the Project or other construction or operations on the site, the term "Contractor" in the Contract Documents in each case shall mean the Contractor who executes each separate Owner-Contractor Agreement.

§ 6.1.3 The Owner shall provide for coordination of the activities of the Owner's own forces and of each separate contractor with the Work of the Contractor, who shall cooperate with them. The Contractor shall participate with other separate contractors and the Owner in reviewing their construction schedules. The Contractor shall make any revisions to the construction schedule deemed necessary after a joint review and mutual agreement. The construction schedules shall then constitute the schedules to be used by the Contractor, separate contractors and the Owner until subsequently revised.

§ 6.1.4 Unless otherwise provided in the Contract Documents, when the Owner performs construction or operations related to the Project with the Owner's own forces, the Owner shall be deemed to be subject to the same obligations and to have the same rights that apply to the Contractor under the Conditions of the Contract, including, without excluding others, those stated in Article 3, this Article 6 and Articles 10, 11 and 12.

§ 6.2 MUTUAL RESPONSIBILITY
§ 6.2.1 The Contractor shall afford the Owner and separate contractors reasonable opportunity for introduction and storage of their materials and equipment and performance of their activities, and shall connect and coordinate the Contractor's construction and operations with theirs as required by the Contract Documents.

§ 6.2.2 If part of the Contractor's Work depends for proper execution or results upon construction or operations by the Owner or a separate contractor, the Contractor shall, prior to proceeding with that portion of the Work, promptly report to the Architect apparent discrepancies or defects in such other construction that would render it unsuitable for such proper execution and results. Failure of the Contractor so to report shall constitute an acknowledgment that the Owner's or separate contractor's completed or partially completed construction is fit and proper to receive the Contractor's Work, except as to defects not then reasonably discoverable.

§ 6.2.3 The Contractor shall reimburse the Owner for costs the Owner incurs that are payable to a separate contractor because of the Contractor's delays, improperly timed activities or defective construction. The Owner shall be responsible to the Contractor for costs the Contractor incurs because of a separate contractor's delays, improperly timed activities, damage to the Work or defective construction.

§ 6.2.4 The Contractor shall promptly remedy damage the Contractor wrongfully causes to completed or partially completed construction or to property of the Owner, separate contractors as provided in Section 10.2.5.

§ 6.2.5 The Owner and each separate contractor shall have the same responsibilities for cutting and patching as are described for the Contractor in Section 3.14.

§ 6.3 OWNER'S RIGHT TO CLEAN UP
If a dispute arises among the Contractor, separate contractors and the Owner as to the responsibility under their respective contracts for maintaining the premises and surrounding area free from waste materials and rubbish, the Owner may clean up and the Architect will allocate the cost among those responsible.

ARTICLE 7   CHANGES IN THE WORK
§ 7.1 GENERAL
§ 7.1.1 Changes in the Work may be accomplished after execution of the Contract, and without invalidating the Contract, by Change Order, Construction Change Directive or order for a minor change in the Work, subject to the limitations stated in this Article 7 and elsewhere in the Contract Documents.

§ 7.1.2 A Change Order shall be based upon agreement among the Owner, Contractor and Architect; a Construction Change Directive requires agreement by the Owner and Architect and may or may not be agreed to by the Contractor; an order for a minor change in the Work may be issued by the Architect alone.

§ 7.1.3 Changes in the Work shall be performed under applicable provisions of the Contract Documents, and the Contractor shall proceed promptly, unless otherwise provided in the Change Order, Construction Change Directive or order for a minor change in the Work.

§ 7.2 CHANGE ORDERS
§ 7.2.1 A Change Order is a written instrument prepared by the Architect and signed by the Owner, Contractor and Architect stating their agreement upon all of the following:
- .1 The change in the Work;
- .2 The amount of the adjustment, if any, in the Contract Sum; and
- .3 The extent of the adjustment, if any, in the Contract Time.

§ 7.3 CONSTRUCTION CHANGE DIRECTIVES
§ 7.3.1 A Construction Change Directive is a written order prepared by the Architect and signed by the Owner and Architect, directing a change in the Work prior to agreement on adjustment, if any, in the Contract Sum or Contract Time, or both. The Owner may by Construction Change Directive, without invalidating the Contract, order changes in the Work within the general scope of the Contract consisting of additions, deletions or other revisions, the Contract Sum and Contract Time being adjusted accordingly.

§ 7.3.2 A Construction Change Directive shall be used in the absence of total agreement on the terms of a Change Order.

§ 7.3.3 If the Construction Change Directive provides for an adjustment to the Contract Sum, the adjustment shall be based on one of the following methods:
- .1 Mutual acceptance of a lump sum properly itemized and supported by sufficient substantiating data to permit evaluation;
- .2 Unit prices stated in the Contract Documents or subsequently agreed upon;
- .3 Cost to be determined in a manner agreed upon by the parties and a mutually acceptable fixed or percentage fee; or
- .4 As provided in Section 7.3.7.

§ 7.3.4 If unit prices are stated in the Contract Documents or subsequently agreed upon, and if quantities originally contemplated are materially changed in a proposed Change Order or Construction Change Directive so that application of such unit prices to quantities of Work proposed will cause substantial inequity to the Owner or Contractor, the applicable unit prices shall be equitably adjusted.

§ 7.3.5 Upon receipt of a Construction Change Directive, the Contractor shall promptly proceed with the change in the Work involved and advise the Architect of the Contractor's agreement or disagreement with the method, if any, provided in the Construction Change Directive for determining the proposed adjustment in the Contract Sum or Contract Time.

§ 7.3.6 A Construction Change Directive signed by the Contractor indicates the Contractor's agreement therewith, including adjustment in Contract Sum and Contract Time or the method for determining them. Such agreement shall be effective immediately and shall be recorded as a Change Order.

§ 7.3.7 If the Contractor does not respond promptly or disagrees with the method for adjustment in the Contract Sum, the Architect shall determine the method and the adjustment on the basis of reasonable expenditures and savings of those performing the Work attributable to the change, including, in case of an increase in the Contract Sum, an amount for overhead and profit as set forth in the Agreement, or if no such amount is set forth in the Agreement, a reasonable amount. In such case, and also under Section 7.3.3.3, the Contractor shall keep and present, in such form as the Architect may prescribe, an itemized accounting together with appropriate supporting data. Unless otherwise provided in the Contract Documents, costs for the purposes of this Section 7.3.7 shall be limited to the following:

.1 Costs of labor, including social security, old age and unemployment insurance, fringe benefits required by agreement or custom, and workers' compensation insurance;
.2 Costs of materials, supplies and equipment, including cost of transportation, whether incorporated or consumed;
.3 Rental costs of machinery and equipment, exclusive of hand tools, whether rented from the Contractor or others;
.4 Costs of premiums for all bonds and insurance, permit fees, and sales, use or similar taxes related to the Work; and
.5 Additional costs of supervision and field office personnel directly attributable to the change.

§ 7.3.8 The amount of credit to be allowed by the Contractor to the Owner for a deletion or change that results in a net decrease in the Contract Sum shall be actual net cost as confirmed by the Architect. When both additions and credits covering related Work or substitutions are involved in a change, the allowance for overhead and profit shall be figured on the basis of net increase, if any, with respect to that change.

§ 7.3.9 Pending final determination of the total cost of a Construction Change Directive to the Owner, the Contractor may request payment for Work completed under the Construction Change Directive in Applications for Payment. The Architect will make an interim determination for purposes of monthly certification for payment for those costs and certify for payment the amount that the Architect determines, in the Architect's professional judgment, to be reasonably justified. The Architect's interim determination of cost shall adjust the Contract Sum on the same basis as a Change Order, subject to the right of either party to disagree and assert a Claim in accordance with Article 15.

§ 7.3.10 When the Owner and Contractor agree with a determination made by the Architect concerning the adjustments in the Contract Sum and Contract Time, or otherwise reach agreement upon the adjustments, such agreement shall be effective immediately and the Architect will prepare a Change Order. Change Orders may be issued for all or any part of a Construction Change Directive.

§ 7.4 MINOR CHANGES IN THE WORK
The Architect has authority to order minor changes in the Work not involving adjustment in the Contract Sum or extension of the Contract Time and not inconsistent with the intent of the Contract Documents. Such changes will be effected by written order signed by the Architect and shall be binding on the Owner and Contractor.

ARTICLE 8 TIME
§ 8.1 DEFINITIONS
§ 8.1.1 Unless otherwise provided, Contract Time is the period of time, including authorized adjustments, allotted in the Contract Documents for Substantial Completion of the Work.

§ 8.1.2 The date of commencement of the Work is the date established in the Agreement.

§ 8.1.3 The date of Substantial Completion is the date certified by the Architect in accordance with Section 9.8.

§ 8.1.4 The term "day" as used in the Contract Documents shall mean calendar day unless otherwise specifically defined.

§ 8.2 PROGRESS AND COMPLETION
§ 8.2.1 Time limits stated in the Contract Documents are of the essence of the Contract. By executing the Agreement the Contractor confirms that the Contract Time is a reasonable period for performing the Work.

§ 8.2.2 The Contractor shall not knowingly, except by agreement or instruction of the Owner in writing, prematurely commence operations on the site or elsewhere prior to the effective date of insurance required by Article 11 to be furnished by the Contractor and Owner. The date of commencement of the Work shall not be changed by the effective date of such insurance.

§ 8.2.3 The Contractor shall proceed expeditiously with adequate forces and shall achieve Substantial Completion within the Contract Time.

§ 8.3 DELAYS AND EXTENSIONS OF TIME
§ 8.3.1 If the Contractor is delayed at any time in the commencement or progress of the Work by an act or neglect of the Owner or Architect, or of an employee of either, or of a separate contractor employed by the Owner; or by changes ordered in the Work; or by labor disputes, fire, unusual delay in deliveries, unavoidable casualties or other causes beyond the Contractor's control; or by delay authorized by the Owner pending mediation and arbitration; or by other causes that the Architect determines may justify delay, then the Contract Time shall be extended by Change Order for such reasonable time as the Architect may determine.

§ 8.3.2 Claims relating to time shall be made in accordance with applicable provisions of Article 15.

§ 8.3.3 This Section 8.3 does not preclude recovery of damages for delay by either party under other provisions of the Contract Documents.

ARTICLE 9   PAYMENTS AND COMPLETION
§ 9.1 CONTRACT SUM
The Contract Sum is stated in the Agreement and, including authorized adjustments, is the total amount payable by the Owner to the Contractor for performance of the Work under the Contract Documents.

§ 9.2 SCHEDULE OF VALUES
Where the Contract is based on a stipulated sum or Guaranteed Maximum Price, the Contractor shall submit to the Architect, before the first Application for Payment, a schedule of values allocating the entire Contract Sum to the various portions of the Work and prepared in such form and supported by such data to substantiate its accuracy as the Architect may require. This schedule, unless objected to by the Architect, shall be used as a basis for reviewing the Contractor's Applications for Payment.

§ 9.3 APPLICATIONS FOR PAYMENT
§ 9.3.1 At least ten days before the date established for each progress payment, the Contractor shall submit to the Architect an itemized Application for Payment prepared in accordance with the schedule of values, if required under Section 9.2., for completed portions of the Work. Such application shall be notarized, if required, and supported by such data substantiating the Contractor's right to payment as the Owner or Architect may require, such as copies of requisitions from Subcontractors and material suppliers, and shall reflect retainage if provided for in the Contract Documents.

§ 9.3.1.1 As provided in Section 7.3.9, such applications may include requests for payment on account of changes in the Work that have been properly authorized by Construction Change Directives, or by interim determinations of the Architect, but not yet included in Change Orders.

§ 9.3.1.2 Applications for Payment shall not include requests for payment for portions of the Work for which the Contractor does not intend to pay a Subcontractor or material supplier, unless such Work has been performed by others whom the Contractor intends to pay.

§ 9.3.2 Unless otherwise provided in the Contract Documents, payments shall be made on account of materials and equipment delivered and suitably stored at the site for subsequent incorporation in the Work. If approved in advance by the Owner, payment may similarly be made for materials and equipment suitably stored off the site at a location agreed upon in writing. Payment for materials and equipment stored on or off the site shall be conditioned upon compliance by the Contractor with procedures satisfactory to the Owner to establish the Owner's title to such materials and equipment or otherwise protect the Owner's interest, and shall include the costs of applicable insurance, storage and transportation to the site for such materials and equipment stored off the site.

§ 9.3.3 The Contractor warrants that title to all Work covered by an Application for Payment will pass to the Owner no later than the time of payment. The Contractor further warrants that upon submittal of an Application for Payment all Work for which Certificates for Payment have been previously issued and payments received from the

Owner shall, to the best of the Contractor's knowledge, information and belief, be free and clear of liens, claims, security interests or encumbrances in favor of the Contractor, Subcontractors, material suppliers, or other persons or entities making a claim by reason of having provided labor, materials and equipment relating to the Work.

§ 9.4 CERTIFICATES FOR PAYMENT

§ 9.4.1 The Architect will, within seven days after receipt of the Contractor's Application for Payment, either issue to the Owner a Certificate for Payment, with a copy to the Contractor, for such amount as the Architect determines is properly due, or notify the Contractor and Owner in writing of the Architect's reasons for withholding certification in whole or in part as provided in Section 9.5.1.

§ 9.4.2 The issuance of a Certificate for Payment will constitute a representation by the Architect to the Owner, based on the Architect's evaluation of the Work and the data comprising the Application for Payment, that, to the best of the Architect's knowledge, information and belief, the Work has progressed to the point indicated and that the quality of the Work is in accordance with the Contract Documents. The foregoing representations are subject to an evaluation of the Work for conformance with the Contract Documents upon Substantial Completion, to results of subsequent tests and inspections, to correction of minor deviations from the Contract Documents prior to completion and to specific qualifications expressed by the Architect. The issuance of a Certificate for Payment will further constitute a representation that the Contractor is entitled to payment in the amount certified. However, the issuance of a Certificate for Payment will not be a representation that the Architect has (1) made exhaustive or continuous on-site inspections to check the quality or quantity of the Work, (2) reviewed construction means, methods, techniques, sequences or procedures, (3) reviewed copies of requisitions received from Subcontractors and material suppliers and other data requested by the Owner to substantiate the Contractor's right to payment, or (4) made examination to ascertain how or for what purpose the Contractor has used money previously paid on account of the Contract Sum.

§ 9.5 DECISIONS TO WITHHOLD CERTIFICATION

§ 9.5.1 The Architect may withhold a Certificate for Payment in whole or in part, to the extent reasonably necessary to protect the Owner, if in the Architect's opinion the representations to the Owner required by Section 9.4.2 cannot be made. If the Architect is unable to certify payment in the amount of the Application, the Architect will notify the Contractor and Owner as provided in Section 9.4.1. If the Contractor and Architect cannot agree on a revised amount, the Architect will promptly issue a Certificate for Payment for the amount for which the Architect is able to make such representations to the Owner. The Architect may also withhold a Certificate for Payment or, because of subsequently discovered evidence, may nullify the whole or a part of a Certificate for Payment previously issued, to such extent as may be necessary in the Architect's opinion to protect the Owner from loss for which the Contractor is responsible, including loss resulting from acts and omissions described in Section 3.3.2, because of

.1 defective Work not remedied;
.2 third party claims filed or reasonable evidence indicating probable filing of such claims unless security acceptable to the Owner is provided by the Contractor;
.3 failure of the Contractor to make payments properly to Subcontractors or for labor, materials or equipment;
.4 reasonable evidence that the Work cannot be completed for the unpaid balance of the Contract Sum;
.5 damage to the Owner or a separate contractor;
.6 reasonable evidence that the Work will not be completed within the Contract Time, and that the unpaid balance would not be adequate to cover actual or liquidated damages for the anticipated delay; or
.7 repeated failure to carry out the Work in accordance with the Contract Documents.

§ 9.5.2 When the above reasons for withholding certification are removed, certification will be made for amounts previously withheld.

§ 9.5.3 If the Architect withholds certification for payment under Section 9.5.1.3, the Owner may, at its sole option, issue joint checks to the Contractor and to any Subcontractor or material or equipment suppliers to whom the Contractor failed to make payment for Work properly performed or material or equipment suitably delivered. If the Owner makes payments by joint check, the Owner shall notify the Architect and the Architect will reflect such payment on the next Certificate for Payment.

§ 9.6 PROGRESS PAYMENTS

§ 9.6.1 After the Architect has issued a Certificate for Payment, the Owner shall make payment in the manner and within the time provided in the Contract Documents, and shall so notify the Architect.

§ 9.6.2 The Contractor shall pay each Subcontractor no later than seven days after receipt of payment from the Owner the amount to which the Subcontractor is entitled, reflecting percentages actually retained from payments to the Contractor on account of the Subcontractor's portion of the Work. The Contractor shall, by appropriate agreement with each Subcontractor, require each Subcontractor to make payments to Sub-subcontractors in a similar manner.

§ 9.6.3 The Architect will, on request, furnish to a Subcontractor, if practicable, information regarding percentages of completion or amounts applied for by the Contractor and action taken thereon by the Architect and Owner on account of portions of the Work done by such Subcontractor.

§ 9.6.4 The Owner has the right to request written evidence from the Contractor that the Contractor has properly paid Subcontractors and material and equipment suppliers amounts paid by the Owner to the Contractor for subcontracted Work. If the Contractor fails to furnish such evidence within seven days, the Owner shall have the right to contact Subcontractors to ascertain whether they have been properly paid. Neither the Owner nor Architect shall have an obligation to pay or to see to the payment of money to a Subcontractor, except as may otherwise be required by law.

§ 9.6.5 Contractor payments to material and equipment suppliers shall be treated in a manner similar to that provided in Sections 9.6.2, 9.6.3 and 9.6.4.

§ 9.6.6 A Certificate for Payment, a progress payment, or partial or entire use or occupancy of the Project by the Owner shall not constitute acceptance of Work not in accordance with the Contract Documents.

§ 9.6.7 Unless the Contractor provides the Owner with a payment bond in the full penal sum of the Contract Sum, payments received by the Contractor for Work properly performed by Subcontractors and suppliers shall be held by the Contractor for those Subcontractors or suppliers who performed Work or furnished materials, or both, under contract with the Contractor for which payment was made by the Owner. Nothing contained herein shall require money to be placed in a separate account and not commingled with money of the Contractor, shall create any fiduciary liability or tort liability on the part of the Contractor for breach of trust or shall entitle any person or entity to an award of punitive damages against the Contractor for breach of the requirements of this provision.

§ 9.7 FAILURE OF PAYMENT
If the Architect does not issue a Certificate for Payment, through no fault of the Contractor, within seven days after receipt of the Contractor's Application for Payment, or if the Owner does not pay the Contractor within seven days after the date established in the Contract Documents the amount certified by the Architect or awarded by binding dispute resolution, then the Contractor may, upon seven additional days' written notice to the Owner and Architect, stop the Work until payment of the amount owing has been received. The Contract Time shall be extended appropriately and the Contract Sum shall be increased by the amount of the Contractor's reasonable costs of shut-down, delay and start-up, plus interest as provided for in the Contract Documents.

§ 9.8 SUBSTANTIAL COMPLETION
§ 9.8.1 Substantial Completion is the stage in the progress of the Work when the Work or designated portion thereof is sufficiently complete in accordance with the Contract Documents so that the Owner can occupy or utilize the Work for its intended use.

§ 9.8.2 When the Contractor considers that the Work, or a portion thereof which the Owner agrees to accept separately, is substantially complete, the Contractor shall prepare and submit to the Architect a comprehensive list of items to be completed or corrected prior to final payment. Failure to include an item on such list does not alter the responsibility of the Contractor to complete all Work in accordance with the Contract Documents.

§ 9.8.3 Upon receipt of the Contractor's list, the Architect will make an inspection to determine whether the Work or designated portion thereof is substantially complete. If the Architect's inspection discloses any item, whether or not included on the Contractor's list, which is not sufficiently complete in accordance with the Contract Documents so that the Owner can occupy or utilize the Work or designated portion thereof for its intended use, the Contractor shall, before issuance of the Certificate of Substantial Completion, complete or correct such item upon notification by the Architect. In such case, the Contractor shall then submit a request for another inspection by the Architect to determine Substantial Completion.

§ 9.8.4 When the Work or designated portion thereof is substantially complete, the Architect will prepare a Certificate of Substantial Completion that shall establish the date of Substantial Completion, shall establish responsibilities of the Owner and Contractor for security, maintenance, heat, utilities, damage to the Work and insurance, and shall fix the time within which the Contractor shall finish all items on the list accompanying the Certificate. Warranties required by the Contract Documents shall commence on the date of Substantial Completion of the Work or designated portion thereof unless otherwise provided in the Certificate of Substantial Completion.

§ 9.8.5 The Certificate of Substantial Completion shall be submitted to the Owner and Contractor for their written acceptance of responsibilities assigned to them in such Certificate. Upon such acceptance and consent of surety, if any, the Owner shall make payment of retainage applying to such Work or designated portion thereof. Such payment shall be adjusted for Work that is incomplete or not in accordance with the requirements of the Contract Documents.

§ 9.9 PARTIAL OCCUPANCY OR USE

§ 9.9.1 The Owner may occupy or use any completed or partially completed portion of the Work at any stage when such portion is designated by separate agreement with the Contractor, provided such occupancy or use is consented to by the insurer as required under Section 11.3.1.5 and authorized by public authorities having jurisdiction over the Project. Such partial occupancy or use may commence whether or not the portion is substantially complete, provided the Owner and Contractor have accepted in writing the responsibilities assigned to each of them for payments, retainage, if any, security, maintenance, heat, utilities, damage to the Work and insurance, and have agreed in writing concerning the period for correction of the Work and commencement of warranties required by the Contract Documents. When the Contractor considers a portion substantially complete, the Contractor shall prepare and submit a list to the Architect as provided under Section 9.8.2. Consent of the Contractor to partial occupancy or use shall not be unreasonably withheld. The stage of the progress of the Work shall be determined by written agreement between the Owner and Contractor or, if no agreement is reached, by decision of the Architect.

§ 9.9.2 Immediately prior to such partial occupancy or use, the Owner, Contractor and Architect shall jointly inspect the area to be occupied or portion of the Work to be used in order to determine and record the condition of the Work.

§ 9.9.3 Unless otherwise agreed upon, partial occupancy or use of a portion or portions of the Work shall not constitute acceptance of Work not complying with the requirements of the Contract Documents.

§ 9.10 FINAL COMPLETION AND FINAL PAYMENT

§ 9.10.1 Upon receipt of the Contractor's written notice that the Work is ready for final inspection and acceptance and upon receipt of a final Application for Payment, the Architect will promptly make such inspection and, when the Architect finds the Work acceptable under the Contract Documents and the Contract fully performed, the Architect will promptly issue a final Certificate for Payment stating that to the best of the Architect's knowledge, information and belief, and on the basis of the Architect's on-site visits and inspections, the Work has been completed in accordance with terms and conditions of the Contract Documents and that the entire balance found to be due the Contractor and noted in the final Certificate is due and payable. The Architect's final Certificate for Payment will constitute a further representation that conditions listed in Section 9.10.2 as precedent to the Contractor's being entitled to final payment have been fulfilled.

§ 9.10.2 Neither final payment nor any remaining retained percentage shall become due until the Contractor submits to the Architect (1) an affidavit that payrolls, bills for materials and equipment, and other indebtedness connected with the Work for which the Owner or the Owner's property might be responsible or encumbered (less amounts withheld by Owner) have been paid or otherwise satisfied, (2) a certificate evidencing that insurance required by the Contract Documents to remain in force after final payment is currently in effect and will not be canceled or allowed to expire until at least 30 days' prior written notice has been given to the Owner, (3) a written statement that the Contractor knows of no substantial reason that the insurance will not be renewable to cover the period required by the Contract Documents, (4) consent of surety, if any, to final payment and (5), if required by the Owner, other data establishing payment or satisfaction of obligations, such as receipts, releases and waivers of liens, claims, security interests or encumbrances arising out of the Contract, to the extent and in such form as may be designated by the Owner. If a Subcontractor refuses to furnish a release or waiver required by the Owner, the Contractor may furnish a bond satisfactory to the Owner to indemnify the Owner against such lien. If such lien remains unsatisfied after payments are made, the Contractor shall refund to the Owner all money that the Owner may be compelled to pay in discharging such lien, including all costs and reasonable attorneys' fees.

§ 9.10.3 If, after Substantial Completion of the Work, final completion thereof is materially delayed through no fault of the Contractor or by issuance of Change Orders affecting final completion, and the Architect so confirms, the Owner shall, upon application by the Contractor and certification by the Architect, and without terminating the Contract, make payment of the balance due for that portion of the Work fully completed and accepted. If the remaining balance for Work not fully completed or corrected is less than retainage stipulated in the Contract Documents, and if bonds have been furnished, the written consent of surety to payment of the balance due for that portion of the Work fully completed and accepted shall be submitted by the Contractor to the Architect prior to certification of such payment. Such payment shall be made under terms and conditions governing final payment, except that it shall not constitute a waiver of claims.

§ 9.10.4 The making of final payment shall constitute a waiver of Claims by the Owner except those arising from
.1 liens, Claims, security interests or encumbrances arising out of the Contract and unsettled;
.2 failure of the Work to comply with the requirements of the Contract Documents; or
.3 terms of special warranties required by the Contract Documents.

§ 9.10.5 Acceptance of final payment by the Contractor, a Subcontractor or material supplier shall constitute a waiver of claims by that payee except those previously made in writing and identified by that payee as unsettled at the time of final Application for Payment.

ARTICLE 10   PROTECTION OF PERSONS AND PROPERTY
§ 10.1 SAFETY PRECAUTIONS AND PROGRAMS
The Contractor shall be responsible for initiating, maintaining and supervising all safety precautions and programs in connection with the performance of the Contract.

§ 10.2 SAFETY OF PERSONS AND PROPERTY
§ 10.2.1 The Contractor shall take reasonable precautions for safety of, and shall provide reasonable protection to prevent damage, injury or loss to
.1 employees on the Work and other persons who may be affected thereby;
.2 the Work and materials and equipment to be incorporated therein, whether in storage on or off the site, under care, custody or control of the Contractor or the Contractor's Subcontractors or Sub-subcontractors; and
.3 other property at the site or adjacent thereto, such as trees, shrubs, lawns, walks, pavements, roadways, structures and utilities not designated for removal, relocation or replacement in the course of construction.

§ 10.2.2 The Contractor shall comply with and give notices required by applicable laws, statutes, ordinances, codes, rules and regulations, and lawful orders of public authorities bearing on safety of persons or property or their protection from damage, injury or loss.

§ 10.2.3 The Contractor shall erect and maintain, as required by existing conditions and performance of the Contract, reasonable safeguards for safety and protection, including posting danger signs and other warnings against hazards, promulgating safety regulations and notifying owners and users of adjacent sites and utilities.

§ 10.2.4 When use or storage of explosives or other hazardous materials or equipment or unusual methods are necessary for execution of the Work, the Contractor shall exercise utmost care and carry on such activities under supervision of properly qualified personnel.

§ 10.2.5 The Contractor shall promptly remedy damage and loss (other than damage or loss insured under property insurance required by the Contract Documents) to property referred to in Sections 10.2.1.2 and 10.2.1.3 caused in whole or in part by the Contractor, a Subcontractor, a Sub-subcontractor, or anyone directly or indirectly employed by any of them, or by anyone for whose acts they may be liable and for which the Contractor is responsible under Sections 10.2.1.2 and 10.2.1.3, except damage or loss attributable to acts or omissions of the Owner or Architect or anyone directly or indirectly employed by either of them, or by anyone for whose acts either of them may be liable, and not attributable to the fault or negligence of the Contractor. The foregoing obligations of the Contractor are in addition to the Contractor's obligations under Section 3.18.

§ 10.2.6 The Contractor shall designate a responsible member of the Contractor's organization at the site whose duty shall be the prevention of accidents. This person shall be the Contractor's superintendent unless otherwise designated by the Contractor in writing to the Owner and Architect.

§ 10.2.7 The Contractor shall not permit any part of the construction or site to be loaded so as to cause damage or create an unsafe condition.

§ 10.2.8 INJURY OR DAMAGE TO PERSON OR PROPERTY
If either party suffers injury or damage to person or property because of an act or omission of the other party, or of others for whose acts such party is legally responsible, written notice of such injury or damage, whether or not insured, shall be given to the other party within a reasonable time not exceeding 21 days after discovery. The notice shall provide sufficient detail to enable the other party to investigate the matter.

§ 10.3 HAZARDOUS MATERIALS
§ 10.3.1 The Contractor is responsible for compliance with any requirements included in the Contract Documents regarding hazardous materials. If the Contractor encounters a hazardous material or substance not addressed in the Contract Documents and if reasonable precautions will be inadequate to prevent foreseeable bodily injury or death to persons resulting from a material or substance, including but not limited to asbestos or polychlorinated biphenyl (PCB), encountered on the site by the Contractor, the Contractor shall, upon recognizing the condition, immediately stop Work in the affected area and report the condition to the Owner and Architect in writing.

§ 10.3.2 Upon receipt of the Contractor's written notice, the Owner shall obtain the services of a licensed laboratory to verify the presence or absence of the material or substance reported by the Contractor and, in the event such material or substance is found to be present, to cause it to be rendered harmless. Unless otherwise required by the Contract Documents, the Owner shall furnish in writing to the Contractor and Architect the names and qualifications of persons or entities who are to perform tests verifying the presence or absence of such material or substance or who are to perform the task of removal or safe containment of such material or substance. The Contractor and the Architect will promptly reply to the Owner in writing stating whether or not either has reasonable objection to the persons or entities proposed by the Owner. If either the Contractor or Architect has an objection to a person or entity proposed by the Owner, the Owner shall propose another to whom the Contractor and the Architect have no reasonable objection. When the material or substance has been rendered harmless, Work in the affected area shall resume upon written agreement of the Owner and Contractor. By Change Order, the Contract Time shall be extended appropriately and the Contract Sum shall be increased in the amount of the Contractor's reasonable additional costs of shut-down, delay and start-up.

§ 10.3.3 To the fullest extent permitted by law, the Owner shall indemnify and hold harmless the Contractor, Subcontractors, Architect, Architect's consultants and agents and employees of any of them from and against claims, damages, losses and expenses, including but not limited to attorneys' fees, arising out of or resulting from performance of the Work in the affected area if in fact the material or substance presents the risk of bodily injury or death as described in Section 10.3.1 and has not been rendered harmless, provided that such claim, damage, loss or expense is attributable to bodily injury, sickness, disease or death, or to injury to or destruction of tangible property (other than the Work itself), except to the extent that such damage, loss or expense is due to the fault or negligence of the party seeking indemnity.

§ 10.3.4 The Owner shall not be responsible under this Section 10.3 for materials or substances the Contractor brings to the site unless such materials or substances are required by the Contract Documents. The Owner shall be responsible for materials or substances required by the Contract Documents, except to the extent of the Contractor's fault or negligence in the use and handling of such materials or substances.

§ 10.3.5 The Contractor shall indemnify the Owner for the cost and expense the Owner incurs (1) for remediation of a material or substance the Contractor brings to the site and negligently handles, or (2) where the Contractor fails to perform its obligations under Section 10.3.1, except to the extent that the cost and expense are due to the Owner's fault or negligence.

§ 10.3.6 If, without negligence on the part of the Contractor, the Contractor is held liable by a government agency for the cost of remediation of a hazardous material or substance solely by reason of performing Work as required by the Contract Documents, the Owner shall indemnify the Contractor for all cost and expense thereby incurred.

§ 10.4 EMERGENCIES
In an emergency affecting safety of persons or property, the Contractor shall act, at the Contractor's discretion, to prevent threatened damage, injury or loss. Additional compensation or extension of time claimed by the Contractor on account of an emergency shall be determined as provided in Article 15 and Article 7.

ARTICLE 11   INSURANCE AND BONDS
§ 11.1 CONTRACTOR'S LIABILITY INSURANCE
§ 11.1.1 The Contractor shall purchase from and maintain in a company or companies lawfully authorized to do business in the jurisdiction in which the Project is located such insurance as will protect the Contractor from claims set forth below which may arise out of or result from the Contractor's operations and completed operations under the Contract and for which the Contractor may be legally liable, whether such operations be by the Contractor or by a Subcontractor or by anyone directly or indirectly employed by any of them, or by anyone for whose acts any of them may be liable:

.1 Claims under workers' compensation, disability benefit and other similar employee benefit acts that are applicable to the Work to be performed;
.2 Claims for damages because of bodily injury, occupational sickness or disease, or death of the Contractor's employees;
.3 Claims for damages because of bodily injury, sickness or disease, or death of any person other than the Contractor's employees;
.4 Claims for damages insured by usual personal injury liability coverage;
.5 Claims for damages, other than to the Work itself, because of injury to or destruction of tangible property, including loss of use resulting therefrom;
.6 Claims for damages because of bodily injury, death of a person or property damage arising out of ownership, maintenance or use of a motor vehicle;
.7 Claims for bodily injury or property damage arising out of completed operations; and
.8 Claims involving contractual liability insurance applicable to the Contractor's obligations under Section 3.18.

§ 11.1.2 The insurance required by Section 11.1.1 shall be written for not less than limits of liability specified in the Contract Documents or required by law, whichever coverage is greater. Coverages, whether written on an occurrence or claims-made basis, shall be maintained without interruption from the date of commencement of the Work until the date of final payment and termination of any coverage required to be maintained after final payment, and, with respect to the Contractor's completed operations coverage, until the expiration of the period for correction of Work or for such other period for maintenance of completed operations coverage as specified in the Contract Documents.

§ 11.1.3 Certificates of insurance acceptable to the Owner shall be filed with the Owner prior to commencement of the Work and thereafter upon renewal or replacement of each required policy of insurance. These certificates and the insurance policies required by this Section 11.1 shall contain a provision that coverages afforded under the policies will not be canceled or allowed to expire until at least 30 days' prior written notice has been given to the Owner. An additional certificate evidencing continuation of liability coverage, including coverage for completed operations, shall be submitted with the final Application for Payment as required by Section 9.10.2 and thereafter upon renewal or replacement of such coverage until the expiration of the time required by Section 11.1.2. Information concerning reduction of coverage on account of revised limits or claims paid under the General Aggregate, or both, shall be furnished by the Contractor with reasonable promptness.

§ 11.1.4 The Contractor shall cause the commercial liability coverage required by the Contract Documents to include (1) the Owner, the Architect and the Architect's Consultants as additional insureds for claims caused in whole or in part by the Contractor's negligent acts or omissions during the Contractor's operations; and (2) the Owner as an additional insured for claims caused in whole or in part by the Contractor's negligent acts or omissions during the Contractor's completed operations.

§ 11.2 OWNER'S LIABILITY INSURANCE
The Owner shall be responsible for purchasing and maintaining the Owner's usual liability insurance.

§ 11.3 PROPERTY INSURANCE
§ 11.3.1 Unless otherwise provided, the Owner shall purchase and maintain, in a company or companies lawfully authorized to do business in the jurisdiction in which the Project is located, property insurance written on a builder's

risk "all-risk" or equivalent policy form in the amount of the initial Contract Sum, plus value of subsequent Contract Modifications and cost of materials supplied or installed by others, comprising total value for the entire Project at the site on a replacement cost basis without optional deductibles. Such property insurance shall be maintained, unless otherwise provided in the Contract Documents or otherwise agreed in writing by all persons and entities who are beneficiaries of such insurance, until final payment has been made as provided in Section 9.10 or until no person or entity other than the Owner has an insurable interest in the property required by this Section 11.3 to be covered, whichever is later. This insurance shall include interests of the Owner, the Contractor, Subcontractors and Sub-subcontractors in the Project.

§ 11.3.1.1 Property insurance shall be on an "all-risk" or equivalent policy form and shall include, without limitation, insurance against the perils of fire (with extended coverage) and physical loss or damage including, without duplication of coverage, theft, vandalism, malicious mischief, collapse, earthquake, flood, windstorm, falsework, testing and startup, temporary buildings and debris removal including demolition occasioned by enforcement of any applicable legal requirements, and shall cover reasonable compensation for Architect's and Contractor's services and expenses required as a result of such insured loss.

§ 11.3.1.2 If the Owner does not intend to purchase such property insurance required by the Contract and with all of the coverages in the amount described above, the Owner shall so inform the Contractor in writing prior to commencement of the Work. The Contractor may then effect insurance that will protect the interests of the Contractor, Subcontractors and Sub-subcontractors in the Work, and by appropriate Change Order the cost thereof shall be charged to the Owner. If the Contractor is damaged by the failure or neglect of the Owner to purchase or maintain insurance as described above, without so notifying the Contractor in writing, then the Owner shall bear all reasonable costs properly attributable thereto.

§ 11.3.1.3 If the property insurance requires deductibles, the Owner shall pay costs not covered because of such deductibles.

§ 11.3.1.4 This property insurance shall cover portions of the Work stored off the site, and also portions of the Work in transit.

§ 11.3.1.5 Partial occupancy or use in accordance with Section 9.9 shall not commence until the insurance company or companies providing property insurance have consented to such partial occupancy or use by endorsement or otherwise. The Owner and the Contractor shall take reasonable steps to obtain consent of the insurance company or companies and shall, without mutual written consent, take no action with respect to partial occupancy or use that would cause cancellation, lapse or reduction of insurance.

§ 11.3.2 BOILER AND MACHINERY INSURANCE
The Owner shall purchase and maintain boiler and machinery insurance required by the Contract Documents or by law, which shall specifically cover such insured objects during installation and until final acceptance by the Owner; this insurance shall include interests of the Owner, Contractor, Subcontractors and Sub-subcontractors in the Work, and the Owner and Contractor shall be named insureds.

§ 11.3.3 LOSS OF USE INSURANCE
The Owner, at the Owner's option, may purchase and maintain such insurance as will insure the Owner against loss of use of the Owner's property due to fire or other hazards, however caused. The Owner waives all rights of action against the Contractor for loss of use of the Owner's property, including consequential losses due to fire or other hazards however caused.

§ 11.3.4 If the Contractor requests in writing that insurance for risks other than those described herein or other special causes of loss be included in the property insurance policy, the Owner shall, if possible, include such insurance, and the cost thereof shall be charged to the Contractor by appropriate Change Order.

§ 11.3.5 If during the Project construction period the Owner insures properties, real or personal or both, at or adjacent to the site by property insurance under policies separate from those insuring the Project, or if after final payment property insurance is to be provided on the completed Project through a policy or policies other than those insuring the Project during the construction period, the Owner shall waive all rights in accordance with the terms of Section 11.3.7 for damages caused by fire or other causes of loss covered by this separate property insurance. All separate policies shall provide this waiver of subrogation by endorsement or otherwise.

§ 11.3.6 Before an exposure to loss may occur, the Owner shall file with the Contractor a copy of each policy that includes insurance coverages required by this Section 11.3. Each policy shall contain all generally applicable conditions, definitions, exclusions and endorsements related to this Project. Each policy shall contain a provision that the policy will not be canceled or allowed to expire, and that its limits will not be reduced, until at least 30 days' prior written notice has been given to the Contractor.

§ 11.3.7 WAIVERS OF SUBROGATION
The Owner and Contractor waive all rights against (1) each other and any of their subcontractors, sub-subcontractors, agents and employees, each of the other, and (2) the Architect, Architect's consultants, separate contractors described in Article 6, if any, and any of their subcontractors, sub-subcontractors, agents and employees, for damages caused by fire or other causes of loss to the extent covered by property insurance obtained pursuant to this Section 11.3 or other property insurance applicable to the Work, except such rights as they have to proceeds of such insurance held by the Owner as fiduciary. The Owner or Contractor, as appropriate, shall require of the Architect, Architect's consultants, separate contractors described in Article 6, if any, and the subcontractors, sub-subcontractors, agents and employees of any of them, by appropriate agreements, written where legally required for validity, similar waivers each in favor of other parties enumerated herein. The policies shall provide such waivers of subrogation by endorsement or otherwise. A waiver of subrogation shall be effective as to a person or entity even though that person or entity would otherwise have a duty of indemnification, contractual or otherwise, did not pay the insurance premium directly or indirectly, and whether or not the person or entity had an insurable interest in the property damaged.

§ 11.3.8 A loss insured under the Owner's property insurance shall be adjusted by the Owner as fiduciary and made payable to the Owner as fiduciary for the insureds, as their interests may appear, subject to requirements of any applicable mortgagee clause and of Section 11.3.10. The Contractor shall pay Subcontractors their just shares of insurance proceeds received by the Contractor, and by appropriate agreements, written where legally required for validity, shall require Subcontractors to make payments to their Sub-subcontractors in similar manner.

§ 11.3.9 If required in writing by a party in interest, the Owner as fiduciary shall, upon occurrence of an insured loss, give bond for proper performance of the Owner's duties. The cost of required bonds shall be charged against proceeds received as fiduciary. The Owner shall deposit in a separate account proceeds so received, which the Owner shall distribute in accordance with such agreement as the parties in interest may reach, or as determined in accordance with the method of binding dispute resolution selected in the Agreement between the Owner and Contractor. If after such loss no other special agreement is made and unless the Owner terminates the Contract for convenience, replacement of damaged property shall be performed by the Contractor after notification of a Change in the Work in accordance with Article 7.

§ 11.3.10 The Owner as fiduciary shall have power to adjust and settle a loss with insurers unless one of the parties in interest shall object in writing within five days after occurrence of loss to the Owner's exercise of this power; if such objection is made, the dispute shall be resolved in the manner selected by the Owner and Contractor as the method of binding dispute resolution in the Agreement. If the Owner and Contractor have selected arbitration as the method of binding dispute resolution, the Owner as fiduciary shall make settlement with insurers or, in the case of a dispute over distribution of insurance proceeds, in accordance with the directions of the arbitrators.

§ 11.4 PERFORMANCE BOND AND PAYMENT BOND
§ 11.4.1 The Owner shall have the right to require the Contractor to furnish bonds covering faithful performance of the Contract and payment of obligations arising thereunder as stipulated in bidding requirements or specifically required in the Contract Documents on the date of execution of the Contract.

§ 11.4.2 Upon the request of any person or entity appearing to be a potential beneficiary of bonds covering payment of obligations arising under the Contract, the Contractor shall promptly furnish a copy of the bonds or shall authorize a copy to be furnished.

ARTICLE 12 UNCOVERING AND CORRECTION OF WORK
§ 12.1 UNCOVERING OF WORK
§ 12.1.1 If a portion of the Work is covered contrary to the Architect's request or to requirements specifically expressed in the Contract Documents, it must, if requested in writing by the Architect, be uncovered for the Architect's examination and be replaced at the Contractor's expense without change in the Contract Time.

§ 12.1.2 If a portion of the Work has been covered that the Architect has not specifically requested to examine prior to its being covered, the Architect may request to see such Work and it shall be uncovered by the Contractor. If such Work is in accordance with the Contract Documents, costs of uncovering and replacement shall, by appropriate Change Order, be at the Owner's expense. If such Work is not in accordance with the Contract Documents, such costs and the cost of correction shall be at the Contractor's expense unless the condition was caused by the Owner or a separate contractor in which event the Owner shall be responsible for payment of such costs.

§ 12.2 CORRECTION OF WORK
§ 12.2.1 BEFORE OR AFTER SUBSTANTIAL COMPLETION
The Contractor shall promptly correct Work rejected by the Architect or failing to conform to the requirements of the Contract Documents, whether discovered before or after Substantial Completion and whether or not fabricated, installed or completed. Costs of correcting such rejected Work, including additional testing and inspections, the cost of uncovering and replacement, and compensation for the Architect's services and expenses made necessary thereby, shall be at the Contractor's expense.

§ 12.2.2 AFTER SUBSTANTIAL COMPLETION
§ 12.2.2.1 In addition to the Contractor's obligations under Section 3.5, if, within one year after the date of Substantial Completion of the Work or designated portion thereof or after the date for commencement of warranties established under Section 9.9.1, or by terms of an applicable special warranty required by the Contract Documents, any of the Work is found to be not in accordance with the requirements of the Contract Documents, the Contractor shall correct it promptly after receipt of written notice from the Owner to do so unless the Owner has previously given the Contractor a written acceptance of such condition. The Owner shall give such notice promptly after discovery of the condition. During the one-year period for correction of Work, if the Owner fails to notify the Contractor and give the Contractor an opportunity to make the correction, the Owner waives the rights to require correction by the Contractor and to make a claim for breach of warranty. If the Contractor fails to correct nonconforming Work within a reasonable time during that period after receipt of notice from the Owner or Architect, the Owner may correct it in accordance with Section 2.4.

§ 12.2.2.2 The one-year period for correction of Work shall be extended with respect to portions of Work first performed after Substantial Completion by the period of time between Substantial Completion and the actual completion of that portion of the Work.

§ 12.2.2.3 The one-year period for correction of Work shall not be extended by corrective Work performed by the Contractor pursuant to this Section 12.2.

§ 12.2.3 The Contractor shall remove from the site portions of the Work that are not in accordance with the requirements of the Contract Documents and are neither corrected by the Contractor nor accepted by the Owner.

§ 12.2.4 The Contractor shall bear the cost of correcting destroyed or damaged construction, whether completed or partially completed, of the Owner or separate contractors caused by the Contractor's correction or removal of Work that is not in accordance with the requirements of the Contract Documents.

§ 12.2.5 Nothing contained in this Section 12.2 shall be construed to establish a period of limitation with respect to other obligations the Contractor has under the Contract Documents. Establishment of the one-year period for correction of Work as described in Section 12.2.2 relates only to the specific obligation of the Contractor to correct the Work, and has no relationship to the time within which the obligation to comply with the Contract Documents may be sought to be enforced, nor to the time within which proceedings may be commenced to establish the Contractor's liability with respect to the Contractor's obligations other than specifically to correct the Work.

§ 12.3 ACCEPTANCE OF NONCONFORMING WORK
If the Owner prefers to accept Work that is not in accordance with the requirements of the Contract Documents, the Owner may do so instead of requiring its removal and correction, in which case the Contract Sum will be reduced as appropriate and equitable. Such adjustment shall be effected whether or not final payment has been made.

ARTICLE 13 MISCELLANEOUS PROVISIONS

§ 13.1 GOVERNING LAW
The Contract shall be governed by the law of the place where the Project is located except that, if the parties have selected arbitration as the method of binding dispute resolution, the Federal Arbitration Act shall govern Section 15.4.

§ 13.2 SUCCESSORS AND ASSIGNS
§ 13.2.1 The Owner and Contractor respectively bind themselves, their partners, successors, assigns and legal representatives to covenants, agreements and obligations contained in the Contract Documents. Except as provided in Section 13.2.2, neither party to the Contract shall assign the Contract as a whole without written consent of the other. If either party attempts to make such an assignment without such consent, that party shall nevertheless remain legally responsible for all obligations under the Contract.

§ 13.2.2 The Owner may, without consent of the Contractor, assign the Contract to a lender providing construction financing for the Project, if the lender assumes the Owner's rights and obligations under the Contract Documents. The Contractor shall execute all consents reasonably required to facilitate such assignment.

§ 13.3 WRITTEN NOTICE
Written notice shall be deemed to have been duly served if delivered in person to the individual, to a member of the firm or entity, or to an officer of the corporation for which it was intended; or if delivered at, or sent by registered or certified mail or by courier service providing proof of delivery to, the last business address known to the party giving notice.

§ 13.4 RIGHTS AND REMEDIES
§ 13.4.1 Duties and obligations imposed by the Contract Documents and rights and remedies available thereunder shall be in addition to and not a limitation of duties, obligations, rights and remedies otherwise imposed or available by law.

§ 13.4.2 No action or failure to act by the Owner, Architect or Contractor shall constitute a waiver of a right or duty afforded them under the Contract, nor shall such action or failure to act constitute approval of or acquiescence in a breach there under, except as may be specifically agreed in writing.

§ 13.5 TESTS AND INSPECTIONS
§ 13.5.1 Tests, inspections and approvals of portions of the Work shall be made as required by the Contract Documents and by applicable laws, statutes, ordinances, codes, rules and regulations or lawful orders of public authorities. Unless otherwise provided, the Contractor shall make arrangements for such tests, inspections and approvals with an independent testing laboratory or entity acceptable to the Owner, or with the appropriate public authority, and shall bear all related costs of tests, inspections and approvals. The Contractor shall give the Architect timely notice of when and where tests and inspections are to be made so that the Architect may be present for such procedures. The Owner shall bear costs of (1) tests, inspections or approvals that do not become requirements until after bids are received or negotiations concluded, and (2) tests, inspections or approvals where building codes or applicable laws or regulations prohibit the Owner from delegating their cost to the Contractor.

§ 13.5.2 If the Architect, Owner or public authorities having jurisdiction determine that portions of the Work require additional testing, inspection or approval not included under Section 13.5.1, the Architect will, upon written authorization from the Owner, instruct the Contractor to make arrangements for such additional testing, inspection or approval by an entity acceptable to the Owner, and the Contractor shall give timely notice to the Architect of when and where tests and inspections are to be made so that the Architect may be present for such procedures. Such costs, except as provided in Section 13.5.3, shall be at the Owner's expense.

§ 13.5.3 If such procedures for testing, inspection or approval under Sections 13.5.1 and 13.5.2 reveal failure of the portions of the Work to comply with requirements established by the Contract Documents, all costs made necessary by such failure including those of repeated procedures and compensation for the Architect's services and expenses shall be at the Contractor's expense.

§ 13.5.4 Required certificates of testing, inspection or approval shall, unless otherwise required by the Contract Documents, be secured by the Contractor and promptly delivered to the Architect.

§ **13.5.5** If the Architect is to observe tests, inspections or approvals required by the Contract Documents, the Architect will do so promptly and, where practicable, at the normal place of testing.

§ **13.5.6** Tests or inspections conducted pursuant to the Contract Documents shall be made promptly to avoid unreasonable delay in the Work.

§ 13.6 INTEREST
Payments due and unpaid under the Contract Documents shall bear interest from the date payment is due at such rate as the parties may agree upon in writing or, in the absence thereof, at the legal rate prevailing from time to time at the place where the Project is located.

§ 13.7 TIME LIMITS ON CLAIMS
The Owner and Contractor shall commence all claims and causes of action, whether in contract, tort, breach of warranty or otherwise, against the other arising out of or related to the Contract in accordance with the requirements of the final dispute resolution method selected in the Agreement within the time period specified by applicable law, but in any case not more than 10 years after the date of Substantial Completion of the Work. The Owner and Contractor waive all claims and causes of action not commenced in accordance with this Section 13.7.

ARTICLE 14   TERMINATION OR SUSPENSION OF THE CONTRACT
§ 14.1 TERMINATION BY THE CONTRACTOR
§ **14.1.1** The Contractor may terminate the Contract if the Work is stopped for a period of 30 consecutive days through no act or fault of the Contractor or a Subcontractor, Sub-subcontractor or their agents or employees or any other persons or entities performing portions of the Work under direct or indirect contract with the Contractor, for any of the following reasons:
  - .1 Issuance of an order of a court or other public authority having jurisdiction that requires all Work to be stopped;
  - .2 An act of government, such as a declaration of national emergency that requires all Work to be stopped;
  - .3 Because the Architect has not issued a Certificate for Payment and has not notified the Contractor of the reason for withholding certification as provided in Section 9.4.1, or because the Owner has not made payment on a Certificate for Payment within the time stated in the Contract Documents; or
  - .4 The Owner has failed to furnish to the Contractor promptly, upon the Contractor's request, reasonable evidence as required by Section 2.2.1.

§ **14.1.2** The Contractor may terminate the Contract if, through no act or fault of the Contractor or a Subcontractor, Sub-subcontractor or their agents or employees or any other persons or entities performing portions of the Work under direct or indirect contract with the Contractor, repeated suspensions, delays or interruptions of the entire Work by the Owner as described in Section 14.3 constitute in the aggregate more than 100 percent of the total number of days scheduled for completion, or 120 days in any 365-day period, whichever is less.

§ **14.1.3** If one of the reasons described in Section 14.1.1 or 14.1.2 exists, the Contractor may, upon seven days' written notice to the Owner and Architect, terminate the Contract and recover from the Owner payment for Work executed, including reasonable overhead and profit, costs incurred by reason of such termination, and damages.

§ **14.1.4** If the Work is stopped for a period of 60 consecutive days through no act or fault of the Contractor or a Subcontractor or their agents or employees or any other persons performing portions of the Work under contract with the Contractor because the Owner has repeatedly failed to fulfill the Owner's obligations under the Contract Documents with respect to matters important to the progress of the Work, the Contractor may, upon seven additional days' written notice to the Owner and the Architect, terminate the Contract and recover from the Owner as provided in Section 14.1.3.

§ 14.2 TERMINATION BY THE OWNER FOR CAUSE
§ **14.2.1** The Owner may terminate the Contract if the Contractor
  - .1 repeatedly refuses or fails to supply enough properly skilled workers or proper materials;
  - .2 fails to make payment to Subcontractors for materials or labor in accordance with the respective agreements between the Contractor and the Subcontractors;
  - .3 repeatedly disregards applicable laws, statutes, ordinances, codes, rules and regulations, or lawful orders of a public authority; or
  - .4 otherwise is guilty of substantial breach of a provision of the Contract Documents.

§ 14.2.2 When any of the above reasons exist, the Owner, upon certification by the Initial Decision Maker that sufficient cause exists to justify such action, may without prejudice to any other rights or remedies of the Owner and after giving the Contractor and the Contractor's surety, if any, seven days' written notice, terminate employment of the Contractor and may, subject to any prior rights of the surety:

.1 Exclude the Contractor from the site and take possession of all materials, equipment, tools, and construction equipment and machinery thereon owned by the Contractor;
.2 Accept assignment of subcontracts pursuant to Section 5.4; and
.3 Finish the Work by whatever reasonable method the Owner may deem expedient. Upon written request of the Contractor, the Owner shall furnish to the Contractor a detailed accounting of the costs incurred by the Owner in finishing the Work.

§ 14.2.3 When the Owner terminates the Contract for one of the reasons stated in Section 14.2.1, the Contractor shall not be entitled to receive further payment until the Work is finished.

§ 14.2.4 If the unpaid balance of the Contract Sum exceeds costs of finishing the Work, including compensation for the Architect's services and expenses made necessary thereby, and other damages incurred by the Owner and not expressly waived, such excess shall be paid to the Contractor. If such costs and damages exceed the unpaid balance, the Contractor shall pay the difference to the Owner. The amount to be paid to the Contractor or Owner, as the case may be, shall be certified by the Initial Decision Maker, upon application, and this obligation for payment shall survive termination of the Contract.

§ 14.3 SUSPENSION BY THE OWNER FOR CONVENIENCE
§ 14.3.1 The Owner may, without cause, order the Contractor in writing to suspend, delay or interrupt the Work in whole or in part for such period of time as the Owner may determine.

§ 14.3.2 The Contract Sum and Contract Time shall be adjusted for increases in the cost and time caused by suspension, delay or interruption as described in Section 14.3.1. Adjustment of the Contract Sum shall include profit. No adjustment shall be made to the extent

.1 that performance is, was or would have been so suspended, delayed or interrupted by another cause for which the Contractor is responsible; or
.2 that an equitable adjustment is made or denied under another provision of the Contract.

§ 14.4 TERMINATION BY THE OWNER FOR CONVENIENCE
§ 14.4.1 The Owner may, at any time, terminate the Contract for the Owner's convenience and without cause.

§ 14.4.2 Upon receipt of written notice from the Owner of such termination for the Owner's convenience, the Contractor shall

.1 cease operations as directed by the Owner in the notice;
.2 take actions necessary, or that the Owner may direct, for the protection and preservation of the Work; and
.3 except for Work directed to be performed prior to the effective date of termination stated in the notice, terminate all existing subcontracts and purchase orders and enter into no further subcontracts and purchase orders.

§ 14.4.3 In case of such termination for the Owner's convenience, the Contractor shall be entitled to receive payment for Work executed, and costs incurred by reason of such termination, along with reasonable overhead and profit on the Work not executed.

ARTICLE 15 CLAIMS AND DISPUTES
§ 15.1 CLAIMS
§ 15.1.1 DEFINITION
A Claim is a demand or assertion by one of the parties seeking, as a matter of right, payment of money, or other relief with respect to the terms of the Contract. The term "Claim" also includes other disputes and matters in question between the Owner and Contractor arising out of or relating to the Contract. The responsibility to substantiate Claims shall rest with the party making the Claim.

§ 15.1.2 NOTICE OF CLAIMS
Claims by either the Owner or Contractor must be initiated by written notice to the other party and to the Initial Decision Maker with a copy sent to the Architect, if the Architect is not serving as the Initial Decision Maker.

Claims by either party must be initiated within 21 days after occurrence of the event giving rise to such Claim or within 21 days after the claimant first recognizes the condition giving rise to the Claim, whichever is later.

§ 15.1.3 CONTINUING CONTRACT PERFORMANCE
Pending final resolution of a Claim, except as otherwise agreed in writing or as provided in Section 9.7 and Article 14, the Contractor shall proceed diligently with performance of the Contract and the Owner shall continue to make payments in accordance with the Contract Documents. The Architect will prepare Change Orders and issue Certificates for Payment in accordance with the decisions of the Initial Decision Maker.

§ 15.1.4 CLAIMS FOR ADDITIONAL COST
If the Contractor wishes to make a Claim for an increase in the Contract Sum, written notice as provided herein shall be given before proceeding to execute the Work. Prior notice is not required for Claims relating to an emergency endangering life or property arising under Section 10.4.

§ 15.1.5 CLAIMS FOR ADDITIONAL TIME
§ 15.1.5.1 If the Contractor wishes to make a Claim for an increase in the Contract Time, written notice as provided herein shall be given. The Contractor's Claim shall include an estimate of cost and of probable effect of delay on progress of the Work. In the case of a continuing delay, only one Claim is necessary.

§ 15.1.5.2 If adverse weather conditions are the basis for a Claim for additional time, such Claim shall be documented by data substantiating that weather conditions were abnormal for the period of time, could not have been reasonably anticipated and had an adverse effect on the scheduled construction.

§ 15.1.6 CLAIMS FOR CONSEQUENTIAL DAMAGES
The Contractor and Owner waive Claims against each other for consequential damages arising out of or relating to this Contract. This mutual waiver includes

.1 damages incurred by the Owner for rental expenses, for losses of use, income, profit, financing, business and reputation, and for loss of management or employee productivity or of the services of such persons; and

.2 damages incurred by the Contractor for principal office expenses including the compensation of personnel stationed there, for losses of financing, business and reputation, and for loss of profit except anticipated profit arising directly from the Work.

This mutual waiver is applicable, without limitation, to all consequential damages due to either party's termination in accordance with Article 14. Nothing contained in this Section 15.1.6 shall be deemed to preclude an award of liquidated damages, when applicable, in accordance with the requirements of the Contract Documents.

§ 15.2 INITIAL DECISION
§ 15.2.1 Claims, excluding those arising under Sections 10.3, 10.4, 11.3.9, and 11.3.10, shall be referred to the Initial Decision Maker for initial decision. The Architect will serve as the Initial Decision Maker, unless otherwise indicated in the Agreement. Except for those Claims excluded by this Section 15.2.1, an initial decision shall be required as a condition precedent to mediation of any Claim arising prior to the date final payment is due, unless 30 days have passed after the Claim has been referred to the Initial Decision Maker with no decision having been rendered. Unless the Initial Decision Maker and all affected parties agree, the Initial Decision Maker will not decide disputes between the Contractor and persons or entities other than the Owner.

§ 15.2.2 The Initial Decision Maker will review Claims and within ten days of the receipt of a Claim take one or more of the following actions: (1) request additional supporting data from the claimant or a response with supporting data from the other party, (2) reject the Claim in whole or in part, (3) approve the Claim, (4) suggest a compromise, or (5) advise the parties that the Initial Decision Maker is unable to resolve the Claim if the Initial Decision Maker lacks sufficient information to evaluate the merits of the Claim or if the Initial Decision Maker concludes that, in the Initial Decision Maker's sole discretion, it would be inappropriate for the Initial Decision Maker to resolve the Claim.

§ 15.2.3 In evaluating Claims, the Initial Decision Maker may, but shall not be obligated to, consult with or seek information from either party or from persons with special knowledge or expertise who may assist the Initial Decision Maker in rendering a decision. The Initial Decision Maker may request the Owner to authorize retention of such persons at the Owner's expense.

§ 15.2.4 If the Initial Decision Maker requests a party to provide a response to a Claim or to furnish additional supporting data, such party shall respond, within ten days after receipt of such request, and shall either (1) provide a response on the requested supporting data, (2) advise the Initial Decision Maker when the response or supporting data will be furnished or (3) advise the Initial Decision Maker that no supporting data will be furnished. Upon receipt of the response or supporting data, if any, the Initial Decision Maker will either reject or approve the Claim in whole or in part.

§ 15.2.5 The Initial Decision Maker will render an initial decision approving or rejecting the Claim, or indicating that the Initial Decision Maker is unable to resolve the Claim. This initial decision shall (1) be in writing; (2) state the reasons therefor; and (3) notify the parties and the Architect, if the Architect is not serving as the Initial Decision Maker, of any change in the Contract Sum or Contract Time or both. The initial decision shall be final and binding on the parties but subject to mediation and, if the parties fail to resolve their dispute through mediation, to binding dispute resolution.

§ 15.2.6 Either party may file for mediation of an initial decision at any time, subject to the terms of Section 15.2.6.1.

§ 15.2.6.1 Either party may, within 30 days from the date of an initial decision, demand in writing that the other party file for mediation within 60 days of the initial decision. If such a demand is made and the party receiving the demand fails to file for mediation within the time required, then both parties waive their rights to mediate or pursue binding dispute resolution proceedings with respect to the initial decision.

§ 15.2.7 In the event of a Claim against the Contractor, the Owner may, but is not obligated to, notify the surety, if any, of the nature and amount of the Claim. If the Claim relates to a possibility of a Contractor's default, the Owner may, but is not obligated to, notify the surety and request the surety's assistance in resolving the controversy.

§ 15.2.8 If a Claim relates to or is the subject of a mechanic's lien, the party asserting such Claim may proceed in accordance with applicable law to comply with the lien notice or filing deadlines.

§ 15.3 MEDIATION
§ 15.3.1 Claims, disputes, or other matters in controversy arising out of or related to the Contract except those waived as provided for in Sections 9.10.4, 9.10.5, and 15.1.6 shall be subject to mediation as a condition precedent to binding dispute resolution.

§ 15.3.2 The parties shall endeavor to resolve their Claims by mediation which, unless the parties mutually agree otherwise, shall be administered by the American Arbitration Association in accordance with its Construction Industry Mediation Procedures in effect on the date of the Agreement. A request for mediation shall be made in writing, delivered to the other party to the Contract, and filed with the person or entity administering the mediation. The request may be made concurrently with the filing of binding dispute resolution proceedings but, in such event, mediation shall proceed in advance of binding dispute resolution proceedings, which shall be stayed pending mediation for a period of 60 days from the date of filing, unless stayed for a longer period by agreement of the parties or court order. If an arbitration is stayed pursuant to this Section 15.3.2, the parties may nonetheless proceed to the selection of the arbitrator(s) and agree upon a schedule for later proceedings.

§ 15.3.3 The parties shall share the mediator's fee and any filing fees equally. The mediation shall be held in the place where the Project is located, unless another location is mutually agreed upon. Agreements reached in mediation shall be enforceable as settlement agreements in any court having jurisdiction thereof.

§ 15.4 ARBITRATION
§ 15.4.1 If the parties have selected arbitration as the method for binding dispute resolution in the Agreement, any Claim subject to, but not resolved by, mediation shall be subject to arbitration which, unless the parties mutually agree otherwise, shall be administered by the American Arbitration Association in accordance with its Construction Industry Arbitration Rules in effect on the date of the Agreement. A demand for arbitration shall be made in writing, delivered to the other party to the Contract, and filed with the person or entity administering the arbitration. The party filing a notice of demand for arbitration must assert in the demand all Claims then known to that party on which arbitration is permitted to be demanded.

§ 15.4.1.1 A demand for arbitration shall be made no earlier than concurrently with the filing of a request for mediation, but in no event shall it be made after the date when the institution of legal or equitable proceedings based on the Claim would be barred by the applicable statute of limitations. For statute of limitations purposes, receipt of a written demand for arbitration by the person or entity administering the arbitration shall constitute the institution of legal or equitable proceedings based on the Claim.

§ 15.4.2 The award rendered by the arbitrator or arbitrators shall be final, and judgment may be entered upon it in accordance with applicable law in any court having jurisdiction thereof.

§ 15.4.3 The foregoing agreement to arbitrate and other agreements to arbitrate with an additional person or entity duly consented to by parties to the Agreement shall be specifically enforceable under applicable law in any court having jurisdiction thereof.

§ 15.4.4 CONSOLIDATION OR JOINDER
§ 15.4.4.1 Either party, at its sole discretion, may consolidate an arbitration conducted under this Agreement with any other arbitration to which it is a party provided that (1) the arbitration agreement governing the other arbitration permits consolidation, (2) the arbitrations to be consolidated substantially involve common questions of law or fact, and (3) the arbitrations employ materially similar procedural rules and methods for selecting arbitrator(s).

§ 15.4.4.2 Either party, at its sole discretion, may include by joinder persons or entities substantially involved in a common question of law or fact whose presence is required if complete relief is to be accorded in arbitration, provided that the party sought to be joined consents in writing to such joinder. Consent to arbitration involving an additional person or entity shall not constitute consent to arbitration of any claim, dispute or other matter in question not described in the written consent.

§ 15.4.4.3 The Owner and Contractor grant to any person or entity made a party to an arbitration conducted under this Section 15.4, whether by joinder or consolidation, the same rights of joinder and consolidation as the Owner and Contractor under this Agreement.

APPENDIX C

Standard General Conditions of the Construction Contract

Engineers Joint Contract Documents Committee (EJCDC) Document C-700 (2007)

Reprinted by permission of the National Society of Professional Engineers. For information on ordering these and other EJCDC documents, call 1-800-417-0348 or visit www.nspe.org.

> This document has important legal consequences; consultation with an attorney is encouraged with respect to its use or modification. This document should be adapted to the particular circumstances of the contemplated Project and the controlling Laws and Regulations.

STANDARD GENERAL CONDITIONS OF THE CONSTRUCTION CONTRACT

Prepared by

ENGINEERS JOINT CONTRACT DOCUMENTS COMMITTEE

and

Issued and Published Jointly by

ACEC
AMERICAN COUNCIL OF ENGINEERING COMPANIES

ASCE American Society of Civil Engineers

National Society of Professional Engineers
Professional Engineers in Private Practice

AMERICAN COUNCIL OF ENGINEERING COMPANIES

ASSOCIATED GENERAL CONTRACTORS OF AMERICA

AMERICAN SOCIETY OF CIVIL ENGINEERS

PROFESSIONAL ENGINEERS IN PRIVATE PRACTICE
A Practice Division of the
NATIONAL SOCIETY OF PROFESSIONAL ENGINEERS

Endorsed by

CONSTRUCTION SPECIFICATIONS INSTITUTE

EJCDC C-700 Standard General Conditions of the Construction Contract
Copyright © 2007 National Society of Professional Engineers for EJCDC. All rights reserved.

Reprinted with permission of the National Society of Professional Engineers (www.nspe.org)

These General Conditions have been prepared for use with the Suggested Forms of Agreement Between Owner and Contractor (EJCDC C-520 or C-525, 2007 Editions). Their provisions are interrelated and a change in one may necessitate a change in the other. Comments concerning their usage are contained in the Narrative Guide to the EJCDC Construction Documents (EJCDC C-001, 2007 Edition). For guidance in the preparation of Supplementary Conditions, see Guide to the Preparation of Supplementary Conditions (EJCDC C-800, 2007 Edition).

Copyright © 2007 National Society of Professional Engineers
1420 King Street, Alexandria, VA 22314-2794
(703) 684-2882
www.nspe.org

American Council of Engineering Companies
1015 15th Street N.W., Washington, DC 20005
(202) 347-7474
www.acec.org

American Society of Civil Engineers
1801 Alexander Bell Drive, Reston, VA 20191-4400
(800) 548-2723
www.asce.org

Associated General Contractors of America
2300 Wilson Boulevard, Suite 400, Arlington, VA 22201-3308
(703) 548-3118
www.agc.org

The copyright for this EJCDC document is owned jointly by the four EJCDC sponsoring organizations and held in trust for their benefit by NSPE.

STANDARD GENERAL CONDITIONS OF THE CONSTRUCTION CONTRACT

TABLE OF CONTENTS

                                                                                                                                Page

Article 1 – Definitions and Terminology 1
    1.01   Defined Terms 1
    1.02   Terminology 5

Article 2 – Preliminary Matters 6
    2.01   Delivery of Bonds and Evidence of Insurance 6
    2.02   Copies of Documents 6
    2.03   Commencement of Contract Times; Notice to Proceed 6
    2.04   Starting the Work 7
    2.05   Before Starting Construction 7
    2.06   Preconstruction Conference; Designation of Authorized Representatives 7
    2.07   Initial Acceptance of Schedules 7

Article 3 – Contract Documents: Intent, Amending, Reuse 8
    3.01   Intent 8
    3.02   Reference Standards 8
    3.03   Reporting and Resolving Discrepancies 8
    3.04   Amending and Supplementing Contract Documents 9
    3.05   Reuse of Documents 10
    3.06   Electronic Data 10

Article 4 – Availability of Lands; Subsurface and Physical Conditions; Hazardous Environmental
        Conditions; Reference Points 10
    4.01   Availability of Lands 10
    4.02   Subsurface and Physical Conditions 11
    4.03   Differing Subsurface or Physical Conditions 11
    4.04   Underground Facilities 13
    4.05   Reference Points 14
    4.06   Hazardous Environmental Condition at Site 14

Article 5 – Bonds and Insurance 16
    5.01   Performance, Payment, and Other Bonds 16
    5.02   Licensed Sureties and Insurers 16
    5.03   Certificates of Insurance 16
    5.04   Contractor's Insurance 17
    5.05   Owner's Liability Insurance 18
    5.06   Property Insurance 18
    5.07   Waiver of Rights 20
    5.08   Receipt and Application of Insurance Proceeds 21

5.09	Acceptance of Bonds and Insurance; Option to Replace	21
5.10	Partial Utilization, Acknowledgment of Property Insurer	21

Article 6 – Contractor's Responsibilities 22
    6.01   Supervision and Superintendence 22
    6.02   Labor; Working Hours 22
    6.03   Services, Materials, and Equipment 22
    6.04   Progress Schedule 23
    6.05   Substitutes and "Or-Equals" 23
    6.06   Concerning Subcontractors, Suppliers, and Others 25
    6.07   Patent Fees and Royalties 27
    6.08   Permits 27
    6.09   Laws and Regulations 27
    6.10   Taxes 28
    6.11   Use of Site and Other Areas 28
    6.12   Record Documents 29
    6.13   Safety and Protection 29
    6.14   Safety Representative 30
    6.15   Hazard Communication Programs 30
    6.16   Emergencies 30
    6.17   Shop Drawings and Samples 30
    6.18   Continuing the Work 32
    6.19   Contractor's General Warranty and Guarantee 32
    6.20   Indemnification 33
    6.21   Delegation of Professional Design Services 34

Article 7 – Other Work at the Site 35
    7.01   Related Work at Site 35
    7.02   Coordination 35
    7.03   Legal Relationships 36

Article 8 – Owner's Responsibilities 36
    8.01   Communications to Contractor 36
    8.02   Replacement of Engineer 36
    8.03   Furnish Data 36
    8.04   Pay When Due 36
    8.05   Lands and Easements; Reports and Tests 36
    8.06   Insurance 36
    8.07   Change Orders 36
    8.08   Inspections, Tests, and Approvals 37
    8.09   Limitations on Owner's Responsibilities 37
    8.10   Undisclosed Hazardous Environmental Condition 37
    8.11   Evidence of Financial Arrangements 37
    8.12   Compliance with Safety Program 37

Article 9 – Engineer's Status During Construction 37
    9.01   Owner's Representative 37
    9.02   Visits to Site 37

9.03 Project Representative 38
9.04 Authorized Variations in Work 38
9.05 Rejecting Defective Work 38
9.06 Shop Drawings, Change Orders and Payments... 38
9.07 Determinations for Unit Price Work 39
9.08 Decisions on Requirements of Contract Documents and Acceptability of Work...... 39
9.09 Limitations on Engineer's Authority and Responsibilities 39
9.10 Compliance with Safety Program 40

Article 10 – Changes in the Work; Claims 40
10.01 Authorized Changes in the Work 40
10.02 Unauthorized Changes in the Work 40
10.03 Execution of Change Orders 41
10.04 Notification to Surety 41
10.05 Claims... 41

Article 11 – Cost of the Work; Allowances; Unit Price Work 42
11.01 Cost of the Work 42
11.02 Allowances 45
11.03 Unit Price Work 45

Article 12 – Change of Contract Price; Change of Contract Times 46
12.01 Change of Contract Price 46
12.02 Change of Contract Times 47
12.03 Delays 47

Article 13 – Tests and Inspections; Correction, Removal or Acceptance of Defective Work 48
13.01 Notice of Defects 48
13.02 Access to Work 48
13.03 Tests and Inspections 48
13.04 Uncovering Work... 49
13.05 Owner May Stop the Work 50
13.06 Correction or Removal of Defective Work 50
13.07 Correction Period 50
13.08 Acceptance of Defective Work 51
13.09 Owner May Correct Defective Work 51

Article 14 – Payments to Contractor and Completion 52
14.01 Schedule of Values 52
14.02 Progress Payments 52
14.03 Contractor's Warranty of Title 55
14.04 Substantial Completion 55
14.05 Partial Utilization 56
14.06 Final Inspection 56
14.07 Final Payment 57
14.08 Final Completion Delayed 58
14.09 Waiver of Claims 58

Article 15 – Suspension of Work and Termination 58
    15.01 Owner May Suspend Work 58
    15.02 Owner May Terminate for Cause 58
    15.03 Owner May Terminate For Convenience 60
    15.04 Contractor May Stop Work or Terminate 60

Article 16 – Dispute Resolution 61
    16.01 Methods and Procedures 61

Article 17 – Miscellaneous 61
    17.01 Giving Notice 61
    17.02 Computation of Times 61
    17.03 Cumulative Remedies 62
    17.04 Survival of Obligations 62
    17.05 Controlling Law 62
    17.06 Headings 62

ARTICLE 1 – DEFINITIONS AND TERMINOLOGY

1.01 *Defined Terms*

A. Wherever used in the Bidding Requirements or Contract Documents and printed with initial capital letters, the terms listed below will have the meanings indicated which are applicable to both the singular and plural thereof. In addition to terms specifically defined, terms with initial capital letters in the Contract Documents include references to identified articles and paragraphs, and the titles of other documents or forms.

1. *Addenda*—Written or graphic instruments issued prior to the opening of Bids which clarify, correct, or change the Bidding Requirements or the proposed Contract Documents.

2. *Agreement*—The written instrument which is evidence of the agreement between Owner and Contractor covering the Work.

3. *Application for Payment*—The form acceptable to Engineer which is to be used by Contractor during the course of the Work in requesting progress or final payments and which is to be accompanied by such supporting documentation as is required by the Contract Documents.

4. *Asbestos*—Any material that contains more than one percent asbestos and is friable or is releasing asbestos fibers into the air above current action levels established by the United States Occupational Safety and Health Administration.

5. *Bid*—The offer or proposal of a Bidder submitted on the prescribed form setting forth the prices for the Work to be performed.

6. *Bidder*—The individual or entity who submits a Bid directly to Owner.

7. *Bidding Documents*—The Bidding Requirements and the proposed Contract Documents (including all Addenda).

8. *Bidding Requirements*—The advertisement or invitation to bid, Instructions to Bidders, Bid security of acceptable form, if any, and the Bid Form with any supplements.

9. *Change Order*—A document recommended by Engineer which is signed by Contractor and Owner and authorizes an addition, deletion, or revision in the Work or an adjustment in the Contract Price or the Contract Times, issued on or after the Effective Date of the Agreement.

10. *Claim*—A demand or assertion by Owner or Contractor seeking an adjustment of Contract Price or Contract Times, or both, or other relief with respect to the terms of the Contract. A demand for money or services by a third party is not a Claim.

11. *Contract*—The entire and integrated written agreement between the Owner and Contractor concerning the Work. The Contract supersedes prior negotiations, representations, or agreements, whether written or oral.

12. *Contract Documents*—Those items so designated in the Agreement. Only printed or hard copies of the items listed in the Agreement are Contract Documents. Approved Shop Drawings, other Contractor submittals, and the reports and drawings of subsurface and physical conditions are not Contract Documents.

13. *Contract Price*—The moneys payable by Owner to Contractor for completion of the Work in accordance with the Contract Documents as stated in the Agreement (subject to the provisions of Paragraph 11.03 in the case of Unit Price Work).

14. *Contract Times*—The number of days or the dates stated in the Agreement to: (i) achieve Milestones, if any; (ii) achieve Substantial Completion; and (iii) complete the Work so that it is ready for final payment as evidenced by Engineer's written recommendation of final payment.

15. *Contractor*—The individual or entity with whom Owner has entered into the Agreement.

16. *Cost of the Work*—See Paragraph 11.01 for definition.

17. *Drawings*—That part of the Contract Documents prepared or approved by Engineer which graphically shows the scope, extent, and character of the Work to be performed by Contractor. Shop Drawings and other Contractor submittals are not Drawings as so defined.

18. *Effective Date of the Agreement*—The date indicated in the Agreement on which it becomes effective, but if no such date is indicated, it means the date on which the Agreement is signed and delivered by the last of the two parties to sign and deliver.

19. *Engineer*—The individual or entity named as such in the Agreement.

20. *Field Order*—A written order issued by Engineer which requires minor changes in the Work but which does not involve a change in the Contract Price or the Contract Times.

21. *General Requirements*—Sections of Division 1 of the Specifications.

22. *Hazardous Environmental Condition*—The presence at the Site of Asbestos, PCBs, Petroleum, Hazardous Waste, or Radioactive Material in such quantities or circumstances that may present a substantial danger to persons or property exposed thereto.

23. *Hazardous Waste*—The term Hazardous Waste shall have the meaning provided in Section 1004 of the Solid Waste Disposal Act (42 USC Section 6903) as amended from time to time.

24. *Laws and Regulations; Laws or Regulations*—Any and all applicable laws, rules, regulations, ordinances, codes, and orders of any and all governmental bodies, agencies, authorities, and courts having jurisdiction.

25. *Liens*—Charges, security interests, or encumbrances upon Project funds, real property, or personal property.

26. *Milestone*—A principal event specified in the Contract Documents relating to an intermediate completion date or time prior to Substantial Completion of all the Work.

27. *Notice of Award*—The written notice by Owner to the Successful Bidder stating that upon timely compliance by the Successful Bidder with the conditions precedent listed therein, Owner will sign and deliver the Agreement.

28. *Notice to Proceed*—A written notice given by Owner to Contractor fixing the date on which the Contract Times will commence to run and on which Contractor shall start to perform the Work under the Contract Documents.

29. *Owner*—The individual or entity with whom Contractor has entered into the Agreement and for whom the Work is to be performed.

30. *PCBs*—Polychlorinated biphenyls.

31. *Petroleum*—Petroleum, including crude oil or any fraction thereof which is liquid at standard conditions of temperature and pressure (60 degrees Fahrenheit and 14.7 pounds per square inch absolute), such as oil, petroleum, fuel oil, oil sludge, oil refuse, gasoline, kerosene, and oil mixed with other non-Hazardous Waste and crude oils.

32. *Progress Schedule*—A schedule, prepared and maintained by Contractor, describing the sequence and duration of the activities comprising the Contractor's plan to accomplish the Work within the Contract Times.

33. *Project*—The total construction of which the Work to be performed under the Contract Documents may be the whole, or a part.

34. *Project Manual*—The bound documentary information prepared for bidding and constructing the Work. A listing of the contents of the Project Manual, which may be bound in one or more volumes, is contained in the table(s) of contents.

35. *Radioactive Material*—Source, special nuclear, or byproduct material as defined by the Atomic Energy Act of 1954 (42 USC Section 2011 et seq.) as amended from time to time.

36. *Resident Project Representative*—The authorized representative of Engineer who may be assigned to the Site or any part thereof.

37. *Samples*—Physical examples of materials, equipment, or workmanship that are representative of some portion of the Work and which establish the standards by which such portion of the Work will be judged.

38. *Schedule of Submittals*—A schedule, prepared and maintained by Contractor, of required submittals and the time requirements to support scheduled performance of related construction activities.

39. *Schedule of Values*—A schedule, prepared and maintained by Contractor, allocating portions of the Contract Price to various portions of the Work and used as the basis for reviewing Contractor's Applications for Payment.

40. *Shop Drawings*—All drawings, diagrams, illustrations, schedules, and other data or information which are specifically prepared or assembled by or for Contractor and submitted by Contractor to illustrate some portion of the Work.

41. *Site*—Lands or areas indicated in the Contract Documents as being furnished by Owner upon which the Work is to be performed, including rights-of-way and easements for access thereto, and such other lands furnished by Owner which are designated for the use of Contractor.

42. *Specifications*—That part of the Contract Documents consisting of written requirements for materials, equipment, systems, standards and workmanship as applied to the Work, and certain administrative requirements and procedural matters applicable thereto.

43. *Subcontractor*—An individual or entity having a direct contract with Contractor or with any other Subcontractor for the performance of a part of the Work at the Site.

44. *Substantial Completion*—The time at which the Work (or a specified part thereof) has progressed to the point where, in the opinion of Engineer, the Work (or a specified part thereof) is sufficiently complete, in accordance with the Contract Documents, so that the Work (or a specified part thereof) can be utilized for the purposes for which it is intended. The terms "substantially complete" and "substantially completed" as applied to all or part of the Work refer to Substantial Completion thereof.

45. *Successful Bidder*—The Bidder submitting a responsive Bid to whom Owner makes an award.

46. *Supplementary Conditions*—That part of the Contract Documents which amends or supplements these General Conditions.

47. *Supplier*—A manufacturer, fabricator, supplier, distributor, materialman, or vendor having a direct contract with Contractor or with any Subcontractor to furnish materials or equipment to be incorporated in the Work by Contractor or Subcontractor.

48. *Underground Facilities*—All underground pipelines, conduits, ducts, cables, wires, manholes, vaults, tanks, tunnels, or other such facilities or attachments, and any encasements containing such facilities, including those that convey electricity, gases, steam, liquid petroleum products, telephone or other communications, cable television, water, wastewater, storm water, other liquids or chemicals, or traffic or other control systems.

49. *Unit Price Work*—Work to be paid for on the basis of unit prices.

50. *Work*—The entire construction or the various separately identifiable parts thereof required to be provided under the Contract Documents. Work includes and is the result of performing or providing all labor, services, and documentation necessary to produce such construction, and furnishing, installing, and incorporating all materials and equipment into such construction, all as required by the Contract Documents.

51. *Work Change Directive*—A written statement to Contractor issued on or after the Effective Date of the Agreement and signed by Owner and recommended by Engineer ordering an

addition, deletion, or revision in the Work, or responding to differing or unforeseen subsurface or physical conditions under which the Work is to be performed or to emergencies. A Work Change Directive will not change the Contract Price or the Contract Times but is evidence that the parties expect that the change ordered or documented by a Work Change Directive will be incorporated in a subsequently issued Change Order following negotiations by the parties as to its effect, if any, on the Contract Price or Contract Times.

1.02 *Terminology*

A. The words and terms discussed in Paragraph 1.02.B through F are not defined but, when used in the Bidding Requirements or Contract Documents, have the indicated meaning.

B. *Intent of Certain Terms or Adjectives:*

1. The Contract Documents include the terms "as allowed," "as approved," "as ordered," "as directed" or terms of like effect or import to authorize an exercise of professional judgment by Engineer. In addition, the adjectives "reasonable," "suitable," "acceptable," "proper," "satisfactory," or adjectives of like effect or import are used to describe an action or determination of Engineer as to the Work. It is intended that such exercise of professional judgment, action, or determination will be solely to evaluate, in general, the Work for compliance with the information in the Contract Documents and with the design concept of the Project as a functioning whole as shown or indicated in the Contract Documents (unless there is a specific statement indicating otherwise). The use of any such term or adjective is not intended to and shall not be effective to assign to Engineer any duty or authority to supervise or direct the performance of the Work, or any duty or authority to undertake responsibility contrary to the provisions of Paragraph 9.09 or any other provision of the Contract Documents.

C. *Day:*

1. The word "day" means a calendar day of 24 hours measured from midnight to the next midnight.

D. *Defective:*

1. The word "defective," when modifying the word "Work," refers to Work that is unsatisfactory, faulty, or deficient in that it:

    a. does not conform to the Contract Documents; or

    b. does not meet the requirements of any applicable inspection, reference standard, test, or approval referred to in the Contract Documents; or

    c. has been damaged prior to Engineer's recommendation of final payment (unless responsibility for the protection thereof has been assumed by Owner at Substantial Completion in accordance with Paragraph 14.04 or 14.05).

E. *Furnish, Install, Perform, Provide:*

1. The word "furnish," when used in connection with services, materials, or equipment, shall mean to supply and deliver said services, materials, or equipment to the Site (or some other specified location) ready for use or installation and in usable or operable condition.

2. The word "install," when used in connection with services, materials, or equipment, shall mean to put into use or place in final position said services, materials, or equipment complete and ready for intended use.

3. The words "perform" or "provide," when used in connection with services, materials, or equipment, shall mean to furnish and install said services, materials, or equipment complete and ready for intended use.

4. When "furnish," "install," "perform," or "provide" is not used in connection with services, materials, or equipment in a context clearly requiring an obligation of Contractor, "provide" is implied.

F. Unless stated otherwise in the Contract Documents, words or phrases that have a well-known technical or construction industry or trade meaning are used in the Contract Documents in accordance with such recognized meaning.

ARTICLE 2 – PRELIMINARY MATTERS

2.01 *Delivery of Bonds and Evidence of Insurance*

A. When Contractor delivers the executed counterparts of the Agreement to Owner, Contractor shall also deliver to Owner such bonds as Contractor may be required to furnish.

B. *Evidence of Insurance.* Before any Work at the Site is started, Contractor and Owner shall each deliver to the other, with copies to each additional insured identified in the Supplementary Conditions, certificates of insurance (and other evidence of insurance which either of them or any additional insured may reasonably request) which Contractor and Owner respectively are required to purchase and maintain in accordance with Article 5.

2.02 *Copies of Documents*

A. Owner shall furnish to Contractor up to ten printed or hard copies of the Drawings and Project Manual. Additional copies will be furnished upon request at the cost of reproduction.

2.03 *Commencement of Contract Times; Notice to Proceed*

A. The Contract Times will commence to run on the thirtieth day after the Effective Date of the Agreement or, if a Notice to Proceed is given, on the day indicated in the Notice to Proceed. A Notice to Proceed may be given at any time within 30 days after the Effective Date of the Agreement. In no event will the Contract Times commence to run later than the sixtieth day after the day of Bid opening or the thirtieth day after the Effective Date of the Agreement, whichever date is earlier.

2.04 *Starting the Work*

   A. Contractor shall start to perform the Work on the date when the Contract Times commence to run. No Work shall be done at the Site prior to the date on which the Contract Times commence to run.

2.05 *Before Starting Construction*

   A. *Preliminary Schedules:* Within 10 days after the Effective Date of the Agreement (unless otherwise specified in the General Requirements), Contractor shall submit to Engineer for timely review:

   1. a preliminary Progress Schedule indicating the times (numbers of days or dates) for starting and completing the various stages of the Work, including any Milestones specified in the Contract Documents;

   2. a preliminary Schedule of Submittals; and

   3. a preliminary Schedule of Values for all of the Work which includes quantities and prices of items which when added together equal the Contract Price and subdivides the Work into component parts in sufficient detail to serve as the basis for progress payments during performance of the Work. Such prices will include an appropriate amount of overhead and profit applicable to each item of Work.

2.06 *Preconstruction Conference; Designation of Authorized Representatives*

   A. Before any Work at the Site is started, a conference attended by Owner, Contractor, Engineer, and others as appropriate will be held to establish a working understanding among the parties as to the Work and to discuss the schedules referred to in Paragraph 2.05.A, procedures for handling Shop Drawings and other submittals, processing Applications for Payment, and maintaining required records.

   B. At this conference Owner and Contractor each shall designate, in writing, a specific individual to act as its authorized representative with respect to the services and responsibilities under the Contract. Such individuals shall have the authority to transmit instructions, receive information, render decisions relative to the Contract, and otherwise act on behalf of each respective party.

2.07 *Initial Acceptance of Schedules*

   A. At least 10 days before submission of the first Application for Payment a conference attended by Contractor, Engineer, and others as appropriate will be held to review for acceptability to Engineer as provided below the schedules submitted in accordance with Paragraph 2.05.A. Contractor shall have an additional 10 days to make corrections and adjustments and to complete and resubmit the schedules. No progress payment shall be made to Contractor until acceptable schedules are submitted to Engineer.

   1. The Progress Schedule will be acceptable to Engineer if it provides an orderly progression of the Work to completion within the Contract Times. Such acceptance will not impose on Engineer responsibility for the Progress Schedule, for sequencing, scheduling, or progress of

the Work, nor interfere with or relieve Contractor from Contractor's full responsibility therefor.

2. Contractor's Schedule of Submittals will be acceptable to Engineer if it provides a workable arrangement for reviewing and processing the required submittals.

3. Contractor's Schedule of Values will be acceptable to Engineer as to form and substance if it provides a reasonable allocation of the Contract Price to component parts of the Work.

ARTICLE 3 – CONTRACT DOCUMENTS: INTENT, AMENDING, REUSE

3.01 *Intent*

A. The Contract Documents are complementary; what is required by one is as binding as if required by all.

B. It is the intent of the Contract Documents to describe a functionally complete project (or part thereof) to be constructed in accordance with the Contract Documents. Any labor, documentation, services, materials, or equipment that reasonably may be inferred from the Contract Documents or from prevailing custom or trade usage as being required to produce the indicated result will be provided whether or not specifically called for, at no additional cost to Owner.

C. Clarifications and interpretations of the Contract Documents shall be issued by Engineer as provided in Article 9.

3.02 *Reference Standards*

A. Standards, Specifications, Codes, Laws, and Regulations

1. Reference to standards, specifications, manuals, or codes of any technical society, organization, or association, or to Laws or Regulations, whether such reference be specific or by implication, shall mean the standard, specification, manual, code, or Laws or Regulations in effect at the time of opening of Bids (or on the Effective Date of the Agreement if there were no Bids), except as may be otherwise specifically stated in the Contract Documents.

2. No provision of any such standard, specification, manual, or code, or any instruction of a Supplier, shall be effective to change the duties or responsibilities of Owner, Contractor, or Engineer, or any of their subcontractors, consultants, agents, or employees, from those set forth in the Contract Documents. No such provision or instruction shall be effective to assign to Owner, Engineer, or any of their officers, directors, members, partners, employees, agents, consultants, or subcontractors, any duty or authority to supervise or direct the performance of the Work or any duty or authority to undertake responsibility inconsistent with the provisions of the Contract Documents.

3.03 *Reporting and Resolving Discrepancies*

A. *Reporting Discrepancies:*

1. *Contractor's Review of Contract Documents Before Starting Work*: Before undertaking each part of the Work, Contractor shall carefully study and compare the Contract Documents and check and verify pertinent figures therein and all applicable field measurements. Contractor shall promptly report in writing to Engineer any conflict, error, ambiguity, or discrepancy which Contractor discovers, or has actual knowledge of, and shall obtain a written interpretation or clarification from Engineer before proceeding with any Work affected thereby.

2. *Contractor's Review of Contract Documents During Performance of Work*: If, during the performance of the Work, Contractor discovers any conflict, error, ambiguity, or discrepancy within the Contract Documents, or between the Contract Documents and (a) any applicable Law or Regulation , (b) any standard, specification, manual, or code, or (c) any instruction of any Supplier, then Contractor shall promptly report it to Engineer in writing. Contractor shall not proceed with the Work affected thereby (except in an emergency as required by Paragraph 6.16.A) until an amendment or supplement to the Contract Documents has been issued by one of the methods indicated in Paragraph 3.04.

3. Contractor shall not be liable to Owner or Engineer for failure to report any conflict, error, ambiguity, or discrepancy in the Contract Documents unless Contractor had actual knowledge thereof.

B. *Resolving Discrepancies*:

1. Except as may be otherwise specifically stated in the Contract Documents, the provisions of the Contract Documents shall take precedence in resolving any conflict, error, ambiguity, or discrepancy between the provisions of the Contract Documents and:

    a. the provisions of any standard, specification, manual, or code, or the instruction of any Supplier (whether or not specifically incorporated by reference in the Contract Documents); or

    b. the provisions of any Laws or Regulations applicable to the performance of the Work (unless such an interpretation of the provisions of the Contract Documents would result in violation of such Law or Regulation).

3.04 *Amending and Supplementing Contract Documents*

A. The Contract Documents may be amended to provide for additions, deletions, and revisions in the Work or to modify the terms and conditions thereof by either a Change Order or a Work Change Directive.

B. The requirements of the Contract Documents may be supplemented, and minor variations and deviations in the Work may be authorized, by one or more of the following ways:

1. A Field Order;

2. Engineer's approval of a Shop Drawing or Sample (subject to the provisions of Paragraph 6.17.D.3); or

3.05 *Reuse of Documents*

A. Contractor and any Subcontractor or Supplier shall not:

1. have or acquire any title to or ownership rights in any of the Drawings, Specifications, or other documents (or copies of any thereof) prepared by or bearing the seal of Engineer or its consultants, including electronic media editions; or

2. reuse any such Drawings, Specifications, other documents, or copies thereof on extensions of the Project or any other project without written consent of Owner and Engineer and specific written verification or adaptation by Engineer.

B. The prohibitions of this Paragraph 3.05 will survive final payment, or termination of the Contract. Nothing herein shall preclude Contractor from retaining copies of the Contract Documents for record purposes.

3.06 *Electronic Data*

A. Unless otherwise stated in the Supplementary Conditions, the data furnished by Owner or Engineer to Contractor, or by Contractor to Owner or Engineer, that may be relied upon are limited to the printed copies (also known as hard copies). Files in electronic media format of text, data, graphics, or other types are furnished only for the convenience of the receiving party. Any conclusion or information obtained or derived from such electronic files will be at the user's sole risk. If there is a discrepancy between the electronic files and the hard copies, the hard copies govern.

B. Because data stored in electronic media format can deteriorate or be modified inadvertently or otherwise without authorization of the data's creator, the party receiving electronic files agrees that it will perform acceptance tests or procedures within 60 days, after which the receiving party shall be deemed to have accepted the data thus transferred. Any errors detected within the 60-day acceptance period will be corrected by the transferring party.

C. When transferring documents in electronic media format, the transferring party makes no representations as to long term compatibility, usability, or readability of documents resulting from the use of software application packages, operating systems, or computer hardware differing from those used by the data's creator.

ARTICLE 4 – AVAILABILITY OF LANDS; SUBSURFACE AND PHYSICAL CONDITIONS; HAZARDOUS ENVIRONMENTAL CONDITIONS; REFERENCE POINTS

4.01 *Availability of Lands*

A. Owner shall furnish the Site. Owner shall notify Contractor of any encumbrances or restrictions not of general application but specifically related to use of the Site with which Contractor must comply in performing the Work. Owner will obtain in a timely manner and pay for easements for permanent structures or permanent changes in existing facilities. If Contractor and Owner are unable to agree on entitlement to or on the amount or extent, if any, of any adjustment in the

Contract Price or Contract Times, or both, as a result of any delay in Owner's furnishing the Site or a part thereof, Contractor may make a Claim therefor as provided in Paragraph 10.05.

B. Upon reasonable written request, Owner shall furnish Contractor with a current statement of record legal title and legal description of the lands upon which the Work is to be performed and Owner's interest therein as necessary for giving notice of or filing a mechanic's or construction lien against such lands in accordance with applicable Laws and Regulations.

C. Contractor shall provide for all additional lands and access thereto that may be required for temporary construction facilities or storage of materials and equipment.

4.02 *Subsurface and Physical Conditions*

A. *Reports and Drawings:* The Supplementary Conditions identify:

1. those reports known to Owner of explorations and tests of subsurface conditions at or contiguous to the Site; and

2. those drawings known to Owner of physical conditions relating to existing surface or subsurface structures at the Site (except Underground Facilities).

B. *Limited Reliance by Contractor on Technical Data Authorized:* Contractor may rely upon the accuracy of the "technical data" contained in such reports and drawings, but such reports and drawings are not Contract Documents. Such "technical data" is identified in the Supplementary Conditions. Except for such reliance on such "technical data," Contractor may not rely upon or make any claim against Owner or Engineer, or any of their officers, directors, members, partners, employees, agents, consultants, or subcontractors with respect to:

1. the completeness of such reports and drawings for Contractor's purposes, including, but not limited to, any aspects of the means, methods, techniques, sequences, and procedures of construction to be employed by Contractor, and safety precautions and programs incident thereto; or

2. other data, interpretations, opinions, and information contained in such reports or shown or indicated in such drawings; or

3. any Contractor interpretation of or conclusion drawn from any "technical data" or any such other data, interpretations, opinions, or information.

4.03 *Differing Subsurface or Physical Conditions*

A. *Notice:* If Contractor believes that any subsurface or physical condition that is uncovered or revealed either:

1. is of such a nature as to establish that any "technical data" on which Contractor is entitled to rely as provided in Paragraph 4.02 is materially inaccurate; or

2. is of such a nature as to require a change in the Contract Documents; or

3. differs materially from that shown or indicated in the Contract Documents; or

4. is of an unusual nature, and differs materially from conditions ordinarily encountered and generally recognized as inherent in work of the character provided for in the Contract Documents;

then Contractor shall, promptly after becoming aware thereof and before further disturbing the subsurface or physical conditions or performing any Work in connection therewith (except in an emergency as required by Paragraph 6.16.A), notify Owner and Engineer in writing about such condition. Contractor shall not further disturb such condition or perform any Work in connection therewith (except as aforesaid) until receipt of written order to do so.

B. *Engineer's Review:* After receipt of written notice as required by Paragraph 4.03.A, Engineer will promptly review the pertinent condition, determine the necessity of Owner's obtaining additional exploration or tests with respect thereto, and advise Owner in writing (with a copy to Contractor) of Engineer's findings and conclusions.

C. *Possible Price and Times Adjustments:*

1. The Contract Price or the Contract Times, or both, will be equitably adjusted to the extent that the existence of such differing subsurface or physical condition causes an increase or decrease in Contractor's cost of, or time required for, performance of the Work; subject, however, to the following:

   a. such condition must meet any one or more of the categories described in Paragraph 4.03.A; and

   b. with respect to Work that is paid for on a unit price basis, any adjustment in Contract Price will be subject to the provisions of Paragraphs 9.07 and 11.03.

2. Contractor shall not be entitled to any adjustment in the Contract Price or Contract Times if:

   a. Contractor knew of the existence of such conditions at the time Contractor made a final commitment to Owner with respect to Contract Price and Contract Times by the submission of a Bid or becoming bound under a negotiated contract; or

   b. the existence of such condition could reasonably have been discovered or revealed as a result of any examination, investigation, exploration, test, or study of the Site and contiguous areas required by the Bidding Requirements or Contract Documents to be conducted by or for Contractor prior to Contractor's making such final commitment; or

   c. Contractor failed to give the written notice as required by Paragraph 4.03.A.

3. If Owner and Contractor are unable to agree on entitlement to or on the amount or extent, if any, of any adjustment in the Contract Price or Contract Times, or both, a Claim may be made therefor as provided in Paragraph 10.05. However, neither Owner or Engineer, or any of their officers, directors, members, partners, employees, agents, consultants, or subcontractors shall be liable to Contractor for any claims, costs, losses, or damages (including but not limited to all fees and charges of engineers, architects, attorneys, and other

professionals and all court or arbitration or other dispute resolution costs) sustained by Contractor on or in connection with any other project or anticipated project.

4.04 *Underground Facilities*

A. *Shown or Indicated:* The information and data shown or indicated in the Contract Documents with respect to existing Underground Facilities at or contiguous to the Site is based on information and data furnished to Owner or Engineer by the owners of such Underground Facilities, including Owner, or by others. Unless it is otherwise expressly provided in the Supplementary Conditions:

1. Owner and Engineer shall not be responsible for the accuracy or completeness of any such information or data provided by others; and

2. the cost of all of the following will be included in the Contract Price, and Contractor shall have full responsibility for:

   a. reviewing and checking all such information and data;

   b. locating all Underground Facilities shown or indicated in the Contract Documents;

   c. coordination of the Work with the owners of such Underground Facilities, including Owner, during construction; and

   d. the safety and protection of all such Underground Facilities and repairing any damage thereto resulting from the Work.

B. *Not Shown or Indicated:*

1. If an Underground Facility is uncovered or revealed at or contiguous to the Site which was not shown or indicated, or not shown or indicated with reasonable accuracy in the Contract Documents, Contractor shall, promptly after becoming aware thereof and before further disturbing conditions affected thereby or performing any Work in connection therewith (except in an emergency as required by Paragraph 6.16.A), identify the owner of such Underground Facility and give written notice to that owner and to Owner and Engineer. Engineer will promptly review the Underground Facility and determine the extent, if any, to which a change is required in the Contract Documents to reflect and document the consequences of the existence or location of the Underground Facility. During such time, Contractor shall be responsible for the safety and protection of such Underground Facility.

2. If Engineer concludes that a change in the Contract Documents is required, a Work Change Directive or a Change Order will be issued to reflect and document such consequences. An equitable adjustment shall be made in the Contract Price or Contract Times, or both, to the extent that they are attributable to the existence or location of any Underground Facility that was not shown or indicated or not shown or indicated with reasonable accuracy in the Contract Documents and that Contractor did not know of and could not reasonably have been expected to be aware of or to have anticipated. If Owner and Contractor are unable to agree on entitlement to or on the amount or extent, if any, of any such adjustment in Contract Price

or Contract Times, Owner or Contractor may make a Claim therefor as provided in Paragraph 10.05.

4.05 *Reference Points*

A. Owner shall provide engineering surveys to establish reference points for construction which in Engineer's judgment are necessary to enable Contractor to proceed with the Work. Contractor shall be responsible for laying out the Work, shall protect and preserve the established reference points and property monuments, and shall make no changes or relocations without the prior written approval of Owner. Contractor shall report to Engineer whenever any reference point or property monument is lost or destroyed or requires relocation because of necessary changes in grades or locations, and shall be responsible for the accurate replacement or relocation of such reference points or property monuments by professionally qualified personnel.

4.06 *Hazardous Environmental Condition at Site*

A. *Reports and Drawings:* The Supplementary Conditions identify those reports and drawings known to Owner relating to Hazardous Environmental Conditions that have been identified at the Site.

B. *Limited Reliance by Contractor on Technical Data Authorized:* Contractor may rely upon the accuracy of the "technical data" contained in such reports and drawings, but such reports and drawings are not Contract Documents. Such "technical data" is identified in the Supplementary Conditions. Except for such reliance on such "technical data," Contractor may not rely upon or make any claim against Owner or Engineer, or any of their officers, directors, members, partners, employees, agents, consultants, or subcontractors with respect to:

1. the completeness of such reports and drawings for Contractor's purposes, including, but not limited to, any aspects of the means, methods, techniques, sequences and procedures of construction to be employed by Contractor and safety precautions and programs incident thereto; or

2. other data, interpretations, opinions and information contained in such reports or shown or indicated in such drawings; or

3. any Contractor interpretation of or conclusion drawn from any "technical data" or any such other data, interpretations, opinions or information.

C. Contractor shall not be responsible for any Hazardous Environmental Condition uncovered or revealed at the Site which was not shown or indicated in Drawings or Specifications or identified in the Contract Documents to be within the scope of the Work. Contractor shall be responsible for a Hazardous Environmental Condition created with any materials brought to the Site by Contractor, Subcontractors, Suppliers, or anyone else for whom Contractor is responsible.

D. If Contractor encounters a Hazardous Environmental Condition or if Contractor or anyone for whom Contractor is responsible creates a Hazardous Environmental Condition, Contractor shall immediately: (i) secure or otherwise isolate such condition; (ii) stop all Work in connection with such condition and in any area affected thereby (except in an emergency as required by

Paragraph 6.16.A); and (iii) notify Owner and Engineer (and promptly thereafter confirm such notice in writing). Owner shall promptly consult with Engineer concerning the necessity for Owner to retain a qualified expert to evaluate such condition or take corrective action, if any. Promptly after consulting with Engineer, Owner shall take such actions as are necessary to permit Owner to timely obtain required permits and provide Contractor the written notice required by Paragraph 4.06.E.

E. Contractor shall not be required to resume Work in connection with such condition or in any affected area until after Owner has obtained any required permits related thereto and delivered written notice to Contractor: (i) specifying that such condition and any affected area is or has been rendered safe for the resumption of Work; or (ii) specifying any special conditions under which such Work may be resumed safely. If Owner and Contractor cannot agree as to entitlement to or on the amount or extent, if any, of any adjustment in Contract Price or Contract Times, or both, as a result of such Work stoppage or such special conditions under which Work is agreed to be resumed by Contractor, either party may make a Claim therefor as provided in Paragraph 10.05.

F. If after receipt of such written notice Contractor does not agree to resume such Work based on a reasonable belief it is unsafe, or does not agree to resume such Work under such special conditions, then Owner may order the portion of the Work that is in the area affected by such condition to be deleted from the Work. If Owner and Contractor cannot agree as to entitlement to or on the amount or extent, if any, of an adjustment in Contract Price or Contract Times as a result of deleting such portion of the Work, then either party may make a Claim therefor as provided in Paragraph 10.05. Owner may have such deleted portion of the Work performed by Owner's own forces or others in accordance with Article 7.

G. To the fullest extent permitted by Laws and Regulations, Owner shall indemnify and hold harmless Contractor, Subcontractors, and Engineer, and the officers, directors, members, partners, employees, agents, consultants, and subcontractors of each and any of them from and against all claims, costs, losses, and damages (including but not limited to all fees and charges of engineers, architects, attorneys, and other professionals and all court or arbitration or other dispute resolution costs) arising out of or relating to a Hazardous Environmental Condition, provided that such Hazardous Environmental Condition: (i) was not shown or indicated in the Drawings or Specifications or identified in the Contract Documents to be included within the scope of the Work, and (ii) was not created by Contractor or by anyone for whom Contractor is responsible. Nothing in this Paragraph 4.06.G shall obligate Owner to indemnify any individual or entity from and against the consequences of that individual's or entity's own negligence.

H. To the fullest extent permitted by Laws and Regulations, Contractor shall indemnify and hold harmless Owner and Engineer, and the officers, directors, members, partners, employees, agents, consultants, and subcontractors of each and any of them from and against all claims, costs, losses, and damages (including but not limited to all fees and charges of engineers, architects, attorneys, and other professionals and all court or arbitration or other dispute resolution costs) arising out of or relating to a Hazardous Environmental Condition created by Contractor or by anyone for whom Contractor is responsible. Nothing in this Paragraph 4.06.H shall obligate Contractor to indemnify any individual or entity from and against the consequences of that individual's or entity's own negligence.

I. The provisions of Paragraphs 4.02, 4.03, and 4.04 do not apply to a Hazardous Environmental Condition uncovered or revealed at the Site.

ARTICLE 5 – BONDS AND INSURANCE

5.01 *Performance, Payment, and Other Bonds*

A. Contractor shall furnish performance and payment bonds, each in an amount at least equal to the Contract Price as security for the faithful performance and payment of all of Contractor's obligations under the Contract Documents. These bonds shall remain in effect until one year after the date when final payment becomes due or until completion of the correction period specified in Paragraph 13.07, whichever is later, except as provided otherwise by Laws or Regulations or by the Contract Documents. Contractor shall also furnish such other bonds as are required by the Contract Documents.

B. All bonds shall be in the form prescribed by the Contract Documents except as provided otherwise by Laws or Regulations, and shall be executed by such sureties as are named in the list of "Companies Holding Certificates of Authority as Acceptable Sureties on Federal Bonds and as Acceptable Reinsuring Companies" as published in Circular 570 (amended) by the Financial Management Service, Surety Bond Branch, U.S. Department of the Treasury. All bonds signed by an agent or attorney-in-fact must be accompanied by a certified copy of that individual's authority to bind the surety. The evidence of authority shall show that it is effective on the date the agent or attorney-in-fact signed each bond.

C. If the surety on any bond furnished by Contractor is declared bankrupt or becomes insolvent or its right to do business is terminated in any state where any part of the Project is located or it ceases to meet the requirements of Paragraph 5.01.B, Contractor shall promptly notify Owner and Engineer and shall, within 20 days after the event giving rise to such notification, provide another bond and surety, both of which shall comply with the requirements of Paragraphs 5.01.B and 5.02.

5.02 *Licensed Sureties and Insurers*

A. All bonds and insurance required by the Contract Documents to be purchased and maintained by Owner or Contractor shall be obtained from surety or insurance companies that are duly licensed or authorized in the jurisdiction in which the Project is located to issue bonds or insurance policies for the limits and coverages so required. Such surety and insurance companies shall also meet such additional requirements and qualifications as may be provided in the Supplementary Conditions.

5.03 *Certificates of Insurance*

A. Contractor shall deliver to Owner, with copies to each additional insured and loss payee identified in the Supplementary Conditions, certificates of insurance (and other evidence of insurance requested by Owner or any other additional insured) which Contractor is required to purchase and maintain.

B. Owner shall deliver to Contractor, with copies to each additional insured and loss payee identified in the Supplementary Conditions, certificates of insurance (and other evidence of insurance requested by Contractor or any other additional insured) which Owner is required to purchase and maintain.

C. Failure of Owner to demand such certificates or other evidence of Contractor's full compliance with these insurance requirements or failure of Owner to identify a deficiency in compliance from the evidence provided shall not be construed as a waiver of Contractor's obligation to maintain such insurance.

D. Owner does not represent that insurance coverage and limits established in this Contract necessarily will be adequate to protect Contractor.

E. The insurance and insurance limits required herein shall not be deemed as a limitation on Contractor's liability under the indemnities granted to Owner in the Contract Documents.

5.04 *Contractor's Insurance*

A. Contractor shall purchase and maintain such insurance as is appropriate for the Work being performed and as will provide protection from claims set forth below which may arise out of or result from Contractor's performance of the Work and Contractor's other obligations under the Contract Documents, whether it is to be performed by Contractor, any Subcontractor or Supplier, or by anyone directly or indirectly employed by any of them to perform any of the Work, or by anyone for whose acts any of them may be liable:

1. claims under workers' compensation, disability benefits, and other similar employee benefit acts;

2. claims for damages because of bodily injury, occupational sickness or disease, or death of Contractor's employees;

3. claims for damages because of bodily injury, sickness or disease, or death of any person other than Contractor's employees;

4. claims for damages insured by reasonably available personal injury liability coverage which are sustained:

    a. by any person as a result of an offense directly or indirectly related to the employment of such person by Contractor, or

    b. by any other person for any other reason;

5. claims for damages, other than to the Work itself, because of injury to or destruction of tangible property wherever located, including loss of use resulting therefrom; and

6. claims for damages because of bodily injury or death of any person or property damage arising out of the ownership, maintenance or use of any motor vehicle.

B. The policies of insurance required by this Paragraph 5.04 shall:

1. with respect to insurance required by Paragraphs 5.04.A.3 through 5.04.A.6 inclusive, be written on an occurrence basis, include as additional insureds (subject to any customary exclusion regarding professional liability) Owner and Engineer, and any other individuals or entities identified in the Supplementary Conditions, all of whom shall be listed as additional insureds, and include coverage for the respective officers, directors, members, partners, employees, agents, consultants, and subcontractors of each and any of all such additional insureds, and the insurance afforded to these additional insureds shall provide primary coverage for all claims covered thereby;

2. include at least the specific coverages and be written for not less than the limits of liability provided in the Supplementary Conditions or required by Laws or Regulations, whichever is greater;

3. include contractual liability insurance covering Contractor's indemnity obligations under Paragraphs 6.11 and 6.20;

4. contain a provision or endorsement that the coverage afforded will not be canceled, materially changed or renewal refused until at least 30 days prior written notice has been given to Owner and Contractor and to each other additional insured identified in the Supplementary Conditions to whom a certificate of insurance has been issued (and the certificates of insurance furnished by the Contractor pursuant to Paragraph 5.03 will so provide);

5. remain in effect at least until final payment and at all times thereafter when Contractor may be correcting, removing, or replacing defective Work in accordance with Paragraph 13.07; and

6. include completed operations coverage:

    a. Such insurance shall remain in effect for two years after final payment.

    b. Contractor shall furnish Owner and each other additional insured identified in the Supplementary Conditions, to whom a certificate of insurance has been issued, evidence satisfactory to Owner and any such additional insured of continuation of such insurance at final payment and one year thereafter.

5.05 *Owner's Liability Insurance*

A. In addition to the insurance required to be provided by Contractor under Paragraph 5.04, Owner, at Owner's option, may purchase and maintain at Owner's expense Owner's own liability insurance as will protect Owner against claims which may arise from operations under the Contract Documents.

5.06 *Property Insurance*

A. Unless otherwise provided in the Supplementary Conditions, Owner shall purchase and maintain property insurance upon the Work at the Site in the amount of the full replacement cost thereof (subject to such deductible amounts as may be provided in the Supplementary Conditions or required by Laws and Regulations). This insurance shall:

1. include the interests of Owner, Contractor, Subcontractors, and Engineer, and any other individuals or entities identified in the Supplementary Conditions, and the officers, directors, members, partners, employees, agents, consultants, and subcontractors of each and any of them, each of whom is deemed to have an insurable interest and shall be listed as a loss payee;

2. be written on a Builder's Risk, "all-risk" policy form that shall at least include insurance for physical loss or damage to the Work, temporary buildings, falsework, and materials and equipment in transit, and shall insure against at least the following perils or causes of loss: fire, lightning, extended coverage, theft, vandalism and malicious mischief, earthquake, collapse, debris removal, demolition occasioned by enforcement of Laws and Regulations, water damage (other than that caused by flood), and such other perils or causes of loss as may be specifically required by the Supplementary Conditions.

3. include expenses incurred in the repair or replacement of any insured property (including but not limited to fees and charges of engineers and architects);

4. cover materials and equipment stored at the Site or at another location that was agreed to in writing by Owner prior to being incorporated in the Work, provided that such materials and equipment have been included in an Application for Payment recommended by Engineer;

5. allow for partial utilization of the Work by Owner;

6. include testing and startup; and

7. be maintained in effect until final payment is made unless otherwise agreed to in writing by Owner, Contractor, and Engineer with 30 days written notice to each other loss payee to whom a certificate of insurance has been issued.

B. Owner shall purchase and maintain such equipment breakdown insurance or additional property insurance as may be required by the Supplementary Conditions or Laws and Regulations which will include the interests of Owner, Contractor, Subcontractors, and Engineer, and any other individuals or entities identified in the Supplementary Conditions, and the officers, directors, members, partners, employees, agents, consultants and subcontractors of each and any of them, each of whom is deemed to have an insurable interest and shall be listed as a loss payee.

C. All the policies of insurance (and the certificates or other evidence thereof) required to be purchased and maintained in accordance with this Paragraph 5.06 will contain a provision or endorsement that the coverage afforded will not be canceled or materially changed or renewal refused until at least 30 days prior written notice has been given to Owner and Contractor and to each other loss payee to whom a certificate of insurance has been issued and will contain waiver provisions in accordance with Paragraph 5.07.

D. Owner shall not be responsible for purchasing and maintaining any property insurance specified in this Paragraph 5.06 to protect the interests of Contractor, Subcontractors, or others in the Work to the extent of any deductible amounts that are identified in the Supplementary Conditions. The risk of loss within such identified deductible amount will be borne by Contractor, Subcontractors, or others suffering any such loss, and if any of them wishes property insurance coverage within the limits of such amounts, each may purchase and maintain it at the purchaser's own expense.

E. If Contractor requests in writing that other special insurance be included in the property insurance policies provided under this Paragraph 5.06, Owner shall, if possible, include such insurance, and the cost thereof will be charged to Contractor by appropriate Change Order. Prior to commencement of the Work at the Site, Owner shall in writing advise Contractor whether or not such other insurance has been procured by Owner.

5.07 *Waiver of Rights*

A. Owner and Contractor intend that all policies purchased in accordance with Paragraph 5.06 will protect Owner, Contractor, Subcontractors, and Engineer, and all other individuals or entities identified in the Supplementary Conditions as loss payees (and the officers, directors, members, partners, employees, agents, consultants, and subcontractors of each and any of them) in such policies and will provide primary coverage for all losses and damages caused by the perils or causes of loss covered thereby. All such policies shall contain provisions to the effect that in the event of payment of any loss or damage the insurers will have no rights of recovery against any of the insureds or loss payees thereunder. Owner and Contractor waive all rights against each other and their respective officers, directors, members, partners, employees, agents, consultants and subcontractors of each and any of them for all losses and damages caused by, arising out of or resulting from any of the perils or causes of loss covered by such policies and any other property insurance applicable to the Work; and, in addition, waive all such rights against Subcontractors and Engineer, and all other individuals or entities identified in the Supplementary Conditions as loss payees (and the officers, directors, members, partners, employees, agents, consultants, and subcontractors of each and any of them) under such policies for losses and damages so caused. None of the above waivers shall extend to the rights that any party making such waiver may have to the proceeds of insurance held by Owner as trustee or otherwise payable under any policy so issued.

B. Owner waives all rights against Contractor, Subcontractors, and Engineer, and the officers, directors, members, partners, employees, agents, consultants and subcontractors of each and any of them for:

1. loss due to business interruption, loss of use, or other consequential loss extending beyond direct physical loss or damage to Owner's property or the Work caused by, arising out of, or resulting from fire or other perils whether or not insured by Owner; and

2. loss or damage to the completed Project or part thereof caused by, arising out of, or resulting from fire or other insured peril or cause of loss covered by any property insurance maintained on the completed Project or part thereof by Owner during partial utilization pursuant to Paragraph 14.05, after Substantial Completion pursuant to Paragraph 14.04, or after final payment pursuant to Paragraph 14.07.

C. Any insurance policy maintained by Owner covering any loss, damage or consequential loss referred to in Paragraph 5.07.B shall contain provisions to the effect that in the event of payment of any such loss, damage, or consequential loss, the insurers will have no rights of recovery against Contractor, Subcontractors, or Engineer, and the officers, directors, members, partners, employees, agents, consultants and subcontractors of each and any of them.

5.08 *Receipt and Application of Insurance Proceeds*

A. Any insured loss under the policies of insurance required by Paragraph 5.06 will be adjusted with Owner and made payable to Owner as fiduciary for the loss payees, as their interests may appear, subject to the requirements of any applicable mortgage clause and of Paragraph 5.08.B. Owner shall deposit in a separate account any money so received and shall distribute it in accordance with such agreement as the parties in interest may reach. If no other special agreement is reached, the damaged Work shall be repaired or replaced, the moneys so received applied on account thereof, and the Work and the cost thereof covered by an appropriate Change Order.

B. Owner as fiduciary shall have power to adjust and settle any loss with the insurers unless one of the parties in interest shall object in writing within 15 days after the occurrence of loss to Owner's exercise of this power. If such objection be made, Owner as fiduciary shall make settlement with the insurers in accordance with such agreement as the parties in interest may reach. If no such agreement among the parties in interest is reached, Owner as fiduciary shall adjust and settle the loss with the insurers and, if required in writing by any party in interest, Owner as fiduciary shall give bond for the proper performance of such duties.

5.09 *Acceptance of Bonds and Insurance; Option to Replace*

A. If either Owner or Contractor has any objection to the coverage afforded by or other provisions of the bonds or insurance required to be purchased and maintained by the other party in accordance with Article 5 on the basis of non-conformance with the Contract Documents, the objecting party shall so notify the other party in writing within 10 days after receipt of the certificates (or other evidence requested) required by Paragraph 2.01.B. Owner and Contractor shall each provide to the other such additional information in respect of insurance provided as the other may reasonably request. If either party does not purchase or maintain all of the bonds and insurance required of such party by the Contract Documents, such party shall notify the other party in writing of such failure to purchase prior to the start of the Work, or of such failure to maintain prior to any change in the required coverage. Without prejudice to any other right or remedy, the other party may elect to obtain equivalent bonds or insurance to protect such other party's interests at the expense of the party who was required to provide such coverage, and a Change Order shall be issued to adjust the Contract Price accordingly.

5.10 *Partial Utilization, Acknowledgment of Property Insurer*

A. If Owner finds it necessary to occupy or use a portion or portions of the Work prior to Substantial Completion of all the Work as provided in Paragraph 14.05, no such use or occupancy shall commence before the insurers providing the property insurance pursuant to Paragraph 5.06 have acknowledged notice thereof and in writing effected any changes in coverage necessitated thereby. The insurers providing the property insurance shall consent by endorsement on the policy or policies, but the property insurance shall not be canceled or permitted to lapse on account of any such partial use or occupancy.

ARTICLE 6 – CONTRACTOR'S RESPONSIBILITIES

6.01   *Supervision and Superintendence*

   A. Contractor shall supervise, inspect, and direct the Work competently and efficiently, devoting such attention thereto and applying such skills and expertise as may be necessary to perform the Work in accordance with the Contract Documents. Contractor shall be solely responsible for the means, methods, techniques, sequences, and procedures of construction. Contractor shall not be responsible for the negligence of Owner or Engineer in the design or specification of a specific means, method, technique, sequence, or procedure of construction which is shown or indicated in and expressly required by the Contract Documents.

   B. At all times during the progress of the Work, Contractor shall assign a competent resident superintendent who shall not be replaced without written notice to Owner and Engineer except under extraordinary circumstances.

6.02   *Labor; Working Hours*

   A. Contractor shall provide competent, suitably qualified personnel to survey and lay out the Work and perform construction as required by the Contract Documents. Contractor shall at all times maintain good discipline and order at the Site.

   B. Except as otherwise required for the safety or protection of persons or the Work or property at the Site or adjacent thereto, and except as otherwise stated in the Contract Documents, all Work at the Site shall be performed during regular working hours. Contractor will not permit the performance of Work on a Saturday, Sunday, or any legal holiday without Owner's written consent (which will not be unreasonably withheld) given after prior written notice to Engineer.

6.03   *Services, Materials, and Equipment*

   A. Unless otherwise specified in the Contract Documents, Contractor shall provide and assume full responsibility for all services, materials, equipment, labor, transportation, construction equipment and machinery, tools, appliances, fuel, power, light, heat, telephone, water, sanitary facilities, temporary facilities, and all other facilities and incidentals necessary for the performance, testing, start-up, and completion of the Work.

   B. All materials and equipment incorporated into the Work shall be as specified or, if not specified, shall be of good quality and new, except as otherwise provided in the Contract Documents. All special warranties and guarantees required by the Specifications shall expressly run to the benefit of Owner. If required by Engineer, Contractor shall furnish satisfactory evidence (including reports of required tests) as to the source, kind, and quality of materials and equipment.

   C. All materials and equipment shall be stored, applied, installed, connected, erected, protected, used, cleaned, and conditioned in accordance with instructions of the applicable Supplier, except as otherwise may be provided in the Contract Documents.

6.04 *Progress Schedule*

A. Contractor shall adhere to the Progress Schedule established in accordance with Paragraph 2.07 as it may be adjusted from time to time as provided below.

1. Contractor shall submit to Engineer for acceptance (to the extent indicated in Paragraph 2.07) proposed adjustments in the Progress Schedule that will not result in changing the Contract Times. Such adjustments will comply with any provisions of the General Requirements applicable thereto.

2. Proposed adjustments in the Progress Schedule that will change the Contract Times shall be submitted in accordance with the requirements of Article 12. Adjustments in Contract Times may only be made by a Change Order.

6.05 *Substitutes and "Or-Equals"*

A. Whenever an item of material or equipment is specified or described in the Contract Documents by using the name of a proprietary item or the name of a particular Supplier, the specification or description is intended to establish the type, function, appearance, and quality required. Unless the specification or description contains or is followed by words reading that no like, equivalent, or "or-equal" item or no substitution is permitted, other items of material or equipment or material or equipment of other Suppliers may be submitted to Engineer for review under the circumstances described below.

1. *"Or-Equal" Items:* If in Engineer's sole discretion an item of material or equipment proposed by Contractor is functionally equal to that named and sufficiently similar so that no change in related Work will be required, it may be considered by Engineer as an "or-equal" item, in which case review and approval of the proposed item may, in Engineer's sole discretion, be accomplished without compliance with some or all of the requirements for approval of proposed substitute items. For the purposes of this Paragraph 6.05.A.1, a proposed item of material or equipment will be considered functionally equal to an item so named if:

   a. in the exercise of reasonable judgment Engineer determines that:

      1) it is at least equal in materials of construction, quality, durability, appearance, strength, and design characteristics;

      2) it will reliably perform at least equally well the function and achieve the results imposed by the design concept of the completed Project as a functioning whole; and

      3) it has a proven record of performance and availability of responsive service.

   b. Contractor certifies that, if approved and incorporated into the Work:

      1) there will be no increase in cost to the Owner or increase in Contract Times; and

      2) it will conform substantially to the detailed requirements of the item named in the Contract Documents.

2. *Substitute Items:*

   a. If in Engineer's sole discretion an item of material or equipment proposed by Contractor does not qualify as an "or-equal" item under Paragraph 6.05.A.1, it will be considered a proposed substitute item.

   b. Contractor shall submit sufficient information as provided below to allow Engineer to determine if the item of material or equipment proposed is essentially equivalent to that named and an acceptable substitute therefor. Requests for review of proposed substitute items of material or equipment will not be accepted by Engineer from anyone other than Contractor.

   c. The requirements for review by Engineer will be as set forth in Paragraph 6.05.A.2.d, as supplemented by the General Requirements, and as Engineer may decide is appropriate under the circumstances.

   d. Contractor shall make written application to Engineer for review of a proposed substitute item of material or equipment that Contractor seeks to furnish or use. The application:

      1) shall certify that the proposed substitute item will:

         a) perform adequately the functions and achieve the results called for by the general design,

         b) be similar in substance to that specified, and

         c) be suited to the same use as that specified;

      2) will state:

         a) the extent, if any, to which the use of the proposed substitute item will prejudice Contractor's achievement of Substantial Completion on time,

         b) whether use of the proposed substitute item in the Work will require a change in any of the Contract Documents (or in the provisions of any other direct contract with Owner for other work on the Project) to adapt the design to the proposed substitute item, and

         c) whether incorporation or use of the proposed substitute item in connection with the Work is subject to payment of any license fee or royalty;

      3) will identify:

         a) all variations of the proposed substitute item from that specified, and

         b) available engineering, sales, maintenance, repair, and replacement services; and

4) shall contain an itemized estimate of all costs or credits that will result directly or indirectly from use of such substitute item, including costs of redesign and claims of other contractors affected by any resulting change.

B. *Substitute Construction Methods or Procedures:* If a specific means, method, technique, sequence, or procedure of construction is expressly required by the Contract Documents, Contractor may furnish or utilize a substitute means, method, technique, sequence, or procedure of construction approved by Engineer. Contractor shall submit sufficient information to allow Engineer, in Engineer's sole discretion, to determine that the substitute proposed is equivalent to that expressly called for by the Contract Documents. The requirements for review by Engineer will be similar to those provided in Paragraph 6.05.A.2.

C. *Engineer's Evaluation:* Engineer will be allowed a reasonable time within which to evaluate each proposal or submittal made pursuant to Paragraphs 6.05.A and 6.05.B. Engineer may require Contractor to furnish additional data about the proposed substitute item. Engineer will be the sole judge of acceptability. No "or equal" or substitute will be ordered, installed or utilized until Engineer's review is complete, which will be evidenced by a Change Order in the case of a substitute and an approved Shop Drawing for an "or equal." Engineer will advise Contractor in writing of any negative determination.

D. *Special Guarantee:* Owner may require Contractor to furnish at Contractor's expense a special performance guarantee or other surety with respect to any substitute.

E. *Engineer's Cost Reimbursement*: Engineer will record Engineer's costs in evaluating a substitute proposed or submitted by Contractor pursuant to Paragraphs 6.05.A.2 and 6.05.B. Whether or not Engineer approves a substitute so proposed or submitted by Contractor, Contractor shall reimburse Owner for the reasonable charges of Engineer for evaluating each such proposed substitute. Contractor shall also reimburse Owner for the reasonable charges of Engineer for making changes in the Contract Documents (or in the provisions of any other direct contract with Owner) resulting from the acceptance of each proposed substitute.

F. *Contractor's Expense*: Contractor shall provide all data in support of any proposed substitute or "or-equal" at Contractor's expense.

6.06 *Concerning Subcontractors, Suppliers, and Others*

A. Contractor shall not employ any Subcontractor, Supplier, or other individual or entity (including those acceptable to Owner as indicated in Paragraph 6.06.B), whether initially or as a replacement, against whom Owner may have reasonable objection. Contractor shall not be required to employ any Subcontractor, Supplier, or other individual or entity to furnish or perform any of the Work against whom Contractor has reasonable objection.

B. If the Supplementary Conditions require the identity of certain Subcontractors, Suppliers, or other individuals or entities to be submitted to Owner in advance for acceptance by Owner by a specified date prior to the Effective Date of the Agreement, and if Contractor has submitted a list thereof in accordance with the Supplementary Conditions, Owner's acceptance (either in writing or by failing to make written objection thereto by the date indicated for acceptance or objection in the Bidding Documents or the Contract Documents) of any such Subcontractor, Supplier, or other individual or entity so identified may be revoked on the basis of reasonable objection after due investigation. Contractor shall submit an acceptable replacement for the rejected Subcontractor, Supplier, or other individual or entity, and the Contract Price will be adjusted by the difference in the cost occasioned by such replacement, and an appropriate Change Order will be issued. No acceptance by Owner of any such Subcontractor, Supplier, or other individual or entity, whether initially or as a replacement, shall constitute a waiver of any right of Owner or Engineer to reject defective Work.

C. Contractor shall be fully responsible to Owner and Engineer for all acts and omissions of the Subcontractors, Suppliers, and other individuals or entities performing or furnishing any of the Work just as Contractor is responsible for Contractor's own acts and omissions. Nothing in the Contract Documents:

1. shall create for the benefit of any such Subcontractor, Supplier, or other individual or entity any contractual relationship between Owner or Engineer and any such Subcontractor, Supplier or other individual or entity; nor

2. shall create any obligation on the part of Owner or Engineer to pay or to see to the payment of any moneys due any such Subcontractor, Supplier, or other individual or entity except as may otherwise be required by Laws and Regulations.

D. Contractor shall be solely responsible for scheduling and coordinating the Work of Subcontractors, Suppliers, and other individuals or entities performing or furnishing any of the Work under a direct or indirect contract with Contractor.

E. Contractor shall require all Subcontractors, Suppliers, and such other individuals or entities performing or furnishing any of the Work to communicate with Engineer through Contractor.

F. The divisions and sections of the Specifications and the identifications of any Drawings shall not control Contractor in dividing the Work among Subcontractors or Suppliers or delineating the Work to be performed by any specific trade.

G. All Work performed for Contractor by a Subcontractor or Supplier will be pursuant to an appropriate agreement between Contractor and the Subcontractor or Supplier which specifically binds the Subcontractor or Supplier to the applicable terms and conditions of the Contract Documents for the benefit of Owner and Engineer. Whenever any such agreement is with a Subcontractor or Supplier who is listed as a loss payee on the property insurance provided in Paragraph 5.06, the agreement between the Contractor and the Subcontractor or Supplier will contain provisions whereby the Subcontractor or Supplier waives all rights against Owner, Contractor, Engineer, and all other individuals or entities identified in the Supplementary Conditions to be listed as insureds or loss payees (and the officers, directors, members, partners, employees, agents, consultants, and subcontractors of each and any of them) for all losses and damages caused by, arising out of, relating to, or resulting from any of the perils or causes of loss covered by such policies and any other property insurance applicable to the Work. If the insurers on any such policies require separate waiver forms to be signed by any Subcontractor or Supplier, Contractor will obtain the same.

6.07 *Patent Fees and Royalties*

　　A. Contractor shall pay all license fees and royalties and assume all costs incident to the use in the performance of the Work or the incorporation in the Work of any invention, design, process, product, or device which is the subject of patent rights or copyrights held by others. If a particular invention, design, process, product, or device is specified in the Contract Documents for use in the performance of the Work and if, to the actual knowledge of Owner or Engineer, its use is subject to patent rights or copyrights calling for the payment of any license fee or royalty to others, the existence of such rights shall be disclosed by Owner in the Contract Documents.

　　B. To the fullest extent permitted by Laws and Regulations, Owner shall indemnify and hold harmless Contractor, and its officers, directors, members, partners, employees, agents, consultants, and subcontractors from and against all claims, costs, losses, and damages (including but not limited to all fees and charges of engineers, architects, attorneys, and other professionals, and all court or arbitration or other dispute resolution costs) arising out of or relating to any infringement of patent rights or copyrights incident to the use in the performance of the Work or resulting from the incorporation in the Work of any invention, design, process, product, or device specified in the Contract Documents, but not identified as being subject to payment of any license fee or royalty to others required by patent rights or copyrights.

　　C. To the fullest extent permitted by Laws and Regulations, Contractor shall indemnify and hold harmless Owner and Engineer, and the officers, directors, members, partners, employees, agents, consultants and subcontractors of each and any of them from and against all claims, costs, losses, and damages (including but not limited to all fees and charges of engineers, architects, attorneys, and other professionals and all court or arbitration or other dispute resolution costs) arising out of or relating to any infringement of patent rights or copyrights incident to the use in the performance of the Work or resulting from the incorporation in the Work of any invention, design, process, product, or device not specified in the Contract Documents.

6.08 *Permits*

　　A. Unless otherwise provided in the Supplementary Conditions, Contractor shall obtain and pay for all construction permits and licenses. Owner shall assist Contractor, when necessary, in obtaining such permits and licenses. Contractor shall pay all governmental charges and inspection fees necessary for the prosecution of the Work which are applicable at the time of opening of Bids, or, if there are no Bids, on the Effective Date of the Agreement. Owner shall pay all charges of utility owners for connections for providing permanent service to the Work.

6.09 *Laws and Regulations*

　　A. Contractor shall give all notices required by and shall comply with all Laws and Regulations applicable to the performance of the Work. Except where otherwise expressly required by applicable Laws and Regulations, neither Owner nor Engineer shall be responsible for monitoring Contractor's compliance with any Laws or Regulations.

　　B. If Contractor performs any Work knowing or having reason to know that it is contrary to Laws or Regulations, Contractor shall bear all claims, costs, losses, and damages (including but not limited to all fees and charges of engineers, architects, attorneys, and other professionals and all court or arbitration or other dispute resolution costs) arising out of or relating to such Work. However, it shall not be Contractor's responsibility to make certain that the Specifications and Drawings are in accordance with Laws and Regulations, but this shall not relieve Contractor of Contractor's obligations under Paragraph 3.03.

C. Changes in Laws or Regulations not known at the time of opening of Bids (or, on the Effective Date of the Agreement if there were no Bids) having an effect on the cost or time of performance of the Work shall be the subject of an adjustment in Contract Price or Contract Times. If Owner and Contractor are unable to agree on entitlement to or on the amount or extent, if any, of any such adjustment, a Claim may be made therefor as provided in Paragraph 10.05.

6.10 *Taxes*

A. Contractor shall pay all sales, consumer, use, and other similar taxes required to be paid by Contractor in accordance with the Laws and Regulations of the place of the Project which are applicable during the performance of the Work.

6.11 *Use of Site and Other Areas*

A. *Limitation on Use of Site and Other Areas:*

1. Contractor shall confine construction equipment, the storage of materials and equipment, and the operations of workers to the Site and other areas permitted by Laws and Regulations, and shall not unreasonably encumber the Site and other areas with construction equipment or other materials or equipment. Contractor shall assume full responsibility for any damage to any such land or area, or to the owner or occupant thereof, or of any adjacent land or areas resulting from the performance of the Work.

2. Should any claim be made by any such owner or occupant because of the performance of the Work, Contractor shall promptly settle with such other party by negotiation or otherwise resolve the claim by arbitration or other dispute resolution proceeding or at law.

3. To the fullest extent permitted by Laws and Regulations, Contractor shall indemnify and hold harmless Owner and Engineer, and the officers, directors, members, partners, employees, agents, consultants and subcontractors of each and any of them from and against all claims, costs, losses, and damages (including but not limited to all fees and charges of engineers, architects, attorneys, and other professionals and all court or arbitration or other dispute resolution costs) arising out of or relating to any claim or action, legal or equitable, brought by any such owner or occupant against Owner, Engineer, or any other party indemnified hereunder to the extent caused by or based upon Contractor's performance of the Work.

B. *Removal of Debris During Performance of the Work:* During the progress of the Work Contractor shall keep the Site and other areas free from accumulations of waste materials, rubbish, and other debris. Removal and disposal of such waste materials, rubbish, and other debris shall conform to applicable Laws and Regulations.

C. *Cleaning:* Prior to Substantial Completion of the Work Contractor shall clean the Site and the Work and make it ready for utilization by Owner. At the completion of the Work Contractor shall remove from the Site all tools, appliances, construction equipment and machinery, and surplus materials and shall restore to original condition all property not designated for alteration by the Contract Documents.

D. *Loading Structures:* Contractor shall not load nor permit any part of any structure to be loaded in any manner that will endanger the structure, nor shall Contractor subject any part of the Work or adjacent property to stresses or pressures that will endanger it.

6.12 *Record Documents*

A. Contractor shall maintain in a safe place at the Site one record copy of all Drawings, Specifications, Addenda, Change Orders, Work Change Directives, Field Orders, and written interpretations and clarifications in good order and annotated to show changes made during construction. These record documents together with all approved Samples and a counterpart of all approved Shop Drawings will be available to Engineer for reference. Upon completion of the Work, these record documents, Samples, and Shop Drawings will be delivered to Engineer for Owner.

6.13 *Safety and Protection*

A. Contractor shall be solely responsible for initiating, maintaining and supervising all safety precautions and programs in connection with the Work. Such responsibility does not relieve Subcontractors of their responsibility for the safety of persons or property in the performance of their work, nor for compliance with applicable safety Laws and Regulations. Contractor shall take all necessary precautions for the safety of, and shall provide the necessary protection to prevent damage, injury or loss to:

1. all persons on the Site or who may be affected by the Work;

2. all the Work and materials and equipment to be incorporated therein, whether in storage on or off the Site; and

3. other property at the Site or adjacent thereto, including trees, shrubs, lawns, walks, pavements, roadways, structures, utilities, and Underground Facilities not designated for removal, relocation, or replacement in the course of construction.

B. Contractor shall comply with all applicable Laws and Regulations relating to the safety of persons or property, or to the protection of persons or property from damage, injury, or loss; and shall erect and maintain all necessary safeguards for such safety and protection. Contractor shall notify owners of adjacent property and of Underground Facilities and other utility owners when prosecution of the Work may affect them, and shall cooperate with them in the protection, removal, relocation, and replacement of their property.

C. Contractor shall comply with the applicable requirements of Owner's safety programs, if any. The Supplementary Conditions identify any Owner's safety programs that are applicable to the Work.

D. Contractor shall inform Owner and Engineer of the specific requirements of Contractor's safety program with which Owner's and Engineer's employees and representatives must comply while at the Site.

E. All damage, injury, or loss to any property referred to in Paragraph 6.13.A.2 or 6.13.A.3 caused, directly or indirectly, in whole or in part, by Contractor, any Subcontractor, Supplier, or any other individual or entity directly or indirectly employed by any of them to perform any of the Work, or anyone for whose acts any of them may be liable, shall be remedied by Contractor (except damage or loss attributable to the fault of Drawings or Specifications or to the acts or omissions of Owner or Engineer or anyone employed by any of them, or anyone for whose acts any of them may be liable, and not attributable, directly or indirectly, in whole or in part, to the fault or negligence of Contractor or any Subcontractor, Supplier, or other individual or entity directly or indirectly employed by any of them).

F. Contractor's duties and responsibilities for safety and for protection of the Work shall continue until such time as all the Work is completed and Engineer has issued a notice to Owner and Contractor in accordance with Paragraph 14.07.B that the Work is acceptable (except as otherwise expressly provided in connection with Substantial Completion).

6.14 *Safety Representative*

A. Contractor shall designate a qualified and experienced safety representative at the Site whose duties and responsibilities shall be the prevention of accidents and the maintaining and supervising of safety precautions and programs.

6.15 *Hazard Communication Programs*

A. Contractor shall be responsible for coordinating any exchange of material safety data sheets or other hazard communication information required to be made available to or exchanged between or among employers at the Site in accordance with Laws or Regulations.

6.16 *Emergencies*

A. In emergencies affecting the safety or protection of persons or the Work or property at the Site or adjacent thereto, Contractor is obligated to act to prevent threatened damage, injury, or loss. Contractor shall give Engineer prompt written notice if Contractor believes that any significant changes in the Work or variations from the Contract Documents have been caused thereby or are required as a result thereof. If Engineer determines that a change in the Contract Documents is required because of the action taken by Contractor in response to such an emergency, a Work Change Directive or Change Order will be issued.

6.17 *Shop Drawings and Samples*

A. Contractor shall submit Shop Drawings and Samples to Engineer for review and approval in accordance with the accepted Schedule of Submittals (as required by Paragraph 2.07). Each submittal will be identified as Engineer may require.

1. *Shop Drawings:*

   a. Submit number of copies specified in the General Requirements.

   b. Data shown on the Shop Drawings will be complete with respect to quantities, dimensions, specified performance and design criteria, materials, and similar data to show Engineer the services, materials, and equipment Contractor proposes to provide and to enable Engineer to review the information for the limited purposes required by Paragraph 6.17.D.

2. *Samples:*

   a. Submit number of Samples specified in the Specifications.

   b. Clearly identify each Sample as to material, Supplier, pertinent data such as catalog numbers, the use for which intended and other data as Engineer may require to enable Engineer to review the submittal for the limited purposes required by Paragraph 6.17.D.

B. Where a Shop Drawing or Sample is required by the Contract Documents or the Schedule of Submittals, any related Work performed prior to Engineer's review and approval of the pertinent submittal will be at the sole expense and responsibility of Contractor.

C. *Submittal Procedures:*

   1. Before submitting each Shop Drawing or Sample, Contractor shall have:

      a. reviewed and coordinated each Shop Drawing or Sample with other Shop Drawings and Samples and with the requirements of the Work and the Contract Documents;

      b. determined and verified all field measurements, quantities, dimensions, specified performance and design criteria, installation requirements, materials, catalog numbers, and similar information with respect thereto;

      c. determined and verified the suitability of all materials offered with respect to the indicated application, fabrication, shipping, handling, storage, assembly, and installation pertaining to the performance of the Work; and

      d. determined and verified all information relative to Contractor's responsibilities for means, methods, techniques, sequences, and procedures of construction, and safety precautions and programs incident thereto.

   2. Each submittal shall bear a stamp or specific written certification that Contractor has satisfied Contractor's obligations under the Contract Documents with respect to Contractor's review and approval of that submittal.

   3. With each submittal, Contractor shall give Engineer specific written notice of any variations that the Shop Drawing or Sample may have from the requirements of the Contract Documents. This notice shall be both a written communication separate from the Shop

Drawings or Sample submittal; and, in addition, by a specific notation made on each Shop Drawing or Sample submitted to Engineer for review and approval of each such variation.

D. *Engineer's Review:*

1. Engineer will provide timely review of Shop Drawings and Samples in accordance with the Schedule of Submittals acceptable to Engineer. Engineer's review and approval will be only to determine if the items covered by the submittals will, after installation or incorporation in the Work, conform to the information given in the Contract Documents and be compatible with the design concept of the completed Project as a functioning whole as indicated by the Contract Documents.

2. Engineer's review and approval will not extend to means, methods, techniques, sequences, or procedures of construction (except where a particular means, method, technique, sequence, or procedure of construction is specifically and expressly called for by the Contract Documents) or to safety precautions or programs incident thereto. The review and approval of a separate item as such will not indicate approval of the assembly in which the item functions.

3. Engineer's review and approval shall not relieve Contractor from responsibility for any variation from the requirements of the Contract Documents unless Contractor has complied with the requirements of Paragraph 6.17.C.3 and Engineer has given written approval of each such variation by specific written notation thereof incorporated in or accompanying the Shop Drawing or Sample. Engineer's review and approval shall not relieve Contractor from responsibility for complying with the requirements of Paragraph 6.17.C.1.

E. *Resubmittal Procedures:*

1. Contractor shall make corrections required by Engineer and shall return the required number of corrected copies of Shop Drawings and submit, as required, new Samples for review and approval. Contractor shall direct specific attention in writing to revisions other than the corrections called for by Engineer on previous submittals.

6.18 *Continuing the Work*

A. Contractor shall carry on the Work and adhere to the Progress Schedule during all disputes or disagreements with Owner. No Work shall be delayed or postponed pending resolution of any disputes or disagreements, except as permitted by Paragraph 15.04 or as Owner and Contractor may otherwise agree in writing.

6.19 *Contractor's General Warranty and Guarantee*

A. Contractor warrants and guarantees to Owner that all Work will be in accordance with the Contract Documents and will not be defective. Engineer and its officers, directors, members, partners, employees, agents, consultants, and subcontractors shall be entitled to rely on representation of Contractor's warranty and guarantee.

B. Contractor's warranty and guarantee hereunder excludes defects or damage caused by:

1. abuse, modification, or improper maintenance or operation by persons other than Contractor, Subcontractors, Suppliers, or any other individual or entity for whom Contractor is responsible; or

2. normal wear and tear under normal usage.

C. Contractor's obligation to perform and complete the Work in accordance with the Contract Documents shall be absolute. None of the following will constitute an acceptance of Work that is not in accordance with the Contract Documents or a release of Contractor's obligation to perform the Work in accordance with the Contract Documents:

1. observations by Engineer;

2. recommendation by Engineer of payment by Owner of any progress or final payment;

3. the issuance of a certificate of Substantial Completion by Engineer or any payment related thereto by Owner;

4. use or occupancy of the Work or any part thereof by Owner;

5. any review and approval of a Shop Drawing or Sample submittal or the issuance of a notice of acceptability by Engineer;

6. any inspection, test, or approval by others; or

7. any correction of defective Work by Owner.

6.20 *Indemnification*

A. To the fullest extent permitted by Laws and Regulations, Contractor shall indemnify and hold harmless Owner and Engineer, and the officers, directors, members, partners, employees, agents, consultants and subcontractors of each and any of them from and against all claims, costs, losses, and damages (including but not limited to all fees and charges of engineers, architects, attorneys, and other professionals and all court or arbitration or other dispute resolution costs) arising out of or relating to the performance of the Work, provided that any such claim, cost, loss, or damage is attributable to bodily injury, sickness, disease, or death, or to injury to or destruction of tangible property (other than the Work itself), including the loss of use resulting therefrom but only to the extent caused by any negligent act or omission of Contractor, any Subcontractor, any Supplier, or any individual or entity directly or indirectly employed by any of them to perform any of the Work or anyone for whose acts any of them may be liable .

B. In any and all claims against Owner or Engineer or any of their officers, directors, members, partners, employees, agents, consultants, or subcontractors by any employee (or the survivor or personal representative of such employee) of Contractor, any Subcontractor, any Supplier, or any individual or entity directly or indirectly employed by any of them to perform any of the Work, or anyone for whose acts any of them may be liable, the indemnification obligation under Paragraph 6.20.A shall not be limited in any way by any limitation on the amount or type of damages, compensation, or benefits payable by or for Contractor or any such Subcontractor,

Supplier, or other individual or entity under workers' compensation acts, disability benefit acts, or other employee benefit acts.

C. The indemnification obligations of Contractor under Paragraph 6.20.A shall not extend to the liability of Engineer and Engineer's officers, directors, members, partners, employees, agents, consultants and subcontractors arising out of:

1. the preparation or approval of, or the failure to prepare or approve maps, Drawings, opinions, reports, surveys, Change Orders, designs, or Specifications; or

2. giving directions or instructions, or failing to give them, if that is the primary cause of the injury or damage.

6.21 *Delegation of Professional Design Services*

A. Contractor will not be required to provide professional design services unless such services are specifically required by the Contract Documents for a portion of the Work or unless such services are required to carry out Contractor's responsibilities for construction means, methods, techniques, sequences and procedures. Contractor shall not be required to provide professional services in violation of applicable law.

B. If professional design services or certifications by a design professional related to systems, materials or equipment are specifically required of Contractor by the Contract Documents, Owner and Engineer will specify all performance and design criteria that such services must satisfy. Contractor shall cause such services or certifications to be provided by a properly licensed professional, whose signature and seal shall appear on all drawings, calculations, specifications, certifications, Shop Drawings and other submittals prepared by such professional. Shop Drawings and other submittals related to the Work designed or certified by such professional, if prepared by others, shall bear such professional's written approval when submitted to Engineer.

C. Owner and Engineer shall be entitled to rely upon the adequacy, accuracy and completeness of the services, certifications or approvals performed by such design professionals, provided Owner and Engineer have specified to Contractor all performance and design criteria that such services must satisfy.

D. Pursuant to this Paragraph 6.21, Engineer's review and approval of design calculations and design drawings will be only for the limited purpose of checking for conformance with performance and design criteria given and the design concept expressed in the Contract Documents. Engineer's review and approval of Shop Drawings and other submittals (except design calculations and design drawings) will be only for the purpose stated in Paragraph 6.17.D.1.

E. Contractor shall not be responsible for the adequacy of the performance or design criteria required by the Contract Documents.

ARTICLE 7 – OTHER WORK AT THE SITE

7.01   *Related Work at Site*

   A. Owner may perform other work related to the Project at the Site with Owner's employees, or through other direct contracts therefor, or have other work performed by utility owners. If such other work is not noted in the Contract Documents, then:

   1. written notice thereof will be given to Contractor prior to starting any such other work; and

   2. if Owner and Contractor are unable to agree on entitlement to or on the amount or extent, if any, of any adjustment in the Contract Price or Contract Times that should be allowed as a result of such other work, a Claim may be made therefor as provided in Paragraph 10.05.

   B. Contractor shall afford each other contractor who is a party to such a direct contract, each utility owner, and Owner, if Owner is performing other work with Owner's employees, proper and safe access to the Site, provide a reasonable opportunity for the introduction and storage of materials and equipment and the execution of such other work, and properly coordinate the Work with theirs. Contractor shall do all cutting, fitting, and patching of the Work that may be required to properly connect or otherwise make its several parts come together and properly integrate with such other work. Contractor shall not endanger any work of others by cutting, excavating, or otherwise altering such work; provided, however, that Contractor may cut or alter others' work with the written consent of Engineer and the others whose work will be affected. The duties and responsibilities of Contractor under this Paragraph are for the benefit of such utility owners and other contractors to the extent that there are comparable provisions for the benefit of Contractor in said direct contracts between Owner and such utility owners and other contractors.

   C. If the proper execution or results of any part of Contractor's Work depends upon work performed by others under this Article 7, Contractor shall inspect such other work and promptly report to Engineer in writing any delays, defects, or deficiencies in such other work that render it unavailable or unsuitable for the proper execution and results of Contractor's Work. Contractor's failure to so report will constitute an acceptance of such other work as fit and proper for integration with Contractor's Work except for latent defects and deficiencies in such other work.

7.02   *Coordination*

   A. If Owner intends to contract with others for the performance of other work on the Project at the Site, the following will be set forth in Supplementary Conditions:

   1. the individual or entity who will have authority and responsibility for coordination of the activities among the various contractors will be identified;

   2. the specific matters to be covered by such authority and responsibility will be itemized; and

   3. the extent of such authority and responsibilities will be provided.

   B. Unless otherwise provided in the Supplementary Conditions, Owner shall have sole authority and responsibility for such coordination.

7.03 *Legal Relationships*

A. Paragraphs 7.01.A and 7.02 are not applicable for utilities not under the control of Owner.

B. Each other direct contract of Owner under Paragraph 7.01.A shall provide that the other contractor is liable to Owner and Contractor for the reasonable direct delay and disruption costs incurred by Contractor as a result of the other contractor's wrongful actions or inactions.

C. Contractor shall be liable to Owner and any other contractor under direct contract to Owner for the reasonable direct delay and disruption costs incurred by such other contractor as a result of Contractor's wrongful action or inactions.

ARTICLE 8 – OWNER'S RESPONSIBILITIES

8.01 *Communications to Contractor*

A. Except as otherwise provided in these General Conditions, Owner shall issue all communications to Contractor through Engineer.

8.02 *Replacement of Engineer*

A. In case of termination of the employment of Engineer, Owner shall appoint an engineer to whom Contractor makes no reasonable objection, whose status under the Contract Documents shall be that of the former Engineer.

8.03 *Furnish Data*

A. Owner shall promptly furnish the data required of Owner under the Contract Documents.

8.04 *Pay When Due*

A. Owner shall make payments to Contractor when they are due as provided in Paragraphs 14.02.C and 14.07.C.

8.05 *Lands and Easements; Reports and Tests*

A. Owner's duties with respect to providing lands and easements and providing engineering surveys to establish reference points are set forth in Paragraphs 4.01 and 4.05. Paragraph 4.02 refers to Owner's identifying and making available to Contractor copies of reports of explorations and tests of subsurface conditions and drawings of physical conditions relating to existing surface or subsurface structures at the Site.

8.06 *Insurance*

A. Owner's responsibilities, if any, with respect to purchasing and maintaining liability and property insurance are set forth in Article 5.

8.07 *Change Orders*

A. Owner is obligated to execute Change Orders as indicated in Paragraph 10.03.

8.08 *Inspections, Tests, and Approvals*

A. Owner's responsibility with respect to certain inspections, tests, and approvals is set forth in Paragraph 13.03.B.

8.09 *Limitations on Owner's Responsibilities*

A. The Owner shall not supervise, direct, or have control or authority over, nor be responsible for, Contractor's means, methods, techniques, sequences, or procedures of construction, or the safety precautions and programs incident thereto, or for any failure of Contractor to comply with Laws and Regulations applicable to the performance of the Work. Owner will not be responsible for Contractor's failure to perform the Work in accordance with the Contract Documents.

8.10 *Undisclosed Hazardous Environmental Condition*

A. Owner's responsibility in respect to an undisclosed Hazardous Environmental Condition is set forth in Paragraph 4.06.

8.11 *Evidence of Financial Arrangements*

A. Upon request of Contractor, Owner shall furnish Contractor reasonable evidence that financial arrangements have been made to satisfy Owner's obligations under the Contract Documents.

8.12 *Compliance with Safety Program*

A. While at the Site, Owner's employees and representatives shall comply with the specific applicable requirements of Contractor's safety programs of which Owner has been informed pursuant to Paragraph 6.13.D.

ARTICLE 9 – ENGINEER'S STATUS DURING CONSTRUCTION

9.01 *Owner's Representative*

A. Engineer will be Owner's representative during the construction period. The duties and responsibilities and the limitations of authority of Engineer as Owner's representative during construction are set forth in the Contract Documents.

9.02 *Visits to Site*

A. Engineer will make visits to the Site at intervals appropriate to the various stages of construction as Engineer deems necessary in order to observe as an experienced and qualified design professional the progress that has been made and the quality of the various aspects of Contractor's executed Work. Based on information obtained during such visits and observations, Engineer, for the benefit of Owner, will determine, in general, if the Work is proceeding in accordance with the Contract Documents. Engineer will not be required to make exhaustive or continuous inspections on the Site to check the quality or quantity of the Work. Engineer's efforts will be directed toward providing for Owner a greater degree of confidence that the completed Work will conform generally to the Contract Documents. On the basis of such visits and observations, Engineer will keep Owner informed of the progress of the Work and will endeavor to guard Owner against defective Work.

B. Engineer's visits and observations are subject to all the limitations on Engineer's authority and responsibility set forth in Paragraph 9.09. Particularly, but without limitation, during or as a result of Engineer's visits or observations of Contractor's Work, Engineer will not supervise, direct, control, or have authority over or be responsible for Contractor's means, methods, techniques, sequences, or procedures of construction, or the safety precautions and programs incident thereto, or for any failure of Contractor to comply with Laws and Regulations applicable to the performance of the Work.

9.03 *Project Representative*

A. If Owner and Engineer agree, Engineer will furnish a Resident Project Representative to assist Engineer in providing more extensive observation of the Work. The authority and responsibilities of any such Resident Project Representative and assistants will be as provided in the Supplementary Conditions, and limitations on the responsibilities thereof will be as provided in Paragraph 9.09. If Owner designates another representative or agent to represent Owner at the Site who is not Engineer's consultant, agent or employee, the responsibilities and authority and limitations thereon of such other individual or entity will be as provided in the Supplementary Conditions.

9.04 *Authorized Variations in Work*

A. Engineer may authorize minor variations in the Work from the requirements of the Contract Documents which do not involve an adjustment in the Contract Price or the Contract Times and are compatible with the design concept of the completed Project as a functioning whole as indicated by the Contract Documents. These may be accomplished by a Field Order and will be binding on Owner and also on Contractor, who shall perform the Work involved promptly. If Owner or Contractor believes that a Field Order justifies an adjustment in the Contract Price or Contract Times, or both, and the parties are unable to agree on entitlement to or on the amount or extent, if any, of any such adjustment, a Claim may be made therefor as provided in Paragraph 10.05.

9.05 *Rejecting Defective Work*

A. Engineer will have authority to reject Work which Engineer believes to be defective, or that Engineer believes will not produce a completed Project that conforms to the Contract Documents or that will prejudice the integrity of the design concept of the completed Project as a functioning whole as indicated by the Contract Documents. Engineer will also have authority to require special inspection or testing of the Work as provided in Paragraph 13.04, whether or not the Work is fabricated, installed, or completed.

9.06 *Shop Drawings, Change Orders and Payments*

A. In connection with Engineer's authority, and limitations thereof, as to Shop Drawings and Samples, see Paragraph 6.17.

B. In connection with Engineer's authority, and limitations thereof, as to design calculations and design drawings submitted in response to a delegation of professional design services, if any, see Paragraph 6.21.

C. In connection with Engineer's authority as to Change Orders, see Articles 10, 11, and 12.

D. In connection with Engineer's authority as to Applications for Payment, see Article 14.

9.07 *Determinations for Unit Price Work*

A. Engineer will determine the actual quantities and classifications of Unit Price Work performed by Contractor. Engineer will review with Contractor the Engineer's preliminary determinations on such matters before rendering a written decision thereon (by recommendation of an Application for Payment or otherwise). Engineer's written decision thereon will be final and binding (except as modified by Engineer to reflect changed factual conditions or more accurate data) upon Owner and Contractor, subject to the provisions of Paragraph 10.05.

9.08 *Decisions on Requirements of Contract Documents and Acceptability of Work*

A. Engineer will be the initial interpreter of the requirements of the Contract Documents and judge of the acceptability of the Work thereunder. All matters in question and other matters between Owner and Contractor arising prior to the date final payment is due relating to the acceptability of the Work, and the interpretation of the requirements of the Contract Documents pertaining to the performance of the Work, will be referred initially to Engineer in writing within 30 days of the event giving rise to the question.

B. Engineer will, with reasonable promptness, render a written decision on the issue referred. If Owner or Contractor believes that any such decision entitles them to an adjustment in the Contract Price or Contract Times or both, a Claim may be made under Paragraph 10.05. The date of Engineer's decision shall be the date of the event giving rise to the issues referenced for the purposes of Paragraph 10.05.B.

C. Engineer's written decision on the issue referred will be final and binding on Owner and Contractor, subject to the provisions of Paragraph 10.05.

D. When functioning as interpreter and judge under this Paragraph 9.08, Engineer will not show partiality to Owner or Contractor and will not be liable in connection with any interpretation or decision rendered in good faith in such capacity.

9.09 *Limitations on Engineer's Authority and Responsibilities*

A. Neither Engineer's authority or responsibility under this Article 9 or under any other provision of the Contract Documents nor any decision made by Engineer in good faith either to exercise or not exercise such authority or responsibility or the undertaking, exercise, or performance of any authority or responsibility by Engineer shall create, impose, or give rise to any duty in contract, tort, or otherwise owed by Engineer to Contractor, any Subcontractor, any Supplier, any other individual or entity, or to any surety for or employee or agent of any of them.

B. Engineer will not supervise, direct, control, or have authority over or be responsible for Contractor's means, methods, techniques, sequences, or procedures of construction, or the safety precautions and programs incident thereto, or for any failure of Contractor to comply with Laws and Regulations applicable to the performance of the Work. Engineer will not be responsible for Contractor's failure to perform the Work in accordance with the Contract Documents.

C. Engineer will not be responsible for the acts or omissions of Contractor or of any Subcontractor, any Supplier, or of any other individual or entity performing any of the Work.

D. Engineer's review of the final Application for Payment and accompanying documentation and all maintenance and operating instructions, schedules, guarantees, bonds, certificates of inspection, tests and approvals, and other documentation required to be delivered by Paragraph 14.07.A will only be to determine generally that their content complies with the requirements of, and in the case of certificates of inspections, tests, and approvals that the results certified indicate compliance with, the Contract Documents.

E. The limitations upon authority and responsibility set forth in this Paragraph 9.09 shall also apply to the Resident Project Representative, if any, and assistants, if any.

9.10 *Compliance with Safety Program*

A. While at the Site, Engineer's employees and representatives shall comply with the specific applicable requirements of Contractor's safety programs of which Engineer has been informed pursuant to Paragraph 6.13.D.

ARTICLE 10 – CHANGES IN THE WORK; CLAIMS

10.01 *Authorized Changes in the Work*

A. Without invalidating the Contract and without notice to any surety, Owner may, at any time or from time to time, order additions, deletions, or revisions in the Work by a Change Order, or a Work Change Directive. Upon receipt of any such document, Contractor shall promptly proceed with the Work involved which will be performed under the applicable conditions of the Contract Documents (except as otherwise specifically provided).

B. If Owner and Contractor are unable to agree on entitlement to, or on the amount or extent, if any, of an adjustment in the Contract Price or Contract Times, or both, that should be allowed as a result of a Work Change Directive, a Claim may be made therefor as provided in Paragraph 10.05.

10.02 *Unauthorized Changes in the Work*

A. Contractor shall not be entitled to an increase in the Contract Price or an extension of the Contract Times with respect to any work performed that is not required by the Contract Documents as amended, modified, or supplemented as provided in Paragraph 3.04, except in the case of an emergency as provided in Paragraph 6.16 or in the case of uncovering Work as provided in Paragraph 13.04.D.

10.03 *Execution of Change Orders*

A. Owner and Contractor shall execute appropriate Change Orders recommended by Engineer covering:

1. changes in the Work which are: (i) ordered by Owner pursuant to Paragraph 10.01.A, (ii) required because of acceptance of defective Work under Paragraph 13.08.A or Owner's correction of defective Work under Paragraph 13.09, or (iii) agreed to by the parties;

2. changes in the Contract Price or Contract Times which are agreed to by the parties, including any undisputed sum or amount of time for Work actually performed in accordance with a Work Change Directive; and

3. changes in the Contract Price or Contract Times which embody the substance of any written decision rendered by Engineer pursuant to Paragraph 10.05; provided that, in lieu of executing any such Change Order, an appeal may be taken from any such decision in accordance with the provisions of the Contract Documents and applicable Laws and Regulations, but during any such appeal, Contractor shall carry on the Work and adhere to the Progress Schedule as provided in Paragraph 6.18.A.

10.04 *Notification to Surety*

A. If the provisions of any bond require notice to be given to a surety of any change affecting the general scope of the Work or the provisions of the Contract Documents (including, but not limited to, Contract Price or Contract Times), the giving of any such notice will be Contractor's responsibility. The amount of each applicable bond will be adjusted to reflect the effect of any such change.

10.05 *Claims*

A. *Engineer's Decision Required*: All Claims, except those waived pursuant to Paragraph 14.09, shall be referred to the Engineer for decision. A decision by Engineer shall be required as a condition precedent to any exercise by Owner or Contractor of any rights or remedies either may otherwise have under the Contract Documents or by Laws and Regulations in respect of such Claims.

B. *Notice:* Written notice stating the general nature of each Claim shall be delivered by the claimant to Engineer and the other party to the Contract promptly (but in no event later than 30 days) after the start of the event giving rise thereto. The responsibility to substantiate a Claim shall rest with the party making the Claim. Notice of the amount or extent of the Claim, with supporting data shall be delivered to the Engineer and the other party to the Contract within 60 days after the start of such event (unless Engineer allows additional time for claimant to submit additional or more accurate data in support of such Claim). A Claim for an adjustment in Contract Price shall be prepared in accordance with the provisions of Paragraph 12.01.B. A Claim for an adjustment in Contract Times shall be prepared in accordance with the provisions of Paragraph 12.02.B. Each Claim shall be accompanied by claimant's written statement that the adjustment claimed is the entire adjustment to which the claimant believes it is entitled as a result of said event. The opposing party shall submit any response to Engineer and the claimant within 30 days after receipt of the claimant's last submittal (unless Engineer allows additional time).

C. *Engineer's Action*: Engineer will review each Claim and, within 30 days after receipt of the last submittal of the claimant or the last submittal of the opposing party, if any, take one of the following actions in writing:

   1. deny the Claim in whole or in part;

   2. approve the Claim; or

   3. notify the parties that the Engineer is unable to resolve the Claim if, in the Engineer's sole discretion, it would be inappropriate for the Engineer to do so. For purposes of further resolution of the Claim, such notice shall be deemed a denial.

D. In the event that Engineer does not take action on a Claim within said 30 days, the Claim shall be deemed denied.

E. Engineer's written action under Paragraph 10.05.C or denial pursuant to Paragraphs 10.05.C.3 or 10.05.D will be final and binding upon Owner and Contractor, unless Owner or Contractor invoke the dispute resolution procedure set forth in Article 16 within 30 days of such action or denial.

F. No Claim for an adjustment in Contract Price or Contract Times will be valid if not submitted in accordance with this Paragraph 10.05.

ARTICLE 11 – COST OF THE WORK; ALLOWANCES; UNIT PRICE WORK

11.01 *Cost of the Work*

   A. *Costs Included:* The term Cost of the Work means the sum of all costs, except those excluded in Paragraph 11.01.B, necessarily incurred and paid by Contractor in the proper performance of the Work. When the value of any Work covered by a Change Order or when a Claim for an adjustment in Contract Price is determined on the basis of Cost of the Work, the costs to be reimbursed to Contractor will be only those additional or incremental costs required because of the change in the Work or because of the event giving rise to the Claim. Except as otherwise may be agreed to in writing by Owner, such costs shall be in amounts no higher than those prevailing in the locality of the Project, shall not include any of the costs itemized in Paragraph 11.01.B, and shall include only the following items:

   1. Payroll costs for employees in the direct employ of Contractor in the performance of the Work under schedules of job classifications agreed upon by Owner and Contractor. Such employees shall include, without limitation, superintendents, foremen, and other personnel employed full time on the Work. Payroll costs for employees not employed full time on the Work shall be apportioned on the basis of their time spent on the Work. Payroll costs shall include, but not be limited to, salaries and wages plus the cost of fringe benefits, which shall include social security contributions, unemployment, excise, and payroll taxes, workers' compensation, health and retirement benefits, bonuses, sick leave, vacation and holiday pay applicable thereto. The expenses of performing Work outside of regular working hours, on

Saturday, Sunday, or legal holidays, shall be included in the above to the extent authorized by Owner.

2. Cost of all materials and equipment furnished and incorporated in the Work, including costs of transportation and storage thereof, and Suppliers' field services required in connection therewith. All cash discounts shall accrue to Contractor unless Owner deposits funds with Contractor with which to make payments, in which case the cash discounts shall accrue to Owner. All trade discounts, rebates and refunds and returns from sale of surplus materials and equipment shall accrue to Owner, and Contractor shall make provisions so that they may be obtained.

3. Payments made by Contractor to Subcontractors for Work performed by Subcontractors. If required by Owner, Contractor shall obtain competitive bids from subcontractors acceptable to Owner and Contractor and shall deliver such bids to Owner, who will then determine, with the advice of Engineer, which bids, if any, will be acceptable. If any subcontract provides that the Subcontractor is to be paid on the basis of Cost of the Work plus a fee, the Subcontractor's Cost of the Work and fee shall be determined in the same manner as Contractor's Cost of the Work and fee as provided in this Paragraph 11.01.

4. Costs of special consultants (including but not limited to engineers, architects, testing laboratories, surveyors, attorneys, and accountants) employed for services specifically related to the Work.

5. Supplemental costs including the following:

   a. The proportion of necessary transportation, travel, and subsistence expenses of Contractor's employees incurred in discharge of duties connected with the Work.

   b. Cost, including transportation and maintenance, of all materials, supplies, equipment, machinery, appliances, office, and temporary facilities at the Site, and hand tools not owned by the workers, which are consumed in the performance of the Work, and cost, less market value, of such items used but not consumed which remain the property of Contractor.

   c. Rentals of all construction equipment and machinery, and the parts thereof whether rented from Contractor or others in accordance with rental agreements approved by Owner with the advice of Engineer, and the costs of transportation, loading, unloading, assembly, dismantling, and removal thereof. All such costs shall be in accordance with the terms of said rental agreements. The rental of any such equipment, machinery, or parts shall cease when the use thereof is no longer necessary for the Work.

   d. Sales, consumer, use, and other similar taxes related to the Work, and for which Contractor is liable, as imposed by Laws and Regulations.

   e. Deposits lost for causes other than negligence of Contractor, any Subcontractor, or anyone directly or indirectly employed by any of them or for whose acts any of them may be liable, and royalty payments and fees for permits and licenses.

f. Losses and damages (and related expenses) caused by damage to the Work, not compensated by insurance or otherwise, sustained by Contractor in connection with the performance of the Work (except losses and damages within the deductible amounts of property insurance established in accordance with Paragraph 5.06.D), provided such losses and damages have resulted from causes other than the negligence of Contractor, any Subcontractor, or anyone directly or indirectly employed by any of them or for whose acts any of them may be liable. Such losses shall include settlements made with the written consent and approval of Owner. No such losses, damages, and expenses shall be included in the Cost of the Work for the purpose of determining Contractor's fee.

g. The cost of utilities, fuel, and sanitary facilities at the Site.

h. Minor expenses such as telegrams, long distance telephone calls, telephone service at the Site, express and courier services, and similar petty cash items in connection with the Work.

i. The costs of premiums for all bonds and insurance Contractor is required by the Contract Documents to purchase and maintain.

B. *Costs Excluded:* The term Cost of the Work shall not include any of the following items:

1. Payroll costs and other compensation of Contractor's officers, executives, principals (of partnerships and sole proprietorships), general managers, safety managers, engineers, architects, estimators, attorneys, auditors, accountants, purchasing and contracting agents, expediters, timekeepers, clerks, and other personnel employed by Contractor, whether at the Site or in Contractor's principal or branch office for general administration of the Work and not specifically included in the agreed upon schedule of job classifications referred to in Paragraph 11.01.A.1 or specifically covered by Paragraph 11.01.A.4, all of which are to be considered administrative costs covered by the Contractor's fee.

2. Expenses of Contractor's principal and branch offices other than Contractor's office at the Site.

3. Any part of Contractor's capital expenses, including interest on Contractor's capital employed for the Work and charges against Contractor for delinquent payments.

4. Costs due to the negligence of Contractor, any Subcontractor, or anyone directly or indirectly employed by any of them or for whose acts any of them may be liable, including but not limited to, the correction of defective Work, disposal of materials or equipment wrongly supplied, and making good any damage to property.

5. Other overhead or general expense costs of any kind and the costs of any item not specifically and expressly included in Paragraphs 11.01.A.

C. *Contractor's Fee:* When all the Work is performed on the basis of cost-plus, Contractor's fee shall be determined as set forth in the Agreement. When the value of any Work covered by a Change Order or when a Claim for an adjustment in Contract Price is determined on the basis of Cost of the Work, Contractor's fee shall be determined as set forth in Paragraph 12.01.C.

D. *Documentation:* Whenever the Cost of the Work for any purpose is to be determined pursuant to Paragraphs 11.01.A and 11.01.B, Contractor will establish and maintain records thereof in accordance with generally accepted accounting practices and submit in a form acceptable to Engineer an itemized cost breakdown together with supporting data.

11.02 *Allowances*

   A. It is understood that Contractor has included in the Contract Price all allowances so named in the Contract Documents and shall cause the Work so covered to be performed for such sums and by such persons or entities as may be acceptable to Owner and Engineer.

   B. *Cash Allowances:*

      1. Contractor agrees that:

         a. the cash allowances include the cost to Contractor (less any applicable trade discounts) of materials and equipment required by the allowances to be delivered at the Site, and all applicable taxes; and

         b. Contractor's costs for unloading and handling on the Site, labor, installation, overhead, profit, and other expenses contemplated for the cash allowances have been included in the Contract Price and not in the allowances, and no demand for additional payment on account of any of the foregoing will be valid.

   C. *Contingency Allowance:*

      1. Contractor agrees that a contingency allowance, if any, is for the sole use of Owner to cover unanticipated costs.

   D. Prior to final payment, an appropriate Change Order will be issued as recommended by Engineer to reflect actual amounts due Contractor on account of Work covered by allowances, and the Contract Price shall be correspondingly adjusted.

11.03 *Unit Price Work*

   A. Where the Contract Documents provide that all or part of the Work is to be Unit Price Work, initially the Contract Price will be deemed to include for all Unit Price Work an amount equal to the sum of the unit price for each separately identified item of Unit Price Work times the estimated quantity of each item as indicated in the Agreement.

   B. The estimated quantities of items of Unit Price Work are not guaranteed and are solely for the purpose of comparison of Bids and determining an initial Contract Price. Determinations of the actual quantities and classifications of Unit Price Work performed by Contractor will be made by Engineer subject to the provisions of Paragraph 9.07.

   C. Each unit price will be deemed to include an amount considered by Contractor to be adequate to cover Contractor's overhead and profit for each separately identified item.

D. Owner or Contractor may make a Claim for an adjustment in the Contract Price in accordance with Paragraph 10.05 if:

1. the quantity of any item of Unit Price Work performed by Contractor differs materially and significantly from the estimated quantity of such item indicated in the Agreement; and

2. there is no corresponding adjustment with respect to any other item of Work; and

3. Contractor believes that Contractor is entitled to an increase in Contract Price as a result of having incurred additional expense or Owner believes that Owner is entitled to a decrease in Contract Price and the parties are unable to agree as to the amount of any such increase or decrease.

ARTICLE 12 – CHANGE OF CONTRACT PRICE; CHANGE OF CONTRACT TIMES

12.01 *Change of Contract Price*

A. The Contract Price may only be changed by a Change Order. Any Claim for an adjustment in the Contract Price shall be based on written notice submitted by the party making the Claim to the Engineer and the other party to the Contract in accordance with the provisions of Paragraph 10.05.

B. The value of any Work covered by a Change Order or of any Claim for an adjustment in the Contract Price will be determined as follows:

1. where the Work involved is covered by unit prices contained in the Contract Documents, by application of such unit prices to the quantities of the items involved (subject to the provisions of Paragraph 11.03); or

2. where the Work involved is not covered by unit prices contained in the Contract Documents, by a mutually agreed lump sum (which may include an allowance for overhead and profit not necessarily in accordance with Paragraph 12.01.C.2); or

3. where the Work involved is not covered by unit prices contained in the Contract Documents and agreement to a lump sum is not reached under Paragraph 12.01.B.2, on the basis of the Cost of the Work (determined as provided in Paragraph 11.01) plus a Contractor's fee for overhead and profit (determined as provided in Paragraph 12.01.C).

C. *Contractor's Fee:* The Contractor's fee for overhead and profit shall be determined as follows:

1. a mutually acceptable fixed fee; or

2. if a fixed fee is not agreed upon, then a fee based on the following percentages of the various portions of the Cost of the Work:

    a. for costs incurred under Paragraphs 11.01.A.1 and 11.01.A.2, the Contractor's fee shall be 15 percent;

    b. for costs incurred under Paragraph 11.01.A.3, the Contractor's fee shall be five percent;

c. where one or more tiers of subcontracts are on the basis of Cost of the Work plus a fee and no fixed fee is agreed upon, the intent of Paragraphs 12.01.C.2.a and 12.01.C.2.b is that the Subcontractor who actually performs the Work, at whatever tier, will be paid a fee of 15 percent of the costs incurred by such Subcontractor under Paragraphs 11.01.A.1 and 11.01.A.2 and that any higher tier Subcontractor and Contractor will each be paid a fee of five percent of the amount paid to the next lower tier Subcontractor;

d. no fee shall be payable on the basis of costs itemized under Paragraphs 11.01.A.4, 11.01.A.5, and 11.01.B;

e. the amount of credit to be allowed by Contractor to Owner for any change which results in a net decrease in cost will be the amount of the actual net decrease in cost plus a deduction in Contractor's fee by an amount equal to five percent of such net decrease; and

f. when both additions and credits are involved in any one change, the adjustment in Contractor's fee shall be computed on the basis of the net change in accordance with Paragraphs 12.01.C.2.a through 12.01.C.2.e, inclusive.

12.02 *Change of Contract Times*

A. The Contract Times may only be changed by a Change Order. Any Claim for an adjustment in the Contract Times shall be based on written notice submitted by the party making the Claim to the Engineer and the other party to the Contract in accordance with the provisions of Paragraph 10.05.

B. Any adjustment of the Contract Times covered by a Change Order or any Claim for an adjustment in the Contract Times will be determined in accordance with the provisions of this Article 12.

12.03 *Delays*

A. Where Contractor is prevented from completing any part of the Work within the Contract Times due to delay beyond the control of Contractor, the Contract Times will be extended in an amount equal to the time lost due to such delay if a Claim is made therefor as provided in Paragraph 12.02.A. Delays beyond the control of Contractor shall include, but not be limited to, acts or neglect by Owner, acts or neglect of utility owners or other contractors performing other work as contemplated by Article 7, fires, floods, epidemics, abnormal weather conditions, or acts of God.

B. If Owner, Engineer, or other contractors or utility owners performing other work for Owner as contemplated by Article 7, or anyone for whom Owner is responsible, delays, disrupts, or interferes with the performance or progress of the Work, then Contractor shall be entitled to an equitable adjustment in the Contract Price or the Contract Times, or both. Contractor's entitlement to an adjustment of the Contract Times is conditioned on such adjustment being essential to Contractor's ability to complete the Work within the Contract Times.

C. If Contractor is delayed in the performance or progress of the Work by fire, flood, epidemic, abnormal weather conditions, acts of God, acts or failures to act of utility owners not under the control of Owner, or other causes not the fault of and beyond control of Owner and Contractor, then Contractor shall be entitled to an equitable adjustment in Contract Times, if such adjustment is essential to Contractor's ability to complete the Work within the Contract Times. Such an adjustment shall be Contractor's sole and exclusive remedy for the delays described in this Paragraph 12.03.C.

D. Owner, Engineer, and their officers, directors, members, partners, employees, agents, consultants, or subcontractors shall not be liable to Contractor for any claims, costs, losses, or damages (including but not limited to all fees and charges of engineers, architects, attorneys, and other professionals and all court or arbitration or other dispute resolution costs) sustained by Contractor on or in connection with any other project or anticipated project.

E. Contractor shall not be entitled to an adjustment in Contract Price or Contract Times for delays within the control of Contractor. Delays attributable to and within the control of a Subcontractor or Supplier shall be deemed to be delays within the control of Contractor.

ARTICLE 13 – TESTS AND INSPECTIONS; CORRECTION, REMOVAL OR ACCEPTANCE OF DEFECTIVE WORK

13.01 *Notice of Defects*

A. Prompt notice of all defective Work of which Owner or Engineer has actual knowledge will be given to Contractor. Defective Work may be rejected, corrected, or accepted as provided in this Article 13.

13.02 *Access to Work*

A. Owner, Engineer, their consultants and other representatives and personnel of Owner, independent testing laboratories, and governmental agencies with jurisdictional interests will have access to the Site and the Work at reasonable times for their observation, inspection, and testing. Contractor shall provide them proper and safe conditions for such access and advise them of Contractor's safety procedures and programs so that they may comply therewith as applicable.

13.03 *Tests and Inspections*

A. Contractor shall give Engineer timely notice of readiness of the Work for all required inspections, tests, or approvals and shall cooperate with inspection and testing personnel to facilitate required inspections or tests.

B. Owner shall employ and pay for the services of an independent testing laboratory to perform all inspections, tests, or approvals required by the Contract Documents except:

1. for inspections, tests, or approvals covered by Paragraphs 13.03.C and 13.03.D below;

2. that costs incurred in connection with tests or inspections conducted pursuant to Paragraph 13.04.B shall be paid as provided in Paragraph 13.04.C; and

3. as otherwise specifically provided in the Contract Documents.

C. If Laws or Regulations of any public body having jurisdiction require any Work (or part thereof) specifically to be inspected, tested, or approved by an employee or other representative of such public body, Contractor shall assume full responsibility for arranging and obtaining such inspections, tests, or approvals, pay all costs in connection therewith, and furnish Engineer the required certificates of inspection or approval.

D. Contractor shall be responsible for arranging and obtaining and shall pay all costs in connection with any inspections, tests, or approvals required for Owner's and Engineer's acceptance of materials or equipment to be incorporated in the Work; or acceptance of materials, mix designs, or equipment submitted for approval prior to Contractor's purchase thereof for incorporation in the Work. Such inspections, tests, or approvals shall be performed by organizations acceptable to Owner and Engineer.

E. If any Work (or the work of others) that is to be inspected, tested, or approved is covered by Contractor without written concurrence of Engineer, Contractor shall, if requested by Engineer, uncover such Work for observation.

F. Uncovering Work as provided in Paragraph 13.03.E shall be at Contractor's expense unless Contractor has given Engineer timely notice of Contractor's intention to cover the same and Engineer has not acted with reasonable promptness in response to such notice.

13.04 *Uncovering Work*

A. If any Work is covered contrary to the written request of Engineer, it must, if requested by Engineer, be uncovered for Engineer's observation and replaced at Contractor's expense.

B. If Engineer considers it necessary or advisable that covered Work be observed by Engineer or inspected or tested by others, Contractor, at Engineer's request, shall uncover, expose, or otherwise make available for observation, inspection, or testing as Engineer may require, that portion of the Work in question, furnishing all necessary labor, material, and equipment.

C. If it is found that the uncovered Work is defective, Contractor shall pay all claims, costs, losses, and damages (including but not limited to all fees and charges of engineers, architects, attorneys, and other professionals and all court or arbitration or other dispute resolution costs) arising out of or relating to such uncovering, exposure, observation, inspection, and testing, and of satisfactory replacement or reconstruction (including but not limited to all costs of repair or replacement of work of others); and Owner shall be entitled to an appropriate decrease in the Contract Price. If the parties are unable to agree as to the amount thereof, Owner may make a Claim therefor as provided in Paragraph 10.05.

D. If the uncovered Work is not found to be defective, Contractor shall be allowed an increase in the Contract Price or an extension of the Contract Times, or both, directly attributable to such uncovering, exposure, observation, inspection, testing, replacement, and reconstruction. If the parties are unable to agree as to the amount or extent thereof, Contractor may make a Claim therefor as provided in Paragraph 10.05.

13.05 *Owner May Stop the Work*

   A. If the Work is defective, or Contractor fails to supply sufficient skilled workers or suitable materials or equipment, or fails to perform the Work in such a way that the completed Work will conform to the Contract Documents, Owner may order Contractor to stop the Work, or any portion thereof, until the cause for such order has been eliminated; however, this right of Owner to stop the Work shall not give rise to any duty on the part of Owner to exercise this right for the benefit of Contractor, any Subcontractor, any Supplier, any other individual or entity, or any surety for, or employee or agent of any of them.

13.06 *Correction or Removal of Defective Work*

   A. Promptly after receipt of written notice, Contractor shall correct all defective Work, whether or not fabricated, installed, or completed, or, if the Work has been rejected by Engineer, remove it from the Project and replace it with Work that is not defective. Contractor shall pay all claims, costs, losses, and damages (including but not limited to all fees and charges of engineers, architects, attorneys, and other professionals and all court or arbitration or other dispute resolution costs) arising out of or relating to such correction or removal (including but not limited to all costs of repair or replacement of work of others).

   B. When correcting defective Work under the terms of this Paragraph 13.06 or Paragraph 13.07, Contractor shall take no action that would void or otherwise impair Owner's special warranty and guarantee, if any, on said Work.

13.07 *Correction Period*

   A. If within one year after the date of Substantial Completion (or such longer period of time as may be prescribed by the terms of any applicable special guarantee required by the Contract Documents) or by any specific provision of the Contract Documents, any Work is found to be defective, or if the repair of any damages to the land or areas made available for Contractor's use by Owner or permitted by Laws and Regulations as contemplated in Paragraph 6.11.A is found to be defective, Contractor shall promptly, without cost to Owner and in accordance with Owner's written instructions:

      1. repair such defective land or areas; or

      2. correct such defective Work; or

      3. if the defective Work has been rejected by Owner, remove it from the Project and replace it with Work that is not defective, and

      4. satisfactorily correct or repair or remove and replace any damage to other Work, to the work of others or other land or areas resulting therefrom.

   B. If Contractor does not promptly comply with the terms of Owner's written instructions, or in an emergency where delay would cause serious risk of loss or damage, Owner may have the defective Work corrected or repaired or may have the rejected Work removed and replaced. All claims, costs, losses, and damages (including but not limited to all fees and charges of engineers, architects, attorneys, and other professionals and all court or arbitration or other dispute resolution costs) arising out of or relating to such correction or repair or such removal and replacement (including but not limited to all costs of repair or replacement of work of others) will be paid by Contractor.

C. In special circumstances where a particular item of equipment is placed in continuous service before Substantial Completion of all the Work, the correction period for that item may start to run from an earlier date if so provided in the Specifications.

D. Where defective Work (and damage to other Work resulting therefrom) has been corrected or removed and replaced under this Paragraph 13.07, the correction period hereunder with respect to such Work will be extended for an additional period of one year after such correction or removal and replacement has been satisfactorily completed.

E. Contractor's obligations under this Paragraph 13.07 are in addition to any other obligation or warranty. The provisions of this Paragraph 13.07 shall not be construed as a substitute for, or a waiver of, the provisions of any applicable statute of limitation or repose.

13.08 *Acceptance of Defective Work*

A. If, instead of requiring correction or removal and replacement of defective Work, Owner (and, prior to Engineer's recommendation of final payment, Engineer) prefers to accept it, Owner may do so. Contractor shall pay all claims, costs, losses, and damages (including but not limited to all fees and charges of engineers, architects, attorneys, and other professionals and all court or arbitration or other dispute resolution costs) attributable to Owner's evaluation of and determination to accept such defective Work (such costs to be approved by Engineer as to reasonableness) and for the diminished value of the Work to the extent not otherwise paid by Contractor pursuant to this sentence. If any such acceptance occurs prior to Engineer's recommendation of final payment, a Change Order will be issued incorporating the necessary revisions in the Contract Documents with respect to the Work, and Owner shall be entitled to an appropriate decrease in the Contract Price, reflecting the diminished value of Work so accepted. If the parties are unable to agree as to the amount thereof, Owner may make a Claim therefor as provided in Paragraph 10.05. If the acceptance occurs after such recommendation, an appropriate amount will be paid by Contractor to Owner.

13.09 *Owner May Correct Defective Work*

A. If Contractor fails within a reasonable time after written notice from Engineer to correct defective Work, or to remove and replace rejected Work as required by Engineer in accordance with Paragraph 13.06.A, or if Contractor fails to perform the Work in accordance with the Contract Documents, or if Contractor fails to comply with any other provision of the Contract Documents, Owner may, after seven days written notice to Contractor, correct, or remedy any such deficiency.

B. In exercising the rights and remedies under this Paragraph 13.09, Owner shall proceed expeditiously. In connection with such corrective or remedial action, Owner may exclude Contractor from all or part of the Site, take possession of all or part of the Work and suspend Contractor's services related thereto, take possession of Contractor's tools, appliances, construction equipment and machinery at the Site, and incorporate in the Work all materials and equipment stored at the Site or for which Owner has paid Contractor but which are stored elsewhere. Contractor shall allow Owner, Owner's representatives, agents and employees, Owner's other contractors, and Engineer and Engineer's consultants access to the Site to enable Owner to exercise the rights and remedies under this Paragraph.

C. All claims, costs, losses, and damages (including but not limited to all fees and charges of engineers, architects, attorneys, and other professionals and all court or arbitration or other dispute resolution costs) incurred or sustained by Owner in exercising the rights and remedies under this Paragraph 13.09 will be charged against Contractor, and a Change Order will be issued incorporating the necessary revisions in the Contract Documents with respect to the Work; and Owner shall be entitled to an appropriate decrease in the Contract Price. If the parties are unable to agree as to the amount of the adjustment, Owner may make a Claim therefor as provided in Paragraph 10.05. Such claims, costs, losses and damages will include but not be limited to all costs of repair, or replacement of work of others destroyed or damaged by correction, removal, or replacement of Contractor's defective Work.

D. Contractor shall not be allowed an extension of the Contract Times because of any delay in the performance of the Work attributable to the exercise by Owner of Owner's rights and remedies under this Paragraph 13.09.

ARTICLE 14 – PAYMENTS TO CONTRACTOR AND COMPLETION

14.01 *Schedule of Values*

A. The Schedule of Values established as provided in Paragraph 2.07.A will serve as the basis for progress payments and will be incorporated into a form of Application for Payment acceptable to Engineer. Progress payments on account of Unit Price Work will be based on the number of units completed.

14.02 *Progress Payments*

A. *Applications for Payments:*

1. At least 20 days before the date established in the Agreement for each progress payment (but not more often than once a month), Contractor shall submit to Engineer for review an Application for Payment filled out and signed by Contractor covering the Work completed as of the date of the Application and accompanied by such supporting documentation as is required by the Contract Documents. If payment is requested on the basis of materials and equipment not incorporated in the Work but delivered and suitably stored at the Site or at another location agreed to in writing, the Application for Payment shall also be accompanied by a bill of sale, invoice, or other documentation warranting that Owner has received the materials and equipment free and clear of all Liens and evidence that the materials and equipment are covered by appropriate property insurance or other arrangements to protect Owner's interest therein, all of which must be satisfactory to Owner.

2. Beginning with the second Application for Payment, each Application shall include an affidavit of Contractor stating that all previous progress payments received on account of the

Work have been applied on account to discharge Contractor's legitimate obligations associated with prior Applications for Payment.

3. The amount of retainage with respect to progress payments will be as stipulated in the Agreement.

B. *Review of Applications:*

1. Engineer will, within 10 days after receipt of each Application for Payment, either indicate in writing a recommendation of payment and present the Application to Owner or return the Application to Contractor indicating in writing Engineer's reasons for refusing to recommend payment. In the latter case, Contractor may make the necessary corrections and resubmit the Application.

2. Engineer's recommendation of any payment requested in an Application for Payment will constitute a representation by Engineer to Owner, based on Engineer's observations of the executed Work as an experienced and qualified design professional, and on Engineer's review of the Application for Payment and the accompanying data and schedules, that to the best of Engineer's knowledge, information and belief:

    a. the Work has progressed to the point indicated;

    b. the quality of the Work is generally in accordance with the Contract Documents (subject to an evaluation of the Work as a functioning whole prior to or upon Substantial Completion, the results of any subsequent tests called for in the Contract Documents, a final determination of quantities and classifications for Unit Price Work under Paragraph 9.07, and any other qualifications stated in the recommendation); and

    c. the conditions precedent to Contractor's being entitled to such payment appear to have been fulfilled in so far as it is Engineer's responsibility to observe the Work.

3. By recommending any such payment Engineer will not thereby be deemed to have represented that:

    a. inspections made to check the quality or the quantity of the Work as it has been performed have been exhaustive, extended to every aspect of the Work in progress, or involved detailed inspections of the Work beyond the responsibilities specifically assigned to Engineer in the Contract Documents; or

    b. there may not be other matters or issues between the parties that might entitle Contractor to be paid additionally by Owner or entitle Owner to withhold payment to Contractor.

4. Neither Engineer's review of Contractor's Work for the purposes of recommending payments nor Engineer's recommendation of any payment, including final payment, will impose responsibility on Engineer:

    a. to supervise, direct, or control the Work, or

b. for the means, methods, techniques, sequences, or procedures of construction, or the safety precautions and programs incident thereto, or

c. for Contractor's failure to comply with Laws and Regulations applicable to Contractor's performance of the Work, or

d. to make any examination to ascertain how or for what purposes Contractor has used the moneys paid on account of the Contract Price, or

e. to determine that title to any of the Work, materials, or equipment has passed to Owner free and clear of any Liens.

5. Engineer may refuse to recommend the whole or any part of any payment if, in Engineer's opinion, it would be incorrect to make the representations to Owner stated in Paragraph 14.02.B.2. Engineer may also refuse to recommend any such payment or, because of subsequently discovered evidence or the results of subsequent inspections or tests, revise or revoke any such payment recommendation previously made, to such extent as may be necessary in Engineer's opinion to protect Owner from loss because:

a. the Work is defective, or completed Work has been damaged, requiring correction or replacement;

b. the Contract Price has been reduced by Change Orders;

c. Owner has been required to correct defective Work or complete Work in accordance with Paragraph 13.09; or

d. Engineer has actual knowledge of the occurrence of any of the events enumerated in Paragraph 15.02.A.

C. *Payment Becomes Due:*

1. Ten days after presentation of the Application for Payment to Owner with Engineer's recommendation, the amount recommended will (subject to the provisions of Paragraph 14.02.D) become due, and when due will be paid by Owner to Contractor.

D. *Reduction in Payment:*

1. Owner may refuse to make payment of the full amount recommended by Engineer because:

a. claims have been made against Owner on account of Contractor's performance or furnishing of the Work;

b. Liens have been filed in connection with the Work, except where Contractor has delivered a specific bond satisfactory to Owner to secure the satisfaction and discharge of such Liens;

c. there are other items entitling Owner to a set-off against the amount recommended; or

d. Owner has actual knowledge of the occurrence of any of the events enumerated in Paragraphs 14.02.B.5.a through 14.02.B.5.c or Paragraph 15.02.A.

2. If Owner refuses to make payment of the full amount recommended by Engineer, Owner will give Contractor immediate written notice (with a copy to Engineer) stating the reasons for such action and promptly pay Contractor any amount remaining after deduction of the amount so withheld. Owner shall promptly pay Contractor the amount so withheld, or any adjustment thereto agreed to by Owner and Contractor, when Contractor remedies the reasons for such action.

3. Upon a subsequent determination that Owner's refusal of payment was not justified, the amount wrongfully withheld shall be treated as an amount due as determined by Paragraph 14.02.C.1 and subject to interest as provided in the Agreement.

14.03 Contractor's Warranty of Title

A. Contractor warrants and guarantees that title to all Work, materials, and equipment covered by any Application for Payment, whether incorporated in the Project or not, will pass to Owner no later than the time of payment free and clear of all Liens.

14.04 Substantial Completion

A. When Contractor considers the entire Work ready for its intended use Contractor shall notify Owner and Engineer in writing that the entire Work is substantially complete (except for items specifically listed by Contractor as incomplete) and request that Engineer issue a certificate of Substantial Completion.

B. Promptly after Contractor's notification, Owner, Contractor, and Engineer shall make an inspection of the Work to determine the status of completion. If Engineer does not consider the Work substantially complete, Engineer will notify Contractor in writing giving the reasons therefor.

C. If Engineer considers the Work substantially complete, Engineer will deliver to Owner a tentative certificate of Substantial Completion which shall fix the date of Substantial Completion. There shall be attached to the certificate a tentative list of items to be completed or corrected before final payment. Owner shall have seven days after receipt of the tentative certificate during which to make written objection to Engineer as to any provisions of the certificate or attached list. If, after considering such objections, Engineer concludes that the Work is not substantially complete, Engineer will, within 14 days after submission of the tentative certificate to Owner, notify Contractor in writing, stating the reasons therefor. If, after consideration of Owner's objections, Engineer considers the Work substantially complete, Engineer will, within said 14 days, execute and deliver to Owner and Contractor a definitive certificate of Substantial Completion (with a revised tentative list of items to be completed or corrected) reflecting such changes from the tentative certificate as Engineer believes justified after consideration of any objections from Owner.

D. At the time of delivery of the tentative certificate of Substantial Completion, Engineer will deliver to Owner and Contractor a written recommendation as to division of responsibilities pending final payment between Owner and Contractor with respect to security, operation, safety, and protection of the Work, maintenance, heat, utilities, insurance, and warranties and guarantees. Unless Owner and Contractor agree otherwise in writing and so inform Engineer in writing prior to Engineer's issuing the definitive certificate of Substantial Completion, Engineer's aforesaid recommendation will be binding on Owner and Contractor until final payment.

E. Owner shall have the right to exclude Contractor from the Site after the date of Substantial Completion subject to allowing Contractor reasonable access to remove its property and complete or correct items on the tentative list.

14.05 *Partial Utilization*

A. Prior to Substantial Completion of all the Work, Owner may use or occupy any substantially completed part of the Work which has specifically been identified in the Contract Documents, or which Owner, Engineer, and Contractor agree constitutes a separately functioning and usable part of the Work that can be used by Owner for its intended purpose without significant interference with Contractor's performance of the remainder of the Work, subject to the following conditions:

1. Owner at any time may request Contractor in writing to permit Owner to use or occupy any such part of the Work which Owner believes to be ready for its intended use and substantially complete. If and when Contractor agrees that such part of the Work is substantially complete, Contractor, Owner, and Engineer will follow the procedures of Paragraph 14.04.A through D for that part of the Work.

2. Contractor at any time may notify Owner and Engineer in writing that Contractor considers any such part of the Work ready for its intended use and substantially complete and request Engineer to issue a certificate of Substantial Completion for that part of the Work.

3. Within a reasonable time after either such request, Owner, Contractor, and Engineer shall make an inspection of that part of the Work to determine its status of completion. If Engineer does not consider that part of the Work to be substantially complete, Engineer will notify Owner and Contractor in writing giving the reasons therefor. If Engineer considers that part of the Work to be substantially complete, the provisions of Paragraph 14.04 will apply with respect to certification of Substantial Completion of that part of the Work and the division of responsibility in respect thereof and access thereto.

4. No use or occupancy or separate operation of part of the Work may occur prior to compliance with the requirements of Paragraph 5.10 regarding property insurance.

14.06 *Final Inspection*

A. Upon written notice from Contractor that the entire Work or an agreed portion thereof is complete, Engineer will promptly make a final inspection with Owner and Contractor and will notify Contractor in writing of all particulars in which this inspection reveals that the Work is incomplete or defective. Contractor shall immediately take such measures as are necessary to complete such Work or remedy such deficiencies.

14.07 *Final Payment*

   A. *Application for Payment:*

   1. After Contractor has, in the opinion of Engineer, satisfactorily completed all corrections identified during the final inspection and has delivered, in accordance with the Contract Documents, all maintenance and operating instructions, schedules, guarantees, bonds, certificates or other evidence of insurance, certificates of inspection, marked-up record documents (as provided in Paragraph 6.12), and other documents, Contractor may make application for final payment following the procedure for progress payments.

   2. The final Application for Payment shall be accompanied (except as previously delivered) by:

      a. all documentation called for in the Contract Documents, including but not limited to the evidence of insurance required by Paragraph 5.04.B.6;

      b. consent of the surety, if any, to final payment;

      c. a list of all Claims against Owner that Contractor believes are unsettled; and

      d. complete and legally effective releases or waivers (satisfactory to Owner) of all Lien rights arising out of or Liens filed in connection with the Work.

   3. In lieu of the releases or waivers of Liens specified in Paragraph 14.07.A.2 and as approved by Owner, Contractor may furnish receipts or releases in full and an affidavit of Contractor that: (i) the releases and receipts include all labor, services, material, and equipment for which a Lien could be filed; and (ii) all payrolls, material and equipment bills, and other indebtedness connected with the Work for which Owner might in any way be responsible, or which might in any way result in liens or other burdens on Owner's property, have been paid or otherwise satisfied. If any Subcontractor or Supplier fails to furnish such a release or receipt in full, Contractor may furnish a bond or other collateral satisfactory to Owner to indemnify Owner against any Lien.

   B. *Engineer's Review of Application and Acceptance:*

   1. If, on the basis of Engineer's observation of the Work during construction and final inspection, and Engineer's review of the final Application for Payment and accompanying documentation as required by the Contract Documents, Engineer is satisfied that the Work has been completed and Contractor's other obligations under the Contract Documents have been fulfilled, Engineer will, within ten days after receipt of the final Application for Payment, indicate in writing Engineer's recommendation of payment and present the Application for Payment to Owner for payment. At the same time Engineer will also give written notice to Owner and Contractor that the Work is acceptable subject to the provisions of Paragraph 14.09. Otherwise, Engineer will return the Application for Payment to Contractor, indicating in writing the reasons for refusing to recommend final payment, in which case Contractor shall make the necessary corrections and resubmit the Application for Payment.

   C. *Payment Becomes Due:*

1. Thirty days after the presentation to Owner of the Application for Payment and accompanying documentation, the amount recommended by Engineer, less any sum Owner is entitled to set off against Engineer's recommendation, including but not limited to liquidated damages, will become due and will be paid by Owner to Contractor.

14.08 *Final Completion Delayed*

A. If, through no fault of Contractor, final completion of the Work is significantly delayed, and if Engineer so confirms, Owner shall, upon receipt of Contractor's final Application for Payment (for Work fully completed and accepted) and recommendation of Engineer, and without terminating the Contract, make payment of the balance due for that portion of the Work fully completed and accepted. If the remaining balance to be held by Owner for Work not fully completed or corrected is less than the retainage stipulated in the Agreement, and if bonds have been furnished as required in Paragraph 5.01, the written consent of the surety to the payment of the balance due for that portion of the Work fully completed and accepted shall be submitted by Contractor to Engineer with the Application for such payment. Such payment shall be made under the terms and conditions governing final payment, except that it shall not constitute a waiver of Claims.

14.09 *Waiver of Claims*

A. The making and acceptance of final payment will constitute:

1. a waiver of all Claims by Owner against Contractor, except Claims arising from unsettled Liens, from defective Work appearing after final inspection pursuant to Paragraph 14.06, from failure to comply with the Contract Documents or the terms of any special guarantees specified therein, or from Contractor's continuing obligations under the Contract Documents; and

2. a waiver of all Claims by Contractor against Owner other than those previously made in accordance with the requirements herein and expressly acknowledged by Owner in writing as still unsettled.

ARTICLE 15 – SUSPENSION OF WORK AND TERMINATION

15.01 *Owner May Suspend Work*

A. At any time and without cause, Owner may suspend the Work or any portion thereof for a period of not more than 90 consecutive days by notice in writing to Contractor and Engineer which will fix the date on which Work will be resumed. Contractor shall resume the Work on the date so fixed. Contractor shall be granted an adjustment in the Contract Price or an extension of the Contract Times, or both, directly attributable to any such suspension if Contractor makes a Claim therefor as provided in Paragraph 10.05.

15.02 *Owner May Terminate for Cause*

A. The occurrence of any one or more of the following events will justify termination for cause:

1. Contractor's persistent failure to perform the Work in accordance with the Contract Documents (including, but not limited to, failure to supply sufficient skilled workers or suitable materials or equipment or failure to adhere to the Progress Schedule established under Paragraph 2.07 as adjusted from time to time pursuant to Paragraph 6.04);

2. Contractor's disregard of Laws or Regulations of any public body having jurisdiction;

3. Contractor's repeated disregard of the authority of Engineer; or

4. Contractor's violation in any substantial way of any provisions of the Contract Documents.

B. If one or more of the events identified in Paragraph 15.02.A occur, Owner may, after giving Contractor (and surety) seven days written notice of its intent to terminate the services of Contractor:

1. exclude Contractor from the Site, and take possession of the Work and of all Contractor's tools, appliances, construction equipment, and machinery at the Site, and use the same to the full extent they could be used by Contractor (without liability to Contractor for trespass or conversion);

2. incorporate in the Work all materials and equipment stored at the Site or for which Owner has paid Contractor but which are stored elsewhere; and

3. complete the Work as Owner may deem expedient.

C. If Owner proceeds as provided in Paragraph 15.02.B, Contractor shall not be entitled to receive any further payment until the Work is completed. If the unpaid balance of the Contract Price exceeds all claims, costs, losses, and damages (including but not limited to all fees and charges of engineers, architects, attorneys, and other professionals and all court or arbitration or other dispute resolution costs) sustained by Owner arising out of or relating to completing the Work, such excess will be paid to Contractor. If such claims, costs, losses, and damages exceed such unpaid balance, Contractor shall pay the difference to Owner. Such claims, costs, losses, and damages incurred by Owner will be reviewed by Engineer as to their reasonableness and, when so approved by Engineer, incorporated in a Change Order. When exercising any rights or remedies under this Paragraph, Owner shall not be required to obtain the lowest price for the Work performed.

D. Notwithstanding Paragraphs 15.02.B and 15.02.C, Contractor's services will not be terminated if Contractor begins within seven days of receipt of notice of intent to terminate to correct its failure to perform and proceeds diligently to cure such failure within no more than 30 days of receipt of said notice.

E. Where Contractor's services have been so terminated by Owner, the termination will not affect any rights or remedies of Owner against Contractor then existing or which may thereafter accrue. Any retention or payment of moneys due Contractor by Owner will not release Contractor from liability.

F. If and to the extent that Contractor has provided a performance bond under the provisions of Paragraph 5.01.A, the termination procedures of that bond shall supersede the provisions of Paragraphs 15.02.B and 15.02.C.

15.03 *Owner May Terminate For Convenience*

A. Upon seven days written notice to Contractor and Engineer, Owner may, without cause and without prejudice to any other right or remedy of Owner, terminate the Contract. In such case, Contractor shall be paid for (without duplication of any items):

1. completed and acceptable Work executed in accordance with the Contract Documents prior to the effective date of termination, including fair and reasonable sums for overhead and profit on such Work;

2. expenses sustained prior to the effective date of termination in performing services and furnishing labor, materials, or equipment as required by the Contract Documents in connection with uncompleted Work, plus fair and reasonable sums for overhead and profit on such expenses;

3. all claims, costs, losses, and damages (including but not limited to all fees and charges of engineers, architects, attorneys, and other professionals and all court or arbitration or other dispute resolution costs) incurred in settlement of terminated contracts with Subcontractors, Suppliers, and others; and

4. reasonable expenses directly attributable to termination.

B. Contractor shall not be paid on account of loss of anticipated profits or revenue or other economic loss arising out of or resulting from such termination.

15.04 *Contractor May Stop Work or Terminate*

A. If, through no act or fault of Contractor, (i) the Work is suspended for more than 90 consecutive days by Owner or under an order of court or other public authority, or (ii) Engineer fails to act on any Application for Payment within 30 days after it is submitted, or (iii) Owner fails for 30 days to pay Contractor any sum finally determined to be due, then Contractor may, upon seven days written notice to Owner and Engineer, and provided Owner or Engineer do not remedy such suspension or failure within that time, terminate the Contract and recover from Owner payment on the same terms as provided in Paragraph 15.03.

B. In lieu of terminating the Contract and without prejudice to any other right or remedy, if Engineer has failed to act on an Application for Payment within 30 days after it is submitted, or Owner has failed for 30 days to pay Contractor any sum finally determined to be due, Contractor may, seven days after written notice to Owner and Engineer, stop the Work until payment is made of all such amounts due Contractor, including interest thereon. The provisions of this Paragraph 15.04 are not intended to preclude Contractor from making a Claim under Paragraph 10.05 for an adjustment in Contract Price or Contract Times or otherwise for expenses or damage directly attributable to Contractor's stopping the Work as permitted by this Paragraph.

ARTICLE 16 – DISPUTE RESOLUTION

16.01 *Methods and Procedures*

A. Either Owner or Contractor may request mediation of any Claim submitted to Engineer for a decision under Paragraph 10.05 before such decision becomes final and binding. The mediation will be governed by the Construction Industry Mediation Rules of the American Arbitration Association in effect as of the Effective Date of the Agreement. The request for mediation shall be submitted in writing to the American Arbitration Association and the other party to the Contract. Timely submission of the request shall stay the effect of Paragraph 10.05.E.

B. Owner and Contractor shall participate in the mediation process in good faith. The process shall be concluded within 60 days of filing of the request. The date of termination of the mediation shall be determined by application of the mediation rules referenced above.

C. If the Claim is not resolved by mediation, Engineer's action under Paragraph 10.05.C or a denial pursuant to Paragraphs 10.05.C.3 or 10.05.D shall become final and binding 30 days after termination of the mediation unless, within that time period, Owner or Contractor:

1. elects in writing to invoke any dispute resolution process provided for in the Supplementary Conditions; or

2. agrees with the other party to submit the Claim to another dispute resolution process; or

3. gives written notice to the other party of the intent to submit the Claim to a court of competent jurisdiction.

ARTICLE 17 – MISCELLANEOUS

17.01 *Giving Notice*

A. Whenever any provision of the Contract Documents requires the giving of written notice, it will be deemed to have been validly given if:

1. delivered in person to the individual or to a member of the firm or to an officer of the corporation for whom it is intended; or

2. delivered at or sent by registered or certified mail, postage prepaid, to the last business address known to the giver of the notice.

17.02 *Computation of Times*

A. When any period of time is referred to in the Contract Documents by days, it will be computed to exclude the first and include the last day of such period. If the last day of any such period falls on a Saturday or Sunday or on a day made a legal holiday by the law of the applicable jurisdiction, such day will be omitted from the computation.

17.03 *Cumulative Remedies*

A. The duties and obligations imposed by these General Conditions and the rights and remedies available hereunder to the parties hereto are in addition to, and are not to be construed in any way as a limitation of, any rights and remedies available to any or all of them which are otherwise imposed or available by Laws or Regulations, by special warranty or guarantee, or by other provisions of the Contract Documents. The provisions of this Paragraph will be as effective as if repeated specifically in the Contract Documents in connection with each particular duty, obligation, right, and remedy to which they apply.

17.04 *Survival of Obligations*

A. All representations, indemnifications, warranties, and guarantees made in, required by, or given in accordance with the Contract Documents, as well as all continuing obligations indicated in the Contract Documents, will survive final payment, completion, and acceptance of the Work or termination or completion of the Contract or termination of the services of Contractor.

17.05 *Controlling Law*

A. This Contract is to be governed by the law of the state in which the Project is located.

17.06 *Headings*

A. Article and paragraph headings are inserted for convenience only and do not constitute parts of these General Conditions.

APPENDIX D

STANDARD FORM OF AGREEMENT BETWEEN OWNER AND ARCHITECT

AIA Document B101–2007

AIA Document B101™ – 2007

Standard Form of Agreement Between Owner and Architect

AGREEMENT made as of the   day of
in the year of
(In words, indicate day, month and year)

BETWEEN the Architect's client identified as the Owner:
(Name, address and other information)

This document has important legal consequences. Consultation with an attorney is encouraged with respect to its completion or modification.

and the Architect:
(Name, address and other information)

for the following Project:
(Name, location and detailed description)

The Owner and Architect agree as follows.

AIA Document B101™ – 2007 (formerly B151™ – 1997). Copyright © 1974, 1978, 1987, 1997 and 2007 by The American Institute of Architects. **All rights reserved. WARNING: This AIA® Document is protected by U.S. Copyright Law and International Treaties. Unauthorized reproduction or distribution of this AIA® Document, or any portion of it, may result in severe civil and criminal penalties, and will be prosecuted to the maximum extent possible under the law.** Purchasers are permitted to reproduce ten (10) copies of this document when completed. To report copyright violations of AIA Contract Documents, e-mail The American Institute of Architects' legal counsel, copyright@aia.org.

Reproduced with permission of The American Institute of Architects, 1735 New York Avenue, NW, Washington, DC 20006.

TABLE OF ARTICLES

1 INITIAL INFORMATION

2 ARCHITECT'S RESPONSIBILITIES

3 SCOPE OF ARCHITECT'S BASIC SERVICES

4 ADDITIONAL SERVICES

5 OWNER'S RESPONSIBILITIES

6 COST OF THE WORK

7 COPYRIGHTS AND LICENSES

8 CLAIMS AND DISPUTES

9 TERMINATION OR SUSPENSION

10 MISCELLANEOUS PROVISIONS

11 COMPENSATION

12 SPECIAL TERMS AND CONDITIONS

13 SCOPE OF THE AGREEMENT

EXHIBIT A INITIAL INFORMATION

ARTICLE 1 INITIAL INFORMATION
§ 1.1 This Agreement is based on the Initial Information set forth in this Article 1 and in optional Exhibit A, Initial Information:
(*Complete Exhibit A, Initial Information, and incorporate it into the Agreement at Section 13.2, or state below Initial Information such as details of the Project's site and program, Owner's contractors and consultants, Architect's consultants, Owner's budget for the Cost of the Work, authorized representatives, anticipated procurement method, and other information relevant to the Project.*)

§ 1.2 The Owner's anticipated dates for commencement of construction and Substantial Completion of the Work are set forth below:
 .1 Commencement of construction date:

 .2 Substantial Completion date:

§ 1.3 The Owner and Architect may rely on the Initial Information. Both parties, however, recognize that such information may materially change and, in that event, the Owner and the Architect shall appropriately adjust the schedule, the Architect's services and the Architect's compensation.

ARTICLE 2 ARCHITECT'S RESPONSIBILITIES
§ 2.1 The Architect shall provide the professional services as set forth in this Agreement.

§ 2.2 The Architect shall perform its services consistent with the professional skill and care ordinarily provided by architects practicing in the same or similar locality under the same or similar circumstances. The Architect shall perform its services as expeditiously as is consistent with such professional skill and care and the orderly progress of the Project.

§ 2.3 The Architect shall identify a representative authorized to act on behalf of the Architect with respect to the Project.

§ 2.4 Except with the Owner's knowledge and consent, the Architect shall not engage in any activity, or accept any employment, interest or contribution that would reasonably appear to compromise the Architect's professional judgment with respect to this Project.

§ 2.5 The Architect shall maintain the following insurance for the duration of this Agreement. If any of the requirements set forth below exceed the types and limits the Architect normally maintains, the Owner shall reimburse the Architect for any additional cost:
(Identify types and limits of insurance coverage, and other insurance requirements applicable to the Agreement, if any.)

- .1 General Liability
- .2 Automobile Liability
- .3 Workers' Compensation
- .4 Professional Liability

ARTICLE 3   SCOPE OF ARCHITECT'S BASIC SERVICES

§ 3.1 The Architect's Basic Services consist of those described in Article 3 and include usual and customary structural, mechanical, and electrical engineering services. Services not set forth in Article 3 are Additional Services.

§ 3.1.1 The Architect shall manage the Architect's services, consult with the Owner, research applicable design criteria, attend Project meetings, communicate with members of the Project team and report progress to the Owner.

§ 3.1.2 The Architect shall coordinate its services with those services provided by the Owner and the Owner's consultants. The Architect shall be entitled to rely on the accuracy and completeness of services and information furnished by the Owner and the Owner's consultants. The Architect shall provide prompt written notice to the Owner if the Architect becomes aware of any error, omission or inconsistency in such services or information.

§ 3.1.3 As soon as practicable after the date of this Agreement, the Architect shall submit for the Owner's approval a schedule for the performance of the Architect's services. The schedule initially shall include anticipated dates for the commencement of construction and for Substantial Completion of the Work as set forth in the Initial Information. The schedule shall include allowances for periods of time required for the Owner's review, for the performance of the Owner's consultants, and for approval of submissions by authorities having jurisdiction over the Project. Once approved by the Owner, time limits established by the schedule shall not, except for reasonable cause, be exceeded by the Architect or Owner. With the Owner's approval, the Architect shall adjust the schedule, if necessary as the Project proceeds until the commencement of construction.

§ 3.1.4 The Architect shall not be responsible for an Owner's directive or substitution made without the Architect's approval.

§ 3.1.5 The Architect shall, at appropriate times, contact the governmental authorities required to approve the Construction Documents and the entities providing utility services to the Project. In designing the Project, the Architect shall respond to applicable design requirements imposed by such governmental authorities and by such entities providing utility services.

§ 3.1.6 The Architect shall assist the Owner in connection with the Owner's responsibility for filing documents required for the approval of governmental authorities having jurisdiction over the Project.

§ 3.2 SCHEMATIC DESIGN PHASE SERVICES
§ 3.2.1 The Architect shall review the program and other information furnished by the Owner, and shall review laws, codes, and regulations applicable to the Architect's services.

§ 3.2.2 The Architect shall prepare a preliminary evaluation of the Owner's program, schedule, budget for the Cost of the Work, Project site, and the proposed procurement or delivery method and other Initial Information, each in terms of the other, to ascertain the requirements of the Project. The Architect shall notify the Owner of (1) any inconsistencies discovered in the information, and (2) other information or consulting services that may be reasonably needed for the Project.

§ 3.2.3 The Architect shall present its preliminary evaluation to the Owner and shall discuss with the Owner alternative approaches to design and construction of the Project, including the feasibility of incorporating environmentally responsible design approaches. The Architect shall reach an understanding with the Owner regarding the requirements of the Project.

§ 3.2.4 Based on the Project's requirements agreed upon with the Owner, the Architect shall prepare and present for the Owner's approval a preliminary design illustrating the scale and relationship of the Project components.

§ 3.2.5 Based on the Owner's approval of the preliminary design, the Architect shall prepare Schematic Design Documents for the Owner's approval. The Schematic Design Documents shall consist of drawings and other documents including a site plan, if appropriate, and preliminary building plans, sections and elevations; and may include some combination of study models, perspective sketches, or digital modeling. Preliminary selections of major building systems and construction materials shall be noted on the drawings or described in writing.

§ 3.2.5.1 The Architect shall consider environmentally responsible design alternatives, such as material choices and building orientation, together with other considerations based on program and aesthetics, in developing a design that is consistent with the Owner's program, schedule and budget for the Cost of the Work. The Owner may obtain other environmentally responsible design services under Article 4.

§ 3.2.5.2 The Architect shall consider the value of alternative materials, building systems and equipment, together with other considerations based on program and aesthetics in developing a design for the Project that is consistent with the Owner's program, schedule and budget for the Cost of the Work.

§ 3.2.6 The Architect shall submit to the Owner an estimate of the Cost of the Work prepared in accordance with Section 6.3.

§ 3.2.7 The Architect shall submit the Schematic Design Documents to the Owner, and request the Owner's approval.

§ 3.3 DESIGN DEVELOPMENT PHASE SERVICES
§ 3.3.1 Based on the Owner's approval of the Schematic Design Documents, and on the Owner's authorization of any adjustments in the Project requirements and the budget for the Cost of the Work, the Architect shall prepare Design Development Documents for the Owner's approval. The Design Development Documents shall illustrate and describe the development of the approved Schematic Design Documents and shall consist of drawings and other documents including plans, sections, elevations, typical construction details, and diagrammatic layouts of building systems to fix and describe the size and character of the Project as to architectural, structural, mechanical and electrical systems, and such other elements as may be appropriate. The Design Development Documents shall also include outline specifications that identify major materials and systems and establish in general their quality levels.

§ 3.3.2 The Architect shall update the estimate of the Cost of the Work.

§ 3.3.3 The Architect shall submit the Design Development documents to the Owner, advise the Owner of any adjustments to the estimate of the Cost of the Work, and request the Owner's approval.

§ 3.4 CONSTRUCTION DOCUMENTS PHASE SERVICES
§ 3.4.1 Based on the Owner's approval of the Design Development Documents, and on the Owner's authorization of any adjustments in the Project requirements and the budget for the Cost of the Work, the Architect shall prepare Construction Documents for the Owner's approval. The Construction Documents shall illustrate and describe the further development of the approved Design Development Documents and shall consist of Drawings and

Specifications setting forth in detail the quality levels of materials and systems and other requirements for the construction of the Work. The Owner and Architect acknowledge that in order to construct the Work the Contractor will provide additional information, including Shop Drawings, Product Data, Samples and other similar submittals, which the Architect shall review in accordance with Section 3.6.4.

§ 3.4.2 The Architect shall incorporate into the Construction Documents the design requirements of governmental authorities having jurisdiction over the Project.

§ 3.4.3 During the development of the Construction Documents, the Architect shall assist the Owner in the development and preparation of (1) bidding and procurement information that describes the time, place and conditions of bidding, including bidding or proposal forms; (2) the form of agreement between the Owner and Contractor; and (3) the Conditions of the Contract for Construction (General, Supplementary and other Conditions). The Architect shall also compile a project manual that includes the Conditions of the Contract for Construction and Specifications and may include bidding requirements and sample forms.

§ 3.4.4 The Architect shall update the estimate for the Cost of the Work.

§ 3.4.5 The Architect shall submit the Construction Documents to the Owner, advise the Owner of any adjustments to the estimate of the Cost of the Work, take any action required under Section 6.5, and request the Owner's approval.

§ 3.5 BIDDING OR NEGOTIATION PHASE SERVICES
§ 3.5.1 GENERAL
The Architect shall assist the Owner in establishing a list of prospective contractors. Following the Owner's approval of the Construction Documents, the Architect shall assist the Owner in (1) obtaining either competitive bids or negotiated proposals; (2) confirming responsiveness of bids or proposals; (3) determining the successful bid or proposal, if any; and, (4) awarding and preparing contracts for construction.

§ 3.5.2 COMPETITIVE BIDDING
§ 3.5.2.1 Bidding Documents shall consist of bidding requirements and proposed Contract Documents.

§ 3.5.2.2 The Architect shall assist the Owner in bidding the Project by
.1 procuring the reproduction of Bidding Documents for distribution to prospective bidders;
.2 distributing the Bidding Documents to prospective bidders, requesting their return upon completion of the bidding process, and maintaining a log of distribution and retrieval and of the amounts of deposits, if any, received from and returned to prospective bidders;
.3 organizing and conducting a pre-bid conference for prospective bidders;
.4 preparing responses to questions from prospective bidders and providing clarifications and interpretations of the Bidding Documents to all prospective bidders in the form of addenda; and
.5 organizing and conducting the opening of the bids, and subsequently documenting and distributing the bidding results, as directed by the Owner.

§ 3.5.2.3 The Architect shall consider requests for substitutions, if the Bidding Documents permit substitutions, and shall prepare and distribute addenda identifying approved substitutions to all prospective bidders.

§ 3.5.3 NEGOTIATED PROPOSALS
§ 3.5.3.1 Proposal Documents shall consist of proposal requirements and proposed Contract Documents.

§ 3.5.3.2 The Architect shall assist the Owner in obtaining proposals by
.1 procuring the reproduction of Proposal Documents for distribution to prospective contractors, and requesting their return upon completion of the negotiation process;
.2 organizing and participating in selection interviews with prospective contractors; and
.3 participating in negotiations with prospective contractors, and subsequently preparing a summary report of the negotiation results, as directed by the Owner.

§ 3.5.3.3 The Architect shall consider requests for substitutions, if the Proposal Documents permit substitutions, and shall prepare and distribute addenda identifying approved substitutions to all prospective contractors.

§ 3.6 CONSTRUCTION PHASE SERVICES
§ 3.6.1 GENERAL

§ 3.6.1.1 The Architect shall provide administration of the Contract between the Owner and the Contractor as set forth below and in AIA Document A201™–2007, General Conditions of the Contract for Construction. If the Owner and Contractor modify AIA Document A201–2007, those modifications shall not affect the Architect's services under this Agreement unless the Owner and the Architect amend this Agreement.

§ 3.6.1.2 The Architect shall advise and consult with the Owner during the Construction Phase Services. The Architect shall have authority to act on behalf of the Owner only to the extent provided in this Agreement. The Architect shall not have control over, charge of, or responsibility for the construction means, methods, techniques, sequences or procedures, or for safety precautions and programs in connection with the Work, nor shall the Architect be responsible for the Contractor's failure to perform the Work in accordance with the requirements of the Contract Documents. The Architect shall be responsible for the Architect's negligent acts or omissions, but shall not have control over or charge of, and shall not be responsible for, acts or omissions of the Contractor or of any other persons or entities performing portions of the Work.

§ 3.6.1.3 Subject to Section 4.3, the Architect's responsibility to provide Construction Phase Services commences with the award of the Contract for Construction and terminates on the date the Architect issues the final Certificate for Payment.

§ 3.6.2 EVALUATIONS OF THE WORK

§ 3.6.2.1 The Architect shall visit the site at intervals appropriate to the stage of construction, or as otherwise required in Section 4.3.3, to become generally familiar with the progress and quality of the portion of the Work completed, and to determine, in general, if the Work observed is being performed in a manner indicating that the Work, when fully completed, will be in accordance with the Contract Documents. However, the Architect shall not be required to make exhaustive or continuous on-site inspections to check the quality or quantity of the Work. On the basis of the site visits, the Architect shall keep the Owner reasonably informed about the progress and quality of the portion of the Work completed, and report to the Owner (1) known deviations from the Contract Documents and from the most recent construction schedule submitted by the Contractor, and (2) defects and deficiencies observed in the Work.

§ 3.6.2.2 The Architect has the authority to reject Work that does not conform to the Contract Documents. Whenever the Architect considers it necessary or advisable, the Architect shall have the authority to require inspection or testing of the Work in accordance with the provisions of the Contract Documents, whether or not such Work is fabricated, installed or completed. However, neither this authority of the Architect nor a decision made in good faith either to exercise or not to exercise such authority shall give rise to a duty or responsibility of the Architect to the Contractor, Subcontractors, material and equipment suppliers, their agents or employees or other persons or entities performing portions of the Work.

§ 3.6.2.3 The Architect shall interpret and decide matters concerning performance under, and requirements of, the Contract Documents on written request of either the Owner or Contractor. The Architect's response to such requests shall be made in writing within any time limits agreed upon or otherwise with reasonable promptness.

§ 3.6.2.4 Interpretations and decisions of the Architect shall be consistent with the intent of and reasonably inferable from the Contract Documents and shall be in writing or in the form of drawings. When making such interpretations and decisions, the Architect shall endeavor to secure faithful performance by both Owner and Contractor, shall not show partiality to either, and shall not be liable for results of interpretations or decisions rendered in good faith. The Architect's decisions on matters relating to aesthetic effect shall be final if consistent with the intent expressed in the Contract Documents.

§ 3.6.2.5 Unless the Owner and Contractor designate another person to serve as an Initial Decision Maker, as that term is defined in AIA Document A201–2007, the Architect shall render initial decisions on Claims between the Owner and Contractor as provided in the Contract Documents.

§ 3.6.3 CERTIFICATES FOR PAYMENT TO CONTRACTOR

§ 3.6.3.1 The Architect shall review and certify the amounts due the Contractor and shall issue certificates in such amounts. The Architect's certification for payment shall constitute a representation to the Owner, based on the Architect's evaluation of the Work as provided in Section 3.6.2 and on the data comprising the Contractor's Application for Payment, that, to the best of the Architect's knowledge, information and belief, the Work has progressed to the point indicated and that the quality of the Work is in accordance with the Contract Documents. The foregoing representations are subject (1) to an evaluation of the Work for conformance with the Contract Documents upon Substantial Completion, (2) to results of subsequent tests and inspections, (3) to correction of minor deviations from the Contract Documents prior to completion, and (4) to specific qualifications expressed by the Architect.

§ 3.6.3.2 The issuance of a Certificate for Payment shall not be a representation that the Architect has (1) made exhaustive or continuous on-site inspections to check the quality or quantity of the Work, (2) reviewed construction means, methods, techniques, sequences or procedures, (3) reviewed copies of requisitions received from Subcontractors and material suppliers and other data requested by the Owner to substantiate the Contractor's right to payment, or (4) ascertained how or for what purpose the Contractor has used money previously paid on account of the Contract Sum.

§ 3.6.3.3 The Architect shall maintain a record of the Applications and Certificates for Payment.

§ 3.6.4 SUBMITTALS
§ 3.6.4.1 The Architect shall review the Contractor's submittal schedule and shall not unreasonably delay or withhold approval. The Architect's action in reviewing submittals shall be taken in accordance with the approved submittal schedule or, in the absence of an approved submittal schedule, with reasonable promptness while allowing sufficient time in the Architect's professional judgment to permit adequate review.

§ 3.6.4.2 In accordance with the Architect-approved submittal schedule, the Architect shall review and approve or take other appropriate action upon the Contractor's submittals such as Shop Drawings, Product Data and Samples, but only for the limited purpose of checking for conformance with information given and the design concept expressed in the Contract Documents. Review of such submittals is not for the purpose of determining the accuracy and completeness of other information such as dimensions, quantities, and installation or performance of equipment or systems, which are the Contractor's responsibility. The Architect's review shall not constitute approval of safety precautions or, unless otherwise specifically stated by the Architect, of any construction means, methods, techniques, sequences or procedures. The Architect's approval of a specific item shall not indicate approval of an assembly of which the item is a component.

§ 3.6.4.3 If the Contract Documents specifically require the Contractor to provide professional design services or certifications by a design professional related to systems, materials or equipment, the Architect shall specify the appropriate performance and design criteria that such services must satisfy. The Architect shall review shop drawings and other submittals related to the Work designed or certified by the design professional retained by the Contractor that bear such professional's seal and signature when submitted to the Architect. The Architect shall be entitled to rely upon the adequacy, accuracy and completeness of the services, certifications and approvals performed or provided by such design professionals.

§ 3.6.4.4 Subject to the provisions of Section 4.3, the Architect shall review and respond to requests for information about the Contract Documents. The Architect shall set forth in the Contract Documents the requirements for requests for information. Requests for information shall include, at a minimum, a detailed written statement that indicates the specific Drawings or Specifications in need of clarification and the nature of the clarification requested. The Architect's response to such requests shall be made in writing within any time limits agreed upon, or otherwise with reasonable promptness. If appropriate, the Architect shall prepare and issue supplemental Drawings and Specifications in response to requests for information.

§ 3.6.4.5 The Architect shall maintain a record of submittals and copies of submittals supplied by the Contractor in accordance with the requirements of the Contract Documents.

§ 3.6.5 CHANGES IN THE WORK
§ 3.6.5.1 The Architect may authorize minor changes in the Work that are consistent with the intent of the Contract Documents and do not involve an adjustment in the Contract Sum or an extension of the Contract Time. Subject to the provisions of Section 4.3, the Architect shall prepare Change Orders and Construction Change Directives for the Owner's approval and execution in accordance with the Contract Documents.

§ 3.6.5.2 The Architect shall maintain records relative to changes in the Work.

§ 3.6.6 PROJECT COMPLETION

§ 3.6.6.1 The Architect shall conduct inspections to determine the date or dates of Substantial Completion and the date of final completion; issue Certificates of Substantial Completion; receive from the Contractor and forward to the Owner, for the Owner's review and records, written warranties and related documents required by the Contract Documents and assembled by the Contractor; and issue a final Certificate for Payment based upon a final inspection indicating the Work complies with the requirements of the Contract Documents.

§ 3.6.6.2 The Architect's inspections shall be conducted with the Owner to check conformance of the Work with the requirements of the Contract Documents and to verify the accuracy and completeness of the list submitted by the Contractor of Work to be completed or corrected.

§ 3.6.6.3 When the Work is found to be substantially complete, the Architect shall inform the Owner about the balance of the Contract Sum remaining to be paid the Contractor, including the amount to be retained from the Contract Sum, if any, for final completion or correction of the Work.

§ 3.6.6.4 The Architect shall forward to the Owner the following information received from the Contractor: (1) consent of surety or sureties, if any, to reduction in or partial release of retainage or the making of final payment; (2) affidavits, receipts, releases and waivers of liens or bonds indemnifying the Owner against liens; and (3) any other documentation required of the Contractor under the Contract Documents.

§ 3.6.6.5 Upon request of the Owner, and prior to the expiration of one year from the date of Substantial Completion, the Architect shall, without additional compensation, conduct a meeting with the Owner to review the facility operations and performance.

ARTICLE 4 ADDITIONAL SERVICES

§ 4.1 Additional Services listed below are not included in Basic Services but may be required for the Project. The Architect shall provide the listed Additional Services only if specifically designated in the table below as the Architect's responsibility, and the Owner shall compensate the Architect as provided in Section 11.2.
(Designate the Additional Services the Architect shall provide in the second column of the table below. In the third column indicate whether the service description is located in Section 4.2 or in an attached exhibit. If in an exhibit, identify the exhibit.)

Additional Services		Responsibility *(Architect, Owner or Not Provided)*	Location of Service Description *(Section 4.2 below or in an exhibit attached to this document and identified below)*
§ 4.1.1	Programming		
§ 4.1.2	Multiple preliminary designs		
§ 4.1.3	Measured drawings		
§ 4.1.4	Existing facilities surveys		
§ 4.1.5	Site Evaluation and Planning (B203™–2007)		
§ 4.1.6	Building information modeling		
§ 4.1.7	Civil engineering		
§ 4.1.8	Landscape design		
§ 4.1.9	Architectural Interior Design (B252™–2007)		
§ 4.1.10	Value Analysis (B204™–2007)		
§ 4.1.11	Detailed cost estimating		
§ 4.1.12	On-site project representation		
§ 4.1.13	Conformed construction documents		
§ 4.1.14	As-designed record drawings		
§ 4.1.15	As-constructed record drawings		
§ 4.1.16	Post occupancy evaluation		
§ 4.1.17	Facility Support Services (B210™–2007)		
§ 4.1.18	Tenant-related services		
§ 4.1.19	Coordination of Owner's consultants		
§ 4.1.20	Telecommunications/data design		

Additional Services		Responsibility *(Architect, Owner or Not Provided)*	Location of Service Description *(Section 4.2 below or in an exhibit attached to this document and identified below)*
§ 4.1.21	Security Evaluation and Planning (B206™–2007)		
§ 4.1.22	Commissioning (B211™–2007)		
§ 4.1.23	Extensive environmentally responsible design		
§ 4.1.24	LEED® Certification (B214™–2007)		
§ 4.1.25	Fast-track design services		
§ 4.1.26	Historic Preservation (B205™–2007)		
§ 4.1.27	Furniture, Finishings, and Equipment Design (B253™–2007)		
§ 4.1.28	Other		

§ 4.2 Insert a description of each Additional Service designated in Section 4.1 as the Architect's responsibility, if not further described in an exhibit attached to this document.

§ 4.3 Additional Services may be provided after execution of this Agreement, without invalidating the Agreement. Except for services required due to the fault of the Architect, any Additional Services provided in accordance with this Section 4.3 shall entitle the Architect to compensation pursuant to Section 11.3 and an appropriate adjustment in the Architect's schedule.

§ 4.3.1 Upon recognizing the need to perform the following Additional Services, the Architect shall notify the Owner with reasonable promptness and explain the facts and circumstances giving rise to the need. The Architect shall not proceed to provide the following services until the Architect receives the Owner's written authorization:
.1 Services necessitated by a change in the Initial Information, previous instructions or approvals given by the Owner, or a material change in the Project including, but not limited to, size, quality, complexity, the Owner's schedule or budget for Cost of the Work, or procurement or delivery method;
.2 Services necessitated by the Owner's request for extensive environmentally responsible design alternatives, such as unique system designs, in-depth material research, energy modeling, or LEED® certification;
.3 Changing or editing previously prepared Instruments of Service necessitated by the enactment or revision of codes, laws or regulations or official interpretations;
.4 Services necessitated by decisions of the Owner not rendered in a timely manner or any other failure of performance on the part of the Owner or the Owner's consultants or contractors;
.5 Preparing digital data for transmission to the Owner's consultants and contractors, or to other Owner authorized recipients;
.6 Preparation of design and documentation for alternate bid or proposal requests proposed by the Owner;
.7 Preparation for, and attendance at, a public presentation, meeting or hearing;
.8 Preparation for, and attendance at a dispute resolution proceeding or legal proceeding, except where the Architect is party thereto;
.9 Evaluation of the qualifications of bidders or persons providing proposals;
.10 Consultation concerning replacement of Work resulting from fire or other cause during construction; or
.11 Assistance to the Initial Decision Maker, if other than the Architect.

§ 4.3.2 To avoid delay in the Construction Phase, the Architect shall provide the following Additional Services, notify the Owner with reasonable promptness, and explain the facts and circumstances giving rise to the need. If the Owner subsequently determines that all or parts of those services are not required, the Owner shall give prompt written notice to the Architect, and the Owner shall have no further obligation to compensate the Architect for those services:

.1 Reviewing a Contractor's submittal out of sequence from the submittal schedule agreed to by the Architect;
.2 Responding to the Contractor's requests for information that are not prepared in accordance with the Contract Documents or where such information is available to the Contractor from a careful study and comparison of the Contract Documents, field conditions, other Owner-provided information, Contractor-prepared coordination drawings, or prior Project correspondence or documentation;
.3 Preparing Change Orders and Construction Change Directives that require evaluation of Contractor's proposals and supporting data, or the preparation or revision of Instruments of Service;
.4 Evaluating an extensive number of Claims as the Initial Decision Maker;
.5 Evaluating substitutions proposed by the Owner or Contractor and making subsequent revisions to Instruments of Service resulting therefrom; or
.6 To the extent the Architect's Basic Services are affected, providing Construction Phase Services 60 days after (1) the date of Substantial Completion of the Work or (2) the anticipated date of Substantial Completion identified in Initial Information, whichever is earlier.

§ 4.3.3 The Architect shall provide Construction Phase Services exceeding the limits set forth below as Additional Services. When the limits below are reached, the Architect shall notify the Owner:

.1 ( ) reviews of each Shop Drawing, Product Data item, sample and similar submittal of the Contractor
.2 ( ) visits to the site by the Architect over the duration of the Project during construction
.3 ( ) inspections for any portion of the Work to determine whether such portion of the Work is substantially complete in accordance with the requirements of the Contract Documents
.4 ( ) inspections for any portion of the Work to determine final completion

§ 4.3.4 If the services covered by this Agreement have not been completed within ( ) months of the date of this Agreement, through no fault of the Architect, extension of the Architect's services beyond that time shall be compensated as Additional Services.

ARTICLE 5 OWNER'S RESPONSIBILITIES

§ 5.1 Unless otherwise provided for under this Agreement, the Owner shall provide information in a timely manner regarding requirements for and limitations on the Project, including a written program which shall set forth the Owner's objectives, schedule, constraints and criteria, including space requirements and relationships, flexibility, expandability, special equipment, systems and site requirements. Within 15 days after receipt of a written request from the Architect, the Owner shall furnish the requested information as necessary and relevant for the Architect to evaluate, give notice of or enforce lien rights.

§ 5.2 The Owner shall establish and periodically update the Owner's budget for the Project, including (1) the budget for the Cost of the Work as defined in Section 6.1; (2) the Owner's other costs; and, (3) reasonable contingencies related to all of these costs. If the Owner significantly increases or decreases the Owner's budget for the Cost of the Work, the Owner shall notify the Architect. The Owner and the Architect shall thereafter agree to a corresponding change in the Project's scope and quality.

§ 5.3 The Owner shall identify a representative authorized to act on the Owner's behalf with respect to the Project. The Owner shall render decisions and approve the Architect's submittals in a timely manner in order to avoid unreasonable delay in the orderly and sequential progress of the Architect's services.

§ 5.4 The Owner shall furnish surveys to describe physical characteristics, legal limitations and utility locations for the site of the Project, and a written legal description of the site. The surveys and legal information shall include, as applicable, grades and lines of streets, alleys, pavements and adjoining property and structures; designated wetlands; adjacent drainage; rights-of-way, restrictions, easements, encroachments, zoning, deed restrictions, boundaries and contours of the site; locations, dimensions and necessary data with respect to existing buildings, other improvements and trees; and information concerning available utility services and lines, both public and private, above and below grade, including inverts and depths. All the information on the survey shall be referenced to a Project benchmark.

§ 5.5 The Owner shall furnish services of geotechnical engineers, which may include but are not limited to test borings, test pits, determinations of soil bearing values, percolation tests, evaluations of hazardous materials, seismic evaluation, ground corrosion tests and resistivity tests, including necessary operations for anticipating subsoil conditions, with written reports and appropriate recommendations.

§ 5.6 The Owner shall coordinate the services of its own consultants with those services provided by the Architect. Upon the Architect's request, the Owner shall furnish copies of the scope of services in the contracts between the Owner and the Owner's consultants. The Owner shall furnish the services of consultants other than those designated in this Agreement, or authorize the Architect to furnish them as an Additional Service, when the Architect requests such services and demonstrates that they are reasonably required by the scope of the Project. The Owner shall require that its consultants maintain professional liability insurance as appropriate to the services provided.

§ 5.7 The Owner shall furnish tests, inspections and reports required by law or the Contract Documents, such as structural, mechanical, and chemical tests, tests for air and water pollution, and tests for hazardous materials.

§ 5.8 The Owner shall furnish all legal, insurance and accounting services, including auditing services, that may be reasonably necessary at any time for the Project to meet the Owner's needs and interests.

§ 5.9 The Owner shall provide prompt written notice to the Architect if the Owner becomes aware of any fault or defect in the Project, including errors, omissions or inconsistencies in the Architect's Instruments of Service.

§ 5.10 Except as otherwise provided in this Agreement, or when direct communications have been specially authorized, the Owner shall endeavor to communicate with the Contractor and the Architect's consultants through the Architect about matters arising out of or relating to the Contract Documents. The Owner shall promptly notify the Architect of any direct communications that may affect the Architect's services.

§ 5.11 Before executing the Contract for Construction, the Owner shall coordinate the Architect's duties and responsibilities set forth in the Contract for Construction with the Architect's services set forth in this Agreement. The Owner shall provide the Architect a copy of the executed agreement between the Owner and Contractor, including the General Conditions of the Contract for Construction.

§ 5.12 The Owner shall provide the Architect access to the Project site prior to commencement of the Work and shall obligate the Contractor to provide the Architect access to the Work wherever it is in preparation or progress.

ARTICLE 6 COST OF THE WORK

§ 6.1 For purposes of this Agreement, the Cost of the Work shall be the total cost to the Owner to construct all elements of the Project designed or specified by the Architect and shall include contractors' general conditions costs, overhead and profit. The Cost of the Work does not include the compensation of the Architect, the costs of the land, rights-of-way, financing, contingencies for changes in the Work or other costs that are the responsibility of the Owner.

§ 6.2 The Owner's budget for the Cost of the Work is provided in Initial Information, and may be adjusted throughout the Project as required under Sections 5.2, 6.4 and 6.5. Evaluations of the Owner's budget for the Cost of the Work, the preliminary estimate of the Cost of the Work and updated estimates of the Cost of the Work prepared by the Architect, represent the Architect's judgment as a design professional. It is recognized, however, that neither the Architect nor the Owner has control over the cost of labor, materials or equipment; the Contractor's methods of determining bid prices; or competitive bidding, market or negotiating conditions. Accordingly, the Architect cannot and does not warrant or represent that bids or negotiated prices will not vary from the Owner's budget for the Cost of the Work or from any estimate of the Cost of the Work or evaluation prepared or agreed to by the Architect.

§ 6.3 In preparing estimates of the Cost of Work, the Architect shall be permitted to include contingencies for design, bidding and price escalation; to determine what materials, equipment, component systems and types of construction are to be included in the Contract Documents; to make reasonable adjustments in the program and scope of the Project; and to include in the Contract Documents alternate bids as may be necessary to adjust the estimated Cost of the Work to meet the Owner's budget for the Cost of the Work. The Architect's estimate of the Cost of the Work shall be based on current area, volume or similar conceptual estimating techniques. If the Owner requests detailed cost estimating services, the Architect shall provide such services as an Additional Service under Article 4.

§ 6.4 If the Bidding or Negotiation Phase has not commenced within 90 days after the Architect submits the Construction Documents to the Owner, through no fault of the Architect, the Owner's budget for the Cost of the Work shall be adjusted to reflect changes in the general level of prices in the applicable construction market.

§ 6.5 If at any time the Architect's estimate of the Cost of the Work exceeds the Owner's budget for the Cost of the Work, the Architect shall make appropriate recommendations to the Owner to adjust the Project's size, quality or budget for the Cost of the Work, and the Owner shall cooperate with the Architect in making such adjustments.

§ 6.6 If the Owner's budget for the Cost of the Work at the conclusion of the Construction Documents Phase Services is exceeded by the lowest bona fide bid or negotiated proposal, the Owner shall
- .1 give written approval of an increase in the budget for the Cost of the Work;
- .2 authorize rebidding or renegotiating of the Project within a reasonable time;
- .3 terminate in accordance with Section 9.5;
- .4 in consultation with the Architect, revise the Project program, scope, or quality as required to reduce the Cost of the Work; or
- .5 implement any other mutually acceptable alternative.

§ 6.7 If the Owner chooses to proceed under Section 6.6.4, the Architect, without additional compensation, shall modify the Construction Documents as necessary to comply with the Owner's budget for the Cost of the Work at the conclusion of the Construction Documents Phase Services, or the budget as adjusted under Section 6.6.1. The Architect's modification of the Construction Documents shall be the limit of the Architect's responsibility under this Article 6.

ARTICLE 7 COPYRIGHTS AND LICENSES

§ 7.1 The Architect and the Owner warrant that in transmitting Instruments of Service, or any other information, the transmitting party is the copyright owner of such information or has permission from the copyright owner to transmit such information for its use on the Project. If the Owner and Architect intend to transmit Instruments of Service or any other information or documentation in digital form, they shall endeavor to establish necessary protocols governing such transmissions.

§ 7.2 The Architect and the Architect's consultants shall be deemed the authors and owners of their respective Instruments of Service, including the Drawings and Specifications, and shall retain all common law, statutory and other reserved rights, including copyrights. Submission or distribution of Instruments of Service to meet official regulatory requirements or for similar purposes in connection with the Project is not to be construed as publication in derogation of the reserved rights of the Architect and the Architect's consultants.

§ 7.3 Upon execution of this Agreement, the Architect grants to the Owner a nonexclusive license to use the Architect's Instruments of Service solely and exclusively for purposes of constructing, using, maintaining, altering and adding to the Project, provided that the Owner substantially performs its obligations, including prompt payment of all sums when due, under this Agreement. The Architect shall obtain similar nonexclusive licenses from the Architect's consultants consistent with this Agreement. The license granted under this section permits the Owner to authorize the Contractor, Subcontractors, Sub-subcontractors, and material or equipment suppliers, as well as the Owner's consultants and separate contractors, to reproduce applicable portions of the Instruments of Service solely and exclusively for use in performing services or construction for the Project. If the Architect rightfully terminates this Agreement for cause as provided in Section 9.4, the license granted in this Section 7.3 shall terminate.

§ 7.3.1 In the event the Owner uses the Instruments of Service without retaining the author of the Instruments of Service, the Owner releases the Architect and Architect's consultant(s) from all claims and causes of action arising from such uses. The Owner, to the extent permitted by law, further agrees to indemnify and hold harmless the Architect and its consultants from all costs and expenses, including the cost of defense, related to claims and causes of action asserted by any third person or entity to the extent such costs and expenses arise from the Owner's use of the Instruments of Service under this Section 7.3.1. The terms of this Section 7.3.1 shall not apply if the Owner rightfully terminates this Agreement for cause under Section 9.4.

§ 7.4 Except for the licenses granted in this Article 7, no other license or right shall be deemed granted or implied under this Agreement. The Owner shall not assign, delegate, sublicense, pledge or otherwise transfer any license granted herein to another party without the prior written agreement of the Architect. Any unauthorized use of the

Instruments of Service shall be at the Owner's sole risk and without liability to the Architect and the Architect's consultants.

ARTICLE 8 CLAIMS AND DISPUTES
§ 8.1 GENERAL
§ 8.1.1 The Owner and Architect shall commence all claims and causes of action, whether in contract, tort, or otherwise, against the other arising out of or related to this Agreement in accordance with the requirements of the method of binding dispute resolution selected in this Agreement within the period specified by applicable law, but in any case not more than 10 years after the date of Substantial Completion of the Work. The Owner and Architect waive all claims and causes of action not commenced in accordance with this Section 8.1.1.

§ 8.1.2 To the extent damages are covered by property insurance, the Owner and Architect waive all rights against each other and against the contractors, consultants, agents and employees of the other for damages, except such rights as they may have to the proceeds of such insurance as set forth in AIA Document A201–2007, General Conditions of the Contract for Construction. The Owner or the Architect, as appropriate, shall require of the contractors, consultants, agents and employees of any of them similar waivers in favor of the other parties enumerated herein.

§ 8.1.3 The Architect and Owner waive consequential damages for claims, disputes or other matters in question arising out of or relating to this Agreement. This mutual waiver is applicable, without limitation, to all consequential damages due to either party's termination of this Agreement, except as specifically provided in Section 9.7.

§ 8.2 MEDIATION
§ 8.2.1 Any claim, dispute or other matter in question arising out of or related to this Agreement shall be subject to mediation as a condition precedent to binding dispute resolution. If such matter relates to or is the subject of a lien arising out of the Architect's services, the Architect may proceed in accordance with applicable law to comply with the lien notice or filing deadlines prior to resolution of the matter by mediation or by binding dispute resolution.

§ 8.2.2 The Owner and Architect shall endeavor to resolve claims, disputes and other matters in question between them by mediation which, unless the parties mutually agree otherwise, shall be administered by the American Arbitration Association in accordance with its Construction Industry Mediation Procedures in effect on the date of the Agreement. A request for mediation shall be made in writing, delivered to the other party to the Agreement, and filed with the person or entity administering the mediation. The request may be made concurrently with the filing of a complaint or other appropriate demand for binding dispute resolution but, in such event, mediation shall proceed in advance of binding dispute resolution proceedings, which shall be stayed pending mediation for a period of 60 days from the date of filing, unless stayed for a longer period by agreement of the parties or court order. If an arbitration proceeding is stayed pursuant to this section, the parties may nonetheless proceed to the selection of the arbitrator(s) and agree upon a schedule for later proceedings.

§ 8.2.3 The parties shall share the mediator's fee and any filing fees equally. The mediation shall be held in the place where the Project is located, unless another location is mutually agreed upon. Agreements reached in mediation shall be enforceable as settlement agreements in any court having jurisdiction thereof.

§ 8.2.4 If the parties do not resolve a dispute through mediation pursuant to this Section 8.2, the method of binding dispute resolution shall be the following:
(Check the appropriate box. If the Owner and Architect do not select a method of binding dispute resolution below, or do not subsequently agree in writing to a binding dispute resolution method other than litigation, the dispute will be resolved in a court of competent jurisdiction.)

☐ Arbitration pursuant to Section 8.3 of this Agreement

☐ Litigation in a court of competent jurisdiction

☐ Other *(Specify)*

§ 8.3 ARBITRATION

§ 8.3.1 If the parties have selected arbitration as the method for binding dispute resolution in this Agreement, any claim, dispute or other matter in question arising out of or related to this Agreement subject to, but not resolved by, mediation shall be subject to arbitration which, unless the parties mutually agree otherwise, shall be administered by the American Arbitration Association in accordance with its Construction Industry Arbitration Rules in effect on the date of this Agreement. A demand for arbitration shall be made in writing, delivered to the other party to this Agreement, and filed with the person or entity administering the arbitration.

§ 8.3.1.1 A demand for arbitration shall be made no earlier than concurrently with the filing of a request for mediation, but in no event shall it be made after the date when the institution of legal or equitable proceedings based on the claim, dispute or other matter in question would be barred by the applicable statute of limitations. For statute of limitations purposes, receipt of a written demand for arbitration by the person or entity administering the arbitration shall constitute the institution of legal or equitable proceedings based on the claim, dispute or other matter in question.

§ 8.3.2 The foregoing agreement to arbitrate and other agreements to arbitrate with an additional person or entity duly consented to by parties to this Agreement shall be specifically enforceable in accordance with applicable law in any court having jurisdiction thereof.

§ 8.3.3 The award rendered by the arbitrator(s) shall be final, and judgment may be entered upon it in accordance with applicable law in any court having jurisdiction thereof.

§ 8.3.4 CONSOLIDATION OR JOINDER

§ 8.3.4.1 Either party, at its sole discretion, may consolidate an arbitration conducted under this Agreement with any other arbitration to which it is a party provided that (1) the arbitration agreement governing the other arbitration permits consolidation; (2) the arbitrations to be consolidated substantially involve common questions of law or fact; and (3) the arbitrations employ materially similar procedural rules and methods for selecting arbitrator(s).

§ 8.3.4.2 Either party, at its sole discretion, may include by joinder persons or entities substantially involved in a common question of law or fact whose presence is required if complete relief is to be accorded in arbitration, provided that the party sought to be joined consents in writing to such joinder. Consent to arbitration involving an additional person or entity shall not constitute consent to arbitration of any claim, dispute or other matter in question not described in the written consent.

§ 8.3.4.3 The Owner and Architect grant to any person or entity made a party to an arbitration conducted under this Section 8.3, whether by joinder or consolidation, the same rights of joinder and consolidation as the Owner and Architect under this Agreement.

ARTICLE 9 TERMINATION OR SUSPENSION

§ 9.1 If the Owner fails to make payments to the Architect in accordance with this Agreement, such failure shall be considered substantial nonperformance and cause for termination or, at the Architect's option, cause for suspension of performance of services under this Agreement. If the Architect elects to suspend services, the Architect shall give seven days' written notice to the Owner before suspending services. In the event of a suspension of services, the Architect shall have no liability to the Owner for delay or damage caused the Owner because of such suspension of services. Before resuming services, the Architect shall be paid all sums due prior to suspension and any expenses incurred in the interruption and resumption of the Architect's services. The Architect's fees for the remaining services and the time schedules shall be equitably adjusted.

§ 9.2 If the Owner suspends the Project, the Architect shall be compensated for services performed prior to notice of such suspension. When the Project is resumed, the Architect shall be compensated for expenses incurred in the interruption and resumption of the Architect's services. The Architect's fees for the remaining services and the time schedules shall be equitably adjusted.

§ 9.3 If the Owner suspends the Project for more than 90 cumulative days for reasons other than the fault of the Architect, the Architect may terminate this Agreement by giving not less than seven days' written notice.

§ 9.4 Either party may terminate this Agreement upon not less than seven days' written notice should the other party fail substantially to perform in accordance with the terms of this Agreement through no fault of the party initiating the termination.

§ 9.5 The Owner may terminate this Agreement upon not less than seven days' written notice to the Architect for the Owner's convenience and without cause.

§ 9.6 In the event of termination not the fault of the Architect, the Architect shall be compensated for services performed prior to termination, together with Reimbursable Expenses then due and all Termination Expenses as defined in Section 9.7.

§ 9.7 Termination Expenses are in addition to compensation for the Architect's services and include expenses directly attributable to termination for which the Architect is not otherwise compensated, plus an amount for the Architect's anticipated profit on the value of the services not performed by the Architect.

§ 9.8 The Owner's rights to use the Architect's Instruments of Service in the event of a termination of this Agreement are set forth in Article 7 and Section 11.9.

ARTICLE 10 MISCELLANEOUS PROVISIONS

§ 10.1 This Agreement shall be governed by the law of the place where the Project is located, except that if the parties have selected arbitration as the method of binding dispute resolution, the Federal Arbitration Act shall govern Section 8.3.

§ 10.2 Terms in this Agreement shall have the same meaning as those in AIA Document A201–2007, General Conditions of the Contract for Construction.

§ 10.3 The Owner and Architect, respectively, bind themselves, their agents, successors, assigns and legal representatives to this Agreement. Neither the Owner nor the Architect shall assign this Agreement without the written consent of the other, except that the Owner may assign this Agreement to a lender providing financing for the Project if the lender agrees to assume the Owner's rights and obligations under this Agreement.

§ 10.4 If the Owner requests the Architect to execute certificates, the proposed language of such certificates shall be submitted to the Architect for review at least 14 days prior to the requested dates of execution. If the Owner requests the Architect to execute consents reasonably required to facilitate assignment to a lender, the Architect shall execute all such consents that are consistent with this Agreement, provided the proposed consent is submitted to the Architect for review at least 14 days prior to execution. The Architect shall not be required to execute certificates or consents that would require knowledge, services or responsibilities beyond the scope of this Agreement.

§ 10.5 Nothing contained in this Agreement shall create a contractual relationship with or a cause of action in favor of a third party against either the Owner or Architect.

§ 10.6 Unless otherwise required in this Agreement, the Architect shall have no responsibility for the discovery, presence, handling, removal or disposal of, or exposure of persons to, hazardous materials or toxic substances in any form at the Project site.

§ 10.7 The Architect shall have the right to include photographic or artistic representations of the design of the Project among the Architect's promotional and professional materials. The Architect shall be given reasonable access to the completed Project to make such representations. However, the Architect's materials shall not include the Owner's confidential or proprietary information if the Owner has previously advised the Architect in writing of the specific information considered by the Owner to be confidential or proprietary. The Owner shall provide professional credit for the Architect in the Owner's promotional materials for the Project.

§ 10.8 If the Architect or Owner receives information specifically designated by the other party as "confidential" or "business proprietary," the receiving party shall keep such information strictly confidential and shall not disclose it to any other person except to (1) its employees, (2) those who need to know the content of such information in order to perform services or construction solely and exclusively for the Project, or (3) its consultants and contractors whose contracts include similar restrictions on the use of confidential information.

ARTICLE 11   COMPENSATION

§ 11.1 For the Architect's Basic Services described under Article 3, the Owner shall compensate the Architect as follows:
(Insert amount of, or basis for, compensation.)

§ 11.2 For Additional Services designated in Section 4.1, the Owner shall compensate the Architect as follows:
(Insert amount of, or basis for, compensation. If necessary, list specific services to which particular methods of compensation apply.)

§ 11.3 For Additional Services that may arise during the course of the Project, including those under Section 4.3, the Owner shall compensate the Architect as follows:
(Insert amount of, or basis for, compensation.)

§ 11.4 Compensation for Additional Services of the Architect's consultants when not included in Section 11.2 or 11.3, shall be the amount invoiced to the Architect plus _____ percent (_____ %), or as otherwise stated below:

§ 11.5 Where compensation for Basic Services is based on a stipulated sum or percentage of the Cost of the Work, the compensation for each phase of services shall be as follows:

Schematic Design Phase:	percent (	%)
Design Development Phase:	percent (	%)
Construction Documents Phase:	percent (	%)
Bidding or Negotiation Phase:	percent (	%)
Construction Phase:	percent (	%)
Total Basic Compensation	one hundred percent (	100.00%)

§ 11.6 When compensation is based on a percentage of the Cost of the Work and any portions of the Project are deleted or otherwise not constructed, compensation for those portions of the Project shall be payable to the extent services are performed on those portions, in accordance with the schedule set forth in Section 11.5 based on (1) the lowest bona fide bid or negotiated proposal, or (2) if no such bid or proposal is received, the most recent estimate of the Cost of the Work for such portions of the Project. The Architect shall be entitled to compensation in accordance with this Agreement for all services performed whether or not the Construction Phase is commenced.

§ 11.7 The hourly billing rates for services of the Architect and the Architect's consultants, if any, are set forth below. The rates shall be adjusted in accordance with the Architect's and Architect's consultants' normal review practices.
(If applicable, attach an exhibit of hourly billing rates or insert them below.)

§ 11.8 COMPENSATION FOR REIMBURSABLE EXPENSES
§ 11.8.1 Reimbursable Expenses are in addition to compensation for Basic and Additional Services and include expenses incurred by the Architect and the Architect's consultants directly related to the Project, as follows:

- .1 Transportation and authorized out-of-town travel and subsistence;
- .2 Long distance services, dedicated data and communication services, teleconferences, Project Web sites, and extranets;
- .3 Fees paid for securing approval of authorities having jurisdiction over the Project;
- .4 Printing, reproductions, plots, standard form documents;
- .5 Postage, handling and delivery;
- .6 Expense of overtime work requiring higher than regular rates, if authorized in advance by the Owner;
- .7 Renderings, models, mock-ups, professional photography, and presentation materials requested by the Owner;
- .8 Architect's Consultant's expense of professional liability insurance dedicated exclusively to this Project, or the expense of additional insurance coverage or limits if the Owner requests such insurance in excess of that normally carried by the Architect's consultants;
- .9 All taxes levied on professional services and on reimbursable expenses;
- .10 Site office expenses; and
- .11 Other similar Project-related expenditures.

§ 11.8.2 For Reimbursable Expenses the compensation shall be the expenses incurred by the Architect and the Architect's consultants plus   percent (%) of the expenses incurred.

§ 11.9 COMPENSATION FOR USE OF ARCHITECT'S INSTRUMENTS OF SERVICE
If the Owner terminates the Architect for its convenience under Section 9.5, or the Architect terminates this Agreement under Section 9.3, the Owner shall pay a licensing fee as compensation for the Owner's continued use of the Architect's Instruments of Service solely for purposes of completing, using and maintaining the Project as follows:

§ 11.10 PAYMENTS TO THE ARCHITECT
§ 11.10.1 An initial payment of   Dollars ($) shall be made upon execution of this Agreement and is the minimum payment under this Agreement. It shall be credited to the Owner's account in the final invoice.

§ 11.10.2 Unless otherwise agreed, payments for services shall be made monthly in proportion to services performed. Payments are due and payable upon presentation of the Architect's invoice. Amounts unpaid (   ) days after the invoice date shall bear interest at the rate entered below, or in the absence thereof at the legal rate prevailing from time to time at the principal place of business of the Architect.
(Insert rate of monthly or annual interest agreed upon.)

§ 11.10.3 The Owner shall not withhold amounts from the Architect's compensation to impose a penalty or liquidated damages on the Architect, or to offset sums requested by or paid to contractors for the cost of changes in the Work unless the Architect agrees or has been found liable for the amounts in a binding dispute resolution proceeding.

§ 11.10.4 Records of Reimbursable Expenses, expenses pertaining to Additional Services, and services performed on the basis of hourly rates shall be available to the Owner at mutually convenient times.

ARTICLE 12 SPECIAL TERMS AND CONDITIONS
Special terms and conditions that modify this Agreement are as follows:

ARTICLE 13 SCOPE OF THE AGREEMENT
§ 13.1 This Agreement represents the entire and integrated agreement between the Owner and the Architect and supersedes all prior negotiations, representations or agreements, either written or oral. This Agreement may be amended only by written instrument signed by both Owner and Architect.

§ 13.2 This Agreement is comprised of the following documents listed below:
- .1 AIA Document B101™–2007, Standard Form Agreement Between Owner and Architect
- .2 AIA Document E201™–2007, Digital Data Protocol Exhibit, if completed, or the following:

- .3 Other documents:
  (List other documents, if any, including Exhibit A, Initial Information, and additional scopes of service, if any, forming part of the Agreement.)

This Agreement entered into as of the day and year first written above.

_____   _____
OWNER *(Signature)*   **ARCHITECT** *(Signature)*

_____   _____
(Printed name and title)   *(Printed name and title)*

CAUTION: You should sign an original AIA Contract Document, on which this text appears in RED. An original assures that changes will not be obscured.

APPENDIX D / STANDARD FORM OF AGREEMENT BETWEEN OWNER AND ARCHITECT

AIA® Document B101™ – 2007 Exhibit A

Initial Information

for the following PROJECT:
(Name and location or address)

THE OWNER:
(Name and address)

This document has important legal consequences. Consultation with an attorney is encouraged with respect to its completion or modification.

THE ARCHITECT:
(Name and address)

This Agreement is based on the following information.
(Note the disposition for the following items by inserting the requested information or a statement such as "not applicable," "unknown at time of execution" or "to be determined later by mutual agreement.")

ARTICLE A.1 PROJECT INFORMATION
§ A.1.1 The Owner's program for the Project:
(Identify documentation or state the manner in which the program will be developed.)

§ A.1.2 The Project's physical characteristics:
(Identify or describe, if appropriate, size, location, dimensions, or other pertinent information, such as geotechnical reports; site, boundary and topographic surveys; traffic and utility studies; availability of public and private utilities and services; legal description of the site; etc.)

§ A.1.3 The Owner's budget for the Cost of the Work, as defined in Section 6.1:
(Provide total, and if known, a line item break down.)

§ A.1.4 The Owner's other anticipated scheduling information, if any, not provided in Section 1.2:

§ A.1.5 The Owner intends the following procurement or delivery method for the Project:
(Identify method such as competitive bid, negotiated contract, or construction management.)

§ A.1.6 Other Project information:
(Identify special characteristics or needs of the Project not provided elsewhere, such as environmentally responsible design or historic preservation requirements.)

ARTICLE A.2   PROJECT TEAM
§ A.2.1 The Owner identifies the following representative in accordance with Section 5.3:
(List name, address and other information.)

§ A.2.2 The persons or entities, in addition to the Owner's representative, who are required to review the Architect's submittals to the Owner are as follows:
(List name, address and other information.)

§ A.2.3 The Owner will retain the following consultants and contractors:
(List discipline and, if known, identify them by name and address.)

§ A.2.4 The Architect identifies the following representative in accordance with Section 2.3:
(List name, address and other information.)

§ A.2.5 The Architect will retain the consultants identified in Sections A.2.5.1 and A.2.5.2.
(List discipline and, if known, identify them by name and address.)

§ A.2.5.1 Consultants retained under Basic Services:
    .1   Structural Engineer

    .2   Mechanical Engineer

    .3   Electrical Engineer

§ A.2.5.2 Consultants retained under Additional Services:

§ A.2.6 Other Initial Information on which the Agreement is based:
(Provide other Initial Information.)

APPENDIX E

2012 Code of Ethics & Professional Conduct (AIA)

FROM THE OFFICE OF GENERAL COUNSEL

2012 Code of Ethics & Professional Conduct

Preamble
Members of The American Institute of Architects are dedicated to the highest standards of professionalism, integrity, and competence. This Code of Ethics and Professional Conduct states guidelines for the conduct of Members in fulfilling those obligations. The Code is arranged in three tiers of statements: Canons, Ethical Standards, and Rules of Conduct:
- Canons are broad principles of conduct.
- Ethical Standards (E.S.) are more specific goals toward which Members should aspire in professional performance and behavior.
- Rules of Conduct (**Rule**) are mandatory; violation of a Rule is grounds for disciplinary action by the Institute. Rules of Conduct, in some instances, implement more than one Canon or Ethical Standard.

The **Code** applies to the professional activities of all classes of Members, wherever they occur. It addresses responsibilities to the public, which the profession serves and enriches; to the clients and users of architecture and in the building industries, who help to shape the built environment; and to the art and science of architecture, that continuum of knowledge and creation which is the heritage and legacy of the profession.

Commentary is provided for some of the Rules of Conduct. That commentary is meant to clarify or elaborate the intent of the rule. The commentary is not part of the **Code**. Enforcement will be determined by application of the Rules of Conduct alone; the commentary will assist those seeking to conform their conduct to the **Code** and those charged with its enforcement.

Statement in Compliance With Antitrust Law
The following practices are not, in themselves, unethical, unprofessional, or contrary to any policy of The American Institute of Architects or any of its components:
(1) submitting, at any time, competitive bids or price quotations, including in circumstances where price is the sole or principal consideration in the selection of an architect;
(2) providing discounts; or
(3) providing free services.

Individual architects or architecture firms, acting alone and not on behalf of the Institute or any of its components, are free to decide for themselves whether or not to engage in any of these practices. Antitrust law permits the Institute, its components, or Members to advocate legislative or other government policies or actions relating to these practices. Finally, architects should continue to consult with state laws or regulations governing the practice of architecture.

CANON I

General Obligations

Members should maintain and advance their knowledge of the art and science of architecture, respect the body of architectural accomplishment, contribute to its growth, thoughtfully consider the social and environmental impact of their professional activities, and exercise learned and uncompromised professional judgment.

E.S. 1.1 Knowledge and Skill:
Members should strive to improve their professional knowledge and skill.

Rule 1.101 In practicing architecture, Members shall demonstrate a consistent pattern of reasonable care and competence, and shall apply the technical knowledge and skill which is ordinarily applied by architects of good standing practicing in the same locality.

Commentary: By requiring a "consistent pattern" of adherence to the common law standard of competence, this rule allows for discipline of a Member who more than infrequently does not achieve that standard. Isolated instances of minor lapses would not provide the basis for discipline.

E.S. 1.2 Standards of Excellence:
Members should continually seek to raise the standards of aesthetic excellence, architectural education, research, training, and practice.

E.S. 1.3 Natural and Cultural Heritage:
Members should respect and help conserve their natural and cultural heritage while striving to improve the environment and the quality of life within it.

E.S. 1.4 Human Rights:
Members should uphold human rights in all their professional endeavors.

Rule 1.401 Members shall not discriminate in their professional activities on the basis of race, religion, gender, national origin, age, disability, or sexual orientation.

E.S. 1.5 Allied Arts & Industries: Members should promote allied arts and contribute to the knowledge and capability of the building industries as a whole.

CANON II

Obligations to the Public

Members should embrace the spirit and letter of the law governing their professional affairs and should promote and serve the public interest in their personal and professional activities.

E.S. 2.1 Conduct: Members should uphold the law in the conduct of their professional activities.

Rule 2.101 Members shall not, in the conduct of their professional practice, knowingly violate the law.

Commentary: The violation of any law, local, state or federal, occurring in the conduct of a Member's professional practice, is made the basis for discipline by this rule. This includes the federal Copyright Act, which prohibits copying architectural works without the permission of the copyright owner. Allegations of violations of this rule must be based on an independent finding of a violation of the law by a court of competent jurisdiction or an administrative or regulatory body.

Rule 2.102 Members shall neither offer nor make any payment or gift to a public official with the intent of influencing the official's judgment in connection with an existing or prospective project in which the Members are interested.

Commentary: This rule does not prohibit campaign contributions made in conformity with applicable campaign financing laws.

Rule 2.103 Members serving in a public capacity shall not accept payments or gifts which are intended to influence their judgment.

Rule 2.104 Members shall not engage in conduct involving fraud or wanton disregard of the rights of others.

Commentary: This rule addresses serious misconduct whether or not related to a Member's professional practice. When an alleged violation of this rule is based on a violation of a law, or of fraud, then its proof must be based on an independent finding of a violation of the law or a finding of fraud by a court of competent jurisdiction or an administrative or regulatory body.

Rule 2.105 If, in the course of their work on a project, the Members become aware of a decision taken by their employer or client which violates any law or regulation and which will, in the Members' judgment, materially affect adversely the safety to the public of the finished project, the Members shall:
(a) advise their employer or client against the decision,
(b) refuse to consent to the decision, and
(c) report the decision to the local building inspector or other public official charged with the enforcement of the applicable laws and regulations, unless the Members are able to cause the matter to be satisfactorily resolved by other means.

Commentary: This rule extends only to violations of the building laws that threaten the public safety. The obligation under this rule applies only to the safety of the finished project, an obligation coextensive with the usual undertaking of an architect.

Rule 2.106 Members shall not counsel or assist a client in conduct that the architect knows, or reasonably should know, is fraudulent or illegal.

E.S. 2.2 Public Interest Services: Members should render public interest professional services, including pro bono services, and encourage their employees to render such services. Pro bono services are those rendered without expecting compensation, including those rendered for indigent persons, after disasters, or in other emergencies.

E.S. 2.3 Civic Responsibility: Members should be involved in civic activities as citizens and professionals, and should strive to improve public appreciation and understanding of architecture and the functions and responsibilities of architects.

Rule 2.301 Members making public statements on architectural issues shall disclose when they are being compensated for making such statements or when they have an economic interest in the issue.

CANON III

Obligations to the Client

Members should serve their clients competently and in a professional manner, and should exercise unprejudiced and unbiased judgment when performing all professional services.

E.S. 3.1 Competence: Members should serve their clients in a timely and competent manner.

Rule 3.101 In performing professional services, Members shall take into account applicable laws and regulations. Members may rely on the advice of other qualified persons as to the intent and meaning of such regulations.

Rule 3.102 Members shall undertake to perform professional services only when they, together with those whom they may engage as consultants, are qualified by education, training, or experience in the specific technical areas involved.

Commentary: This rule is meant to ensure that Members not undertake projects that are beyond their professional capacity. Members venturing into areas that require expertise they do not possess may obtain that expertise by additional education, training, or through the retention of consultants with the necessary expertise.

Rule 3.103 Members shall not materially alter the scope or objectives of a project without the client's consent.

E.S. 3.2 Conflict of Interest:
Members should avoid conflicts of interest in their professional practices and fully disclose all unavoidable conflicts as they arise.

Rule 3.201 A Member shall not render professional services if the Member's professional judgment could be affected by responsibilities to another project or person, or by the Member's own interests, unless all those who rely on the Member's judgment consent after full disclosure.

Commentary: This rule is intended to embrace the full range of situations that may present a Member with a conflict between his interests or responsibilities and the interest of others. Those who are entitled to disclosure may include a client, owner, employer, contractor, or others who rely on or are affected by the Member's professional decisions. A Member who cannot appropriately communicate about a conflict directly with an affected person must take steps to ensure that disclosure is made by other means.

Rule 3.202 When acting by agreement of the parties as the independent interpreter of building contract documents and the judge of contract performance, Members shall render decisions impartially.

Commentary: This rule applies when the Member, though paid by the owner and owing the owner loyalty, is nonetheless required to act with impartiality in fulfilling the architect's professional responsibilities.

E.S. 3.3 Candor and Truthfulness:
Members should be candid and truthful in their professional communications and keep their clients reasonably informed about the clients' projects.

Rule 3.301 Members shall not intentionally or recklessly mislead existing or prospective clients about the results that can be achieved through the use of the Members' services, nor shall the Members state that they can achieve results by means that violate applicable law or this **Code**.

Commentary: This rule is meant to preclude dishonest, reckless, or illegal representations by a Member either in the course of soliciting a client or during performance.

E.S. 3.4 Confidentiality:
Members should safeguard the trust placed in them by their clients.

Rule 3.401 Members shall not knowingly disclose information that would adversely affect their client or that they have been asked to maintain in confidence, except as otherwise allowed or required by this **Code** or applicable law.

*Commentary: To encourage the full and open exchange of information necessary for a successful professional relationship, Members must recognize and respect the sensitive nature of confidential client communications. Because the law does not recognize an architect-client privilege, however, the rule permits a Member to reveal a confidence when a failure to do so would be unlawful or contrary to another ethical duty imposed by this **Code**.*

CANON IV

Obligations to the Profession

Members should uphold the integrity and dignity of the profession.

E.S. 4.1 Honesty and Fairness:
Members should pursue their professional activities with honesty and fairness.

Rule 4.101 Members having substantial information which leads to a reasonable belief that another Member has committed a violation of this **Code** which raises a serious question as to that Member's honesty, trustworthiness, or fitness as a Member, shall file a complaint with the National Ethics Council.

Commentary: Often, only an architect can recognize that the behavior of another architect poses a serious question as to that other's professional integrity. In those circumstances, the duty to the professional's calling requires that a complaint be filed. In most jurisdictions, a complaint that invokes professional standards is protected from a libel or slander action if the complaint was made in good faith. If in doubt, a Member should seek counsel before reporting on another under this rule.

Rule 4.102 Members shall not sign or seal drawings, specifications, reports, or other professional work for which they do not have responsible control.

Commentary: Responsible control means the degree of knowledge and supervision ordinarily required by the professional standard of care. With respect to the work of licensed consultants, Members may sign or seal such work if they have reviewed it, coordinated its preparation, or intend to be responsible for its adequacy.

Rule 4.103 Members speaking in their professional capacity shall not knowingly make false statements of material fact.

Commentary: This rule applies to statements in all professional contexts, including applications for licensure and AIA membership.

E.S. 4.2 Dignity and Integrity:
Members should strive, through their actions, to promote the dignity and integrity of the profession, and to ensure that their representatives and employees conform their conduct to this **Code**.

Rule 4.201 Members shall not make misleading, deceptive, or false statements or claims about their professional qualifications, experience, or performance and shall accurately state the scope and nature of their responsibilities in connection with work for which they are claiming credit.

Commentary: This rule is meant to prevent Members from claiming or implying credit for work which they did not do, misleading others, and denying other participants in a project their proper share of credit.

Rule 4.202 Members shall make reasonable efforts to ensure that those over whom they have supervisory authority conform their conduct to this **Code**.

Commentary: What constitutes "reasonable efforts" under this rule is a common sense matter. As it makes sense to ensure that those over whom the architect exercises supervision be made generally aware of the **Code***, it can also make sense to bring a particular provision to the attention of a particular employee when a situation is present which might give rise to violation.*

CANON V

Obligations to Colleagues

Members should respect the rights and acknowledge the professional aspirations and contributions of their colleagues.

E.S. 5.1 Professional Environment: Members should provide their associates and employees with a suitable working environment, compensate them fairly, and facilitate their professional development.

E.S. 5.2 Intern and Professional Development: Members should recognize and fulfill their obligation to nurture fellow professionals as they progress through all stages of their career, beginning with professional education in the academy, progressing through internship and continuing throughout their career.

Rule 5.201 Members who have agreed to work with individuals engaged in an architectural internship program or an experience requirement for licensure shall reasonably assist in proper and timely documentation in accordance with that program.

E.S. 5.3 Professional Recognition: Members should build their professional reputation on the merits of their own service and performance and should recognize and give credit to others for the professional work they have performed.

Rule 5.301 Members shall recognize and respect the professional contributions of their employees, employers, professional colleagues, and business associates.

Rule 5.302 Members leaving a firm shall not, without the permission of their employer or partner, take designs, drawings, data, reports, notes, or other materials relating to the firm's work, whether or not performed by the Member.

Rule 5.303 A Member shall not unreasonably withhold permission from a departing employee or partner to take copies of designs, drawings, data, reports, notes, or other materials relating to work performed by the employee or partner that are not confidential.

Commentary: A Member may impose reasonable conditions, such as the payment of copying costs, on the right of departing persons to take copies of their work.

CANON VI

Obligations to the Environment

Members should promote sustainable design and development principles in their professional activities.

E.S. 6.1 Sustainable Design: In performing design work, Members should be environmentally responsible and advocate sustainable building and site design.

E.S. 6.2 Sustainable Development: In performing professional services, Members should advocate the design, construction, and operation of sustainable buildings and communities.

E.S. 6.3 Sustainable Practices: Members should use sustainable practices within their firms and professional organizations, and they should encourage their clients to do the same.

RULES OF APPLICATION, ENFORCEMENT, AND AMENDMENT

Application

The **Code of Ethics and Professional Conduct** applies to the professional activities of all members of the AIA.

Enforcement

The Bylaws of the Institute state procedures for the enforcement of the **Code of Ethics and Professional Conduct**. Such procedures provide that:

(1) Enforcement of the **Code** is administered through a National Ethics Council, appointed by the AIA Board of Directors.

(2) Formal charges are filed directly with the National Ethics Council by Members, components, or anyone directly aggrieved by the conduct of the Members.

(3) Penalties that may be imposed by the National Ethics Council are:
   (a) Admonition
   (b) Censure
   (c) Suspension of membership for a period of time
   (d) Termination of membership.

(4) Appeal procedures are available.

(5) All proceedings are confidential, as is the imposition of an admonishment; however, all other penalties shall be made public.

Enforcement of Rules 4.101 and 4.202 refer to and support enforcement of other Rules. A violation of Rules 4.101 or 4.202 cannot be established without proof of a pertinent violation of at least one other Rule.

Amendment

The **Code of Ethics and Professional Conduct** may be amended by the convention of the Institute under the same procedures as are necessary to amend the Institute's Bylaws. The **Code** may also be amended by the AIA Board of Directors upon a two-thirds vote of the entire Board.

2012 Edition. This copy of the **Code of Ethics is current as of September 2012. Contact the General Counsel's Office for further information at (202) 626-7348.*

APPENDIX F

Code of Ethics for Engineers (NSPE 2007)

Code of Ethics for Engineers

Preamble
Engineering is an important and learned profession. As members of this profession, engineers are expected to exhibit the highest standards of honesty and integrity. Engineering has a direct and vital impact on the quality of life for all people. Accordingly, the services provided by engineers require honesty, impartiality, fairness, and equity, and must be dedicated to the protection of the public health, safety, and welfare. Engineers must perform under a standard of professional behavior that requires adherence to the highest principles of ethical conduct.

I. Fundamental Canons
Engineers, in the fulfillment of their professional duties, shall:
1. Hold paramount the safety, health, and welfare of the public.
2. Perform services only in areas of their competence.
3. Issue public statements only in an objective and truthful manner.
4. Act for each employer or client as faithful agents or trustees.
5. Avoid deceptive acts.
6. Conduct themselves honorably, responsibly, ethically, and lawfully so as to enhance the honor, reputation, and usefulness of the profession.

II. Rules of Practice
1. Engineers shall hold paramount the safety, health, and welfare of the public.
   a. If engineers' judgment is overruled under circumstances that endanger life or property, they shall notify their employer or client and such other authority as may be appropriate.
   b. Engineers shall approve only those engineering documents that are in conformity with applicable standards.
   c. Engineers shall not reveal facts, data, or information without the prior consent of the client or employer except as authorized or required by law or this Code.
   d. Engineers shall not permit the use of their name or associate in business ventures with any person or firm that they believe is engaged in fraudulent or dishonest enterprise.
   e. Engineers shall not aid or abet the unlawful practice of engineering by a person or firm.
   f. Engineers having knowledge of any alleged violation of this Code shall report thereon to appropriate professional bodies and, when relevant, also to public authorities, and cooperate with the proper authorities in furnishing such information or assistance as may be required.
2. Engineers shall perform services only in the areas of their competence.
   a. Engineers shall undertake assignments only when qualified by education or experience in the specific technical fields involved.
   b. Engineers shall not affix their signatures to any plans or documents dealing with subject matter in which they lack competence, nor to any plan or document not prepared under their direction and control.
   c. Engineers may accept assignments and assume responsibility for coordination of an entire project and sign and seal the engineering documents for the entire project, provided that each technical segment is signed and sealed only by the qualified engineers who prepared the segment.
3. Engineers shall issue public statements only in an objective and truthful manner.
   a. Engineers shall be objective and truthful in professional reports, statements, or testimony. They shall include all relevant and pertinent information in such reports, statements, or testimony, which should bear the date indicating when it was current.
   b. Engineers may express publicly technical opinions that are founded upon knowledge of the facts and competence in the subject matter.
   c. Engineers shall issue no statements, criticisms, or arguments on technical matters that are inspired or paid for by interested parties, unless they have prefaced their comments by explicitly identifying the interested parties on whose behalf they are speaking, and by revealing the existence of any interest the engineers may have in the matters.
4. Engineers shall act for each employer or client as faithful agents or trustees.
   a. Engineers shall disclose all known or potential conflicts of interest that could influence or appear to influence their judgment or the quality of their services.
   b. Engineers shall not accept compensation, financial or otherwise, from more than one party for services on the same project, or for services pertaining to the same project, unless the circumstances are fully disclosed and agreed to by all interested parties.
   c. Engineers shall not solicit or accept financial or other valuable consideration, directly or indirectly, from outside agents in connection with the work for which they are responsible.
   d. Engineers in public service as members, advisors, or employees of a governmental or quasi-governmental body or department shall not participate in decisions with respect to services solicited or provided by them or their organizations in private or public engineering practice.
   e. Engineers shall not solicit or accept a contract from a governmental body on which a principal or officer of their organization serves as a member.
5. Engineers shall avoid deceptive acts.
   a. Engineers shall not falsify their qualifications or permit misrepresentation of their or their associates' qualifications. They shall not misrepresent or exaggerate their responsibility in or for the subject matter of prior assignments. Brochures or other presentations incident to the solicitation of employment shall not misrepresent pertinent facts concerning employers, employees, associates, joint venturers, or past accomplishments.
   b. Engineers shall not offer, give, solicit, or receive, either directly or indirectly, any contribution to influence the award of a contract by public authority, or which may be reasonably construed by the public as having the effect or intent of influencing the awarding of a contract. They shall not offer any gift or other valuable consideration in order to secure work. They shall not pay a commission, percentage, or brokerage fee in order to secure work, except to a bona fide employee or bona fide established commercial or marketing agencies retained by them.

III. Professional Obligations
1. Engineers shall be guided in all their relations by the highest standards of honesty and integrity.
   a. Engineers shall acknowledge their errors and shall not distort or alter the facts.
   b. Engineers shall advise their clients or employers when they believe a project will not be successful.
   c. Engineers shall not accept outside employment to the detriment of their regular work or interest. Before accepting any outside engineering employment, they will notify their employers.
   d. Engineers shall not attempt to attract an engineer from another employer by false or misleading pretenses.
   e. Engineers shall not promote their own interest at the expense of the dignity and integrity of the profession.
2. Engineers shall at all times strive to serve the public interest.
   a. Engineers are encouraged to participate in civic affairs; career guidance for youths; and work for the advancement of the safety, health, and well-being of their community.
   b. Engineers shall not complete, sign, or seal plans and/or specifications that are not in conformity with applicable engineering standards. If the client or employer insists on such unprofessional conduct, they shall notify the proper authorities and withdraw from further service on the project.
   c. Engineers are encouraged to extend public knowledge and appreciation of engineering and its achievements.
   d. Engineers are encouraged to adhere to the principles of sustainable development[1] in order to protect the environment for future generations.

3. Engineers shall avoid all conduct or practice that deceives the public.
   a. Engineers shall avoid the use of statements containing a material misrepresentation of fact or omitting a material fact.
   b. Consistent with the foregoing, engineers may advertise for recruitment of personnel.
   c. Consistent with the foregoing, engineers may prepare articles for the lay or technical press, but such articles shall not imply credit to the author for work performed by others.
4. Engineers shall not disclose, without consent, confidential information concerning the business affairs or technical processes of any present or former client or employer, or public body on which they serve.
   a. Engineers shall not, without the consent of all interested parties, promote or arrange for new employment or practice in connection with a specific project for which the engineer has gained particular and specialized knowledge.
   b. Engineers shall not, without the consent of all interested parties, participate in or represent an adversary interest in connection with a specific project or proceeding in which the engineer has gained particular specialized knowledge on behalf of a former client or employer.
5. Engineers shall not be influenced in their professional duties by conflicting interests.
   a. Engineers shall not accept financial or other considerations, including free engineering designs, from material or equipment suppliers for specifying their product.
   b. Engineers shall not accept commissions or allowances, directly or indirectly, from contractors or other parties dealing with clients or employers of the engineer in connection with work for which the engineer is responsible.
6. Engineers shall not attempt to obtain employment or advancement or professional engagements by untruthfully criticizing other engineers, or by other improper or questionable methods.
   a. Engineers shall not request, propose, or accept a commission on a contingent basis under circumstances in which their judgment may be compromised.
   b. Engineers in salaried positions shall accept part-time engineering work only to the extent consistent with policies of the employer and in accordance with ethical considerations.
   c. Engineers shall not, without consent, use equipment, supplies, laboratory, or office facilities of an employer to carry on outside private practice.
7. Engineers shall not attempt to injure, maliciously or falsely, directly or indirectly, the professional reputation, prospects, practice, or employment of other engineers. Engineers who believe others are guilty of unethical or illegal practice shall present such information to the proper authority for action.
   a. Engineers in private practice shall not review the work of another engineer for the same client, except with the knowledge of such engineer, or unless the connection of such engineer with the work has been terminated.
   b. Engineers in governmental, industrial, or educational employ are entitled to review and evaluate the work of other engineers when so required by their employment duties.
   c. Engineers in sales or industrial employ are entitled to make engineering comparisons of represented products with products of other suppliers.
8. Engineers shall accept personal responsibility for their professional activities, provided, however, that engineers may seek indemnification for services arising out of their practice for other than gross negligence, where the engineer's interests cannot otherwise be protected.
   a. Engineers shall conform with state registration laws in the practice of engineering.
   b. Engineers shall not use association with a nonengineer, a corporation, or partnership as a "cloak" for unethical acts.
9. Engineers shall give credit for engineering work to those to whom credit is due, and will recognize the proprietary interests of others.
   a. Engineers shall, whenever possible, name the person or persons who may be individually responsible for designs, inventions, writings, or other accomplishments.
   b. Engineers using designs supplied by a client recognize that the designs remain the property of the client and may not be duplicated by the engineer for others without express permission.
   c. Engineers, before undertaking work for others in connection with which the engineer may make improvements, plans, designs, inventions, or other records that may justify copyrights or patents, should enter into a positive agreement regarding ownership.
   d. Engineers' designs, data, records, and notes referring exclusively to an employer's work are the employer's property. The employer should indemnify the engineer for use of the information for any purpose other than the original purpose.
   e. Engineers shall continue their professional development throughout their careers and should keep current in their specialty fields by engaging in professional practice, participating in continuing education courses, reading in the technical literature, and attending professional meetings and seminars.

Footnote 1 "Sustainable development" is the challenge of meeting human needs for natural resources, industrial products, energy, food, transportation, shelter, and effective waste management while conserving and protecting environmental quality and the natural resource base essential for future development.

As Revised July 2007

"By order of the United States District Court for the District of Columbia, former Section 11(c) of the NSPE Code of Ethics prohibiting competitive bidding, and all policy statements, opinions, rulings or other guidelines interpreting its scope, have been rescinded as unlawfully interfering with the legal right of engineers, protected under the antitrust laws, to provide price information to prospective clients; accordingly, nothing contained in the NSPE Code of Ethics, policy statements, opinions, rulings or other guidelines prohibits the submission of price quotations or competitive bids for engineering services at any time or in any amount."

Statement by NSPE Executive Committee
In order to correct misunderstandings which have been indicated in some instances since the issuance of the Supreme Court decision and the entry of the Final Judgment, it is noted that in its decision of April 25, 1978, the Supreme Court of the United States declared: "The Sherman Act does not require competitive bidding."

It is further noted that as made clear in the Supreme Court decision:
1. Engineers and firms may individually refuse to bid for engineering services.
2. Clients are not required to seek bids for engineering services.
3. Federal, state, and local laws governing procedures to procure engineering services are not affected, and remain in full force and effect.
4. State societies and local chapters are free to actively and aggressively seek legislation for professional selection and negotiation procedures by public agencies.
5. State registration board rules of professional conduct, including rules prohibiting competitive bidding for engineering services, are not affected and remain in full force and effect. State registration boards with authority to adopt rules of professional conduct may adopt rules governing procedures to obtain engineering services.
6. As noted by the Supreme Court, "nothing in the judgment prevents NSPE and its members from attempting to influence governmental action . . ."

Note: In regard to the question of application of the Code to corporations vis-a-vis real persons, business form or type should not negate nor influence conformance of individuals to the Code. The Code deals with professional services, which services must be performed by real persons. Real persons in turn establish and implement policies within business structures. The Code is clearly written to apply to the Engineer, and it is incumbent on members of NSPE to endeavor to live up to its provisions. This applies to all pertinent sections of the Code.

1420 King Street
Alexandria, Virginia 22314-2794
703/684-2800 • Fax:703/836-4875
www.nspe.org
Publication date as revised: July 2007 • Publication #1102

Appendix G

DBIA Code of Professional Conduct (2008)

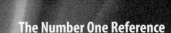

Preamble: Members of the Design-Build Institute of America are dedicated to the highest level of ethical conduct, professionalism, and service to clients. The following principles are guidelines for the conduct of DBIA members in fulfilling these obligations. They address responsibilities to the public to owners, and to members of our industry, all of who help to shape the built environment.

The principles articulated herein are limited to those which are particularly relevant to the design-build process. Although not specifically stated, it should be manifestly understood that members subscribe to the fundamental principles of honesty, fairness and lawfulness which are long and well established in the professions of architecture, engineering and construction.

Obligations to the Public:

- Members must respect their obligations to protect the safety, health and welfare of the public, and to disclose any knowledge of abuses affecting the public's interest.

- Members should seek to extend the public's knowledge and appreciation of the design-build industry and its achievements.

Obligations to the Owner:

- Members should accept responsibility for understanding and meeting the owner's broad design and construction objectives.

- Members should seek to consistently fulfill the inherent advantages of design-build delivery: efficiency, singularity of responsibility, economy, timeliness, and quality.

- Members shall not misrepresent their experiences and abilities, and shall undertake design and construction only when their knowledge and experience in the work involved, or that of their members of the design-build team, qualify them to do so.

- Members shall be responsible for the quality of the completed project, reflective of the design-builder's responsibility for performance of both the design and construction.

- Members must be open and honest in their contractual relationships with others, informing owners of any circumstances that could lead to a conflict of interest, or the perception of a conflict of interest.

Obligations to Members of the Design-Build Team:

- Members should think always as designers and builders, and respect the talents and points of view which each teammate brings to the design-build process.

- Members must respect the obligations which all design professionals have to protect the safety, health and welfare of the public.

- Members shall select design professionals, contractors and the other principal teammates on the basis of qualifications and best value.

DBIA Manual of Practice Document Number 305
DBIA Professional Code of Conduct
Second Edition, November 2008 Design-Build Institute of America Washington, D.C.
© Copyright 2008 by the Design-Build Institute of America. All rights reserved. Printed in the United States of America. No part of this publication may be reproduced, stored in a retrieval system, or transmitted, in any form or by any means, electronic, mechanical, photocopying, recording or otherwise, without the prior written permission of the DBIA.
This publication should not be understood to offer legal or other professional service. If legal advice or other expert assistance is required, the services of a competent professional person should be sought.
Design-Build Institute of America 1100 H Street, Suite 500 NW Washington, D.C. 20005 (202) 682-0110
First Edition, November 2004 Design-Build Institute of America Washington, D.C.
Design-Build Institute of America

Glossary

- **Acceleration**: A contractor speeding up performance, either voluntarily to finish the project early or because the owner has delayed the project but refused to grant the contractor a time extension.
- **Acceptance doctrine**: A defense to a contractor's liability for physical harm caused by building defects, but only if that harm occurred after the completed building had been accepted by the owner.
- **Act**: A statute; e.g., the Davis-Bacon Act or a prompt payment act.
- **Act or omission**: Another way of saying "negligence"; a defendant may have either acted in a negligent manner or negligently failed to act when it had a duty to do so.
- **Action**: (1) Lawsuit (same as "legal action"); (2) act (opposite of "omission").
- **ADR**: *See* Alternative dispute resolution.
- **Adhesion contract**: Contract offered on a take-it-or-leave-it basis, with terms favoring the economically stronger party, often involving mass-produced standardized contracts.
- **Administrative law judge (ALJ)**: A government agency employee who presides over hearings to resolve disputes between the agency and someone affected by a decision or action of the agency. The ALJ is the initial trier of fact and decision maker, whose decision may be appealed to a court. Also referred to as a "hearing officer."
- **Agent**: One who works for another (the principal or employer) and has the authority to bind the principal to a third party. Opposite of an independent contractor.
- **Alliance**: *See* Project alliance.
- **Alternative dispute resolution (ADR)**: A dispute-resolution process not involving litigation; arbitration is the most popular.
- **Answer**: The defendant's responsive pleading (document) to the plaintiff's complaint.
- **Appellant**: The party who appeals a trial court's decision, seeking to overturn the decision.
- **Appellee**: The party who seeks to uphold a trial court decision being appealed by the appellant. Same as respondent.
- **Application for payment**: Request by contractor to be paid for work performed.
- **Arbitration**: A dispute-resolution process created by agreement of the parties, in which a neutral party (the arbitrator), chosen by the parties to resolve disputes that may arise, issues an award resolving the dispute.
- **Assignment**: Transfer of a right from the initial party (for example, a party to a contract) to a third party.
- **Award**: (1) A money judgment made by a judge (2) an arbitrator's decision.
- **Benefit of the bargain**: The basic measure of compensatory damages for breach of contract.
- **"Betterment" rule**: Legal doctrine protecting a design professional from an owner's claim of liability for the cost of building the structure in compliance with the building codes, where the original design did not comply with the code.
- **Bid bond**: A bond guaranteeing that the principal will enter into a contract awarded to it.
- **Board of contract appeals (BCA)**: An administrative tribunal addressing monetary claims on federal government contracts; appeal is to the U.S. Court of Appeals for the Federal Circuit.
- **Bona fide**: Latin for "in good faith."
- **Bond**: An agreement in which a surety guarantees one party's (the principal's) conduct in favor of a beneficiary (the obligee). Using the terminology of the Restatement (Third) of Suretyship and Guaranty (1996), the principal is the "primary obligor," the surety is the "secondary obligor," and the term "obligee" is unchanged.
- **Boring**: *See* Soil boring.
- **Bridging**: A hybrid design-build project delivery method, under which the owner's design team (not the contractor's) creates the preliminary design.
- **Build-operate-transfer (BOT)**: Usually large infrastructure projects in third-world countries, under which a private contractor designs and builds, is paid through operation of the facility, then transfers title to the government.
- **Building codes**: Statutes and regulations that establish minimum design and construction standards to protect against structural failures, fire, and unsanitary conditions in the improvement.
- **Building information modeling (BIM)**: A technology-driven organizational model, centered around an information-rich computer model of the project.
- **Condition**: In contract law, an event that must occur before one contracting party is obligated to the other contracting party.
- *Contra proferentem*: A canon (or guide) for interpreting contracts, which provides that an ambiguity will be interpreted against the drafter of the contract (the one who supplied the contract language).

- **Covenant**: Promise.
- **Certificate of merit statutes**: A procedural device for weeding out frivolous claims against design professionals, requiring the plaintiff to file together with the complaint an affidavit from a design professional stating that the plaintiff's complaint has merit.
- **Certificate for payment**: Issued by the design professional in response to a contractor's application for payment, authorizing the owner to pay all or part of the application. *See also* Application for payment.
- **Change**: A mechanism, usually expressly authorized by contract, allowing the owner to unilaterally modify the contract work without the owner having to obtain the contractor's consent. *See also* Change order and Construction change directive.
- **Change order (CO)**: A written agreement as to project work to be changed, including any adjustment as to price and time. *See also* Change.
- **Changed conditions clause**: *See* Differing site condition clause.
- **Claims court**: *See* Court of claims.
- **Clause**: A part (such as a paragraph) of a contract or statute devoted to a specified topic; for example, an arbitration clause or an indemnity clause; also called a provision.
- **Codes**: *See* Building codes.
- **Common law**: The body of legal rules made by the courts (i.e., by judges) when issuing decisions.
- **Comparative negligence**: When a plaintiff obtains a damages award against multiple tortfeasors, each defendant is liable to the plaintiff only for that defendant's percentage of fault in having caused plaintiff's injury.
- **Compensatory damages**: A money award meant to make the plaintiff "whole." In contract law, this is measured by the "benefit of the bargain" lost by the defendant's breach.
- **Complaint**: The document used by the plaintiff to begin the lawsuit; contains allegations of facts and lists legal theories under which the defendant would be liable.
- **Conduit clause**: *See* Flow-through clause.
- **Consideration**: In contract law, a bargained-for exchange between two parties, which converts their agreement into a binding contract.
- **Construction change directive (CCD)**: Under the AIA A201-2007, an order to perform changed project work, even though the contractor has not agreed as to an adjustment in the contract price or time. *See also* Change.
- **Continuous treatment doctrine**: Temporary suspension of the running of the statute of limitations based on a client's reasonable reliance upon a professional. For example, an owner's reliance on the advice of his architect may suspend running of the limitations period for a claim against the architect.
- **Contra proferentem**: A guide for contract language interpretation under which ambiguity is construed against the drafter of the contract.
- **Contract**: A binding agreement, such that one party's breach of the contract will (absent a defense) subject it to damages owed to the other party.

- **Contribution**: A method of damages allocation among multiple defendants who had tortiously caused the plaintiff a single, indivisible physical injury.
- **Copyright**: The right of an author to bar the unauthorized reproducing, preparing derivative works, or distributing copies of copyrighted work. For a design professional, plans and drawings, as well as the completed structure, may be subject to copyright protection. Violation of the copyright is called an "infringement."
- **Corporation**: A business which is a legal entity separate from its shareholders (or owners); the company name is followed by the designation: 'Inc.,' 'Corp.' or 'Co.'
- **Court of claims**: A special trial court which hears claims against the government. Also called a "claims court."
- **Critical path method (CPM)**: A schedule describing the sequence and timing of the different project activities. The late start of an activity which will result in late completion of the entire project is on the "critical" path. *See also* Float.
- **Damages**: Money sought or received by a plaintiff to address a legal wrong caused by a defendant or defendants.
- **DB**: *See* Design-build.
- **DBB**: *See* Design-bid-build.
- **Declaratory judgment**: A court decision establishing the rights of the parties, usually based on stipulated facts; the purpose of a declaratory judgment is to preclude further litigation.
- **Defendant**: The party against whom a lawsuit is brought (by the plaintiff).
- **Demurrer**: Pre-trial dismissal of a lawsuit on the ground that, even if the facts as alleged by the plaintiff are true, the plaintiff has no valid legal claim.
- **Design-bid-build (DBB)**: The traditional project delivery method, in which the owner separately hires a designer and builder; first the designer creates the design, then the owner uses the completed design to solicit a contractor.
- **Design-build (DB)**: A project delivery method under which a single entity promises to both design and construct the project.
- **Design delegation**: Process by which the owner shifts to the contractor creation of part of the design, usually through the submittal process.
- **Design professional**: An architect or professional engineer.
- **Design specifications**: Specify in detail the materials to furnish and work to perform, giving the contractor little discretion in how to create the project.
- **Differing site condition (DSC) clause**: Contract language shifting to the owner costs associated with unanticipated subsurface conditions. Previously called a "changed conditions" clause.
- **Disclaimer**: Contract language purporting to shield one party from liability for certain actions.
- **Discovery**: Pre-trial gathering of information by the litigants.
- **Discovery rule**: Tolls (delays) running of the statute of limitations for latent defects until the time the claimant discovered or should have discovered the defect.

- **Economic loss rule**: Judicial doctrine that prohibits the recovery of solely economic (financial or pecuniary) losses under a theory of non-intentional tort.
- **Emotional distress damages**: Compensation for the claimant's mental anguish or emotional distress, usually recoverable only in tort, not for breach of contract.
- **Employee**: One who works for an employer who (1) has the right to control the worker's actions and (2) withholds employment-related taxes (such as social security) from the worker's compensation.
- **Ethics**: *See* Professional ethics.
- **Exculpatory clause**: Contract provision that shields the protected party from the consequences of its own negligence. *See also* Disclaimer.
- **Express warranty**: *See* Warranty.
- **Fast-track**: *See* Phased construction.
- **Fiduciary relationship**: A relationship based on trust and loyalty where one person relies on the integrity and fidelity of the other and the latter must not take unfair advantage of the trust in them by benefiting at the expense of the former.
- **Final completion**: The point at which the project is ready for acceptance, entitling the contractor to final payment.
- **Float**: Slack time in a project schedule; delay of an activity during the float period will not result in delayed completion of the entire project. *See also* Critical path method.
- **Flow-through clause**: A subcontract clause which imposes on the subcontractor the same obligations (and sometimes also the same benefits) owed to the prime contractor that the contractor has toward the owner; also called a "conduit" clause.
- *Force majeure* **clause**: Relieves the contractor from the risk of rare events over which it has no control, such as "acts of God."
- **Forensic scheduling**: Use of expert witnesses to testify as to responsibility for project delays.
- **Fraud**: An intentional misrepresentation of fact; a fraud claim also requires proof of reasonable reliance by the plaintiff and a legal injury.
- **Good faith and fair dealing**: An implied promise (or covenant or obligation) that each party to a contract will not only avoid deliberate and willful frustration of the other party's expectations, but will also extend a helping hand to achieve those expectations, where to do so would not be unreasonably burdensome.
- **Good Samaritan laws**: These immunize design professionals from potential liability when they provide services in emergencies.
- **Guaranteed maximum price (GMP)**: Most commonly a price cap on a "cost" contract entered into by an at-risk construction manager, sometimes agreed to before the design is entirely complete.
- **Habitability**: *See* Warranty of habitability.
- **Hold harmless**: *See* Indemnity.
- **Homeowner association**: In a planned development (such as a condominium), an association elected by and representative of the unit owners, which can sue to protect the common properties of the development.
- **Implied**: Imposed by operation of law, in contrast to agreed-upon expressly by the parties.
- **Implied warranty**: *See* Warranty.
- **Incorporation by reference**: Contracting-writing device by which a document or writing is designated as part of a contract, as if reproduced in the contract itself.
- **Indemnitee**: The recipient of indemnification.
- **Indemnitor**: The party providing indemnification.
- **Indemnity (or indemnification)**: A device—either arising from common law principles of unjust enrichment or specified by contract—in which the indemnitor pays for the loss owed the indemnitee to the plaintiff. The indemnitor "holds harmless" the indemnitee's liability to the plaintiff.
- **Independent contractor**: An independent business, hired to perform a job, who retains control over the manner in which the work is performed and is paid on a per-project basis.
- **Independent contractor rule**: A legal presumption that the hirer of an independent contractor is not vicariously liable for the torts of the contractor.
- **Infringement**: *See* Copyright.
- **Initial decision maker (IDM)**: A new position created by the AIA in A201-2007, designating the person to make an initial decision as to claims between the owner and contractor; if no separate person is designated as the IDM, the architect is the IDM.
- **Injunction**: Court order either prohibiting a person from performing certain acts or ordering a person to perform an act (the latter called "specific performance").
- **Instruments of service**: The physical plans and specifications created by a design professional.
- **Insurance policy**: The contract between the insurance company and the insured (or policyholder).
- **Integrated project delivery (IPD)**: *See* Project alliance.
- **Integration clause**: A contract provision stating that the written contract is the entire agreement of the parties and so cannot be modified by alleged oral agreements. *See also* Parol evidence rule.
- **Interpleader action**: Where various claims are made against the contract balance held by the owner, the owner deposits the disputed funds with the court, files a lawsuit in which it names as parties all claimants to the funds, and withdraws from the fray; the court then unscrambles the claims.
- **Joint and several liability**: Each one of multiple defendants is liable to the claimant for the entirety of the judgment or debt.
- **Joint check rule**: A check written for at least two parties in the format of "A and B," so that both must sign the check for it to be cashed. Each party is deemed to have been fully paid the amount from the check due them, regardless of how they divide the proceeds between themselves.
- **Joint venture**: Created by two or more separate entities who associate, usually to engage in one specific project or transaction.
- **Judgment**: The decision made by a trial court resolving the litigation by specifying which party (if any) is liable and the amount of damages due (if any).

- **Jurisdiction**: A court's authority to hear a particular dispute.
- **Lean project**: A project delivery method which applies the business principles of the Toyota Motor Company (such as just-in-time delivery of materials) to the construction process.
- **License**: Authorization for a limited purpose; for example, a design professional may grant the owner a license to use the plans and specifications for maintenance of the completed project without having to obtain prior consent from the designer.
- **License bond**: Acquired as a condition to receiving a license; the bond guarantees for the benefit of the contractee (such as a homeowner) that the licen*see* (such as a home builder) will comply with certain laws or regulations.
- **Licensing**: Regulation of persons (including business entities) authorized by the state to engage in professional design (also called registration) or construction activities.
- **Lien**: *See* Mechanics' lien.
- **Lien release bond**: Posted by an owner whose property is subjected to a mechanics' lien; the lien is then transferred from the property to the bond.
- **Lien waiver**: A subcontractor or supplier giving up the statutory right to file a mechanics' lien.
- **Limitation of liability clause**: Contract provision, usually found in design agreements, limiting the designer's liability to a specific sum or the designer's fee, whichever is greater.
- **Limitations**: *See* Statute of limitations.
- **Limited liability company (LLC)/limited liability partnership (LLP)**: A business form which is a separate legal entity (akin to a corporation), such that the shareholders or partners have limited legal liability, but with different tax treatment than for a corporation.
- **Liquidated damages clause**: Contract remedy to compensate an owner for the contractor's unexcused delay, consisting of a specified amount of money for each day of delay after the contract's completion date.
- **Litigation**: The dispute-resolution process that uses the court system.
- **Mechanics' lien**: A security interest (or right) in improved real property, held by those who provide labor or materials for the improvement.
- **Mediation**: A dispute-resolution mechanism in which the mediator does not decide issues but instead helps the parties to settle.
- **Material**: (1) As a noun: supplies used to build a building. (2) As an adjective: significant or major; for example, a material breach of a contract authorizes the nonbreaching party to cease performance and seek damages.
- **Modification**: A two-party agreement in which (for example) the owner and contractor mutually agree to change portions of the work.
- **Motion**: A request by a party to the litigation, made to the trial court.
- **Multi-prime project**: The owner directly contracts with the trade contractors, rather than one prime contractor who then hires the subcontractors.
- **Negligence**: A tort, defined as conduct which is below the societal standard of care.
- **Negligence *per se***: Conduct which is conclusively negligent without need of further proof; usually involves violation of a statute.
- **No-damages-for-delay clause**: Contract clause which denies the contractor delay damages and limits the contractor's remedy to time extensions.
- **Obligee**: *See* bond.
- **Omission**: A failure to act; may be negligence if the person had a duty to act.
- **Opinion**: A court decision.
- **Parol evidence rule**: Bars the introduction of evidence (prevents a party from testifying) about alleged oral agreements (1) made prior to or contemporaneous with the signing of a contract (2) which contradict or vary the terms of a written contract. *See also* Integration clause.
- **Partnering**: A team-approach project delivery system in which the owner, design team and major constructors agree to work together to accomplish the project.
- **Partnership**: An association of two or more persons to carry on a business for profit as co-owners; the partnership is not a separate legal entity.
- **Party**: A legal entity (whether a person or business). E.g.: a party to a lawsuit, or a party to a contract.
- **Pay when paid clause**: A contract provision which attempts to make the prime contractor's obligation to pay a subcontractor contingent upon the prime contractor first having been paid by the owner for that work.
- **Payment bond**: An agreement in which a surety guarantees the principal's payment obligations owed to the obligee.
- **Penal sum**: The amount of a bond.
- **Per diem**: Per day; for each day.
- **Performance bond**: An agreement in which a surety guarantees the principal's performance owed to the obligee.
- **Performance specifications**: Specifying only the characteristics of the completed project, giving the contractor discretion in how to achieve that result.
- **Phased construction**: Project delivery method under which construction begins while the design is still being worked out; also known as fast-track.
- **Plaintiff**: The party that commences a lawsuit (by bringing a claim against the defendant(s)).
- **Pleading**: A document filed in a lawsuit (e.g., complaint and answer).
- **Policy**: *See* Insurance policy.
- **Precedent**: The binding nature of a legal rule established by a higher court on a lower court.
- **Premises liability**: Tort law governing the responsibility of a land owner or possessor to one who is injured while on the property because of a defect in the condition of the building or land.
- **Primary obligor**: *See* bond.
- **Principal**: A noun: (1) the party whose obligations are guaranteed by a surety; *see* bond (2) a person with authority to bind a business entity, such as an owner, officer or director.

- **Principal debtor**: The Restatement terminology for a principal. *See* Principal.
- **Privity**: A contract relationship.
- **Products liability**: Liability of a manufacturer, distributor or retailer of a product. Depending on the type of defect, may be in strict liability or require proof of negligence.
- **Professional ethics**: Rules of conduct created by business or professional associations, which apply to members of that association.
- **Professional standard**: The negligence standard of care as applied to professionals; usually must be established using expert testimony.
- **Program management**: A program manager acts as the owner's advisor and organizer: analyzing the owner's needs, prequalifying the design and construction team, and administering the construction process.
- **Project alliance**: A project delivery method based on teamwork between the owner, design team and major constructors, in which a party is financially rewarded or penalized based on meeting predetermined goals, and there is a general release of liability among the parties.
- **Promissory estoppel**: Justifiable reliance by one party (the promisee) in response to a representation of another (the promisor) such that the promise becomes binding on the promisor.
- **Provision**: *See* Clause.
- **Public-private partnership (PPP or P3)**: The private sector finances, builds, and/or operates public-sector activities in exchange for a contractually specified stream of future income.
- **Punch list**: List of minor defects that need to be repaired or completed before the project is complete.
- **Punitive damages**: A money award meant to punish the defendant, rather than compensate the plaintiff.
- *Quantum meruit*: *See* Restitution.
- **Registration**: *See* Licensing.
- **Regulation**: (1) A rule issued by an administrative agency, (2) more generally, a body of rules governing a particular activity; e.g., regulation of who may legally may perform construction or design activities is done through the licensing statutes.
- **Repose**: *See* Statute of repose.
- *Respondeat superior*: A doctrine of vicarious (strict) liability imposed upon employers (or principals) for the torts of their employees (or agents).
- **Respondent**: The party seeking to uphold the trial court's decision, which is being appealed by the appellant. Same as the appellee.
- **Restatement of the Law**: A compilation of legal rules in an area of the law (Restatement of Agency; Restatement of Torts, etc.) which may be adopted by a state's courts. Restatements are created by the American Law Institute, a private organization made up of lawyers, judges, and legal scholars.
- **Restitution**: An award of damages intended to restore the status quo that existed before the contract; the plaintiff is awarded both the loss to itself and the benefit conferred on the defendant. Sometimes phrased as a claim for *quantum meruit* or unjust enrichment.
- **Retainage**: A contractually created security system under which the owner keeps a specified portion of earned progress payments to secure itself against the risks of unfinished or defective work or unpaid subcontractor claims.
- **Schedule of values**: Agreed valuation of designated portions of the work, usually providing the basis for progress payments; the aggregate of the schedule should be the contract price.
- **Secondary obligor**: *See* bond.
- **Security interest**: A lender's or creditor's right to property in the event a loan or debt is not repaid; usually, the property is sold and the lender or creditor is paid from the sale proceeds. *See also* Mechanics' lien.
- **Sole proprietorship**: A one-person business; the proprietorship is not a separate legal entity.
- **Soil boring**: A metal tube is drilled into the soil to a specified depth, then the soil content of the core is examined to determine the nature of the subsurface soil.
- **Specific performance**: Court order requiring a person to perform a certain act.
- **Standard of conduct**: The standard of prudent conduct of a reasonable person; conduct below this standard (less prudent conduct) establishes negligence.
- **Statute**: A law passed by the legislative branch and signed by the executive (or passed over the executive's veto).
- **Statute of Frauds**: A statute requiring evidence of a writing for certain agreements to be enforceable.
- **Statute of limitations**: A statute specifying how long the plaintiff has to begin a lawsuit.
- **Strict liability**: Liability imposed without proof of fault. *See also* Products liability.
- **Subdivision bond**: A bond guaranteeing public improvements (streets, sewers, etc.) promised by a developer building a housing subdivision.
- **Submittal**: Information provided by the contractor which illustrates in detail portions of the work; for examples, shop drawings, product data and physical samples.
- **Subrogation**: A type of lawsuit; in the construction industry, most commonly occurring when: (1) a contractor damages the building under construction, (2) the owner's property insurer pays the owner to repair the damage, and (3) the insurer "steps into the shoes" of the owner and sues the contractor (but in the owner's name) to recoup that payment to the owner; the insurer's lawsuit against the contractor brought in the name of the owner is a "subrogation" action.
- **Substantial completion**: The point at which the project may be occupied and used for its intended purpose; not the same as project acceptance.
- **Substantial performance doctrine**: Common law doctrine intended to prevent forfeiture or economic waste: if the project is unfinished yet sufficiently complete that the owner may use it, the contractor is entitled to the contract balance, less the cost of repairs or completion.

- **Summary judgment**: Pre-trial dismissal of a lawsuit on the ground that, based on the facts as developed in discovery, the plaintiff has no valid legal claim.
- **Surety**: A person or commercial entity that promises to perform another's obligations; the agreement containing the surety's promise is called a bond.
- **Teaming agreement**: In federal procurement, a project delivery method by which two business entities (usually with different areas of expertise) team up as the prime contractor on the job, although the entities formally divide between prime contractor and subcontractor.
- **Toll**: Temporary suspension of a defense based on the passage of time, such as a statute of limitations.
- **Tort**: A civil wrong other than a breach of contract, for which the law will grant a remedy, typically a money award.; includes negligence, misrepresentation, and strict products liability.
- **Tortfeasor**: A person who commits a tort.
- **Trust**: A financial arrangement under which one person (the trustee) holds money or property (the *res*) for the benefit of another person (the beneficiary).
- **Turnkey contract**: A project delivery method subject to great variation; at its simplest, the owner gives the builder general directions as to the completed project, which the contractor then designs, builds and furnishes; so that the owner can simply turn the key and take over.
- **Unbalanced bid**: Bid based on unit pricing, in which the contractor bids high for unit work that will be performed early, and low for later work.
- **Unconscionability**: A defense to enforcement of a contract based upon the twin elements of lack of freedom of choice by the weaker party and unfair terms benefitting the stronger party.
- **Unjust enrichment**: *See* Restitution.
- **Uniform Commercial Code (UCC)**: A set of suggested laws relating to the commercial sale of goods (personal property) between merchants; the UCC has been adopted in whole or in part by all states and the District of Columbia.
- **Value Engineering Change Proposal (VECP)**: A contract incentive, under which a contractor who proposes to the owner a change in the design to achieve a cost saving, shares with the owner in that saving.
- **Verdict**: The decision made by a jury.
- **Vicarious liability**: Strict liability, usually imposed on a hirer or employer for the negligence of an employee, agent or even independent contractor.
- **Waiver**: The voluntary giving-up of a known right; may be demonstrated by words or by conduct when a party performs notwithstanding a condition excusing that performance.
- **Warranty**: An assurance by one party, relied on by the other party, that a particular outcome will be achieved by the warrantor. A warranty is "express" if written in a contract. An "implied" warranty is imposed by operation of law, independent of the parties' intentions, and can extend to parties not in privity.
- **Warranty of habitability**: An implied warranty, usually from a builder-vendor of a residence to the purchaser (and sometimes extended to subsequent purchasers) that the house is structurally sound and its basic components (plumbing, electrical, etc.) function properly.
- **Warranty of workmanlike conduct**: An implied warranty imposed on contractors to build in a non-negligent manner.
- **Work for hire doctrine**: Under copyright law, a hirer (who possesses the copyright) is the author of (1) a work prepared by an employee or (2) a work specially ordered or commissioned when done by an agent (not an independent contractor).
- **Workers' compensation**: A form of insurance procured by employers promising compensation (payment) for employees injured or killed in the course of their employment; insurance coverage exists as specified by statute and without the employee having to prove the employer was negligent.
- **Zoning**: The division of land into distinct districts and the regulation of the uses and developments within those districts.

Reproduced Case Index

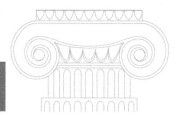

Bilt-Rite Contractors, Inc. v. The Architectural Studio 239

City of Mounds View v. Walijarvi 226

Duncan v. Missouri Board for Architects, Professional Engineers and Land Surveyors 140

Griswold and Rauma, Architects, Inc. v. Aesculapius Corp. 195

SME Industries, Inc. v. Thompson, Ventulett, Stainback & Assocs., Inc. 241

United States v. Spearin 324

Watson Lumber Co. v. Guennewig 361

Watson, Watson, Rutland/Architects, Inc. v. Montgomery County Board of Education 204

* The full Case Index is found on the book's website.

Subject Index

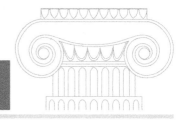

ACCEPTANCE DOCTRINE
Prime contractor liability, 124
Third-party injury from construction defects, 124

ACT OF GOD
See Force Majeure Clause

ADHESION CONTRACTS
Formation of, 37
Interpretation of, 47–48

AGENCY
Agent roles, 23, 25–26
Authority, 26
    Actual, 26
    Apparent, 26
Business associations, 23–27
Concept of, 24–25
Design professional role in, 23–25
Disputes, 26–27
Enforcement of regulations, 7
Policies of, 25
Principal–agent disputes, 26
Principal–third party disputes, 26
Third-party relationships, 26–27

AIA DOCUMENTS
A101-2007
    § 3.1, p. 391
    § 3.3, p. 405
A102-2007
    Article 7, p. 282
    Article 8, p. 282
A132-2009
    Generally, 126
A133-2009
    Generally, 126, 281
A134-2009
    Generally, 126
A201-1987
    ¶ 4.5.1, p. 493
    ¶ 1.6.1, pp. 214–215
    ¶ 4.5, p. 493
    ¶ 13.7, p. 258
A201-2007
    Article 6, p. 292
    Article 7, p. 360
    Article 14, p. 433
    Article 15, pp. 169, 383
    Generally, pp. 58, 336
    § 1.1.1, p. 383
    § 1.1.2, pp. 235, 259
    § 1.1.8, p. 493
    § 1.3.1, p. 365
    § 1.5.1, p. 214

§ 2.2.1, pp. 49, 333, 413
§ 3.2.1, pp. 329, 383
§ 3.2.2, pp. 323, 329
§ 3.2.3, pp. 45, 105
§ 3.5, pp. 322, 330
§ 3.7.1, p. 104
§ 3.7.3, pp. 45, 105
§ 3.7.4, pp. 382–383
§ 3.10.1, p. 395
§ 3.12.1, p. 209
§ 3.12.2, p. 209
§ 3.12.3, p. 209
§ 3.12.10, p. 210
§ 3.18, pp. 453, 455
§ 3.18.1, pp. 453–454
§ 4.2.1, p. 181
§ 4.2.2, p. 493
§ 4.2.7, p. 211
§ 4.2.12, pp. 347, 495–496
§ 4.2.13, p. 498
§ 5.2.1, pp. 416–417
§ 5.2.2, p. 417
§ 5.3, pp. 412–413
§ 6.2, p. 493
§ 6.2.3, p. 292
§ 7.2.1, p. 365
§ 7.3.1, pp. 364–365
§ 7.3.3, p. 360
§ 7.3.4, pp. 282, 366
§ 7.3.5, p. 366
§ 7.3.6, p. 366
§ 7.3.7, pp., 366–367
§ 7.3.10, p. 367
§ 7.4, pp. 285, 365
§ 8.1.4, p. 286
§ 8.2.2, p. 391
§ 8.3.1, pp., 393, 399
§ 8.3.3, p. 399
§ 9.1.7, p. 383
§ 9.7, pp. 350, 435
§ 9.7.1, p. 49
§ 9.8, p. 391
§ 9.10.4, p. 252
§ 10.4, p. 366
§ 11.3, pp. 454
§ 12.1.1, p. 497
§ 12.1.2, p. 497
§ 12.2.2.1, p. 330
§ 12.2.5, p. 330
§ 13.4.1, p. 350
§ 13.5.3, p. 497
§ 13.7, p. 259
§ 14.1–14.4, p. 434
§ 14.1.1, p. 434

§ 14.1.3, p. 435
§ 14.1.4, p. 435
§ 14.1.1.3, p. 350
§ 14.1.1.4, p. 333
§ 14.2, p., 433
§ 14.2.1, p., 434
§ 14.2.1.1, p., 395
§ 14.2.2, pp. 244, 436, 438
§ 14.4, p. 436
§ 14.4.3, p. 436
§ 15.1.2, p. 286
§ 15.1.3, p. 497
§ 15.1.4, p., 368
§ 15.1.5.2, p. 393
§ 15.1.6, p. 58
§ 15.1.6.2, p. 63
§ 15.2, p. 493
§ 15.2.1, p. 497
§ 15.2.5, p. 496
§ 15.2.6.1, p. 498
§ 15.3.1, p. 493
§ 15.4.1, p. 493
§ 15.4.4, p. 506
A232-2009
    Article 4, p. 129
    Article 9, p. 130
    Generally, p. 126
    § 3.7.1, p. 127
    § 3.7.4, p. 130
    § 4.2, p. 129
    § 4.2.2, p. 129
    § 4.2.3, p. 129
    § 4.2.5, p. 129
    § 4.2.6, p. 129
    § 4.2.7, p. 130
    § 4.2.8, p. 130
    § 4.2.9, p. 130
    § 4.2.10, p. 130
    § 9.5.1, p. 130
A312–1984
    ¶ 6.1, p. 485
A312-2010
    § 1, pp. 476, 479
    § 2, pp. 479, 484
    § 3, p. 476
    § 5, p. 480
    § 6, p. 481
    § 7.1, p. 485
    § 7.3, pp. 482, 485
    § 10, p. 482
A401-2007
    § 4.7, p. 418
    § 11.3, p. 418
    § 12.1, p. 418

SUBJECT INDEX

B101-2007
   Article 3, pp. 168–169, 173
   Article 4, pp. 173, 175, 178
   Article 6, p. 201
   Generally, p. 58, 168–178, 201
   § 2.2, pp. 230, 256
   § 2.4, p. 171
   § 3.2, pp. 168, 233
   § 3.3, p. 168
   § 3.4, p. 168
   § 3.4.3, p. 193
   § 3.6, p. 168
   § 3.6.1.3, p. 181
   § 3.6.2.1, p. 208
   § 3.6.3.2, p. 345
   § 4.1, pp. 174, 175
   § 4.1.5, p. 174
   § 4.1.6, p. 302
   § 4.1.18, p. 175
   § 4.1.23, p. 233
   § 4.1.24, p. 233
   § 4.1.26, p. 175
   § 4.2, p. 175
   § 4.3, p. 175
   § 4.3.1, p. 175
   § 4.3.1.7, p., 192
   § 4.3.2, p. 175
   § 4.3.2.6, p. 181
   § 5.1, p. 181
   § 5.2, p. 181
   § 5.3, p. 181
   § 5.4, p. 181
   § 5.5, p. 181
   § 5.7, p. 181
   § 5.8, pp. 181, 193
   § 6.1, p. 173
   § 6.2, p. 201
   § 6.7, pp. 199, 201
   § 7.1, p. 214
   § 7.2, p. 213
   § 7.3, p. 213
   § 8.1.1, p. 259
   § 8.1.3, pp. 58, 187
   § 8.3.4, p. 506
   § 9.1, p. 182
   § 9.2, p. 182
   § 9.3, pp. 180, 182
   § 9.4, p. 183
   § 9.5, p. 183
   § 9.6, pp. 180, 183
   § 9.7, p. 183
   § 10.3, p. 185
   § 10.5, p. 235
   § 11.5, p. 179
   § 11.8, p. 177
   § 11.8.1, p. 177
   § 11.8.2, p. 177
   § 11.8.1.9, p. 177
   § 11.8.1.10, p. 177
   § 11.8.1.11, p. 177
   § 11.10.2, pp. 179, 180
   § 11.10.3, p. 178
   § 11.10.4, p. 176
   § 13.1, p. 44
B132-2009
   Generally, 126

B141-1997
   ¶ 1.3.7.3, p. 258
   ¶ 1.3.9.4, p. 175
   ¶ 2.8.3.6, p. 192
   Changes, 168–173
   Generally, p. 175
B141a-1977,
   Generally, 179
B144 ARCH-CM-1993,
   Generally, 126
B214-2007,
   Generally, 233
E202-2008,
   Generally, 302
J330-1958
   Standard 9, p. 265
   Standard 10, p. 265

ALTERNATIVE DISPUTE RESOLUTION (ADR)
See also Arbitration
Arbitration method for, 506–507
Contract Dispute Act (CDA) for, 508

AMERICAN ARBITRATION ASSOCIATION (AAA)
See also CIA Rules
Arbitration process, 503–506
Arbitrator neutrality, 504–505
Awards, 506
Consolidation, 505–506
Construction Industry Arbitration (CIA) rules, 503–504
Discovery, 504
Joinder, 505–506
Multiple-party arbitration, 505–506
Selection of arbitrators, 504–505
Waiver of arbitration, 503

AMERICAN INSTITUTE OF ARCHITECTS (AIA)
See also AIA Documents
Arbitration, dispute resolution provisions, 493
Architect contracts, 129–130, 168–169
Breach of contract remedies, 58
Building information modeling (BIM) documents, 302
Business role of, 23
Changes clauses, 360, 364–367
Code of Ethics & Professional Conduct, 264–268, E-1–E-5
Code of Ethics, 264–268
   Rule 2.105, p. 266
   Rule 3.202, p. 267
   Rule 3.401, p. 267
   Rule 4.102, p. 267
   Rule 4.103, pp. 267, 269
Collaboration of documents and designs, 329
Compensation of design professionals, 172–178
Construction contract documents, 115, 117, 129–130, 391
Construction management (CM) contracts, 126, 129–130, 289
Contract language interpretation, 201
Contract-specified remedies, 58
Contractor rights in event of nonpayment, 350

Contractual fee arrangements, 175
Contractual time limitations, 336
Cost prediction disclaimers, 201
Default termination provisions, 433–435
Design-build (DB) standardized document, 168–169, 294
Design defect awareness, 323
Design professional services, 168–169, 175, 203
Disclaimers, 203
Drawing ownership: License to use, 213–214
Economic feasibility studies, 192
Electronic data and protection, 214–215
Ethics canons, 264–268
Fiduciary relationships, 169–172
Guaranteed maximum price (GMP), 126, 281
General Conditions of the Contract for Construction, A201-2007, pp. 49, 58, 323, B-1–B-38
Generally accepted accounting principles (GAAP), 176
"Green" design and LEED certification documents, 233
Indemnity clauses, 453–454
Legal view of standardized contracts, 13
Notice to proceed (NTP), 391
Objective standards, 208
Ownership of documents clause, 215
Payment timing, 179–180
Performance and Payment Bonds, A312-2010, pp. 476, 479–482, 484–485
Professional standards, 228, 256
Project delivery alternative methods, 289
Scheduling approaches, 395
Separate contract systems, 292
Standard Form of Agreement Between Contractor and Subcontractor, A401-2007, pp. 412
Standard Form of Agreement Between Owner and Architect, (B101–2007), pp. 168–169, D-1–D-22
Standard Form of Agreement Between Owner and Contractor, A101-2007, pp. A-1–A-8
Standardized contracts of, 13
Statute of limitations provisions, 258–259
Subjective standards, 208–209
Subsurface problems, risk allocation approach to, 382–383

AMERICAN INSTITUTE OF CONSTRUCTORS (AIC)
Constructor Code of Conduct, 269–270

AMERICANS WITH DISABILITIES ACT (ADA) OF 1990
Building standards and codes, 228
Design professional liability and standards, 228

APPEALS
See Judicial System

APPORTIONMENT OF LIABILITY
See also Loss shifting
Comparative negligence and, 451
Contribution, 450–451
Design professional fault defense, 255
First instance vs. ultimate responsibility, 446–447
Multiple defendants of, 446, 447–448
Third-party action of, 448
Tort law and, 450

ARBITRATION

Abuse of, 500–501
Adjuncts of judicial system, 507
Administrative Dispute Resolution Act (ADRA), 508
Administrative review for, 492
Alternative dispute resolution (ADR), 492–493, 506–507
AIA documents for, 491, 503–506
   Arbitration process, 503–506
   Documents for arbitration, 491, 503–504
Appeal time, 498
Arbitrator selection, 504–505
Awards, 506
Claims resolution, 492–493
Common law defenses to, 501–503
   Conditions precedent, 504
   Termination of contract, 504
   Mutuality, 503
   Unconscionability, 502
   Waiver of arbitration, 503
Construction Industry Arbitration (CIA) rules, 503–504
Contract Dispute Act (CDA), 508
Discovery, 504
Dispute resolution methods, 492
EJCDC specifications, 493
Federal Arbitration Act (FAA), 499–501
   Arbitration authorization and award enforcement, 499–500
   Arbitration agreement requirements, § 2, p. 508
   Preemption, 508
Federal preemption, 501
Federal procurement contracts, 508
Immunity of arbitrators, 495
Imposed resolution by third parties, 492
Initial decision maker (IDM) position, 169, 493, 496–499
Initial decisions, 493–498
   Costs, 496–497
   Design professional as decision maker, 493–494
   Elemental fairness requirements, 494–495
   Finality of, 497–498
   Forms of communication, 496
   Payment certificate issuance, 498
   Review process, 498
   Standard of interpretation, 495–496
   Subject matter of dispute, 498
International, 508–509
Introduction process, 503–504
Judicial review, 499–500
Mediation, 492
Multiple party, 505–506
Mutuality, 503
Precedent conditions, 503
Process of, 503–506
   Arbitrator neutrality, 504–505
   Consolidation, 505–506
   Construction Industry Arbitration (CIA) rules, 503
   Discovery, 504
   Joinder, 505–506
   Multiple-party arbitration, 505–506
   Selection of arbitrators, 504–505

Public contracts, 507–508
Revised Uniform Arbitration Act (RUAA), 504–505
State and local contracts, 508
State judicial regulation, 500–501
State statutes, 501
Statutory framework, 499–500
Summary jury trials, 507
Termination of contract affected by, 503
Unconscionability, 502
Uniform Arbitration Act (UAA), 499–500
Voluntary resolution, 492
Waiver of, 503

ARCHITECTS

See also American Institute of Architects (AIA); Design-build (DB); Design Professionals; Design Professional Liability; Design Professional Services
Authority conflicts, 285
American Institute of Architects (AIA) specifications, 168–169, 264–268
Architectural Works Copyright Protection Act (AWCPA) of 1990, p. 216
Breach of contract, 185–187
   "Betterment" rule defense, 185–187
   Claims against client, 187
Compensation for services, 172–180
   Fee methods, 172–178
   Payment before building, 180–181
   Timing of payment, 179–180
Construction manager (CM) relationship, 129–130
Drawings and specifications, 169, 209–221
Ethics of, 264–268
Initial decision maker (IDM), as, 169
Licensing laws, 148–155, 160–162
   Application of seal to another's work, 151–152
   "Holding out" statutes, 148–149
   Liability problems, 162
   Out-of-state practice, 153–155
   Possessor of license, 152–153
   "Practicing" statutes, 149
   Regulation differences, 149–151
   Statutory exemptions, 152
   Unlicensed, 160–162
Owner relationships, 166–188
   Conflict of interest, 170–171
   Confidentiality, 170–171
   Contract completion and, 181–182
   Death or unavailability of design professional, 184–185
   Fiduciary relationships, 169–172
   Involving compensation, 172–180
   Not involving compensation, 181
   Suspension of performance, 182
   Termination of contract, 182–184
Project administration, 169, 191–193
   Assistance in obtaining financing, 191–192
   Economic feasibility studies, 192
   Legal services, 193
   Obtain public approval, 192–193
   Project delivery and, 285
Professional service contracts, 168–169

ASSOCIATED GENERAL CONTRACTORS (AGC) OF AMERICA

Building information modeling (BIM) documents, 302
Business role of, 23
Construction management (CM) contracts, 126, 289
Design–build standardized documents, 294
Document applications:
   ConsensusDOC 300, p. 299
   ConsensusDOC 301, p. 302
   ConsensusDOC 310, p. 233
   ConsensusDOC 500, p. 126
   ConsensusDOC 800, p. 299
   ConsensusDOC 801, p. 126
   ConsensusDOC 803, p. 126
"Green" Building Addendum, 233
Legal view of standardized contracts, 13
Program management (PM) documents, 299
Project delivery alternative methods, 289
Sustainable building design and construction, 233
Tri-Party Collaborative Agreement, 299

ATTORNEYS

See Judicial System

AWARDS

See Arbitration, Compensation; Money Awards; Remedies; Recovery

BAD FAITH

Surety claims, 486–487

BANKRUPTCY

Contractors: obligation to release surety's, 487

BIDDING

See also Public Contracts
Award of, 312
Brooks Act, 310
Competitive process, 311–313
Corruption of, 310–311
Design-award-build (DAB) contract compared to, 286
Design-bid-build (DBB), 286
Federal Acquisitions Regulations (FAR), 311
Injunction barring award, 313
Invitation for bids (IFB), 311–312
Opening, 312
Protests, 312–313
Public contracts and, 306–316
Public works specifications and, 309–310
Rejection of all, 312
Remedies for rejected bidders, 313
Request for information (RFI), 312
Responsibility of bidder, 312–313
Responsiveness of, 312
Subcontractors, 413–416
   Bargaining (shopping and peddling), 414–416
   Drennan rule, 414–416
   Irrevocable sub-bids, 414
   Postaward negotiations, 415
   Process of bidding for, 413–414
   Promissory estoppel and, 414
   Sub-bid "quotations", 416

Submission deadlines, 112
Unit pricing and, 282–283

BONDS
See also Payment Bonds; Performance Bonds; Sureties
Ancillary bonds, 478
Bad-faith claims, 486–487
Bid bonds, 478
Construction industry need for, 477
Construction process surety, 132
International contracts and, 487
License bonds, 478
Lien release bonds, 478
Miller Act, 422, 485
Payment bonds, 132, 422, 483–485
Penal sum, 476, 486
Performance bonds, 132, 370, 478–483
Process of obtaining, 476
Subdivision bonds, 478
Sureties, 438–439, 474–489
Terminology, 476
Time requirements for claims, 485

BREACH OF CONTRACT
Attorney's fees, 56
Anticipatory repudiation, 49
Client claims against design professionals, 232
Consumer protection and written defects, 42–43
Cost conditions, 199
    Breach of condition and, 199
    Breach of promise and, 199
Design professional services, 185–187
    "Betterment" rule defense, 185–187
    Claims against client, 187
    Consequential damages, 187
    Defective design, 185–186
    Unexcused delay, 186
Discovery rule, 232
Factors of, 48–49
Future, 49
Material breach, 48–49, 183
Recovery, 56–68
    Attorneys' fees, 56
    Causation, 57
    Certainty, 57
    Construction-specific damages, 59–67
    Contractor vs. owner claims, 59–64
    Contract-specified remedies, 58
    Forseeability (freak events), 57
    Limits on, 57–59
    Lost profits, 58
    Mitigation (avoidable consequences), 57–58
    Multiple defendant claims, 66–67
    Noneconomic losses (emotional distress), 56–57, 232
    Owner vs. contractor claims, 64–66
Remedies, 52–68
    Construction-specific damages, 59–67
    Declaratory judgments, 54–55
    Injunction, 54
    Money awards, 55–57
    Punitive damages, 56–57, 232
    Recovery limits, 57–59
    Single recovery rule, 66–67
    Specific performance, 54–55

Restitution, 56
Substantial performance doctrine, 352–353
Time limits for claims, 232
Warranty and, 329–330

BROOKS ACT
See Design Professionals

BUILD-OPERATE-TRANSFER (BOT)
Production organization and delivery, 300–301

BUILDING CODES
Contract requirements, 45
California "Green" Building Standards Code (CALGreen), 105
Housing codes, 228
International Building Code (IBC), 105
Land use regulation, 104–105

BUILDING INFORMATION MODELING (BIM)
Computer-aided design (CAD) compared to, 301
Project organization and delivery, 301–302
Standardized documents for, 302

BUILDING STANDARDS
See also Construction Process; Project Delivery
Americans with Disabilities Act (ADA), 228
Building codes, 104–105, 227–228
Building inspections, 105–106
California "Green" Building Standards Code (CALGreen), 105
Design and construction process regulations, 104–106
Design professional liability for, 227–228
International Building Code (IBC), 105
Land use regulations, 104–108
New York Scaffold Act, 106–107
Occupational Safety and Health (OSH) laws, 107–108
Permits, 104
Safety legislation, 106–108
Self-certification, 104
State safe workplace statutes, 106–107
Tort laws, 108
Violation of, 227–228

BUSINESS ASSOCIATIONS
Agency relationships, 24–27
Agents, 23–24
Authority of private ownership, 115, 117–119
Corporations, 20–21, 118–119
Design professional use of, 18–25
Employees, 23–24
Employment relationships, 23–24, 27–30
Independent contractors, 23–24, 30–31
Integrated Agreement, 298–299
Joint ventures, 22
Limited liability companies (LLC), 21
Limited liability partnerships (LLP), 21
Partnerships, 18–20, 118, 297–298, 300
Professional associations, 23
Project organization and delivery, 297–298, 300
    Build-operate-transfer (BOT), 300–301
    Integrated Project Delivery (IPD), 299
    Lean project delivery, 298–299

Partnering, 297–298
    Project Alliance Agreement (PAA), 299
    Public-private partnerships (PPP), 300
    Teaming agreements, 298
Relevance of, 18
Share-office arrangements, 22–23
Sole proprietors, 18, 118
Spouse and cohabitant ownership, 119
Unincorporated associations, 22, 119

CALIFORNIA BUSINESS AND PROFESSIONS CODE
§ 5537, p. 152
§ 7031, pp. 158–159
§ 7031(a), pp. 157–158

CALIFORNIA CODE OF CIVIL PROCEDURE
§ 411.35, p. 254

CERTAINTY
Breach of contract recovery, 57

CHANGES CLAUSE
Absence of, 359
Authority to order change, 364–366
Bargaining power shifts, 360–361
Cardinal change, 369
Case scenario of, 361–363
Change order (CO), 360
Common law for, 359
Construction change directive (CCD), 360, 366
Constructive change, 369–370
Contractor "changes" claims, 368–370
Contractor's perspective, 358–359
Deductive change (Deletions), 367
Initial decision maker (IDM) role, 358, 369–370
Legal issues, 361–364
Limitations on power of, 364–365
    Public contracts, 365
    Time of contract performance, 365
    Work performance, 364–365
Misrepresentation of authority, 366
Owner's perspective, 358
Performance bonds, effect of changes on, 370
Standard documents for, 360
Waivers excusing formal requirements, 367–366

CIA RULES
See also American Arbitration Association (AAA)
Fast Track Rules (Rules F-1 through F-13), p. 503
Large/Complex Track Rules (Rules L-1 through L-4), p. 503
Regular Track Rules (Rules R-1 through R-55), p. 503
R-3, p. 504
R-7, p. 506
R-20(b), p. 504
R-43, p. 506

CLAIMS
See also Construction-Related Claims; Professional Liability; Third-Party Claims
"Betterment" rule defense, 185–187

SUBJECT INDEX

Change clauses, 368–370
  Cardinal change, 369
  Constructive change, 369–370
Completed projects, 60–64
  Actual cost and recordkeeping, 61–62
  Correction cost, 64–65
  Defective performance, 64–66
  Delay, 66
  Diminished value, 65–66
  Economic waste, 65–66
  Eichleay formula, 62–63
  Extended home office overhead, 62–63
  Jury verdicts, 64
  Productivity loss preferred formulas, 63
  Site chaos and productivity, 60–61
  Total cost, 63–64
Contract law, 320
Contractor claims, 59–64, 323–327, 369–370
  Case scenario of, 324–325
  "Changes" clauses, 368–370
  Completed projects, 60–64,
  Damages, formulas for, 59–60
  Defective specifications, 326–327
  Design specifications, 326
  Eichleay formula, 62–63
  Extended home office overhead damages, 62–63
  Failure to commence project, 59
  Injury, 327
  Jury verdicts, 64
  Misrepresentation, 326–327
  Owner nonpayment, 327
  Partially completed projects, 59–60
  Productivity loss preferred formulas, 63
  Spearin doctrine, 326–327
  Total cost, 63–64
Critical path method (CPM) for, 398
Design professionals:
  Breach of cost condition, 199
  Breach of promise, 199
  Claims against client, 187
  Contractual language and, 201, 208–209
  Cost predictions, 199, 201
  Defenses for, 201, 208–209, 250–261
  Discovery rule for, 232
  "Green" or sustainable design and, 232–234
  Immunity from decision making, 258
  Negligence, 234–247
  Liability for claims against, 199, 201, 208–209, 231–234
  Site visit negligence, 208–209
  Third party claims against, 234–247
  Tort versus contract law, 231–232
Federal False Claims Act (FCA), 314–315
Forfeiture of Fraudulent Claims Act, 314
Ideal decision maker (IDM) role in, 358, 369–370
Misrepresentation of subsurface conditions, 376
Performance disputes, 320–329
  Contract law, 320
  Defects, 321–322, 323
  Design control, 321
  Express warranty, 320
  Good faith and fair dealing principles of, 322–323

Implied warranty, 320
  Legal liability, 320–321
  Statutory, 320–321
  Theories of liability, 321–323
  Tort law, 320
Public entities (against), 214–315
Subcontractors, 417–425
  Liquidating agreements, 424–425
  Mechanics' liens for, 418–422
  Nonstatutory disputes, 423
  Owner disputes, 423–425
  Pass-through claims, 424–425
  "Pay when paid" clause, 417–418
  Payment bonds for, 422
  Payment claims against contractor, 417–418
  Performance-related claims, 423–424
  Prime contractor disputes, 423–424
  Stop notices and, 422
  Third-party disputes, 421

CLIENTS
See Owners

COMMON LAW
Arbitration defenses, 501–503
Changes clause, 359
Contract termination, 435, 438–439
  Anticipatory[repudiation, 439
  Material breach, 438–439
  Wrongful termination for default, 435
Contractor rights to payment, 349–350
Court system for, 7
Delays risk allocation, 392
Design professional liability, 245–246
Duty to disclose (warn), 87, 122–123
Economic loss rule, 89–93
Effect of termination of contract, 503
Forseeability, 57
Indemnification regulations, 456–457
Implied warranty, 87, 122–132
Intellectual property and, 212–219
  Copyright protection, 219
  Ownership, 212–215, 217
  "Work for hire" doctrine, 216–217
Late and nonpayment during performance, 349–350
Mutuality, 503
Payment rules, 342
Precedent, 7, 503
Privity, 91, 237
Public land use controls, 98–99
Safety of workplace, 245–246
Soil condition liability, 122–123
Substantial performance doctrine, 352–353
Subsurface problem rules, 122–125, 375
Termination of contract applications, 435, 438–439, 503
Unconscionability, 502
Waiver of arbitration, 503

COMPENSATION
See Payment

COMPETITIVE BIDDING
See Bidding

COMPREHENSIVE ENVIRONMENTAL RESPONSE, COMPENSATION, AND LIABILITY ACT (CERCLA)
See also Environmental Protection Movement
Brownfields Act (2001), 103
Construction industry participation, 103
Hazardous waste cleanup, 102–103
Land use controls, 102–104
Potentially responsible parties (PRPs), § 107, pp. 102–103
State laws modeled after, 104
Toxic Mold Protection Act, 104

CONCENSUSDOCS
See Associated General Contractors of America

CONSEQUENTIAL DAMAGES
Breach of contract, 187
Exclusion of, 256

CONSTITUTIONAL LAW (UNITED STATES)
Land use regulation, 100–101
Legislative role of, 5
Professional licensing and registration, 148
Punitive damage remedies, 89
Zoning limitations (takings), 100–101

CONSTRUCTION CONTRACTS
Conditions, 180–181
Construction manager (CM) role in awards of, 127–128
Cost-plus, 200–201
Design professional role, 169, 181–185
  Bidding and negotiations, 169
  Obligations on death or unavailability, 184–185
  Suspension of performance, 182
  Termination of contract, 182–184
  Upon completion, 181–182
Drawings and specifications, 209
Federal Acquisition Regulations (FAR), 433
Flow-through (conduit) clauses, 412–413
"Green" design and LEED certification, 233
Incorporation by reference, 412
No-lien contracts, 421
Payment before building, 180–181
Project delivery and, 278–282, 290–293
  Cost contracts, 280–282
  Fixed-price contracts, 278–280
  Separate for multiple prime contractors, 290–293
  Turnkey contracts, 293
Subcontractor rights and duties, 412–413, 439
Termination of, 430–441
  Agreement of, 433
  Anticipatory repudiation, 439
  Common law and, 435, 438–439
  Completion date reinstatement, 437
  Contractor suspension, 435–436
  Contractual power to terminate, 433
  Default, 433–435
  Design professional role in, 436–437

Termination of (*Continued*)
   Material breach, 438–439
   Notice of, 434–435, 437–438
   Owner suspension and convenience, 435–436
   Waiver of, 437
   Wrongful termination for default, 435

CONSTRUCTION MANAGEMENT ASSOCIATION OF AMERICA (CMMA)

Business role of, 23
Code of Professional Ethics, 271–272
Construction manager (CM) documents, 124, 126, 289
Contract documents
   CMMA A-1 (2013), 126
   CMMA CMAR-1 (2013), 126
Design-build standardized documents, 294
Legal view of standardized contracts, 13
Project delivery alternative methods, 289

CONSTRUCTION MANAGER (CM)

Agent (CMa), 125–126, 288
American Institute of Architects (AIA) contracts, 126, 289
Architect relationship, 129–130
Associated General Construction (AGC) contracts, 126, 289
Authority conflicts, 284
Code of Professional Ethics, 271–272
Commercial general liability (CGL) coverage, 129
Contract awards by, 127–128
Construction Management Association of America (CMMA), 124, 126, 271–272, 289
Constructor (CMc), 125–126
Cost contracts, 281
Cost prediction calculations, 127–128, 194
Design-build (DB) responsibilities, 288–289
Design professional use of, 127–128
Differing site conditions (DSC), 127
Ethics of, 271–272
Guaranteed maximum price (GMP), 127–128, 281
Legal issues, 126–129
   Liability and rights of, 128–129
   Owner relationship, 127–128
Licensing laws, 159–160
Project delivery responsibilities, 288–289
Project organization and delivery, 281, 284, 288–289
Public contracts, 313–314
Reasons for, 125
Role in construction process, 124–130, 368–369

CONSTRUCTION PROCESS

See also Construction Contracts; Design-Build; Land Regulation; Project Delivery
Acceleration of performance, 404–406
   Bonus/penalty clauses, 405–406
   Constructive acceleration, 404–405
   Early completion, 405
Acceptance doctrine and, 124
Business associations, 297–301
Build-operate-transfer (BOT), 300–301
Building standards, 103–106
Change clauses effects on work, 365
Delays, 391–392, 394–395, 398–404
   Allocation of risks, 391–392

Causes of, 391–392
Common law allocation of, 392
Concurrent causes, 398
Remedies of, 392
Schedules and, 394–398
Time extensions, 399
Design professional responsibilities, 119–120
"Green" Building Facilitator (GBF), 233
Hazardous material removal, 102–103
Independent contractor status, 121–122
Insurers, 132
International building code (IBC), 103
Land use regulations, 96–110
   Building codes, 104–105
   Building inspections, 105–106
   Environmental laws, 102–104
   Permits, 104
   Zoning, 99–101
Legal framework of, 1–110
   American legal system, 3–14
   Business associations, 16–32
   Contracts, 34–50
   Disputes, 52–68
   Land regulation, 96–110
   Torts, 70–94
Lenders responsibilities, 130–132
Organization of, 129–130
Owner responsibilities, 114–119
Payment bonds, 132
Performance bonds, 132
Prime contractor responsibilities, 120–124
Program manager (PM) responsibilities, 299
Project commencement, 390–391
Project completion, 181, 350–352, 391
   Date of, 391
   Final completion, 351–352
   Notification and payment, 350–352
   Punch list for, 181, 351
   Reinstatement of date after contract termination, 437
   Substantial completion, 350–351
Records and notices, 61–62, 382
   Actual cost and recordkeeping, 61–62
   Differing Sites Condition (DSC), 382
   Notice to proceed (NTP), 390–391
Safety legislation, 106–108
   Occupational Safety and Health (OSH) laws, 107–108
   State safe workplace statutes, 106–107
   Tort laws for, 108
Self-certification, 104
Site visits, 202–209
   Active role of design professional, 203
   Case scenario of, 204–209
   Certification of progress payments, 209
   Contractual specifications, 207–209
   Obligations of design professional, 203–204
   Observation of construction, 203–209
   Passive role of design professional, 203
Soil conditions, 122–123
Standard contracts for, 114, 117
Subcontractor responsibilities, 123–124
Sureties, 132
Time and, 388–408
   Acceleration of contractor performance, 404–405

Commencement, 390–391
Completion, 391
Concurrent causes of delay, 398
Contract clauses for, 392–394, 400–406
Delays, 391–392, 394–395, 398–404
Early completion, 405
Extensions, 399
Fault, 392–394
Legal view of, 390
Schedules, 394–398
Weather conditions, 393–394
Tort law, 72, 108
   Independent contractor rule, 108
   Premises liability, 108
   Relevance to, 72
   Site safety and liability, 108

CONSTRUCTION-RELATED CLAIMS

See also Breach of Contract; Labor Disputes
Breach of contract damages, 59–67
Completed projects, 60–64
   Actual cost and recordkeeping, 61–62
   Correction cost, 64–65
   Defective performance, 64–66
   Delay, 66
   Diminished value, 65–66
   Economic waste, 65–66
   Eichleay formula, 62–63
   Extended home office overhead, 62–63
   Jury verdicts, 64
   Productivity loss preferred formulas, 63
   Site chaos and productivity, 60–61
   Total cost, 63–64
Contractor vs. owner 59–64
   Completed projects, 60–64,
   Damages, formulas for, 59–60
   Eichleay formula, 62–63
   Extended home office overhead damages, 62–63
   Failure to commence project, 59
   Jury verdicts, 64
   Partially completed projects, 59–60
   Productivity loss preferred formulas, 63
   Total cost, 63–64
Defective performance, 60–61, 64–65
Delay, 400–404
   Contractor caused, 400–401
   Liquidated damages clauses, 400–401
   No-damages-for-delay clauses, 402–403
   Owner caused, 401–404
   Records for, 404
   Subcontractor claims, 404
Lost profits, 58
Misrepresentation, 237–240

CONSTRUCTION-RELATED CLAIMS

Mitigation (avoidable consequences), 57–58
Multiple parties, 66–67
Owner vs. contractor, 64–65
   Correction costs, 64–66
   Delay, 66
   Diminished value, 64–66
   Economic waste, 65–66
   Failure to commence project, 64
   Partially completed projects, 64

Partially completed projects, 59–60, 64
  Damages formulas, 59–60
  Restitution to contractor, 60
Performance disputes, 320–329
  Contract law, 320
  Defects, 321–322, 323
  Design control, 321
  Good faith and fair dealing principles of, 322–323
  Legal liability, 320–321
  Statutory, 320–321
  Theories of liability, 321–323
  Tort law, 320
Spearin doctrine, 324–327
Subcontractors, 417–425
  Liquidating agreements, 424–425
  Mechanics' liens for, 418–422
  Nonstatutory disputes, 423
  Owner disputes, 423–425
  Pass-through claims, 424–425
  Payment bonds for, 422
  Payment claims against contractor, 417–418
  Performance-related claims, 423–424
  Prime contractor disputes, 423–424
  Stop notices and, 422
  Third-party disputes, 421
Subrogation (project destruction), 334–335
Unconscionability, 41, 334
Weather related risks and delays, 392–393

CONSULTANTS
Outside of professional organization, 212
Within professional organization, 211–212

CONSUMER PROTECTION
  See also Fraud; Licensing Laws
Breach of contract, 42–43
Homeowner claims, 331–332
Homeowner written contracts, 42–43
Magnuson-Moss Warranty Act, 332
New home warranty acts (NHWA), 332
Professional liability and, 231
Right to repair acts, 332

CONTRACT CLAUSES
  See also Changes Clauses; Different Site Conditions (DSC) Clause; Indemnity Clauses; Warranties
Bonus/penalty, 405–406
Change order (CO), 360
Consequential damages, waiver of, 256
Construction change directive (CCD), 360, 366
Contractor's warranty, 329–330
  Express, 329–330
  Implied, 329
Design professional contracts, 199, 256–259
Differing site conditions (DSC), 377–382
Dispute resolution, 259
Exclusion of consequential damages, 256
Exculpatory, 47
Flow-through (conduit), 412–413
Force majeure, 392–393
Indemnity, 451–455
Immunity from decision making, 258
Integration, 199
Limitation of liability, 256–258, 452

Liquidated damages, 400–401, 452
Modifications, 360
No-damages-for-delay, 492–493
"Pay when paid", 4174–418
Risk management and, 256–259
Scope of services, 256
Standard of performance, 256
Statutes of limitations, 258–259
Third-party claims, 259
"Time is of the essence", 393–394

CONTRACT DAMAGES
  See also Breach of Contract; Payment
Attorneys' fees, 56
Benefit of the bargain, 55
Betterment (Added value) rule, 185–187
Compensatory damages, 55
Consequential, 187
Contractor versus owner, 59–64
Defective design, 185–186
Lost profits, 58
Money awards, 55–57
Owner versus contractor, 64–66
Punitive, 56–57
Reimbursement, 55
Restitution, 56
Unexcused delay, 186

CONTRACT DISPUTES ACT (CDA)
Alternate dispute resolution (ADR) and, 508
CDA § 7103(h), 508
Public contracts and, 314

CONTRACTORS
  See Independent Contractor; Prime Contractors; Subcontractors

CONTRACTS
  See also Changes Clauses; Construction Contracts; Interpretation of Contracts; Professional Service Contracts; Public Contracts; Termination of Contract
Adhesion, 37, 47–48
Breach, 42–43, 48–49
  Anticipatory repudiation, 49, 439
  Compensation, 55
  Construction disputes, 59–67
  Declaratory judgments, 55
  Future, 49
  Material, 48–49, 183
  Money awards, 55–57
  Punitive damages, 56–57
  Recovery limits, 57–58
  Remedies for, 54–68
  Single recovery rule, 66–67
  Specific decrees, 55
Building code requirements, 45
Consideration, 37, 39
Consumer protection and, 42–43
Cost contracts, 280–283
Custom, 45
Defects, 40–43
  Duty to Disclose, 40
  Economic duress, 41
  Fraud, 40
  Homeowner consumer protection, 42–43

Intent of parties, 41
Mistake, 40–41
Mutual assent affects from, 40–41
Restitution from, 56
Statute of frauds, 42
Unconscionability, 41
Writing requirements and, 41–43
Design professional, 182–184, 230–231
  Material breach, 183–184
  Measure of conduct, 230–231
  Notice period, 183–184
  Termination of, 182–184
Dispute resolution, 259
Enforcing, 36–37
Fixed-price contracts, 278–280
Formation of, 37–40
Freedom of contract, 36–37
Good faith and fair dealing, 45
Homeowner, 42–43
Implied terms, 44
Interpretation, 46–48, 208–209
  Canons of, 47
  Contra proferentem guide, 47
  Exculpatory clauses, 47
  Intention of the parties, 46
  Language, 46–48, 208–209
International, 13
Law, 8–9, 43–48
  Contracting parties, 12
  Court systems for, 8–9
  International, 13
  Interpretation, 46–48
  Judicial terms for, 44–46
  Parole evidence rule, 43–44
  Precedent, 7
  Private sources of, 12–13
  Public sources of, 4–9
  Restatements of, 13
  Standardized publications and, 13
Mutual assent, 37–41
  Defects in formation, 40–41
  Objective theory of contracts, 37–38
  Offers and acceptance of, 38–39
Negotiated, 47
Negotiations in, 44–46
Offeree, 37
Offeror, 37
Promisee, 37
Promisor, 37
Promissory estoppel, 39
Prospective advantage, 82
Reasonable certainty of terms, 40
Relevance of, 36
Requirements of, 37
Risk management, 256–259
  Contractual provision for, 256–259
  Dispute resolution, 259
  Exclusion of consequential damages, 256
  Force majeure clauses, 392–393
  Immunity from decision making, 258
  Limitation of liability clauses, 256–258
  Scope of services, 256
  Standard of performance, 256
  Statutes of limitations, 258–259
  Third-party claims, 259

Summary, 46
Writing requirements, 41–43
   Consumer protection of, 42–43
   Intent of parties, 41
   Memorandums, 41, 50
   `Statute of frauds and, 42

CONTRIBUTION
Apportionment, 450–451
Tort law for, 450

COPYRIGHT
Architectural Works Copyright Protection Act (AWCPA) of 1990, p. 216
Categories subjected to protection, 215–216
Common law and, 216–217
Copyright Act of 1976, pp. 215–219
   § 101, p. 216
   § 102(a), p. 216
   § 201, p. 216
   § 204(a), pp. 217, 219
   § 411(a), p. 218
   § 412, p. 218
   § 504, p. 219
Design professional relevance, 215–216
Duration of protection, 219
Infringement of copyright, 218–219
   Registration, 218
   Remedies for, 218–219
   Substantial similarity, 218
Intellectual property rights from, 212–213
Nonexclusive license, 217
Notice requirements, 219
Obtaining a copyright, 219
Owner use of copyrighted work, 217, 219
Ownership of drawings and specifications, 212–213, 219
Preemption of, 215
Recovery for claims, 218–219
   Injunctive relief, 218–219
   Damages (actual and statutory), 218–219
Tangible manifestation protection, 216
Transfer of ownership, 217, 219
"Work for hire" doctrine, 216–217

CORPORATIONS
See also Business Associations
Authority problems, 118–119
Business association role of, 20–22
Construction process and owner authority, 118–119
Design professional and, 20–22
Joint ventures, 22
Legal aspects of, 20–22
Liability of, 20–21
Licensing, 155
Limited Liability Companies (LLC), 21

COST PREDICTIONS
AIA (B141-1997) contracts and disclaimers, 201
Construction manager (CM) calculations, 127–128, 194
Cost estimates compared to, 194, 198
Design professional responsibilities, 193–202
   Accuracy/inaccuracy of, 193–194
   Breach of condition and, 199

Breach of promise and, 199
Case scenario of, 194–199
Client damages, 199
Cost-plus contracts, 200–201
Danger of excessive costs, 202
Legal issues, 199–201
Parole evidence rule, 199
Refined method for, 194
Traditional method for, 194
Tolerance, 199–200
Waiving of cost condition, 200
EJCDC contracts and disclaimers, 201
Guaranteed maximum price (GMP), 127–128, 194

COSTS
See Payment; Project Pricing Variations; Project Delivery

COURTS
See Judicial System

CRITICAL PATH METHOD (CPM)
Construction schedules, 395–398
Delay claim defense using, 398
Description of, 395–396
Float (slack time) and, 396–397

DAMAGES
See also Contract Damages; Delays
Actual value formula, 400
Attorney's fees, 56
Benefit of the bargain, 55
Breach of condition, 199
Breach of promise, 199
Breach of contract remedies, 55–57, 232
Compensatory damages, 55
Consequential, 256
Cost predictions, 199
Emotional distress, 55–56
Exclusion of, 256
Interest, 55
Liquidated damages clauses, 400–401
Litigation costs, 56
No-damages-for delay clauses, 402–403
Punitive damages, 56–57, 232
   Breach of contract compensation, 56–57
   Client claims against design professionals, 232
   Compensation/recovery for, 56–57, 232
   Constitutionality of, 89
Quantum meruit, 56, 353
Reimbursement, 55
Restitution, 56, 353
Single-recovery rule (multiple defendants), 66–67
Tort law, 88–89
   Compensatory, 88
   Punitive, 89
Unjust enrichment, 56
Work not substantially completed, 353

DECISIONS
Architect as Initial decision maker (IDM), 169
Arbitration, 493–498
   Design professional as judge, 493–494
   Finality of, 497–498
   Initial, 493–498

   Forms of communication, 496
   Payment certificate issuance, 498
   Review process, 498
   Standard of interpretation, 495–496
   Subject matter of dispute, 498
Contractual, 169
Contractor constructive change claims, 369–370
Design professional as IDM, 169, 369–370
Initial decision maker (IDM) role in, 369–370
Termination of construction contract, 432–433

DEFECTS
See also Subrogation
Construction, 328–329
Consumer protection, 42–43, 331–332
Consumer warranties, 332
Contract formation, 40–43
Contractor awareness of design defects, 323
Contractor claims, 323–327
Deductive changes, 322–323
Defective materials, 321–322
Design control, 321
Design professional liability, 246–247
Duty to disclose, 40
Duty to warn, 246–247
Formal contract requirements (writing), 41–43
Good faith and fair dealing, 322–323
Homeowner claims, 42–43, 331–332
Implied warranty, 331
Misrepresentation, 323–327
Multiple causation, 328
Mutual assent and, 40–41
Mutual mistakes, 333–334
New home warranty acts (NHWA), 332
Owner claims, 327–329
Performance disputes and, 321–334
Restitution for, 56
Right to repair acts, 332
Shared responsibility, 328–329
   Design delegation, 328–329
   Owner and contractor, 328
Spearin doctrine, 324–327
Statute of frauds, 42
Unconscionability of, 41, 334

DEFENSES TO CLAIMS
See also Immunity; Statutes of limitations
Anti–indemnity statutes, 456
Certificates of merit, 254–255
Design professionals, 106–107, 254–255
"Good Samaritan" laws, 255
Indemnification regulations, 455–456
Limitation of liability clauses, 256–258
Negligence, 80
Premises liability, 84–85
Products liability, 87–88
State safe workplace, 106–107
Superior knowledge, 84–85
Worker's compensation, 27–28, 255

DELAYS
Breach of contract damages, 66, 186
Categorization of, 391–392
Causes, 391–392

Common law allocation of, 392
Construction process and, 394–404
    Concurrent causes, 398
    Impact on scheduling, 394–398
    Notice to proceed (NTP), 390–391
    Schedules and, 394–398
    Time extensions, 399
Continuous treatment doctrine, 253
Contract allocation of, 392–394
Contractor-caused, 400–401
Critical path method (CPM) used for claim defense, 398
Damage formula for, 400
Damages and, 400–404
    Contractor-caused, 400–401
    Eichleay formula, 62–63
    Owner-caused, 401–404
    Records for, 404
    Subcontractor claims, 404
Design professional services and, 186
Force majeure clauses for, 392–393
Late payments, 180
Liquid damages clauses, 400–401
No-damages-for-delay clauses, 402–403
Per diem rate, 400–401
Remedies for, 392
"Time is of the essence" clauses, 393–394
Time-related, 391–392, 394–404
Unexcused, 186

DESIGN-BID-BUILD (DBB)
Advantages, 286
Contract bidding, 286
Definition, 286
Design-award-build (DAB) contract compared to, 286
Subsurface problem risk allocation, 374–375
Weaknesses, 287–288

DESIGN BUILD INSTITUTE OF AMERICA (DBIA)
Code of Professional Conduct, 270–271, G-1–G-4
Design-build standardized contracts, 294

DESIGN-BUILD (DB)
Advantages/disadvantages, 296–297
Bridging method for, 294–295
Construction managers (CM) used for, 288–289
Design-award-build (DAB) process, 286
Design-bid-build (DBB) process, 286
Design Build Institute of America (DBIA), 270–271
Design-build project delivery, 294–295
Ethics of, 270–271
Insurance, 296
Joint ventures, 22
Licensing laws and, 295–296
Owner-designer relationships, 294–295
Project delivery, 293–297
Public contracts for, 313–314
Reasons for, 293–294
Standardized documents for, 168–169, 294
Subsurface problem risk allocation, 374–375

DESIGN PROFESSIONAL LIABILITY
See also Cost Predictions; Design Service Contracts; Risk Management
Americans with Disabilities Act (ADA) standards, 228
Building code compliance and violations, 227–228
Client claims, 231–234
    "Green" or sustainable design, 232–234
    Tort versus contract law, 231–232
Consumer protection statutes, 231
Contractual vs. professional standards, 230–231
"Green" design and LEED certification, 232–234
Interference with contract, 243–244
Intentional torts, 243–244
Legal defenses, 201, 208–209, 250–261
    Acceptance of project, 252
    Apportionment of fault, 255
    Certificates of merit, 254–255
    Consequential damages exclusion, 256
    Contract language and, 201, 208–209, 256–259
    Contractual risk control, 256–259
    Dispute resolution, 259
    "Good Samaritan" laws, 255
    Immunity from decision making, 253–254, 258
    Limitation of liability clauses, 256–258
    Statutes of limitations, 252–253, 258–259
    Statutory, 254–255
    Third-party claims, 259
    Workers' compensation, 255
Measure of conduct, 230–231
    Contractual standards, 230–231
    Consumer protection statutes, 231
Negligence, 227–228, 237–247
    Adviser's privilege, 244
    Americans with Disabilities Act (ADA) and, 228
    Case scenarios of, 239–240, 241–243
    Code compliance and, 227–228
    Duty of care, 228, 235–237
    Duty to warn, 246–247
    Economic loss rule, 237–243
    Industry vs. professional standards, 228
    Misrepresentation, 237–240
    Privity and, 237
    Professional ethics and, 228, 241–243
    Safety and, 244–247
    Third-party claims, 234–247
Professional standards, 225–230, 256
    Case scenario of, 226–227
    Contractual risk control, 256
    Expert testimony, 229–230
    Industry standards and, 228
    Measure of reasonable care, 225–227
    Professional ethics, 228
    Violation of statutes and administrative regulations, 227–228
Prospective advantage, 243–244
Risk management, 256–259
    Consequential damages, 256
    Contractual provisions for, 256–259
    Exclusion of consequential damages, 256
    Force majeure clauses, 392–393
    Good faith decisions, 258
    Immunity from decision making, 258
    Limitations of liability clauses, 256–258
    Scope of services, 256
    Standard of performance, 256
    Statute of limitations, 258–259
Safety of workplace, 244–246
    Common law and, 245–246
    Occupational Safety and Health Act (OSHA), 246
Third-party claims, 234–247, 259
    Contract duty for benefit of, 235
    Duty of care, 235–237
    Duty to warn, 246–247
    Economic loss rule, 237–243
    Negligent misrepresentation, 237–240
    Potential of, 234–235
    Professional negligence, 241–243
    Risk management, 259
    Safety of workplace, 244–246
Tort law and, 231–232, 235–237, 243–244

DESIGN PROFESSIONAL SERVICES
See also Design Professional Liability; Dispute Resolution; Payment; Professional Service Contracts; Termination of Contract
Additional services, 169
AIA specifications, 168–187
    Compensation methods, 172–173
    Drawing and specification ownership, 212–215
    Fiduciary relationships, 169–172
    Electronic data ownership, 214–215
    Generally accepted accounting principles (GAAP), 176
    Payment timing, 179–180
    Traditional phases of service, 168–169
    Standard Form of Agreement Between Owner and Architect, (B101-2007), 168
AIA standards for, 168–169, 308
Architect-owner relationships, 166–168
Basic services, 169
Bidding/negotiations for contractor, 169
Breach of contract remedies, 185–187
    "Betterment" rule defense, 185–187
    Claims against client, 187
    Consequential damages, 187
    Defective design, 185–186
    Unexcused delay, 186
Compensation for services, 172–180
    Adjustment of fee, 178
    AIA computation methods, 175
    Basic vs. additional services, 174–175
    Daily or hourly rates, 175–176
    Deductions from fee, 178
    EJCDC computation methods, 175
    Fee ceilings, 178
    Fixed fees, 176–177
    Generally accepted accounting principles (GAAP), 176
    Limitations of liability from fee, 178
    Methods of, 172–178
    Monthly billing, 179
    Percentage of construction costs, 173–175
    Personnel expenses (daily or hourly), 175–176
    Professional fee plus expenses, 176
    Reasonable value of services, 177
    Reimbursable expenses, 177

Construction responsibilities, 119–120
Consultant use and responsibility, 211–212
    Outside of professional organization, 212
    Within professional organization, 211–212
Contractor submittal review, 209–211
Cost predictions, 193–202
    Accuracy of, 193–194
    AIA (B141-1997) contracts and disclaimers, 201
    Breach of condition and, 199
    Breach of promise and, 199
    Case scenario of, 194–199
    Client damages, 199
    Construction manager (CM) calculations, 194
    Cost estimates compared to, 194, 198
    Cost-plus contracts, 200–201
    Danger of excessive costs, 202
    EJCDC contracts and disclaimers, 201
    Inaccuracy of, 193–194
    Legal issues, 199–201
    Parole evidence rule, 199
    Refined method for, 194
    Traditional method for, 194
    Tolerance, 199–200
    Waiving of cost condition, 200
Drawings and specifications, 209–221
    AIA and EJCDC approaches for use of, 213–215
    Copyright, 215–221
    Design-build (DB) ownership, 215
    Determining ownership of, 212–215
    Electronic data ownership, 214–215
    Infringement of copyright, 218–219
    Obtaining a copyright, 219
    Owner use of copyrighted work, 217
    Ownership of, 212–215, 219
    Preparation of, 168–169
    Relevance of, 215–216
    Review of contractor submittals, 209–211
    "Work for hire" doctrine, 216–217
EJCDC specifications, 201, 214–215
Initial decision maker (IDM) responsibilities, 169, 358
Notices and duration of time extensions, 399
Owner relationships, 166–188
    Confidentiality, 170–171
    Conflict of interest, 170–171
    Contract completion and, 181–182
    Death or unavailability of design professional, 184–185
    Deterioration due to excessive costs, 202
    Fiduciary relationships, 169–172
    Good faith and fair dealing, 169
    Involving compensation, 172–180
    Not involving compensation, 181
    Suspension of performance, 182
    Termination of contract, 182–1484
Project administration, 169, 191–193
    Assistance in obtaining financing, 191–192
    Economic feasibility studies, 192
    Legal services, 193
    Obtain public approval, 192–193
Site visits, 202–209
    Active role in, 203
    AIA standards for, 208–209
    Case scenario of, 204–209
    Certification of progress payments, 209
    Contract disclaimers for, 203
    Contractual specifications, 207–209
    Obligations to, 203–204
    Observation of construction, 203–209
    Passive role in, 203
Termination of contract, 182–184
    Material breach, 183–184
    Notice period, 183–184
Timing of payment, 179–180
    Conditions when project not built, 180–181
    Interim fees, 179
    Late payments, 180
    Monthly billing, 179
    "Right to be paid" on performance, 179

DESIGN PROFESSIONALS

See also Design Professional Liability; Duty; Risk Management

Brooks Act, 308
Business associations, 18–25
    Agencies, 24–25
    Joint ventures, 22
    Limited Liability Companies (LLC), 21
    Limited Liability Partnerships (LLP), 21
    Partnerships, 18–20
    Professional corporations, 21
    Share-office arrangements, 22–23
    Sole proprietorship, 18
    Unincorporated, 22
Defenses to claims, 201, 208–209, 250–261
    Acceptance of project, 252
    Apportionment of fault, 255
    Certificates of merit, 254–255
    Consequential damages exclusion, 256
    Contract language and, 201, 208–209, 256–259
    Contractual risk control, 256–259
    Dispute resolution, 259, 259
    "Good Samaritan" laws, 255
    Immunity from decision making, 253–254, 258
    Limitation of liability clauses, 256–258
    Statutes of limitations, 252–253, 258–259
    Statutory, 254–255
    Third-party claims, 259
    Workers' compensation, 255
Economic loss rule applied to, 91–92
Insurance claims, preparation for, 470–471
Liability, 224–249
    Adviser's privilege, 244
    Case scenarios of, 226–227, 239–240, 241–243
    Client claims, 231–234
    Common law, 245–246
    Consumer protection statutes, 231
    Contractual standards, 230–231
    Duty of care, 235–237
    Duty to warn, 246–247
    Expert testimony and, 229–230
    "Green" or sustainable design, 232–234
    Intentional torts, 243–244
    Interference with contract, 243–244
    Measure of conduct,. 230–231
    Measure of reasonable care, 225–227
    Negligence, 237–243
    Professional standards, 225–230
    Prospective advantage, 243–244
    Safety of workplace and, 244–246
    Third-party claims, 234–247
    Tort versus contract law, 231–232
Licensing, 148–159, 160–162
    Business organization and, 152–153
    Constitutionality of, 148
    "Holding out" statutes, 148–149
    Out-of-state practice, 153–155
    Possessor of license, 152–153
    "Practicing" statutes, 149
    Regulation differences, 149–151
    Statutory exemptions, 152
Moonlighting, 160–162
    Ethical and legal responsibility, 161
    Liability problems, 162
    Third-party claims, 162
    Unlicensed vs. licensed, 160–161
Public contracts, 308–313
    Brooks Act, 308
    Competitive bidding, 310–313
    Hiring qualifications, 308
    Public works specifications, 309–310
Recovery for services, 156–159, 161
Risk management, 256–259
Standards of, 225–231
Subcontractor nonstatutory claims against, 423
Unlicensed work, 156–159
    Criminal sanctions, 156
    Payment reimbursement, 158–159
    Right to compensation statutes, 157–158
    Substantial compliance doctrine, 159

DIFFERING SITE CONDITIONS (DSC) CLAUSE

See also Subsurface Problems

Construction manager (CM) and, 127
Contractor liability protection, 377–380
Disclaimer system compared to, 377–378
Federal approach, 379–382
    Notice requirements, 382
    Type I clause, 380–381
    Type II clause, 381
Subsurface problem risk allocation, 377–382

DISCLAIMERS

Contractor responsibility for subsurface conditions, 378–379
Design professional responsibility for site visits, 203, 208
Subsurface risk allocation, 377–379

DISCOVERY

See Judicial System

DISPUTE RESOLUTION

See also Arbitration

Alternative dispute resolution (ADR), 492–493, 506–507
Agency business associations, 26–27
Agent-third party relationships, 26–27
Principle-third party, 26
Administrative Dispute Resolution Act (ADRA), 508
Arbitration methods of, 492–493

Breach of contract, 52–68
    Attorney's fees, 56
    Contract-specified remedies, 54–55, 58
    Damages specific to construction, 59–67
    Money awards, 55–57
    Limits on recovery, 57–58
Contract Dispute Act (CDA) for, 508
Contractual control of, 259
Court systems, 8–9
Design professionals and, 259
Discovery, 11
EJCDC specifications, 493
Initial administrative review, 492
Imposed by third parties, 492
Judgment, 8, 11–12
Judicial system and, 8–12
Litigation process, 10–12
No initial review for, 493
Mediator, 492
Public contracts (claims court), 9
Respondent, 8
Trial courts, 8, 11–12
Verdict, 8
Voluntary resolution, 492

DISPUTES
See Performance Disputes

DRAWINGS AND SPECIFICATIONS
AIA and EJCDC approaches to ownership, 213–215
    Electronic data and protection, 214–215
    License to use, 213–214
Building information modeling (BIM), 301–302
Collaborative review of, 329
Construction contract provisions for, 209
    Outside of professional organization, 212
    Within professional organization, 211–212
Contractor submittals, 209–211
Copyright protection, 212–213, 215–221
Defective, 326–327
Design-build (DB) ownership, 215
Design professional use and responsibilities, 169, 209–221
    Collaboration for, 211
    Contractual rights as instrument of service, 212–213
    Liability and, 209–211
    Review of submittals, 169, 209–211
Design specifications, 283–284
"Design" vs. "Performance" decisions, 283–284
"Impossibility" of construction, 333–334
Infringement of copyright, 218–219
Obtaining a copyright, 219
Owner use of copyrighted work, 217
Ownership of, 213–215
    Contractual rights, 212, 213–215
    Copyright protection, 212–213, 219
    Instruments of service, 212–213
Performance disputes and, 326–327, 329, 333–334
Performance specifications, 264
Preparation of, 168–169
Project delivery and, 283–284
Public works specifications, 309–310
Relevance of, 215–216
Shop drawings, 209
Spearin doctrine applications, 326–327
Warranty of, 326
"Work for hire" doctrine, 216–217

DRENNAN RULE
Avoidance of, 416
Promissory estoppel and, 414
Restatement (Second) of Contracts, 414
Subcontractor bargaining and, 414–415

DUTY
See also Tort Law
Concept of, 75
Contractual process: Duty to disclose, 40, 134–135, 416–417
Forseeability doctrine, 77–78
Hazardous conditions: Duty to warn, 246–247
Negligence:
    Duty of care, 77–79, 228, 235–237
    Limits on, 77–79
    Professional ethics and, 228
    Tests for risky behavior, 77
    Third party claims, 235–237
    Tort law and, 77–79
Premises liability:
    Defenses to, 84–85
    Duty of care limitation, 78–79
    Duty of possessor of land, 82–85
    Tort law for, 82–85
Product liability: Duty to warn, 87
Professional liability:
    Duty of care, 235–237
    Duty to warn, 246–247
Soil condition: Duty to disclose, 122–123
Third-party claims, 235–237, 246–247

EASEMENTS
Light, air, and water, 99

ECONOMIC DURESS
Contract failure from, 41

ECONOMIC LOSS RULE
See also Design Professional Liability
Common law and, 89–90
Design professional applications, 91–92, 237–243
Development of, 90–91
Hazardous defect and prophylactic remediation, 92
Negligence liability and, 78–79, 237–243
    Duty of care limitations and, 78–79
    Negligent misrepresentation, 237–241
    Professional negligence, 241–243
Pecuniary loss limitations, 89–93
Permutations of, 91–93
Property damage applications, 92
Statutory developments, 93
Tests for, 91
Tort law and, 78–79, 89–93

EICHLEAY FORMULA
Home office overhead claims, 62–63

EJCDC DOCUMENTS
C–525, p. 282
C–700
    Article 10 p. 360
    Article 15, p. 434
    Article 16, p. 360
    ¶2.03, p. 391
    ¶2.05(A), p. 395
    ¶2.06, p. 395
    ¶2.07, p. 395
    ¶4.02, p. 384
    ¶4.02(A), p. 384
    ¶4.02(B), p. 384
    ¶4.03(A), p. 384
    ¶4.03(B), p. 384
    ¶4.03(C), p. 384
    ¶4.04(A)(1), p. 384
    ¶4.04(B)(2), p. 384
    ¶6.06(A), p. 417
    ¶6.06(B), p. 417
    ¶6.09(B), p. 45
    ¶7.03, p. 292
    ¶9.08(D), p. 347
    ¶10.05, p. 384
    ¶12.03(A), p.393
    ¶14.02(B)(4)(d), p. 346
    ¶ 14.04, p. 391
    ¶15.01–15.03, p. 434
    ¶15.03, p. 436
    ¶15.04, p. 435
    ¶15.04(B), p. 436
E–500
    Exhibit A, ¶A2.01, p. 175
    Exhibit A, ¶A2.01A(1), p. 192
    Exhibit A, ¶A2.02, p. 175
    Exhibit A, ¶A2.02(6), p.496
    Exhibit F, ¶F5.02, p. 201
    Exhibit I, ¶I A(1), p. 257
    ¶5.01, p. 201
    ¶6.01A, p. 230
    ¶6.03, p. 214
    ¶6.03B, p. 214
    ¶6.03C, p. 214
    ¶6.03D, p. 214
    ¶6.03E, p. 214
    ¶6.03F, p. 214

EMPLOYMENT RELATIONSHIPS
Agency and, 23–27
Business associations, 23–24, 27–30
Employee leasing companies, 28
Independent contractors, 23–24, 30–31
Labor disputes, 28–30
    Picket lines, 29
    Project Labor Agreements (PLA), 29–30
    Right-to-work laws, 30
    Secondary boycotts, 29
    Union involvement and, 28–29
Respondeat superior, 27
Workers' compensation, 27–28

ENGINEERING PROFESSION
See also Design Professionals
Application of seal to another's work, 151–152
Ethics, 268–269
Geotechnical role for soil reports, 374
"Holding out" statutes, 148–149
Liability problems, 162
Limitations of liability clauses for, 256
National Society of Professional Engineers (NSPE), 23, 268–269

Out-of-state practice, 153–155
Possessor of license, 152–153
"Practicing" statutes, 149
Regulation differences, 149–151
Statutory exemptions, 152
Unlicensed, 160–162

ENGINEERS JOINT
CONTRACT DOCUMENTS
COMMITTEE (EJCDC)
 See also EJCDC Documents
Arbitration, dispute resolution provisions, 493
Change clause specifications, 360
Construction contracts, 115, 117, 282
Contract language interpretation, 201
Cost prediction disclaimers, 201
Design-build standardized documents, 294
Differing site conditions (DSC) document approach, 384
Drawing ownership: License to use, 213–214
Electronic data and protection, 214–215
Legal view of standardized contracts, 13
Notice to proceed (NTP), 391
Payment bonds, document C-615 (2010), 476
Performance bonds, document C-610 (2010), 476
Program management (PM) documents, 299
Scheduling approaches, 395
Separate contract systems, 292
Standard General Conditions of the Construction Contract, C-1–C-69
Standardized contracts of, 13
Subrogation waivers, 335

ENVIRONMENTAL LAW
 See also Comprehensive Environment Response, Compensation, and Liability Act (CERCLA); Hazardous Materials; Land Use Regulation
Administrative use of regulations, 7
Brownfields Act, 103–104
California "Green" Building Standards Code (CALGreen"), 105
Comprehensive Environment Response, Compensation, and Liability Act (CERCLA), 103–104
Environmental impact reports (EIR), 103
Environmental impact statements (EIS), 102
Environmental Protection Agency (EPA) enforcement, 7, 102, 103
Land use control, 102–104
National Environmental Policy Act (NEPA), 102
Potentially responsible parties (PRPs), liability on, 102–103
Sick building, 104
State law, 103–104
Superfund Act, 102–103
Toxic Mold Protection Act, 104

ETHICS
Architects, 264–268
Concepts of, 263–264
Construction managers, 271–272
Contractors, 269–270
Design/builders, 270–271
Design professionals, 228, 241–243
Engineers, 268–269

Morality compared to, 263
Negligence and, 228, 241–243
Professional, 228, 241–243, 262–274

EXCULPATION
Exculpatory clauses, 47
Indemnification compared to, 451–452
Limitation of liability compared to, 257
Risk management and, 257

EXPERT TESTIMONY
Design professional standards and, 229–230
Liability of witnesses, 230
Witness immunity, 230

EXTENSIONS
 See also Time
Duration and notice of, 399

FAST-TRACKING
Advantages of, 289
Disadvantages of, 289–290
Project delivery and, 289–290

FEDERAL ACQUISITION
REGULATION (FAR)
Administrative use of regulations, 7
Invitation for bids (IFB), 311
Public contracts, 311

FEDERAL FALSE
CLAIMS ACT (FCA)
Claims against public entities, 314–315

FEDERAL SYSTEM
Administrative agencies, 7
Appeals to, 8–9
Constitutions, 5
Contracting parties, 12–13
Courts, 7–9
Execution of, 4–5, 7–12
Executive branch, 6–7
Government functions, 4–5
Importance of laws, 4
Jurisdiction of, 8
Legislation, 5–6
Private contracts, 12–13
Public contracts 9, 425–426
Standardized documents, 13
Statutes, 5–6

FEDERAL TORT CLAIMS
ACT (FTCA) OF 1946
Negligence claims and, 80–81

FEDERATION INTERNATIONALE
DES INGENIEURS-CONSEILS
(INTERNATIONAL FEDERATION OF
CONSULTING ENGINEERS: FIDIC)
International contracts, 384–385
Red Book (1999), 384–385
 Subclause 4.10, p. 385
 Subclause 4.11, p. 385
 Subclause 4.12, p. 385
Subsurface problem risk allocation approach, 384–385

FEES
 See Attorneys; Compensation; Pricing Variations

FIDUCIARY RELATIONSHIPS
American Institute of Architects (AIA) standards for, 169–172
Limited Liability Partnerships (LLP), 21
Partnerships, 18–20
Professional design-owner relationship, 169–172

FINANCING
Assignment of payment, 347
Design professional assistance in obtaining, 191–192
Economic feasibility studies, 192
Lender financing, 130–132
Lender liability, 347–348
Loan to contractor, 347
Loan to owner, 347–348
Progress payments, 347–348

FLOW-THROUGH CLAUSES
Construction (subcontractor) contracts, 412–413

FORCE *MAJEURE* CLAUSE
Construction delay justification, 392–393

FORSEEABILITY
Breach of contract recovery and, 57
Duty of care limitations, 77–78

FRAUD
 See also Statute of Fraud
Contract defects from, 40

GOOD FAITH AND FAIR DEALING
Contract negotiations, 45
Contractor awareness of design defects, 323
Deductive changes, 322–323
Immunity from decision making, 258
Performance disputes and, 322–323
Risk management and, 258

"GREEN" BUILDING DESIGN
Certification and claims, 234
Commentator speculation and, 232–233
Design professional liability and, 232–234
Energy-savings and claims, 234
"Green" building, 234
"Green" Building Facilitator (GBF), 233
High-performance building, 234
Leadership in Energy and Environmental Design (LEED), 233–234
Standardized contracts for, 233
United States "Green" Building Council (USGBC), 233

GUARANTEED MAXIMUM
PRICE (GMP)
Construction manager (CM) fee calculation, 127–128, 194, 281
Cost contracts using, 281
Cost predictions and,127–128, 194
Project organization and delivery, 281

HAZARDOUS MATERIALS
Brownfields Act, 103
Comprehensive Environment Response, Compensation, and Liability Act (CERCLA), 102–103
Construction industry participation, 103
Economic loss rule applied to, 92
Environmental protection policies for, 102–104
Land use regulation, 102–104
Prophylactic remediation, 92
State legislation for, 103–104

HOMEOWNERS
See Consumer Protection; Owners
Associations for private land regulation, 98
Consumer protection, 42–43
Statutory protections, 331–332
Written contracts, 42–43

HOUSING DEVELOPMENT
Euclidian zones, 99
Homeowner associations, 98
Land use regulation, 98–101
Restrictive covenants, 98
Smart growth, 100
Takings, 100–101
Zoning, 99–101

IMMUNITY
Arbitration decision-making position, 495
Contractual risk control, 258
Decisions in good faith and, 258
Design professional legal defenses, 255, 258
Federal Tort Claims Act (FTCA), 80–81
Workers' compensation, 255

IMPLIED TERMS
See Construction Contracts

IMPLIED WARRANTY
Common law and, 122–123
Contractor's workmanship, 331
Defects and, 326–327
Design specifications, 326
Injury and, 327
Homeowner claims, 331
Legal liability and, 320
Misrepresentation claims, 326–327
Performance disputes and, 320, 326, 327, 329, 331
Product liability, 87
Soil condition liability, 122–123
Spearin doctrine, 122, 326–327

INDEMNIFICATION
Anti-indemnity statutes, 456
Contractual, 449, 451–452
Exculpation compared to, 451–452
Indemnitee, 449, 453
Indemnitor, 449, 453
Indemnity clauses, 453–457
   AIA document provision, 453–454
   Common law regulation of, 456–457
   Functions of, 454–455
   Statutory regulation of, 455–456
Insurance requirements of indemnitor, 457
Limitation of liability clauses compared to, 452

Liquidated damages compared to, 452
Loss shifting, 444–459
Non-contractual, 449–450

INDEPENDENT CONTRACTOR
Business associations, 23–24, 30–31
Contractor responsibility, 108
Liability of, 85–86
Role of, 85
Rule and exceptions of, 85–86, 108
Site safety and, 86, 108
Tort law and, 85–86, 108
Vicarious liability and, 86
Workman's compensation for, 86

INITIAL DECISION MAKER (IDM)
Administrative (judge) responsibilities, 493–498
Arbitration, 493–499
Architect as, 169
Change clause responsibilities, 369–370
Communication of decisions, 496
Contractor constructive claims and, 369–370
Creation of position, 498–499
Design professional as, 169, 493–498
Elemental fairness responsibilities, 494–495
Finality of decision, 498–499
Payment of costs incurred by, 496–497
Selection of, 493–494
Standards of interpretation, 495–496

INJUNCTIONS
Bid protests (barring award), 313
Breach of contract remedies, 54–55
Copyright infringement (injunctive relief), 218–219

INSURANCE
See also Subrogation
Commercial general liability (CGL), 129, 462, 466–467, 471
   Construction manager (CM) coverage, 129
   Coverage, 462, 466–467
   Defective construction claims, 466–467
   Project manager (PM) coverage, 471
Conflict of interest, 465
Construction insurers' role, 132
Construction project types of, 462–463
Deductible policies, 464
Design-build construction, 296
Destruction of project (subrogation), 335
Duty to defend, 465
Elements of risk and, 462–463
Employer's liability, 462
Insurance Services Office, Inc. (ISO), 463
Liability, 462
Notice of claim: cooperation, 464–465
Performance disputes, 335
Policies, 463–466
Policy limits, 464
Premiums, 464
Professional liability, 462, 467–471
   Coverage, 462, 468–470
   Design professional preparation for claims, 470–471
   Exclusions, 470
   Policy types, 468
   Requirement of, 467–468

Project delivery and, 471
Property insurance, 335, 462, 466
   Coverage, 462, 466
   Subrogation waivers, 335
   Total destruction of project, 466
Regulation, 463–464
Risk management and, 461–462
Standard policies for, 463
Sureties compared to, 477–478
Workers' compensation, 462

INTEREST
Breach of contract recovery, 56

INTERNATIONAL BUILDING CODE (IBC)
Design and construction regulation, 103

INTERNATIONAL CONTRACTS
Arbitration and, 508–509
Federation Internationale des Ingenieurs-Conseil (FIDIC), 384–385, 509
Legal systems and, 13
Sureties, 487

INTERPRETATION OF CONTRACTS
See also Changes Clauses; Indemnity Clauses
Adhesion contracts, 47–48
Canons of interpretation, 47
Language, 46–48, 201, 208–209
   American Institute of Architects (AIA), 201, 208–209
   Design professional claim defense, 201, 208–209
   Contra proferentem, 47
   Contract disclaimers, 201, 208
   Cost prediction and, 201
   Engineers Joint Contract Documents Committee (EJCDC), 201
   Exculpatory clauses, 47
   Intention of the parties, 46
   Objective standards, 208
   Subjective standards, 208–209
Negotiated contracts, 47

INVITATION FOR BIDS (IFB)
Public contracts, 311–312

JOINT VENTURES
See Partnerships

JUDGES
See Judicial System

JUDICIAL SYSTEM
See also Arbitration; Dispute Resolution
Appeals, 8–9
Appellant, 8
Attorneys, 10
   American rule, 56
   Breach of contract remedies, 56
   Fees, 10, 56
   Hiring and compensation of, 10
Courts,
   Appellate, 8–9
   Common law of, 7
   Federal, 8

Courts (*Continued*)
    Judgments, 8, 11–12
    Jurisdiction, 8
    Legislative process of, 7–9
    Litigation, 8
    Precedent in, 7
    Public contracts (claims), 9
    State, 8–9
    Trial, 8, 11–12
    Verdict, 8
Discovery, 11, 504
Dismissal, 11
Expert testimony, 229–230
Federal court system, 8
Judges, 8–9
Judgments, 8, 11–12
    Breach of contract, 54–55
    Declaratory, 54–55
    Enforcement of, 12
    Summary, 11
    Trial court, 8
Jurisdiction of courts, 8
Jury verdict, 64
Litigation, 8, 10–12
    Attorney role and compensation for, 10
    Breach of contract remedies, 56
    Complaint, 10–11
    Cost of, 12, 56
    Defendant, 10
    Discovery, 11
    Judgment enforcement, 12
    Motions, 11
    Parties to, 10
    Plaintiff, 10
    Pleadings, 10–11
    Pretrial dismissal, 11
    Summary judgment, 11
    Trial process, 11–12
Mediation, 492
Parties, 10, 12
Pleadings, 10–11
State court systems, 8
Statute of limitations, 258–259, 335–336
Statute of repose and, 335–336
Summary jury trails, 507
Trials, 8, 11–12

JUNK SCIENCE DEBATE
Expert testimony and, 230

LABOR DISPUTES
Agency and employment relationships, 28–30
Picket lines, 29
Project Labor Agreements (PLA), 29–30
Right-to-work laws, 30
Secondary boycotts, 29
Union involvement and, 28–29

LAND POSSESSION
See Premises Liability

LAND REGULATION
See also Comprehensive Environment Response, Compensation, and Liability Act (CERCLA); Restrictive Covenants; Zoning
Constitutional limitations on, 100–101
Construction process and, 96–110
Design and construction processes, 104–106
    Building codes, 104–105
    Building inspections, 105–106
    Permits, 104
Environmental protection movement, 99–100, 102–104
    Brownfields Act, 103
    Comprehensive Environment Response, Compensation, and Liability Act (CERCLA), 102–103
    Environmental impact reports (EIR), 103
    Environmental impact statements (EIS), 102
    Environmental Policy Act (EPA), 102
    National Environmental Policy Act (NEPA), 102
    Potentially contaminated urban sites, 103
    Potentially responsible parties (PRP), 102–103
    State laws, 103–104
    Toxic Mold Protection Act, 104
    Zoning concerns, 99–100
Homeowner associations, 98
Housing development, 98
Legal limitations of, 97–98
Neighboring landowner protection, 98–99
    Drainage, 98–99
    Easements for light, air, and view, 99
    Soil support, 98
    Surface waters, 98–99
Private action limits, 97–98
Public land use controls, 97, 98–108
    Common law doctrines, 98–99
    Environmental law, 102–104
    Private restrictions compared to, 97
    Zoning, 99–101
Restrictive covenants, 98
Safety, 106–108
    Occupational Safety and Health Administration (OSHA), 107–108
    Tort law, 108
    Workplace statutes, 106–107
Subdivision development, 98
Takings, 100–101

LAW
See Common Law; Federal System; Judicial System; Legislation; Torts

LEAN PROJECT DELIVERY
Integrated Agreement, 298–299
Project organization and delivery, 298–299

LEADERSHIP IN ENERGY AND ENVIRONMENTAL DESIGN (LEED)
"Green" sustainable building design and certification, 233–234

LEGAL SYSTEMS
Administrative agencies, 7
American, 3–14
Civil law, 4
Constitutions, 5
Contracting parties and, 12–13
Courts, 7–9
Criminal law, 4
Executive branch, 6–7
Federal, 4–5
International contracts, 13
Legislation, 5–6
Litigation process, 10–12
Private sources, 12–13
Public sources, 4–9
Regulation of construction industry, 4
Restatements of the law, 13
Standardized documents and, 13

LEGISLATION
Arbitration, 499–502, 508
    Contract regulation, 508
    Judicial regulation, 501–502
    Judicial review enforcement, 499–500
Certificates of merit, 254–255
Consumer protection, 331–332
Design professional claim defenses, 254–255
Economic loss rule developments, 93
"Good Samaritan" laws, 255
Insurance regulation, 463–464
Land use, 100–101, 103–104
    Constitutionality of takings, 100–101
    Hazardous materials, 103–104
    State environmental regulation, 103–104
Lien laws, 419
New home warranty acts (NHWA), 332
Per se rule, 76
Process of, 5–6
Sick building, 104
Statutes, role of, 5–6
Trust funds, 422
Uniform Commercial Code (UCC), 6
Workers' compensation, 255

LENDERS
See Financing; Payment

LIABILITY
See also Duty; Professional Liability; Risk Management; Tort Law
Acceptance doctrine, 124
Active, 449
Common law rule, 122–123
Apportionment, 446, 449–451
    Comparative negligence and, 451
    Contribution, 450–451
    Third-party action of, 448
Causal connection and, 79
Construction manager (CM), 128–129
Corporations, 20–21
Duty to disclose, 40, 122–123
First instance vs. ultimate responsibility, 446–447
Direct vs. third-party action, 448
Employment relationships, 27–30
Express warranty, 320
Foreseeability and, 77–78
Implied warranty, 87, 122–123, 320
Independent contractor rule, 85–86
Labor disputes, 28–30
Limited Liability Companies (LLC), 21
Limited Liability Partnerships (LLP), 21
Loss shifting, 444–459
    Common law regulation, 456–457
    Direct action of, 448
    Exculpation vs. contractual indemnity, 451–452

Indemnity clauses, 453–455
Limit of liability clause vs. indemnity clause, 452
Noncontractual indemnity, 449–450
Negligence and, 78–79
Partnerships, 19–20
Passive, 449
Performance disputes, 320–321
Premises, 78–79, 82–85, 108
Prime contractor, 122–124
Product, 86–88
Privity doctrine, 237
Respondeat superior, 27
Statutes and, 320
Strict liability, 80–81, 86–87
Tort law and, 80–81, 86–87, 320
Unincorporated associations, 22
Vicarious, 27, 86
Worker's compensation, 27–28

LICENSING
Administration of, 139–148
   Postadmission discipline, 139–148
   State requirements, 139
Architects and engineers, 149–152
   Application of seal to another's work, 151–152
   Regulation differences, 149–151
Construction managers (CM), 159–160
Contractors, 155–159
   Business organization and, 155
   Forfeiture for unlicensed work, 156–159
   Requirements for, 155
   Responsible managing employee (RME), 155
   Responsible managing officer (RMO), 155
   Right to compensation statutes, 157–158
Criminal and quasi-criminal sanctions, 156
Criticism of need for, 138–139
Design-build construction, 295–296
Design professionals, 148–159, 160–162
   Business organization and, 152–153
   Constitutionality of, 148
   Ethical and legal responsibility, 161
   "Holding out" statutes, 148–149
   Liability problems, 162
   Moonlighting, 160–162
   Out-of-state practice, 153–155
   Possessor of license, 152–153
   "Practicing" statutes, 149
   Recovery for services, 156–159, 161
   Regulation differences, 149–151
   Statutory exemptions, 152
   Unlicensed vs. licensed, 160–161
Federal vs. state, 138
Justification for, 138–139
Malum in se and *prohibitum*, 156–157
Public regulation of, 138–139
Registration compared to, 138
Self-certification, 104
Unlicensed work, 156–159, 160–162
   Criminal sanctions, 156
   Moonlighting, 160–162
   Payment reimbursement, 158–159
   Recovery for work, 156–159
   Right to compensation statutes, 157–158
   Substantial compliance doctrine, 159

LIENS
See Mechanics' Liens

LIMITATION OF LIABILITY CLAUSES
Contractual risk management, 256–258
Engineer liability and, 256
Exculpation compared to, 257
Indemnification and, 257
Third-party claims, 257–258

LIMITATIONS
See Statute of Limitations

LIMITED LIABILITY COMPANIES
See Business Associations

LIQUIDATED DAMAGES CLAUSE
Contractor-caused delay remedies, 400–401
Indemnification compared to, 452
Per diem rate, 400–401

LITIGATION
See also Arbitration; Judicial System

LOSS SHIFTING
Common law regulation, 456–457
Direct action of, 448
Exculpation vs. contractual indemnity, 451–452
First instance vs. ultimate responsibility, 446–447
Indemnification and, 446
Indemnity clauses, 453–455
Insurance indemnification requirement, 457
Limit of liability clause vs. indemnity clause, 452
Noncontractual indemnity, 449–450

MAGNUSON-MOSS WARRANTY ACT
Consumer warranty regulation, 332

MASSACHUSETTS SMART GROWTH ZONING AND HOUSING PROTECTION ACT
Land use regulation, 100

MATERIAL BREACH
Design professional contracts termination, 183
Factors of, 48–49, 438–439

MECHANICAL CONTRACTORS OF AMERICA (MCAA) BULLETIN
Productivity factors, 63

MECHANICS' LIENS
Claimants and lienable work, 419–420
Criticism of, 421–422
Entitlement to compensation, 421
Legal complexity, 418–419
Lien laws, 419
No-lien contracts, 421
Priority of, 420–421
Stop notices, 422
Subcontractor payment claims, 418–422
Unjust enrichment justification of, 419
Subcontractor payment by, 418–422

MEDIATION
See Arbitration

MILLER ACT
See also Bonds
Payment bonds and, 422
Surety time requirements, 485

MISREPRESENTATION
Defects and, 326–327
Design specifications and, 324–327
Implied warranty and, 326
Intentional, 81
Personal loss distinction, 81
Negligent, 81, 237–240
Opinion versus representation, 81
Performance disputes claims, 326–327
Reliance, 81
Spearin doctrine, 324–327
Subsurface conditions, 376
Third-party claims, 82, 237–240
Tort law for, 81–82

MISSOURI STATUTES
§ 327.411, p. 151

MISTAKES
Contracts, restitution for, 56
Contractual defects from, 40–41

MITIGATION
Avoidable consequences, 57–58
Breach of contract recovery, 57–58

MOLD
See Environmental Law

MONEY AWARDS
See Contract Damages; Damages

MOONLIGHTING
See Design Professionals

MUTUAL ASSENT
Arbitration mutuality, 503
Contractual formation and, 37–39
   Closing the deal, 38
   Objective theory of contracts, 37–38
   Offer and acceptance, 38
Defects in contractual process, 40–43
   Duty to disclose, 40
   Economic duress, 41
   Effects from, 40–41
   Fraud, 40
   Mistake, 40–41
   Restitution from, 56
   Unconscionability, 41

NATIONAL ASSOCIATION OF PROFESSIONAL ENGINEERS (NSPE)
Business role of, 23
Code of Ethics for Engineers, 268–269, F-1–F-3
   Rule 5, p. 269

NATIONAL ENVIRONMENTAL POLICY ACT (NEPA) of 1969
Environmental impact report (EIR), 103
Environmental impact statement (EIS), 102
Land use regulation, 102–104
State laws modeled after, 103–104

NEGLIGENCE
See also Misrepresentation; Tort Law
Apportionment of fault, 255, 451
Assumption of the risk, 80
Classification of, 74
Comparative negligence, 80, 451
Concept of, 74, 75
Contributory negligence, 80
Definition, 75
Design professional liability and, 227–228, 234–247
  Adviser's privilege, 244
  Americans with Disabilities Act (ADA) and, 228, 254–257
  Case scenarios of, 239–240, 241–243
  Code compliance and, 227–228, 241–243
  Duty of care, 228, 235–237
  Economic loss rule, 237–243
  Industry vs. professional standards, 228
  Misrepresentation, 237–240
  Professional ethics and, 228, 241–243
Duty of care, 77–79, 228, 235–237
  Design professional liability and, 235–237
  Limits on, 77–79
  Professional ethics and, 228
  Tests for, 77
Economic loss rule, 78–79
Emotional distress, 79–80
Federal Tort Claims Act (FTCA), 80–81
Forseeability doctrine, 77–79
Government defendants, claims against, 80–81
Liability and, 78–79
Noneconomic losses and, 79–80
Occupational Safety and Health (OSH) Act and, 76, 246
Privity and, 237
Protected Interests, 79–80
Standard of conduct, 75–77
  Community (objective), 75–76
  Factual vs. legal issues, 76–77
  Per se doctrine, 76
  Statutory violations, 76
Third-party claims, 234–247
  Duty of care, 235–237
  Duty to warn, 246–247
  Economic loss rule, 237–243
  Negligent misrepresentation, 237–240
  Professional negligence, 241–243
  Privity and, 237
  Safety of workplace, 244–246

NEW HOME WARRANTY ACTS (NHWA)
Performance dispute claims, 332

NEW YORK SCAFFOLD ACT OF 1885
Construction safety, 106–107

NONECONOMIC LOSSES
Breach of contract recovery for, 56–57
Emotional distress, 56–57, 79–80
Negligence and, 79–80
Tort law and, 79–80

NONPAYMENT
See also Payment
AIA document regulations for, 350
Common law rights, 349–350
Contractor performance dispute claim, 327
During performance, 349–350
Surety withholding, 348–349

NOTICE TO PROCEED (NTP)
Construction projects, 390–391

OCCUPATIONAL SAFETY AND HEALTH (OSH) ACT
See also Safety
Administrative use of regulations, 7
Construction safety standards, 107–180
Negligence and, 76
Occupational Safety and Health Administration (OSHA), 73, 76, 107–108
Occupational Safety and Health Review Commission (OSHRC), 107
Tort law and, 73, 76

OWNERS
See also Business Associations; Public Contracts
Authority for projects, 117–119, 364–366
Claims against contractor, 64–65
  Correction costs, 64–66
  Delay, 66
  Diminished value, 64–66
  Economic waste, 65–66
  Failure to commence project, 64
  Partially completed projects, 64
Claims against design professionals, 199, 201, 208–209, 231–234
  Breach of cost condition, 199
  Breach of promise, 199
  Contractual language and, 201, 208–209, 256–259
  Cost predictions, 199, 201
  Discovery rule for, 232
  "Green" or sustainable design and, 232–234
  Good faith decisions, 258
  Immunity from decision making, 258
  Limitation of liability clauses, 256–258
  Risk management, 256–259
  Site visit negligence, 208–209
  Tort versus contract law, 231–232
Comparison of public vs. private, 115
Contract clauses, 358, 364–366, 404–406
  Acceleration of performance using, 404–405
  Authority to order, 364–366
  Bonus/penalty, 405–406
  Changes, 358, 364–366, 404
  Early completion, 405
  Perspective of, 358
Construction manager (CM) relationship, 125–128, 288
  Agent (CMa), 125–126, 288
  Authority of, 127–128
  Constructor (CMc), 125–126
Construction responsibilities, 114–119
Contract termination by, 433–436
  Convenience, 436
  Default, 433–434
  Suspension of work, 435–436
Cost predictions, dispensing with, 200
Delays caused by, 401–404
  Contract defenses, 401–402
  No-damages-for-delay clauses, 402–403
  Records for, 404
  Sources of, 401–402
  Subcontractor claims, 404
Design professional relationship, 166–188,
  Confidentiality, 170–171
  Conflict of interest, 170–171
  Contract completion and, 181–182
  Death or unavailability of design professional, 184–185
  Design-build (DB) drawing ownership, 215
  Deterioration due to excessive costs, 202
  Fiduciary relationships, 169–172
  Good faith and fair dealing, 169
  Involving compensation, 172–180
  Not involving compensation, 181
  Suspension of performance, 182
  Termination of contract, 182–1484
Drawing and specification ownership and use, 212–215
Experience of, 116–117
Financial problems, 333
Foreign ownership, 119
Nonpayment, 349–350
Nonstatutory claims against, 423
Pass-through claims, 424–425
Private owners, 115, 117–119, 416–417
  Authority problems of, 117–119
  Corporations as, 118–119
  Partnerships of, 118
  Sole proprietors, 118
  Spouses or cohabitants of, 119
  Subcontractor use, 416–417
  Unincorporated associations as, 119
Public owners, 115, 119
Project organization and delivery, 278, 280–286
  Administration costs, 281–282
  Authority: delegation of representatives, 284–285
  Choices for, 278
  Communication and provision of information, 285–286
  Cost contracts and, 280–282
  "Design" vs. "performance" specification decisions, 283–284
  Unit pricing and, 282–283
  Value Engineering Change Proposal (VECP), 283
Subcontractor claims against, 423–425
Subcontractor selection and approval, 416–417
Subsurface information furnished by, 375–376
  Misrepresentation claims and, 376
  Types of reports for, 375–376
Use of copyrighted work, 217

PAROL EVIDENCE RULE
Cost predictions and, 199
Integration clause, 199
Definition, 43–44

PARTNERSHIPS
Authority of partners, 19, 118
Business associations, 18–21
Construction process and owner authority, 118
Design professionals in, 18–21
Fiduciary duties, 19
Joint ventures, 22
Legal aspects of, 18–21
Liability of, 19–20
Licensing, 155
Limited Liability Partnerships (LLP), 21
Project delivery partnering, 297–298, 300
Public-private partnership (PPP), 300
Uniform Partnership Act (UPA), 18–19

PAYMENT
See also Bonds; Mechanics' Liens; Payment Bonds; Sureties
Arbitration costs, 496–497
Common law rules, 342
Construction process, 130–132
Design professional responsibilities, 342–343, 345–347
   Application of payment, 342–343
   Certificate of payment, 343
   Defenses to liability, 346–347
   Liability of work completed, 346
   Observation of work, 345–346
   Progress payment mechanisms, 342–342
   Schedule of values, 342
Design professional services, 179–181
   Adjustment of fee, 178
   AIA computation methods, 175
   Basic vs. additional services, 174–175
   Claims against the client, 187
   Conditions when project not built, 180–181
   Daily or hourly rates, 175–176
   Deductions from fee, 178
   EJCDC computation methods, 175
   Fee ceilings, 178
   Fixed fees, 176–177
   Generally accepted accounting principles (GAAP), 176
   Interim fees, 179
   Interim fee payment, 179
   Late payments, 180
   Limitations of liability from fee, 178
   Methods of, 172–178
   Monthly billing, 179
   Percentage of construction costs, 173–175
   Personnel expenses (daily or hourly), 175–176
   Professional fee plus expenses, 176
   Reasonable value of services, 177
   Reimbursable expenses, 177
   "Right to be paid" on performance, 179
Initial decision maker (IDM) costs, 496–497
Insurance policy limits, 464
Interpleader action, 344
Joint checks, 344–345, 423
Late payments, 180, 349–340

Lenders responsibilities, 130–132, 347–348
   Assignment, 347
   Loan to contractor, 347
   Loan to owner, 347–348
   Project funding, 130–132
   Subcontractor claims, 348
Lien waivers, 344
Mechanics liens, 418–422
   Claimants and lienable work, 419–420
   Criticism of, 421–422
   Entitlement to compensation, 421
   Legal complexity, 418–419
   Lien laws, 419
   No-lien contracts, 421
   Priority of, 420–421
   Subcontractor payment claims, 418–422
   Unjust enrichment justification of, 419
Miller Act, 422
Nonpayment during performance, 349–350
Prime contractors, 340–355
   Acceptance: effects on future claims, 352
   Common law rights, 349–350
   Final completion and, 351
   Late and nonpayment during performance, 349–350
   Payment to subcontractors and suppliers, 344–345
   Progress payments, 342–343
   Retainage, 343–344
   Substantial completion and, 350–351
   Uncompleted (substantially) work and, 353
   Withholding due to surety requests, 348–349
   Work not substantially complete, 353
Project completion process and, 350–352
Prompt payment acts, 349
Quantum meruit recovery, 353
Subcontractors, 344–345, 348–349, 417–423
   "Pay when paid" clause, 417–418
   Bonds, 422
   Lender duty to shield from misconduct, 348
   Mechanics liens, 418–422
   Prime contractor payment of, 344–345
   Prime contractor, claims against, 417–418
   Stop notices, 422
   Surety requests for withholding from prime, 348–349
   Third parties, claims against, 421
   Trust fund legislation, 422
Substantial performance doctrine, 352–353
Sureties, 132
Timing of, 179–181, 342
Unlicensed contractors, 157–159

PAYMENT BONDS
American Institute of Architects (AIA) document A312-2010, pp. 476, 484–485
Claimants coverage issues, 484
Construction process and, 132
Engineers Joint Contract Documents Committee (EJCDC) document C-615, p. 476
Function of, 483
Liability, 484

PERFORMANCE BONDS
American Institute of Architects (AIA) document A312-2010, pp. 476, 479–482
Changes clauses, effect of on, 370
Construction process and, 132
Engineers Joint Contract Documents Committee (EJCDC) document C-610, p. 476
Liability to owner, 482–483
Owner initial responsibilities, 479–480
   Conference, 479
   Declaration of default, 479–480
   Tender contract balance, 480
Surety claim defenses, 482
Surety options, 480–481
Surety promise to owner, 478–479

PERFORMANCE DISPUTES
Claims, 320–329
   Contract law, 320
   Defects, 321–322, 323
   Design control, 321
   Good faith and fair dealing principles of, 322–323
   Legal liability, 320–321
   Statutory, 320–321
   Theories of liability, 321–323
   Tort law, 320
Construction changes, 356–371
Contractor claims, 323–327
   Case scenario of, 324–325
   Defective specifications, 326–327
   Design specifications, 326
   Injury, 327
   Misrepresentation, 326–327
   Owner nonpayment, 327
   Spearin doctrine, 326–327
Contract payment, 340–355
Contractor defenses, 333–336
   Commercial impracticability, 334
   Contractual limitations, 336
   Design defects, 334
   Frustration, 334
   "Impossible" specifications, 333–334
   Laches, 336
   Mutual mistake, 333
   Statutes of limitation, 335–336
   Statutes of repose, 335–336
   Subrogation waivers, 334–335
   Unconscionability, 334
Delay, 390–408
Financial problems and, 333
Owner claims, 327–329
   Design responsibilities, 328–329
   Multiple causation of defects, 328
   Shared responsibility for defects, 328–329
Project scheduling, 390–408
Subcontracting process, 410–441
Subsurface problems, 372–387
Warranty, 320, 329–332
   Consumer, 332
   Contractor's clause, 329–330
   Design specifications, 326
   Express, 320, 329, 330
   Homeowner claims, 331–332
   Implied, 320, 326, 327, 329, 331

Warranty (*Continued*)
   Injury and, 327
   New home warranty acts (NHWA), 332
   Right to repair acts, 332
   Statutory protection and, 331–332
   Workmanship liability and, 331

PHASED CONSTRUCTION
See Fast-Tracking

PLEADINGS
See Judicial System

PREMISES LIABILITY
See Tort Law

PRICING VARIATIONS
Administration costs, 281–282
Construction management (CM) fees, 281
Cost contracts, 280–283
Cost plus fee contract, 281
Fixed-price contracts, 278–280
Guaranteed maximum price (GMP), 281
Owner concerns and protections, 280–281
Pricing changed work, 366–367
Project organization and delivery, 278–284
Unit pricing, 282–283
Value Engineering Change Proposal (VECP), 283

PRIME CONTRACTORS
See also Construction Contracts; Payment; Termination of Contract
Acceleration of performance, 404–406
   Bonus/penalty clauses, 405–406
   Changes Claus, 404
   Constructive acceleration, 404–405
   Early completion, 405
Acceptance doctrine and liability, 124
American Institute of Constructors (AIC), 269–270
Changes claims, 368–370
   Cardinal change, 369
   Constructive change, 369–370
Claims against owner, 59–64
   Completed projects, 60–64,
   Damages, formulas for, 59–60
   Eichleay formula, 62–63
   Extended home office overhead damages, 62–63
   Failure to commence project, 59
   Jury verdicts, 64
   Partially completed projects, 59–60
   Productivity loss preferred formulas, 63
   Total cost, 63–64
Common law rights to payment, 349–350
Construction industry expectations, 120–121
Construction process responsibilities, 120–124
Contract termination by, 434–435
   Default, 434–435
   Suspension of work, 435
Design-award-build (DAB) contract, 286
Design-bid-build (DBB) contract, 286
Differing site condition (DSC) clauses for, 377–382
Ethics of, 269–270
Financial problems, 333

Independent contractor status, 121–122
Licensing laws, 155–159
   Business organization and, 155
   Forfeiture for unlicensed work, 156–157
   Requirements for, 155
   Responsible managing employee (RME), 155
   Responsible managing officer (RMO), 155
   Right to compensation statutes, 157–158
Payment, 340–355
   Common law rules, 342
   Completion process and, 350–352
   Design professional certification of work for, 345–347
   Late and nonpayment during performance, 349–350
   Lender's interest in process of, 347–348
   Progress payments, 342–343
   Retainage, 343–344
   Subcontractors and suppliers, 344–345
   Substantial performance doctrine, 352–353
   Surety requests for withholding, 348–349
   Uncompleted (substantially) work and, 353
Performance dispute claims, 323–329
   Case scenario of, 324–325
   Defective specifications, 326–327
   Design responsibilities, 328–329
   Design specifications, 326
   Injury, 327
   Misrepresentation, 326–327
   Multiple causation of defects, 328
   Owner claims against, 327–329
   Owner nonpayment, 327
   Shared responsibility for defects, 328–329
   Spearin doctrine for, 326–327
Performance dispute defenses, 333–336
   Commercial impracticability, 334
   Contractual limitations, 336
   Design defects, 334
   Frustration, 334
   "Impossible" specifications, 333–334
   Laches, 336
   Mutual mistake, 333
   Statutes of limitation, 335–336
   Statutes of repose, 335–336
   Subrogation waivers, 334–335
   Unconscionability, 334
Project organization and delivery, 278–284, 290–291
   Compensation for work, 276
   Cost contracts, 280–282
   "Design" vs. "performance" specification decisions, 283–284
   Fixed-price contracts, 278–280
   Multiple contractors, 290–291
   Separate contracts for, 290–291
   Unit pricing, 282–283
   Value engineering change proposal (VECP), 283
Shop drawings, 209
Soil condition liability, 122–123
Subcontractor obligations, 123–124
Submittals, 209–211
Subsurface problems, responsibility of, 378–379
Sureties, 479–480, 487
   Bankruptcy of, 487
   Declaration of default, 479–480

Unlicensed work, 156–159
   Criminal sanctions, 156
   Malum prohibitum violation, 156–157
   Payment reimbursement, 158–159
   Recovery for work, 156–159
   Right to compensation statutes, 157–158
   Substantial compliance doctrine, 159
Warranty clause, 329–330
   Express, 320, 329, 330
   Implied, 320, 329, 331

PRIVITY REQUIREMENT
Economic loss rule and, 91
Negligence and, 237

PRODUCTIVITY LOSS PREFERRED FORMULAS
Industry productivity studies, 63
Measured mile method, 63

PRODUCTS LIABILITY
See Tort Law

PROFESSIONAL ETHICS
See Ethics

PROFESSIONAL LIABILITY
See Design Professional Liability

PROFESSIONAL SERVICE CONTRACTS
Brooks Act, 308
Code of Ethics and, 308
Cost prediction agreements, 201
Language interpretation, 201, 208–209
   Disclaimers, 201, 208–209
   Objective standards, 208
   Subjective standards, 208–209
Public contracts, 308–310
Site visit liability, 208–209
Termination of services, 182–184
   Material breach, 183–184
   Notice period, 183–184

PROJECT ALLIANCE AGREEMENT (PAA)
See Project Delivery

PROJECT DELIVERY
Administrative problems, 284–286
   Architect authority, 285
   Authority conflicts, 284–285
   Communication, 285–286
   Owner site representatives, 284
Building information modeling (BIM), 301–302
Build-operate-transfer (BOT), 300–301
Compensating contractor work, 278
Construction manager role in, 281, 284, 288–289
Contract documents for modern methods, 289
Cost contracts, 280–282
Design-award-build (DAB) contract, 286
Design-bid-build (DBB), 286–288
   Advantages, 286
   Contract bidding, 286
   Weaknesses, 287–288

Design-build (DB), 293–297
    Advantages and disadvantages, 296–297
    Insurance, 296
    Licensing, 295–296
    Owner-designer relationships, 294–295
    Reasons for, 293–294
    Standardized contracts for, 294
"Design" vs. "Performance" specifications, 283–284
Fixed-price contracts, 278–280
Insurance coverage and, 471
Lean project delivery, 298–299
Modern variations in, 288–289
Multiple-prime contractors, 290–293
Organization and, 276–304
Owner's choices, 278
Partnering, 297–298, 300
Phased construction (fast-tracking), 289–290
Program manager (PM), 299
Project Alliance Agreement (PAA), 299
Project planning, 278
Public contracts, 306–316
Public-private partnerships (PPP), 300
Separate contract systems, 290–293
Teaming agreements, 298
Turnkey contracts, 293
Unit pricing, 282–283
Value engineering change proposal (VECP), 283

PROJECT MANAGER
Authority conflicts, 284
Commercial general liability (CGL) coverage for, 471
Project organization and delivery, 299

PROMISSORY ESTOPPEL
Contract consideration and, 39
Subcontractor irrevocable sub-bids, 414

PROMPT PAYMENT ACT
Late and nonpayment during performance, 349

PROPERTY DAMAGE
Economic loss rule applications, 92

PUBLIC CONTRACTS
    See also Bidding
Arbitration, 507–508
    Federal procurement, 508
    State and local contracts, 508
Brooks Act, 308
Change clauses effects on, 365
Competitive bidding, 310–313
Construction Managers (CM), 313–314
Contract Disputes Act (CDA), 314, 508
Court system, 9
Design-build (DB) projects and, 313–314
Design professionals, 308–310
    Hiring qualifications, 308
    Public works specifications, 309–310
Federal False Claims Act (FCA), 314–315
Liquidated damages clause, 400–401
Listing laws, 425
Subcontractors, 425–426

PUBLIC-PRIVATE COLLABORATION
Build-operate transfer (BOT), 300–301
Public-private partnerships (PPP), 300

PUNITIVE DAMAGES
See Damages

QUANTUM MERUIT
See Damages

RECOVERY
    See also Claims; Compensation; Restitution
Breach of contract, 52–68
Attorneys' fees, 56
Causation, 57
Certainty, 57
Client claims against design professionals, 232
Contractor work performed, 156–159
Copyright infringement, 218–219
Damages, 59, 218–219
Declaratory judgments, 54–55
Design professional services, 156–159, 161
Forseeability (freak events), 57
Infringement, 218–219
Injunction (injunctive relief), 54–55, 218–219
Limits on, 57–59
Lost profits, 58
Mitigation (avoidable consequences), 57–58
Money awards, 55–57
Moonlighting services, 161
Noneconomic losses (emotional distress), 79–80, 232
Punitive damages, 56–57, 232
Tort law, 79–80, 88–89
    Compensatory damages, 88
    Punitive damages, 89
Unlicensed work, 156–159
    Payment reimbursement, 158–159
    Stature bars right to compensation, 157–158
    Statute silent on compensation, 157
    Statute specifies right to compensation, 157

RELIANCE
Misrepresentation and, 81
Tort laws, 81

REMEDIES
    See also Damages
Bid protests, 313
Breach of contract, 54–57
Construction delay, 392
Contract-specified, 59–67
    Contractor vs. owner claims, 59–64
    Multiple defendant claims, 66–67
    Owner vs. contractor claims, 64–66
Copyright infringement, 218–219
Declaratory judgments, 54–55
Injunction, 54–55, 218–219, 313
Money awards, 55–57, 64
Tort law recovery, 88–89, 232
    Compensatory damages, 88
    Punitive damages, 89, 232

RESTATEMENT OF CONTRACTS (SECOND)
§ 90, p. 414
§ 241, p. 48
§ 251, p. 49

RESTATEMENT OF LAWS
Description, 13

RESTATEMENT OF LAWS AND TORTS
Duty of care, 77–78
Duty of possessor of land, 83–84
Economic loss rule, 92, 238
    Negligence and, 238
    Property damage, 92
Independent contractor liability, 85–86
Negligence, 76, 77–78
Premises liability 83–84
Product liability, 86–88
    Defenses to, 87–88
    Duty to warn, 87
    Services contracts, 88
    Strict liability, 86–87
Trespassing children, liability of, 83

RESTATEMENT OF TORTS (SECOND)
§ 324A, pp. 246–247
§ 339, p. 83
§ 409, p. 85
§ 552, pp. 238, 241
§ 552(1) p. 238
§ 772, p. 244

RESTATEMENT OF TORTS (THIRD)
Liability for Physical and Emotional Harm (2005)
    § 3, p. 76
    § 7, p. 84
    § 7(a), p. 77
    § 7(b), pp. 78–79
    § 26, p. 79
    § 51, p. 84
Products Liability (1998)
    § 2, p. 87
    § 19(b), p. 88
    § 21(c), p. 92

RESTITUTION
See Damages

RESTRICTIVE COVENANTS
Homeowner associations, 98
Land use regulations, 97–98, 100–101
Private land use controls, 97
Public land use compared to private, 97
Takings, 100–101

RIGHT TO REPAIR ACTS
Performance dispute claims, 332

RISK MANAGEMENT
    See also Apportionment of Liability; Indemnification; Insurance; Loss Shifting; Professional Liability Insurance
Apportionment of liability, 446–451
Consequential damages, 256
Contractual provision for, 256–259
    Dispute resolution, 259
    Exclusion of consequential damages, 256
    Force majeure provision, 392–393

Contractual provision for (*Continued*)
    Good faith decisions, 258
    Immunity of decision making, 328
    Limitation of liability clauses, 256–258
    Scope of services, 256
    Standard of performance, 256
    Statutes of limitations, 258–259
    Third-party claims, 259
Contribution, 450–451
Design professional defense to claims, 256–259
First instance responsibility, 447
Indemnification, 444–459
Insurance, 460–473
Loss shifting, 444–459
Subsurface problem allocation, 374–379
    Construction use of, 376–378
    Contractors and, 378–379
    Design-bid (DB) vs. design-bid-build (DBB) risks, 374–375
    Differing Site Conditions (DSC) clause, 377–378
    Disclaimers for, 377–379
Ultimate responsibility, 447

SAFETY
See also Defects; Negligence; Strict Liability; Third-party Claims; Torts
Acceptance doctrine, 124
Building code violation, 227–228
Construction processes, 124
Design professional liability, 244–247
    Common law liability, 245–246
    Duty of care, 235–237
    Duty to warn, 246–247
    Third-party claims, 244–247
    Violation of codes, 227–228
    Workplace responsibility, 244–246
Independent contractor rule, 85–86, 108
Land use regulation, 106–108
Occupational Safety and Health (OSH) laws, 107–108
Occupational Safety and Health Act (OSHA) liability, 246
Premises liability, 78–79, 82–85, 108
Third-party claims, 124, 244–247
Tort law for, 108, 235–237
Workplace statutes, 106–107

SCHEDULES
Bar (Gantt) chart, 394–395
Critical path method (CPM), 395–399
Delay and, 394–398
Float (slack) time, 397
Measuring delay using, 394–395
Preliminary, 395
Standard document approaches, 395

SEPARATE CONTRACTS
Multiple prime contractors and project delivery, 290–291

SICK BUILDING
Toxic Mold Protection Act, 104

SITE VISITS
AIA standards for, 208–209
Contract disclaimers for, 203

Design professionals, 202–209
    Active role of, 203
    Case scenario of, 204–209
    Certification of progress payments, 209
    Contractual specifications, 207–209
    Obligations of, 203–204
    Observation of construction, 203–209
    Passive role of, 203
"Green" Building Facilitator (GBF), 233

SOIL CONDITIONS
Common law liability, 122–123
Support for neighboring land, 98

SOLE PROPRIETORS
Business associations, 18
Construction process and owner authority, 118

SPEARIN DOCTRINE
Defective specifications, 326–327
Implied warranty, 122, 326–327
Injury, 327
Misrepresentation and, 324–327
Performance disputes and, 324–327
Soil condition liability, 122
Warranty of design specifications, 326

SPECIFICATIONS
See Drawings and Specifications

STANDARD OF CONDUCT
Community (objective), 75–76
Factual vs. legal issues, 76–77
Negligence *per se* doctrine, 76
Statutory violations, 76
Tort law and negligence, 75–77

STATUTES OF FRAUD
Contract writing requirements, 42
Transactions subject to, 42
Uniform Commercial Code (UCC) for, 42

STATUTES OF LIMITATIONS
Continuous treatment doctrine, 253
Contractor performance dispute defense, 335–336
Contractual limitations, 258–259
Design professional claim defense, 252–253, 258–259
Passage of time defense, 252–253
Project delivery and, 286
Statute of repose and, 335–336
Written communication and, 286

STOP NOTICES
See Mechanics' Liens

STRICT LIABILITY
Federal Tort Claims Act (FTCA), 80–81
Historical development of, 86
Implied warranty replaced with, 86–87
Negligence and, 80–81
Product liability and, 86–87
Tort law for, 80–81, 86–87

SUBCONTRACTORS
Bidding, 413–416
    Bargaining (shopping and peddling), 414–416

    Drennan rule, 414–416
    Irrevocable sub-bids, 414
    Postaward negotiations, 415
    Process of, 413–414
    Promissory estoppel and, 414
    Sub-bid "quotations", 416
Construction process responsibilities, 123–124
Construction-related claims, 417–425
    Liquidating agreements, 424–425
    Mechanics' liens for, 418–422
    Nonstatutory disputes, 423
    Owner disputes, 423–425
    Pass-through claims, 424–425
    Payment bonds for, 422
    Payment claims against contractor, 417–418
    Performance-related claims, 423–424
    Prime contractor disputes, 417–418, 423–424
    Third-party disputes, 421
Contract rights and duties of, 412–413
Delay claims by, 404
Flow-through (conduit) clauses, 412–413
Listing laws, 416–417, 425
Payment, 344–345, 348–349
    Lender duty to shield from misconduct, 348
    Surety requests for withholding from prime, 348–349
    Prime contractor responsibilities, 344–345
Payment claims, 417–423
    Bonds, 422
    Entitlement to compensation, 421
    Joint checks, 423
    Mechanics liens, 418–422
    Miller Act, 422
    No-lien contracts, 421
    "Pay when paid" clause, 417–418
    Prime contractor, against, 417–418
    Stop notices, 422
    Third parties, against, 421
    Trust fund legislation, 422
Prime contractor liability for, 123–124
Public contracts and, 425–426
Selection and approval of, 416–417
Standard documents for, 413, 416–418
Termination of contract rights, 439

SUBMITTALS
See Drawings and Specifications

SUBROGATION
Performance disputes and, 334–335
Property insurance claims and, 335
Waivers, 335

SUBSTANTIAL COMPLIANCE DOCTRINE
Unlicensed work recovery, 159

SUBSURFACE PROBLEMS
See also Differing Site Conditions (DSC) Clause
Common law rules for, 122–123, 375
Concealed (AIA) approach, 382–383
Contractual (Federal) approaches, 379–382
Federation Internationale des Ingenieurs-Conseil (FIDIC) approach, 384–385
Misrepresentation claims and, 376
Notice requirements, 382

Owner responsibility for information, 375–376
Risk allocation, 376–379
    Benefits and drawbacks of, 376–378
    Construction use of, 376–378
    Contractors and, 378–379
    Differing Site Conditions (DSC) clause, 377–378
    Disclaimer for, 377–379
Standard documents for, 384
Unforeseen conditions, 374–375
    Design-bid (DB) vs. design-bid-build (DBB) risks, 374–375
    Effect on performance, 374
    Geotechnical engineer role, 374

SUPERFUND
See Comprehensive Environmental Response, Compensation, and Liability Act (CERCLA)

SURETIES
See also Bonds; Payment Bonds; Performance Bonds
AIA documents for, 479–482, 484–485
Asserting claims, 485
Bad faith claims, 486–487
Bankruptcy of contractor, 487
Bonds, 438–439, 474–489
    Ancillary, 478
    Construction industry need for, 477
    Payment, 132, 422, 483–485
    Performance, 132, 478–483
    Process of obtaining, 476
Construction process requirements, 132
Declaration of default, 479–480
Defense from owner actions, 482
Discharge of, 482
Insurance compared to, 477–478
International contracts, 487
Miller Act, 422
Obligee (creditor), 476
Penal sum, 476, 486
Principal (debtor), 476
Reimbursement, 485–486
Terminology, 476
Time requirements for claims, 485
Withholding payment requests, 348–349

TAKINGS
Land use regulations, 100–101

TEAMING AGREEMENTS
See Project Delivery

TECHNOLOGICAL ADVANCES
Building information modeling (BIM), 301–302
Electronic data and protection, 214–215

TENNESSEE STATUTES
§ 62-6-103(b), p. 157

TERMINATION OF CONTRACT
Anticipatory repudiation, 43
Arbitration defense of effects of, 503
Architect-owner relationships, 182–184
Common law applications, 435, 438–439, 503
Construction parties, 430–441
    Agreement of, 433
    Completion date reinstatement, 437
    Contractor suspension, 435–436
    Contractual power to terminate, 433
    Owner suspension and convenience, 435–436
    Waiver of, 437
Default, 433–435
    Contractor termination, 434–435
    Owner termination, 433–434
    Wrongful termination, 435
Design professionals, 182–184
Difficulty of decision, 432–433
Federal Acquisition Regulations (FAR), 433
Material breach, 183–184, 438–439
Notice of, 183–184, 437–438
Subcontractors and, 439

THIRD-PARTY CLAIMS
See also Construction-Related Claims
Acceptance doctrine, 124
Agency relationships, 26–27
    Agent disputes, 26–27
    Apparent authority, 26
    Principal disputes, 26
Apportionment for, 448, 450–451
Construction defects, 124
Limitations of liability clauses for, 257–258
Moonlighting and, 162
Professional liability for, 234–237
    Adviser's privilege, 244
    Contract duty for benefit of, 235
    Duty of care, 235–237
    Duty to warn, 246–247
    Economic loss rule, 237–243
    Intentional tort claims, 243–244
    Negligent misrepresentation, 237–240
    Potential of, 234–235
    Professional negligence, 241–243
    Risk management, 259
    Safety of workplace, 244–246
Subcontractor nonstatutory claims against, 423
Tort law for, 235–247

TIME
See also Delays; Extensions; Schedules
Arbitration appeal, 498
Arbitration award, 506
Bid submission deadlines, 312
Breach of contract claim limits, 232
Change clauses effects, 365
Client claims against design professionals, 232
Common law payment rules, 342
Communication, notification and response, 286
Construction process and, 388–408
    Acceleration of contractor performance, 404–405
    Bonus/penalty clauses and, 405–406
    Commencement, 390–391
    Completion, 391
    Concurrent causes of delay, 398
    Contract clauses for, 392–394, 400–406
    Delays, 391–392, 394–404
    Early completion, 405
    Extensions, 399
    Fault, 392–394
    Legal view of, 390
    Schedules, 394–398
    Weather conditions, 393–394
Contractual control of, 259
Contractual limitations, 336
Copyrights, duration of protection, 219
Design professional claim defense, 252–253, 258–259
Differing site condition (DSC) notice requirements, 382
Discovery rule, 232, 504
Float (slack) time, 397
Laches, 336
Listing laws, 416–417, 425
Miller Act specification of, 485
Notices and duration of extension, 399
Notice of termination, 434–435, 437–438
Notice to proceed (NTP), 390–391
Payments, 179–181
    Conditions when project not built, 180–181
    Interim fees, 179
    Late payments, 180, 349–350
    Monthly billing, 179
    Prompt payment acts, 349
    "Right to be paid" on performance, 179
Project completion notification, 350–352
    Final completion, 351–352
    Substantial completion, 350–351
Project organization and delivery, 286
Statutes of limitations, 252–253, 258–259, 335–336
    Continuous treatment doctrine, 253
    Contractual provision for, 258–259
    Passage of time defense, 252–253
    Performance disputes, 335–336
Statutes of repose, 335–336
Subcontractor selection and approval, 416–417
Surety claim requirements, 485
"Time is of the essence" clause, 393–394
Tort claim limits, 232
Unexcused delay, 186

TORT LAW
See also Negligence; Restatement of Laws and Torts; Restatement of Torts
Breach of contract compared to, 73
Client claims against design professionals, 231–232
Construction process, 72, 108
    Relevance to, 72
    Site safety and liability, 108
Contract law compared to, 231–232
Contribution apportions, 450
Damages
    Compensatory, 88
    Punitive, 89
Defined, 72–73
Design professional liability, 231–232, 235–237
Duty:
    Concept of, 75
    Hazardous conditions (to warn), 246–247
    Limits on, 77–79
    Negligence (duty of care), 77–79, 235–237
    Possessor of land, 82–85
    Product liability (to warn), 87
    Restatement tests for, 77
    Third-party protection, 235–237, 246–247

Economic loss rule, 78–79, 89–93
   Common law and, 89–90
   Development of, 90–91
   Negligence liability and, 78–79
   Pecuniary loss limitations, 89–93
   Permutations of, 91–93
Federal Tort Claims Act (FTCA), 80
Forseeability, 77–79
Function of, 73–74
Independent contractors, 85–86, 108
Intentional claims, 243–244
   Interference with contract, 243–244
   Prospective advantage, 243–244
Intention, 81–82
Misrepresentation, 81–82
Negligence, 75–81
Noneconomic losses, 79–80
Occupational Safety and Health Administration (OSHA) and, 73, 76
Premises liability, 78–79, 82–85, 108
   Construction site safety and liability, 108
   Defenses for, 84–85
   Duty of care limitation, 78–79
   Duty of possessor of land, 82–85
   Negligence and, 78–79
   Relevance, 82–83
   Restatement (Third) of Torts for, 84
   Status of entrant, 83–84
   Superior knowledge defense, 84–85
Products liability, 86–88
   Defects and, 87
   Defenses to, 87–88
   Duty to warn, 87
   Restatement (Third) of Torts, 86–87
   Services contracts, 87
   Strict liability, 86
Professional liability claims, 231–232, 235–237
Prospective advantage, 82
Relevance to construction process, 72
Reliance, 81
Recovery remedies, 88–89
   Compensatory damages, 88
   Punitive damages, 89, 232
Restatements, 77–78, 83–84, 86–87
Strict liability, 80–81, 86
Third-party claims, 235–247
Threshold classifications, 74
Time period for claims, 232
Vicarious liability, 27, 86
Workers' compensation, 86

TOXIC MOLD PROTECTION ACT OF 2001
Sick building legislation, 104

TRIALS
See Arbitration; Judicial System

TRUST FUND LEGISLATION
Subcontractor payment using, 422

TURNKEY CONTRACTS
Project delivery, 293

UNCONSCIONABILITY
Arbitration defense of, 502
Contract defects from, 41
Contractor performance dispute defense, 334
Defects of construction and, 334
Procedural, 502
Substantive, 502

UNIFORM COMMERCIAL CODE (UCC)
Contract applications of, 38, 41, 42, 45–46
Description, 6
Good faith and fair dealing doctrine, 45–46
Statute of Frauds and, 42
U.C.C. Article 2, p. 6
U.C.C. § 2-201, pp. 42
U.C.C. § 2-207, p. 38
U.C.C. § 2-302, p. 41
Unconscionability, 41

UNIFORM PARTNERSHIP ACT (UPA)
See Partnerships

UNINCORPORATED ASSOCIATIONS
See Business Associations

UNIONS
See Labor Disputes

UNITED STATES "GREEN" BUILDING COUNCIL (USGBC)
"Green" or sustainable design, 233

UNJUST ENRICHMENT
See Damages

VALUE ENGINEERING CHANGE PROPOSAL (VECP)
Project delivery and, 283

VICARIOUS LIABILITY
See Tort Law

WAIVERS
Arbitration, 503
Changes clauses and, 367–366
Formal contract requirements excused by, 367–366
Subrogation, 335
Termination of contract, 437

WARRANTIES
See also Spearin Doctrine
Breach of contract and, 329–330
Consumer protection acts, 331–332
Consumer warranty, 332
Contractor's clause, 329–330
Express warranty, 320, 329–330
Homeowner claims, 331–332
Implied warranty, 320, 329, 331
Injury and, 327
Magnuson–Moss Warranty Act, 332
New home warranty acts (NHWA), 332
Performance disputes and, 320, 326, 327, 329–332
Right to repair acts, 332
Sale of homes, 331
Statutory protection and, 331–332

WEATHER-RELATED ISSUES
Construction delays, 393
Force majeure clauses for, 392–39

WORKERS' COMPENSATION
Employment relationships, 27–28
Immunity and, 255, 462
Independent contractor's, 86
Insurance, 462
Professional liability and, 255
Statutory employment, 27–28
Tort law for, 86

ZONING
Constitutional limitations, 100–101
Euclidian, 99
Land use control, 99–101
Societal and environmental concerns, 99–100
Smart growth, 100
Takings, 100–101

TABLE OF ABBREVIATIONS *(continued from front endsheets)*

Ky.L.J.	Kentucky Law Journal[6]	N.D.	North Dakota Reports[1]
Ky.Rev.Stat.	Kentucky Revised Statutes[8]	N.D.W.Va	Northern District of West Virginia (federal trial court)
LDA	Local development authority	N.E.	North Eastern Reporter[3]
L.R–H.L.	Law Reports–House of Lords (England)	N.E.2d	North Eastern Reporter, Second Series[3]
La.	Louisiana Reports[1]	NEPA	National Environment Policy Act (federal)
La.App.	Louisiana Court of Appeals	N.H.	New Hampshire Reports[1]
Law & Contemp.Probs.	Law and Contemporary Problems[6]	N.H.Rev.Stat.Ann.	New Hampshire Revised Statutes Annotated[8]
Law Div.	Law Division (New Jersey trial court)	N.J.	New Jersey Reports[1]
Loy.L.A.L.Rev.	Loyola (Los Angeles) Law Review[6]	N.J.L.	New Jersey Law Reports
Loy.L.Rev.	Loyola Law Review[6]	N.J.Stat.Ann	New Jersey Statutes Annotated[8]
MBE	Minority Business Enterprise	N.J.Super.	New Jersey Superior Court Reports
M.D. Tenn.	Middle District of Tennessee (federal trial court)	NLRB	National Labor Relations Board (federal)
MLDC	Model Land Development Code	N.M.	New Mexico Reports[1]
Marq.L.Rev.	Marquette Law Review[6]	N.M.Stat.Ann.	New Mexico Statutes Annotated[8]
Mass.	Massachusetts Reports[1]	NSPE	National Society of Professional Engineers
Mass.App.Ct.	Massachusetts Appeals Court Reports	NTP	notice to proceed
Mass.Gen.Laws Ann.	Massachusetts General Laws Annotated[8]	N.W.	North Western Reporter[3]
Md.	Maryland Reports[1]	N.W.2d	North Western Reporter, Second Series[3]
Md.App.	Maryland Appellate Reports	N.Y.	New York Reports[1]
Md.L.Rev.	Maryland Law Review[6]	N.Y.2d	New York Reports, Second Series[1]
Md.Real Prop.Ann.Code	Maryland Real Property Annotated Code[8]	N.Y.Gen.Obl.Law	New York General Obligations Act[8]
Me.	Maine Reports[1]	N.Y.–McKinney's Lien Law	New York McKinney's Lien Law
Mich.	Michigan Reports[1]	N.Y.S.	New York Supplement[4]
Mich.App.	Michigan Appeals Reports	N.Y.S.2d	New York Supplement, Second Series[4]
Mich.Comp.Laws.Ann.	Michigan Compiled Laws Annotated[8]	Neb.	Nebraska Reports[1]
Minn.	Minnesota Law Review[6]	Neb.L.Rev.	Nebraska Law Review[6]
Minn.Stat.	Minnesota Statutes[8]	Neb.Rev.Stat.	Nebraska Revised Statutes[8]
Misc.	Miscellaneous New York Reports	Nev.	Nevada Reports[1]
Misc.2d	Miscellaneous New York Reports, Second Series	O.C.G.A.	Official Code of Georgia Annotated
Miss.	Mississippi Reports[1]	OFPP	Office of Federal Procurement Policy
Miss.L.J.	Mississippi Law Journal[6]	OSHA	Occupational Safety and Health Act (federal)
Mo.Ct.App.	Missouri Appeal Reports	OSHComm'n	Occupational Safety and Health Commission (federal)
Mo.L.Rev.	Missouri Law Review[6]	O.S.H.Cas.	Occupational Safety & Health Cases (federal)
Mo.Rev.Stat.	Missouri Revised Statutes[8]	Ohio App.2d	Ohio Appellate Reports, Second Series
Mont.	Montana Reports[1]	Ohio App.3d	Ohio Appellate Reports, Third Series
Mont.CodeAnn.	Montana Code Annotated[8]	Ohio Misc.	Ohio Miscellaneous Reports
N.C.	North Carolina Reports[1]	Ohio N.U.L.Rev.	Ohio Northern University Law Review[6]
N.C.App.	North Carolina Court of Appeals Reports	Ohio St.	Ohio State Reports[1]
N.C.Gen.Stat.	North Carolina General Statutes[8]	Ohio St.2d	Ohio State Reports, Second Series[1]
		Ohio St.3d	Ohio State Reports, Third Series[1]
		Okla.Stat.	Oklahoma Statutes[8]
		Op.Cal.Att'y.Gen.	Opinions of the California Attorney General
		Op.Colo.Att'yGen.	Opinions of the Colorado Attorney General
		Or.	Oregon Reports[1]
		Or.App.	Oregon Court of Appeals Reports
		Or.L.Rev.	Oregon Law Review[6]
		Or.Rev.Stat.	Oregon Revised Statutes[8]
		P.	Pacific Reporter[3]
		P.2d	Pacific Reporter, Second Series[3]
		P.3d	Pacific Reporter, Third Series[3]

Key/Footnotes:
[1] Highest Court of State
[2] Highest Court of United States
[3] Regional Reporter
[4] All New York cases
[5] All California cases
[6] Periodical
[7] Annotation to important cases
[8] Statute
[9] Encyclopedia
[10] Federal Agency Appeals Board

PSBCA	Postal Services Board of Contract Appeals[10]	Tex.Ct.App.	Texas Court of Appeals
Pa.	Pennsylvania State Reports[1]	Tex.Prop.Code.Ann.	Texas Property Code Annotated[8]
Pac.L.J.	Pacific Law Journal[6]	Tex.Tech.L.Rev.	Texas Tech. Law Review[6]
Pa.Commw.	Pennsylvania Commonwealth Court Reports	Tul.L.Rev.	Tulane Law Review[6]
Pa.Super.	Pennsylvania Superior Court Reports	U.Bridgeport L.Rev.	University of Bridgeport Law Review[6]
P.L.	Public Law (U.S.)[8]	UCC	Uniform Commercial Code
Procur.Lawyer	Procurement Lawyer[6]	UC Davis L.Rev.	University of California Davis Law Review[6]
Pub.Cont.L.J.	Public Contract Law Journal[6]	UCLA L.Rev.	University of California, Los Angeles Law Review[6]
Pub.L.	Public Law (U.S.)[8]	U.Pa.L.Rev.	University of Pennsylvania Law Review[6]
RCRA	Resource Conservation and Recovery Act (federal)	U.S.	United States Supreme Court Reports[2]
R.I.	Rhode Island Reports[1]	U.S.C.	United States Code[8]
Real Est.L.J.	Real Estate Law Journal[6]	USCFC	United States Court of Federal Claims
		Univ.Cin.L.Rev.	University of Cincinnati Law Review[6]
S.	Senate Bill (federal)	Utah 2d	Utah Reports, Second Series[1]
S.C.	South Carolina Reports[1]		
S.C.L.Rev.	South Carolina Law Review[6]	Va.	Virginia Reports[1]
S.Ct.	Supreme Court Reporter (U.S.)[2]	Va.Code Ann.	Virginia Code Annotated[8]
S.D.	South Dakota Reports[1]	Vill.L.Rev.	Villanova Law Review[6]
S.D.N.Y.	Southern District of New York (federal trial court)	Vt.	Vermont Reports[1]
		Vt.L.Rev.	Vermont Law Review[6]
S.E.2d	South Eastern Reporter, Second Series[3]		
S.W.	South Western Reporter	WBE	Women's Business Enterprise
S.W.2d	South Western Reporter, Second Series[3]	W.D. PA	Western District of Pennsylvania (federal trial court)
Seton Hall Legisl.J.	Seton Hall Journal of Legislation[6]		
So.	Southern Reporter[3]	W.Va.	West Virginia Reports[1]
So.2d	Southern Reporter, Second Series[3]	Wake Forest L.Rev.	Wake Forest Law Review[8]
Stan.L.Rev.	Stanford Law Review[6]	Wash.	Washington Reports[1]
Stat.	Statutes-at-Large (U.S.)[8]	Wash.2d	Washington Reports, Second Series[1]
		Wash.App.	Washington Appellate Reports
Temp.L.Rev.	Temple Law Review[6]	Wash. & Lee L.Rev.	Washington & Lee Law Review
Tenn.	Tennessee Reports[1]	Wash.L.Rev.	Washington Law Review[6]
Tenn.Code Ann.	Tennessee Code Annotated[8]	Wash.Rev.Code Ann.	Washington Revised Code Annotated
Tenn.Ct.App.	Tennessee Court of Appeals	Wis.	Wisconsin Reports[1]
Tex.	Texas Supreme Court[1]	Wis.2d	Wisconsin Reports, Second Series[1]
Tex. Civ. Proc. & Rem.Code	Texas Civil Procedure and Remedies Code[8]	Wis.L.Rev.	Wisconsin Law Review[6]
		Wis.Stat.	Wisconsin Statutes
Tex.Civ.Stat.Ann.	Texas Civil Statutes Annotated[8]	Wm.Mitchell L.Rev.	William Mitchell Law Review[6]